RNA Technologies

More information about this series at http://www.springer.com/series/8619

Nikolaus Rajewsky • Stefan Jurga •
Jan Barciszewski
Editors

Plant Epigenetics

Springer

Editors
Nikolaus Rajewsky
Max Delbrück Center for Molecular Medicine
Berlin Institute for Medical Systems Biology
Berlin-Buch, Berlin
Germany

Stefan Jurga
Nanobiomedical Center
Adam Mickiewicz University
Poznań, Poland

Jan Barciszewski
Polish Academy of Sciences
Institute of Bioorganic Chemistry
Poznań, Poland

ISSN 2197-9731 ISSN 2197-9758 (electronic)
RNA Technologies
ISBN 978-3-319-55519-5 ISBN 978-3-319-55520-1 (eBook)
DOI 10.1007/978-3-319-55520-1

Library of Congress Control Number: 2017937907

Printed on acid-free paper

This Springer imprint is published by Springer Nature
The registered company is Springer International Publishing AG
The registered company address is: Gewerbestrasse 11, 6330 Cham, Switzerland

Preface

Plant Epigenetics: From Genotype to Phenotype

The Last Unicellular Common Ancestor (LUCA) has existed more than 1 billion years ago. During that time, the plant and animal kingdoms have evolved separately and adopted a multicellular system, with sophisticated pathways of development and capability for perfect adaptation to the environment. Today, in the era of genomics it is known that many developmental processes of plants and animals are similar, although they have evolved independently. The carriers of the logic in these two major lineages are different and show a complicated network of ancient protein and nucleic acid domains, but at the same time a very high conservation and similarity of chromatin proteins and regulatory mechanisms is observed. This, however, does not exclude differences of structure and functions of chromatin that exist between plants and animals. They have evolved very efficient and flexible but different adaptation mechanisms to the local environment in order to ensure survival and reproduction. The specific differences connected to lineage-specific features may provide strong information on the general mechanisms underlying the complexity and regulatory and integratory role of chromatin in all eukaryotes. During a movement towards their final differentiated states, various changes occur in cells due to genetic and environmental factors. Resulted altered properties of the cells have been memorized after each cell division.

Recent technological advances allow genome-wide analysis of DNA and histone modifications, which affect their structures, and have the potential to reveal the regulation mechanisms in plants on the level above nucleotide sequence. Those chemical changes allow the manifestation of multiple phenotypes encoded in the same DNA sequence. In this way, chromatin modifications contribute to variation at multiple levels, ranging from the expression of individual genes, to the differentiation of cell types, to population-level phenotypic diversity. In other words, that is epigenetics.

Formally, the term epigenetics is a combination of two words ‘epigenesis’ and ‘genetics’ and has been coined 75 years ago (Brilliant Jubilee) in 1942 by Conrad H. Waddington. He proposed epigenetics as the branch of biology that studies the causal interaction genes and their products, which brings the phenotype into being, and proposed the concept of the epigenetic landscape as a metaphor for cell differentiation. Currently, epigenetics is interpreted as the study of mitotically and/or meiotically heritable changes in patterns of gene expression that occur without alterations in DNA sequence. Generally, epigenetic studies are focused on chemical modifications of chromatin and their roles in transcriptional silencing. Epigenetic modifications contribute to phenotypic variation at multiple levels, from gene regulation to development, stress response, and population level phenotypic diversity and evolution. A lot of epigenomic features have been comprehensively profiled in health and disease across cell types, tissues and individuals.

Plant development particularly depends on epigenetics. They integrate various environmental signals into different phenotypic or growth responses. Therefore, an understanding of mechanisms of how epigenetic modifications affect the expression of genotype into phenotype in plants is of prime interest.

There are a number of epigenetic phenomena discovered in plants: (i) paramutation which describes the heritable change in expression status of an allele upon its exposure to an allele with the same sequence but displays a different expression status, (ii) nucleolar dominance that is a selective silencing of the ribosomal RNA genes inherited from one progenitor of a genetic hybrid, (iii) imprinting which is characterized by selective expression of genes inherited from only the maternal or the paternal parent, (iv) vernalization which induces flowering in plants in response to cold, (v) RNA-mediated homology-dependent technologies that have important contributions for plant genetic engineering, (vi) RNA-mediated DNA methylation that leads to gene downregulation and (vii) RNA-mediated mRNA degradation or inactivation.

Nowadays, genome sequences for Arabidopsis, rice, poplar, maize and many other plants are known and thus facilitate genome-wide analyses of DNA methylation and histone modifications and their relationships to coding as well as short (miRNAs, siRNAs) and long noncoding RNAs, which can function as epigenetic marks of transcriptional gene silencing and also a defence against transposable elements and viruses. Thus, plants are good model systems and stay as first line of discoveries in the fields of epigenetics.

To deeply discuss and present the frontiers of plant epigenetics, we brought together a diverse group of experts from academia, who working both from the bottom (mechanism) up and top (phenotype) down. We believe that these complementary approaches enable high-impact science.

In the book, there are 26 chapters, which present the current state of epigenomic profiling, and how functional information can be indirectly inferred is discussed. New approaches that promise functional answers, collectively referred to as epigenome editing, are also described. The book highlights the latest important advances in our understanding of the functions of plant epigenomics or new technologies for the study of epigenomic marks and mechanisms in plants. Topics include the deposition or removal of chromatin modifications and histone variants, the role of epigenetics in development and response to environmental signals, natural variation and ecology, and applications for epigenetics in crop improvement. The chapters in this book are variable in nature, ranging from the complex regulation of stress and heterosis to the precise mechanisms of DNA and histone modifications, providing breakthroughs in the explanation of complex phenotypic phenomena. We hope that the chapters in this book present outstanding significance and will capture broad interest.

Berlin — Nikolaus Rajewsky
Poznań — Stefan Jurga
Poznań — Jan Barciszewski
January 2017

Contents

Conservation, Divergence, and Abundance of MiRNAs and Their Effect in Plants

Flor de Fátima Rosas-Cárdenas and Stefan de Folter

Abstract More than 80% of our food comes from seed, flower, and fruit parts. Therefore, understanding the genetic networks that regulate how these organs are formed is important. Transcription factors are one of the main regulators. In addition, it has been demonstrated that another level of regulation includes epigenetic mechanisms. Epigenetic mechanisms include changes in DNA methylation, histone modifications, and noncoding RNAs. Noncoding RNAs include miRNAs that regulate gene expression, at the transcriptional or posttranscriptional level. Next-generation sequencing has demonstrated the expression, conservation, and divergence of miRNAs. Furthermore, functional studies of miRNAs have allowed elucidating their important roles in many developmental processes, including in flower and fruit development, providing potential applications of the use of miRNAs in crop improvement. In this chapter, we describe the conservation and divergence of miRNAs in plants and the advance in the elucidation of their functions.

Keywords miRNA conservation • miRNA abundance • miRNA function • Gene expression • Gene regulation • Plants

Contents

F. de F. Rosas-Cárdenas
Centro de Investigación en Biotecnología Aplicada del Instituto Politécnico Nacional (CIBA-IPN), Ex-Hacienda San Juan Molino, Carretera Estatal Tecuexcomac-Tepetitla Km 1.5, 90700 Tlaxcala, México

S. de Folter (✉)
Unidad de Genómica Avanzada, Laboratorio Nacional de Genómica para la Biodiversidad (LANGEBIO), Centro de Investigación y de Estudios Avanzados del Instituto Politécnico Nacional (CINVESTAV-IPN), Km. 9.6 Libramiento Norte, Carretera Irapuato-León, 36821 Irapuato, Guanajuato, México
e-mail: stefan.defolter@cinvestav.mx

N. Rajewsky et al. (eds.), *Plant Epigenetics*, RNA Technologies,
DOI 10.1007/978-3-319-55520-1_1

1 Introduction

Plants have involved several processes that allowed them to survive their sessile life, which is observed in the gene expression when exposed to different environments (Zhu et al. 2011). Changes in gene expression patterns are widely believed to underlie many of the phenotypic differences within and between plant species (Chen and Rajewsky 2007). Gene expression is regulated at many levels and involves transcription factors and epigenetic mechanisms that include noncoding RNAs (Chen and Rajewsky 2007; Nonogaki 2010; Cech and Steitz 2014; Holoch and Moazed 2015). Noncoding RNAs include small RNAs and are known to silence genes post-transcriptionally by guiding target mRNAs for degradation or repressing translation (Sunkar et al. 2007). Next-generation sequencing has revealed the sRNAs present in diverse organisms (e.g., Chen et al. 2010; Chávez Montes et al. 2014). On the one hand, variation in size and abundance of these molecules has been shown, but the conservation and specificity between species has also been shown, with special attention to a subgroup of sRNAs denominated miRNAs. MiRNAs add an extensive layer of gene control, affecting transcription, stability, localization, and translation (Kidner 2010; Cech and Steitz 2014).

2 General Aspects of MiRNAs

MiRNAs are small sequences of RNA of 20–26 nucleotides (nt), being of 21 nt the most abundant (Jones-Rhoades et al. 2006; Källman et al. 2013; Chávez Montes et al. 2014; Liu et al. 2014a, c). MiRNAs represent a diminutive fraction (<10%) of the total number of sRNAs in plants (Lu et al. 2005; Chávez Montes et al. 2014). Although miRNAs comprise a tiny portion of the genome, they are the best characterized sRNAs (Jones-Rhoades et al. 2006). MiRNAs are derived from a process that starts in the nucleus and finishes in the cytoplasm, where they perform their action. MiRNAs are transcribed from MIRNA loci by RNA polymerase II (Xie et al. 2010), processed of the primary transcript by the DICER-LIKE1 (DCL1) complex, and finally form the effector complex that includes the mature miRNA and an ARGONAUTE (AGO) protein. MiRNAs are key components in complex gene regulatory pathways and they affect transcription, stability, localization, and translation of genes (Kidner 2010); they target mRNAs for cleavage or translational repression (Mallory and Bouché 2008; Voinnet 2009; Cech and Steitz 2014). MiRNAs are grouped into families based on nucleotide sequence of the mature miRNA. These mature miRNA sequences are roughly the same, however, the precursors (sequences outside the mature miRNA) can be highly variable

(Ha et al. 2008). Several miRNA genes can exist for a mature miRNA; for example, in *Arabidopsis thaliana* three different precursors exist for miR164. MiRNAs are classically multigene families, allowing for subtlety and complexity of control, and perhaps a fast response to evolution (Maher et al. 2006; Kidner 2010).

Most plants have more than 100 miRNA genes (*MIR*) (Jones-Rhoades and Bartel 2004; Axtell et al. 2007; Joshi et al. 2010; Nozawa et al. 2012; Chávez Montes et al. 2014), found mainly in intergenic regions and widely distributed in the genome. Although, intronic miRNAs (called "mirtrons"), exonic miRNAs (transcribed from the exons of protein-coding genes), and miRNAs generated from transposable elements, also have been described in plants (Reinhart et al. 2002; Piriyapongsa and Jordan 2008; Voinnet 2009; Xie et al. 2010; Nozawa et al. 2012; Rogers and Chen 2013). MiRNA genes in plants, as well as in animals, are sometimes found in clusters in the genome (Jones-Rhoades and Bartel 2004; Griffiths-Jones et al. 2008; Zhu et al. 2008; Merchan et al. 2009; Liu et al. 2014c). Although miRNA genes in plants are generally monocistronic, some miRNAs originate from polycistronic MIRNA loci (such as miR156, miR166, miR395), i.e., that a single pri-miRNA can form two or more hairpins, each containing a distinct mature miRNA species (Wang et al. 2007; Griffiths-Jones et al. 2008; Merchan et al. 2009; Zhang et al. 2009; Guo et al. 2012; Baldrich et al. 2016). MiRNAs play an essential role in post-transcriptional gene regulation, but transcriptional regulation by miRNAs has also been reported (Khraiwesh et al. 2010). Plant miRNAs are highly complementary to its RNA target, allowing a fast and reliable bioinformatics identification of their targets, which are mostly transcription factors. MiRNAs are expressed frequently in a temporally and spatially regulated manner; their expression and abundance vary widely, depending on the tissue, organ, or developmental stage of the plant (Axtell and Bartel 2005; Jones-Rhoades et al. 2006; Xie et al. 2010; Rosas-Cárdenas et al. 2014).

3 Biogenesis and Action of MiRNAs

Currently, many reviews exist that describe the biogenesis of plant miRNAs and all the proteins that have been identified that participate in this process (Chen 2009; Voinnet 2009; Axtell et al. 2011; Rogers and Chen 2013; Budak and Akpinar 2015; Reis et al. 2015). In general, miRNA biogenesis in plants includes the transcription of the *MIR* gene, processing, modification, and transference to the RISC complex (RNA-induced silencing complex), which determines the level of mature miRNAs found in the cell (Rogers and Chen 2013). Plant miRNAs are processed of MIRNA loci. In MIRNA loci of *A. thaliana*, core promoter elements including TATA box and transcription initiator (INR) elements have been identified, indicating a multicomponent mode of regulation of *MIR* transcription (Zhou et al. 2007; Xie et al. 2010; Rogers and Chen 2013). A *MIR* gene is transcribed by DNA-dependent RNA polymerase II, to form the pri-miRNA (Lee et al. 2004). Many pri-miRNAs are subjected to similar post-transcriptional modifications as mRNAs

(7-methylguanosine cap at the 5′ end and a poly(A) tail at the 3′) (Jones-Rhoades and Bartel 2004; Xie et al. 2005; Rogers and Chen 2013). The pri-miRNA is stabilized by DAWDLE (DDL, RNA-binding protein) (Yu et al. 2008) and is processed by a protein complex by combinatorial action of DCL1 (RNase III family enzyme Dicer-like 1), HYPONASTIC LEAVES (HYL1 a protein binding to double-stranded RNA) (Vazquez et al. 2004), SERRATE (SE, a C_2H_2 Zinc Finger protein) (Lobbes et al. 2006; Yang et al. 2006), and CBC (Cap-binding protein) (Kim et al. 2008). The pri-miRNA is processed into a precursor miRNA (pre-miRNA), with stem-loop-shaped secondary structures, by the endonuclease Dicer-like 1 protein (DCL1) (Kurihara and Watanabe 2004). DCL1 also carries out the subsequent cleavage of pre-miRNA to release the miRNA/miRNA* duplex (Axtell et al. 2011). The generation of the miRNA/miRNA* duplex occurs inside the nucleus. The miRNA/miRNA* is modified at the 3′ terminus by methylation by Hua Enhancer 1 (HEN1) and is then exported to the cytoplasm, possibly through HST (HASTY, a nuclear exportin) (Axtell et al. 2011). In the cytoplasm, the miRNA is loaded into the RNA-induced silencing complex (RISC) through binding with Argonaute (AGO) proteins and then binds to its target through sequence complementarity to regulate the target mRNA (Rogers and Chen 2013; Budak and Akpinar 2015).

Plant miRNAs are highly complementary to their mRNA targets; perfect or near perfect pairing of miRNA and its target site supports endonucleolytic cleavage of the mRNA by AGO (Llave et al. 2002). Different mechanisms exist how a miRNA can regulate its target mRNA such as mRNA cleavage (mostly in the middle of the duplex mRNA–miRNA between nucleotide 10 and 11 of 5′ end of the miRNA) (Reinhart et al. 2002; German et al. 2008), translational inhibition of the mRNA target, causing reduced levels of protein but not mRNA (Jones-Rhoades et al. 2006; Brodersen et al. 2008), and mRNA deadenylation or alteration of mRNA stability (Chen and Rajewsky 2007). These processes eventually result in the decrease of the product of the target gene (Meyers et al. 2006). It has been shown that several miRNAs (e.g., miR156 and miR164) can regulate the expression of multiple genes, while multiple miRNAs may control a single gene (e.g., *GhMYB2D* is targeted by miR828 and miR858) (Karlova et al. 2013; Jin et al. 2013; Guan et al. 2014). However, target genes may have a perfectly complementary site to the miRNA, they may not be targeted by the miRNA due to nonoverlapping expression patterns, because of different promoter activities (Tang 2010). However, it has also been shown that plant miRNAs can move from one tissue to another (e.g., miR165 and miR166 can move within the root) (Carlsbecker et al. 2010) and through the vascular system (e.g., miR399 can move between shoot and root) (Pant et al. 2008; Sun 2012). Moreover, it has been suggested that a very small number of copies of a specific miRNA per cell can regulate many transcripts (Voinnet 2005, 2009).

Functionally, plant miRNAs are involved in many fundamental biological processes, and their conservation across the plant kingdom suggests that these molecules have played key roles in plant developmental processes since the earliest stages of their evolution (Zhang et al. 2006; Jasinski et al. 2010; Chávez Montes et al. 2014).

4 Classification, Conservation, Divergence, and Abundance of MiRNAs in Plants

With the development of new technologies such as next-generation sequencing (NGS), and with genomic information available, it has been possible to explore the presence, conservation, abundance, and divergence of miRNAs in many plant species (e.g., Zhang et al. 2006; Cuperus et al. 2011; Pareek et al. 2011; Xu et al. 2013; Chávez Montes et al. 2014; Evers et al. 2015). This information has been increasing exponentially in recent years (Kozomara and Griffiths-Jones 2014). Based on the database version 20, 24,521 miRNA loci from 206 species are reported, which are processed to produce 30,424 mature miRNAs in plants, which have been added to miRBase (miRNA Registry Database v20, http://microrna.sanger.ac.uk/) (Kozomara and Griffiths-Jones 2014). Currently, based on the database version 21, 7057 plant miRNA loci have been identified in 73 plant species.

Based on the nucleotide sequence, the mature miRNAs are classified into families, with identical or very similar sequences grouped into the same family (Chávez Montes et al. 2014). MiRNAs are also classified in respect to their length; most plant miRNAs are 21 nt in length (Chávez Montes et al. 2014), generated by DCL1 or DCL4. DCL2 and DCL3 proteins, on the other hand, tend to generate miRNAs that are 22 and 24 nt long, respectively (Liu et al. 2009; Rogers and Chen 2013). MiRNAs processed by DCL1 are considered canonical miRNAs (Budak and Akpinar 2015). However, variants of canonical miRNAs are considered non-canonical and have been termed “isomiRs” (Lee et al. 2010; Cloonan et al. 2011; Guo et al. 2012). The names/identifiers in the database are of the form ath-miR-156 (mature sequence) and ath-MIR156 (precursor). The first three letters refer to the organism; the mature miRNA is designated miR-156. Distinct precursor sequences and genomic loci that express identical mature sequences get names of the form ath-miR-156-1 and ath-miR-156-2. Lettered suffixes denote closely related mature sequences—for example ath-miR-156a and ath-miR-156b would be expressed from precursor ath-MIR-156a and ath-MIR-156b, respectively (Ambros et al. 2003).

The database that contains the libraries of miRNAs, permitted to compare the frequency of diverse members sequenced (e.g., Nozawa et al. 2012; Chávez Montes et al. 2014; Kozomara and Griffiths-Jones 2014), and endorsed the identification and classification of miRNAs based on their conservation and divergence. MiRNAs can be classified as conserved, non-conserved, or specific (Allen et al. 2004; Cuperus et al. 2011). Conserved miRNAs are considered those miRNAs that are present in at least two different species, and these can be grouped as highly, moderately, or lowly conserved, based on the number of plants in which each family of miRNA is predicted (Zhang et al. 2006). The analysis of miRNA conservation has identified a small number of highly conserved miRNAs across the plant kingdom, from ferns to Solanaceae (Fig. 1), such as miR156, miR160, miR164, miR166, miR167, miR168, miR169, miR172, and miR396 (Fig. 1). This suggests that these conserved miRNAs already existed since the early stages of vascular land plant evolution (425 million years ago), and that these miRNAs may play the same function in different species (Zhang et al. 2006; Axtell and Bowman 2008; Chávez Montes et al. 2014). Beyond moderately or lowly conserved families, there are also miRNAs that are distributed across species with diverse lineage enrichment, such as miR536, miR1083, and miR1314, which are predominant in Gymnosperms (Fig. 1) (Chávez Montes et al. 2014). On the other hand, the non-conserved miRNAs are considered specific when detected in only one or a few phylogenetically related species (Chávez Montes et al. 2014), which suggests that these miRNA loci have emerged recently. Most miRNAs are specific or lowly conserved, indicating that most miRNA loci evolved recently (Nozawa et al. 2012; Chávez Montes et al. 2014). In most cases, conserved plant miRNAs regulate homologous targets, at identical target sites in every species in which they exist;

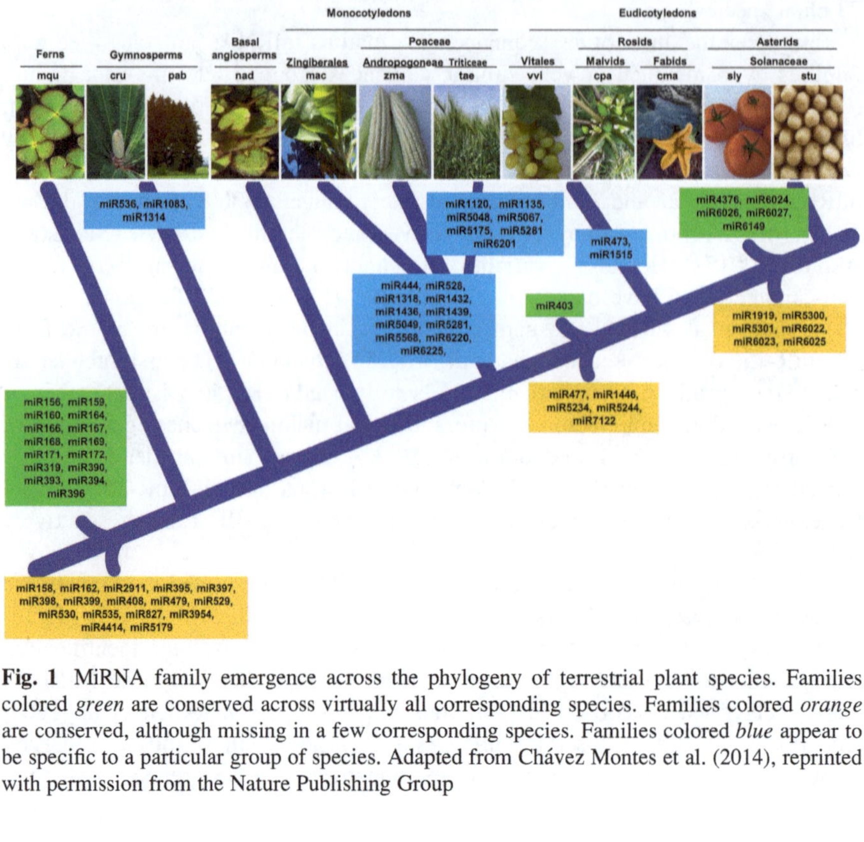

Fig. 1 MiRNA family emergence across the phylogeny of terrestrial plant species. Families colored *green* are conserved across virtually all corresponding species. Families colored *orange* are conserved, although missing in a few corresponding species. Families colored *blue* appear to be specific to a particular group of species. Adapted from Chávez Montes et al. (2014), reprinted with permission from the Nature Publishing Group

many conserved miRNAs regulate genes that encode transcription factors (Ha et al. 2008). While non-conserved miRNAs target a more heterogeneous set of gene families, which are not enriched in transcription factors, likely a reflection of the infrequent transition to an evolutionary conserved function (Kasschau et al. 2007; Axtell and Bowman 2008).

Additionally, the abundance of the same or different miRNA families varies drastically over the plant kingdom, ranging from 1 to more than 300,000 reads of the same nucleotide sequence in a different species (Chávez Montes et al. 2014). Interestingly, the abundance and conservation of miRNAs are correlated; miRNA abundance increases as the conservation of the sequence increases (Fig. 1) (Chávez Montes et al. 2014). Moreover, like the expression of transcription factors, the expression of miRNAs is thought to underlie many of the phenotypic differences within and between species (Chen and Rajewsky 2007). Although, the transcriptional regulation of miRNAs has not been extensively examined, it has been suggested that the expression of a miRNA could be affected by the spatiotemporal expression pattern of proteins that are involved in pri-miRNA processing and generating mature miRNAs (Xie et al. 2010). On the other hand, despite wide variation in the level of expression of miRNAs, studies are required to understand better what is the minimal concentration necessary to produce an effect by a miRNA (Chen and Rajewsky 2007; Voinnet 2009).

5 MiRNA Functions in Plants

The correct regulation of gene expression in response to developmental and environmental factors is essential during the plant life cycle. Many miRNAs have been identified by sequencing and bioinformatic analysis (e.g., Chávez Montes et al. 2014). The elucidation of the biogenesis and mechanism of action, the progress in sequencing technologies, the identification and validation of miRNAs, the quantification of their accumulation, the validation of miRNA–target interactions (Fig. 2), and the overexpression and silencing of miRNAs in model and non-model plants (Table 1) have facilitated the elucidation and understanding of miRNA-mediated regulatory networks and their role in plants. All these analyses have demonstrated that the regulatory role of miRNAs is very extensive. Currently, it is known that miRNAs have versatile functions and play an important role in diverse biological processes. Some miRNAs participate in diverse developmental process such as leaf development (e.g., Kidner 2010; Sarwat et al. 2013), flower development (e.g., Spanudakis and Jackson 2014; Hong and Jackson 2015), and fruit development and ripening (e.g., Rosas-Cárdenas et al. 2014; Silva et al. 2014; Gao et al. 2015; Ripoll et al. 2015), among others. For example, miR156/157, miR164, miR165/166, and miR168 are involved in gene expression regulation during most stages of plant development. MiR156 and miR390 regulate plant developmental timing; miR159, miR167, and miR171 are required for reproductive development; miR164 and miR166 control meristem formation (Table 1). Likewise, miRNAs could play

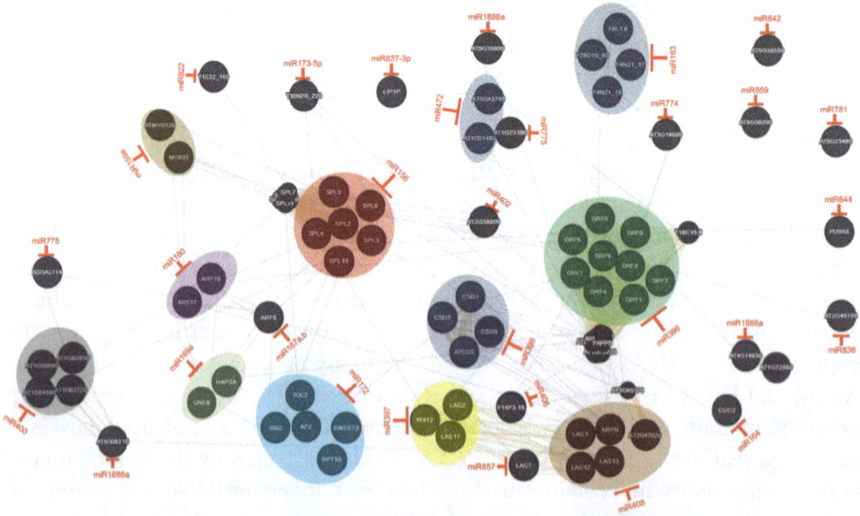

Fig. 2 Schematic representation of miRNA–target interactions. The interactions of miRNAs and target genes are illustrated based on information of Arabidopsis in miRTarBase (The experimentally validated microRNA–target interaction database, http://mirtarbase.mbc.nctu.edu.tw/) (Chou et al. 2016). The list of validated Arabidopsis target genes was downloaded from miRTarBase and imported in GeneMania (http://www.genemania.org/), to visualize interactions among miRNAs and their target genes. Finally, manual grouping of genes was performed. The *purple lines* indicate co-expression of genes, genes targeted by a miRNA are indicated in *black circles*, and those who belong to the same family and targeted by the same miRNA are grouped in a colored *big circle*

different roles in the regulation of gene expression at different times in development, such as miR156 (Xu et al. 2016). Moreover, phylogenetic analyses suggest that miRNA acquisition could play a role in phenotypic innovation (Jovelin 2013).

On the other hand, it has been demonstrated that communication exists between miRNAs, and between miRNAs and hormones (e.g., Nonogaki 2010; Rubio-Somoza and Weigel 2011; Jin et al. 2013; Curaba et al. 2014; Xue et al. 2014). For example, miR156 and miR172 exhibit a complementary temporal expression pattern; in Arabidopsis and maize, the vegetative phase change is controlled by miR156 and miR172, the first is highly abundant in seedlings and decreases during the juvenile to adult transition, while miR172 has an opposite expression pattern (Wu et al. 2010). Various studies have validated target genes, demonstrating that some miRNAs regulate several genes of the same family (Table 1). This can also be observed in Fig. 2, where we analyzed the interaction between miRNAs and targets validated in Arabidopsis based on the information available in miRTarBase (http://mirtarbase.mbc.nctu.edu.tw/) (Chou et al. 2016) (Fig. 2). Liu and Chen (2009) proposed a model that provides the intersection between miRNA pathways and phytohormone responses; different phytohormones regulate several miRNAs, and several miRNAs appear to respond to multiple plant hormones. An example of an

Table 1 Phenotypic effects caused by overexpressing of highly conserved miRNAs in different plant species

Precursor	Transgenic plant	Targets validated	Effect caused by overexpressed miRNA	References
MIR156a	Arabidopsis	SPL3, 4, 5	Extended juvenile traits and delayed flowering	Wu and Poethig (2006)
MIR156	Maize	Not-validated	Increased starch content, biomass and tiller number, prevented flowering, and generated dwarfism	Chuck et al. (2007, 2011)
Sly-MIR156a	Tomato	SGN-U345132, U313540, SlSPL2,3,6a,6b,15, CNR	More leaves and adventitious roots, generated smaller fruits, dwarfism, a "bush-like" structure, shorter plastochron, delayed flowering, and decreased fruit yield and fruit less red	Zhang et al. (2011b)
Osa-MIR156b	Switchgrass	PvSPL1,2, 3,6	Increased biomass, saccharification efficiency, and forage digestibility	Fu et al. (2012)
Stu-MIR156a	Potato	StSPL3,6,9,13, StLIGULELESS1	Reduced stomatal density in leaves and levels of tuberization, delayed flowering, increased branching, and higher number of leaves with reduced leaflets	Bhogale et al. (2014)
Ath-MIR156b	Tomato	Not-validated	Increased axillary shoots and branching index, delayed flowering, reduced number of fruits and seeds, variable number of locules, fruit-like structures emerged from the stylar end of the main fruit, abnormal growth of flower, flowers with a squashed appearance, and higher number of small and pale-green leaves	Silva et al. (2014)
Brp-MIR156a	Cabbage	BrpSPL9-2	Delayed time of leaf folding, concomitant with prolongation of the seedling and rosette stages	Wang et al. (2014b)

(continued)

Table 1 (continued)

Precursor	Transgenic plant	Targets validated	Effect caused by overexpressed miRNA	References
MIR156	Alfalfa	SPL6,12,13	Reduced internode length and stem thickness, enhanced shoot branching, increased trichome density, delayed flowering, and elevated biomass production	Aung et al. (2015)
Lju-MIR156a	Lotus	AU089181, TC70253, TC57859	Enhanced branching, delayed flowering, reduced nodulation, smaller organs, dwarf phenotype, lateral shoots developed vigorously, and emerged from almost every leaf axil	Wang and Wang (2015)
MIR156	Tobacco	NtSPL2,4,9	Delayed juvenile-to-adult transition and flowering, increased number of lateral shoots, very node developed a lateral shoot, increased tillers or branches, smaller flowers, and leaves pale green	Zhang et al. (2015)
Ath-MIR156A	Tobacco	Not-validated	Decreased stomatal density and content chlorophyll, increased stomata and cell number, and produced many adventitious roots on stems during development	Feng et al. (2016)
MIR156d	Arabidopsis	Not-validated	Increased in Rhizosphere acidification capacity	Lei et al. (2016)
MIR159a	Arabidopsis	MYB33, LEAFY	Delayed floral transition in short days, small cauline leaves, short and sterile siliques, increased anther size, darkening of anthers, and a failure to release pollen	Achard et al. (2004)
Ta-MIR159	Rice	Not-validated	Increased sensitive to heat stress	Wang et al. (2012)
Gma-MIR160	Soybean	Not-validated	Enhanced inhibition of root growth, roots hypersensitive to auxin and hyposensitive to cytokinin, and decreased nodulation	Turner et al. (2013)

(continued)

Table 1 (continued)

Precursor	Transgenic plant	Targets validated	Effect caused by overexpressed miRNA	References
MIR164b	Arabidopsis	CUC1,2, NAC1, ORE1, At5g07680, At5g61430	Organ fusion and negatively regulated aging-induced cell death and senescence	Mallory et al. (2004), Kim et al. (2009)
Ath-MIR164a,c	Arabidopsis	Not-validated	Decreased leaf serration	Nikovics et al. (2006)
Ath-MIR164b	Tomato	GOBLET	Simple leafless	Berger et al. (2009)
Ath-MIR164a	Arabidopsis	ORE1	Delayed leaf senescence	Li et al. (2013)
MIR167b	Arabidopsis	Not-validated	Increased in hypocotyls' size, sterile and smaller flowers	Ru et al. (2006)
Ath-MIR167a,b, c,d	Arabidopsis	ARF6,8	Affected ovule and anther development	Wu et al. (2006)
Ath-MIR167a	Tomato	ARF6,8a,8b	Decreased leaf size, internode length and petals, stamens, and styles, flowers did not open and arrested, and female sterility	Liu et al. (2014b)
MIR167c	Soybean roots	Not-validated	Increased in lateral root number and lateral root length, and reduced sensitivity to auxin	Wang et al. (2015a)
Ath-MIR168a	Arabidopsis	Not-validated	Displayed ABA and salt hypersensitivity and enhanced drought tolerance, delayed flowering, and decreased root and plant size	Li et al. (2012)
Sly-MIR169c	Tomato	Not-validated	Enhanced drought tolerance, reduced stomatal opening, decreased transpiration rate, decreased leaf water loss, and enhanced drought tolerance	Zhang et al. (2011a)
Ath-MIR169d	Arabidopsis	Not-validated	Accelerated flowering	Xu et al. (2014)
Hvu-MIR171a	Barley	SCL6-like	Delayed flowering. Increased number of short vegetative phytomers, defected in branching	Curaba et al. (2013)

(continued)

Table 1 (continued)

Precursor	Transgenic plant	Targets validated	Effect caused by overexpressed miRNA	References
MIR171a,b,c	Arabidopsis	Not-validated	Decreased trichome density on stems and floral organs, dark-green, and narrower leaves	Xue et al. (2014)
Stu-MIR172b	Potato	Not-validated	Promoted flowering, accelerated tuberization under moderately inductive photoperiods, and triggered tuber formation under long days	Martin et al. (2009)
Osa-MIR172a,b,c,d	Rice	Not-validated	Smaller plant and flowered early, panicle architecture significantly altered, reduced primary branches and total spikelets, developed several bract-like structures at the base of the spikelets, and most spikelets continuously generated bracts and terminated with distorted leaf-like hulls that lacked any floral organs	Lee and An (2012)
Stu-MIR172b	Potato	Not-validated	Promoted flowering, accelerated tuberization under moderately inductive photoperiods, and triggered tuber formation under long days	Martin et al. (2009)
Gma-MIR172c	Arabidopsis	Glyma01g39520	Reduced leaf water loss and increased survival rate under stress conditions, conferred tolerance to water deficit and salt stress, increased in ABA sensitivity, and accelerated flowering	Li et al. (2016)
Gma-MIR172a	Arabidopsis	Glyma03g33470	Accelerated flowering both in long and short day conditions	Wang et al. (2016)
MIR172	Apple	Not-validated	Flowers consisted of carpel tissue only, failed to produce fruit, and reduced fruit weight	Yao et al. (2016)

(continued)

Table 1 (continued)

Precursor	Transgenic plant	Targets validated	Effect caused by overexpressed miRNA	References
MIR172	Tomato	Not-validated	Carpel-only flowers, which developed into parthenocarpic fruit, alteration in flower and fruit development, flowers showing yellowish sepals and underdeveloped stamens, flowers showing fully formed ovary and incompletely developed style and stigma, seedless fruit development, produced ectopic ovaries inside the fruit	Yao et al. (2016)
MIR319/ JAW	Arabidopsis	Not-validated	Aberrant curling and serration of the leaves	Palatnik et al. (2003)
MIR319	Arabidopsis	Not-validated	Delayed in leaf senescence	Schommer et al. (2008)
Osa-MIR319a	Creeping bentgrass	Not-validated	Enhanced drought and salt tolerance associated with increased leaf wax content and water retention but reduced sodium uptake, greater leaf expansion, and thicker, increased stem diameter	Zhou et al. (2013)
MIR319		Not-validated	Enhanced cold tolerance, affected leave morphogenesis	Yang et al. (2013)
Osa-MIR319b	Rice	Not-validated	Delayed development and increased leaf width, and enhanced tolerance to cold stress	Wang et al. (2014a)
MIR390c	*Physcomitrella patens*	Not-validated	Repressed bud and leafy gametophore formation, and impeded in developmental transition	Cho et al. (2012)
Osa-MIR390	Rice	Not-validated	Enhanced sensitivity to Cd stress, and delayed seedling growth under Cd stress conditions	Ding et al. (2016)
Ath-MIR393a,b	Arabidopsis	TIR1, AFB2,3	Increased bacterial resistance	Navarro et al. (2005)

(continued)

Table 1 (continued)

Precursor	Transgenic plant	Targets validated	Effect caused by overexpressed miRNA	References
Osa-MIR393	Rice	Not-validated	Increased tillers and early flowering, and reduced tolerance to salt and drought, hyposensitivity to auxin	Xia et al. (2012)
Gma-MIR394a	Arabidopsis	Glyma08g11030	Decreased leaf water loss and enhanced drought tolerance	Ni et al. (2012)
Gma-MIR394a/b	Arabidopsis	Not-validated	Increased sensitive to salt stress and ABA, accumulated higher levels of ABA-induced hydrogen peroxide and superoxide anion radicals	Song et al. (2013)
Osa-MIR396c	Rice, Arabidopsis	Not-validated	Reduced salt and alkali stress tolerance	Gao et al. (2010)

Note: Origin of the miRNA precursor indicated by species three-letter code when known. Furthermore, the references given in the table is not an exhaustive list
MiRNAs conserved across virtually all plant species (based on Fig. 1) are shown

miRNA regulated by hormones is miR319; it is repressed by cytokinin (CK), abscisic acid (ABA), and gibberellic acid (GA) and activated by jasmonic acid (JA) and auxins (AUX) (Liu and Chen 2009; Curaba et al. 2014).

Functional studies of miRNAs have focused mainly on highly conserved and abundant miRNAs such as miR156, miR166, miR172, among others. For example, miR156, one of the most highly conserved and ubiquitous expressed miRNA in plants, is found throughout the complete plant kingdom (Chávez Montes et al. 2014), and miR156 participates in vegetative development, is highly abundant in seedlings, and decreases during the juvenile-to-adult transition. In general, miR156 regulates age-related processes (Wu and Poethig 2006). The overexpression of miR156 prolongs the juvenile phase (Table 1). However, other functions can be observed for miR156 when overexpressed in transgenic plants (Table 1), for instance, miR156 overexpression also delays flowering (Cao et al. 2015), represses nodulation, and causes morphological and developmental changes in *Lotus japonicus* (Wang et al. 2015b). Moreover, in Arabidopsis it also has been found that miR156 modulates rhizosphere acidification in response to phosphate limitation (Lei et al. 2016). In tomato, overexpression of miR156 negatively affects yield and quality of the fruit (Zhang et al. 2011b; Silva et al. 2014). More studies suggest that miRNAs could regulate characteristics of agronomic interest such as plant tolerance to abiotic stress (Zhang 2015), using miR319, which is involved in cold resistance and drought tolerance (Wang et al. 2014a; Zhou and Luo 2014), or regulation of senescence, using miR164, which is involved in senescence regulation (Kim et al. 2009, 2014).

Besides studies using mutants for miRNAs, studies using overexpression of miRNAs have demonstrated the function and targets of miRNAs. Despite the high conservation of some miRNAs in plants, each miRNA may have other targets and species-specific functions. In Table 1, several examples are given of overexpression analyses of highly conserved miRNAs in different species (Fig. 1). Several studies have shown that miRNAs may generate severe effects in one plant species but less in another. In the article by Zhang et al. (2015), the authors mentioned that the genotype-dependent response of miRNAs to abiotic stresses is not only different among plant species but also varies among genotypes of the same species. Also, the level of expression of a miRNA in overexpressing plants may drastically affect the plant phenotype (Berger et al. 2009; Fu et al. 2012). Not surprisingly, though, it suggests that the abundance of a miRNA is important for its function and, thereby, for the severity of a generated phenotype in overexpression lines. All previous mentioned evidence suggests that plant miRNAs can be manipulated to enhance characteristics of agronomic interest. Moreover, although several studies have shown that miRNAs are not conserved between plants and animals, recently it has been shown that it is possible that a plant-derived miRNA, through dietary intake, could have functions in mammals, apparently, miR159 inhibits growth of breast tumors in mice (Chin et al. 2016). It will be interesting to study if this phenomenon is more widespread.

In summary, it is of great interest to know the role of each miRNA and their effects in different plant species.

6 Pleiotropic Effects of MiRNAs

Analyses of miRNAs have shown that many miRNAs are present in many plant species, and that the level of expression of miRNAs changes with respect to the tissue type, developmental stage, etc. It has also been revealed that specific miRNAs may regulate several target genes (Fig. 2), resulting that a single miRNA can participate in different processes. Pleiotropic developmental defects are caused by mutations in the genes required for the functioning of miRNAs, supporting the crucial roles for the components of the miRNA pathway (Wu 2013). However, overexpression (Table 1) or silencing of miRNAs has also shown that some miRNAs can cause pleiotropic effects in plants. For example, the overexpression of miR156, which has at least five in Arabidopsis and eight target genes in tomato validated, generated a bushy phenotype and delayed flowering in different plant species (Table 1). Moreover, in tomato, the overexpression of miR156 caused more vigorously developed axillary shoots, and almost every leaf axil formed a new shoot, increasing the branching index. Furthermore, abnormal growth of flowers was observed, and plants had flowers with a squashed appearance. Also, it affects tomato fruit development, like reduced fruit number, fruit-like structures emerged from the stylar end of the main fruit, and a more variable number of locules and less seed is observed (Zhang et al. 2011b; Silva et al.

2014). Another example is miR164, which is encoded by three loci in Arabidopsis (MIR164A, B, C) (Mallory et al. 2004; Baker et al. 2005; Sieber et al. 2007) and regulates six target genes in Arabidopsis and tomato. The effect observed by overexpression of this miRNA is organ fusion. Based on expression studies in prickly pear cactus, it has been suggested that miR164 may play different roles during fruit development (Rosas-Cárdenas et al. 2014). All these studies suggest that the number and abundance of targets and spatiotemporal expression of both miRNAs and targets can contribute to generate pleiotropic effects when miRNAs are overexpressed. Moreover, we could suggest that the expression level of a miRNA may also contribute to pleiotropic effects.

7 Conclusions and Future Prospects

Currently, sequencing efforts have identified many miRNAs present in many species spread over the plant kingdom, allowing understanding miRNA diversity, conservation, and knowing their abundance. Moreover, the integration of strategies and methods allows to analyze the function of these molecules in both model and non-model plants, many of them of agronomic interest (de Rosas-Cárdenas et al. 2015). However, still many studies will be necessary to understand the role of all existing miRNAs in plants. Nevertheless, the potential to use miRNAs to improve desirable traits in crop plants already exists.

Acknowledgements We thank the Mexican National Council of Science and Technology (CONACyT) for the grant CB2012-177739 to SdF and grant CB2013-221522 to FFRC. We also thank the Instituto Politécnico Nacional for the grant SIP-20160215 to FFRC.

References

Achard P, Herr A, Baulcombe DC et al (2004) Modulation of floral development by a gibberellin-regulated microRNA. Development 131:3357–3365

Allen E, Xie Z, Gustafson AM et al (2004) Evolution of microRNA genes by inverted duplication of target gene sequences in *Arabidopsis thaliana*. Nat Genet 36:1282–1290

Ambros V, Bartel B, Bartel DP et al (2003) A uniform system for microRNA annotation. RNA 9:277–279

Aung B, Gruber MY, Amyot L et al (2015) MicroRNA156 as a promising tool for alfalfa improvement. Plant Biotechnol J 156:1–12

Axtell MJ, Bartel DP (2005) Antiquity of microRNAs and their targets in land plants. Plant Cell 17:1658–1673

Axtell MJ, Bowman JL (2008) Evolution of plant microRNAs and their targets. Trends Plant Sci 13:343–349

Axtell MJ, Snyder JA, Bartel DP (2007) Common functions for diverse small RNAs of land plants. Plant Cell 19:1750–1769

Axtell MJ, Westholm JO, Lai EC (2011) Vive la différence: biogenesis and evolution of microRNAs in plants and animals. Genome Biol 12:1–13

Baker CC, Sieber P, Wellmer F et al (2005) The *early extra petals1* mutant uncovers a role for microRNA miR164c in regulating petal number in Arabidopsis. Curr Biol 15:303–315
Baldrich P, Hsing Y-IC, San Segundo B (2016) Genome-wide analysis of polycistronic micrornas in cultivated and wild rice. Genome Biol Evol 8:1104–1114
Berger Y, Harpaz-Saad S, Brand A et al (2009) The NAC-domain transcription factor GOBLET specifies leaflet boundaries in compound tomato leaves. Development 136:823–832
Bhogale S, Mahajan AS, Natarajan B et al (2014) MicroRNA156: a potential graft-transmissible microRNA that modulates plant architecture and tuberization in Solanum tuberosum ssp. andigena. Plant Physiol 164:1011–1027
Brodersen P, Sakvarelidze-Achard L, Bruun-Rasmussen M et al (2008) Widespread translational inhibition by plant miRNAs and siRNAs. Science 320:1185–1190
Budak H, Akpinar BA (2015) Plant miRNAs: biogenesis, organization and origins. Funct Integr Genomics 15:523–531
Cao D, Li Y, Wang J et al (2015) GmmiR156b overexpression delays flowering time in soybean. Plant Mol Biol 89:353–363
Carlsbecker A, Ji-Young L, Roberts C et al (2010) Cell signalling by microRNA165/6 directs gene dose-dependent root cell fate. Nature 465:316–321
Cech TR, Steitz JA (2014) The noncoding RNA revolution – trashing old rules to forge new ones. Cell 157:77–94
Chávez Montes RA, de Rosas-Cárdenas FF, De Paoli E et al (2014) Sample sequencing of vascular plants demonstrates widespread conservation and divergence of microRNAs. Nat Commun 5:3722
Chen X (2009) Small RNAs and their roles in plant development. Annu Rev Cell Dev Biol 25:21–44
Chen K, Rajewsky N (2007) The evolution of gene regulation by transcription factors and microRNAs. Nat Rev Genet 8:93–103
Chen M, Meng Y, Gu H, Chen D (2010) Functional characterization of plant small RNAs based on next-generation sequencing data. Comput Biol Chem 34:308–312
Chin AR, Fong MY, Somlo G et al (2016) Cross-kingdom inhibition of breast cancer growth by plant miR159. Cell Res 26:1–12
Cho SH, Coruh C, Axtell MJ (2012) miR156 and miR390 regulate tasiRNA accumulation and developmental timing in Physcomitrella patens. Plant Cell 24:4837–4849
Chou CH, Chang NW, Shrestha S et al (2016) miRTarBase 2016: updates to the experimentally validated miRNA-target interactions database. Nucleic Acids Res 44:D239–D247
Chuck G, Cigan AM, Saeteurn K et al (2007) The heterochronic maize mutant Corngrass1 results from overexpression of a tandem microRNA. Nat Genet 39:544–549
Chuck GS, Tobias C, Sun L et al (2011) Overexpression of the maize Corngrass1 microRNA prevents flowering, improves digestibility, and increases starch content of switchgrass. Proc Natl Acad Sci USA 108:17550–17555
Cloonan N, Wani S, Xu Q et al (2011) MicroRNAs and their isomiRs function cooperatively to target common biological pathways. Genome Biol 12:R126
Cuperus JT, Fahlgren N, Carrington JC (2011) Evolution and functional diversification of MIRNA genes. Plant Cell 23:431–442
Curaba J, Talbot M, Li Z et al (2013) Over-expression of microRNA171 affects phase transitions and floral meristem determinancy in barley. BMC Plant Biol 13:6
Curaba J, Singh MB, Bhalla PL (2014) MiRNAs in the crosstalk between phytohormone signalling pathways. J Exp Bot 65:1425–1438
de Rosas-Cárdenas FF, Escobar-Guzmán R, Cruz-Hernández A et al (2015) An efficient method for miRNA detection and localization in crop plants. Front Plant Sci 6:99
Ding Y, Ye Y, Jiang Z et al (2016) MicroRNA390 is involved in cadmium tolerance and accumulation in rice. Front Plant Sci 7:1–10
Evers M, Huttner M, Dueck A et al (2015) miRA: adaptable novel miRNA identification in plants using small RNA sequencing data. BMC Bioinformatics 16:370

Feng S, Xu Y, Guo C et al (2016) Modulation of miR156 to identify traits associated with vegetative phase change in tobacco (*Nicotiana tabacum*). J Exp Bot 67:1493–1504

Fu C, Sunkar R, Zhou C et al (2012) Overexpression of miR156 in switchgrass (*Panicum virgatum* L.) results in various morphological alterations and leads to improved biomass production. Plant Biotechnol J 10:443–452

Gao P, Bai X, Yang L et al (2010) Overexpression of osa-MIR396c decreases salt and alkali stress tolerance. Planta 231:991–1001

Gao C, Ju Z, Cao D et al (2015) MicroRNA profiling analysis throughout tomato fruit development and ripening reveals potential regulatory role of RIN on microRNAs accumulation. Plant Biotechnol J 13:370–382

German MA, Pillay M, Jeong D-H et al (2008) Global identification of microRNA-target RNA pairs by parallel analysis of RNA ends. Nat Biotechnol 26:941–946

Griffiths-Jones S, Saini HK, van Dongen S, Enright AJ (2008) miRBase: tools for microRNA genomics. Nucleic Acids Res 36:D154–D158

Guan X, Pang M, Nah G et al (2014) miR828 and miR858 regulate homoeologous MYB2 gene functions in Arabidopsis trichome and cotton fibre development. Nat Commun 5:3050

Guo L, Li H, Liang T et al (2012) Consistent isomiR expression patterns and 3' addition events in miRNA gene clusters and families implicate functional and evolutionary relationships. Mol Biol Rep 39:6699–6706

Ha M, Pang M, Agarwal V, Chen ZJ (2008) Interspecies regulation of microRNAs and their targets. Biochim Biophys Acta 1779:735–742

Holoch D, Moazed D (2015) RNA-mediated epigenetic regulation of gene expression. Nat Rev Genet 16:71–84

Hong Y, Jackson S (2015) Floral induction and flower formation-the role and potential applications of miRNAs. Plant Biotechnol J 13:282–292

Jasinski S, Vialette-Guiraud ACM, Scutt CP (2010) The evolutionary-developmental analysis of plant microRNAs. Philos Trans R Soc Lond Ser B Biol Sci 365:469–476

Jin D, Wang Y, Zhao Y, Chen M (2013) MicroRNAs and their cross-talks in plant development. J Genet Genomics 40:161–170

Jones-Rhoades MW, Bartel DP (2004) Computational identification of plant microRNAs and their targets, including a stress-induced miRNA. Mol Cell 14:787–799

Jones-Rhoades MW, Bartel DP, Bartel B (2006) MicroRNAs and their regulatory roles in plants. Annu Rev Plant Biol 57:19–53

Joshi T, Yan Z, Libault M et al (2010) Prediction of novel miRNAs and associated target genes in Glycine max. BMC Bioinformatics 11(Suppl 1):S14

Jovelin R (2013) Pleiotropic constraints, expression level, and the evolution of miRNA sequences. J Mol Evol 77:206–220

Källman T, Chen J, Gyllenstrand N, Lagercrantz U (2013) A significant fraction of 21-nucleotide small RNA originates from phased degradation of resistance genes in several perennial species. Plant Physiol 162:741–754

Karlova R, van Haarst JC, Maliepaard C et al (2013) Identification of microRNA targets in tomato fruit development using high-throughput sequencing and degradome analysis. J Exp Bot 64:1863–1878

Kasschau KD, Fahlgren N, Chapman EJ et al (2007) Genome-wide profiling and analysis of Arabidopsis siRNAs. PLoS Biol 5:e57

Khraiwesh B, Arif MA, Seumel GI et al (2010) Transcriptional control of gene expression by microRNAs. Cell 140:111–122

Kidner CA (2010) The many roles of small RNAs in leaf development. J Genet Genomics 37:13–21

Kim S, Yang J-Y, Xu J et al (2008) Two cap-binding proteins CBP20 and CBP80 are involved in processing primary MicroRNAs. Plant Cell Physiol 49:1634–1644

Kim J, Woo H, Kim J et al (2009) Trifurcate feed-forward regulation of age-dependent cell death involving miR164 in Arabidopsis. Science 323:1053–1057

Kim HJ, Hong SH, Kim YW et al (2014) Gene regulatory cascade of senescence-associated NAC transcription factors activated by ETHYLENE-INSENSITIVE2- mediated leaf senescence signalling in Arabidopsis. J Exp Bot 65:4023–4036

Kozomara A, Griffiths-Jones S (2014) MiRBase: annotating high confidence microRNAs using deep sequencing data. Nucleic Acids Res 42:68–73

Kurihara Y, Watanabe Y (2004) Arabidopsis micro-RNA biogenesis through Dicer-like 1 protein functions. Proc Natl Acad Sci USA 101:12753–12758

Lee DY, An G (2012) Two AP2 family genes, SUPERNUMERARY BRACT (SNB) and OsINDETERMINATE SPIKELET 1 (OsIDS1), synergistically control inflorescence architecture and floral meristem establishment in rice. Plant J 69:445–461

Lee Y, Kim M, Han J et al (2004) MicroRNA genes are transcribed by RNA polymerase II. EMBO J 23:4051–4060

Lee LW, Zhang S, Etheridge A et al (2010) Complexity of the microRNA repertoire revealed by next-generation sequencing. RNA 16:2170–2180

Lei KJ, Lin YM, An GY (2016) miR156 modulates rhizosphere acidification in response to phosphate limitation in Arabidopsis. J Plant Res 129:275–284

Li W, Cui X, Meng Z et al (2012) Transcriptional regulation of Arabidopsis MIR168a and ARGONAUTE1 homeostasis in abscisic acid and abiotic stress responses. Plant Physiol 158:1279–1292

Li Z, Peng J, Wen X, Guo H (2013) ETHYLENE-INSENSITIVE3 is a senescence-associated gene that accelerates age-dependent leaf senescence by directly repressing mir164 transcription in Arabidopsis. Plant Cell 25:3311–3328

Li W, Wang T, Zhang Y, Li Y (2016) Overexpression of soybean miR172c confers tolerance to water deficit and salt stress, but increases ABA sensitivity in transgenic *Arabidopsis thaliana*. J Exp Bot 67:175–194

Liu Q, Chen Y-Q (2009) Insights into the mechanism of plant development: interactions of miRNAs pathway with phytohormone response. Biochem Biophys Res Commun 384:1–5

Liu Q, Feng Y, Zhu Z (2009) Dicer-like (DCL) proteins in plants. Funct Integr Genomics 9:277–286

Liu H, Qin C, Chen Z et al (2014a) Identification of miRNAs and their target genes in developing maize ears by combined small RNA and degradome sequencing. BMC Genomics 15:25

Liu N, Wu S, Van Houten J et al (2014b) Down-regulation of AUXIN RESPONSE FACTORS 6 and 8 by microRNA 167 leads to floral development defects and female sterility in tomato. J Exp Bot 65:2507–2520

Liu YX, Wang M, Wang XJ (2014c) Endogenous small RNA clusters in plants. Genomics Proteomics Bioinformatics 12:64–71

Llave C, Kasschau KD, Rector MA, Carrington JC (2002) Endogenous and silencing-associated small RNAs in plants. Plant Cell 14:1605–1619

Lobbes D, Rallapalli G, Schmidt DD et al (2006) SERRATE: a new player on the plant microRNA scene. EMBO Rep 7:1052–1058

Lu C, Tej SS, Luo S et al (2005) Elucidation of the small RNA component of the transcriptome. Science 309:1567–1569

Maher C, Stein L, Ware D (2006) Evolution of Arabidopsis microRNA families through duplication events. Genome Res 16:510–519

Mallory AC, Bouché N (2008) MicroRNA-directed regulation: to cleave or not to cleave. Trends Plant Sci 13:359–367

Mallory AC, Dugas DV, Bartel DP, Bartel B (2004) MicroRNA regulation of NAC-domain targets is required for proper formation and separation of adjacent embryonic, vegetative, and floral organs. Curr Biol 14:1035–1046

Martin A, Adam H, Díaz-Mendoza M et al (2009) Graft-transmissible induction of potato tuberization by the microRNA miR172. Development 136:2873–2881

Merchan F, Boualem A, Crespi M et al (2009) Plant polycistronic precursors containing non-homologous microRNAs target transcripts encoding functionally related proteins. Genome Biol 10:R136
Meyers BC, Souret FF, Lu C et al (2006) Sweating the small stuff: microRNA discovery in plants. Curr Opin Biotechnol 17:139–146
Navarro L, Dunoyer P, Jay F et al (2005) A plant miRNA contributes to antibacterial resistance by repressing auxin signaling. Science 312:436–439
Ni Z, Hu Z, Jiang Q et al (2012) Overexpression of gma-MIR394a confers tolerance to drought in transgenic *Arabidopsis thaliana*. Biochem Biophys Res Commun 427:330–335
Nikovics K, Blein T, Peaucelle A et al (2006) The balance between the MIR164A and CUC2 genes controls leaf margin serration in Arabidopsis. Plant Cell 18:2929–2945
Nonogaki H (2010) MicroRNA gene regulation cascades during early stages of plant development. Plant Cell Physiol 51:1840–1846
Nozawa M, Miura S, Nei M (2012) Origins and evolution of microRNA genes in plant species. Genome Biol Evol 4:230–239
Palatnik JF, Allen E, Wu X et al (2003) Control of leaf morphogenesis by microRNAs. Nature 425:257–263
Pant BD, Buhtz A, Kehr J et al (2008) MicroRNA399 is a long-distance signal for the regulation of plant phosphate homeostasis. Plant J 53:731–738
Pareek CS, Smoczynski R, Tretyn A (2011) Sequencing technologies and genome sequencing. J Appl Genet 52:413–435
Piriyapongsa J, Jordan IK (2008) Dual coding of siRNAs and miRNAs by plant transposable elements. RNA 14:814–821
Reinhart BJ, Weinstein EG, Rhoades MW et al (2002) MicroRNAs in plants. Genes Dev 16:1616–1626
Reis RS, Eamens AL, Waterhouse PM (2015) Missing pieces in the puzzle of plant microRNAs. Trends Plant Sci 20:721–728
Ripoll JJ, Bailey LJ, Mai QA et al (2015) MicroRNA regulation of fruit growth. Nat Plants 1:15036
Rogers K, Chen X (2013) Biogenesis, turnover, and mode of action of plant microRNAs. Plant Cell 25:2383–2399
Rosas-Cárdenas FDF, Caballero-Pérez J, Gutiérrez-Ramos X et al (2014) miRNA expression during prickly pear cactus fruit development. Planta 241:435–448
Ru P, Xu L, Ma H et al (2006) Plant fertility defects induced by the enhanced expression of microRNA167. Cell Res 16:457–465
Rubio-Somoza I, Weigel D (2011) MicroRNA networks and developmental plasticity in plants. Trends Plant Sci 16:258–264
Sarwat M, Naqvi AR, Ahmad P et al (2013) Phytohormones and microRNAs as sensors and regulators of leaf senescence: assigning macro roles to small molecules. Biotechnol Adv 31:1153–1171
Schommer C, Palatnik JF, Aggarwal P et al (2008) Control of jasmonate biosynthesis and senescence by miR319 targets. PLoS Biol 6:e230
Sieber P, Wellmer F, Gheyselinck J et al (2007) Redundancy and specialization among plant microRNAs: role of the MIR164 family in developmental robustness. Development 134:1051–1060
Silva GFFE, Silva EM, Da Silva AM et al (2014) MicroRNA156-targeted SPL/SBP box transcription factors regulate tomato ovary and fruit development. Plant J 78:604–618
Song JB, Gao S, Sun D et al (2013) miR394 and LCR are involved in Arabidopsis salt and drought stress responses in an abscisic acid-dependent manner. BMC Plant Biol 13:210
Spanudakis E, Jackson S (2014) The role of microRNAs in the control of flowering time. J Exp Bot 65:365–380
Sun G (2012) MicroRNAs and their diverse functions in plants. Plant Mol Biol 80:17–36

Sunkar R, Chinnusamy V, Zhu J, Zhu J-K (2007) Small RNAs as big players in plant abiotic stress responses and nutrient deprivation. Trends Plant Sci 12:301–309

Tang G (2010) Plant microRNAs: an insight into their gene structures and evolution. Semin Cell Dev Biol 21:782–789

Turner M, Nizampatnam NR, Baron M et al (2013) Ectopic expression of miR160 results in auxin hypersensitivity, cytokinin hyposensitivity, and inhibition of symbiotic nodule development in soybean. Plant Physiol 162:2042–2055

Vazquez F, Gasciolli V, Crété P et al (2004) The nuclear dsRNA binding protein HYL1 is required for microRNA accumulation and plant development, but not posttranscriptional transgene silencing. Curr Biol 14:346–351

Voinnet O (2005) Non-cell autonomous RNA silencing. FEBS Lett 579:5858–5871

Voinnet O (2009) Origin, biogenesis, and activity of plant microRNAs. Cell 136:669–687

Wang H, Wang H (2015) The miR156/SPL module, a regulatory hub and versatile toolbox, gears up crops for enhanced agronomic traits. Mol Plant 8:677–688

Wang S, Zhu QH, Guo X et al (2007) Molecular evolution and selection of a gene encoding two tandem microRNAs in rice. FEBS Lett 581:4789–4793

Wang Y, Sun F, Cao H et al (2012) TamiR159 directed wheat TaGAMYB cleavage and its involvement in anther development and heat response. PLoS One 7:e48445

Wang ST, Sun XL, Hoshino Y et al (2014a) MicroRNA319 positively regulates cold tolerance by targeting OsPCF6 and OsTCP21 in rice (*Oryza sativa* L.) PLoS One 9:1–12

Wang Y, Wu F, Bai J et al (2014b) BrpSPL9 (Brassica rapa ssp. pekinensis SPL9) controls the earliness of heading time in Chinese cabbage. Plant Biotechnol J 12:312–321

Wang Y, Li K, Chen L et al (2015a) MicroRNA167-directed regulation of the auxin response factors GmARF8a and GmARF8b is required for soybean nodulation and lateral root development. Plant Physiol 168:984–999

Wang Y, Wang Z, Amyot L et al (2015b) Ectopic expression of miR156 represses nodulation and causes morphological and developmental changes in *Lotus japonicus*. Mol Gen Genomics 290:471–484

Wang T, Sun M-Y, Wang X-S et al (2016) Over-expression of GmGIa-regulated soybean miR172a confers early flowering in transgenic *Arabidopsis thaliana*. Int J Mol Sci 17:645

Wu G (2013) Plant microRNAs and development. J Genet Genomics 40:217–230

Wu G, Poethig RS (2006) Temporal regulation of shoot development in *Arabidopsis thaliana* by miR156 and its target SPL3. Development 133:3539–3547

Wu M-F, Tian Q, Reed JW (2006) Arabidopsis microRNA167 controls patterns of ARF6 and ARF8 expression, and regulates both female and male reproduction. Development 133:4211–4218

Wu G, Park MY, Conway SR et al (2010) The sequential action of miR156 and miR172 regulates developmental timing in Arabidopsis. Cell 138:750–759

Xia K, Wang R, Ou X et al (2012) OsTIR1 and OsAFB2 downregulation via OsmiR393 overexpression leads to more tillers, early flowering and less tolerance to salt and drought in rice. PLoS One 7:1–10

Xie Z, Allen E, Fahlgren N et al (2005) Expression of Arabidopsis MIRNA genes. Plant Physiol 138:2145–2154

Xie Z, Khanna K, Ruan S (2010) Expression of microRNAs and its regulation in plants. Semin Cell Dev Biol 21:790–797

Xu X, Yin L, Ying Q et al (2013) High-throughput sequencing and degradome analysis identify mirnas and their targets involved in fruit senescence of Fragaria ananassa. PLoS One 8:e70959

Xu MY, Zhang L, Li WW et al (2014) Stress-induced early flowering is mediated by miR169 in *Arabidopsis thaliana*. J Exp Bot 65:89–101

Xu M, Hu T, Zhao J et al (2016) Developmental functions of miR156-regulated squamosa promoter binding protein-like (SPL) genes in *Arabidopsis thaliana*. PLoS Genet 12:e1006263

Xue XY, Zhao B, Chao LM et al (2014) Interaction between two timing micrornas controls trichome distribution in Arabidopsis. PLoS Genet 10:e1004266

Yang L, Liu Z, Lu F et al (2006) SERRATE is a novel nuclear regulator in primary microRNA processing in Arabidopsis. Plant J 47:841–850

Yang C, Li D, Mao D et al (2013) Overexpression of microRNA319 impacts leaf morphogenesis and leads to enhanced cold tolerance in rice (*Oryza sativa* L.) Plant Cell Environ 36:2207–2218

Yao JL, Tomes S, Xu J, Gleave AP (2016) How microRNA172 affects fruit growth in different species is dependent on fruit type. Plant Signal Behav 11:e1156833

Yu B, Bi L, Zheng B et al (2008) The FHA domain proteins DAWDLE in Arabidopsis and SNIP1 in humans act in small RNA biogenesis. Proc Natl Acad Sci USA 105:10073–10078

Zhang B (2015) MicroRNA: a new target for improving plant tolerance to abiotic stress. J Exp Bot 66:1749–1761

Zhang B, Pan X, Cannon CH et al (2006) Conservation and divergence of plant microRNA genes. Plant J 46:243–259

Zhang L, Chia JM, Kumari S et al (2009) A genome-wide characterization of microRNA genes in maize. PLoS Genet 5:e1000716

Zhang X, Zou Z, Gong P et al (2011a) Over-expression of microRNA169 confers enhanced drought tolerance to tomato. Biotechnol Lett 33:403–409

Zhang X, Zou Z, Zhang J et al (2011b) Over-expression of sly-miR156a in tomato results in multiple vegetative and reproductive trait alterations and partial phenocopy of the sft mutant. FEBS Lett 585:435–439

Zhang TQ, Wang JW, Zhou CM (2015) The role of miR156 in developmental transitions in *Nicotiana tabacum*. Sci China Life Sci 58:253–260

Zhou M, Luo H (2014) Role of microRNA319 in creeping bentgrass salinity and drought stress response. Plant Signal Behav 9:e28700

Zhou X, Ruan J, Wang G, Zhang W (2007) Characterization and identification of microRNA core promoters in four model species. PLoS Comput Biol 3:0412–0423

Zhou M, Li D, Li Z et al (2013) Constitutive expression of a miR319 gene alters plant development and enhances salt and drought tolerance in transgenic creeping bentgrass. Plant Physiol 3 (161):1375–1391

Zhu Q-H, Spriggs A, Matthew L et al (2008) A diverse set of microRNAs and microRNA-like small RNAs in developing rice grains. Genome Res 18:1456–1465

Zhu C, Ding Y, Liu H (2011) MiR398 and plant stress responses. Physiol Plant 143:1–9

The Role of MiRNAs in Auxin Signaling and Regulation During Plant Development

Clelia De-la-Peña, Geovanny I. Nic-Can, Johny Avilez-Montalvo, José E. Cetz-Chel, and Víctor M. Loyola-Vargas

Abstract Auxins are involved in almost every aspect of plant physiology. For instance, auxins play a central role in the differentiation process during the development of plants. Furthermore, the homeostasis of auxins involves biosynthesis and degradation as well as their conjugation with amino acids and carbohydrates, and the hydrolysis of some of these conjugates liberates indole-3-acetic acid (IAA). The balance in the IAA concentration triggers its own signal transduction pathway and produces a molecular and biochemical response. This response begins with the sensing of the IAA concentration through the construction of a co-receptor complex that includes an F-box protein from the TRANSPORT INHIBITOR RESPONSE 1 (TIR1)/AUXIN SIGNALING F-BOX PROTEIN (AFB) family and a member of the AUXIN/IAA-INDUCIBLE (AUX/IAA) family of transcriptional repressors. This complex allows the expression of auxin response genes. Most of the auxin-regulated processes are tightly regulated. Several differentially expressed miRNAs, which alter the auxin response, have been identified in *Arabidopsis thaliana* somatic embryogenesis development. Also, during the stress response in soybean roots, auxin-responsive *cis*-elements in the promoters of many salt-responsive miRNAs have been found. These findings suggest that miRNAs may be regulated by auxins. In this chapter, we analyze developing research related to the interaction between auxins and miRNAs.

Keywords Auxins • Development • miRNAs

C. De-la-Peña
Unidad Biotecnología, Centro de Investigación Científica de Yucatán, Calle 43 No. 130 × 32 y 34 Col. Chuburná de Hidalgo, 97205 Mérida, Yucatán, México

G.I. Nic-Can
CONACYT Research Fellow-Campus de Ciencias Exactas e Ingeniería, Universidad Autónoma de Yucatán, Periférico Norte. Km 33.5, Tablaje catastral 13615 Col. Chuburná de Hidalgo Inn, 97203 Mérida, Yucatán, México

J. Avilez-Montalvo • J.E. Cetz-Chel • V.M. Loyola-Vargas (✉)
Unidad de Bioquímica y Biología Molecular de Plantas, Centro de Investigación Científica de Yucatán, Calle 43 No. 130 × 32 y 34 Col. Chuburná de Hidalgo, 97205 Mérida, Yucatán, México
e-mail: vmloyola@cicy.mx

N. Rajewsky et al. (eds.), *Plant Epigenetics*, RNA Technologies,
DOI 10.1007/978-3-319-55520-1_2

Contents

1 Introduction

1.1 Auxins

As sessile organisms, plants have sophisticated development mechanisms to overcome challenges of growth in a hostile environment. Central to this operation are the substances known as plant growth regulators (PGRs), which can trigger multiple responses to influence specific physiological responses. PGRs include cytokinins, ethylene, abscisic acid, gibberellins, auxins, and others and are involved in all phases of plant development, from the response and adaption to recurring biotic and abiotic stresses to seed-to-seed signaling (Weijers et al. 2006; Zhu and Lee 2015). Among PGRs, auxins were the first to be isolated and are probably the most studied. Early studies on the phototropic curvature of coleoptiles by Charles Darwin and his son Francis suggested the existence of a mobile signal that controls cell elongation; this signal was later named auxin and identified as indole-3-acetic acid (IAA) [reviewed in Su et al. (2015)]. Subsequent studies have also demonstrated that auxins are involved in many biological processes, such as embryogenesis, organogenesis, vascular tissue differentiation, hypocotyl and root elongation and apical dominance, among other important processes during the development of plants (Berleth et al. 2000; Leyser 2005; Woodward and Bartel 2005; Benjamins and Scheres 2008; Robert and Friml 2009; Zhao 2010; Ayil-Gutiérrez et al. 2013; Pacurar et al. 2014).

Further studies determined that IAA is mainly synthesized in young leaves, cotyledons, expanding leaves, and root tissues (Ljung et al. 2001; Ljung 2013) by two major pathways: the Trp-dependent and the Trp-independent (Zazimalová and Napier 2003; Woodward and Bartel 2005; Normanly 2010). The IAA biosynthesis pathways are highly conserved throughout the plant kingdom, and four Trp-dependent pathways have been described, including 3-acetaldoximine (IAOx), indole-3-pyruvic acid (IPA), indole-3-acetamide (IAM), and tryptamine (TAM). The conversion of Trp to indole-3-pyruvate and subsequently into IAA by the TRYPTOPHAN AMINOTRANSFERASE and the YUCCA (YUC) family of flavin monooxygenases, respectively, is the predominant Trp-dependent pathway (Zhao et al. 2001; Stepanova et al. 2008; Tao et al. 2008; Mano and Nemoto 2012). In contrast, evidence exists for a Trp-independent pathway that involves the conversion of indole-3-glycerol phosphate to IAA (Strader and Bartel 2011; Nonhebel 2015; Wang et al. 2015).

Once auxins are synthesized, they have to be differentially distributed throughout the different tissues of the plant. Auxins can move through the phloem or via the cell-to-cell transport system (Habets and Offringa 2014). This distribution is required for local auxin accumulation and the generation of gradients during crucial stages of growth and development (Geisler et al. 2014; Soriano et al. 2014). It has been shown that IAA can move through the plasma membrane by passive diffusion in its uncharged form (IAAH), whereas in its anionic form (IAA-) it requires specific auxin efflux and influx carriers (Petrášek and Friml 2009). Efflux carriers such as AUX1 and its related LIKE AUX1 (AUX/LAX) proteins seem to be crucial, especially when the auxin efflux is high, although PIN-FORMED (PIN) proteins, which also act as efflux carriers, contribute to polar distribution and high directionally of auxins according to growth responses (Friml 2003; Petrášek and Friml 2009). Another family group of proteins known as ATP-binding cassette subgroup B can act as efflux and influx carriers. In this way, development in plant cells largely depends on auxin accumulation in the right place at the right moment (Friml et al. 2003). IAA levels are tightly regulated through its conjugation with amino acids or carbohydrates, and the hydrolysis of some of these conjugates liberates IAA according to internal cellular requirements. Therefore, IAA perception inside the cell must be able to change transcriptional events before it produces a biological response.

The regulation of auxin response genes are principally mediated by two families of transcription factors, the auxin response factor (ARF) and the auxin/indole-3-acetic acid (AUX/IAA) (Guilfoyle 2015). At low levels of auxin, the AUX/IAA proteins interact with ARF family proteins, which are targeted to auxin response promoter elements (AuxRES) in several auxin-regulated genes. Thus, to repress the ARF function, these are kept away from their target promoter by AUX/IAA proteins. Additionally, the effects of auxin on transcription also involve changes in chromatin structure and histone modifications. For instance, TOPLESS (TPL) and TPL-related proteins (TRP), other transcriptional repressors, downregulate gene expression through diverse transcriptional regulators. It has been shown that TPL/TRP can recruit histone deacetylases to promote heterochromatin generation, and thus block the ARF function [reviewed in Perrot-Rechenmann (2014) and Retzer et al. (2014)].

In contrast, when the auxin level increases, the auxin binds to TIR1, a component of the SCF (SKP1, CUL1, and F-box protein) ubiquitin ligase complex. This complex promotes AUX/IAA degradation using the 26S proteasome (Gray et al. 1999; Jing et al. 2015). Control of AUX/IAA protein degradation is important for activating the plant cells' response to auxin. Due to AUX/IAA forming heterodimers with ARF, auxin-induced degradation of AUX/IAA reactivates the

ARF protein's function, and thus activates the transcription of primary genes (Weijers 2015; Dinesh et al. 2016).

Independently of the TIR1 auxin receptor, it is also proposed that the perception of extracellular auxin is mediated by auxin-binding protein 1 (ABP1), through interaction with a transmembrane kinase (TMK). ABP1 and TMK form a cell surface auxin perception complex, where the auxin that activates the Rho-like guanosine triphosphatases (ROP) signaling pathway regulates a plethora of plasma membrane or cytoplasmic responses more than it regulates transcriptional activity (Xu et al. 2014).

Taken together, auxin controls most, if not all, aspects of plant growth and development. Although key genes involved in biosynthesis, transport, response, and degradation have been identified, it is still not known how the different levels of this molecular signal are individually perceived by plant cells to generate a molecular or physiological response.

1.2 MiRNAs

Ribonucleic acid (RNA) is a wonderful macromolecule, performing a set of essential functions in living organisms. Most genes use mRNA as an intermediate for protein production. However, there are genes whose final products are RNA that do not code for protein. Such non-protein-coding RNAs (ncRNA) range from the transfer and ribosomal RNAs, which are involved in protein-synthesizing machinery, to the more recently discovered regulatory small RNAs (sRNAs). There are several kinds of sRNAs that can be classified into three categories: microRNAs (miRNAs), small interfering RNAs (siRNAs), and Piwi-interacting RNAs (piRNAs). There are also new types of sRNAs under investigation, such as small temporal RNAs (stRNAs), small modular RNAs (smRNAs), tiny non-coding RNAs (tncRNAs), trans-acting siRNAs (Ta-siRNAs), repeat-associated siRNAs (Ra-siRNAs), and natural-antisense transcript-derived siRNAs (Nat-siRNAs) (Boopathi 2015).

Mature plant miRNAs are small (20–24 bp in length), and they are produced from longer RNA precursors, which contain a stem loop or hairpin structure with imperfect base pairing in the stem region. MiRNAs are able to regulate gene expression at the post-transcriptional level through specific base pairing with cognate target mRNAs. The recognition of the miRNA by its targeted mRNA produces a cleavage, translation inhibition, or both in the mRNA. The vascular tissue of the plants is the medium through which the miRNAs can move from one tissue to another (Sun 2012).

Since the first reported miRNAs in plants (Llave et al. 2002; Mette et al. 2002; Park et al. 2002; Reinhart et al. 2002), the techniques of massive sequencing and the improvement of the bioinformatics programs have increased the numbers of entries in the microRNA database (http://www.mirbase.org; release 21; verified on 28th September, 2016). There are a little more than 7000 miRNAs from plants belonging

Table 1 Presence of miRNAs in plants

	microRNAs	
	Precursors	Mature
Chlorophyta	50	86
Coniferophyta	108	110
Magnoliophyta	124	129
Araliaceae	29	32
Asteraceae	84	94
Brassicaceae	726	1071
Caricaceae	79	81
Cucurbitaceae	120	120
Euphorbiaceae	247	247
Fabaceae	1379	1545
Lamiales	65	71
Linaceae	124	124
Malvaceae	458	460
Ranunculaceae	45	45
Rhizophoraceae	8	8
Rosaceae	386	421
Rutaceae	75	79
Salicaceae	356	405
Solanaceae	463	617
Vitaceae	163	186
Monocotyledons	1616	2221
Total	6992	8496

to 73 species (Table 1), which represents 32% of the total entries in the database. An important segment of the total number of miRNAs come from economically important crops and is highly conserved. MiRNAs participate in the gene regulation of several developmental processes in plants. Among these processes are the response to external environmental stimuli, organogenesis, plant immunity, plant-pathogen interactions, cell proliferation, signaling, and cell death (Boopathi 2015; Shivaprasad et al. 2012).

Even before the discovery of miRNAs in plants, a link among miRNAs and PGRs was established. The hyponastic leaves (*HYL1-1)* mutant exhibited diminished responses to auxin and cytokinin, and hypersensitivity to abscisic acid (ABA) (Lu and Fedoroff 2000). Since then, there has been increasing evidence for the interaction between PGRs and miRNAs, and many of them target mRNAs implicated in auxin responses (Table 2). Auxins are a central player in the regulation of cell division and differentiation, particularly through the interaction with other PGRs, mainly cytokinins (Zhang et al. 2013; Schaller et al. 2015; Xu et al. 2015). However, this interaction goes further and involves brassinosteroids (Zhou et al. 2013) and gibberellins (Liu et al. 2016). Several of these interactions are mediated by miRNAs. Treatment of *Oryza sativa* with PGRs led to the discovery of 22 conserved miRNAs (Liu et al. 2009). Eleven of these were deregulated by one or more

Table 2 Representative examples of auxins' miRNA targets

Species	microRNA	Target	References
Arabidopsis thaliana	160	ARF10	Qiao et al. (2012)
		ARF10, ARF16, ARF17	Kasschau et al. (2003), Mallory et al. (2005), Rhoades et al. (2002), Wang et al. (2005)
	164	NAC1	Guo et al. (2005b), Rhoades et al. (2002)
	166	ARF6, ARF8	Gutierrez et al. (2009)
	167	ARF8	Kasschau et al. (2003), Park et al. (2002), Rhoades et al. (2002)
		ARF6, ARF8	Rhoades et al. (2002), Su et al. (2016)
		ARF10, ARF17	Sorin et al. (2005)
		ARF10, ARF16, ARF17	Mallory et al. (2005)
	319	SAUR IAA3/ SHY2	Koyama et al. (2010)
	390	tasiRNAs ARF3, ARF4	Allen et al. (2005), Marin et al. (2010), Yoon et al. (2010)
	393	F-box protein family	Iglesias et al. (2014), Navarro et al. (2006), Si-Ammour et al. (2011), Sunkar and Zhu (2004), Wang et al. (2004), Windels and Vazquez (2011)
	847	Aux/IAA	Wang and Guo (2015)
	10515	SUR1	Kong et al. (2015)
Glycine max	156g/j, 172f, 390e, 399a/b, 1511, 2111b/c/f, Gly03, Gly04, Gly16a/b, Gly20, Gly13		Sun et al. (2016)
Raphanus sativus	160, 161	ARF16, ARF17	Zhai et al. (2016)
	167	ARF8	Zhai et al. (2016)
Solanum lycopersicum	160, 167	ARF6, ARF8, ARF10, ARF16	Liu et al. (2016)
	166	ARF6, ARF8	Fan et al. (2015)
Zingiber officinale	167	ARF	Singh et al. (2016)
Coffea spp.	167	ARF8	Chaves et al. (2015)

(continued)

Table 2 (continued)

Species	microRNA	Target	References
Zea mays	160, 167	ARFs	Shen et al. (2013)
	393	F-box protein family	
Oryza sativa	167	ARF8	Yang et al. (2006)

ARF auxin response factor, *CUC1* cup-shaped cotyledons 1, *IAA3/SHY2* INDOLE-3-ACETIC ACID3/SHORT HYPOCOTYL2, *SAUR* SMALL AUXIN UP RNA, *SUR1* SUPERROOT1, *tasiRNAs* TAS3-derived trans-acting short-interfering RNAs

PGR treatments. The expression of miR159 and miR394 is regulated by ethylene, while miR167 and miR413 are regulated by ABA (Liu et al. 2009). For auxins specifically, it has been found that miR164 mediates cleavage of *NAM/ATAF/CUC (NAC)* domain-encoding mRNAs, in particular *NAC1*, producing an auxin-mediated induction of adventitious and lateral root formation (Guo et al. 2005b). Several of the auxin response factors (ARF), a key component of the auxin signaling cascade, are regulated by various miRNAs. miR160 targets three *ARF*, in particular *ARF17*, modifying the expression of auxin-inducible *GH3* genes, which encode auxin-conjugating proteins (Mallory et al. 2005). On the other hand, *ARF6* and *ARF8* are targeted by miR167 (Gutierrez et al. 2009). Another miRNA that plays an important role in auxin homeostasis is miR393 (Windels and Vazquez 2011; Eckardt 2012). MiR393 downregulates four auxin receptor family F-box protein (*TAAR*) genes (Si-Ammour et al. 2011; Windels and Vazquez 2011).

2 Biogenesis and Function of MiRNAs in Plants

Genetic screening of plant mutants affected in developmental processes, including PGR signaling, has helped to identify many of the genes involved in miRNA biogenesis (Rubio-Somoza and Weigel 2011; Khraiwesh et al. 2012). There are many proteins and enzymes involved in miRNA biogenesis, such as HYL1 (HYPONASTIC LEAVES 1), SE (C2H2 Zn-finger protein SERRATE), DCL1 (RNase III DICER-LIKE 1 enzyme), DDL (RNA-binding protein DAWDLE), HEN1 (HUA ENHANCER 1), HASTY (homolog of exportin 5), and AGO1 (ARGONAUTE1) (Ha and Kim 2014).

DNA-dependent RNA Pol II transcribes miRNAs from the *MIR* genes. miRNAs originate from a hairpin or stem-loop precursor (Bartel 2004; Jones-Rhoades et al. 2006; Cuperus et al. 2011). First, the primary miRNA transcript, also called Pri-miRNA, is cleaved by DCL1 to form the stem-loop intermediate precursor pre-miRNA (Tang et al. 2003). The resulting miRNA duplex is then processed by DCL1, the main function of which is to cleave double strand (ds) pri-miRNA or ds pre-miRNA to produce mature miRNA, together with the protein HYL1 in the nucleus (Bartel 2004). The methyltransferase HEN1 incorporates a methyl group on the 2′ OH of the 3′ last nucleotide of the mature duplex miRNA. These methylations

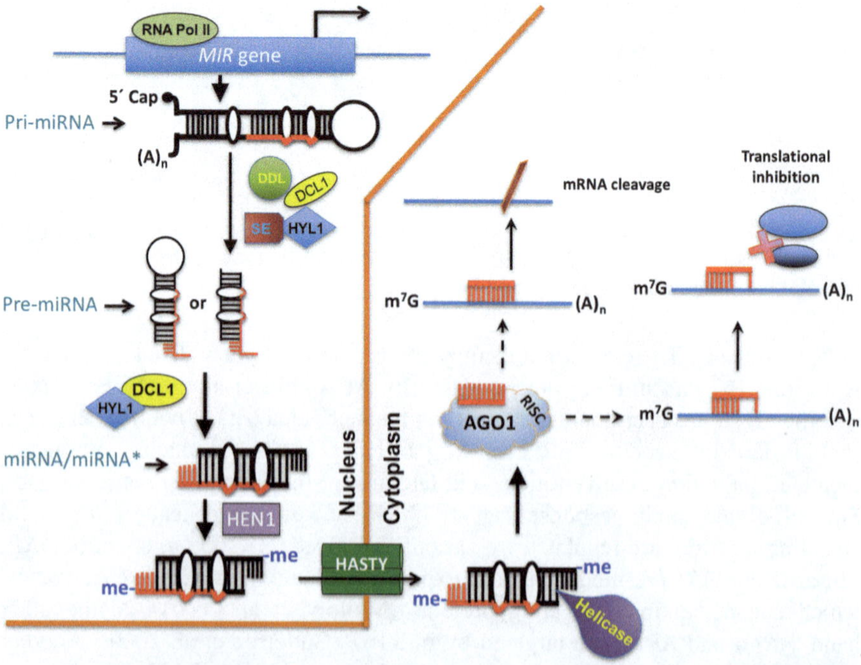

Fig. 1 miRNA biogenesis and silencing mechanisms in plants. MiRNAs are transcribed by DNA-dependent RNA Pol II from the MIR genes. Pri-miRNA is cleaved by RNase III DICER-LIKE 1 enzyme (DCL1). The resulting miRNA duplex is then processed by DCL1. The methyltransferase HUA ENHANCER 1 (HEN1) incorporates a methyl group on the 2′OH of the 3′ last nucleotide of the mature duplex miRNA. Then, a homolog of exporting 5 (HASTY) exports the miRNA from the nucleus to the cytoplasm. The mature miRNA is recognized by ARGONAUTE1 (AGO1). There are two types of silencing via miRNAs: one that cleaves the mRNA and one that represses the translation by binding stably to the mRNA targets. Other important players in miRNA biogenesis are DDL (RNA-binding protein DAWDLE), HYL1 (HYPONASTIC LEAVES 1), and SE (C2H2 Zn-finger protein SERRATE). See text for more details

are recognized by HASTY, which exports the miRNA from the nucleus to the cytoplasm. In the cytoplasm, the methyl groups are removed, and a helicase unwinds the ds to produce a single-strand mature miRNA, which is recognized by AGO1 (Bartel 2004). AGO1 is responsible for recruiting all the parts of the RNA-induced silencing complex (RISC) that recognizes the mRNA targets where the mature miRNA has its regulating function (Bartel 2004; Chen 2009). Furthermore, AGO1 is involved in the translational repression of the target mRNAs in the endoplasmic reticulum with the association of AMP1 (ALTERED MERISTEM PROGRAM1) (Li et al. 2013; Iwakawa and Tomari 2013). The resulting biogenesis can generate two types of miRNAs: those which perfectly complement their mRNA targets and those with mismatches to their targets. MiRNAs with perfect complementarity to their target mRNA tend to induce mRNA cleavage by silencing. On the other hand, miRNAs with mismatches tend to repress translation by binding stably to the mRNA targets (Jones-Rhoades et al. 2006; Axtell et al. 2011; Cuperus et al. 2011) (Fig. 1).

The importance of the proteins involved in the miRNA biogenesis during plant development was shown using *A. thaliana* mutants that present phenotypic alterations (Bohmert et al. 1998; Schauer et al. 2002; Mallory and Vaucheret 2006; Chen 2009). For instance, the *dcl1* mutant develops embryo lethality, suggesting that miRNAs are needed for plant viability (Schauer et al. 2002). The *ago1* mutant maintained viability although with dramatic phenotypic changes (Bohmert et al. 1998). An interesting discovery about miRNAs biogenesis is that it can be regulated itself by two miRNAs, miR162 and miR168, targeting *DCL1* and *AGO1*, respectively (Xie et al. 2003; Vaucheret et al. 2004).

3 Evolution of Plant MicroRNA Genes

From the evolutionary point of view, miRNAs can be divided into two groups. In one of the groups are the miRNA families that are highly conserved, from ferns and mosses to higher plants, with only one or two different nucleotides (Llave et al. 2002; Reinhart et al. 2002; Bonnet et al. 2004; Floyd and Bowman 2004; Jones-Rhoades and Bartel 2004; Sunkar and Zhu 2004; Wang et al. 2004; Adai et al. 2005; Axtell and Bartel 2005; Zhang et al. 2005, 2006a, b; Cuperus et al. 2011). In the other group, the non-conserved miRNAs play specific roles in different tissues or plant species (Lu et al. 2005; Zhang et al. 2005; Sanan-Mishra et al. 2009). Certainly, most of the miRNAs present in plants seem to be distinct to that species, and there are many miRNAs that are present only in a base family or are species specific (Axtell 2008, 2013; Axtell and Bowman 2008; Cuperus et al. 2011). Phylogenetic analysis of embryophytes identified eight miRNA families as their common ancestor (Cuperus et al. 2011). On the other hand, the unicellular green alga *Chlamydomonas reinhardtii* does not have any miRNA family in common with the embryophyte plants (Molnár et al. 2007; Zhao et al. 2007). These facts suggest three main hypotheses: (1) that during eukaryotic evolution miRNAs arose at least twice from an ancestral small RNA (Axtell 2008); (2) that individual miRNA families have to be conserved for a long time, remaining basically unchanged since before the appearance of angiosperms (Axtell and Bartel 2005; Axtell 2013); and (3) that most of the known miRNA genes could have arisen relatively recently (Cuperus et al. 2011).

The miRNAs, miR160, miR166, and miR390, involved in response to auxins, are present in all the Embryophyta (Cuperus et al. 2011). Analyzing the miR167 from 20 plant species, we found that all of the miR167-5p do not show any variation among them (Fig. 2). On the other hand, the miR167-3p shows a high level of variability. Half of the species have members in both groups. The other half, including *Brassica rapa*, *B. napus*, *Carica papaya*, *Gossypium hirsutum*, *Nicotiana tabacum*, *Citrus sinensis*, *Vitis vinifera*, *Theobroma cacao*, *Triticum aestivum*, *Sorghum bicolor*, only have members of the conserved family.

An analysis of the presence of miRNAs related to auxins shows that seven of them, 156, 160, 164, 166, 167, 172, and 399, are present in most of the families

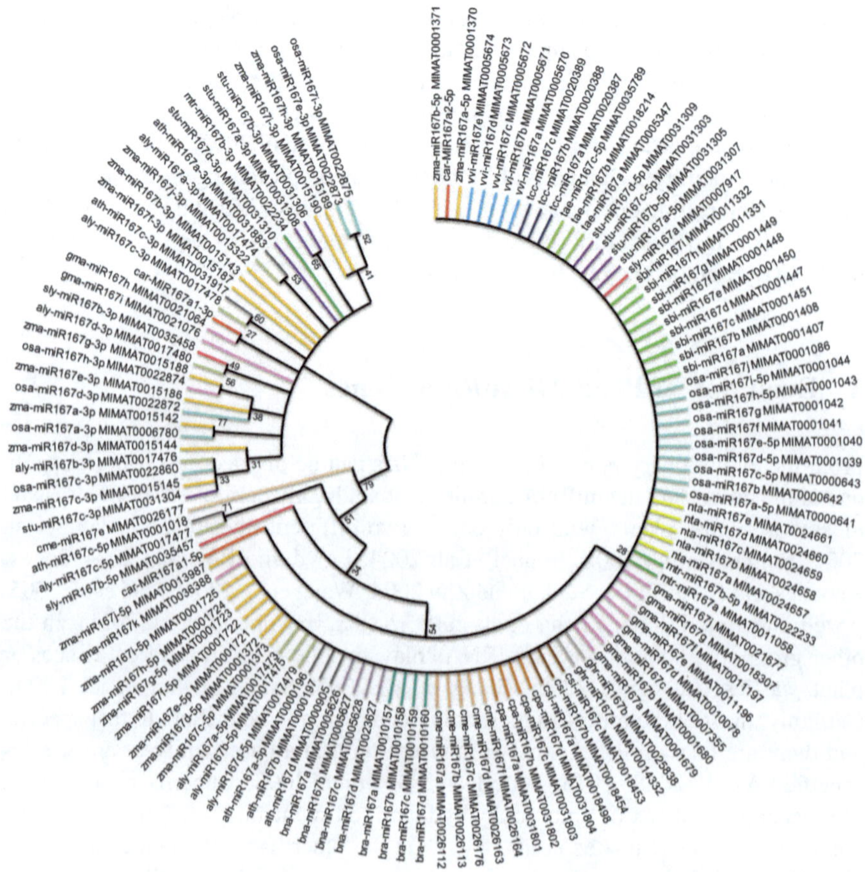

Fig. 2 Phylogenetic relationships of miRNA167 from several species. The tree was constructed using the neighbor joining method using the MEGA program v 6.0. Bootstrap values from 5000 replicates are indicated at each branch. Each *color* represents one species. Abbreviations: car *Coffea arabica*, osa *Oryza sativa*, ath *Arabidopsis thaliana*, aly *Arabidopsis lyrata*, zma *Zea mays*, cme Cucumismelo, cca *Carica papaya*, bna *Brassica napus*, bra *Brassica rapa*, mtr *Medicago truncatula*, nta *Nicotiana tabacum*, sly *Solanum lycopersicum*, stu *Solanum tuberosum*, vvi *Vitis vinifera*, sbi *Sorghum bicolor*, gma *Glycine max*, csi *Citrus sinensis*, tcc *Theobroma cacao*, ghi *Gossypium hirsitum*, tae *Triticum aestivum*

studied (Fig. 3). On the other hand, miRNA161 is present only in the Panicoideae and Brassicaceae, while miR2111 is present in only three families: Fabaceae, Vitaceae, and Brassicaceae. This suggests a clear division among highly conserved miRNAs that are shared among most of the plant families, and very specific miRNAs present in only a few families.

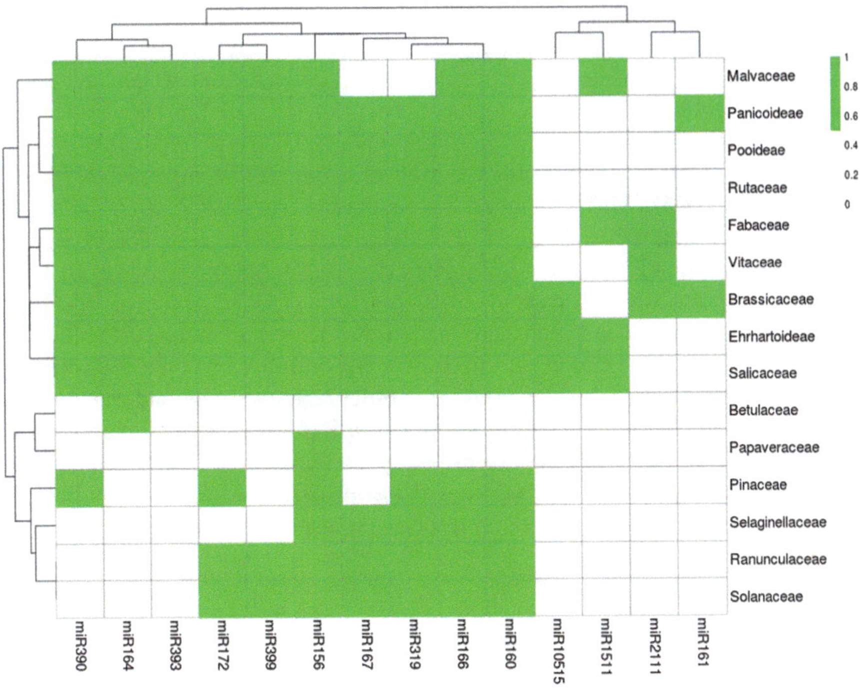

Fig. 3 microRNAs related to auxin metabolism found in the families studied. Families that have at least one microRNA, according to the plant microRNA database, were included. Clustering was performed according to the presence or absence of the microRNAs

4 Gene Regulation by MicroRNAs in Plants

MiRNAs were first discovered in *Caenorhabditis elegans* by Lee et al. (1993). Since then, miRNAs have become one of the most important and studied topics in biology (Cifuentes et al. 2010; Olmedo-Monfil et al. 2010; Cuperus et al. 2011; Djuranovic et al. 2012; Khraiwesh et al. 2012; Manohar et al. 2013). Although much of the available information about the role of miRNAs has been done in *A. thaliana*, many miRNAs and their complementary sites in the targeted genes are conserved among angiosperms and gymnosperms (Table 1) (Floyd and Bowman 2004).

Although miRNAs are very small in size, they have big regulatory roles, e.g., regulating cell homeostasis during differentiation, organ development, cell death, and plant growth during normal or in vitro conditions (Bartel 2004, 2009; Jones-Rhoades and Bartel 2004; Mallory and Vaucheret 2004; Mallory and Vaucheret 2006). Bioinformatics tools have helped to discover that at least 30% of miRNA gene families, including miR156, 157, 158, 159, 163, 165, 166, 168, and 319, are present in approximately 10 different plant species and participate very actively in different plant developmental processes (Jung and Park 2007; Reyes and Chua 2007; Sunkar and Zhu 2007; Li et al. 2012; Su et al. 2016).

Most of the miRNAs negatively regulate transcription factors, which have major roles in morphogenesis and development. For instance, miR156 targets the gene *SPL* (*Squamosa Promoter binding-Like*) involved in the regulation of the vegetative-to-reproductive transition as well as organ size (Wang et al. 2008; Wu and Poethig 2006). MiR164 targets the expression of *CUC1* (*CUP-SHAPED COTYLEDON*) and *NAC* (*NAM, ATAF1/2, and CUC2*) (Guo et al. 2005b; Raman et al. 2008) while miR165 and 166 regulate class III *HD-ZIP* genes (Bao et al. 2004; Carlsbecker et al. 2010; Furuta et al. 2012). *MYB* and *GAMYB*, which participate in the flowering process and seed germination, are targets of miR159 (Millar and Gubler 2005; Alonso-Peral et al. 2010). Another miRNA involved in flowering, specifically targeted *AP2 (APETALA2)*, is miR172 (Aukerman and Sakai 2003; Chen 2004). On the other hand, miR160 and miR167 regulate the expression of *ARF* (Wang et al. 2005), one of the most important transcription factor in regulating the expression of auxin response genes (Li et al. 2016). MiRNAs have become important signal molecules for auxin response, transport, and regulation.

5 MiRNAs in Auxins Signaling and Homeostasis

The precise mechanism of how auxins modulate plant growth and development is not fully understood. However, great progress has been made in the understanding of the signaling and transport of this PGR (McSteen 2010; Zhao 2010) (Table 3).

Strict control of auxin homeostasis and the maintenance of an appropriate level of IAA is important for normal growth and development. IAA is transported to the growing regions of a plant, and high IAA content correlates with intense cell division (Ljung et al. 2001; Tanaka et al. 2006). Proteins such as GRETCHEN HAGEN 3 (GH3), TRYPTOPHAN SYNTHASE β (TRP2), YUCCA (YUC), and others are important players in auxin homeostasis (Table 3).

In order to have a strict balance of IAA concentration in the cells, IAA is first distributed via phloem in a slow transport method from cell to cell, which is highly regulated by specific transport proteins such as AUXIN-RESISTANT 1/LIKE AUXIN-RESISTANT (AUX1/LAX), the ATP-BINDING CASSETTE SUBFAMILY B TRANSPORTER (ABCB), and the PIN-FORMED (PIN) (Petrášek et al. 2006; Cho et al. 2007; Petrášek and Friml 2009) (Table 3). PIN proteins are important components of auxin efflux, and its subcellular localization guides the flow of auxins (Tanaka et al. 2006). The polar movement of IAA allows a differential distribution, or gradients, of auxin within the plant tissues, and these gradients are dynamic during different developmental processes (Tanaka et al. 2006). The chemiosmotic transport of auxins is based on the differential pH between the apoplast (pH 5.5) and the cytoplasm (pH 7.0). IAAH is diffused throughout the plasmatic membrane or carried by the influx transport AUX1/LAX1 into the cell. Inside the cytoplasm, IAAH is dissociated to form IAA, which can be exported

Table 3 Proteins and genes required during the biosynthesis, transport and signaling of auxins

Proteins	*Genes*	References
Biosynthesis		
Tryptophan synthase β	*TRP2*	Bartel (1997)
Amidase	*AMI1*	Mano et al. (2010)
Tryptophan aminotransferase	*TAA1 TAR1, 2*	Won et al. (2011)
Aldehyde oxidase	*AtAO1*	Ljung et al. (2002)
Flavin monoxygenase	*YUC1-11*	Zhao et al. (2001)
Cytochrome P450	*CYP79B2/3*	Zhao et al. (2002)
Nitrilase	*NIT1-4*	Cohen et al. (2003)
Transport		
Auxin influx transporter	*AUX1 LAX*	Kramer (2004), Zazimalová et al. (2010)
Auxin efflux carrier	*PIN1*	Gälweiler et al. (1998), Kramer (2004)
Serine threonine kinase	*PID*	Christensen et al. (2000)
ABC transporter	*ABCB1,19*	Zazimalová et al. (2010)
Signaling		
Aux/IAA transcription factor	*IAA1-25*	Hagen and Guilfoyle (2002)
Auxin response factor	*ARF1-23*	Guilfoyle and Hagen (2001, 2007)
F-box	*TIR1, AFB*	Dharmasiri et al. (2005), Parry et al. (2009)
Small Auxin Up RNA	*SAUR*	Hagen and Guilfoyle (2002)
Conjugation		
Gretchen Hagen 3	*GH3*	Hagen and Guilfoyle (2002)

from the cell by the efflux carriers PGP or PIN (Petrášek and Friml 2009; Robert and Friml 2009).

Auxins increase the early transcription of several genes such as *AUX/IAA*, *GH3*, and *SMALL AUXIN UP RNA (SAUR)* (Table 3). These genes can regulate plant physiology by modulating the interaction between transcription factors and the auxin response elements (AuxREs) of the involved genes (Abel and Theologis 1996) that normally are activated in the 2–20 min response (Guilfoyle et al. 1998). It is known that, under low nuclear concentration of auxin, the transcriptional repressors Aux/IAAs associates with the C-terminal domain of the ARF proteins, a class of transcriptional regulators that mediate the auxin-dependent response by binding directly to TGTCTC sequence of the auxin-responsive element (auxREs) found in the promoters of auxin-inducible genes (Kim et al. 1997; Reed 2001; Hagen and Guilfoyle 2002; Liscum and Reed 2002; Vernoux et al. 2011). When the nuclear concentration of auxin increases, AUX/IAA repressor interacts with TIR1/AUXIN SIGNALING F-BOX PROTEIN (AFB) required for recognition by CULLIN scaffold-type E3 ligases (SCF E3), and it is targeted for degradation (Dharmasiri et al. 2005; Kepinski and Leyser 2005; Tan et al. 2007; Vernoux et al. 2011; Dinesh et al. 2016) (Table 3). The degradation of Aux/IAA induces the reactivation of the ARF function and the expression of the targeted genes involved in early auxin response.

Despite the fact that auxins have been studied biochemically, molecularly, and physiologically, it is still unknown how their regulation is coordinated in plant development. A small but important piece of evidence has proposed that the small RNAs are one of the responsible elements in the regulation of auxin homeostasis, transport, and signaling (Axtell 2013). There are many studies indicating that miRNAs have a major role in auxin genes related to homeostasis and signaling (Navarro et al. 2006; Gutierrez et al. 2009; Marin et al. 2010; Yoon et al. 2010; Si-Ammour et al. 2011; Chen et al. 2012; Kinoshita et al. 2012; Iglesias et al. 2014; Hrtyan et al. 2015), but there are still missing pieces in the picture of the role of miRNAs in plant development.

5.1 Auxin Homeostasis and MiRNAs

The homeostasis of auxins involves biosynthesis and degradation, as well as their conjugation with amino acids and carbohydrates. The hydrolysis of some of these conjugates liberates IAA (Ljung et al. 2002; Ljung 2013). The level of IAA concentration releases its signal transduction pathway and produces a molecular and biochemical response. This response begins with the sensing of IAA concentration through the assembly of a co-receptor complex that includes an F-box protein from the TIR1/AUXIN SIGNALING F-BOX PROTEIN (AFB) family and a member of the AUXIN/IAA-INDUCIBLE (AUX/IAA) family of transcriptional repressors. The homeostasis of IAA is mainly regulated by the GH3 proteins, which catalyze the conjugation of IAA with several amino acids to keep the auxin levels as normal as possible for each biological function (Bajguz and Piotrowska 2009; Chen et al. 2010).

It has been found that *GH3* genes are regulated by ARF8, ARF19, and ARF17 (Yang et al. 2006; Ding et al. 2013; Zhang et al. 2015). It is interesting to note that these ARF are targeted by miRNAs 167, 847, and 160, respectively (Table 2) (Mallory et al. 2005; Sorin et al. 2005; Yang et al. 2006; Ding et al. 2013; Wang and Guo 2015; Zhang et al. 2015; Li and Zhang 2016). Therefore, it seems that the final function of GH3 is a two-player regulation that depends not only on the ARF directly but also on the miRNAs that cleave specific ARFs' mRNAs (Fig. 4). For instance, ARF8 is involved in the regulation of the gene *GH3* during lateral root formation and hypocotyl elongation in Arabidopsis (Tian et al. 2004). ARF19 regulates the expression of *GH3* to mediate auxin homeostasis in lateral root formation (De Rybel et al. 2010). On the other hand, ARF17, which alters the expression of *GH3*, has been actively participating in embryonic, root, and floral development (Mallory et al. 2005) (Fig.4).

Although auxin transport is an important part of auxin regulation, the evidence for their involvement is very scarce. In a multicellular computational model, Muraro et al. (2014) found that miR165/166 can act as a regulatory mechanism for vascular patterning by targeting genes involved in auxin transport such as PIN1 (Muraro et al. 2014).

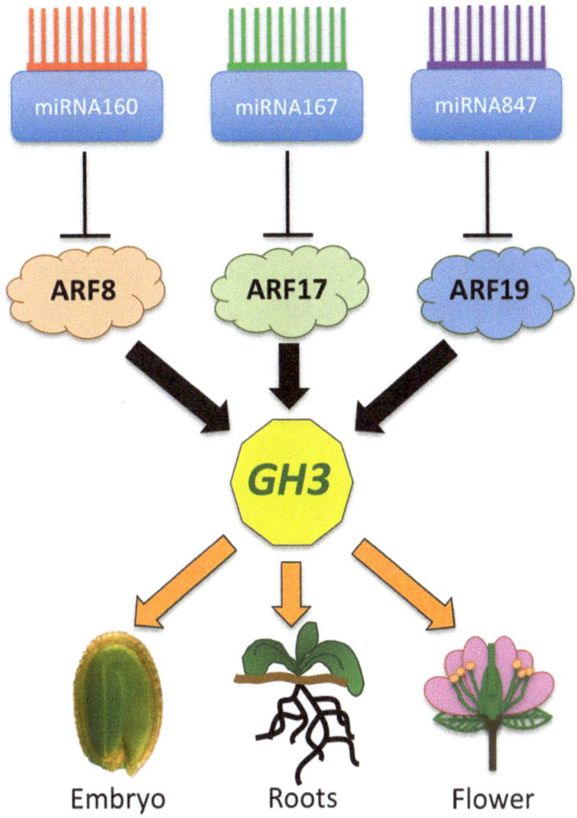

Fig. 4 microRNAs involved in ARF regulation for GH3 expression during embryo, root, and flower development. *ARF* AUXIN RESPONSE FACTORS, *GH3* GRETCHEN HAGEN 3

5.2 Auxin Signaling and MiRNAs

MiRNAs have also regulated the signaling pathway in response to auxin. While miRNA390 targets the expression of *TAS3* and this regulates ARF2, 3, and 4 - (Barrera-Figueroa et al. 2011), miR393 can target five different genes that belong to the TIR1 family (Sunkar and Zhu 2004), a positive regulator of auxin signaling. TIR1 targets the ARF transcriptional repressor, AUXIN/INDOLE-3-ACETIC ACID (Aux/IAA), for degradation (Quint and Gray 2006; Tan et al. 2007).

Several miRNAs can downregulate some of the most studied classes of the ARF genes and direct their mRNA cleavage (Jones-Rhoades and Bartel 2004; Rhoades et al. 2002). For instance, miR167 targets *ARF6* and *ARF8*, while miR160 targets *ARF10*, *ARF16*, and *ARF17* (Rhoades et al. 2002; Kasschau et al. 2003; Allen et al. 2005; Mallory et al. 2005). Disrupting miRNA-mediated regulation of specific *ARF* *s*can alters the normal phenotype. Regulation of *ARF17* mRNA levels by miR160 control is necessary for a proper transcriptional regulation of *GH3*-like early auxin response, which encodes IAA-amino acid-conjugating proteins (Mallory et al. 2005; Staswick et al. 2005; Park et al. 2007). Plants with five silent mutations within the miR160-complementary domain of an ARF17 genomic clone, named 5mARF17, show an increase in the levels of expression of ARF17, reduced accumulation of GH3.5, and several floral defects (Mallory et al. 2005). Examples of these floral defects include accelerated flowering time, rosette serration, reduced petal size, abnormal stamen structure, and reduced fertility.

6 Role of MiRNA in Plant Growth and Development Mediated by Auxins

Recently, miRNAs appear to be key regulators that help to integrate diverse biological responses mediated by PGRs (Sanan-Mishra et al. 2013; Liu et al. 2016). Auxin signaling is highly regulated by miRNAs and appears to be conserved among different plant species, including *A. thaliana*, *Oryza sativa*, *Solanum lycopersicum*, and others (Rhoades et al. 2002; Eckardt, 2005; Hendelman et al. 2012; Sanan-Mishra et al. 2013). The first efforts led to the discovery that several *ARF* family members (*ARF8, ARF10, ARF16, ARF17,* and others) may be regulated by both miR160 and miR167 during early development in plants (Rhoades et al. 2002; Jones-Rhoades and Bartel 2004; Wang et al. 2005). Other crucial components of auxin signaling, such as *TIR1* and F-box auxin transcripts, have also been determined to be regulated by miR393 and miR394 (Allen et al. 2005; Jones-Rhoades and Bartel 2004). Interestingly, miR393 negatively regulates *TIR1*, *AFB2,* and *AFB3* transcripts to repress auxin signaling, thus increasing antibacterial resistance (Navarro et al. 2006) or helps to regulate auxin-related development of leaves by initiating the biogenesis of small interference RNAs to regulate the expression of *TIR* and *AFB2* (Si-Ammour et al. 2011). This suggests that miRNAs, and auxin levels, might regulate plant responses. Sorin et al. (2005) showed that Arabidopsis plants lacking ARGONAUTE1 (AGO1), a key player in the miRNA pathway, generate a super root, but are impaired in adventitious root formation. This impairment was connected to a defect in the regulation of auxin homeostasis. For instance, a study determined that a decrease in free IAA and its conjugates was correlated with downregulation of *GH3* genes that encode to acyl amidosynthetases (Hagen et al. 1991), which are also presumed to be targets of ARF17. Jain et al. (2006) found that high levels of *ARF17* mRNA are due to its resistance to cleavage by miR160; the same ARF that promotes the accumulation of *GH3* transcripts, including *GH3.2, GH3.3, GH3.5,* and *GH3.6*, are involved in the conjugation of IAA to amino acids (Jain et al. 2006).Therefore, a decrease of active IAA levels in the cell leads to dramatic pleiotropic defects, such as deformed embryos, abnormal stamens, and sterility, among others (Mallory et al. 2005). Many of these defects resemble phenotypes observed previously in Arabidopsis plants with mutations in *DCL1*, *AGO1*, *HYL,* and *HEN1* genes.

Five NAM/ATAF/CUC (NAC) domain transcription factor families are predicted to be under regulation by miR164. NAC1 acts downstream of TIR1, transmitting auxin signals by promoting proper lateral root development. Guo et al. (2005a) showed that miR164 guides the cleavage of endogenous and transgenic *NAC1*. However, when *NAC1* is mutated to avoid the cleavage induced by miR164, NAC1 messengers accumulate in the plant, allowing the production of more lateral roots (Guo et al. 2005b). Wang et al. (2005) determined that ARF10 and ARF16 are key controllers of root cap cell differentiation. They showed that miR160 overproduction in Pro35S:MIR160 plants and the ARF10-2 ARF16-2 double mutants display the same root tip defects such as uncontrolled cell division, impairment of cell differentiation, and loss of gravity sensing. This suggests that in the root auxin response, miR160 regulates the expression of ARF10 and ARF16 to promote the columnella proper cell differentiation (Wang et al. 2005). A similar effect was found in miR167, which regulates both female and male reproduction through the post-transcriptional control of ARF6 and ARF8. The overexpression of miR167 and the *ARF6ARF8* double mutants led to observation of similar and dramatic defects during the development of ovules and anthers. Therefore, miR167 is essential for regulating *ARF6* and *ARF8* during gynoecium and stamen development (Wu et al. 2006). Also, it has been found that transcripts of *ARF6* and *ARF8*, as well as *ARF17*, require additional levels of post-transcriptional regulation during the generation of adventitious roots. Gutierrez et al. (2009) showed that ARF6 positively regulates the amounts of both miR160 and miR167. ARF8 negatively regulates levels of miR167, whereas ARF17 regulates miR160 negatively, but positively affects the levels of miR167. Therefore, a complex regulatory mechanism apparently contributes to the regulation of adventitious root development, through a feedback regulation of miRNA homeostasis through direct and non-direct target TFs. A similar mechanism has been proposed between miR390 and *ARF4* during lateral rooting in Arabidopsis (Marin et al. 2010). In that model, miR390 expression senses external auxin concentration and directs the cleavage of the non-coding *TAS3* transcripts to affect the production of tasiRNA-*ARF* production. *ARF4* expression is critical for lateral root development, but interestingly inhibits miR390 expression through a negative regulation between the tasiRNA-ARF pathway and ARF4, which allows the spatiotemporal expression of *ARF4* (Yoon et al. 2010). In another study, it was shown that there is a positive and negative feedback regulation of miR390 by ARF2, ARF3, and ARF4, which regulate the miR390 expression pattern. This regulatory network allows the maintenance of optimal levels of the transcripts of *ARF* and thus specifies the timing of lateral root growth (Marin et al. 2010).

On the other hand, the regulation of shoot regeneration in vitro also requires the participation of miRNAs. By using Arabidopsis calli, it was found that non-totipotent cells contain more miR160 transcripts than a totipotent line. Taken into account that *ARF10* is a target of miR160, Qiao et al. (2012) showed that transgenic plants with an miR160-resistant form of *ARF10* increased shoot regeneration up to fivefold. In contrast, shoot regeneration was practically null when miR160 was overexpressed. More recently, deep small RNA sequencing during embryo differentiation in maize allowed the detection of more than 100 known miRNAs belonging to 23 miRNA families (Shen et al. 2013), where several of them could be involved in the establishment of embryogenic calli. It appears that the initial callus formation requires a major transcriptional regulation due to the high levels of differentially expressed miRNAs found in the callus (Shen et al. 2013). Moreover, the participation of miR393, which targets *TIR1*, and both miR160 and miR167, which target ARF messengers, seem to be important during the early stages of somatic embryogenesis. For instance, the overexpression of miR167 inhibits somatic embryo generation by altering the auxin response and auxin transport in the embryogenic calli (Su et al. 2016). This study also found that *ARF6* and *ARF8* transcripts regulated by miR167 are necessary for SE since both TFs are required in auxin signaling pathways and can mediate auxin-induced gene activation during the SE induction.

Taken together, these results clearly show that miRNAs play a crucial role during plant growth and that their participation in the auxin signaling pathway is necessary to help to coordinate plant development by controlling the auxin response genes properly.

7 Concluding Remarks

Since auxins are involved in almost every aspect of the physiology of plants, the balance of auxin concentration is crucial in producing such a diversity of responses. Most of these responses are regulated by miRNAs. The miRNA databases grow daily, and more and more of their roles are discovered. However, a deeper understanding of their function is needed for a complete model of action of the miRNAs. The highly networked regulation that miRNAs exhibit during root development, embryo formation, and in auxin signaling generally throughout the plant increases the complexity of our understanding of PGR. There are a couple of miRNAs, such as 160 and 167, that can regulate many ARFs. However, a more detailed knowledge of the miRNAs involved in specific target degradation in particular processes would help to understand the communication and balance among all players. For instance, it is necessary to understand the role of miRNAs in the epigenetic regulation of differentiation and somatic embryogenesis in order to outline a signaling map to increase the efficiency of the process in some recalcitrant species.

Acknowledgements The work from the VMLV laboratory was supported by a grant from the National Council of Science and Technology (CONACyT; Grant No. 257436), and a scholarship from CONACyT to JAM.

References

Abel S, Theologis A (1996) Early genes and auxin action. Plant Physiol 111:9–17

Adai A, Johnson C, Mlotshwa S et al (2005) Computational prediction of miRNAs in *Arabidopsis thaliana*. Genome Res 15:78–91

Allen E, Xie Z, Gustafson AM et al (2005) microRNA-directed phasing during trans-acting siRNA biogenesis in plants. Cell 121:207–221

Alonso-Peral MM, Li J, Li Y et al (2010) The microRNA159-regulated GAMYB-like genes inhibit growth and promote programmed cell death in Arabidopsis. Plant Physiol 154:757–771

Aukerman MJ, Sakai H (2003) Regulation of flowering time and floral organ identity by a microRNA and its APETALA2-like target genes. Plant Cell 15:2730–2741

Axtell MJ (2008) Evolution of microRNAs and their targets: are all microRNAs biologically relevant? BBA-Gene Regul Mech 1779:725–734

Axtell MJ (2013) Classification and comparison of small RNAs from plants. Annu Rev Plant Biol 64:137–159

Axtell MJ, Bartel DP (2005) Antiquity of microRNAs and their targets in land plants. Plant Cell 17:1658–1673

Axtell MJ, Bowman JL (2008) Evolution of plant microRNAs and their targets. Trends Plant Sci 13:343–349

Axtell MJ, Westholm JO, Lai EC (2011) Vive la différence: biogenesis and evolution of microRNAs in plants and animals. Genome Biol 12:221

Ayil-Gutiérrez BA, Galaz-Ávalos RM, Peña-Cabrera E et al (2013) Dynamics of the concentration of IAA and some of its conjugates during the induction of somatic embryogenesis in *Coffea canephora*. Plant Signal Behav 8:e26998

Bajguz A, Piotrowska A (2009) Conjugates of auxin and cytokinin. Phytochemistry 70:957–969

Bao N, Lye KW, Barton MK (2004) MicroRNA binding sites in Arabidopsis class III HD-ZIP mRNAs are required for methylation of the template chromosome. Dev Cell 7:653–662

Barrera-Figueroa BE, Gao L, Diop NN et al (2011) Identification and comparative analysis of drought-associated microRNAs in two cowpea genotypes. BMC Plant Biol 11:127

Bartel B (1997) Auxin biosynthesis. Annu Rev Plant Physiol Plant Mol Biol 48:51–66. doi:10.1146/annurev.arplant.48.1.51

Bartel DP (2004) MicroRNAs: genomics, biogenesis, mechanism, and function. Cell 116:281–297

Bartel DP (2009) MicroRNAs: target recognition and regulatory functions. Cell 136:215–233

Benjamins R, Scheres B (2008) Auxin: the looping star in plant development. Annu Rev Plant Biol 59:443–465

Berleth T, Mattsson J, Hardtke CS (2000) Vascular continuity and auxin signals. Trends Plant Sci 5:387–393

Bohmert K, Camus I, Bellini C et al (1998) *AGO1* defines a novel locus of *Arabidopsis* controlling leaf development. EMBO J 17:170–180

Bonnet E, Wuyts J, Rouzé P et al (2004) Detection of 91 potential conserved plant microRNAs in *Arabidopsis thaliana* and *Oryza sativa* identifies important target genes. Proc Natl Acad Sci USA 101:11511–11516

Boopathi NM (2015) Plant miRNomics: novel insights in gene expression and regulation. In: Barh D, Khan MS, Davies E (eds) Plant omics: the omics of plant science. Springer, New Delhi, pp 181–211

Carlsbecker A, Lee JY, Roberts CJ et al (2010) Cell signalling by microRNA165/6 directs gene dose-dependent root cell fate. Nature 465:316–321

Chaves SS, Fernandes-Brum CN, Silva GFF et al (2015) New insights on coffea miRNAs: features and evolutionary conservation. Appl Biochem Biotechnol 177:879–908

Chen X (2004) A microRNA as a translational repressor of *APETALA2* in *Arabidopsis* flower development. Science 303:2022–2025

Chen X (2009) Small RNAs and their roles in plant development. Annu Rev Cell Dev Biol 25:21–44

Chen Q, Westfall CS, Hicks LM et al (2010) Kinetic basis for the conjugation of auxin by a GH3 family indole-acetic acid-amido synthetase. J Biol Chem 285:29780–29786

Chen H, Li Z, Xiong L (2012) A plant microRNA regulates the adaptation of roots to drought stress. FEBS Lett 586:1742–1747

Cho M, Lee SH, Cho HT (2007) P-Glycoprotein 4 displays auxin efflux transporter-like action in Arabidopsis root hair cells and tobacco cells. Plant Cell 19:3930–3943

Christensen SK, Dagenais N, Chory J et al (2000) Regulation of auxin response by the protein kinase PINOID. Cell 100:469–478

Cifuentes D, Xue H, Taylor DW et al (2010) A novel miRNA processing pathway independent of dicer requires argonaute2 catalytic activity. Science 328:1694–1698

Cohen JD, Slovin JP, Hendrickson AM (2003) Two genetically discrete pathways convert tryptophan to auxin: more redundancy in auxin biosynthesis. Trends Plant Sci 8:197–199

Cuperus JT, Fahlgren N, Carrington JC (2011) Evolution and functional diversification of *MIRNA* genes. Plant Cell 23:431–442

De Rybel B, Vassileva V, Parizot B et al (2010) A novel Aux/IAA28 signaling cascade activates GATA23-dependent specification of lateral root founder cell identity. Curr Biol 20:1697–1706

Dharmasiri N, Dharmasiri S, Estelle M (2005) The F-box protein TIR1 is an auxin receptor. Nature 435:441–445

Dinesh DC, Villalobos LI, Abel S (2016) Structural biology of nuclear auxin action. Trends Plant Sci 21:302–316

Ding Y, Tao Y, Zhu C (2013) Emerging roles of microRNAs in the mediation of drought stress response in plants. J Exp Bot 64:3077–3086

Djuranovic S, Nahvi A, Green R (2012) miRNA-mediated gene silencing by translational repression followed by mRNA deadenylation and decay. Science 336:237–240

Eckardt NA (2005) MicroRNAs regulate auxin homeostasis and plant development. Plant Cell 17:1335–1338

Eckardt NA (2012) A microRNA cascade in plant defense. Plant Cell 24:840

Fan S, Li Q, Guo G et al (2015) Identification of microRNAs in two species of tomato, *Solanum lycopersicum* and *Solanum habrochaites*, by deep sequencing. J Integr Agric 14:42–49

Floyd SK, Bowman JL (2004) Gene regulation: ancient microRNA target sequences in plants. Nature 428:485–486

Friml J (2003) Auxin transport – shaping the plant. Curr Opin Plant Biol 6:7–12

Friml J, Vieten A, Sauer M et al (2003) Efflux-dependent auxin gradients establish the apical-basal axis of Arabidopsis. Nature 426:147–153

Furuta K, Lichtenberger R, Helariutta Y (2012) The role of mobile small RNA species during root growth and development. Curr Opin Cell Biol 24:211–216

Gälweiler L, Guan C, Müller A et al (1998) Regulation of polar auxin transport by *At*PIN1 in Arabidopsis vascular tissue. Science 282:2226–2230

Geisler M, Wang B, Zhu J (2014) Auxin transport during root gravitropism: transporters and techniques. Plant Biol 16:50–57

Gray WM, del Pozo JC, Walker L et al (1999) Identification of an SCF ubiquitin-ligase complex required for auxin response in *Arabidopsis thaliana*. Genes Dev 13:1678–1691

Guilfoyle TJ (2015) The PB1 domain in auxin response factor and Aux/IAA proteins: a versatile protein interaction module in the auxin response. Plant Cell 27:33–43

Guilfoyle TJ, Hagen G (2001) Auxin response factors. J Plant Growth Regul 20:281–291

Guilfoyle TJ, Hagen G (2007) Auxin response factors. Curr Opin Plant Biol 10:453–460

Guilfoyle T, Hagen G, Ulmasov T et al (1998) How does auxin turn on genes? Plant Physiol 118:341–347

Guo A, He K, Liu D et al (2005a) DATF: a database of Arabidopsis transcription factors. Bioinformatics 21:2568–2569

Guo HS, Xie Q, Fei JF et al (2005b) MicroRNA directs mRNA cleavage of the transcription factor *NAC1* to downregulate auxin signals for Arabidopsis lateral root development. Plant Cell 17:1376–1386

Gutierrez L, Bussell JD, Pacurar DI et al (2009) Phenotypic plasticity of adventitious rooting in Arabidopsis is controlled by complex regulation of AUXIN RESPONSE FACTOR transcripts and microRNA abundance. Plant Cell 21:3119–3132

Ha M, Kim VN (2014) Regulation of microRNA biogenesis. Nat Rev Mol Cell Biol 15:509–524

Habets MEJ, Offringa R (2014) PIN-driven polar auxin transport in plant developmental plasticity: a key target for environmental and endogenous signals. New Phytol 203:362–377

Hagen G, Guilfoyle T (2002) Auxin-responsive gene expression: genes, promoters and regulatory factors. Plant Mol Biol 49:373–385

Hagen G, Martin G, Li Y et al (1991) Auxin-induced expression of the soybean GH3 promoter in transgenic tobacco plants. Plant Mol Biol 17:567–579

Hendelman A, Buxdorf K, Stav R et al (2012) Inhibition of lamina outgrowth following *Solanum lycopersicum* auxin response factor 10 (*SlARF10*) derepression. Plant Mol Biol 78:561–576

Hrtyan M, Šliková E, Hejátko J et al (2015) RNA processing in auxin and cytokinin pathways. J Exp Bot 66:4897–4912

Iglesias MJ, Terrile MC, Windels D et al (2014) MiR393 regulation of auxin signaling and redox-related components during acclimation to salinity in Arabidopsis. PLoS One 9:e107678

Iwakawa H, Tomari Y (2013) Molecular insights into microRNA-mediated translational repression in plants. Mol Cell 52:591–601

Jain M, Kaur N, Tyagi AK et al (2006) The auxin-responsive GH3 gene family in rice (*Oryza sativa*). Funct Integr Genomics 6:36–46

Jing H, Yang X, Zhang J et al (2015) Peptidyl-prolyl isomerization targets rice Aux/IAAs for proteasomal degradation during auxin signalling. Nat Commun 6:7395

Jones-Rhoades MW, Bartel DP (2004) Computational identification of plant microRNAs and their targets, including a stress-induced miRNA. Mol Cell 14:787–799

Jones-Rhoades MW, Bartel DP, Bartel B (2006) MicroRNAs and their regulatory roles in plants. Annu Rev Plant Biol 57:19–53

Jung JH, Park CM (2007) MIR166/165 genes exhibit dynamic expression patterns in regulating shoot apical meristem and floral development in Arabidopsis. Planta 225:1327–1338

Kasschau KD, Xie Z, Allen E et al (2003) P1/HC-Pro, a viral suppressor of RNA silencing, interferes with Arabidopsis development and miRNA function. Dev Cell 4:205–217

Kepinski S, Leyser O (2005) The Arabidopsis F-box protein TIR1 is an auxin receptor. Nature 435:446–451

Khraiwesh B, Zhu JK, Zhu J (2012) Role of miRNAs and siRNAs in biotic and abiotic stress responses of plants. BBA-Gene Regul Mech 1819:137–148

Kim J, Harter K, Theologis A (1997) Protein-protein interactions among the Aux/IAA proteins. Proc Natl Acad Sci USA 94:11786–11791

Kinoshita N, Wang H, Kasahara H et al (2012) *IAA-Ala resistant3*, an evolutionarily conserved target of miR167, mediates Arabidopsis root architecture changes during high osmotic stress. Plant Cell 24:3590–3602

Kong W, Li Y, Zhang M et al (2015) A novel Arabidopsis microRNA promotes IAA biosynthesis via the indole-3-acetaldoxime pathway by suppressing superroot1. Plant Cell Physiol 56:715–726

Koyama T, Mitsuda N, Seki M et al (2010) TCP transcription factors regulate the activities of asymmetric leaves1 and miR164, as well as the auxin response, during differentiation of leaves in Arabidopsis. Plant Cell 22:3574–3588

Kramer EM (2004) PIN and AUX/LAX proteins: their role in auxin accumulation. Trends Plant Sci 9:578–582

Lee RC, Feinbaum RL, Ambros V (1993) The *C. elegans* heterochronic gene lin-4 encodes small RNAs with antisense complementarity to lin-14. Cell 75:843–854

Leyser O (2005) Auxin distribution and plant pattern formation: how many angels can dance on the point of PIN? Cell 121:819–822

Li C, Zhang B (2016) MicroRNAs in control of plant development. J Cell Physiol 231:303–313

Li W, Cui X, Meng Z et al (2012) Transcriptional regulation of Arabidopsis MIR168a and argonaute1 homeostasis in abscisic acid and abiotic stress responses. Plant Physiol 158:1279–1292

Li S, Liu L, Zhuang X et al (2013) MicroRNAs inhibit the translation of target mRNAs on the endoplasmic reticulum in Arabidopsis. Cell 153:562–574

Li S-B, Xie ZZ, Hu CG et al (2016) A review of auxin response factors (ARF) in plants. Front Plant Sci 7:47

Liscum E, Reed JW (2002) Genetics of Aux/IAA and ARF action in plant growth and development. Plant Mol Biol 49:387–400

Liu Q, Zhang YC, Wang CY et al (2009) Expression analysis of phytohormone-regulated microRNAs in rice, implying their regulation roles in plant hormone signaling. FEBS Lett 583:723–728

Liu X, Xu T, Dong X et al (2016) The role of gibberellins and auxin on the tomato cell layers in pericarp via the expression of ARFs regulated by miRNAs in fruit set. Acta Physiol Plant 38:1–11

Ljung K (2013) Auxin metabolism and homeostasis during plant development. Development 140:943–950

Ljung K, Bhalerao RP, Sandberg G (2001) Sites and homeostatic control of auxin biosynthesis in Arabidopsis during vegetative growth. Plant J 28:465–474

Ljung K, Hull AK, Kowalczyk M et al (2002) Biosynthesis, conjugation, catabolism and homeostasis of indole-3-acetic acid in *Arabidopsis thaliana*. Plant Mol Biol 50:309–332

Llave C, Xie Z, Kasschau KD et al (2002) Cleavage of scarecrow-like mRNA targets directed by a class of Arabidopsis miRNA. Science 297:2053–2056

Lu C, Fedoroff N (2000) A mutation in the Arabidopsis HYL1 gene encoding a dsRNA binding protein affects responses to abscisic acid, auxin, and cytokinin. Plant Cell 12:2351–2365

Lu S, Sun YH, Shi R et al (2005) Novel and mechanical stress-responsive microRNAs in *Populus trichocarpa* that are absent from Arabidopsis. Plant Cell 17:2186–2203

Mallory AC, Vaucheret H (2004) MicroRNAs: something important between the genes. Curr Opin Plant Biol 7:120–125

Mallory AC, Vaucheret H (2006) Functions of microRNAs and related small RNAs in plants. Nat Genet 38:S31–S36

Mallory AC, Bartel DP, Bartel B (2005) MicroRNA-directed regulation of Arabidopsis auxin response factor17 is essential for proper development and modulates expression of early auxin response genes. Plant Cell 17:1360–1375

Mano Y, Nemoto K (2012) The pathway of auxin biosynthesis in plants. J Exp Bot 63:2853–2872

Mano Y, Nemoto K, Suzuki M et al (2010) The AMI1 gene family: indole-3-acetamide hydrolase functions in auxin biosynthesis in plants. J Exp Bot 61:25–32

Manohar S, Jagadeeswaran G, Nimmakayala P et al (2013) Dynamic regulation of novel and conserved miRNAs across various tissues of diverse cucurbit species. Plant Mol Biol Rep 31:335–343

Marin E, Jouannet V, Herz A et al (2010) miR390, Arabidopsis TAS3 tasiRNAs, and their auxin response factor targets define an autoregulatory network quantitatively regulating lateral root growth. Plant Cell 22:1104–1117

McSteen P (2010) Auxin and monocot development. Cold Spring Harb Perspect Biol 2:a001479

Mette MF, van der Winden J, Matzke M et al (2002) Short RNAs can identify new candidate transposable element families in Arabidopsis. Plant Physiol 130:6–9

Millar AA, Gubler F (2005) The Arabidopsis GAMYB-like genes, MYB33 and MYB65, are microRNA-regulated genes that redundantly facilitate anther development. Plant Cell 17:705–721

Molnár A, Schwach F, Studholme DJ et al (2007) miRNAs control gene expression in the single-cell alga *Chlamydomonas reinhardtii*. Nature 447:1126–1129

Muraro D, Mellor N, Pound MP et al (2014) Integration of hormonal signaling networks and mobile microRNAs is required for vascular patterning in Arabidopsis roots. Proc Natl Acad Sci USA 111:857–862

Navarro L, Dunoyer P, Jay F et al (2006) A plant miRNA contributes to antibacterial resistance by repressing auxin signaling. Science 312:436–439

Nonhebel HM (2015) Tryptophan-independent IAA synthesis: critical evaluation of the evidence. Plant Physiol 169:1001–1005

Normanly J (2010) Approaching cellular and molecular resolution of auxin biosynthesis and metabolism. Cold Spring Harb Perspect Biol 2:a001594

Olmedo-Monfil V, Duran-Figueroa N, Arteaga-Vazquez M et al (2010) Control of female gamete formation by a small RNA pathway in Arabidopsis. Nature 464:628–632

Pacurar DI, Perrone I, Bellini C (2014) Auxin is a central player in the hormone cross-talks that control adventitious rooting. Physiol Plant 151:83–96

Park W, Li J, Song R et al (2002) CARPEL FACTORY, a dicer homolog, and HEN1, a novel protein, act in microRNA metabolism in *Arabidopsis thaliana*. Curr Biol 12:1484–1495

Park JE, Park JY, Kim YS et al (2007) GH3-mediated auxin homeostasis links growth regulation with stress adaptation response in Arabidopsis. J Biol Chem 282:10036–10046

Parry G, Calderon-Villalobos LI, Prigge M et al (2009) Complex regulation of the TIR1/AFB family of auxin receptors. Proc Natl Acad Sci USA 106:22540–22545

Perrot-Rechenmann C (2014) Auxin signaling in plants. In: Howell SH (ed) Molecular biology. The plant sciences, vol 2. Springer, New York, pp 245–268

Petrášek J, Friml J (2009) Auxin transport routes in plant development. Development 136:2675–2688

Petrášek J, Mravec J, Bouchard R et al (2006) PIN proteins perform a rate-limiting function in cellular auxin efflux. Science 312:914–918

Qiao M, Zhao Z, Song Y et al (2012) Proper regeneration from *in vitro* cultured *Arabidopsis thaliana* requires the microRNA-directed action of an auxin response factor. Plant J 71:14–22

Quint M, Gray WM (2006) Auxin signaling. Curr Opin Plant Biol 9:448–453

Raman S, Greb T, Peaucelle A et al (2008) Interplay of miR164, cup-shaped cotyledon genes and lateral suppressor controls axillary meristem formation in *Arabidopsis thaliana*. Plant J 55:65–76

Reed JW (2001) Roles and activities of Aux/IAA proteins in Arabidopsis. Trends Plant Sci 6:420–425

Reinhart BJ, Weinstein EG, Rhoades MW et al (2002) MicroRNAs in plants. Genes Dev 16:1616–1626

Retzer K, Butt H, Korbei B et al (2014) The far side of auxin signaling: fundamental cellular activities and their contribution to a defined growth response in plants. Protoplasma 251:731–746

Reyes JL, Chua NH (2007) ABA induction of miR159 controls transcript levels of two MYB factors during Arabidopsis seed germination. Plant J 49:592–606

Rhoades MW, Reinhart BJ, Lim LP et al (2002) Prediction of plant microRNA targets. Cell 110:513–520

Robert HS, Friml J (2009) Auxin and other signals on the move in plants. Nat Chem Biol 5:325–332

Rubio-Somoza I, Weigel D (2011) MicroRNA networks and developmental plasticity in plants. Trends Plant Sci 16:258–264

Sanan-Mishra N, Kumar V, Sopory SK et al (2009) Cloning and validation of novel miRNA from basmati rice indicates cross talk between abiotic and biotic stresses. Mol Genet Genomics 282:463–474

Sanan-Mishra N, Varanasi SP, Mukherjee S (2013) Micro-regulators of auxin action. Plant Cell Rep 32:733–740

Schaller GE, Bishopp A, Kieber JJ (2015) The Yin-Yang of hormones: cytokinin and auxin interactions in plant development. Plant Cell 27:44–63

Schauer SE, Jacobsen SE, Meinke DW et al (2002) *DICER-LIKE1*: blind men and elephants in Arabidopsis development. Trends Plant Sci 7:487–491

Shen Y, Jiang Z, Lu S et al (2013) Combined small RNA and degradome sequencing reveals microRNA regulation during immature maize embryo dedifferentiation. Biochem Biophys Res Commun 441:425–430

Shivaprasad PV, Chen HM, Patel K et al (2012) A microRNA superfamily regulates nucleotide binding site-leucine-rich repeats and other mRNAs. Plant Cell 24:859–874

Si-Ammour A, Windels D, Arn-Bouldoires E et al (2011) miR393 and secondary siRNAs regulate expression of the TIR1/AFB2 auxin receptor clade and auxin-related development of Arabidopsis leaves. Plant Physiol 157:683–691

Singh N, Srivastava S, Sharma A (2016) Identification and analysis of miRNAs and their targets in ginger using bioinformatics approach. Gene 575:570–576

Soriano M, Li H, Jacquard C et al (2014) Plasticity in cell division patterns and auxin transport dependency during *in vitro* embryogenesis in *Brassica napus*. Plant Cell 26:2568–2581

Sorin C, Bussell JD, Camus I et al (2005) Auxin and light control of adventitious rooting in Arabidopsis require ARGONAUTE1. Plant Cell 17:1343–1359

Staswick PE, Serban B, Rowe M et al (2005) Characterization of an Arabidopsis enzyme family that conjugates amino acids to indole-3-acetic acid. Plant Cell 17:616–627

Stepanova AN, Robertson-Hoyt J, Yun J et al (2008) TAA1-mediated auxin biosynthesis is essential for hormone crosstalk and plant development. Cell 133:177–191

Strader LC, Bartel B (2011) Transport and metabolism of the endogenous auxin precursor indole-3-butyric acid. Mol Plant 4:477–486

Su SH, Gray WM, Masson PH (2015) Auxin: shape matters. Nat Plants 1:15097

Su Y, Liu Y, Zhou C et al (2016) The microRNA167 controls somatic embryogenesis in Arabidopsis through regulating its target genes ARF6 and ARF8. Plant Cell Tiss Org 124:405–417

Sun G (2012) MicroRNAs and their diverse functions in plants. Plant Mol Biol 80:17–36

Sun Z, Wang Y, Mou F et al (2016) Genome-wide small RNA analysis of soybean reveals auxin-responsive microRNAs that are differentially expressed in response to salt stress in root apex. Front Plant Sci 6:1273

Sunkar R, Zhu JK (2004) Novel and stress-regulated microRNAs and other small RNAs from Arabidopsis. Plant Cell 16:2001–2019

Sunkar R, Zhu JK (2007) Micro RNAs and short-interfering RNAs in plants. J Integr Plant Biol 49:817–826

Tan X, Calderon-Villalobos LI, Sharon M et al (2007) Mechanism of auxin perception by the TIR1 ubiquitin ligase. Nature 446:640–645

Tanaka H, Dhonukshe P, Brewer PB et al (2006) Spatiotemporal asymmetric auxin distribution: a means to coordinate plant development. Cell Mol Life Sci 63:2738–2754

Tang G, Reinhart BJ, Bartel DP et al (2003) A biochemical framework for RNA silencing in plants. Genes Dev 17:49–63

Tao Y, Ferrer JL, Ljung K et al (2008) Rapid synthesis of auxin via a new tryptophan-dependent pathway is required for shade avoidance in plants. Cell 133:164–176

Tian C, Muto H, Higuchi K et al (2004) Disruption and overexpression of auxin response factor 8 gene of Arabidopsis affect hypocotyl elongation and root growth habit, indicating its possible involvement in auxin homeostasis in light condition. Plant J 40:333–343

Vaucheret H, Vazquez F, Crété P et al (2004) The action of ARGONAUTE1 in the miRNA pathway and its regulation by the miRNA pathway are crucial for plant development. Genes Dev 18:1187–1197

Vernoux T, Brunoud G, Farcot E et al (2011) The auxin signalling network translates dynamic input into robust patterning at the shoot apex. Mol Syst Biol 7:508

Wang JJ, Guo HS (2015) Cleavage of *INDOLE-3-ACETIC ACID INDUCIBLE28* mRNA by microRNA847 upregulates auxin signaling to modulate cell proliferation and lateral organ growth in Arabidopsis. Plant Cell 27:574–590

Wang XJ, Reyes JL, Chua NH et al (2004) Prediction and identification of *Arabidopsis thaliana* microRNAs and their mRNA targets. Genome Biol 5:R65
Wang JW, Wang LJ, Mao YB et al (2005) Control of root cap formation by microRNA-targeted auxin response factors in Arabidopsis. Plant Cell 17:2204–2216
Wang JW, Schwab R, Czech B et al (2008) Dual effects of miR156-targeted SPL genes and CYP78A5/KLUH on plastochron length and organ size in *Arabidopsis thaliana*. Plant Cell 20:1231–1243
Wang B, Chu J, Yu T et al (2015) Tryptophan-independent auxin biosynthesis contributes to early embryogenesis in *Arabidopsis*. Proc Natl Acad Sci USA 112:4821–4826
Weijers D (2015) Auxin: harnessing a loose cannon. Nat Plants 1:15024
Weijers D, Schlereth A, Ehrismann JS et al (2006) Auxin triggers transient local signaling for cell specification in Arabidopsis embryogenesis. Dev Cell 10:265–270
Windels D, Vazquez F (2011) miR393: integrator of environmental cues in auxin signaling? Plant Signal Behav 6:1672–1675
Won C, Shen X, Mashiguchi K et al (2011) Conversion of tryptophan to indole-3-acetic acid by tryptophan aminotransferases of arabidopsis and YUCCAs in Arabidopsis. Proc Natl Acad Sci USA 108:18518–18523
Woodward AW, Bartel B (2005) Auxin: regulation, action, and interaction. Ann Bot 95:707–735
Wu G, Poethig RS (2006) Temporal regulation of shoot development in *Arabidopsis thaliana* by miR156 and its target SPL3. Development 133:3539–3547
Wu MF, Tian Q, Reed JW (2006) Arabidopsis microRNA167 controls patterns of ARF6 and ARF8 expression, and regulates both female and male reproduction. Development 133:4211–4218
Xie Z, Kasschau KD, Carrington JC (2003) Negative feedback regulation of dicer-like1 in Arabidopsis by microRNA-guided mRNA degradation. Curr Biol 13:784–789
Xu T, Dai N, Chen J et al (2014) Cell surface ABP1-TMK auxin-sensing complex activates ROP GTPase signaling. Science 343:1025–1028
Xu J, Zha M, Li Y et al (2015) The interaction between nitrogen availability and auxin, cytokinin, and strigolactone in the control of shoot branching in rice (*Oryza sativa* L.) Plant Cell Rep 34:1647–1662
Yang JH, Han SJ, Yoon EK et al (2006) Evidence of an auxin signal pathway, microRNA167-ARF8-GH3, and its response to exogenous auxin in cultured rice cells. Nucleic Acids Res 34:1892–1899
Yoon EK, Yang JH, Lim J et al (2010) Auxin regulation of the microRNA390-dependent transacting small interfering RNA pathway in Arabidopsis lateral root development. Nucleic Acids Res 38:1382–1391
Zazimalová E, Napier RM (2003) Points of regulation for auxin action. Plant Cell Rep 21:625–634
Zazimalová E, Murphy AS, Yang H et al (2010) Auxin transporters – why so many? Cold Spring Harb Perspect Biol 2:a001552
Zhai L, Xu L, Wang Y et al (2016) Transcriptional identification and characterization of differentially expressed genes associated with embryogenesis in radish (*Raphanus sativus* L.) Sci Rep 6:21652
Zhang BH, PAN XP, Wang QL et al (2005) Identification and characterization of new plant microRNAs using EST analysis. Cell Res 15:336–360
Zhang B, Pan X, Anderson TA (2006a) Identification of 188 conserved maize microRNAs and their targets. FEBS Lett 580:3753–3762
Zhang B, Pan X, Cannon CH et al (2006b) Conservation and divergence of plant microRNA genes. Plant J 46:243–259
Zhang W, Swarup R, Bennett M et al (2013) Cytokinin induces cell division in the quiescent center of the Arabidopsis root apical meristem. Curr Biol 23:1979–1989
Zhang S, Wang S, Xu Y et al (2015) The auxin response factor, OsARF19, controls rice leaf angles through positively regulating *OsGH3-5* and *OsBRI1*. Plant Cell Environ 38:638–654
Zhao Y (2010) Auxin biosynthesis and its role in plant development. Annu Rev Plant Biol 61:49–64

Zhao Y, Christensen SK, Fankhauser C et al (2001) A role for flavin monooxygenase-like enzymes in auxin biosynthesis. Science 291:306–309

Zhao Y, Hull AK, Gupta NR et al (2002) Trp-dependent auxin biosynthesis in *Arabidopsis*: involvement of cytochrome P450s CYP79B2 and CYP79B3. Genes Dev 16:3100–3112

Zhao T, Li G, Mi S et al (2007) A complex system of small RNAs in the unicellular green alga *Chlamydomonas reinhardtii*. Genes Dev 21:1190–1203

Zhou XY, Song L, Xue HW (2013) Brassinosteroids regulate the differential growth of Arabidopsis hypocotyls through auxin signaling components IAA19 and ARF7. Mol Plant 6:887–904

Zhu Z, Lee B (2015) Friends or foes: new insights in jasmonate and ethylene co-actions. Plant Cell Physiol 56:414–420

Growing Diversity of Plant MicroRNAs and *MIR*-Derived Small RNAs

Mariyana Gozmanova, Vesselin Baev, Elena Apostolova, Gaurav Sablok, and Galina Yahubyan

Abstract Small, noncoding RNAs are essential regulatory molecules of plant genome. Small RNAs (sRNAs) have been classified on the basis of their biogenesis and mode of action in two major types—microRNAs (miRNAs) and small interfering RNAs (siRNAs). Plant miRNAs are typically 21 nucleotides in length and derive from unique genetic loci (*MIR* genes). Next-generation sequencing approaches have increased significantly the number of known plant miRNAs and have revealed that *MIR* genes frequently produce sRNAs, known as miRNA variants, isoforms or isomiRs, which exhibit differences from their corresponding "reference" mature sequences. The main mechanism of action of canonical miRNAs is sequence-specific repression of gene expression on posttranscriptional level. Recent studies have revealed that noncanonical miRNAs and *MIR*-derived siRNAs (a particular subset of isomiRs) can act as well in sequence-specific transcriptional silencing thus influencing genome function through DNA methylation. Moreover, miRNAs can be regulated by epigenetic alteration such as DNA methylation and histone modifications of *MIR* genes. Having profound role in genetic and epigenetic control, plant miRNAs, and *MIR*-derived siRNAs can potentially participate in most developmental processes, plant stress response, and adaptation. In this chapter, we discuss the biogenesis of miRNAs and *MIR*-derived sRNAs and their regulatory impact on plant gene expression.

The original version of this chapter was revised. An erratum to this chapter can be found at DOI 10.1007/978-3-319-55520-1_26.

M. Gozmanova (✉) • V. Baev • G. Yahubyan
Department of Plant Physiology and Molecular Biology, University of Plovdiv, 24 Tsar Assen Str, Plovdiv, Bulgaria

Institute of Molecular Biology and Biotechnologies, 105, Ruski Blvd., 4000 Plovdiv, Bulgaria
e-mail: mgozmanova@gmail.com

E. Apostolova
Department of Plant Physiology and Molecular Biology, University of Plovdiv, 24 Tsar Assen Str, Plovdiv, Bulgaria

G. Sablok
Plant Functional Biology and Climate Change Cluster (C3), University of Technology, Sydney, PO Box 123, Broadway, NSW 2007, Australia

N. Rajewsky et al. (eds.), *Plant Epigenetics*, RNA Technologies,
DOI 10.1007/978-3-319-55520-1_3

The algorithms for accurate annotation of novel miRNAs and isomiR sequences are still a challenging task, requiring integration of experimental and computational approaches. Here, we shortly present some recent tools which have been developed to facilitate this task by providing friendly user interface, without requiring computing skills for the purpose of the analysis.

Keywords miRNAs • *MIR*-derived siRNAs • Isomirs • Long miRNAs • DNA methylation • Histone modifications

Contents

1 Introduction

RNA-mediated processes together with DNA methylation and histone modifications are considered as the main molecular mechanisms constructing the epigenetic regulatory network which has an essential role in plant developmental programs, stress response and adaptation, transposon silencing, signaling pathways, and various non-Mendelian patterns of inheritance (Grant-Downton and Dickinson 2005; Avramova 2011).

2 Micro RNAs in the Plant Small RNA World

Plant small RNAs (sRNAs), which range in size from approximately 20 to 30 nucleotides (nt), can be distinguished both by their biogenesis and mode of action. They include microRNAs (miRNAs) and small interfering RNAs (siRNAs) such as repeat-associated siRNAs (ra-siRNAs), trans-acting siRNAs (ta-siRNAs), and natural antisense transcript siRNAs (nat-siRNAs) (Vaucheret 2006). Each type of sRNAs is unique with respect to the molecular size, the plant DICER-like (DCL) RNaseIII enzyme which is involved in its biogenesis, and the ARGONAUTE (AGO) protein which is directed by the sRNA for silencing of the target gene expression. Most siRNAs target the same locus they are derived from, except for miRNAs and ta-siRNAs, which target mRNAs produced from different loci. Plant sRNA molecules function as negative regulators of gene expression on a transcriptional or posttranscriptional level (Ruiz-Ferrer and Voinnet 2009).

In general, ra-siRNAs (or heterochromatic siRNAs, hc-siRNAs) are 21–24 nt in length, and their biogenesis involves DCL2, DCL3, and DCL4 (Kasschau et al. 2007; Xie et al. 2004). These sRNAs are mainly loaded into AGO3 and are known to be involved in DNA and histone methylation (Xie et al. 2004; Zheng et al. 2007; Zilberman et al. 2003). ta-siRNAs are phased 21 nt RNA molecules whose production involves only DCL4 and is triggered by miRNA-directed cleavage of the TAS transcripts. Further ta-siRNAs are loaded into AGO1 and, like miRNAs, promote sequence-specific cleavage of their targeted gene transcripts (Allen et al. 2005, 2006). The biogenesis of 21 and 24 ntnat-siRNAs involves one of DCL1, DCL2, or DCL3, and a subgroup of nat-siRNAs is dependent on RDR2 and PolIV (Borsani et al. 2005; Katiyar-Agarwal et al. 2006; Zhang et al. 2012). A category of sRNAs ranging from 30 to 40 nt in size, referred to as long siRNAs (lsiRNAs), has been subsequently identified (Katiyar-Agarwal et al. 2007). siRNAs act as transcriptional repressors of a subset of transposons and genes through triggering de novo methylation of homologous DNA in the process of RNA-dependent DNA methylation (RdDM) (Wassenegger et al. 1994; Law and Jacobsen 2010).

First plant sRNAs with miRNA characteristics were recognized in *Arabidopsis* in 2002 (Llave et al. 2002; Mette et al. 2002; Park et al. 2002; Reinhart et al. 2002). Canonical miRNAs are typically 21 nt in length, and their precursors (pri-miRNAs) are transcribed by RNA polymerase II (PolII) (occasionally by RNA Polymerase III) from miRNA (*MIR*) genes (Jones-Rhoades et al. 2006). Some precursors were found produced from the spliced introns of the gene transcripts in *Arabidopsis* and rice (Meng and Shao 2012). The single-stranded precursor forms a hairpin structure which is cut out by DCL1 releasing double-stranded miRNA with an approximately 2-nt 3′-end overhang. Since other DCL proteins (DCL2, DCL3, or DCL4) may recognize and process miRNA precursors, miRNAs of diverse sizes ranging from 20- to 24-nt could be generated (Margis et al. 2006). The double-stranded miRNA is subsequently 2′-*O*-methylated by the methyl transferase Hua enhancer1 (HEN1) and protected by RNA degradation (Yu et al. 2005; Molnár et al. 2007; Abe et al. 2010). Only one strand of the duplex is selected to become the functional miRNA (guide strand, mature miRNA) while the other strand is degraded (passenger strand, miRNA*). In some cases, the passenger strand can be differentially expressed in different tissues and developmental stages and can be functionally active (Okamura et al. 2008). In *Arabidopsis*, miR396a-5p is the mature miRNA which is preferentially expressed in root to target growth-regulating factors (GRFs) family (Jones-Rhoades and Bartel 2004), while miR396a-3p is preferentially expressed in flower (Jeong et al. 2013). There are examples of multiple mature miRNA production from a single precursor as in the case of closely related miR161.1, miR161.2, and miR161.3 targeting genes encoding pentatricopeptide repeat proteins (PPRs) in *Arabidopsis* (Allen et al. 2004; Jeong et al. 2013).

Mature miRNA strand, or occasionally miRNA* strand, is loaded into an AGO protein to form a RNA-induced silencing complex (RISC). Most plant miRNAs possess 5′ uridine, which serves as a sign for association with AGO1 (Mi et al. 2008; Montgomery et al. 2008; Takeda et al. 2008). MiRNA directs RISC to target mRNAs in a sequence-dependent manner. Plant miRNAs share perfect or nearly perfect complementarity with their targets (Rhoades et al. 2002) and bind to complementary sequences in the 3′UTRs of target mRNAs. While miRNAs with near perfect complementarity repress predominantly translation machinery, those with perfect complementarity induce transcript cleavage (Hutvágner and Zamore 2002; Llave et al. 2002; Bartel 2004).

Though miRNA and miRNA* are the most predominant species from a precursor, high-throughput sequencing (HTS) data show that there are often low-frequency positional and length variations from miRNA hairpin structures (Meyers et al. 2008; Morin et al. 2008). These length and sequence variants of canonical miRNAs have been widely demonstrated as isoforms or isomiRs. Based on the homology assessment methods, they have been primarily classified into templated (miRNAs length variants having homology to parent genes) or non-templated (nucleotide additions and/or posttranscriptional RNA edits resulting in no homology to parent genes) isomiRs (Neilsen et al. 2012; Rogans and Rey 2016). Although the precise mechanism by which these isomiRs originate has not yet been established, it has been laid forth that the isomiR biogenesis is due to the imprecise cleavage activity of DCL1 (Bartel 2004), 3′ uridylation by nucleotidyl transferases such as UTP:RNA uridylyltransferase (URT1) and HEN1 suppressor1 (HESO1) (Tu et al. 2015; Wang et al. 2015), or 3′-5′ exoribinuclease activity of Small RNA Degrading Nucleases (SDNs) (Ramachandran and Chen 2008), producing isomiRs with 5′-, 3′-nt additions or deletions. The regulatory roles of isomiRs have already been established in tissue specificity, development, leaf senescence, and stress response (Colaiacovo et al. 2012; Hackenberg et al. 2013; Xu et al. 2014). In *Arabidopsis*, the comparative assessment of the genome-wide sRNA profiles induced by temperature stress displayed differential expression of a specific subset of mature miRNAs and a variety of isomiRs (Baev et al. 2014). For example, the miR160c precursor gave rise to numbers of isomiRs which were found to be differentially expressed upon high- and low-temperature treatment (Fig. 1). The isomiR with the highest copy number was the most considerably upregulated during stress exposure implying its likely functional significance for plant stress response (Baev et al. 2014).

Many miRNAs come in families, in which *MIR* loci are often closely related and occasionally produce identical miRNAs. Some of annotated miRNA families are conserved across vast phylogenetic scales, while others are family- or species-specific (Cuperus et al. 2011). Like miRNAs, numerous isomiRs have been found conserved across species and are likely to participate in regulation of important biological processes (Ameres and Zamore 2013).

3 MiRNA-Mediated DNA Methylation

The increasing evidences of isomiRs complexity in plants have prompted searching for biologically significant variants amongst the large number of *MIR*-derived sRNAs (Sablok et al. 2015). The well-recognized mechanism of action of miRNAs

```
>ath-_MIR160c_MI0000192
GTTATGCCTGGCTCCCTGTATGCCACGAGTGGATACCGATTTTGGTTTTAAAATCGGCTGCCGGTGGCGTACAAGGAGTCAAGCATGAC
((((((.((((((.((((((((((..(..(...(((((((((....)))))))))..)..)..)))))))))).)))))).))))))
                                                                                           Sample   Copy number
...ATGCCTGGCTCCCTGTATGCC.................................................................  LT       515
...ATGCCTGGCTCCCTGTATGCC.................................................................  HT       335
...ATGCCTGGCTCCCTGTATGCC.................................................................  NT       456
...ATGCCTGGCTCCCTGTATGCCA................................................................  LT       2
...ATGCCTGGCTCCCTGTATGCCA................................................................  HT       10
...ATGCCTGGCTCCCTGTATGCCA................................................................  NT       3
....TGCCTGGCTCCCTGTATGCCA................................................................  LT       451
....TGCCTGGCTCCCTGTATGCCA................................................................  HT       741
....TGCCTGGCTCCCTGTATGCCA................................................................  NT       713
..................................................................GCGTACAAGGAGTCAAGC.....  NT       2
..................................................................GCGTACAAGGAGTCAAGCA....  LT       12
..................................................................GCGTACAAGGAGTCAAGCA....  HT       7
..................................................................GCGTACAAGGAGTCAAGCA....  NT       8
..................................................................GCGTACAAGGAGTCAAGCAT...  LT       74
..................................................................GCGTACAAGGAGTCAAGCAT...  HT       222
..................................................................GCGTACAAGGAGTCAAGCAT...  NT       52
..................................................................GCGTACAAGGAGTCAAGCATG..  LT       4331
..................................................................GCGTACAAGGAGTCAAGCATG..  HT       7208
..................................................................GCGTACAAGGAGTCAAGCATG..  NT       2739
..................................................................GCGTACAAGGAGTCAAGCATGA.  HT       4
..................................................................GCGTACAAGGAGTCAAGCATGA.  NT       5
...................................................................CGTACAAGGAGTCAAGCATG..  LT       51
...................................................................CGTACAAGGAGTCAAGCATG..  HT       50
...................................................................CGTACAAGGAGTCAAGCATG..  NT       37
...................................................................CGTACAAGGAGTCAAGCATGA.  LT       295
...................................................................CGTACAAGGAGTCAAGCATGA.  HT       96
...................................................................CGTACAAGGAGTCAAGCATGA.  NT       306
```

Fig. 1 miR160c and its isomiRs in *Arabidopsis*. The profile was generated from NGS datasets derived from temperature conditional responses (*NT* normal temperature, *LT* low temperature, *HT* high temperature). The mature miRNA indexed in miRBase is shown in bold; the isomiR having the highest copy number is underlined

is sequence-specific repression of gene expression on posttranscriptional level. Recent findings have revealed that newly identified miRNAs and *MIR*-derived sRNAs can act as well in sequence-specific transcriptional silencing thus influencing genome function through DNA methylation (Bao et al. 2004; Chellappan et al. 2010; Wu et al. 2010; Khraiwesh et al. 2010).

3.1 First Evidences for an Indirect Link Between MiRNAs and DNA Methylation

The first evidence for involvement of miRNAs in DNA methylation came from the study of Bao et al. (2004) in *Arabidopsis* that intended to answer whether the miRNA complementary site in the target mRNA affected the chromatin state of corresponding gene. For that, they compared DNA methylation between wild-type plants and two mutant lines—*phb-1d* and *phv-1d* in which dominant mutations in the *PHABULOSA* (*PHB*) and PHAVOLUTA (*PHV*) genes disrupted the complementarity site of miRNAs 165 and 166 in *PHB* and *PHV* mRNAs (McConnell et al. 2001; Emery et al. 2003). In the two genes, the miR165/166 complementary site is split by an intron, and the downstream exons were found heavy methylated in the most cells of wild type. Decreased methylation at the *PHB* and *PHV* loci in *phb-1d* and *phv-1d*, respectively, revealed indirect interaction between miR165/166 and the *PHB* and *PHV* templates. *It was hypothesized that the miRNA binds to the complementary site of the processed, nascent mRNA and recruits in trans a chromatin-modifying complex to the closely located template locus* (Bao et al. 2004).

Intriguingly, the *PHB* and *PHV* methylation was not affected in *dcl1* and *ago1* mutants (Bao et al. 2004) that raised the question of whether other players, different but still related to miRNAs, could mediate DNA methylation.

The hypothesis proposed by Bao and collaborators (2004) has been confirmed in moss *Physcomitrella patens* through analyzing the interaction between miR166 and its target mRNAs—*PpC3HDZIP1* and *PpHB10* (Khraiwesh et al. 2010). Similarly to the *PHB* and *PHV* genes in *Arabidopsis*, the miRNA binding site contains an intron and is reconstituted upon splicing of the *PpC3HDZIP1* and *PpHB10* primary transcripts. It was shown in moss mutants without DCL1b that, although the miRNA level was unchanged, the target transcripts were significantly downregulated without being cleaved. Evidences were obtained for accumulation of stable miRNA:target mRNA complexes and for hypermethylation of target loci in these mutants, and was proposed that the ratio between miRNAs and their targets was determining for recruitment *in trans* of the DNA-methylation effector molecules to template loci (Khraiwesh et al. 2010).

3.2 MIR-Derived sRNAs: The Real Players in MiRNA-Mediated DNA Methylation

MIR-Derived sRNA Diversity and Biogenesis In *Arabidopsis*, a novel class of 23- to 27-nt sRNAs was identified in *Arabidopsis* that was produced together with the canonical 20- to 22-nt miRNAs from number of *MIR* genes (Vazquez et al. 2008; Chellappan et al. 2010). Unlike canonical miRNAs which originate from ancient, highly conserved *MIR* genes, the long *MIR*-derived sRNAs originate from recently evolved *MIR* genes (Vazquez et al. 2008). These studies demonstrate that canonical miRNAs and *MIR*-derived sRNAs can be generated independently from the same hairpins by DCL1 and DCL3, respectively. Mutational analysis revealed that the accumulation of *MIR*-derived sRNAs was dependent not only on DCL3 but also on RDR2 and PolIV in *Arabidopsis* (Chellappan et al. 2010). The involvement of PolIV, RDR2, and DCL3, which are the main components of the small interfering RNA (siRNA) biogenesis, in the processing of *MIR*-derived sRNAs is the reason to refer to these sRNAs as *MIR*-derived siRNAs (Chellappan et al. 2010). The question what is the activity of PolIV and RDR2 in *MIR*-derived siRNAs biogenesis in *Arabidopsis* remains still open.

Unlike *Arabidopsis*, numbers of *MIR* genes were found to produce both canonical miRNAs and 24 ntsRNAs, or only 24 ntsRNAs in rice (Zhu et al. 2008; Wu et al. 2009). The two types of sRNA can be processed from the same hairpins by the cooperative action of DCL1 and DCL3, or, as in some cases, a hairpin precursors can be processed by DCL3 giving rise to only 24-nt long sRNA species called long miRNAs or lmiRNAs (Wu et al. 2010). 31 of the 54 lmiRNAs identified by Wu and coworkers in rice (2010) were observed later to be located in the intronic regions of protein-coding genes (Tong et al. 2013). Contrasting to the *Arabidopsis MIR*-

derived siRNAs, there is no evidence to suggest that the formation of lmiRNAs needs RDR2 and PolIV. In tomato, 10 loci encoding putative 24 ntlmiRNAs were predicted, and the expression of four of them was proved to be DCL3-dependent confirming their identity as lmiRNAs (Kravchik et al. 2014). In addition, the expression profiles of two of the tomato lmiRNAs in different organs showed their involvement in tomato reproductive development.

MIR-Derived sRNA Effector Complexes and DNA Methylation Canonical miRNAs and MIR-derived sRNA are loaded on functionally different argonaute complexes—canonical miRNAs associate specifically with AGO1 while MIR-derived siRNAs are sorted in AGO4 clade proteins. The reduced levels of *MIR*-derived siRNAs in *ago4-1* mutant and AGO4-coimmunoprecipitation assay have revealed that these sRNAs could associate with AGO4 in *Arabidopsis* (Chellappan et al. 2010). The observations of Wu et al. (2010) in rice suggest that lmiRNAs initiated with adenine sort into AGO4a, AGO4b, and AGO16, and those beginning with uracil are loaded on AGO4b.The association of *MIR*-derived siRNAs with AGO4 clade proteins suggested their involvement in DNA methylation.

There are increasing evidences that some *MIR*-derived siRNAs are indispensable for DNA methylation of their target loci *in trans* and/or of their own *MIR* loci *in cis*. In the *Arabidopsis* mutant *nrpd1-3* lacking the Pol IV largest subunit, where 23- to 26-nt sRNAs were absent, reduced DNA methylation of the putative target mRNAs At4g16580 and At5g08490 for the recently evolved miR2328 and miR2831-5P, respectively, correlated with upregulated mRNA expression (Chellappan et al. 2010). In the same mutant, the DNA methylation of *SPL2* (squamosa-promoter binding protein-like), a target of the canonical miR156, was found reduced up- and downstream of the miR target site compared to wild type. To examine the DNA methylation of lmiRNA-producing and target loci in rice, two mutant lines *dcl3a-17* and *rdr2-2* were subjected to bisulfite sequencing (Wu et al. 2010). Since the lmiRNAs biosynthesis requires DCL3, but not RDR2, the comparison of methylation profiles of these mutants would allow for differentiation between siRNA- and lmiRNA-dependent DNA methylation. In that way, it was demonstrated that miR1873 directed the methylation at its own locus as well as that miR1863, miR820.2, miR1873.1, and miR1876 mediated the methylation of their target genes (Wu et al. 2010). Unlike siRNA-dependent DNA methylation spreading in the 3′-direction only, lmiRNAs induce DNA methylation bidirectionally from the miRNA-binding site in target genesas exemplified by bisulfite sequencing of the target genes of miR1862c, miR1863b, miR1867, miR2121b, miR5150, and miR5831 in rice (Hu et al. 2014) and of miR160 and miR166 in moss (Kravchik et al. 2014).

The described variety of plant canonical and non-canonical miRNAs and *MIR*-derived siRNAs, with the particularities of their biogenesis, AGO incorporation, and mode of action, is schematically presented in Fig. 2.

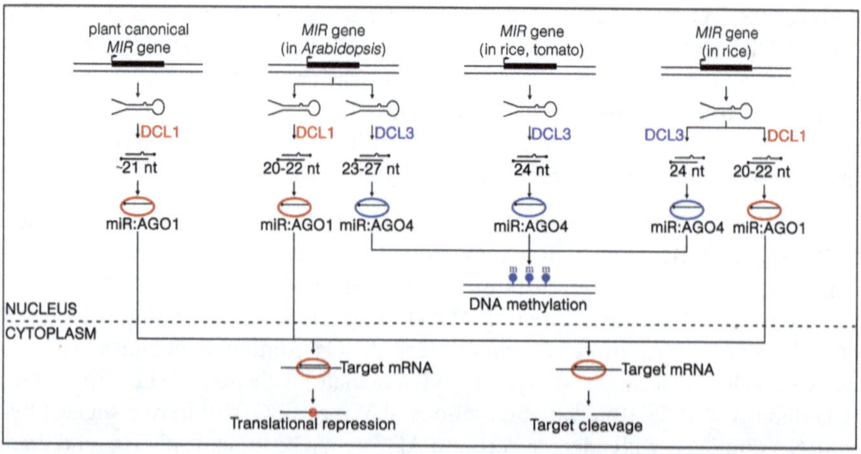

Fig. 2 Diversity of plant *MIR*-derived sRNAs. The hairpin precursors, transcribed from most plant *MIR* genes, are cut out by DCL1 to produce canonical mature ~21 nt miRNAs which associate with AGO1 and mediate target mRNA cleavage or translational repression. In addition to this classical pathway, some plant *MIR* genes can generate sRNA species that differ from the canonical miRNAs. In *Arabidopsis*, two sRNA species—canonical miRNAs and *MIR*-derived siRNAs (23–27 nt)—can be generated independently from different molecules of the same hairpin population by DCL1 and DCL3, respectively (Chellappan et al. 2010). In rice and tomato, some *MIR* genes produce only 24 ntlmiRNAs using DCL3, while other *MIR* genes can produce canonical miRNA and lmiRNA species simultaneously by coordinate activities of DCL1 and DCL3 on the same molecule (Wu et al. 2010; Kravchik et al. 2014). *MIR*-derived siRNAs and lmiRNAs associate predominantly with AGO4 and mediate DNA methylation of target genes or their own *MIR* genes. The DCL1/AGO1 pathway is depicted in *red* color, while the DCL3/AGO4 pathway is depicted in *blue* color

4 Epigenetic Control of *MIR* Genes

Plant miRNA expression might be regulated on transcriptional level by chromatin remodeling due to histone modification or DNA methylation of the corresponding *MIR* loci.

4.1 Impact of Histone Modifications of MIR Loci on MiRNA Expression

The studies of Kim et al. (2009) performed with the *Arabidopsis* mutants, *gcn5-1* and *gcn 5-2*, revealed that acetylation of histone H3 lysine 14 at numbers of miRNA loci interfered with miRNA production. The analysis of pri-miRNA/mature miRNA accumulation levels suggested a role of GCN5 (histone acetyltransferase) at both pri-miRNAs maturation and miRNA expression.

Briefly, the reduced expression of miRNA processing genes—DCL1, SE, HYL, and AGO1, correlated with low accumulation of pri-miRNAs and high levels of mature miRNAs in the wild type. In *gcn5* mutants, the expression of these components of miRNA processing pathway was upregulated and correlated with increased levels of miRNAs and decreased levels of corresponding pre-miRNAs. No direct interaction between GCN5 and DCL1, SE, HYL, and AGO1 genes was identified. These data inspired the hypothesis of the presence of a common repressor of these genes, the activity of which was influenced by GCN5.

Furthermore, a direct interaction was described between GCN5 and four *MIR* genes (miR165a, miR172a, miR395e, and miR399d) in the wild-type *Arabidopsis*. Based on studies of *gcn 5-2* mutant, the specific miRNA-protein interaction was mapped to the bromodomain of GCN5. Moreover, some histone deacetylases (HDA9 and HDA19) were also seen as factors regulating miRNA accumulation.

Nosaka et al. (2012) have found that miR820 cleaves *OsDRM2* mRNA and induce DNA methylation at the miR820 target site in *OsDRM2* gene. On the other hand, it became clear that miR820 is encoded by CACTA TEs (five copies, located on different chromosomes) (Nosaka et al. 2013). The heterochromatic mark (H3K9 dimethylation) is detected in CACTA copies carrying miR820 genes suggesting repression of miR820. However, in one of the CACTA copies, that resides in chromosome 7 and produces transcripts, low levels of active histone marks (H3K4 di/tri methylation and H3K9 acetylation) correlate with high level of asymmetric cytosine methylation (CHH) in the same region, but still the transcription of miR820 from the this locus is allowed.

In *Brachypodium distachyon*, a representative of Pooideae plants, a newly evolved, species-specific miRNA-miR5200 was identified to target the mRNAs of two florigen genes, FTL1 and FTL2 (Wu et al. 2013a, b). The miR5200 expression was found changed under different day lengths being upregulated under short day (SD) and downregulated under long day (LD). The authors claimed that differential expression of miR5200 might due to changes in chromatin modifications at the *MIR* genes, MIR5200a and MIR5200b. A repressive histone mark H3K27tri-methylated was enriched in the MIR5200 genes under LD conditions. Thus, epigenetics control of miR5200 expression was found to participate in the photoperiodic regulation of the transition from vegetative toward reproductive stage in *B. distachyon* (Wu et al. 2013a, b).

4.2 DNA Methylation of MIR Genes Affects MiRNA Expression

DNA methylation status at CG, CHG, and CHH contexts has been explored in both promoter region and gene body of *MIR* genes in rice. These parameters were compared between conserved and species-specific miRNA (Hu et al. 2014). The highest rate of methylation was determined in CG context of both promoters and gene bodies of non-conserved *MIR* genes. It was found that the genes of majority of highly expressed, conserved miRNAs were constitutively hypomethylated, while the genes of recently evolved miRNAs were hypermethylated at both promoters and gene bodies. The authors suggested that a strong control orchestrated by DNA methylation had been established in plant evolution as a means for repression of newly evolved species-specific miRNAs.

DNA methylation landscape was monitored in five different rice tissues – embryo, endosperm, root, shoot, and mature leaves (Hu et al. 2014). The promoters and gene bodies of non-conserved *MIR* genes showed the same trend in their methylation profile in CG context among the studied tissues. Hypomethylation was detected in all contexts in endosperm. A higher rate of methylation was detected only in CHH context in embryo and mature leaves. DNA methylation of promoters and gene bodies of conserved miRNAs didn't show significant differences in all tissues except for the endosperm where reduced methylation in CHG and CHH context was observed.

In poplar, a relationship between DNA methylation and gene expression was described for both long noncoding RNAs (lncRNA) >200 bp and miRNAs (Song et al. 2016).The forth exon of lncRNA00268512 gene was observed on the complementary strand of the first exon of the protein-coding gene Potri.018G127000, in which miR396e was located. Due to this substantial overlap, it was proposed that the lncRNA00268512 and miR396e may interact (Song et al. 2016). Moreover, a stress-specific differentially methylated region (SDMR 162) was identified in the first exon of Potri.018G127000. The decreased DNA methylation of SDMR 162 correlated with increased levels of lncRNA upon cold and osmotic stress. It was suggested that the excess lncRNA molecules could captured miRNA396e-3p and caused its lower abundance in response to stress.

4.3 Link Between MIR Gene DNA Methylation and Plant Stress Response

Abiotic stress-responsive DNA methylation was observed in five conserved *MIR* genes (miR167-3p, miR6445a, miRNA319c, miR156f, and miR472a) and eleven non-conserved *MIR* genes in *Populus simonii* (Song et al. 2016). A long-term impact of DNA methylation on gene expression of miRNAs was seen in response to short-term abiotic stress in ~15% of de novo methylated sites. Stress-induced DNA demethylation was reported at two *MIR* genes encoding miR156f and miR472b transcripts. Approximately, 11% of demethylated sites that responded to abiotic stress were found preserved 6 months later (Song et al. 2016).

Ci et al. (2015) identified 1066 stress-specific DNA methylated sites in poplarin response to heat and cold stress. Analysis of DNA methylation levels showed 150 stress-specific DNA methylated sites per each type of stress and 100 DNA methylated sites common to both stress types. Seven *MIR* genes (miR156i, miR156j, miR167h, miR390c, miR393a, miR396e, and miR396g) were found to be differentially methylated in response to temperature stress. Moreover, their expression was influenced by the cytosine methylation pattern. Most temperature responsive *MIR* genes that carry CNG methylation pattern displayed higher expression than those having CG methylation.

5 Computational Tools for Plant MiRNA Analysis from NGS Datasets

The identification of the entire repertoire of sRNAs and miRNAs has been made possible by the next-generation sequencing (NGS) techniques in an efficient and cost-effective fashion. Due to the big data output of these methods, the computational approaches are a necessary step to depict miRNAs in outgrowing datasets.

A typical bioinformatics miRNA identification algorithm based on NGS datasets involves several steps including, but not limited to (1) quality filtering and adapter trimming; (2) mapping of sRNA reads to identify corresponding genomic loci; (3) estimating the miRNA expression based on copy number; (4) exploring the 2D RNA structure of the loci for identification of pre-miRNA; etc. sRNA sequencing libraries usually also contain other non-miRNA RNA molecules, such as other sRNAs, degradation reads from protein-coding genes, rRNA reads, etc. The computational miRNA identification also compels filtering these non-miRNAs as much as possible from the reads in the library to make analysis more accurate and the "background" to be discarded. Each step in the process may result in a change of the output of the analysis, but perhaps the most crucial part is how the mapping stage is done. The software pipelines available so far vary in terms of user interface, user control, parameters, the input format data, reference databases, and how they adopt each of the above analysis stages.

The key step in NGS data analysis is the mapping the huge amount of short reads to a given genome. Several algorithms and software modules have been specifically designed for dealing with the alignment of millions of reads. Some of the most used tools for the alignment to the reference genome are Bowtie (http://bowtie.cbcb.umd.edu/) (Langmead et al. 2009), BWA (http://maq.sourceforge.net/) (Li and Durbin 2009), MAQ (http://maq.sourceforge.net/) (Li et al. 2008), and SOAP (http://soap.genomics.org.cn/) (Li et al. 2008).

The standard approach for identification of plant miRNAs involves cross-species discovery of the conserved miRNAs in the sample due to the fact that large population of miRNAs has orthologues in the plant kingdom. However, this is limited to organisms where known reference genome and miRNAs genes are available.

In such comparative analysis, the NGS short reads are aligned to a known reference database. These miRNA databases are the source for known miRNA sequence (mature and hairpin sequences) and annotation information. They also are essential for expression profiling of miRNAs. The most popular miRNA databases include, but not limited to: miRBase (Griffiths-Jones 2004), deep Base (Yang and Qu 2012), microRNA.org (Betel et al. 2008), miRGen (Megraw et al. 2007), miRNAMap (Hsu et al. 2006), miRNEST (Szcześniak et al. 2012), and PMRD (Zhang et al. 2010); among them the most comprehensive ones for plant miRNAs are miRBase and PMRD.

The more challenging analysis is to try to identify novel mature and precursor miRNA sequences directly from NGS read data without any dependence upon homologous references of known conserved plant miRNAs. Such methods usually include deep investigation of the mapping results and the clustering of the sRNAs produced from miRNA/miRNA* regions. To discover the mature miRNA among such clusters of sRNAs expressed from the precursor region, only those reads which fit a specific criteria (e.g., could form duplex, maximum with four mismatches, observed 3′ overhangs, duplex length stayed within the range of 18–24 bp, and high copy number) are considered as miRNA candidates (Meyers et al. 2008). However, such rules have limitations and necessitate experimental validation over large amount of datasets. Based on PCR experiments which aimed to validate de novo predicted miRNA candidates, some authors reported that 40% of them were false positive (Wei et al. 2009). All these studies have urged the researchers in the recent years to try to optimize the existing and develop new methods to decrease the false positive rate of the output.

A variety of web-based and stand-alone software have been developed for analyses of plant miRNA data. The list of some of the available software that are designed specifically for plant miRNAs or can be used with plant miRNAs datasets can be found in Table 1.

The recent discovery that miRNAs can both regulate and be regulated by target interactions has a key role for understanding their roles in gene regulation (Salmena et al. 2011). A unique and remaining task in the field is the capacity to identify miRNA targets with high confidence. Moreover, finding the real functional miRNA targets is still puzzling even though the biological rules of miRNA targeting have been shown experimentally and computationally.

The classical way of identifying a plant miRNA target lays on the complementarity between itself and its mRNA site defined by the stability of the duplex which has been utilized widely as a main feature in the analysis step by computational tools. Some of the tools that utilize these algorithms are psRNATarget (Dai and Zhao 2011) and imiRTP (Ding et al. 2012), which predict the functional type of miRNA based on the complementary at the central region of the miRNA:target pair.

Recently in the era of NGS, degradome datasets have been used to find evidence of cleaved miRNA targets without relying on computational RNA folding predictions. Experimental methods have shown that miRNA AGO-mediated cleavage of mRNA happens exactly between the 10th and 11th nucleotide of miRNA–mRNA duplex. The subsequent upstream molecule of the cleaved target is degraded, but the downstream fragment is shown to be stable (Llave et al. 2002). And therefore, NGS techniques involving the capturing of these downstream fragments (Addo-Quaye et al. 2008; German et al. 2008; Gregory et al. 2008) and

Table 1 Bioinformatics tools for plant miRNA analysis from NGS datasets

Tool name	Reference	Web page
PsRobot	Wu et al. (2012)	http://omicslab.genetics.ac.cn/psRobot/
Semirna	Muñoz-Mérida et al. (2012)	http://www.bioinfocabd.upo.es/semirna/
miRDeepFinder	Xie et al. (2012)	http://www.leonxie.com/DeepFinder.php
psRNATarget	Dai and Zhao (2011)	http://plantgrn.noble.org/psRNATarget/
miRDeep-P	Yang and Li. (2011)	http://faculty.virginia.edu/lilab/miRDP/
miRA	Evers et al. (2015)	https://github.com/mhuttner/miRA
plantDARIO	Patra et al. (2014)	https://github.com/mhuttner/miRA
CAP-miRSeq	Sun et al. (2014)	http://bioinformaticstools.mayo.edu/research/cap-mirseq/
miREvo	Wen et al. (2012)	http://evolution.sysu.edu.cn/software/mirevo.htm
miRanalyser	Hackenberg et al. (2011)	http://bioinfo2.ugr.es/miRanalyzer/miRanalyzer.php
miR-PREFeR	Lei and Sun. (2014)	https://github.com/hangelwen/miR-PREFeR
miRPlant	An et al. (2014)	http://sourceforge.net/projects/mirplant/
mirTools 2.0	Wu et al. (2013a, b)	http://centre.bioinformatics.zj.cn/mr2_dev/index.php
sRNAtoolbox	Rueda et al. (2015)	http://bioinfo5.ugr.es/srnatoolbox/index
UEA sRNA Workbench	Stocks et al. (2012)	http://srna-workbench.cmp.uea.ac.uk/
ShortStack	Axtell (2013)	http://sites.psu.edu/axtell/software/shortstack/
CleaveLand	Addo-Quaye et al. (2009)	http://www.bio.psu.edu/people/faculty/Axtell/AxtellLab/Software.html
PAREsnip	Folkes et al. (2012)	http://srna-workbench.cmp.uea.ac.uk/tools/analysis-tools/paresnip/
PatMaN	Prüfer et al. (2008)	http://bioinf.eva.mpg.de/patman
SoMART	Li et al. (2012)	http://bakerlab.berkeley.edu/somart-webserver-mirna-sirna-analysis
StarScan	Liu et al. (2015)	http://mirlab.sysu.edu.cn/starscan/

the following bioinformatics analysis of such datasets can be used to identify the miRNA targets. Recently, several tools have been developing that use degradome data in order to predict miRNA targets. Among the most popular software are CleaveLand (Addo-Quaye et al. 2009), SeqTar (Zheng et al. 2012), PAREsnip (Folkes et al. 2012), PatMaN (Prüfer et al. 2008), SoMART (Li et al. 2012), and StarScan (Liu et al. 2015).

In the past few years, cohorts of tools have been developed to identify isomiRs, which are either stand-alone or web-based (Table 2). Among the tools used to profile isomiRs in plants are SeqCluster (Pantano et al. 2011), miRSeqNovel (Qian et al. 2012), isomiRID (de Oliveira et al. 2013), sRNAtoolbox (Rueda et al. 2015), isomiRex (Sablok et al. 2013), and isomiRage (Muller et al. 2014). Although these tools allow the identification of the isomiRs, they suffer some omissions such as isomiR profiling that would take into account the sequencing artifacts, expression-

Table 2 Recently developed classification tools for identifying isomiRs

Tool name	Reference	Web page
SeqCluster	Pantano et al. (2011)	https://github.com/lpantano/seqbuster
miRSeqNovel	Qian et al. (2012)	http://sourceforge.net/projects/mirseq/files
isomiRID	de Oliveira et al. (2013)	http://www.ufrgs.br/RNAi/isomiRID/
isomiRex	Sablok et al. (2013)	bioinfo1.uni-plovdiv.bg/isomiRex/
miRspring	Humphreys and Suter (2013)	http://mirspring.victorchang.edu.au
IsomiRage	Muller et al. (2014)	http://cru.genomics.iit.it/Isomirage/
sRNAtoolbox	Rueda et al. (2015)	http://bioinfo5.ugr.es/srnatoolbox
isomiR-SEA	Urgese et al. (2016)	http://eda.polito.it/isomir-sea
miR-isomiRExp	Guo et al. (2016)	http://mirisomirexp.aliapp.com
DeAnnoIso	Zhang et al. (2016)	http://mcg.ustc.edu.cn/bsc/deanniso/
IsomiRbank	Zhang et al. (2016)	http://mcg.ustc.edu.cn/bsc/isomir/

based read support, visualization of the isomiRs with respect to read depth and mapping, target predictions, and functional enrichment. Some tools such as isomiRex (Sablok et al. 2013) are web-based, and without allowing the identification of novel miRNAs, provide support for isomiRs visualization based on read depth, whereas stand-alone tools lack for isomiRs visualization, but support the PARE-Seq-based targets predictions. Although the recently published isomiR Detection tool DeAnnoIso (Zhang et al. 2016) allows for the detection of isomiRs, it can be only accessible through the web interface and lacks to detect the isomiRs across a wide range of plant species (only four plant species are supported). IsomiRage can group the functionally relevant isomiRs according to the adenylation, uridylation, and other respective events in response to biological context (Muller et al. 2014). Other tools such as miR-isomiRExp (Guo et al. 2016) and isomiR-SEA (Urgese et al. 2016) are yet to be assessed for their efficiency in plant isomiR profiling. In terms of the pre-analyzed isomiRs, only one such database tool isomiRBank (Zhang et al. 2016) exists, providing the pre-compiled set of isomiRs across four plant species.

Conclusions The use of genome-wide technologies has enabled the identification of novel miRNAs and a large number of isomiRs that transform the plant miRNAome in an increasingly complex world. Though some of the emerging *MIR*-derived siRNAs resemble canonical miRNAs, they show deviation from the conventional view of miRNA biogenesis, AGO incorporation, mode of action, and regulatory effect on gene expression. LmiRNAs are an excellent example of the biological significance of such *MIR*-derived siRNAs, that being mediators in a noncanonical RdDM pathway, lead to the fine-tuning of epigenetic control in plants. Furthermore, the variability in DNA methylation patterns of genes encoding novel, species-specific miRNAs, induced by different types of stress, upgrades our view about the regulatory networks associated with plant response, adaptability, and tolerance.

References

Abe M, Yoshikawa T, Nosaka M et al (2010) WAVY LEAF1, an ortholog of arabidopsis HEN1, regulates shoot development by maintaining microRNA and trans-acting small interfering RNA accumulation in rice. Plant Physiol 154:1335–1346

Addo-Quaye C, Eshoo TW, Bartel DP et al (2008) Endogenous siRNA and microRNA targets identified by sequencing of the Arabidopsis degradome. Curr Biol 18:758–762

Addo-Quaye C, Miller W, Axtell MJ (2009) CleaveLand: a pipeline for using degradome data to find cleaved small RNA targets. Bioinformatics 25:130–131

Allen E, Xie Z, Gustafson AM et al (2004) Evolution of microRNA genes by inverted duplication of target gene sequences in Arabidopsis thaliana. Nat Genet 36:1282–1290

Allen E, Xie Z, Gustafson AM et al (2005) microRNA-directed phasing during trans-acting siRNA biogenesis in plants. Cell 121:207–221

Ameres SL, Zamore PD (2013) Diversifying microRNA sequence and function. Nat Rev Mol Cell Biol 14:475–488

An J, Lai J, Sajjanhar A et al (2014) miRPlant: an integrated tool for identification of plant miRNA from RNA sequencing data. BMC Bioinformatics 15:275

Avramova Z (2011) Handbook of epigenetics. In: Tollefsbol T (ed) The new molecular and medical genetics. Elsevier, Amsterdam, pp 251–278

Axtell MJ (2013) ShortStack: comprehensive annotation and quantification of small RNA genes. RNA 19:740–751

Axtell MJ, Jan C, Rajagopalan R et al (2006) A two-hit trigger for siRNA biogenesis in plants. Cell 127:565–577

Baev V, Milev I, Naydenov M et al (2014) Insight into small RNA abundance and expression in high- and low-temperature stress response using deep sequencing in Arabidopsis. Plant Physiol Biochem 84:105–114

Bao N, Lye K-W, Barton MK (2004) MicroRNA binding sites in Arabidopsis class III HD-ZIP mRNAs are required for methylation of the template chromosome. Dev Cell 7:653–662

Bartel DP (2004) MicroRNAs: genomics, biogenesis, mechanism, and function. Cell 116:281–297

Betel D, Wilson M, Gabow A et al (2008) The microRNA.org resource: targets and expression. Nucleic Acids Res 36:D149–D153

Borsani O, Zhu J, Verslues PE et al (2005) Endogenous siRNAs derived from a pair of natural cis-antisense transcripts regulate salt tolerance in Arabidopsis. Cell 123:1279–1291

Chellappan P, Xia J, Zhou X et al (2010) siRNAs from miRNA sites mediate DNA methylation of target genes. Nucleic Acids Res 38:6883–6894

Ci D, Song Y, Tian M et al (2015) Methylation of miRNA genes in the response to temperature stress in Populussimonii. Front Plant Sci 6:921

Colaiacovo M, Bernardo L, Centomani I et al (2012) A survey of MicroRNA length variants contributing to miRNome complexity in peach (PrunusPersica L.) front. Plant Sci 3:165

Cuperus JT, Fahlgren N, Carrington JC (2011) Evolution and functional diversification of MIRNA genes. Plant Cell 23:431–442

Dai X, Zhao PX (2011) psRNATarget: a plant small RNA target analysis server. Nucleic Acids Res 39:W155–W159

de Oliveira LF, Christoff AP, Margis R (2013) isomiRID: a framework to identify microRNA isoforms. Bioinformatics 29:2521–2523

Ding J, Li D, Ohler U et al (2012) Genome-wide search for miRNA-target interactions in Arabidopsis thaliana with an integrated approach. BMC Genomics 13:S3–S3

Emery JF, Floyd SK, Alvarez J et al (2003) Radial patterning of Arabidopsis shoots by class III HD-ZIP and KANADI genes. Curr Biol 13:1768–1774

Evers M, Huttner M, Dueck A et al (2015) miRA: adaptable novel miRNA identification in plants using small RNA sequencing data. BMC Bioinformatics 16:370

Folkes L, Moxon S, Woolfenden HC et al (2012) PAREsnip: a tool for rapid genome-wide discovery of small RNA/target interactions evidenced through degradome sequencing. Nucleic Acids Res 40:e103–e103

German MA, Pillay M, Jeong D-H et al (2008) Global identification of microRNA-target RNA pairs by parallel analysis of RNA ends. Nat Biotech 26:941–946

Grant-Downton RT, Dickinson HG (2005) Epigenetics and its implications for plant biology. 1. The epigenetic network in plants. Ann Bot 96:143–1164

Gregory BD, O'Malley RC, Lister R, Urich MA, Tonti-Filippini J, Chen H, Millar AH, Ecker JR (2008) A link between RNA metabolism and silencing affecting Arabidopsis development. Develop Cell 14:854–866

Griffiths-Jones S (2004) The microRNA registry. Nucleic Acids Res 32:D109–D111

Guo L, Yu J, Liang T et al (2016) miR-isomiRExp: a web-server for the analysis of expression of miRNA at the miRNA/isomiR levels. Scientific Rep 6:23700

Hackenberg M, Rodríguez-Ezpeleta N, Aransay AM (2011) miRanalyzer: an update on the detection and analysis of microRNAs in high-throughput sequencing experiments. Nucleic Acids Res 39:W132–W138

Hackenberg M, Shi B-J, Gustafson P et al (2013) Characterization of phosphorus-regulated miR399 and miR827 and their isomirs in barley under phosphorus-sufficient and phosphorus-deficient conditions. BMC Plant Biol 13:214–214

Hsu PWC, Huang H-D, Hsu S-D et al (2006) miRNAMap: genomic maps of microRNA genes and their target genes in mammalian genomes. Nucleic Acids Res 34:D135–D139

Hu W, Wang T, Xu J et al (2014) MicroRNA mediates DNA methylation of target genes. Biochem Biophys Res Commun 444:676–681

Humphreys DT, Suter CM (2013) miRspring: a compact standalone research tool for analyzing miRNA-seq data. Nucleic Acids Res 41:e147

Hutvágner G, Zamore PD (2002) A microRNA in a multiple-turnover RNAi enzyme complex. Science 297:2056

Jeong DH, Thatcher SR, Brown RS et al (2013) Comprehensive investigation of microRNAs enhanced by analysis of sequence variants, expression patterns, ARGONAUTE loading, and target cleavage. Plant Physiol 162:1225–1245

Jones-Rhoades Matthew W, Bartel David P (2004) Computational identification of plant microRNAs and their targets, including a stress-induced miRNA. Mol Cell 14:787–799

Jones-Rhoades MW, Bartel DP, Bartel B (2006) MicroRNAs and their regulatory roles in plants. Annu Rev Plant Biol 57:19–53

Kasschau KD, Fahlgren N, Chapman EJ et al (2007) Genome-wide profiling and analysis of Arabidopsis siRNAs. PLoS Biol 5:e57

Katiyar-Agarwal S, Morgan R, Dahlbeck D et al (2006) A pathogen-inducible endogenous siRNA in plant immunity. Proc Nat Acad Sci U S A 103:18002–18007

Katiyar-Agarwal S, Gao S, Vivian-Smith A et al (2007) A novel class of bacteria-induced small RNAs in Arabidopsis. Genes Dev 21:3123–3134

Khraiwesh B, Arif MA, Seumel GI et al (2010) Transcriptional control of gene expression by microRNAs. Cell 140:111–122

Kim W, Benhamed M, Servet C et al (2009) Histone acetyltransferase GCN5 interferes with the miRNA pathway in Arabidopsis. Cell Res 19:899–909

Kravchik M, Damodharan S, Stav R et al (2014) Generation and characterization of a tomato DCL3-silencing mutant. Plant Sci 221–222:81–89

Langmead B, Trapnell C, Pop M et al (2009) Ultrafast and memory-efficient alignment of short DNA sequences to the human genome. Genome Biol 10:R25–R25

Law JA, Jacobsen SE (2010) Establishing, maintaining and modifying DNA methylation patterns in plants and animals. Nat Rev Genet 11:204–220

Lei J, Sun Y (2014) miR-PREFeR: an accurate, fast and easy-to-use plant miRNA prediction tool using small RNA-Seq data. Bioinformatics 30:2837–2839

Li H, Durbin R (2009) Fast and accurate short read alignment with Burrows–Wheeler transform. Bioinformatics 25:1754–1760

Li R, Li Y, Kristiansen K et al (2008) SOAP: short oligonucleotide alignment program. Bioinformatics 24:713–714

Li F, Orban R, Baker B (2012) SoMART: a web server for plant miRNA, tasiRNA and target gene analysis. Plant J 70:891–901

Liu X, Hao L, Li D et al (2015) Long non-coding RNAs and their biological roles in plants. Genomics Proteomics Bioinformatics 13:137–147

Llave C, Xie Z, Kasschau KD et al (2002) Cleavage of Scarecrow-like mRNA targets directed by a class of Arabidopsis miRNA. Science 297:2053–2056

Margis R, Fusaro AF, Smith NA et al (2006) The evolution and diversification of Dicers in plants. FEBS Lett 580:2442–2450

McConnell JR, Emery J, Eshed Y et al (2001) Role of PHABULOSA and PHAVOLUTA in determining radial patterning in shoots. Nature 411:709–713

Megraw M, Sethupathy P, Corda B et al (2007) miRGen: a database for the study of animal microRNA genomic organization and function. Nucleic Acids Res 35:D149–D155

Meng Y, Shao C (2012) Large-scale identification of mirtrons in Arabidopsis and rice. PLoS ONE 7:e31163

Mette MF, van der Winden J, Matzke M et al (2002) Short RNAs can identify new candidate transposable element families in Arabidopsis. Plant Physiol 130:6–9

Meyers BC, Axtell MJ, Bartel B et al (2008) Criteria for annotation of plant microRNAs. Plant Cell 20:3186–3190

Mi S, Cai T, Hu Y et al (2008) Sorting of small RNAs into Arabidopsis argonaute complexes is directed by the 5′ terminal nucleotide. Cell 133:116–127

Molnár A, Schwach F, Studholme DJ et al (2007) miRNAs control gene expression in the single-cell alga Chlamydomonasreinhardtii. Nature 447:1126–1129

Montgomery TA, Howell MD, Cuperus JT et al (2008) Specificity of ARGONAUTE7-miR390 interaction and dual functionality in TAS3 trans-acting siRNA formation. Cell 133:128–141

Morin RD, O’Connor MD, Griffith M et al (2008) Application of massively parallel sequencing to microRNA profiling and discovery in human embryonic stem cells. Genome Res 18:610–621

Muller H, Marzi MJ, Nicassio F (2014) IsomiRage: from functional classification to differential expression of miRNA isoforms. Front Bioeng Biotechnol 2:38

Muñoz-Mérida A, Perkins JR, Viguera E et al (2012) Semirna: searching for plant miRNAs using target sequences. OMICS 16:168–177

Neilsen CT, Goodall GJ, Bracken CP (2012) IsomiRs—the overlooked repertoire in the dynamic microRNAome. Trends Genet 28:544–549

Nosaka M, Itoh J-I, Nagato Y et al (2012) Role of transposon-derived small RNAs in the interplay between genomes and parasitic DNA in rice. PLoS Genet 8:e1002953

Nosaka M, Ono A, Ishiwata A et al (2013) Expression of the rice microRNA miR820 is associated with epigenetic modifications at its own locus. Genes Genet Syst 88:105–112

Okamura K, Phillips MD, Tyler DM et al (2008) The regulatory activity of microRNA* species has substantial influence on microRNA and 3′ UTR evolution. Nat Struct Mol Biol 15:354–363

Pantano L, Estivill X, Martí E (2011) A non-biased framework for the annotation and classification of the non-miRNA small RNA transcriptome. Bioinformatics 27:3202–3203

Park W, Li J, Song R et al (2002) CARPEL FACTORY, a dicer homolog, and HEN1, a novel protein, act in microRNA metabolism in Arabidopsis thaliana. Curr Biol 12:1484–1495

Patra D, Fasold M, Langenberger D et al (2014) plantDARIO: web based quantitative and qualitative analysis of small RNA-seq data in plants. Front Plant Sci 5:708

Prüfer K, Stenzel U, Dannemann M et al (2008) PatMaN: rapid alignment of short sequences to large databases. Bioinformatics 24:1530–1531

Qian K, Auvinen E, Greco D et al (2012) miRSeqNovel: An R based workflow for analyzing miRNA sequencing data. Mol Cell Probes 26:208–2011

Ramachandran V, Chen X (2008) Degradation of microRNAs by a family of exoribonucleases in Arabidopsis. Science 321:490–1492

Reinhart BJ, Weinstein EG, Rhoades MW et al (2002) MicroRNAs in plants. Genes Develop 16:1616–1626

Rhoades MW, Reinhart BJ, Lim LP et al (2002) Prediction of plant microRNA targets. Cell 110:513–520

Rogans SJ, Rey C (2016) Unveiling the micronome of cassava (Manihot esculenta Crantz). PLoS One 11:e0147251

Rueda A, Barturen G, Lebrón R et al (2015) sRNAtoolbox: an integrated collection of small RNA research tools. Nucleic Acids Res 43:W467–W473

Ruiz-Ferrer V, Voinnet O (2009) Roles of plant small RNAs in biotic stress responses. Annu Rev Plant Biol 60:485–510

Sablok G, Milev I, Minkov G et al (2013) isomiRex: web-based identification of microRNAs, isomiR variations and differential expression using next-generation sequencing datasets. FEBS Lett 587:2629–2634

Sablok G, Srivastva AK, Suprasanna P et al (2015) isomiRs: increasing evidences of isomiRs complexity in plant stress functional biology. Front Plant Sci 6:949–949

Salmena L, Poliseno L, Tay Y et al (2011) AceRNA hypothesis: the Rosetta stone of a hidden RNA language? Cell 146:353–358

Song Y, Ci D, Tian M et al (2016) Stable methylation of a non-coding RNA gene regulates gene expression in response to abiotic stress in Populussimonii. J Exp Bot 67:1477–1492

Stocks MB, Moxon S, Mapleson D et al (2012) The UEA sRNA workbench: a suite of tools for analysing and visualizing next generation sequencing microRNA and small RNA datasets. Bioinformatics 28:2059–2061

Sun Z, Evans J, Bhagwate A et al (2014) CAP-miRSeq: a comprehensive analysis pipeline for microRNA sequencing data. BMC Genomics 15:423

Szcześniak MW, Deorowicz S, Gapski J et al (2012) miRNEST database: an integrative approach in microRNA search and annotation. Nucleic Acids Res 40:D198–D204

Takeda A, Iwasaki S, Watanabe T et al (2008) The mechanism selecting the guide strand from small RNA duplexes is different among argonaute proteins. Plant Cell Physiol 49:493–500

Tong Y-A, Peng H, Zhan C et al (2013) Genome-wide analysis reveals diversity of rice intronic miRNAs in sequence structure, biogenesis and function. PLoS One 8:63938

Tu B, Liu L, Xu C et al (2015) Distinct and cooperative activities of HESO1 and URT1 nucleotidyl transferases in microRNA turnover in Arabidopsis. PLoS Genet 11:e1005119

Urgese G, Paciello G, Acquaviva A et al (2016) isomiR-SEA: an RNA-Seq analysis tool for miRNAs/isomiRs expression level profiling and miRNA-mRNA interaction sites evaluation. BMC Bioinformatics 17:148

Vaucheret H (2006) Post-transcriptional small RNA pathways in plants: mechanisms and regulations. Genes Develop 20:759–771

Vazquez F, Blevins T, Ailhas J et al (2008) Evolution of Arabidopsis MIR genes generates novel microRNA classes. Nucleic Acids Res 36:6429–6438

Wang X, Zhang S, Dou Y et al (2015) Synergistic and independent actions of multiple terminal nucleotidyl transferases in the 3′ tailing of small RNAs in Arabidopsis. PLoS Genet 11: e1005091

Wassenegger M, Heimes S, Riedel L et al (1994) RNA-directed de novo methylation of genomic sequences in plants. Cell 76:567–576

Wei L, Zhang D, Xiang F et al (2009) Differentially expressed miRNAs potentially involved in the regulation of defense mechanism to drought stress in maize seedlings. Int J Plant Sci 170:979–989

Wen M, Shen Y, Shi S et al (2012) miREvo: an integrative microRNA evolutionary analysis platform for next-generation sequencing experiments. BMC Bioinformatics 13:140

Wu L, Zhang Q, Zhou H et al (2009) Rice microRNA effector complexes and targets. Plant Cell 21:3421–3435

Wu L, Zhou H, Zhang Q et al (2010) DNA methylation mediated by a microRNA pathway. Mol Cell 38:465–475

Wu H-J, Ma Y-K, Chen T et al (2012) PsRobot: a web-based plant small RNA meta-analysis toolbox. Nucleic Acids Res 40:W22–W28

Wu J, Liu Q, Wang X et al (2013a) mirTools 2.0 for non-coding RNA discovery, profiling, and functional annotation based on high-throughput sequencing. RNA Biol 10:1087–1092

Wu L, Liu D, Wu J et al (2013b) Regulation of FLOWERING LOCUS T by a microRNA in Brachypodium distachyon. Plant Cell 25:4363–4377

Xie Z, Johansen LK, Gustafson AM et al (2004) Genetic and functional diversification of small RNA pathways in plants. PLoS Biol 2:e104

Xie F, Xiao P, Chen D et al (2012) miRDeepFinder: a miRNA analysis tool for deep sequencing of plant small RNAs. Plant Mol Biol 80:75–84

Xu X, Bai H, Liu C, Chen E, Chen Q, Zhuang J et al (2014) Genome-wide analysis of microRNAs and their target genes related to leaf senescence of rice. PLoS One 9(12):e114313

Yang X, Li L (2011) miRDeep-P: a computational tool for analyzing the microRNA transcriptome in plants. Bioinformatics 27:2614–2615

Yang J-H, Qu L-H (2012) Deepbase: annotation and discovery of microRNAs and other noncoding RNAs from deep-sequencing data. In: Fan J-B (ed) Next-generation microRNA expression profiling technology: methods and protocols. Humana Press, Totowa, NJ, pp 233–248

Yu B, Yang Z, Li J et al (2005) Methylation as a crucial step in plant microRNA biogenesis. Science 307:932

Zhang Z, Yu J, Li D et al (2010) PMRD: plant microRNA database. Nucleic Acids Res 38:D806–D813

Zhang X, Xia J, Lii YE et al (2012) Genome-wide analysis of plant nat-siRNAs reveals insights into their distribution, biogenesis and function. Genome Biol 13:R20

Zhang Y, Zang Q, Zhang H et al (2016) DeAnnIso: a tool for online detection and annotation of isomiRs from small RNA sequencing data. Nucleic Acids Res 44:W166–W175

Zheng X, Zhu J, Kapoor A et al (2007) Role of Arabidopsis AGO6 in siRNA accumulation, DNA methylation and transcriptional gene silencing. EMBO J 26:1691–1701

Zheng Y, Li Y-F, Sunkar R et al (2012) SeqTar: an effective method for identifying microRNA guided cleavage sites from degradome of polyadenylated transcripts in plants. Nucleic Acids Res 40:e28

Zhu Q-H, Spriggs A, Matthew L et al (2008) A diverse set of microRNAs and microRNA-like small RNAs in developing rice grains. Genome Res 18:1456–1465

Zilberman D, Cao X, Jacobsen SE (2003) Control of locus-specific siRNA accumulation and DNA and histone methylation. Science 299:716

An Evolutionary View of the Biogenesis and Function of Rice Small RNAs

Tian Tang, Ming Wen, Pei Lin, and Yushuai Wang

Abstract Small RNAs are key players in transcriptional and posttranscriptional gene silencing. The rice genus *Oryza* comprises two domesticated species and 22 wild species. Using deep-sequencing technology, a variety of small RNAs, such as microRNAs (miRNAs) and small interfering RNAs (siRNAs), have been characterized in Asian-cultivated rice (*Oryza sativa*) and its wild relatives. These RNA species are processed by different types of DICER-LIKE proteins (DCL) and/or RNA-dependent RNA polymerases (RDRs) and are loaded into distinct types of ARGONAUTE (AGO) family members. Compared with *Arabidopsis*, rice has an expanded number of core genes in RNA-silencing pathways, resulting from multiple gene duplication events, and functional diversification of these genes remains largely unexplored. Rice also has an evolutionarily dynamic small RNA repertoire, with several special classes of small RNAs unique to rice or to monocots. While miRNAs can serve as a driving force for rice domestication, knowledge about evolutionary trajectories and specialized functions of rice small RNAs is still lacking to a large extent. In this chapter, we summarize our current understanding of the evolution of biogenesis and functional diversity of rice small RNAs.

Keywords RNAi pathway genes • miRNAs • siRNAs • Rice • Positive selection

Contents

T. Tang (✉) • P. Lin • Y. Wang
State Key Laboratory of Biocontrol and Guangdong Key Laboratory of Plant Resources, Sun Yat-sen University, Xingangxi Road No. 135, Guangzhou 510275, Guangdong, China
e-mail: lsstt@mail.sysu.edu.cn

M. Wen
Program in Emerging Infectious Diseases, Duke–National University of Singapore Graduate Medical School, Singapore, 169857, Singapore

N. Rajewsky et al. (eds.), *Plant Epigenetics*, RNA Technologies,
DOI 10.1007/978-3-319-55520-1_4

1 Introduction

Small RNAs are key regulators in eukaryotes that guide gene silencing at the transcriptional or posttranscriptional level (Carrington and Ambros 2003; Carthew and Sontheimer 2009; Simon and Meyers 2011). The evolution of the small RNA repertoire is important for the emergence of phenotypic novelty and diversification (Axtell and Bowman 2008; Axtell et al. 2011; Berezikov 2011; Cuperus et al. 2011). In plants, microRNAs (miRNAs) and small interfering RNAs (siRNAs) are two major classes of small RNAs distinguished by their mode of biogenesis and mechanisms of action. MiRNAs, typically 21–22 nt in length, are processed from hairpin-structured single-stranded RNA precursors that are transcribed from MIRNA genes by RNA polymerase II. Plant miRNAs direct sequence-specific mRNA cleavage or translation inhibition (Voinnet 2009) or in some rare cases mediate DNA methylation (Wu et al. 2010). Unlike miRNAs, siRNAs arise from perfectly complementary double-stranded RNAs (dsRNAs) that are typically converted from single-stranded RNAs (ssRNAs) by RNA-dependent RNA polymerase (RDR or RdRP). Plant siRNAs can be further classified into at least two major classes, including phased siRNAs (phasiRNAs) and heterochromatic siRNAs (hc-siRNAs) (Axtell 2013). Some phased siRNAs that function in *trans*-regulation of target gene expression are called trans-acting siRNAs (tasiRNAs) (Allen et al. 2005). PhasiRNAs commonly trigger cleavage and decay of target transcripts, while hc-siRNAs suppress the activities of transposable elements via DNA methylation and chromatin modifications (Slotkin and Martienssen 2007; Matzke and Mosher 2014). Plant small RNAs play critical roles in diverse biological processes, including organ morphogenesis and polarity, developmental phase transition, and biotic and abiotic stress responses (Sunkar et al. 2007; Chen 2009; Ruiz-Ferrer and Voinnet 2009; Chuck and O'Connor 2010; Lelandais-Briere et al. 2010; Chitwood and Sinha 2014), and have become promising targets in bioengineering for crop improvement (Zhang et al. 2013; Kong et al. 2014).

Rice (*Oryza sativa* L.) is one of the world's most important staple food crops, feeding more than half of the world's population. The rice genus *Oryza* comprises 24 species and 9 recognized genome types, including both diploid and tetraploid (AA, BB, CC, EE, FF, GG, BBCC, CCDD, and HHJJ) genomes (http://www.gramene.org). The AA genome species complex comprises two cultivated species, *Oryza sativa* L. and *Oryza glaberrima* Steud., and their six wild relatives, which show different biogeographic distribution and extensive adaptations. The Asian-cultivated rice *O. sativa* contains two subspecies, *O. sativa* ssp. *indica* and *O. sativa* ssp. *japonica*, which originated from the wild progenitor *Oryza rufipogon* Griff. (Oka 1988; Khush 1997). Introgressive hybridization between rice cultivars and other wild relatives of *Oryza* has greatly broadened the gene pool of *O. sativa* (Oka 1988; Khush 1997; Harushima et al. 2002; Tang et al. 2006; He et al. 2011; Huang et al. 2012). However, the African-cultivated rice *O. glaberrima* was domesticated independently of *O. sativa* in Africa from the wild progenitor *Oryza barthii* A. Chev (Wang et al. 2014). The greater tolerance of African rice to salinity, drought, and flooding has attracted increasing attention in breeding (Meyer et al. 2016).

With the fascinating history of domestication and speciation, rice represents an excellent model system of monocots for evolutionary biology and cereal functional genomics. Whole genome sequences of the eight AA genome rice (Goff et al. 2002; Yu et al. 2002, 2005; Project 2005; Wang et al. 2014; Chen et al. 2013a; Zhang et al. 2014), population genome sequencing of the two types of cultivated rice and their wild progenitors (Tang et al. 2006; He et al. 2011; Huang et al. 2012), and the partial sequencing of other genome types of *Oryza* species (Wing et al. 2005; Jacquemin et al. 2013) have provided an unprecedented opportunity to illuminate the evolution of rice genes and genomes and to probe the coding and noncoding genes that are favored by artificial selection for agriculturally important traits during rice domestication and improvement (Yu et al. 2009; Huang et al. 2012; Meyer et al. 2016). In the last decade, exhaustive efforts of high-throughput sequencing have identified numerous silencing small RNAs and elucidated their targets in cultivated and wild rice species. This chapter focuses on recent literature that sheds light on the evolution of small RNA biogenesis, diversity, and function in *Oryza* species, with an emphasis on the effect of selection.

2 Evolution of Core RNA Interference (RNAi) Pathway Genes in Rice

Plant small RNAs are invariably dependent on several major classes of proteins for their biogenesis or activities (Fig. 1a). RNase III endonuclease DICER-LIKE proteins (DCL) produce all known classes of plant-silencing RNAs via their dsRNA-processing activities, whereas RDRs act as an amplificatory component necessary for the production of siRNAs. Plant miRNAs and siRNAs are stabilized via 2′-O-methylation mediated by the RNA methyltransferase HUA ENHANCER 1 (HEN1) and are loaded into the ARGONAUTE (AGO) effector proteins to carry out their molecular functions via the RNA-induced silencing complex (RISC). Core genes in the RNAi pathway underwent losses or extensive expansion events during eukaryotic evolution (Burroughs et al. 2014). Positive selection plays a significant role in

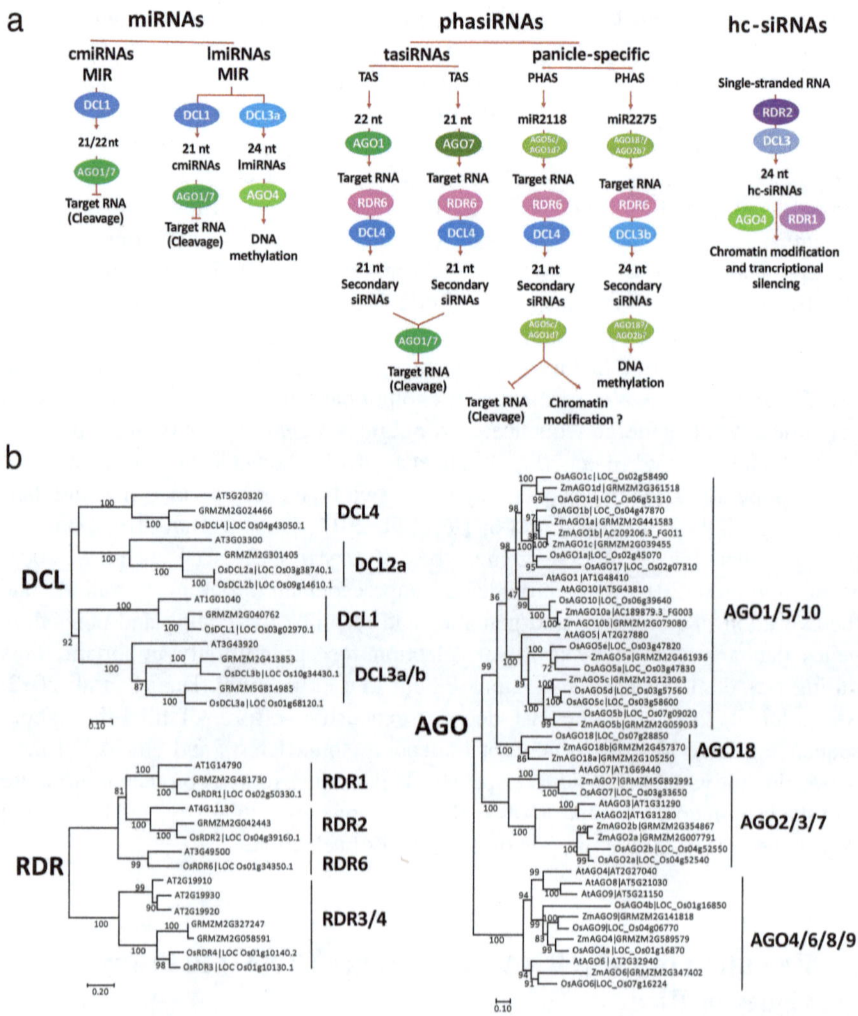

Fig. 1 RNA-silencing pathways in rice. (**a**) A schematic overview of biogenesis pathways and functional modes of rice small RNAs. (**b**) Maximum likelihood phylogenetic tree of *DCL*, *RDR*, and *AGO* coding DNA sequences (CDSs) from *Arabidopsis thaliana*, *Zea mays*, and *O. sativa*. CDSs for each gene were aligned in a codon-based manner using TranslatorX (http://pc16141.mncn.csic.es/index_v4.html), and the trees were constructed using FastTree with default parameters. Numbers above branches indicate bootstrap values greater than 50% based on 1000 replicates. *nt* nucleotides

driving diversification of the core RNAi components; consequently, it promotes the diversity of their associated small RNA populations (Kolaczkowski et al. 2011; Obbard et al. 2011; Mukherjee et al. 2013).

2.1 Dicer-Like

Rice and *Arabidopsis* genomes encode six and four *DCL* genes, respectively (Margis et al. 2006; Garcia-Ruiz et al. 2010), which form four distinct groups resulting from a rapid four-way duplication early in plant evolution (Mukherjee et al. 2013) (Fig. 1b). Two gene duplication events lead to two copies of both *DCL2* and *DCL3* in rice. The *DCL2* duplication is specific to the *Oryza* lineages, and the *DCL3* duplication occurred before the split of *O. sativa* and maize (*Zea mays*) but after the monocot–dicot divergence (Mukherjee et al. 2013). Rice DCLs have distinct functions: OsDCL1 is mainly responsible for processing miRNAs and triggering the generation of phasiRNAs (Liu et al. 2005); OsDCL2s participate in the maintenance of endogenous dsRNA virus, probably functioning antagonistically against other DCLs (Takanashi et al. 2011); OsDCL3s are involved in the production of 24-nt long miRNAs (lmiRNAs) and siRNAs (Itoh et al. 2000; Wu et al. 2010; Song et al. 2012a); and OsDCL4 is required for the biogenesis of 21-nt phasiRNAs including tasiRNAs (Liu et al. 2007; Nagasaki et al. 2007; Song et al. 2012a).

Knockdown of *OsDCL2* leads to unstable maintenance of the highly symbiotic endornavirus, which is common in multiple strains of *O. sativa* and the wild rice *O. rufipogon* (Urayama et al. 2010). It is unclear whether the duplication of *OsDCL2* confers any dosage advantage or promotes functional specialization. By contrast, functional divergence has occurred between the *OsDCL3* paralogs, both of which generate 24-nt small RNAs. *OsDCL3a* produces transposable element (TE)-derived lmiRNAs and unphased hc-siRNAs (Itoh et al. 2000; Song et al. 2012a), while *OsDCL3b* has evolved a panicle-preferential expression pattern to produce stamen-specific phasiRNAs (Kapoor et al. 2008). Such differences in function are accompanied with rapid sequence evolution of their double-stranded RNA binding (DSRM) domains. Using the branch-site model, Mukherjee et al. showed that positive selection specific to the *OsDCL3b* lineage is likely to be the driving force for the diversification of the *OsDCL3* paralogs (Mukherjee et al. 2013). DSRM domains typically function in coordinating the hand-off of the RNA template from Dicer to an AGO protein (Parker et al. 2008). The divergence in DSRM of *OsDCL3* paralogs probably plays a role in defining distinct RNAi pathways. Although which AGO is the downstream component of the OsDCL3b-dependent 24-nt phasiRNA pathway remains unknown, it is likely to be different from AGO4, which is the effector of the OsDCL3a-derived lmiRNAs or hc-siRNA (Wu et al. 2010). In maize, the two *DCL3* paralogs are mainly divergent in their helicase domains (Margis et al. 2006), suggesting different evolutionary trajectories of *DCL3s* across monocots.

As is the case with plant DCL4 throughout monocots and dicots, adaptive evolution has also targeted the PAZ domain of the antiviral OsDCL4, probably due to the long-term "arms race" associated with the recognition of viral RNA. Dicer PAZ typically contains a positively charged pocket that binds the RNA-end structure primarily through electrostatic interactions (Ma et al. 2004). However, the electrostatic charge of the DCL4 PAZ-RNA binding pocket has reversed to primarily negative in monocots (Mukherjee et al. 2013). Rice has specific adaptive substitutions in the PAZ-RNA binding loop that distinguish *OsDCL4* from its orthologs in *Z. mays* or *Sorghum bicolor*; one position of these adaptive substitutions is extremely conserved across other Dicer proteins and forms a critical RNA contact in *Giardia* Dicer (Simon and Meyers 2011). These observations suggest OsDCL4 may exhibit different RNA-binding properties compared with its orthologs and other paralogous DCL proteins.

2.2 RNA-Dependent RNA Polymerases

Rice has five *RDR* genes: *OsRDR1*, *OsRDR2*, and *OsRDR6* originate from the *RDRα* clade that are present in all three kingdoms (plants, animals, and fungi); *OsRDR3* and *OsRDR4* are rice tandem duplicates from the *RDRγ* clade that are unique to plants and fungi (Zong et al. 2009). In comparison with rice, *Arabidopsis* has one more *RDR* resulting from the triplicated *RDR3/4/5* family (Willmann et al. 2011) (Fig. 1b). *RDRα* clade members are functionally diverse in distinct RNAi pathways, while the *RDR* orthologs between rice and *Arabidopsis* have largely conserved molecular functions. RDR1 is a key component of the antiviral RNAi pathway (Garcia-Ruiz et al. 2010; Wang et al. 2010b) and participates in the 21-nt siRNA-dependent DNA methylation (Pontier et al. 2012; Stroud et al. 2013); RDR2 acts upstream of DCL3 action to generate 24-nt hc-siRNAs involved in directing DNA methylation (Havecker et al. 2010) and histone modifications; RDR6 is required for the generation of 21-nt and 24-nt phasiRNAs including tasiRNAs (Song et al. 2012b). RDR1, RDR2, and RDR6 play a role in the biogenesis of siRNAs from plant viruses (Diaz-Pendon et al. 2007; Donaire et al. 2008). Despite functional conservation with *Arabidopsis*, *OsRDRs* may have evolved distinct expression regulatory networks. For example, the activation of *OsRDR1* in rice antiviral response is under the direct control of a monocot-specific microRNA, miR444, and its MADS box targets (Wang et al. 2016).

2.3 Hua Enhancer 1

The rice *HEN1* ortholog is *WAF1* (*WAVY LEAF1*). Stabilization of small RNAs by *WAF1* is indispensable for rice development, especially for shoot apical meristem (SAM) maintenance, leaf morphogenesis, and floral development governed by the tasiRNA pathway (Abe et al. 2010). Interestingly, a recent comparative genomic study of six AA genome rice species detected signs of positive selection on *WAF1*, probably due to its role in maintaining floral meristem identity (Zhang et al. 2014).

2.4 Argonaute

As the main RNA-silencing effectors, *AGO* genes have diverged rapidly in plants, with 19 members in rice (Kapoor et al. 2008), 17 in maize (Qian et al. 2011; Zhai et al. 2014), and 10 in *Arabidopsis* (Chen 2009). Phylogenic analyses group plant *AGOs* into three major clades: *AGO1/5/10*, *AGO2/3/7*, and *AGO4/6/8/9* (Vaucheret 2008) (Fig. 1b). Studies in *Arabidopsis* have experimentally demonstrated functional diversification between different *AGO* subfamilies. AGO1, AGO2, AGO7, and AGO10 mediate posttranscriptional gene silencing (PTGS), whereas AGO4 mediates transcriptional gene silencing (TGS). AGO4, AGO6, and AGO9 associate mostly with 24-nt siRNAs, whereas AGO1, AGO2, AGO5, AGO7, and AGO10 bind 21/22-nt molecules. AGO1, AGO2, and AGO5 preferentially bind small RNAs exhibiting a 5′-end uridine, adenosine, or cytosine, respectively, while AGO7 and AGO10 are associated almost exclusively with miR390 and miR165/166, respectively (reviewed in Bologna and Voinnet 2014; Fang and Qi 2016).

Like other monocots, rice has an expanded *AGO1/5/10* clade and an extra subclade, *AGO18*, which is a sister group of the clade *AGO1/5/10* (Fig. 1b). Such expansion is most prominent for the *AGO1* and *AGO5* subfamilies (Fig. 1b). Although *Arabidopsis* has only one copy each of *AGO1* and *AGO5* genes, rice has four *AGO1* and five *AGO5* homologs. The *OsAGO1* subfamily comprises the segmentally duplicated pair *OsAGO1a* and *OsAGO1b* and their sister clades *OsAGO1c* and *OsAGO1d*, while the *OsAGO5* subfamily consists of the tandem duplicated pair *OsAGO5a* and *OsAGO5e*, the segmentally duplicated pair *OsAGO5b* and *OsAGO5c*, and *OsAGO5d* duplicated from *OsAGO5c* (Kapoor et al. 2008). Rice also has a lineage-specific *AGO17*, which was probably duplicated from *OsAGO1a* after the split of rice and maize (Fig. 1b). Experiments have elucidated molecular functions for the members of OsAGO1 (Wu et al. 2009), OsAGO5c (MEL) (Nonomura et al. 2007; Komiya et al. 2014), OsAGO10 (PNH1) (Nishimura et al. 2002), OsAGO7 (SHL4) (Nagasaki et al. 2007), and OsAGO18 (Wu et al. 2015).

The *AGO* orthologs between rice and *Arabidopsis* have largely conserved functions (Bologna and Voinnet 2014; Zhang et al. 2015), whereas paralogous *OsAGOs* within the same subfamily may have maintained, gained, or lost function after recent duplications. The *OsAGO1* subfamily is a good case in point. Deep sequencing of rice AGO1 immunoprecipitates revealed that OsAGO1a, OsAGO1b, and OsAGO1 all have a strong preference for binding small RNAs with 5′-uridine (Wu et al. 2009), the same as the AtAGO1 ortholog in *Arabidopsis* (Mi et al. 2008). While most of the miRNAs were evenly distributed in the three OsAGO1 complexes, a subset of miRNAs were specifically incorporated into or excluded from one of the OsAGO1s (Wu et al. 2009), suggesting a potential role of subfunctionalization in the maintenance of these *OsAGO1* genes. The binding specificity of OsAGO1d is unknown, but *OsAGO1d* shows the same expression pattern as *OsAGO5c/MEL* (see below) and was thus hypothesized to play a role in the biogenesis of phasiRNAs in rice anthers (Fei et al. 2016), representing a candidate of neofunctionalization. In contrast, *OsAGO17* shows high sequence divergence from *OsAGO1a* and exhibits a disrupted Asp-Asp-His (DDH) motif (Kapoor et al. 2008), which is common in an active site critical for AGO proteins carrying on endonuclease activity (Rivas et al. 2005), representing a candidate for nonfunctionalization or neofunctionalization. The validation of the hypothesis of subfunctionalization or neofunctionalization requires the detection of positive selection in one or both lineages after gene duplication (Assis and Bachtrog 2013), albeit differential expression was used to infer the evolutionary fates of *OsAGOs* (Yang et al. 2013).

In the *OsAGO5* subfamily, OsAGO5c, also known as MEL1, was demonstrated to regulate the cell division of the premeiotic germ cell in rice anthers (Nonomura et al. 2007) and to preferentially bind 21-nt phasiRNAs therein (Komiya et al. 2014). While MEL1-like AtAGO5 binds to small RNAs with a 5′ cytosine preferentially, this preference was not observed in the total 21-nt phasiRNA population of rice or maize (Komiya et al. 2014; Zhai et al. 2015), suggesting MEL1 may not be the only AGO protein that binds 21-nt phasiRNAs. It is unclear whether OsAGO5s may have evolved redundant molecular functions but specialized binding preference as OsAGO1s did, or if other OsAGOs, such as OsAGO1d, participate in the biogenesis of 21-nt phasiRNAs in rice anthers. In addition, *MEL1* null mutants display severely disrupted large-scale meiotic chromosome reprogramming and hypermethylation of H3K9 in pollen mother cells (Nonomura et al. 2007; Liu and Nonomura 2016), suggesting a novel function of MEL1 and its associated 21-nt phasiRNAs in chromatin modification in addition to target cleavage (Fei et al. 2016).

Interestingly, the monocot-specific AGO18 cooperates with AGO1 in controlling the virus resistance in rice (Wu et al. 2015). The *OsAGO18* mutant rice is more sensitive to viral infection, while the overexpression of *OsAGO18* confers broad-spectrum virus resistance (Wu et al. 2015). Genetic studies reveal that the antiviral function of OsAGO18 depends on its activity to sequester microRNA168 (miR168) to alleviate repression of rice *AGO1* essential for antiviral RNAi (Wu et al. 2015). While the sequestering miR168 to regulate AGO1 homeostasis represents an evolutionary novelty of OsAGO18 specific to rice, the expression pattern of *OsAGO18* suggests that it may have a role in the biogenesis of 24-nt phasiRNAs (Fei et al. 2016), conserved with its maize homolog *ZmAGO18b* (Zhai et al. 2015).

3 Evolution of Rice Small RNAs and Their Targets

3.1 *Canonical miRNAs*

3.1.1 Evolution of miRNAs in AA Genome Oryza Species

The evolution of miRNA genes is a birth-and-death process (Fahlgren et al. 2007; Nozawa et al. 2012). While the overall number of miRNA families increases, a high

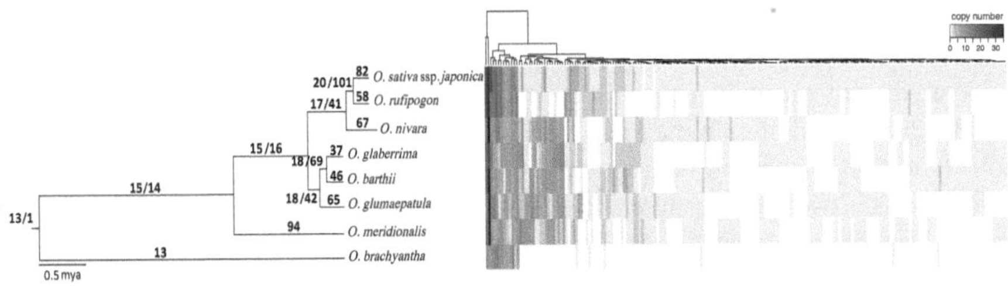

Fig. 2 The evolution of miRNA families in the AA genome *Oryza* species. The heat map shows the clustering of miRNA families based on their copy numbers. The numbers of miRNA families in six AA genome wild rice were retrieved from Table 43 in Zhang et al. (2014). As for *O. rufipogon* and *O. baranchyantha*, sequences of known rice miRNA genes were retrieved from miRBase v. 21 (http://mirbase.org/) and aligned to their genomes using Blast, respectively. Putative miRNA precursors with sequence identity >80% were further evaluated using miRCheck program (Jones-Rhoades et al. 2004) and grouped into miRNA families. Branch lengths of the phylogenetic tree of AA genome *Oryza* species reflect evolutionary divergence times in millions of year (Mya) as determined in Zhu and Ge (2005). Numbers on each internal branch show highly conserved and lowly conserved miRNA families (before and after slash), and the number on each terminal branch shows species-specific miRNA families. miRNA families present in both monocots and eudicots are defined as highly conserved (Cuperus et al. 2011)

turnover rate exists for miRNA families in plant species (Xie et al. 2016). The net gain rate for the recent terminal branch leading to *O. sativa* was estimated to be 7.3 families Myr^{-1}, much higher than *Arabidopsis thaliana* (1.64 families Myr^{-1}), *Populus trichocarpa* (1.64 families Myr^{-1}), *Brachypodium distachyon* (3.75 families Myr^{-1}), and maize (0 families Myr^{-1}) (Xie et al. 2016). The high net gain rate in rice may result from the high birth rate of novel miRNA families that often present in single or low copy numbers (Fig. 2). Comparative analyses of miRNA annotation in AA genome rice species revealed that the overall number of miRNA families did not vary substantially (Fig. 2), whereas miRNA gene families related to flower development exhibited a wide range of variation in copy number among rice species (Zhang et al. 2014). Lineage-specific expansion of the flowering-associated miRNAs accompanied with the expansion of the MADS-box gene family may partially account for the characteristic differences of flower development and reproduction (Zhang et al. 2014). In addition, rice miRNAs can exist as polycistronic clusters, similar to animal miRNAs, and many polycistronic miRNAs are conserved among AA genome rice species, presumably acting as versatile regulators of gene expression in different biological processes (Baldrich et al. 2016).

3.1.2 MiRNA Genes Under Positive Selection in Cultivated Rice

The vast majority of miRNA genes are highly conserved across the AA genome rice species (Zhang et al. 2014), whereas a small proportion bear signatures of positive selection in the cultivated rice, likely to be important for agriculturally important traits in rice domestication and improvement (Wang et al. 2010a; Liu et al. 2013a, 2015).

Using population genetics analyses, independent studies have identified tens of miRNAs exhibiting significantly reduced nucleotide diversity in the cultivated rice compared with the wild progenitor *O. rufipogon* (Wang et al. 2010a; Liu et al. 2013a, 2015). Although reduction of nucleotide diversity is an indicator of positive selection, bottleneck and demographic effects, such as population expansion, can also lead to low genetic diversity. Some candidates of the positively selected miRNAs genes are located in the previously identified genomic regions associated with rice domestication or show differential expression between the cultivated and wild rice, suggesting they are more than products of neutral evolution (Liu et al. 2013a, 2015). Among them, *MIR390* and *TAS3a2* are responsible for the biogenesis of ta-siRNAs involved in the Auxin regulatory pathway (Allen et al. 2005; Wang et al. 2010a), while *MIR1436*, *MIR156b/c/i*, *MIR160f*, and *MIR167i* are related to flowering in common wild rice (Chen et al. 2013b; Liu et al. 2015). Selection could target miRNA genes involved in the same regulatory pathway for enhanced changes in the developmental process.

In addition to sequence evolution, changes in miRNA expression can also be subject to positive selection during rice domestication and/or improvement. Wang et al. (2012) reported differential expression of miRNAs between cultivated and wild rice species. The *miR164* family is upregulated in both *indica* and *japonica* cultivars compared with *O. rufipogon,* and neutrality tests such as Fay and Wu's *H* (Fay and Wu 2000) and Tajima's D (Tajima 1989) detect signatures of selective sweeps in *MIR164c, MIR164d*, and *MIR164e* in the cultivated population but not in the wild rice population. Recently, we studied the expression variation of miRNAs and mRNAs between *indica* and *japonica* rice (Wen et al. 2016). While 196 out of 272 miRNAs have an expression pattern compatible with the evolutionary mode of stabilizing selection, 29 miRNAs exhibit low within-subspecies expression polymorphism and high between-subspecies expression divergence, indicating a role of directional selection in shaping miRNA expression during rice domestication (Wen et al. 2016). Interestingly, lowly conserved miRNAs confer most of the expression variation between rice subspecies, whereas highly conserved miRNAs that are common between eudicots and monocots show stronger negative correlation with their targets (Wen et al. 2016) (Fig. 3), indicating distinct evolutionary trajectories of the two categories of miRNAs.

In contrast to deeply conserved miRNA families, young miRNA genes are often weakly expressed, processed imprecisely, and tend to lack targets, suggesting that most are neutrally evolving, evolutionarily transient loci (Fahlgren et al. 2007; Cuperus et al. 2011). Nevertheless, expression levels of more than half of the rice-specific miRNAs are under directional selection (Wen et al. 2016), suggesting that young miRNAs could be sources of novel regulatory variations when occasionally integrated into the existing regulatory networks. One well-studied case is *osa-MIR7695*, a novel miRNA specifically produced in *japonica* but not in *indica* rice (Campo et al. 2013). *osa-miR7695* targets an alternatively spliced transcript of the *Natural resistance-associated macrophage protein 6* (*Nramp6*) gene, and its overexpression contributes to increased rice blast resistance (Campo et al. 2013).

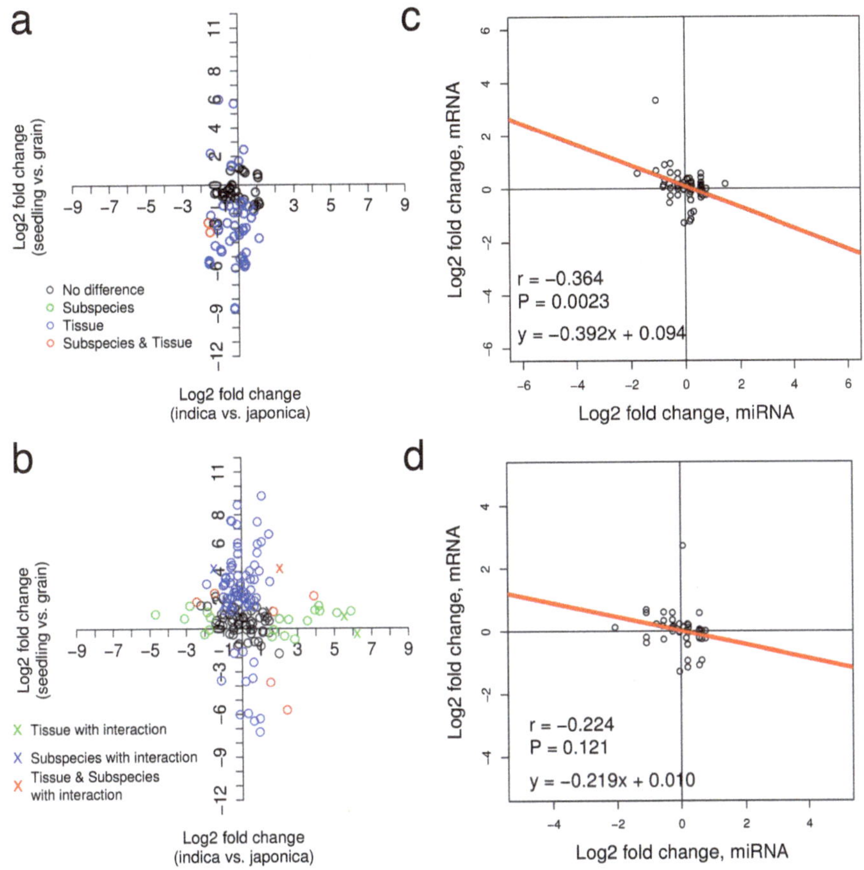

Fig. 3 Expression variations of known miRNAs and their targets in rice. (**a**, **b**) Differential expression of highly conserved (**a**) and lowly conserved (**b**) miRNAs between seedlings and grains of three *indica* and three *japonica* rice cultivars. Mature miRNAs that show no differential expression (*black*) or show significant differential expression between subspecies (*green*), developmental stage (*blue*), and both (*red*) are indicated by *circles* in different colors, while miRNAs with differential expression for the additional factor of development-by-subspecies interaction are indicated by crosses with the same color setting. (**c**, **d**) Correlation between the expression of highly conserved miRNAs and lowly conserved miRNAs and their targets in seedlings: (**c**) highly conserved miRNAs and their predicted targets (390 pairs) and (**d**) lowly conserved miRNAs and their predicted targets (219 pairs). Adapted from Wen et al. (2016)

The processing of *osa-MIR7695* is imperfect in *O. glaberrima*, where the pattern resembles that in wild rice accessions rather than in *O. sativajaponica* subspecies, indicating that selection favors the improvement of *osa-MIR7695* precursor processing during the domestication of *japonica* rice.

3.1.3 MiRNA Targets Under Positive Selection in Cultivated Rice

Like miRNA genes, miRNA target sites are mainly under strong purifying selection (Guo et al. 2008). The gain or loss of miRNA target sites could occasionally be beneficial to agriculturally important traits. For example, a point mutation in the SBP-box-containing transcription factor *OsSPL14* perturbs miR156-directed regulation of *OsSPL14*, generating an "ideal" rice plant with a reduced tiller number, increased lodging resistance, and enhanced grain yield (Jiao et al. 2010; Miura et al. 2010). Coincidently, both *osa-MIR156i* and its target gene *OsSPL14* were found to bear signatures of positive selection (Liu et al. 2015), suggesting that coevolution between miRNAs and their target sequences could happen in cultivated rice (Guo et al. 2008). Using population genetic analyses, Liu et al. (2015) detected significant negative Tajima's *D* (Tajima 1989) and/or Fu and Li's D^* and F^* (Fu and Li 1993) values in 12 out of 68 miRNA target genes in rice cultivars (Liu et al. 2015), suggesting an excess of low-frequency alleles. Because such departures of neutrality could be caused by many factors, such as bottleneck or demography, further study using coalescent simulation would be necessary to validate the candidates of positively selected miRNAs targets in rice.

3.2 Long miRNAs

The 24-nt lmiRNAs are processed by DCL3, sorted into AGO4 clade proteins, and mediate DNA methylation at their target loci *in trans*, which is very different from cmiRNAs that are generated by DCL1 and specifically loaded into AGO1 clade proteins to direct the cleavage of their target mRNAs (Wu et al. 2010) (Fig. 1a). In *Arabidopsis*, lmiRNAs are consistently generated by the recently evolved *MIRNA* genes, while ancient *MIRNA* genes give rise predominantly to canonical miRNAs (Vazquez et al. 2008). Most rice miRNA precursors give rise to a single cmiRNA or lmiRNA, while a subset of rice pri-miRNAs can produce both cmiRNAs and lmiRNAs through the coordinated actions of DCL1 and DCL3a (Wu et al. 2010). These results suggest that evolution of *MIRNA* genes is associated with gradual, overlapping changes in DCL usage. cmiRNA and lmiRNA, derived from the same pri-miRNA, are sorted into distinct effector proteins directing either mRNA cleavage or DNA methylation, indicating that these pri-miRNAs can have dual functions (Wu et al. 2010). miR820 involved in the evolutionary "arms race" between a host and its parasite DNA represents such a special class of bifunctional miRNA genes.

miR820 is a rice-specific miRNA family that has five copies in the *japonica* cv. Nipponbare genome but none in *Arabidopsis* or maize. MiR820 family members are located within CACTA DNA transposons and target the de novo DNA methyltransferase *OsDRM2*, a component of the host's silencing machinery. Pri-miR820 produces two miRNA variants: canonical 21-nt miR820 regulates *OsDRM2* expression primarily through mRNA cleavage, while 24-nt miR820 directs DNA methylation of miR820 loci *in cis* as well as at *OsDRM2 in trans* (Wu et al. 2010). As a result, the negative regulation of *OsDRM2* by miR820 activates TE expression, providing a mechanism by which parasitic elements in the genome escape the host's defense machinery (Nosaka et al. 2012). Interestingly, the nucleotide sequence of miR820 and its target site in BB/BBCC *Oryza* species have coevolved to maintain their base-pairing ability, and the coevolution has led to the proliferation of TE carrying pre-miR820 in BB genome species (Nosaka et al. 2012).

TEs can serve as a driving force for the evolution of miRNAs by providing raw materials for RNA-hairpin formation and the subsequent siRNA–miRNA transition (Piriyapongsa and Jordan 2008; Zhang et al. 2011; Ou-Yang et al. 2013). The miR820-mediated relaxation of TE repression may have contributed to the massive origination of novel miRNAs and the high proportion of TE-related miRNAs in rice (Zhang et al. 2011).

3.3 Phased siRNAs

The biogenesis of phasiRNAs, including tasiRNAs, is triggered by miRNA targeting on either noncoding or protein-coding genes, known as phasiRNA- or tasiRNA-producing loci (*PHAS* or *TAS* genes), respectively [reviewed in Fei et al. (2013)]. The rice genome encodes a *TAS3* family that is targeted by miR390 to produce tasiRNAs (Axtell et al. 2006). *TAS3*-derived tasiRNAs target three *Auxin Responsive Factor* (*ARF*) genes, *ARF2*, *ARF3*/*ETT*, and *ARF4* (Allen et al. 2005), regulating shoot apical meristem (SAM) formation during embryogenesis and the adaxial–abaxial polarity of floral organs in rice (Nagasaki et al. 2007; Toriba et al. 2010). While the miR390-TAS3 pathway is deeply conserved from moss to higher plants (Allen et al. 2005; Fei et al. 2013), *TAS* and miRNA loci involved in phasiRNA biogenesis, such as *TAS3a2* and *MIR390*, have experienced direct selection during Asian rice domestication (Wang et al. 2010a; Liu et al. 2013a, b, 2015).

Other phasiRNAs in rice, mainly derived from long noncoding RNAs (lncRNAs), are preferentially expressed in panicles (Liu et al. 2013b; Komiya et al. 2014). Two 22-nt miRNAs, miR2118 and miR2275, specifically trigger the production of 21-nt (by miR2118) and 24-nt (by miR2275) phasiRNAs in rice panicles, respectively (Johnson et al. 2009; Song et al. 2012a). The functions of reproductive 21-nt and 24-nt phasiRNAs are largely unknown, but their preferential expression in panicles, especially anthers, is conserved in monocots, including *Brachypodium distachyon* and maize (International Brachypodium 2010; Zhai et al. 2015). In maize, each phasiRNA type exhibits independent spatiotemporal regulation, with 21-nt premeiotic phasiRNAs dependent on stamen epidermal differentiation and 24-nt meiotic phasiRNAs dependent on tapetal cell differentiation (Zhai et al. 2015). In rice, the reproductive phasiRNAs also display stage-specific expression patterns during early stages of another development, suggesting that the timing of phasiRNA biogenesis is crucial in rice microsporogenesis (Fei et al. 2016). Recently, Ta et al. reported a delayed expression of the 21-nt phasiRNAs as well as their lncRNA precursors and regulators (miR2118 and *MEL1* gene) during the panicle development of *O. glaberrima* compared to that of *O. barthii*. Such a differential expression of the miR2118-triggered 21-nt phasiRNA pathway reflects differential rates of determinate fate acquisition of panicle meristems between the wild and domesticated African rice species (Ta et al. 2016).

Interestingly, the miR484/miR2118-triggered phasiRNAs in eudicots, such as legumes and Solanaceae, are mainly generated from genes encoding the nucleotide binding site (NBS)- and leucine-rich repeat (LRR)-containing disease resistance proteins (Zhai et al. 2011; Li et al. 2012; Shivaprasad et al. 2012; Arikit et al. 2014; Fei et al. 2015). The link between NBS-LRRs and their regulation by miRNAs can be traced back to >100 million years after the origin of NBS-LRR genes in early land plants such as mosses and spike mosses (Xia et al. 2015). However, the genomes of Poaceae and Brassicaceae essentially lack the widespread targeting of NBS-LRRs by miR482/2118, and species from these families are distinguished by high heterogeneity of their NBS-LRRs (Zhang et al. 2016a). How miR2118 lost the regulation of NBS-LRRs and gained the new targets of lncRNA clusters accompanied with the panicle-specific expression in grasses would be an intriguing question.

3.4 Heterochromatic siRNAs

The 24-nt heterochromatic siRNAs (hc-siRNAs) or repeat-associated siRNAs (rasiRNAs) are the most abundant small RNAs identified in rice (Jeong et al. 2011). These siRNAs originated from TEs or DNA repeats usually guiding DNA methylation, an activity known as RNA-directed DNA methylation (RdDM), and histone modifications. While hc-siRNAs mainly function in the maintenance of genome integrity by silencing TEs (Chen 2009), they may also silence endogenous non-TE genes serving as an important component of the host gene regulatory network. For example, the TE-derived siR815 in rice contributes to the natural variation of the two alleles of transcription factor *WRKY45*, and siR815-induced suppression of *ST1* by RdDM results in the negative role of *WRKY45-1* but in the positive role of *WRKY45-2* in regulating disease resistance (Zhang et al. 2016b).

In rice, OsDCL3a is primarily responsible for TE-associated 24-nt siRNA processing (Wei et al. 2014). Reduction of the OsDCL3a function reduced the 24-nt siRNAs predominately from miniature inverted-repeat TEs (MITEs), a class of short (<600 bp) nonautonomous DNA transposons and elevated expression of nearby genes, resulting in phenotypes such as dwarfism and enlarged flag leaf angle (Wei et al. 2014). These results indicate that OsDCL3a-dependent 24-nt siRNAs derived from MITEs are broadly functioning regulators for fine-tuning gene expression. MITEs generate 23.5% (183,837 of 781,885) of all small RNAs identified from rice (Lu et al. 2012). Considering the insertional polymorphism of MITEs between rice cultivars, MITE-derived small RNAs downregulate gene expression of different sets of genes in different genotypes, which may provide considerable phenotypic diversity to *O. sativa* (Lu et al. 2012).

4 Conclusions and Future Prospects

Our review of the literature yields two main conclusions. First, rice has expanded core protein families in the RNAi pathways, but their functional diversification and the underlying evolutionary force remain unclear, especially for AGO proteins. Second, rice has a large number of novel small RNAs and uses a variety of distinct small RNA types, such as lmRNAs and panicle-specific phasiRNAs. Although these small RNAs are supposed to be selected during the evolution of regulatory pathways, little is known about their evolution and function in the domestication and speciation of *Oryza* species. Considering the differential composition of TEs among *Oryza* species (Zhang et al. 2014), the interplay between TE-derived small RNAs and the parasite DNA and how such interplay may have contributed to the rice genome and gene expression evolution would be a fascinating question. The biogenesis, function, and evolution of rice reproductive phasiRNAs are another mystery. Moreover, young miRNAs involved in rice immunity and flowering may have undergone rapid evolution and play a role in phenotypic evolution during rice domestication and improvement. The recent advance of genome editing technologies, such as the CRISPR/Cas system (Belhaj et al. 2013; Shan et al. 2014) and anti-miRNA oligonucleotide technologies (He et al. 2016), allows the manipulation of RNAi pathway genes and small RNAs individually or in combination in different plant species. With genomes available from multiple *Oryza* species, a real opportunity exists to dissect the regulatory roles of rice small RNAs and to understand the evolution of small RNA-mediated regulatory network in cereals in the near future.

Acknowledgements This work was supported by National Science Foundation of China (31170308 and 91231117 to T.T.), the Science Foundation for Outstanding Young Teachers in Higher Education of Guangdong (Yq2013005) and the Fundamental Research Funds for the Central Universities (16lgjc75) to T. T., and the General Financial Grant from the China Postdoctoral Science Foundation (2013M531981) to M.W.

References

Abe M, Yoshikawa T, Nosaka M et al (2010) WAVY LEAF1, an ortholog of Arabidopsis HEN1, regulates shoot development by maintaining microRNA and trans-acting small interfering RNA accumulation in rice. Plant Physiol 154:1335–1346

Allen E, Xie Z, Gustafson AM et al (2005) microRNA-directed phasing during trans-acting siRNA biogenesis in plants. Cell 121:207–221

Arikit S, Xia R, Kakrana A et al (2014) An atlas of soybean small RNAs identifies phased siRNAs from hundreds of coding genes. Plant Cell 26:4584–4601

Assis R, Bachtrog D (2013) Neofunctionalization of young duplicate genes in drosophila. Proc Natl Acad Sci USA 110:17409–17414

Axtell MJ (2013) Classification and comparison of small RNAs from plants. Annu Rev Plant Biol 64:137–159

Axtell MJ, Bowman JL (2008) Evolution of plant microRNAs and their targets. Trends Plant Sci 13:343–349

Axtell MJ, Jan C, Rajagopalan R et al (2006) A two-hit trigger for siRNA biogenesis in plants. Cell 127:565–577

Axtell MJ, Westholm JO, Lai EC (2011) Vive la difference: biogenesis and evolution of microRNAs in plants and animals. Genome Biol 12:221

Baldrich P, Hsing YI, San Segundo B (2016) Genome-wide analysis of polycistronic microRNAs in cultivated and wild rice. Genome Biol Evol 8:1104–1114

Belhaj K, Chaparro-Garcia A, Kamoun S et al (2013) Plant genome editing made easy: targeted mutagenesis in model and crop plants using the CRISPR/Cas system. Plant Methods 9:39

Berezikov E (2011) Evolution of microRNA diversity and regulation in animals. Nat Rev Genet 12:846–860

Bologna NG, Voinnet O (2014) The diversity, biogenesis, and activities of endogenous silencing small RNAs in Arabidopsis. Annu Rev Plant Biol 65:473–503

Burroughs AM, Ando Y, Aravind L (2014) New perspectives on the diversification of the RNA interference system: insights from comparative genomics and small RNA sequencing. Wiley Interdiscip Rev RNA 5:141–181

Campo S, Peris-Peris C, Sire C et al (2013) Identification of a novel microRNA (miRNA) from rice that targets an alternatively spliced transcript of the Nramp6 (natural resistance-associated macrophage protein 6) gene involved in pathogen resistance. New Phytol 199:212–227

Carrington JC, Ambros V (2003) Role of microRNAs in plant and animal development. Science 301:336–338

Carthew RW, Sontheimer EJ (2009) Origins and mechanisms of miRNAs and siRNAs. Cell 136:642–655

Chen X (2009) Small RNAs and their roles in plant development. Annu Rev Cell Dev Biol 25:21–44

Chen JF, Huang QF, Gao DY et al (2013a) Whole-genome sequencing of Oryza Brachyantha reveals mechanisms underlying Oryza genome evolution. Nat Commun 4:1595

Chen Z, Li F, Yang S et al (2013b) Identification and functional analysis of flowering related microRNAs in common wild rice (*Oryza rufipogon* Griff.) PLoS One 8:e82844

Chitwood DH, Sinha NR (2014) Plant development: small RNAs and the metamorphosis of leaves. Curr Biol 24:R1087–R1089

Chuck G, O'Connor D (2010) Small RNAs going the distance during plant development. Curr Opin Plant Biol 13:40–45

Cuperus JT, Fahlgren N, Carrington JC (2011) Evolution and functional diversification of MIRNA genes. Plant Cell 23:431–442

Diaz-Pendon JA, Li F, Li WX et al (2007) Suppression of antiviral silencing by cucumber mosaic virus 2b protein in Arabidopsis is associated with drastically reduced accumulation of three classes of viral small interfering RNAs. Plant Cell 19:2053–2063

Donaire L, Barajas D, Martinez-Garcia B et al (2008) Structural and genetic requirements for the biogenesis of tobacco rattle virus-derived small interfering RNAs. J Virol 82:5167–5177

Fahlgren N, Howell MD, Kasschau KD et al (2007) High-throughput sequencing of Arabidopsis microRNAs: evidence for frequent birth and death of MIRNA genes. PLoS One 2:e219

Fang X, Qi Y (2016) RNAi in plants: an Argonaute-centered view. Plant Cell 28:272–285

Fay JC, Wu CI (2000) Hitchhiking under positive Darwinian selection. Genetics 155:1405–1413

Fei Q, Xia R, Meyers BC (2013) Phased, secondary, small interfering RNAs in posttranscriptional regulatory networks. Plant Cell 25:2400–2415

Fei Q, Li P, Teng C et al (2015) Secondary siRNAs from medicago NB-LRRs modulated via miRNA-target interactions and their abundances. Plant J 83:451–465

Fei Q, Yang L, Liang W et al (2016) Dynamic changes of small RNAs in rice spikelet development reveal specialized reproductive phasiRNA pathways. J Exp Bot 67:6037–6049
Fu YX, Li WH (1993) Statistical tests of neutrality of mutations. Genetics 133:693–709
Garcia-Ruiz H, Takeda A, Chapman EJ et al (2010) Arabidopsis RNA-dependent RNA polymerases and Dicer-like proteins in antiviral defense and small interfering RNA biogenesis during turnip mosaic virus infection. Plant Cell 22:481–496
Goff SA, Ricke D, Lan TH et al (2002) A draft sequence of the rice genome (*Oryza sativa* L. ssp. japonica). Science 296:92–100
Guo X, Gui Y, Wang Y et al (2008) Selection and mutation on microRNA target sequences during rice evolution. BMC Genomics 9:454
Harushima Y, Nakagahra M, Yano M et al (2002) Diverse variation of reproductive barriers in three intraspecific rice crosses. Genetics 160:313–322
Havecker ER, Wallbridge LM, Hardcastle TJ et al (2010) The Arabidopsis RNA-directed DNA methylation Argonautes functionally diverge based on their expression and interaction with target loci. Plant Cell 22:321–334
He L, Xie M, Huang J et al (2016) Efficient and specific inhibition of plant microRNA function by anti-microRNA oligonucleotides (AMOs) in vitro and in vivo. Plant Cell Rep 35:933–945
He Z, Zhai W, Wen H et al (2011) Two evolutionary histories in the genome of rice: the roles of domestication genes. PLoS Genet 7:e1002100
Huang X, Kurata N, Wei X et al (2012) A map of rice genome variation reveals the origin of cultivated rice. Nature 490:497–501
International Brachypodium I (2010) Genome sequencing and analysis of the model grass Brachypodium distachyon. Nature 463:763–768
Itoh JI, Kitano H, Matsuoka M et al (2000) Shoot organization genes regulate shoot apical meristem organization and the pattern of leaf primordium initiation in rice. Plant Cell 12: 2161–2174
Jacquemin J, Bhatia D, Singh K et al (2013) The international Oryza map alignment project: development of a genus-wide comparative genomics platform to help solve the 9 billion-people question. Curr Opin Plant Biol 16:147–156
Jeong DH, Park S, Zhai J et al (2011) Massive analysis of rice small RNAs: mechanistic implications of regulated microRNAs and variants for differential target RNA cleavage. Plant Cell 23:4185–4207
Jiao Y, Wang Y, Xue D et al (2010) Regulation of OsSPL14 by OsmiR156 defines ideal plant architecture in rice. Nat Genet 42:541–544
Johnson C, Kasprzewska A, Tennessen K et al (2009) Clusters and superclusters of phased small RNAs in the developing inflorescence of rice. Genome Res 19:1429–1440
Jones-Rhoades MW, Bartel DP, Bartel B (2004) MicroRNAsand their regulatory roles in plants. Annu Rev Plant Biol 57:19–53
Kapoor M, Arora R, Lama T et al (2008) Genome-wide identification, organization and phylogenetic analysis of Dicer-like, Argonaute and RNA-dependent RNA polymerase gene families and their expression analysis during reproductive development and stress in rice. BMC Genomics 9:451
Khush GS (1997) Origin, dispersal, cultivation and variation of rice. Plant Mol Biol 35:25–34
Kolaczkowski B, Hupalo DN, Kern AD (2011) Recurrent adaptation in RNA interference genes across the drosophila phylogeny. Mol Biol Evol 28:1033–1042
Komiya R, Ohyanagi H, Niihama M et al (2014) Rice germline-specific Argonaute MEL1 protein binds to phasiRNAs generated from more than 700 lincRNAs. Plant J 78:385–397
Kong X, Zhang M, Xu X et al (2014) System analysis of microRNAs in the development and aluminium stress responses of the maize root system. Plant Biotechnol J 12:1108–1121
Lelandais-Briere C, Sorin C, Declerck M et al (2010) Small RNA diversity in plants and its impact in development. Curr Genomics 11:14–23
Li F, Pignatta D, Bendix C et al (2012) MicroRNA regulation of plant innate immune receptors. Proc Natl Acad Sci USA 109:1790–1795

Liu H, Nonomura KI (2016) A wide reprogramming of histone H3 modifications during male meiosis I in rice is dependent on the Argonaute protein MEL1. J Cell Sci 129:3553–3561
Liu B, Li P, Li X et al (2005) Loss of function of OsDCL1 affects microRNA accumulation and causes developmental defects in rice. Plant Physiol 139:296–305
Liu B, Chen Z, Song X et al (2007) Oryza sativa Dicer-like4 reveals a key role for small interfering RNA silencing in plant development. Plant Cell 19:2705–2718
Liu Q, Wang H, Zhu L et al (2013a) Genome-wide identification and analysis of miRNA-related single nucleotide polymorphisms (SNPs) in rice. Rice (NY) 6:10
Liu Y, Wang Y, Zhu QH et al (2013b) Identification of phasiRNAs in wild rice (*Oryza rufipogon*). Plant Signal Behav 8:e25079
Liu Q, Wang H, Hu H et al (2015) Genome-wide identification and evolutionary analysis of positively selected miRNA genes in domesticated rice. Mol Genet Genomics 290:593–602
Lu C, Chen J, Zhang Y et al (2012) Miniature inverted-repeat transposable elements (MITEs) have been accumulated through amplification bursts and play important roles in gene expression and species diversity in *Oryza sativa*. Mol Biol Evol 29:1005–1017
Ma JB, Ye K, Patel DJ (2004) Structural basis for overhang-specific small interfering RNA recognition by the PAZ domain. Nature 429:318–322
Margis R, Fusaro AF, Smith NA et al (2006) The evolution and diversification of Dicers in plants. FEBS Lett 580:2442–2450
Matzke MA, Mosher RA (2014) RNA-directed DNA methylation: an epigenetic pathway of increasing complexity. Nat Rev Genet 15:394–408
Meyer RS, Choi JY, Sanches M et al (2016) Domestication history and geographical adaptation inferred from a SNP map of African rice. Nat Genet 48:1083–1088
Mi S, Cai T, Hu Y et al (2008) Sorting of small RNAs into Arabidopsis Argonaute complexes is directed by the 5′ terminal nucleotide. Cell 133:116–127
Miura K, Ikeda M, Matsubara A et al (2010) OsSPL14 promotes panicle branching and higher grain productivity in rice. Nat Genet 42:545–549
Mukherjee K, Campos H, Kolaczkowski B (2013) Evolution of animal and plant Dicers: early parallel duplications and recurrent adaptation of antiviral RNA binding in plants. Mol Biol Evol 30:627–641
Nagasaki H, Itoh J, Hayashi K et al (2007) The small interfering RNA production pathway is required for shoot meristem initiation in rice. Proc Natl Acad Sci USA 104:14867–14871
Nishimura A, Ito M, Kamiya N et al (2002) OsPNH1 regulates leaf development and maintenance of the shoot apical meristem in rice. Plant J 30:189–201
Nonomura K, Morohoshi A, Nakano M et al (2007) A germ cell specific gene of the Argonaute family is essential for the progression of premeiotic mitosis and meiosis during sporogenesis in rice. Plant Cell 19:2583–2594
Nosaka M, Itoh J, Nagato Y et al (2012) Role of transposon-derived small RNAs in the interplay between genomes and parasitic DNA in rice. PLoS Genet 8:e1002953
Nozawa M, Miura S, Nei M (2012) Origins and evolution of microRNA genes in plant species. Genome Biol Evol 4:230–239
Obbard DJ, Jiggins FM, Bradshaw NJ et al (2011) Recent and recurrent selective sweeps of the antiviral RNAi gene Argonaute-2 in three species of drosophila. Mol Biol Evol 28:1043–1056
Oka H-I (1988) Origin of cultivated rice. Elsevier/JSSP, Tokyo
Ou-Yang F, Luo QJ, Zhang Y et al (2013) Transposable element-associated microRNA hairpins produce 21-nt sRNAs integrated into typical microRNA pathways in rice. Funct Integr Genomics 13:207–216
Parker GS, Maity TS, Bass BL (2008) dsRNA binding properties of RDE-4 and TRBP reflect their distinct roles in RNAi. J Mol Biol 384:967–979
Piriyapongsa J, Jordan IK (2008) Dual coding of siRNAs and miRNAs by plant transposable elements. RNA 14:814–821
Pontier D, Picart C, Roudier F et al (2012) NERD, a plant-specific GW protein, defines an additional RNAi-dependent chromatin-based pathway in Arabidopsis. Mol Cell 48:121–132

Project I R G S (2005) The map-based sequence of the rice genome. Nature 436:793–800

Qian Y, Cheng Y, Cheng X et al (2011) Identification and characterization of Dicer-like, Argonaute and RNA-dependent RNA polymerase gene families in maize. Plant Cell Rep 30:1347–1363

Rivas FV, Tolia NH, Song JJ et al (2005) Purified Argonaute2 and an siRNA form recombinant human RISC. Nat Struct Mol Biol 12:340–349

Ruiz-Ferrer V, Voinnet O (2009) Roles of plant small RNAs in biotic stress responses. Annu Rev Plant Biol 60:485–510

Shan Q, Wang Y, Li J et al (2014) Genome editing in rice and wheat using the CRISPR/Cas system. Nat Protoc 9:2395–2410

Shivaprasad PV, Chen HM, Patel K et al (2012) A microRNA superfamily regulates nucleotide binding site-leucine-rich repeats and other mRNAs. Plant Cell 24:859–874

Simon SA, Meyers BC (2011) Small RNA-mediated epigenetic modifications in plants. Curr Opin Plant Biol 14:148–155

Slotkin RK, Martienssen R (2007) Transposable elements and the epigenetic regulation of the genome. Nat Rev Genet 8:272–285

Song X, Li P, Zhai J et al (2012a) Roles of DCL4 and DCL3b in rice phased small RNA biogenesis. Plant J 69:462–474

Song X, Wang D, Ma L et al (2012b) Rice RNA-dependent RNA polymerase 6 acts in small RNA biogenesis and spikelet development. Plant J 71:378–389

Stroud H, Greenberg MV, Feng S et al (2013) Comprehensive analysis of silencing mutants reveals complex regulation of the Arabidopsis methylome. Cell 152:352–364

Sunkar R, Chinnusamy V, Zhu J et al (2007) Small RNAs as big players in plant abiotic stress responses and nutrient deprivation. Trends Plant Sci 12:301–309

Ta KN, Sabot F, Adam H et al (2016) miR2118-triggered phased siRNAs are differentially expressed during the panicle development of wild and domesticated African rice species. Rice (NY) 9:10

Tajima F (1989) Statistical method for testing the neutral mutation hypothesis by DNA polymorphism. Genetics 123:585–595

Takanashi H, Ohnishi T, Mogi M et al (2011) DCL2 is highly expressed in the egg cell in both rice and Arabidopsis. Plant Signal Behav 6:604–606

Tang T, Lu J, Huang J et al (2006) Genomic variation in rice: genesis of highly polymorphic linkage blocks during domestication. PLoS Genet 2:e199

Toriba T, Suzaki T, Yamaguchi T et al (2010) Distinct regulation of adaxial-abaxial polarity in anther patterning in rice. Plant Cell 22:1452–1462

Urayama S, Moriyama H, Aoki N et al (2010) Knock-down of OsDCL2 in rice negatively affects maintenance of the endogenous dsRNA virus, Oryza sativa endornavirus. Plant Cell Physiol 51:58–67

Vaucheret H (2008) Plant Argonautes. Trends Plant Sci 13:350–358

Vazquez F, Blevins T, Ailhas J et al (2008) Evolution of Arabidopsis MIR genes generates novel microRNA classes. Nucleic Acids Res 36:6429–6438

Voinnet O (2009) Origin, biogenesis, and activity of plant microRNAs. Cell 136:669–687

Wang Y, Shen D, Bo S et al (2010a) Sequence variation and selection of small RNAs in domesticated rice. BMC Evol Biol 10:119

Wang XB, Wu Q, Ito T et al (2010b) RNAi-mediated viral immunity requires amplification of virus-derived siRNAs in Arabidopsis thaliana. Proc Natl Acad Sci USA 107:484–489

Wang Y, Bai XF, Yan CH et al (2012) Genomic dissection of small RNAs in wild rice (Oryza rufipogon): lessons for rice domestication. New Phytol 196:914–925

Wang MH, Yu Y, Haberer G et al (2014) The genome sequence of African rice (Oryza glaberrima) and evidence for independent domestication. Nat Genet 46:982–988

Wang H, Jiao X, Kong X et al (2016) A signaling cascade from miR444 to RDR1 in rice antiviral RNA silencing pathway. Plant Physiol 170:2365–2377

Wei L, Gu L, Song X et al (2014) Dicer-like 3 produces transposable element-associated 24-nt siRNAs that control agricultural traits in rice. Proc Natl Acad Sci USA 111:3877–3882

Wen M, Xie M, He L et al (2016) Expression variations of miRNAs and mRNAs in rice (*Oryza sativa*). Genome Biol Evol 8(11):3529–3544. doi:10.1093/gbe/evw252

Willmann MR, Endres MW, Cook RT et al (2011) The functions of RNA-dependent RNA polymerases in Arabidopsis. Arabidopsis Book 9:e0146

Wing RA, Ammiraju JS, Luo M et al (2005) The Oryza Map Alignment Project: the golden path to unlocking the genetic potential of wild rice species. Plant Mol Biol 59:53–62

Wu L, Zhang Q, Zhou H et al (2009) Rice microRNA effector complexes and targets. Plant Cell 21:3421–3435

Wu L, Zhou H, Zhang Q et al (2010) DNA methylation mediated by a microRNA pathway. Mol Cell 38:465–475

Wu J, Yang Z, Wang Y et al (2015) Viral-inducible Argonaute18 confers broad-spectrum virus resistance in rice by sequestering a host microRNA. Elife 4. doi: 10.7554/eLife.05733

Xia R, Xu J, Arikit S et al (2015) Extensive families of miRNAs and PHAS loci in Norway spruce demonstrate the origins of complex phasiRNA networks in seed plants. Mol Biol Evol 32:2905–2918

Xie J, Yang X, Song Y et al (2016) Adaptive evolution and functional innovation of Populus-specific recently evolved microRNAs. New Phytol 213:206–219

Yang Y, Zhong J, Ouyang YD et al (2013) The integrative expression and co-expression analysis of the AGO gene family in rice. Gene 528:221–235

Yu J, Hu S, Wang J et al (2002) A draft sequence of the rice genome (*Oryza sativa* L. ssp. indica). Science 296:79–92

Yu J, Wang J, Lin W et al (2005) The genomes of *Oryza sativa*: a history of duplications. PLoS Biol 3:e38

Yu YC, Tang T, Qian Q et al (2009) Independent losses of function in a polyphenol oxidase in rice: differentiation in grain discoloration between subspecies and the role of positive selection under domestication (vol 20, pp 2946, 2008). Plant Cell 21:1020–1020

Zhai J, Jeong DH, De Paoli E et al (2011) MicroRNAs as master regulators of the plant NB-LRR defense gene family via the production of phased, trans-acting siRNAs. Genes Dev 25: 2540–2553

Zhai L, Sun W, Zhang K et al (2014) Identification and characterization of Argonaute gene family and meiosis-enriched Argonaute during sporogenesis in maize. J Integr Plant Biol 56:1042–1052

Zhai J, Zhang H, Arikit S et al (2015) Spatiotemporally dynamic, cell-type-dependent premeiotic and meiotic phasiRNAs in maize anthers. Proc Natl Acad Sci USA 112:3146–3151

Zhang Y, Jiang WK, Gao LZ (2011) Evolution of microRNA genes in *Oryza sativa* and *Arabidopsis thaliana*: an update of the inverted duplication model. PLoS One 6:e28073

Zhang YC, Yu Y, Wang CY et al (2013) Overexpression of microRNA OsmiR397 improves rice yield by increasing grain size and promoting panicle branching. Nat Biotech 31: 848–852

Zhang QJ, Zhu T, Xia EH et al (2014) Rapid diversification of five Oryza AA genomes associated with rice adaptation. Proc Natl Acad Sci USA 111:E4954–E4962

Zhang H, Xia R, Meyers BC et al (2015) Evolution, functions, and mysteries of plant Argonaute proteins. Curr Opin Plant Biol 27:84–90

Zhang Y, Xia R, Kuang H et al (2016a) The diversification of plant NBS-LRR defense genes directs the evolution of microRNAs that target them. Mol Biol Evol 33:2692–2705

Zhang H, Tao Z, Hong H et al (2016b) Transposon-derived small RNA is responsible for modified function of WRKY45 locus. Nat Plants 2:16016

Zhu Q, Ge S (2005) Phylogenetic relationships among A-genome species of the genus Oryza revealed by intron sequences of four nuclear genes. New Phytol 167:249–265

Zong J, Yao X, Yin J et al (2009) Evolution of the RNA-dependent RNA polymerase (RdRP) genes: duplications and possible losses before and after the divergence of major eukaryotic groups. Gene 447:29–39

Small RNAs: Master Regulators of Epigenetic Silencing in Plants

Sarma Rajeev Kumar, Safia, and Ramalingam Sathishkumar

Abstract From fairly simple beginnings, research on epigenetic silencing in plants has revealed a highly complex epigenetic pathway. In the last two decades, several interesting phenomena associated with epigenetic regulation in plants were dissected giving insights into the biological significance of epigenetic marks and the role it plays in an organism's life cycle by controlling different physiological processes like plant development, morphogenesis, reproduction, and stress response. Epigenetics refers to either heritable or reversible genetic modifications in DNA or histone proteins that maintain the nucleosome structure in a dynamic manner or those mediated by small RNAs (sRNAs) that in turn modulate gene expression. Plants are equipped with intricate regulatory mechanism to elicit highly sequence-specific chromatin-based gene silencing. Diverse classes of RNAs like small interfering RNA (siRNA), microRNAs (miRNAs), and long noncoding RNAs (lnc RNAs) have emerged as key regulators of gene expression along with several accessory proteins. sRNAs are widespread in various eukaryotes and are specifically involved in the maintenance of chromatin modifications in plants. These sRNAs regulate gene expression in different ways including post-transcriptional gene silencing (PTGS) in cytosol by targeting complementary transcripts for degradation, thereby repressing protein synthesis. In nucleus, sRNAs are responsible for transcriptional gene silencing (TGS) by directing epigenetic modifications like cytosine or histone methylation to homologous regions of the genome. This chapter gives an overview of the role of small RNAs in PTGS and TGS.

Keywords DNA methylation • miRNAs • Post-transcriptional gene silencing • sRNA • Transcriptional gene silencing • Transposon silencing

S. Rajeev Kumar
Molecular Plant Biology and Biotechnology Laboratory, CSIR—Central Institute of Medicinal and Aromatic Plants Research Centre, Bangalore, India

Safia • R. Sathishkumar (✉)
Plant Genetic Engineering Laboratory, Department of Biotechnology, Bharathiar University, Coimbatore, India
e-mail: rsathish@buc.edu.in

N. Rajewsky et al. (eds.), *Plant Epigenetics*, RNA Technologies,
DOI 10.1007/978-3-319-55520-1_5

Contents

1 Introduction

Small RNAs (sRNA) play distinct role and are involved in diverse physiological and metabolic processes in plants including developmental, reproduction, genome reprogramming, and phenotypic plasticity (Borges and Martienssen 2015). sRNAs are also a part of cellular defense mechanism against invading RNA viruses and transposable or retro elements. In addition, lately, their roles in regulating endogenous gene expression including epigenetic silencing and silencing of transgenes have been reported, which is summarized in Tables 1 and 2 (Simon and Meyers 2011; Holoch and Moazed 2015; Rajeev kumar et al. 2015). The major epigenetic mechanisms deciphered in eukaryotes to date are using plants as a model system. Epigenetic silencing is a highly regulated and tightly controlled process involving sRNA-directed transcript cleavage in cytosol or DNA methylation, histone, and chromatin modifications, ultimately controlling gene expression in nucleus. Deep understanding of sRNA-directed DNA methylation gave insights into highly complex and distinct pathways involved in different physiological processes in plants. In addition, DNA methylation mediated by sRNA also plays an important role in the regulation of genes although most of methylated DNA sequences found in genomes belong to class of transposable elements and DNA repeats (Chan et al. 2005).

In plants, sRNAs are produced as 21–24-nucleotide (nt) molecules by DICER-LIKE proteins, using double-stranded RNA (dsRNA) intermediates from hairpin precursors, or from the synthesis of dsRNA from single-stranded RNA (ssRNA) by RNA-dependent RNA polymerases (RdRps). The sRNA thus produced is loaded onto ARGONAUTE proteins (AGOs) to target coding or noncoding RNAs by sequence complementarity. Further, depending on the nature of the target gene and AGO involved, this process led to cleavage and degradation of transcripts and translational repression leading to gene silencing.

There are several kinds of endogenous sRNAs reported in plants like microRNAs (miRNAs), hairpin-derived siRNAs (hp-siRNAs), trans-acting siRNA (tasiRNA), natural antisense siRNAs (natsiRNAs), secondary siRNAs, repeat-associated siRNA (rasiRNA), and heterochromatic siRNAs (hetsiRNAs) (Borges and Martienssen 2015). To confer the stability and protection from 3-′-uridylation, these sRNAs are often modified at the 3′-end by 2′-*O*-methylation. In this chapter, we have briefed the involvement of sRNA in gene silencing by PTGS and TGS.

Table 1 Role of miRNA/sRNA induced physiological response in plants

S no	Host plant	RNA type	Role/physiological process involved in plant/trait associated	References
01	*N. tabacum*	mi	Viral infection	Li et al. (2012)
02	*Arabidopsis*	mi	Organ initiation from meristematic tissue	Guleria et al. (2011)
03	*Arabidopsis* and *N. benthamiana*	mi	Bacterial infection	Zhang et al. (2011)
04	*Populus trichocarpa*	mi	Tree development	Lelandais-Briere et al. (2010)
05	*Glycine max*	mi	Nodulation	Li et al. (2010)
06	*Arabidopsis*	mi	Bacterial infection	Li et al. (2010), Jagadeeswaran et al. (2009), Navarro et al. (2006)
07	*Triticum aestivum*	mi	Regulate salt stress	Yao et al. (2010)
08	*Oryza sativa*	mi	Regulate water stress in rice	Zhou et al. (2010)
09	*Medicago truncatula*	mi	Root and nodule development	Lelandais-Briere et al. (2009)
10	*M. truncatula*	mi	Root vascular tissue patterning	Boualem et al. (2008)
11	Tomato	mi	Assist fruit ripening	Moxon et al. (2008)
12	*Zea mays*	mi	Associated with plant development and organ formation	Zhang et al. (2008)
13	*Z. mays*	mi	Associated with sugar metabolism	Zhang et al. (2008)
14	*O. sativa*	mi	Modulate grain maturation	Zhu et al. (2008)
15	*Petunia hybrida, Antirrhinum majus*	mi	Floral organ identity	Cartolano et al. (2007)
16	*Z. mays*	mi	Floral and vegetative phase transitions	Chuck et al. (2007)
17	*Arabidopsis*	mi	Regulate ABA, dehydration, and salt stress	Jung and Kang (2007)
18	*O. sativa*	mi	Shoot meristem initiation	Nagasaki et al. (2007)
19	*Arabidopsis*	mi	miRNAs are involved in maintaining sulfur and phosphate homeostasis	Sunkar et al. (2007)
20	*Arabidopsis*	mi	Regulation of copper homeostasis	Yamasaki et al. (2007)
21	*Arabidopsis*	mi	Involved in regulation of light stress	Zhou et al. (2007)
22	*M. truncatula*	mi	Associated with nodule differentiation	Combier et al. (2006)
23	*Arabidopsis*	mi	Oxidative stress tolerance	Sunkar et al. (2006)
24	*Arabidopsis*	mi	Ovule and anther development	Wu et al. (2006)

(continued)

Table 1 (continued)

S no	Host plant	RNA type	Role/physiological process involved in plant/trait associated	References
25	*Arabidopsis*	mi	Involved in regulation of petal number	Baker et al. (2005)
26	*Arabidopsis*	mi	Lateral root development	Guo et al. (2005)
27	*Arabidopsis*	mi	Root cap development	Wang et al. (2005)
28	*Arabidopsis*	mi	Shoot vascular development	Williams et al. (2005)
29	*Z. mays*	mi	Leaf polarity	Juarez et al. (2004)
30	*Arabidopsis*	mi	Regulate cold, salt, and dehydration stress	Sunkar and Zhu (2004)
31	*Arabidopsis*	mi	siRNA, miRNA processing	Vaucheret et al. (2004)
32	*Arabidopsis*	mi	Promote flowering	Aukerman and Sakai (2003)
33	*Arabidopsis*	mi	miRNA maturation	Bartel and Bartel (2003)
34	*Arabidopsis*	mi	Leaf morphogenesis	Palatnik et al. (2003)
35	*Arabidopsis*	mi	Floral and vegetative phase transitions	Park et al. (2002)
36	*Arabidopsis*	mi	Leaf polarity	Rhoades et al. (2002)
37	*Arabidopsis*	si—RNA	Regulate hypoxia or survival under anaerobic conditions	Moldovan et al. (2010)
38	*T. aestivum*	si—RNA	Regulate cold, heat, salt, dehydration stress	Yao et al. (2010)
39	*Craterostigma plantagineum*	si—RNA	Dehydration and ABA stress tolerance	Hilbricht et al. (2008)
40	*Arabidopsis*	si—RNA	Bacterial infection	Katiyar-Agarwal et al. (2006)
41	*Arabidopsis*	ta-siRNA	Regulate hypoxia or survival under anaerobic conditions	Moldovan et al. (2010)
42	*Arabidopsis*	nat-siRNA	Bacterial infection	Katiyar-Agarwal et al. (2006)
43	*Arabidopsis*	nat-siRNA	Regulate salt stress	Borsani et al. (2005)

2 Nuclear sRNA-Dependent Gene Silencing

Similar to *Caenorhabditis elegans*, RNA silencing in plants is also initiated after generation of dsRNA. The dsRNA-silencing mechanism results in production of 21–23-nt RNAs. These small RNAs are referred to as sRNAs that were first reported in plants (Hamilton and Baulcombe 1999). Later, the coincidence of RNA silencing and simultaneous sRNA accumulation have been reported in many plants, and lately these small RNA molecules have attracted a lot of attention

Table 2 Role of sRNA induced methylation or associated mechanism in plants

S no	Host plant	Species/ process involved	Role/physiological process involved in plant/trait associated	References
01	*Arabidopsis*	Methylation	Associated with plant-induced immune response	Yu et al. (2013)
02	*Z. mays*	RdDM	Paramutation	Castel and Martienssen (2013), Bond and Baulcombe (2014)
03	*Arabidopsis*, Tomato	Methylation	Increased insect resistance	Rasmann et al. (2012)
04	*Arabidopsis*	Methylation	Mediates plant growth rates and responses to defense hormones	Latzel et al. (2012)
05	*Mimulus guttatus*	Methylation	Enhanced physical and chemical defense, increase in glandular leaf trichome	Scoville et al. (2011)
06	*Arabidopsis*	Methylation	Phenotypic plasticity in response to different nutrient	Bossdorf et al. (2010)
07	*Arabidopsis*	Methylation	Pathogen resistance (viral, bacterial, fungal)	Kathiria et al. (2010)
08	*Arabidopsis*	siRNA	Methylation in meristematic root stem cells	Melnyk et al. (2011)
09	*Arabidopsis*	Si RNA via RdDM	Control specification of gametic egg cells by transposon silencing in the germ cells	Olmedo-Monfil et al. (2010)
10	*Arabidopsis*	Methylation via SiRNA	Cytokinesis during endosperm development	Gerald et al. (2009)
11	*Arabidopsis*	Methylation	Regulate height and flowering time	Johannes et al. (2009)
12	*Arabidopsis*	Methylation via siRNA	Regulation of flowering time	Jullien et al. (2006), Kinoshita et al. (2004)
13	*Arabidopsis*	Methylation mediated by siRNA	Seed development and maintenance of genomic imprinting	Kinoshita et al. (1999), Vielle-Calzada et al. (1999)

that are now considered to be integral part of RNA silencing machinery. sRNA-dependent pathways can be either localized in nucleus mediated by transcriptional gene silencing (TGS) or can be cytosolic through post- transcriptional gene silencing (PTGS). In both situations, a similar set of steps is followed from activation of the pathway to the sRNA-dependent silencing of gene expression (Fig. 1). sRNA-mediated silencing pathways are activated by double-stranded RNAs that are processed into sRNAs by DICER, a type III ribonuclease. DICERs belong to a novel class of specific RNase III family of proteins with dual catalytic domains, dsRNA-binding domains and helicase and PAZ motifs (Nicholson 2014). These proteins are evolutionarily conserved across the species including worms, flies, plants, fungi, and mammals, supporting the fact that DICER and its orthologues play critical role in the initiation step of the RNA-silencing mechanism. sRNAs are then incorporated into RNAi effector complexes that use the sRNA sequence as a specificity determinant to bind only with complementary RNA. The configuration of generated sRNAs is very crucial for incorporation into RNA-Induced Silencing Complex (RISC). In general, ds sRNAs generated by DICER are 2-nucleotide-long with 3′ overhangs and 5′ phosphate and 3′ hydroxyl ends that are effective inducers of RNA silencing. However, synthetic or artificial duplex siRNAs or siRNAs with extensive 2′-deoxy modifications did not mediate RNA silencing efficiently and that was implicated due to misincorporation into the RISC machinery (Depicker et al. 2005). RISC and RNAi-Induced Transcriptional Silencing (RITS) complex mediate, respectively, sRNA-dependent PTGS and TGS. Both types of complexes contain an AGO protein, but conversely to RISC, RITS appears to localize exclusively in nucleus. Once a perfect match occurs between the siRNA and a targeted RNA, target transcript is cleaved. RNA-directed RNA polymerases (RdRp) amplify this silencing response using the target RNA as a template to synthesize more dsRNAs that fuel the sRNA pathway and amplify the sRNA-mediated silencing (Pikaard and Mittelsten Scheid 2014). Besides, RdRP also efficiently synthesizes dsRNA from target mRNAs, and these dsRNAs become the source for the production of secondary sRNAs. Once RISC is activated, a helicase function separates two sRNA strands leading to sequence-specific cleavage of the single-stranded complementary target (endogenous) RNA by AGO proteins.

In addition, silencing of endogenous transcript by miRNA also regulates gene expression in plants. miRNA biogenesis in plants is a complex process and involves several steps to form mature miRNAs from miRNA genes (Bartel 2004). Initially, the transcript encoding miRNA is transcribed to a primary miRNA (pri-miRNA) by RNA polymerase II (pol II), and in most of the cases, it is longer than several hundred nucleotides (Bartel 2004; Lee et al. 2004). Subsequently, the pri-miRNA formed is cleaved to a stem loop/hairpin intermediate by DICER-1 enzyme (DCL1) in plants called pre-miRNA (Tang et al. 2003; Kurihara and Watanabe 2004). Further, miRNAs are degraded to miRNA:miRNA* duplex by DCL1 in the nucleus and the duplex is transported to cytoplasm by HASTY, a plant exportin 5 protein (Park et al. 2005). Once in cytosol, miRNAs are converted to single strand mature miRNAs by helicase (Bartel 2004). Finally, the mature miRNAs are coupled to RISC complex and regulate the expression of target gene (Bartel 2004). In general, mature miRNAs are 21–22 nt ssRNAs that regulate gene expression by pairing to specific transcripts leading to either RNA cleavage or inhibition of protein translation (Baulcombe 2004). Also, these sRNA can trigger a long-term and heritable gene silencing which is maintained in an epigenetic fashion.

In the RISC complex, miRNAs bind to target transcript and repress gene expression through perfect or in some cases through near-perfect complementarity between miRNA and target mRNA (Bartel 2004). In plants, most target mRNAs contain a single miRNA complementary site, and the corresponding miRNAs perfectly complement to these sites leading to the degradation of target mRNAs (Bartel 2004). The unique mechanism with plant miRNA is that complementary sites can exist at any position of the target mRNA rather than in the 3′UTR (in animals, the complementary sites are mostly in 3′UTR regions). It has been

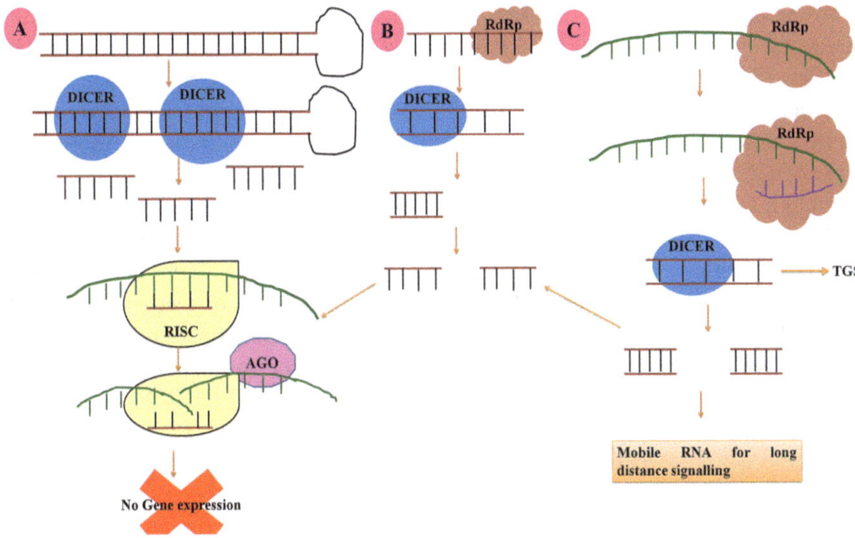

Fig. 1 RNA silencing in plants. (**a**) Hairpin dsRNAs are precursors for generation of miRNA. These dsRNA are cleaved by DICER (*blue* circles) into small ssRNA. The ssRNA thus generated are incorporated into RISC complex (*yellow* teardrop) where target mRNA is cleaved by AGO proteins (*pink* dodecagon) leading to PTGS. (**b**) ss viral RNAs are transcribed by RNA-dependent RNA polymerase (RdRp) (*pink* cloud) and the dsRNAs generated are cleaved by DICER and subsequently incorporated into RISC. (**c**) Aberrant RNAs generated in plants are also targets of RdRp which transcribes a complementary strand to form dsRNA which is further cleaved by DICER and incorporated into RISC complex. The dsRNA generated can also act as mobile signals to regulate gene expression. Alternatively, they can induce TGS by methylation of DNA (Modified with permission from Volkov et al. 2006)

also reported that some of the miRNAs regulate gene expression by repressing translation by interfering with ribosome movement (Aukerman and Sakai 2003; Chen 2004). Schwab et al. (2005) reported that miR172 efficiently guides cleavage of its target transcripts implicating that complex mechanisms are involved in gene silencing mediated by miRNA. Further, it is also evident that plant miRNAs regulate gene expression at PTGS, not only by repression of mRNA translation but also by direct cleavage of target transcripts (Zhang et al. 2006).

3 Small RNA-Directed DNA Methylation in Plants

RNA-directed DNA methylation (RdDM) is one of the well-characterized and completely deciphered epigenetic mechanisms in plants. It is required for many critical cellular functions including transposon silencing, genome stability, cell identity maintenance, and defense against exogenous DNAs or invading viruses (Zhang et al. 2013a). DNA methylation is an important epigenetic modification in the genomes of all eukaryotes that involves addition of a methyl group to the 5th position of the cytosine (m5C) by DNA methyltransferase. Unlike in animals, where cytosine is methylated only at CpG sequences (Bird 1986), in plants methylation occurs in all sequence contexts (CpG, CpHpG, and CpHpH, where H is adenine, cytosine, or thymine) that are known as symmetric and asymmetric methylation, respectively. In plants including *Arabidopsis thaliana*, more than 25% of CG, 7% of CHG, and 2% of CHH are methylated (Cokus et al. 2008). Although lower levels of DNA methylation are reported in euchromatin part with coding gene (Zhang et al. 2006), methylation mainly occurs in transposon-rich heterochromatic regions (Zilberman and Henikoff 2007). To date, different enzymes have been reported to be involved in the process of DNA methylation, its maintenance, and subsequent demethylation. In plants, maintenance of DNA methylation state is regulated by the activity of three methyltransferases: The DOMAIN REARRANGED METHYLTRANSFERASE 2 (DRM2), which is involved in *de novo* DNA methylation (Zilberman and Henikoff 2007; Takuno and Gaut 2012); METHLYTRANSFERASE1 (MET1) is crucial for maintaining CG methylation (Zilberman and Henikoff 2007); and finally, CHROMOMETHYLASE3 (CMT3), involved in CHG methylation maintenance (Cao and Jacobsen 2002a, b). A family of 5-methylcytosine DNA glycosylases that catalyzes active DNA demethylation in plants has also been reported, and process of demethylation is equally important as DNA methylation (Barreto et al. 2007).

As mentioned above, several types of endogenous sRNAs (trans-acting siRNAs, natural antisense siRNAs, miRNA, lc RNA, and heterochromatic siRNAs) exist in plants. Among them, heterochromatic siRNAs and miRNAs (in some cases) are responsible for mediating gene silencing by directing DNA methylation machineries (Fig. 2) (Xie et al. 2004; Zhang et al. 2007; Mosher et al. 2008; Jia et al. 2011). In general, siRNAs are 21–24 nt long that are generated from different RNA precursors. siRNA generation is initiated by a plant-specific DNA-dependent RNA polymerase known as polymerase IV (pol IV), which produces ssRNA from a transposon or repeat-containing region (Sidorenko et al. 2009; Havecker et al. 2010). This process is assisted by other proteins like SHH1and DTF1 that aids in recruiting pol IV to target loci (Law et al. 2013; Zhang et al. 2013b). The ssRNA thus generated are converted by RNA-dependent RNA polymerase 2 (RDR2) to dsRNAs. Subsequently, these dsRNA species are cleaved to smaller fragments by DICER-LIKE3 (DCL3), followed by HEN1 methylation (Xie et al. 2004; Alleman et al. 2006). siRNAs thus generated are further loaded to AGO4 and are recruited back to homologous genomic loci from which the RNA was generated (Ye et al. 2012). The AGO-bound 24-nt siRNAs complex can serve as sequence-specific guide for methylation by pairing with complementary target DNA or nascent scaffold RNA (Wierzbicki et al. 2009). The methylation signal can also be transmitted to long-distance regions. However, it is important to mention here that heterochromatic silencing generally requires two rounds of siRNA production. The secondary siRNAs produced from methylated loci are often associated with short repetitive sequences in intergenic regions (Lee et al. 2012) and are involved in spreading of methylation (Pontier et al. 2005; Matzke et al. 2009; Kanno et al. 2010). During the second round of siRNA generation, RNA-DIRECTED DNA

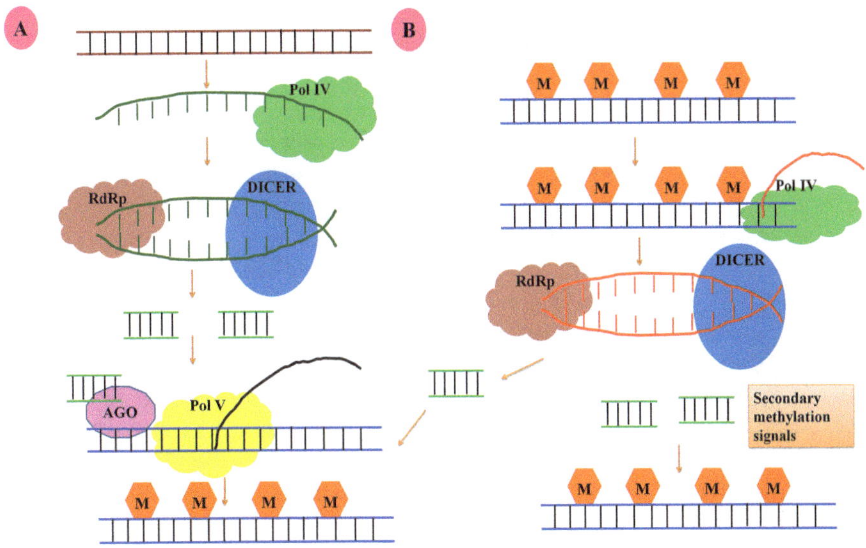

Fig. 2 RNA-induced DNA methylation in plants. (**a**) ssRNA is transcribed using transposons, repeat DNA, or DNA from heterochromatin region by polymerase IV (*green* clouds). The RdRP (*rose* clouds) makes a complementary RNA strand and the dsRNA is cleaved by DICER (*blue* circle) to generate 21–24 nt dsRNA. The dsRNA is coupled with AGO (*pink* wavy dodecagon) and transported to target DNA loci where pol V generates another nascent RNA (*black* line) that further recruits methylation machinery leading to methylation in target loci (orange hexagon). (**b**) Alternatively, aberrant transcripts (*red* line) are made from methylated DNA by pol V. RdRP uses these ssRNAs to make a complementary strand and DICER cleaves the dsRNA to small 24 nt sRNA. These sRNA can either direct methylation as mentioned above or can also act as secondary signals for methylation of target loci (Modified with permission from Simon and Meyers 2011)

METHYLATION (RDM1) protein binds to single-stranded methylated DNA and aids in recruitment of silencing complex to methylated DNA generated from the previous process (Gao et al. 2010). AGO4 in the complex cleaves pol II and pol V transcripts that are complementary to siRNAs. Subsequently, RDR2 transcribes the cleaved transcript fragments into dsRNAs. Finally, DCL3 cleavage of dsRNAs leads to the production of secondary siRNAs at the methylated loci (Gao et al. 2010).

It is worth mentioning here that DNA and histone demethylation requires AGO-siRNA complex and chromatin methylation enzymes transported or recruited to target loci. Long noncoding RNAs (lncRNAs) act as scaffold RNAs that are involved in this recruitment, and their biogenesis is independent of siRNA biogenesis (Wierzbicki et al. 2008). Polymerase V (pol V) plays critical role in scaffold RNAs production, and it has been also reported that pol II is also required at certain loci (Wierzbicki 2012; Wierzbicki et al. 2008). A GHKL-type ATPase, and associated DMS11, is essential for production of the pol V transcript (Lorkovic et al. 2012). Further, pol V-dependent scaffold RNA production requires several proteins including DDR complex containing a chromatin remodeling protein DRD1, a chromosome hinge domain protein DMS3, and RDM1 that binds ss methylated DNA to facilitate pol V transcription (Kanno et al. 2005, 2008; Law and Jacobsen 2010).

Although the initial steps of DNA methylation are completely known, the final steps involved in silencing mechanism including process involved in recruiting DNA methyltransferases target loci are not fully understood. It has been proposed that this process is assisted by complementary base pairing between 24-nt siRNAs and nascent scaffold RNAs generated from lnc RNAs. One possible mode of mechanism proposed is that 24-nt siRNAs incorporated into AGO4 interact with scaffold RNAs by base pairing, and IDN2 protein binds to dsRNA with 5′ overhangs (Zhang et al. 2012) to stabilize interactions between an AGO-bound siRNA and a scaffold RNA (Finke et al. 2012). This complex then directs the silencing machinery, including the *de novo* cytosine methyltransferase DRM2 activities to the target loci. Alternatively, the scaffold RNAs may stabilize siRNA–DNA interactions by binding directly to AGO4 and activates subsequent methylation. Another proposed model is that scaffold RNAs enhance the interaction efficiency of AGO4 with target loci sequence by altering structural features of heterochromatin (Wierzbicki et al. 2008). Ultimately, any of the above methods results in recruitment of silencing machinery including DNA methyltransferases to the target sequences leading to silencing of specific genomic loci.

There are some minor differences between methylation mediated by siRNA and miRNA. Key distinctions can be made between these two types with respect to their biogenesis and the targets with which they interact (Jia et al. 2011). Unlike siRNAs that are produced from a long dsRNA, the production of miRNA-derived siRNAs is site specific, and their structures are almost identical to the miRNAs. Another distinction is that the length of miRNA-derived siRNAs is 23–27 nt and that of long miRNAs (lmiRNAs) is 24 nt. Pol II catalyzes the precursors of both miRNA-derived siRNAs and lmiRNAs. While the synthesis of lmiRNAs in rice does not require RDR2 activity, its activity is inevitable for miRNA-derived siRNAs in *Arabidopsis* (Chellappan et al. 2010). lmiRNA-directed DNA methylation in rice is limited to as close as 80 nt region around the lmiRNA and target loci (Wu et al. 2010). Another striking difference within miRNA is while lmiRNA may direct methylation signals both in cis and trans loci, miRNA-derived siRNAs mainly guide DNA methylation at their target site in trans and rarely mediate DNA methylation in cis (Chellappan et al. 2010; Wu et al. 2010). Taken together, these studies suggest distinct pathways involving siRNA- and miRNA-mediated DNA methylation exists in plants.

4 Mechanism of Transposon Repression by sRNAs and Silencing of Transposons

The involvement of DNA methylation in transposable elements and subsequent silencing in plants was reported two decades back (Chandler and Walbot 1986). An important factor of the successful communication between transposable elements and its host is epigenetic modifications including DNA methylation mediated by sRNAs (Ito 2012). A critical role of RNA silencing at genome level is mainly to protect the genome from damage caused by active transposons (mobile elements) and by extreme amplification generated by repetitive elements resulting in the formation of heterochromatin (Lippman and Martienssen 2004). In most cases, transposons are silent in their host plants; however, certain genomic shocks like environmental stress or a hybridization event trigger transposon activation. Since transposons can also affect the regulatory machinery of host genes, it has been lately reported that transposons could have coevolved as an important mechanism for plant growth, development, and adaptation. It has been recently reported that sRNAs suppress transgenerational transposition of transposable elements making sRNAs critically important toward maintaining genome stability in plants.

In all eukaryotes, AGO proteins mediate transposon repression by sRNAs. Their specificity is conferred by sRNAs that identify silencing targets through complementary base pairing. The molecular mechanisms by which AGO-small RNA complexes repress transposon expression vary across species. In general, mechanisms include small RNA-directed mRNA cleavage, TGS through DNA and histone methylation, and excision of transposon sequences from the host genome. In plants, AGO4 binds to 24-nt sRNAs and is essential for non-CG methylation of transposons (Zilberman et al. 2003) and for *de novo* methylation of the repeat-containing FWA (Flowering Wageningen) locus (Fujimoto et al. 2008). The role of sRNA silencing and transposon silencing is reported in detail by Ito (2012).

Progress in understanding the silencing of transposable elements in *A. thaliana* has revealed a close association between DNA and histone methylation and sRNAs. Although DNA and histone methylation are essential to maintain silencing, RNA-mediated mechanism is inevitable to initiate as well as to maintain silencing (Zilberman and Henikoff 2004). It was reported in *A. thaliana* that inheritance of epigenetic gene silencing for several generations could function as a transgenerational genome defense mechanism against movement of transposons (Kato et al. 2004). Further, it was reported that a silent CACTA is mobilized by the DNA hypomethylation mutation ddm1. However, CACTA triggered by the ddm1 mutation was mobile in the presence of the WT *DDM1* gene, implicating that *de novo* silencing alone is not efficient for genome defense against CACTA movement. Nevertheless, defense depends on maintenance of transposon silencing over many generations. It was shown that heterochromatin region in *Arabidopsis* genome is determined by the presence of transposable elements and associated tandem repeats, under the control of the chromatin remodeling ATPase DDM1 proteins, and sRNAs correspond to these sequences that are involved in guiding

DDM1 for methylation (Zilberman et al. 2003, 2004). Transposable elements can regulate genes epigenetically; however, it happens only when they are inserted within or very close to them.

sRNAs derived from transposon can introduce DNA methylation of nearby genes through RdDM and regulate gene expression. One well best example is FLOWERING LOCUS C (FLC), a major repressor of flowering; FLC is negatively regulated by vernalization leading to variation in flowering behavior among different *Arabidopsis* accessions (Boss et al. 2004). The variation is associated with a 1.2 kb nonautonomous mutator-like transposon insertion in the first intron of *FLC* gene (Gazzani et al. 2003; Michaels et al. 2003) acting in cis to reduce expression of the *FLC* allele in the accession Landsberg erecta (Ler) (Liu et al. 2004). The imprinted gene FLOWERING WAGENINGEN (FWA) is specifically expressed in the endosperm of *A. thaliana* but is silent in vegetative tissues of the plant. The tissue-specific expression of FWA is regulated by DNA demethylation of the FWA promoter, which comprised of two direct repeats containing a sequence related to a short interspersed nuclear (SINE) retrotransposon (Kinoshita et al. 2007). When the FWA promoter is methylated, localized heterochromatin was established leading to transcriptional silencing and further resulted in generation of sRNA from SINE-related tandem repeats (Lippman et al. 2004; Chan et al. 2005). Similar to the inserted transposon of FLC, these siRNAs can also target the RdDM activity to establish *de novo* silencing at unmethylated FWA transgenes (Cao and Jacobsen 2002b; Chan et al. 2005).

5 Conclusion and Future Perspectives

Since past two decades, explosion of information and generation of huge data related to proteins, RNAs, and chemical modifications that regulate epigenetic control mediated by different sRNA was reported. However, still gaps remain in the nature of epigenetic inheritance and the role of epigenetics as a source of variation contributing different morphological traits in plants. Plants possess a highly complex network of diverse sRNA pathway involved in various physiological functions. These sRNA pathways vary with respect to sRNA biogenesis and subsequent sRNA recruitment to different effector complexes. The 21-nt and 24-nt sRNAs generated are highly mobile and have the ability to move not only from cell to cell but over long distances to direct mRNA cleavage and induce DNA methylation in recipient cells. These sRNAs have also role in epigenetic modifications associated with genome defense and protection from transposon and inverted repeats. Lately, it has been reported that sRNA signals can move even to developing seeds or pollens and can induce epigenetic changes that ultimately initiate transgenerational effects.

Application of sRNA-based silencing in crop plants has opened new avenues in improvement of crop productivity and quality. Targeted sRNA-based gene silencing can be used to modulate metabolic pathways in specific tissues to accumulate various pharmacologically important metabolites that are not otherwise possible by traditional breeding. Although still several gaps exist in molecular processes and roles of sRNA-mediated silencing in plants, current understanding has already provided new platforms for designing molecular tools for functional genomic studies and crop improvements exploiting innate sRNA-based methods. It can be expected in the near future that sRNA-based silencing technologies will help to face the challenges of productive agriculture in unfavorable environmental conditions that can improve food safety.

References

Alleman M, Sidorenko L, McGinnis K et al (2006) An RNA-dependent RNA polymerase is required for paramutation in maize. Nature 442:295–298

Aukerman MJ, Sakai H (2003) Regulation of flowering time and floral organ identity by a microRNA and its APETALA2-like target genes. Plant Cell 11:2730–2741

Baker CC, Sieber P, Wellmer F et al (2005) The early extra petals1 mutant uncovers a role for microRNA miR164c in regulating petal number in *Arabidopsis*. Curr Biol 15:303–315

Barreto G, Schafer A, Marhold J et al (2007) Gadd45a promotes epigenetic gene activation by repair-mediated DNA demethylation. Nature 445:671–675

Bartel DP (2004) MicroRNAs: genomics, biogenesis, mechanism, and function. Cell 116:281–297

Bartel B, Bartel DP (2003) MicroRNAs: at the root of plant development. Plant Physiol 132:709–717

Baulcombe D (2004) RNA silencing in plants. Nature 431:356–363

Bird AP (1986) CpG-rich islands and the function of DNA methylation. Nature 321:209–213

Bond DM, Baulcombe DC (2014) Small RNAs and heritable epigenetic variation in plants. Trends Cell Biol 24:100–107

Borges F, Martienssen RA (2015) The expanding world of small RNAs in plants. Nat Rev Mol Cell Biol 16:727–741

Borsani O, Zhu J, Verslues PE et al (2005) Endogenous siRNAs derived from a pair of natural cis-antisense transcripts regulate salt tolerance in *Arabidopsis*. Cell 123:1279–1291

Boss PK, Bastow RM, Mylne JS et al (2004) Multiple pathways in the decision to flower: enabling, promoting, and resetting. Plant Cell 16(Suppl):S18–S31

Bossdorf O, Arcuri D, Richards CL et al (2010) Experimental alteration of DNA methylation affects the phenotypic plasticity of ecologically relevant traits in *Arabidopsis thaliana*. Evol Ecol 24:541–553

Boualem A, Laporte P, Jovanovic M et al (2008) MicroRNA166 controls root and nodule development in *Medicago truncatula*. Plant J 54:876–887

Cao XF, Jacobsen SE (2002a) Locus-specific control of asymmetric and CpNpG methylation by the DRM and CMT3 methyltransferase genes. Proc Natl Acad Sci U S A 99(suppl 4):16491–16498

Cao X, Jacobsen SE (2002b) Role of the *Arabidopsis* DRM methyltransferases in de novo DNA methylation and gene silencing. Curr Biol 12:1138–1144

Cartolano M, Castillo R, Efremova N et al (2007) A conserved microRNA module exerts homeotic control over *Petunia hybrida* and *Antirrhinum majus* floral organ identity. Nat Genet 39:901–905

Castel SE, Martienssen RA (2013) RNA interference in the nucleus: roles for small RNAs in transcription, epigenetics and beyond. Nat Rev Genet 14:100–112

Chan SW, Henderson IR, Jacobsen SE (2005) Gardening the genome: DNA methylation in *Arabidopsis thaliana*. Nat Rev Genet 6:351–360

Chandler VL, Walbot V (1986) DNA modification of a maize transposable element correlates with loss of activity. Proc Natl Acad Sci U S A 83:1767–1771

Chellappan P, Xia J, Zhou X et al (2010) siRNAs from miRNA sites mediate DNA methylation of target genes. Nucleic Acids Res 38:6883–6894

Chen X (2004) A microRNA as a translational represser of APETALA2 in *Arabidopsis* flower development. Science 303:2022–2025

Chuck G, Cigan AM, Saeteurn K et al (2007) The heterochronic maize mutant corngrass1 results from overexpression of a tandem microRNA. Nat Genet 39:544–549

Cokus SJ, Feng S, Zhang X et al (2008) Shotgun bisulphite sequencing of the *Arabidopsis* genome reveals DNA methylation patterning. Nature 452:215–219

Combier JP, Frugier F, De Billy F et al (2006) MtHAP2-1 is a key transcriptional regulator of symbiotic nodule development regulated by microRNA169 in *Medicago truncatula*. Genes Dev 20:3084–3088

Depicker A, Sanders M, Meyer P (2005) Transgene silencing. Plant Epigenetics 19:1–32

Finke A, Kuhlmann M, Mette MF (2012) IDN2 has a role downstream of siRNA formation in RNA directed DNA methylation. Epigenetics 7:950–960

Fujimoto R, Kinoshita Y, Kawabe A et al (2008) Evolution and control of imprinted FWA genes in the genus *Arabidopsis*. PLoS Genet 4:e1000048

Gao Z, Liu HL, Daxinger L et al (2010) An RNA polymerase II- and AGO4-associated protein acts in RNA-directed DNA methylation. Nature 465:106–109

Gazzani S, Gendall AR, Lister C et al (2003) Analysis of the molecular basis of flowering time variation in *Arabidopsis* accessions. Plant Physiol 132:1107–1114

Gerald JNF, Hui PS, Berger F (2009) Polycomb group-dependent imprinting of the actin regulator AtFH5 regulates morphogenesis in *Arabidopsis thaliana*. Development 136:3399–3404

Guleria P, Mahajan M, Bhardwaj J et al (2011) Plant small RNAs: biogenesis, mode of action and their roles in abiotic stresses. Genomics Proteomics Bioinformatics 9:183–199

Guo HS, Xie Q, Fei JF et al (2005) Micro RNA directs mRNA cleavage of the transcription factor NAC1 to down regulate auxin signals for *Arabidopsis* lateral root development. Plant Cell 17:1376–1386

Hamilton AJ, Baulcombe DC (1999) A species of small antisense RNA in post transcriptional gene silencing in plants. Science 286:950–952

Havecker ER, Wallbridge LM, Hardcastle TJ et al (2010) The Arabidopsis RNA-directed DNA methylation argonautes functionally diverge based on their expression and interaction with target loci. Plant Cell 22:321–334

Hilbricht T, Varotto S, Sgaramella V et al (2008) Retrotransposons and siRNA have a role in the evolution of desiccation tolerance leading to resurrection of the plant *Craterostigma plantagineum*. New Phytol 179:877–887

Holoch D, Moazed D (2015) RNA-mediated epigenetic regulation of gene expression. Nat Rev Genet 16:71–84

Ito H (2012) Small RNAs and transposon silencing in plants. Dev Growth Differ 54:100–107

Jagadeeswaran G, Saini A, Sunkar R (2009) Biotic and abiotic stress down-regulate miR398 expression in *Arabidopsis*. Planta 229:1009–1014

Jia X, Yan J, Tang G (2011) Micro RNA-mediated DNA methylation in plants. Front Biol 6:133–139

Johannes F, Porcher E, Teixeira FK et al (2009) Assessing the impact of transgenerational epigenetic variation on complex traits. PLoS Genet 5:1000530

Juarez MT, Kui JS, Thomas J et al (2004) MicroRNA-mediated repression of rolled leaf specifies maize leaf polarity. Nature 428:84–88

Jullien PE, Kinoshita T, Ohad N et al (2006) Maintenance of DNA methylation during the *Arabidopsis* life cycle is essential for parental imprinting. Plant Cell 18:1360–1372

Jung HJ, Kang H (2007) Expression and functional analyses of microRNA417 in *Arabidopsis* thaliana under stress conditions. Plant Physiol Biochem 45:805–811

Kanno T, Huettel B, Mette MF et al (2005) A typical RNA polymerase subunits required for RNA-directed DNA methylation. Nat Genet 37:761–765

Kanno T, Bucher E, Daxinger L et al (2008) A structural maintenance of chromosome hinge domain-containing protein is required for RNA-directed DNA methylation. Nat Genet 40:670–675

Kanno T, Bucher E, Daxinger L et al (2010) RNA-directed DNA methylation and plant development require an IWR1-type transcription factor. EMBO Rep 11:65–71

Kathiria P, Sidler C, Golubov A et al (2010) Tobacco mosaic virus infection results in an increase in recombination frequency and resistance to viral, bacterial, and fungal pathogens in the progeny of infected tobacco plants. Plant Physiol 153:1859–1870

Katiyar-Agarwal S, Morgan R, Dahlbeck D et al (2006) A pathogen-inducible endogenous siRNA in plant immunity. Proc Natl Acad Sci U S A 103:18002–18007

Kato M, Takashima K, Kakutani T (2004) Epigenetic control of CACTA transposon mobility in Arabidopsis. Genetics 168:961–969

Kinoshita T, Yadegari R, Harada JJ et al (1999) Imprinting of the MEDEA polycomb gene in the *Arabidopsis* endosperm. Plant Cell 11:1945–1952

Kinoshita T, Miura A, Choi Y et al (2004) One-way control of FWA imprinting in *Arabidopsis* endosperm by DNA methylation. Science 303:521–523

Kinoshita Y, Saze H, Kinoshita T et al (2007) Control of FWA gene silencing in *Arabidopsis thaliana* by SINE-related direct repeats. Plant J 49:38–45

Kurihara Y, Watanabe Y (2004) *Arabidopsis* micro-RNA biogenesis through Dicer-like 1 protein functions. Proc Natl Acad Sci U S A 101:12753–12758

Latzel V, Zhang Y, Moritz K et al (2012) Epigenetic variation in plant responses to defence hormones. Ann Bot 110:1423–1428

Law JA, Jacobsen SE (2010) Establishing, maintaining and modifying DNA methylation patterns in plants and animals. Nat Rev Genet 11:204–220

Law JA, Du J, Hale CJ et al (2013) Polymerase-IV occupancy at RNA-directed DNA methylation sites requires SHH1. Nature 498:385–389

Lee Y, Kim M, Han J et al (2004) MicroRNA genes are transcribed by RNA polymerase II. EMBO J 23:4051–4060

Lee TF, Gurazada SG, Zhai J et al (2012) RNA polymerase V-dependent small RNAs in Arabidopsis originate from small, intergenic loci including most SINE repeats. Epigenetics 7:781–795

Lelandais-Briere C, Naya L, Sallet E et al (2009) Genome-wide *Medicago truncatula* small RNA analysis revealed novel microRNAs and isoforms differentially regulated in roots and nodules. Plant Cell 21:2780–2796

Lelandais-Briere C, Sorin C, Declerck M et al (2010) Small RNA diversity in plants and its impact in development. Curr Genomics 11:14–23

Li Y, Zhang Q, Zhang J et al (2010) Identification of microRNAs involved in pathogen-associated molecular pattern-triggered plant innate immunity. Plant Physiol 152:2222–2231

Li F, Pignatta D, Bendix C et al (2012) MicroRNA regulation of plant innate immune receptors. Proc Natl Acad Sci U S A 109:1790–1795

Lippman Z, Martienssen R (2004) The role of RNA interference in heterochromatic silencing. Nature 431:364–370

Lippman Z, Gendrel AV, Black M et al (2004) Role of transposable elements in heterochromatin and epigenetic control. Nature 430:471–476

Liu J, He Y, Amasino R et al (2004) siRNAs targeting an intronic transposon in the regulation of natural flowering behavior in *Arabidopsis*. Genes Dev 18:2873–2878

Lorković ZJ, Naumann U, Matzke AJM et al (2012) Involvement of a GHKL ATPase in RNA directed DNA methylation in *Arabidopsis thaliana*. Curr Biol 22:933–938

Matzke M, Kanno T, Daxinger L et al (2009) RNA-mediated chromatin-based silencing in plants. Curr Opin Cell Biol 21:367–376

Melnyk CW, Molnar A, Bassett A et al (2011) Mobile 24 nt small RNAs direct transcriptional gene silencing in the root meristems of *Arabidopsis thaliana*. Curr Biol 21:1678–1683

Michaels SD, He YH, Scortecci KC et al (2003) Attenuation of FLOWERING LOCUS C activity as a mechanism for the evolution of summer-annual flowering behavior in *Arabidopsis*. Proc Natl Acad Sci U S A 100:10102–10107

Moldovan D, Spriggs A, Yang J et al (2010) Hypoxia-responsive microRNAs and trans-acting small interfering RNAs in *Arabidopsis*. J Exp Bot 61:165–177

Mosher RA, Schwach F, Studholme D et al (2008) PolIVb influences RNA-directed DNA methylation independently of its role in siRNA biogenesis. Proc Natl Acad Sci U S A 105:3145–3150

Moxon S, Jing R, Szittya G et al (2008) Deep sequencing of tomato short RNAs identifies microRNAs targeting genes involved in fruit ripening. Genome Res 18:1602–1609

Nagasaki H, Itoh JI, Hayashi K et al (2007) The small interfering RNA production pathway is required for shoot meristem initiation in rice. Proc Natl Acad Sci U S A 104:14867–14871

Navarro L, Dunoyer P, Jay F et al (2006) A plant miRNA contributes to antibacterial resistance by repressing auxin signaling. Science 312:436–439

Nicholson AW (2014) Ribonuclease III mechanisms of double-stranded RNA cleavage. Wiley Interdiscip Rev RNA 5:31–48

Olmedo-Monfil V, Durán-Figueroa N, Arteaga-Vázquez M et al (2010) Control of female gamete formation by a small RNA pathway in *Arabidopsis*. Nature 464:628–632

Palatnik JF, Allen E, Wu X et al (2003) Control of leaf morphogenesis by microRNAs. Nature 425:257–263

Park W, Li J, Song R et al (2002) CARPEL FACTORY, a Dicer homolog, and HEN1, a novel protein, act in microRNA metabolism in *Arabidopsis thaliana*. Curr Biol 21:1484–1495

Park MY, Wu G, Gonzalez-Sulser A et al (2005) Nuclear processing and export of microRNAs in *Arabidopsis*. Proc Natl Acad Sci U S A 8:3691–3696

Pikaard CS, Mittelsten Scheid O (2014) Epigenetic regulation in plants. Cold Spring Harb Perspect Biol 6:a019315

Pontier D, Yahubyan G, Vega D et al (2005) Reinforcement of silencing at transposons and highly repeated sequences requires the concerted action of two distinct RNA polymerases IV in *Arabidopsis*. Genes Dev 19:2030–2040

Rajeevkumar S, Anunanthini P, Sathishkumar R (2015) Epigenetic silencing in transgenic plants. Front Plant Sci 6:693

Rasmann S, De Vos M, Casteel CL et al (2012) Herbivory in the previous generation primes plants for enhanced insect resistance. Plant Physiol 158:854–863

Rhoades MW, Reinhart BJ, Lim LP et al (2002) Prediction of plant microRNA targets. Cell 110:513–520

Schwab R, Palatnik JF, Riester M et al (2005) Specific effects of microRNAs on the plant transcriptome. Dev Cell 8:517–527

Scoville AG, Barnett LL, Bodbyl-Roels S et al (2011) Differential regulation of a MYB transcription factor is correlated with transgenerational epigenetic inheritance of trichome density in Mimulus Guttatus. New Phytol 191:251–263

Sidorenko L, Dorweiler JE, Cigan AM et al (2009) A dominant mutation in mediator of paramutation2, one of three second-largest subunits of a plant-specific RNA polymerase, disrupts multiple siRNA silencing processes. PLoS Genet 5:e1000725

Simon SA, Meyers BC (2011) Small RNA-mediated epigenetic modifications in plants. Curr Opin Plant Biol 14:148–155

Sunkar R, Zhu JK (2004) Novel and stress-regulated microRNAs and other small RNAs from *Arabidopsis*. Plant Cell 16:2001–2019

Sunkar R, Kapoor A, Zhu JK (2006) Posttranscriptional induction of two Cu/Zn superoxide dismutase genes in Arabidopsis is mediated by downregulation of miR398 and important for oxidative stress tolerance. Plant Cell 18:2051–2065

Sunkar R, Chinnusamy V, Zhu J et al (2007) Small RNAs as big players in plant abiotic stress responses and nutrient deprivation. Trends Plant Sci 12:301–309

Takuno S, Gaut BS (2012) Body-methylated genes in Arabidopsis thaliana are functionally important and evolve slowly. Mol Biol Evol 29:219–227

Tang G, Reinhart BJ, Bartel DP et al (2003) A biochemical framework for RNA silencing in plants. Genes Dev 17:49–63

Vaucheret H, Vazquez F, Crété P et al (2004) The action of ARGONAUTE1 in the miRNA pathway and its regulation by the miRNA pathway are crucial for plant development. Genes Dev 18:1187–1197

Vielle-Calzada JP, Thomas J, Spillane C et al (1999) Maintenance of genomic imprinting at the *Arabidopsis* Medea locus requires zygotic DDM1 activity. Genes Dev 13:2971–2982

Volkov RA, Komarova NY, Zentgraf U et al (2006) Molecular cell biology: epigenetic gene silencing in plants. Prog Bot 67:101–133

Wang JW, Wang LJ, Mao YB et al (2005) Control of root cap formation by microRNA-targeted auxin response factors in *Arabidopsis*. Plant Cell 17:2204–2216

Wierzbicki AT (2012) The role of long non-coding RNAs in transcriptional gene silencing. Curr Opin Plant Biol 15:517–522

Wierzbicki AT, Haag JR, Pikaard CS (2008) Noncoding transcription by RNA polymerase Pol IVb/Pol V mediates transcriptional silencing of overlapping and adjacent genes. Cell 135:635–648

Wierzbicki AT, Ream TS, Haag JR et al (2009) RNA polymerase V transcription guides ARGONAUTE4 to chromatin. Nat Genet 41:630–634

Williams L, Grigg SP, Xie M et al (2005) Regulation of *Arabidopsis* shoot apical meristem and lateral organ formation by microRNA miR166g and its AtHD-ZIP target genes. Development 132:3657–3668

Wu MF, Tian Q, Reed JW (2006) Arabidopsis microRNA167 controls patterns of ARF6 and ARF8 expression, and regulates both female and male reproduction. Development 133:4211–4218

Wu L, Zhou H, Zhang Q et al (2010) DNA methylation mediated by a microRNA pathway. Mol Cell 38:465–475

Xie Z, Johansen LK, Gustafson AM et al (2004) Genetic and functional diversification of small RNA pathways in plants. PLoS Biol 2:E104

Yamasaki H, Abdel-Ghany SE, Cohu CM et al (2007) Regulation of copper homeostasis by micro-RNA in *Arabidopsis*. J Biol Chem 282:16369–16378

Yao Y, Ni Z, Peng H et al (2010) Non-coding small RNAs responsive to abiotic stress in wheat (*Triticum aestivum* L.) Funct Integr Genomics 10:187–190

Ye R, Wang W, Iki T et al (2012) Cytoplasmic assembly and selective nuclear import of *Arabidopsis* ARGONAUTE4/siRNA complexes. Mol Cell 46:859–870

Yu A, Lepère G, Jay F et al (2013) Dynamics and biological relevance of DNA demethylation in *Arabidopsis* antibacterial defense. Proc Natl Acad Sci U S A 110:2389–2394

Zhang B, Pan X, Cobb GP et al (2006) Plant microRNA: a small regulatory molecule with big impact. Dev Biol 289:3–16

Zhang X, Henderson IR, Lu C et al (2007) Role of RNA polymerase IV in plant small RNA metabolism. Proc Natl Acad Sci U S A 104:4536–4541

Zhang Z, Wei L, Zou X et al (2008) Submergence-responsive microRNAs are potentially involved in the regulation of morphological and metabolic adaptations in maize root cells. Ann Bot 102:509–519

Zhang X, Zhao H, Gao S et al (2011) *Arabidopsis* Argonaute 2 regulates innate immunity via miRNA393 mediated silencing of a Golgi-localized snare gene, MEMB12. Mol Cell 42:356–366

Zhang CJ, Ning YQ, Zhang SW et al (2012) IDN2 and its paralogs form a complex required for RNA-directed DNA methylation. PLoS Genet 8:e1002693

Zhang H, He X, Zhu JK (2013a) RNA-directed DNA methylation in plants: where to start? RNA Biol 10:1593–1596

Zhang H, Ma ZY, Zeng L et al (2013b) DTF1 is a core component of RNA-directed DNA methylation and may assist in the recruitment of pol IV. Proc Natl Acad Sci U S A 110:8290–8295
Zhou X, Wang G, Zhang W (2007) UV-B responsive microRNA genes in *Arabidopsis* thaliana. Mol Syst Biol 3:103
Zhou L, Liu Y, Liu Z et al (2010) Genome-wide identification and analysis of drought-responsive microRNAs in *Oryza sativa*. J Exp Bot 61:4157–4168
Zhu QH, Spriggs A, Matthew L et al (2008) A diverse set of microRNAs and microRNA-like small RNAs in developing rice grains. Genome Res 18:1456–1465
Zilberman D, Henikoff S (2004) Silencing of transposons in plant genomes: kick them when they're down. Genome Biol 5:249
Zilberman D, Henikoff S (2007) Genome-wide analysis of DNA methylation patterns. Development 134:3959–3965
Zilberman D, Cao X, Jacobsen SE (2003) ARGONAUTE4 control of locus-specific siRNA accumulation and DNA and histone methylation. Science 299:716–719
Zilberman D, Cao X, Johansen LK et al (2004) Role of *Arabidopsis* ARGONAUTE4 in RNA-directed DNA methylation triggered by inverted repeats. Curr Biol 14:1214–1220

Small RNA Biogenesis and Degradation in Plants

Qiming Yu, Yaling Liu, Mu Li, and Bin Yu

Abstract Small RNAs (sRNAs), ~20–25 nucleotide (nt) in size, regulate various biological processes in plants through directing sequence-specific gene silencing. sRNAs are derived from either single- or double-stranded precursor RNAs. Proper levels of sRNAs are crucial for plant growth, development, genomic stability, and adaptation to abiotic and biotic stresses. Studies have identified the machineries controlling sRNA levels through biogenesis and degradation. This chapter covers recent progresses related to mechanisms governing small RNA biogenesis and degradation.

Keywords Plants • miRNAs • ta-siRNAs • pha-siRNAs • nat-siRNAs • ra-siRNAs • Biogenesis • Degradation

Contents

Q. Yu
Jinan Foreign Language School, Jinan 250001, Shandong, China

Y. Liu
Center for Plant Science Innovation, University of Nebraska-Lincoln, Lincoln, NE 68588–0666, USA

College of Life Science, Shanxi Agricultural University, Taigu 030801, Shanxi, P. R. China

M. Li • B. Yu (✉)
Center for Plant Science Innovation, University of Nebraska-Lincoln, Lincoln, NE 68588–0666, USA
e-mail: byu3@unl.edu

N. Rajewsky et al. (eds.), *Plant Epigenetics*, RNA Technologies,
DOI 10.1007/978-3-319-55520-1_6

1 Introduction

Small RNAs (sRNAs) are repressors of gene expression and play essential roles in various biological processes (Baulcombe 2004; Bologna and Voinnet 2014; Borges and Martienssen 2015). Upon production, sRNAs are sorted into their effector protein called ARGONUATE (AGO) and guide it to recognize target RNAs or DNA loci through sequence complementarity. AGO then silences gene expression at transcriptional levels through directing DNA methylation or histone modification and/or at post-transcriptional levels through target RNA cleavage or translational inhibition. microRNAs (miRNAs) and small interfering RNAs (siRNAs) are two major classes of sRNAs. They are chemically identical, but disguisable at their origin and biogenesis (Chen 2009). miRNAs are derived from primary miRNA transcripts (pri-miRNAs), which contain one or more miRNA-residing imperfect step-loops, while siRNAs are produced from long perfect double-stranded RNAs (dsRNAs) (Chen 2009). Depending on their origin, biogenesis and acting model, endogenous siRNAs can be further divided into several classes: trans-acting siRNAs (ta-siRNAs), phased siRNAs (pha-siRNAs), repeated DNA-derived siRNAs (ra-siRNAs), and natural *cis*-antisense siRNAs (nat-siRNAs) (Baulcombe 2004; Bologna and Voinnet 2014; Borges and Martienssen 2015). sRNAs are also subject to degradation (Xie et al. 2015). Turnover is also critical for proper function of sRNAs because excess amount of sRNA impairs plant development. Here, we review our current knowledge related to sRNA biogenesis and degradation.

2 miRNA Biogenesis in Plants

Most pri-miRNA-encoding genes (*MIR*) are localized at intergenic regions and transcribed as independent units by the DNA-dependent RNA polymerase II (Pol II) (Coruh et al. 2014; Nozawa et al. 2012; Xie et al. 2005). However, some *MIRs* are co-transcribed with host genes, some of which are transposons, as intronic or exonic sequences (Piriyapongsa and Jordan 2008; Yang et al. 2012).Upon transcription, pri-miRNAs are cut by the RNAseIII enzyme called DICER-LIKE 1 (DCL1) at least two times to release the miRNA/miRNA* (passenger strand) duplexes in the nucleus (Fig. 1). HUA1 ENHANCER 1 (HEN1), a small RNA methylase, then deposits a methyl group at the 3′ end of the miRNA duplex to stabilize them (Fig. 1) (Xie et al. 2015; Yu et al. 2005).

Many additional factors contribute to miRNA biogenesis by modulating pri-miRNA transcription, processing, and/or stability. Similar to mRNA-coding

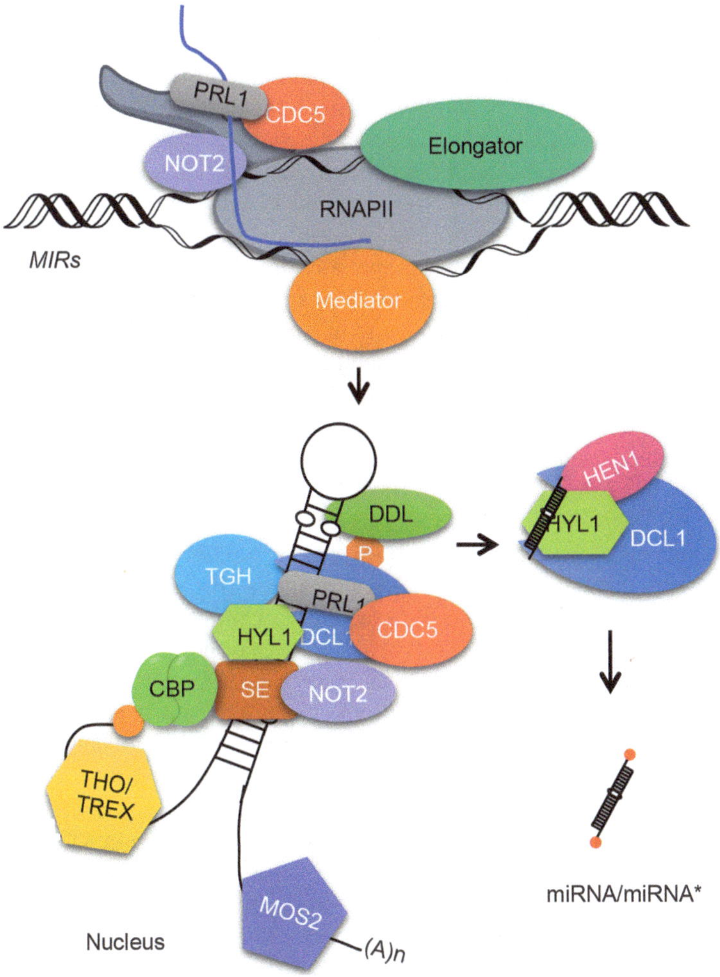

Fig. 1 The miRNA biogenesis pathway. Most *MIRs* are transcribed by Pol II to produce pri-miRNAs. CDC5, NOT2, and Mediator interact with Pol II and *MIR* promoters and are required for *MIR* transcription. After transcription, DDL-binding and PRL1-binding stabilize pri-miRNAs. Pri-miRNAs are co-transcriptionally processed, which requires the elongator complex. SE and HYL1 form a complex with DCL1 to precisely and efficiently process pri-miRNAs in the nucleus. Additional factors such as TGH, CDC5, PRL1, NOT2, CBP20/80, and DDL associate with the DCL1 complex to facilitate its activity. NOT2, MOS2, and TGH also promote the recruitment of pri-miRNAs to the DCL1 complex. After processing, HEN1, which interacts with HYL1 and DCL1, methylates the miRNA duplex

genes, *MIR* transcription requires general and specific transcription factors. NOT2a and its homolog NOT2b, which are core subunits of the conserved CARBON CATABOLITE REPRESSION4 (CCR4)-NOT complex, CDC5, which is an atypical MYB transcription factor, and Mediator (a transcription factor) interact with

Pol II and *MIR* promoters, and positively regulate transcription of many *MIR* genes (Fig. 1) (Kim et al. 2011; Wang et al. 2013; Zhang et al. 2013b). Besides them, the cycling DOF transcription factor (CDF2) binds a subset of *MIR* promoters to promote or repress their activities (Sun et al. 2015). In addition, the ATP-dependent SWR1 chromatin-remodeling complex (SWR1-C) also positively contributes to the expression of *MIRs* through changings of the nucleosome dynamics (Choi et al. 2016). Notably, transcription of some *MIRs* is temporally and spatially regulated. For instance, the transcription factors SCARECROW (SCR) and SHORT ROOT activate the expression of *MIR166* in root endodermis (Carlsbecker et al. 2010). After transcription, a 5′ 7-methylguanosine cap and a 3′ polyadenylated tail (poly-A) are added to pri-miRNAs (Jones-Rhoades and Bartel 2004; Xie et al. 2005). 5′ cap likely stabilizes pri-miRNAs since defection in 5′ capping reduces pri-miRNA accumulation (Hajheidari et al. 2012). Besides 5′ cap, two proteins, DAWDLE (DDL) and PLEIOTROPIC REGULATORY LOCUS 1 (PRL1), bind and stabilize pri-miRNAs following transcription (Fig. 1) (Yu et al. 2008; Zhang et al. 2014, 2015).

Pri-miRNAs are co-transcriptionally processed by DCL1, which is evidenced by the involvement of the elongator complex, which is required for the elongation of Pol II-dependent transcripts, in miRNA biogenesis (Fig. 1) (Fang et al. 2015a). Elongator interacts with DCL1 and is required for the association of DCL1 with *MIR* loci. This observation suggests that DCL1 may be recruited to the nascent pri-miRNAs during transcript elongation (Fang et al. 2015a). The efficient cleavage of miRNA/miRNA duplex from pri-miRNAs by DCL1 requires HYL1 (and sRNA-binding protein), TOUGH (TGH; an RNA-binding protein), and SERRATE (SE; a zinc-finger protein) (Fig. 1) (Dong et al. 2008; Fang and Spector 2007; Fujioka et al. 2007; Ren et al. 2012b; Ren and Yu 2012). HYL1 and SE are also required for precise cleavage of miRNA/miRNA* from pri-miRNAs (Dong et al. 2008), while TGH also modulates the interaction between pri-miRNAs and the processing complex (Ren et al. 2012b). It has been proposed that DCL1, SE, and HYL1 form a Dicing body (D-body) (Fang and Spector 2007), whose formation requires MOS2, an RNA-binding protein (Wu et al. 2013). In addition, NOT2, CDC5, PRL1, and CDF2 also interact with DCL1 and SE to promote pri-miRNA processing (Fig.1) (Sun et al. 2015; Wang et al. 2013; Zhang et al. 2014, 2013b). As these proteins also associate with Pol II, it is possible that they play a role in the co-transcriptional recruitment of DCL1 to pri-miRNAs. Interestingly, the CAP-binding proteins 80 (CBP 80) and CBP20 also associate with the DCL1 complex and are required for miRNA accumulation (Fig. 1) (Gregory et al. 2008; Kim et al. 2008; Laubinger et al. 2008). The recruitment of pri-miRNA to the DCL1 complex involves the THO/TREX complex that functions in the transport of nascent mRNAs from the nucleus towards the cytoplasm (Fig. 1) (Francisco-Mangilet et al. 2015). Furthermore, several additional proteins participate in miRNA biogenesis through the interaction with the accessory factors of DCL1. RECEPTOR FOR ACTIVATED C KINASE 1 (RACK1) that serves as a scaffold for protein bindings interacts with SE to promote pri-miRNA processing (Speth et al. 2013), whereas SICKLE (SIC), a proline-rich protein, co-localizes with HYL1 and is required for the accumulation of a subset of miRNAs (Zhan et al. 2012). Interestingly, GRP7, a homology of human hnRNP A1 involved in splicing, binds a subset of pri-miRNAs to repress their processing (Koster et al. 2014). Notably, REGULATOR OF CBF GENE EXPRESSION 3 (RCF3, also known as HOS5 and SHI1) can bind a subset of pri-miRNAs to regulate their processing in a tissue-specific manner (Chen et al. 2015; Karlsson et al. 2015).

The miRNA biogenesis machinery itself is regulated at both transcriptional and post-transcription levels. The histone acetyltransferase GCN5 promotes the transcription of both *DCL1* and *HYL1* (Kim et al. 2009). Optimal *DCL*1 transcription also requires the STA1, a splicing factor, and CAM33/XAP CIRCADIAN TIME-KEEPER (XCT, a nuclear localized protein) (Ben Chaabane et al. 2013; Fang et al. 2015b). Notably, SE and DCL1 are targets of miR863 and miR162, respectively (Niu et al. 2016; Rajagopalan et al. 2006). This suggests that *DCL1* and *SE* transcripts subject to feedback regulation. Furthermore, both DCL1 and HYL1 activities are regulated by protein phosphorylation. HYL1 is phosphorylated by the MITOGEN-ACTIVATED PROTEIN KINASE 3 (MPK3), which inhibits miRNA biogenesis (Raghuram et al. 2015). To counteract MPK3 activity, C-TERMINAL DOMAIN PHOSPHATASE-LIKE 1 and 2 (CPL1 and CPL2) dephosphorylate HYL1 to enhance miRNA biogenesis in an SE-dependent manner (Manavella et al. 2012). Besides HYL1, CPL1 and CPL2 also recognize RCF3 to positively impact its function in miRNA biogenesis (Chen et al. 2015; Karlsson et al. 2015). DCL1 interacts with the forkhead domain (FHA) of DDL, which mediates protein–protein interactions by targeting phospho-threonine containing motifs (Machida and Yuan 2013). The phospho-threonine binding cleft of FHA interacts with the helicase domain of DCL1 that contains potential phospho-threonine motifs, suggesting that DCL1 may be phosphorylated for its optimal activity (Machida and Yuan 2013). Interestingly, ubiquitination also plays a role in miRNA biogenesis. For instance, Constitutive Photomorphogenic 1 (COP1), an E3 ubiquitin ligase, has been recently shown to block an activity degrading HYL1 in light via unknown mechanism (Cho et al. 2014).

3 The Biogenesis of ta-siRNAs and pha-siRNAs

ta-siRNAs refer to a class siRNAs that act on targets other than the genes that derive ta-siRNAs (Allen et al. 2005; Peragine et al. 2004). This distinguishes ta-siRNAs from other endogenous siRNAs, which mostly silence genes that are the same as or homologs to the genes from which they derive. The production of ta-siRNAs requires miRNA-directed cleavage of primary ta-siRNA transcripts (*TASs*) that have the same structures as mRNAs (Fig. 2a) (Allen et al. 2005; Axtell et al. 2006; Yoshikawa et al. 2005). Two models for ta-siRNA production have been proposed. In the one-hit model, a 22-nt miRNA first directs AGO to cleave *TASs* (Fig. 2a) (Axtell et al. 2006). The 3′ cleavage products are then used as a template to synthesize dsRNAs by RDR6 (Fig. 2a). In the two-hit model, two 21-nt miRNAs

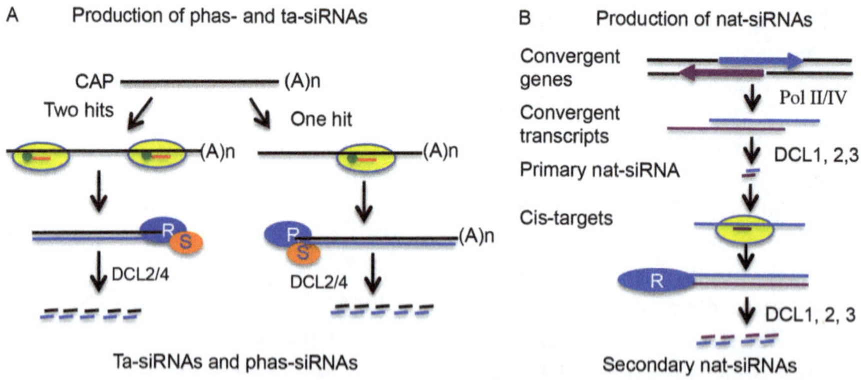

Fig. 2 The biogenesis pathways for ta-/phas-siRNAs and nat-siRNAs. (**a**) Proposed biogenesis pathways for phas- and ta-siRNAs. The precursor RNAs of ta- and phas-siRNAs are first targeted by miRNAs through the one-hit or two-hit model, which triggers the production of dsRNAs by RDR6 (R) with the assistance of SGS3 (S). The resulting dsRNAs will be cleaved by DCLs to produce phased siRNAs. *Yellow oval* indicates AGOs associated with miRNAs. (**b**) Proposed biogenesis pathways for nat-siRNA production. Pol II- or Pol IV-dependent transcription of convergent genes results in dsRNAs. The dsRNAs will be processed by one or more DCLs to produce primary nat-siRNAs. Primary nat-siRNAs will be loaded into AGO (*Yellow oval*) to recognize one of original transcripts. This leads to the production of RDR-dependent dsRNAs, which will be further processed by DCLs to generate secondary nat-siRNAs

recognize *TASs* at independent target sites along the transcript (Fig. 2a) (Axtell et al. 2006). RDR6 is then recruited to the cleavage products to generate dsRNAs. In both scenarios, the dsRNAs are cleaved by DCL4 or DCL2 every 21 or 22 nt from the initial miRNA cleavage point, resulting in a phased production of secondary siRNAs (Fig. 2a) (Fei et al. 2013; Ronemus et al. 2006). ta-siRNA biogenesis requires the assistance of SGS3 and DRB4 (Fig. 2a). DRB4 is a HYL1 homolog and interacts with DCL4 (Adenot et al. 2006), while SGS3 binds dsRNAs with 5′ overhang and partners with RDR6 (Fukunaga and Doudna 2009). In Arabidopsis, miR173 and miR828, 22 nt in size, induce the production of ta-siRNAs from *TAS1*/*TAS2* and *TAS4*, respectively, whereas miR390, 21 nt in size binds AGO7 to target TAS3 (Allen et al. 2005; Axtell et al. 2006; Yoshikawa et al. 2005). The miR390-AGO7-TAS3 combination appears to be conserved among moss, rice, maize, and gymnosperms, suggesting that the two-hit model may be ancestral to the one-hit model (Fei et al. 2013).

Unlike the Brassica plants that only encode few *TAS* loci, most non-brassica plants contain larger number of loci that produce pha-siRNAs (Arikit et al. 2014; Johnson et al. 2009; Shivaprasad et al. 2012; Zhai et al. 2015b). These pha-siRNAs are derived from many mRNAs and long noncoding RNAs called *PHAS* ncRNAs that are transcribed by Pol II, capped and polyadenylated, resembling mRNAs. However, some pha-siRNAs act in *cis* rather in *trans*. The production of pha-siRNAs resembles that of ta-siRNAs and is triggered by miRNA-directed cleavage (Fig. 2a). It has been proposed that RDR6 together with SGS3 use the 3′ cleavage fragments of *PHAS* transcripts as templates to synthesize dsRNA from the poly-A tail to the cleavage site (Fig. 2a) (Song et al. 2012b). DCL4 and DCL5/DCL3b subsequently process the dsRNAs to generate 21- and 24-nucleotide pha-siRNAs, respectively (Song et al. 2012a). In dicots, there is a conserved miR2118-482 superfamily that triggers the production of pha-siRNAs from transcripts encoding nucleotide-binding/leucine-rich repeat (NB-LRR) proteins (Arikit et al. 2014; Shivaprasad et al. 2012; Zhai et al. 2011). NB-LRR-derived pha-siRNAs act both in *cis* and in *trans* and thereby regulate additional members of NB-LRR family (Arikit et al. 2014; Shivaprasad et al. 2012; Zhai et al. 2011). This pha-siRNAs seem to benefit the plant resistance to bacterial infection, as NB-LRRs play essential role in plant immunity. However, NB-LRR pha-siRNAs are lost in grass genomes, which possess anther-specific pha-siRNAs instead (Arikit et al. 2014; Shivaprasad et al. 2012; Zhai et al. 2011).

4 The Biogenesis of Natural *cis*-antisense siRNAs (nat-siRNAs)

nat-siRNAs are derived from dsRNAs that are formed by convergent bidirectional transcripts generated from two partially overlapping genes (Fig. 2b) (Borsani et al. 2005; Katiyar-Agarwal et al. 2006; Ron et al. 2010; Zhang et al. 2012; Zubko and Meyer 2007). These *cis*-antisense transcripts are common in plant genome. Notably, nat-siRNAs are often induced by various stresses or at specific developmental stage or tissues and appear to be required for plant immunity and development (Borsani et al. 2005; Katiyar-Agarwal et al. 2006; Ron et al. 2010; Zhang et al. 2012; Zubko and Meyer 2007). DCL1, DCL2, and/or DCL3 cleave the dsRNA formed by the *cis*-antisense transcripts to initiate the production of primary nat-siRNAs (Fig. 2b) (Borsani et al. 2005; Katiyar-Agarwal et al. 2006; Ron et al. 2010; Zhang et al. 2012; Zubko and Meyer 2007). The primary nat-siRNAs will then guide the cleavage of the complementary transcripts (Fig. 2b). The resulting cleavage products will be used as templates for RDRs to produce dsRNAs, which will be further processed by DCLs into secondary siRNAs, leading to the reinforcement phase (Fig. 2b) (Borsani et al. 2005; Katiyar-Agarwal et al. 2006; Ron et al. 2010; Zhang et al. 2012; Zubko and Meyer 2007). The second phase resembles the production of pha-siRNAs involving RDRs and SGS3. The production of DCL3-dependent nat-siRNAs also requires the RNA-dependent RNA polymerase 2 (RDR2) and plant-specific RNA polymerase IV (Pol IV) (Zhang et al. 2012). In contrast, not all DCL1-dependent nat-siRNAs require RDR6/RDR2 and Pol IV for production (Zhang et al. 2012).

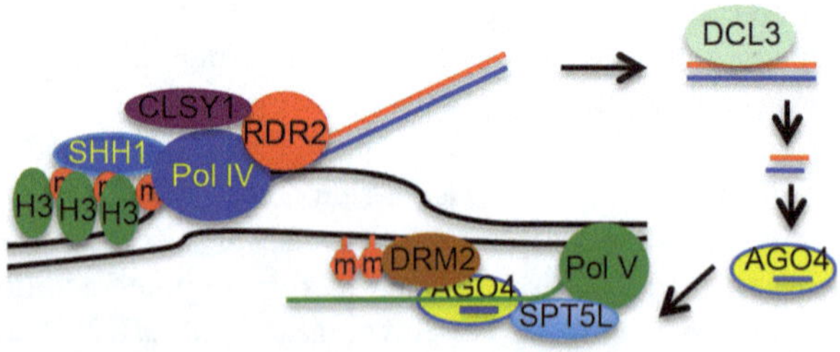

Fig. 3 The biogenesis of canonical ra-siRNAs. SHH1 recognizes the K9 dimethylation (m in *red circle*) of H3 (shown in *green oval*) and recruits Pol IV-RDR2 to the RdDM loci, leading to the production of P4R2 transcripts that are converted to dsRNAs by RdR2. CLSY1 helps the correct localization of Pol IV and RDR2. The resulting dsRNAs are processed by DCL3 to produce 24-nt siRNAs, which are loaded into AGO4 (*Yellow oval*). The AGO4-ra-siRNAs are recruited to chromatin by the Pol V–AGO 4 interaction and the base-pairing between ra-siRNAs and Pol V-dependent transcripts flanking the RdDM loci. AGO4 recruits DRM2 to catalyze the de novo methylation (*Red hexagon*) of RdDM loci

5 The Production of sRNAs Involved in RNA–Direct DNA Methylation (RdDM)

RdDM is a conserved process to silence transposable elements, to direct gene imprinting, and to maintain genome stability in plants and many metazoans (Castel and Martienssen 2013; Law and Jacobsen 2010; Matzke and Mosher 2014). Studies have revealed the presence of both canonical and non-canonical RdDM mechanisms in plants. In canonical RdDM, 24-nt siRNAs derived from repeated DNAs or heterochromatic regions (ra-siRNAs) direct the DNA de novo methyl transferase DOMAINS REARRANGED METHYLTRANSFERASE 2 (DRM2) to recognize RdDM target loci and to catalyze the methylation of previously unmodified cytosine in any sequence context (Fig. 3) (Castel and Martienssen 2013; Law and Jacobsen 2010; Matzke and Mosher 2014). In contrast, in non-canonical RdDM, miRNAs, ta-siRNAs, and 21-nt siRNAs partner with AGO to direct DNA methylation (Xie and Yu 2015). The mechanisms leading to the production of sRNAs involved in canonical and non-canonical RdDM have been proposed.

5.1 The Biogenesis of Canonical ra-siRNAs

The biogenesis of ra-siRNAs starts with Pol IV-dependent transcription from the RdDM loci (Herr et al. 2005; Onodera et al. 2005). Pol IV is a plant-specific DNA-dependent RNA polymerase. It composes of twelve subunits, which are either identical or paralog to subunit of Pol II (Haag et al. 2014; Ream et al. 2009). Pol IV interacts with RDR2 (Haag et al. 2012; Law et al. 2011), which is also required for the production of Pol IV-dependent transcripts (Fig. 3) (P4R2 RNAs). After transcription, RDR2 converts the Pol IV-dependent transcripts into dsRNAs in the nucleus (Fig. 3). DCL3 then cuts the dsRNAs into 24-nt siRNAs (Fig. 3). When DCL3 is defective, DCL2 and DCL4 can process the dsRNAs into 22-nt and 21-nt siRNAs, respectively. 24-nt siRNAs are then loaded into AGO4 or AGO6 and recruited to the RdDM loci through base-pairing with Pol V-dependent transcripts and the interaction between AGO4 and Pol V (Fig. 3). AGO4/AGO6 subsequently recruits DRM2 to catalyze de novo DNA methylation (Fig. 3). Consequently, DNA methylation causes histone deacetylation, histone H3 lysine 9 methylation, and histone H3 lysine 4 demethylation, which repress gene transcription and cause heterochromatin formation.

The nature of Pol IV-dependent transcripts (P4R2 RNAs) has been mystery since the discovery of Pol IV. Recently, several groups identified P4R2 RNAs through sequencing RNAs in the mutant deficient in DCL3 (Blevins et al. 2015; Li et al. 2015; Yang et al. 2016; Ye et al. 2016; Zhai et al. 2015a). The sizes of P4R2 RNAs are ranging from ~26 to 45 nt although long P4R2 RNAs may also exist (Li et al. 2015). Unlike Pol II transcripts, P4R2 RNAs are not polyadenylated and capped (Blevins et al. 2015; Li et al. 2015; Yang et al. 2016; Ye et al. 2016; Zhai et al. 2015a). They often start with a purine (A or G) at their 5′ ends and contain one or two untemplated nucleotide (s) at the 3′ end. Notably, P4R2 RNAs can guide DNA methylation without being diced into siRNAs (Yang et al. 2016; Ye et al. 2016), raising the possibility that ra-siRNAs may not be required for RdDM. Pol IV activity may not require the promoters since Pol IV-associated regions do not contain consensus sequences (Law et al. 2013). Instead, Pol IV needs chromatin marks to define the transcript sites (Blevins et al. 2014; Law et al. 2013; Law et al. 2011; Zhang et al. 2013a). Histone deacetylation, maintenance DNA methylation, and histone H3K9 dimethylation (H3K9me2) have been shown to recruit Pol IV to the chromatin (Blevins et al. 2014; Law et al. 2011, 2013; Zhang et al. 2013a). Consistent with these observations, SAWADEE HOMEODOMAIN HOMOLOGUE 1 (SHH1), which interacts with Pol IV and binds H3K9me2 and unmethylated H3K4 through its unique tandem Tudor-like fold, is required for the recruitment of Pol IV to the chromatin (Fig. 3) (Law et al. 2011; Zhang et al. 2013a). In addition, CLSY1, a putative chromatin-remodeling factor, is required for Pol IV-RDR2 localization and activity (Fig. 3) (Smith et al. 2007). Interestingly, Pol II also assists the recruitment of Pol IV to chromatin at some RdDM loci presumably through its transcription activity, which suggest the interplay among different polymerase may be required for ra-siRNA production (Zheng et al. 2009).

5.2 *The Biogenesis of Non-canonical sRNAs Involved in RdDM*

Plants also use non-canonical RdDM pathways to defend transposons. In non-canonical RdDM pathways, sRNAs are produced differently from those in canonical pathways. Some Pol II-derived hairpin transcripts and pri-miRNAs can be processed by DCL3 to generate 24-nt siRNAs or miRNAs (Chellappan et al. 2010; Dunoyer et al. 2010; Khraiwesh et al. 2010; Slotkin et al. 2005; Wu et al. 2010). These 24-nt siRNAs and miRNAs are then fed into the canonical RdDM pathway to direct DNA methylation either in *trans* or in *cis*. RDR6 also converts TE mRNAs into dsRNAs if these TEs are highly expressed (Gasciolli et al. 2005; Mari-Ordonez et al. 2013). When DCL2 and DCL4 become saturated, DCL3 cuts TE-derived, RDR6-dependent dsRNAs, which are not typically DCL3 substrates, into 24-nt siRNAs (Gasciolli et al. 2005; Mari-Ordonez et al. 2013). This pathway relies on the hierarchical activity of DCLs. Interestingly, RDR6-dependent 21- and 22-nt siRNAs can also participate in RdDM, which is evidenced by the fact that ta-siRNAs can be loaded into AGO4 or AGO6 to direct DNA methylation at *TAS* loci (McCue et al. 2015; Wu et al. 2012). Notably, this pathway is independent of Pol IV, RDR2, and DCL3 although it requires Pol V and DRM2 (Wu et al. 2012). Subsequent studies show that similar mechanisms are employed to initiate and establish the silencing of transcriptionally active TEs, which are long and autonomous (Nuthikattu et al. 2013). Like ta-siRNAs, the production of TE-derived 21-/22-nt siRNAs depends on Pol II, SGS3, DCL2, and DCL4 (McCue et al. 2015; Wu et al. 2012). Recent studies also uncovered a non-canonical RdDM pathway that targets a subset of non-conserved genomic loci (Garcia et al. 2012; Pontier et al. 2012). This pathway depends on Pol IV, Pol V, AGO2, and a protein named Needed for RDR2-independent DNA methylation (NERD), which contains GW repeats and a PHD finger domain and interacts with Pol V and AGO2 (Garcia et al. 2012; Pontier et al. 2012). The biogenesis pathway of siRNAs that act in NERD-dependent RdDM is not well established. However, it seems that siRNA production in this pathway requires the combination of canonical and non-canonical RdDM proteins since siRNA accumulation is reduced in *rdr1*, *rdr6*, *sde3*, *dcl2*, *dcl3*, and *nrpd1a* (a Pol IV mutant) (Garcia et al. 2012; Pontier et al. 2012).

6 Methylation Stabilizes miRNAs and siRNAs

In plants, the 3′ termini of miRNAs and siRNAs harbor a 2′-*O*-methyl group (Fig. 1) (Yu et al. 2005). This modification is added by HEN1 following the release of the miRNA or siRNA duplexes (Fig. 1) (Yang et al. 2006; Yu et al. 2005). HEN1 exists in both eukaryotes and prokaryotes (Huang et al. 2009). Plant HEN1 contains several protein domains, including two dsRNA-binding domains (dsRBD1 and dsRBD2), a La-motif containing domain (LCD), and a methyltransferase domain (MTase) (Huang et al. 2009). Among these protein domains, the dsRBD1 and dsRBD2 domains enable HEN1 to recognize dsRNAs, while the distance between the LCD domain and the MTase domain helps HEN1 to determine its substrate length (Huang et al. 2009). As a result, plant HEN1 specifically deposits a methyl group to the 2′ OH position of the 3′ end in each strand of 21–24 base-pair (bp) dsRNAs with 2-nt overhangs (Huang et al. 2009; Yang et al. 2006), which

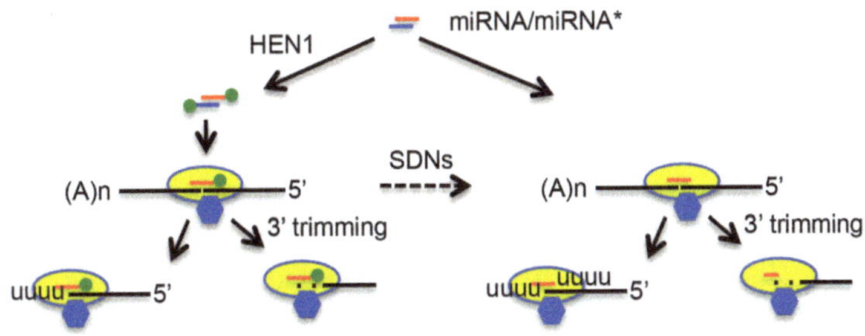

Fig. 4 Proposed model for miRNA methylation and degradation. After production, the miRNA/miRNA* duplexes are methylated (*Green* cycle) by HEN1. Methylated miRNAs are then loaded into AGO1 (*Yellow oval*) and direct AGO1 to cleave targets. The resulted 5′ cleavage products (5′ CP) are degraded through HESO1/URT1 (*Blue hexagon*)-mediated uridylation or 3′ trimming. The 2′-*O*-methylation protects miRNAs from uridylation and 3′ trimming. When methylation is lacking, miRNAs will be attacked by the AGO1-associated uridylation and 3′ trimming activities, leading to degradation. SDN1 may target AGO1-bound miRNAs with the 2′-*O*-methyl group and lead to their uridylation and degradation

are typical features of the miRNA and siRNA duplexes. HEN1 interacts with HYL1 and DCL1 (Fig. 1) (Baranauske et al. 2015), suggesting that miRNA production and methylation are a coupled process. In contrast, HEN1s from metazoans and bacteria lack the dsRNA-binding domain, and therefore, act on ssRNAs (Chan et al. 2009a; Horwich et al. 2007; Kirino and Mourelatos 2007). Consistent with this, metazoan HEN1 recognizes AGO-bound sRNAs (Ohara et al. 2007; Saito et al. 2007), whereas bacterial HEN1 modifies transfer RNAs (tRNAs) (Chan et al. 2009b). Interestingly, fly HEN1 interacts with PIWI (an AGO protein) (Ohara et al. 2007; Saito et al. 2007), indicating a potential role of AGO in determining substrate specificity for HEN1 in metazoans.

In plant *hen1*, sRNAs are reduced in abundance and become heterogeneity in size that is caused by untemplated uridine addition at 3′ termini (uridylation) and 3′-to-5′ exonucleolytic trimming activity (Fig. 4) (Abe et al. 2010; Li et al. 2005; Zhai et al. 2013). Consistent with the observation in plant *hen1*, sRNAs are also subjected to 3′-to-5′ trimming and uridylation in metazoan *hen1* (Billi et al. 2012; Horwich et al. 2007; Kamminga et al. 2010; Kamminga et al. 2012; Montgomery et al. 2012). These results demonstrate that methylation is a conserved mechanism to protect sRNAs from degradation and uridylation.

7 Uridylation Triggers the Degradation of siRNAs and miRNAs

In plants, sRNAs become uridylated globally when HEN1 is lacking (Ren et al. 2014a). In Arabidopsis, a terminal uridyl transferase (TUTase) named HEN1 SUPPRESSOR1 (HESO1) catalyzes the uridylation of miRNAs and siRNAs (Ren et al. 2012a; Zhao et al. 2012). HESO1 acts progressively on RNAs *in vitro*, but its activity is blocked by 2′-*O*-methylation (Fig. 4) (Ren et al. 2012a; Zhao et al. 2012). Besides HESO1, UTP: RNA uridylyltransferase (URT1), which has been shown to uridylate some mRNAs, also act on miRNAs (Fig. 4) (Tu et al. 2015; Wang et al. 2015). However, unlike HESO1, URT1 does not recognize ra-siRNAs, likely due to its exclusive localization in the cytoplasm (Wang et al. 2015). In addition, URT1 appears to add short U-tail to miRNAs in vivo and seems to have a different preference to 3′ end nucleotides with HESO1 (Tu et al. 2015).

In *hen1*, *heso1* reduces the U-tail length, resulting in increased abundance of most normal-sized, 3′ trimmed, and/or short-tailed sRNAs (Ren et al. 2012a; Zhao et al. 2012), whereas *urt1* only affects a few miRNAs (Tu et al. 2015; Wang et al. 2015). However, when both HESO1 and URT1 are lacking, miRNA uridylation is globally abolished in *hen1*, resulting in elevated abundance of miRNAs and an extensive increase of 3′-to-5′ trimming (Wang et al. 2015). Furthermore, overexpression of HESO1 in *hen1* further reduced the accumulation of normal-sized and 3′ trimmed miRNAs (Ren et al. 2012a). Taken together, these results reveal that HESO1 and URT1 synergistically and independently act on miRNAs and that uridylation triggers miRNA degradation and competes with the 3′-to-5′ trimming activity for substrates (Fig. 4) (Tu et al. 2015; Wang et al. 2015). In Chlamydomonas, the MUT68 nucleotidyltransferase uridylates miRNAs and siRNAs to trigger their degradation by the exosome components (Ibrahim et al. 2010). Bedside triggering degradation, uridylation may also block miRNA activity, which is evidenced by the fact that tailing of AGO1-bound miRNA165/6 greatly reduced its cleavage activity on their targets (Tu et al. 2015). Intriguingly, a single U addition in miR171 in *hen1 urt1* enables miR171 to trigger the production of secondary siRNAs from its targets (Tu et al. 2015), suggesting that U addition may also alter miRNA activity. sRNA uridylation also exist in metazoans (Burroughs et al. 2010; Wyman et al. 2011). Like in plants, uridylation affects stability and function of metazoan sRNAs. Interestingly, many TUTases act on sRNAs and some of them act on miRNAs in a sequence-specific manner in metazoans (Burroughs et al. 2010; Wyman et al. 2011).

In addition to miRNAs, HESO1 and URT1 also uridylates the 5′ RNA fragments (5′ CP) generated by AGO1 cleavage of target RNAs (Fig. 4) (Ren et al. 2014b; Wang et al. 2015). Uridylation of 5′ CP is also a conserved process in both metazoans and plants (Shen and Goodman 2004). Similar to its effect on sRNAs, uridylation triggers degradation of 5′ CP, but competes with 3′-to-5′ exonucleolytic trimming activity (Fig. 4) (Ren et al. 2014b). Interestingly, MUT68 from Chlamydomonas also acts on 5′ CP (Ibrahim et al. 2006). These results suggest the presence of a common mechanism, by which TUTase recognize both sRNAs and 5′ CP. Indeed, both HESO1 and URT1 interact with AGO1 and add U-tails to AGO1-bound miRNAs (Ren et al. 2014b; Tu et al. 2015; Wang et al. 2015). Furthermore, defection of AGO1 abolishes uridylation of miRNAs in HEN1 (Ren et al. 2014b; Zhai et al. 2013). Thus, it is likely that TUTases recognizes its substrates in the AGO1 complex. These results also answer the question why plant miRNAs, but metazoan miRNAs, require methylation for stability (Ren et al. 2014a). In plants, miRNAs majorly direct target cleavage. The cleavage products need to be further eliminated. Otherwise, they cause lethality of plants. To ensure the rapid degradation of 5′ CP, uridylation and other degradation activities are associated with the AGO1 complex (Fig. 4). However, base-pairing plant miRNAs and their targets may expose miRNA 3′ end to these AGO1-associated activities due to the extensive complementarity (Ren et al. 2014a). Thus, plant miRNAs may need methylation to protect them from such activities. In contrast, metazoan miRNAs are less complementary to their targets and majorly inhibit translation. They consequently may not be exposed to uridylation/degradation activities when meeting with their targets. Consistent with this notion, the high complementarity between aritificial target RNAs and endogenous miRNAs. triggers miRNA tailing and trimming in metazoans (Ameres et al. 2010, 2011).

Similar to uridylation, a common mechanism may exist to degrade both uridylated miRNAs and 5′ CPs, which transiently associate with AGO. Indeed, the exosome has been shown to degrade both miRNAs and 5′ CP in Chlamydomonas (Ibrahim et al. 2006, 2010). However, such enzymes remain to be identified in higher plants. In metazoans, Dis3l2, a paralog of RRP44 that is a core component of the exosome, degrades uridylated precursor of let-7 miRNA (pre-let-7) (Chang et al. 2013; Ustianenko et al. 2013). By analog, plant homologs of Dis3l2 such as SUPPRESSOR OF VARICOSE and RRP44A may act on uridylated sRNAs and 5′ CPs (Ren et al. 2014a). Alternatively, U-tail may disassociate miRNAs and 5′ CPs from the AGO1 complex, causing their rapid degradation. This is supported by the observation that long tails can be added to AGO1-bound miRNAs in vitro (Ren et al. 2014b), but no long-tailed miRNAs/5′ CP can be detected in vivo (Tu et al. 2015; Wang et al. 2015). It is also possible that some unusual exoribonucleases may recognize the 3′ U-overhang of miRNA-5′ CP duplex to trigger their degradation.

8 Exoribonucleases Degrading sRNAs in Plants

In *Arabidopsis*, a family of 3′-to-5′ exoribonucleases including SMALL RNA DEGRADING NUCLEASE 1 (SDN1), 2, and 3 have been shown to degrade mature miRNAs (Ramachandran and Chen 2008). SDNs appear to act on short single-stranded RNAs, but not sRNA duplexes or pre-miRNAs (Ramachandran and Chen 2008). Furthermore, SDNs act on methylated, but not uridylated miRNAs (Ramachandran and Chen 2008), suggesting that SDNs may function coordinately with HESO1/URT1 to regulate miRNA abundance (Fig. 4). Lack of SDNs increases the abundance of miRNAs and causes pleiotropic development defects, demonstrating that turnover is an essential mechanism to maintain proper miRNA activities (Ramachandran and Chen 2008). Recent studies show that target mimicry, which blocks miRNA-mediate target cleavage, can induce the degradation of miRNAs by SDNs (Yan et al. 2012). This result further reinforces that the function of SDNs is to eliminate the unnecessary miRNAs. AGO10 has been shown to decoy miR165/166 from AGO1, and AGO10 binding also seems to trigger degradation of miR165/166 (Zhu et al. 2011). It is possible that SDNs may play a role in miR165/166 degradation caused by AGO10-decoy.

Unmethylated miRNAs also subjects to 3′-to-5′ trimming activity in higher plants. However, the enzyme remains to be identified. In Chlamydomonas, the RRP6 that possesses 3′-to-5′ exoribonuclease activity can degrade miRNAs (Ibrahim et al. 2006; Ibrahim et al. 2010). Arabidopsis encode three RRP6-LIKE (RRP6L) proteins (Lange et al. 2008). It is possible that RRP6L can act on unmethylated miRNAs in higher plants. In fly, a 3′-to-5′ exoribonuclease named Nibbler, binds AGO to trim miRNAs (Han et al. 2011; Liu et al. 2011). It has two homologs in Arabidopsis (Xie et al. 2015). It is reasonable to speculate that these two enzymes may act on unmethylated miRNAs in Arabidopsis.

9 Perspective

In the past decades, the framework for sRNA biogenesis has been established and progresses have been made towards to sRNA degradation. In addition, factors which play regulatory roles in miRNA biogenesis also have been isolated. These advances have resulted in a better understanding of biological processes involving sRNAs and improved our ability to apply related technologies. However, challenges still exist. Plants employ multiple mechanisms to regulate gene expression and genome stability. How sRNAs are coordinated with other regulatory mechanisms in various biological processes is not well known. In addition, the functional mechanisms of accessory proteins involved in sRNA biogenesis are mostly not well defined. Notably, many of these protein factors also function in transcription, RNA processing, splicing, and RNA decay. Thus, sRNA biogenesis may have interconnections with other RNA metabolisms. Further elucidation of these interconnections is still an obstacle to our understanding of various sRNA pathways. miRNA biogenesis is regulated through transcription, processing, and stability. It remains poorly understood how plants coordinate these processes to ensure proper miRNA levels in response to development and physiological signals. In higher plants, the enzymes degrading modified and unmodified sRNAs are largely unknown, which have greatly limited our understanding of sRNA turn over. Furthermore, the factors regulating siRNA production are largely unknown. Consequently, it is not clear how related biological processes such as DNA methylation are regulated at various developmental stages and in response to biotic and abiotic stresses. Finally, a practical challenge is the optimization of sRNA-based technology and related application used to improve agricultural trait of crops.

Acknowledgement This work was supported by Nebraska Soybean Board (Award 16R-05-3/3 #1706 to B.Y) and National Science Foundation (OIA-1557417 to B.Y).

Note Due to the comprehensive nature of the review topic and limited space allowed for the paper, we apologize to the scientists whose work is not cited here.

No conflict interest declared.

References

Abe M, Yoshikawa T, Nosaka M et al (2010) WAVY LEAF1, an ortholog of Arabidopsis HEN1, regulates shoot development by maintaining microRNA and trans-acting small interfering RNA accumulation in rice. Plant Physiol 154:1335–1346

Adenot X, Elmayan T, Lauressergues D et al (2006) DRB4-dependent TAS3 trans-acting siRNAs control leaf morphology through AGO7. Curr Biol 16:927–932

Allen E, Xie Z, Gustafson AM et al (2005) microRNA-directed phasing during trans-acting siRNA biogenesis in plants. Cell 121:207–221

Ameres SL, Horwich MD, Hung JH et al (2010) Target RNA-directed trimming and tailing of small silencing RNAs. Science 328:1534–1539

Ameres SL, Hung JH, Xu J et al (2011) Target RNA-directed tailing and trimming purifies the sorting of endo-siRNAs between the two Drosophila Argonaute proteins. RNA 17:54–63

Arikit S, Xia R, Kakrana A et al (2014) An atlas of soybean small RNAs identifies phased siRNAs from hundreds of coding genes. Plant Cell 26:4584–4601

Axtell MJ, Jan C, Rajagopalan R et al (2006) A two-hit trigger for siRNA biogenesis in plants. Cell 127:565–577

Baranauske S, Mickute M, Plotnikova A et al (2015) Functional mapping of the plant small RNA methyltransferase: HEN1 physically interacts with HYL1 and DICER-LIKE 1 proteins. Nucleic Acids Res 43:2802–2812

Baulcombe D (2004) RNA silencing in plants. Nature 431:356–363

Ben Chaabane S, Liu R, Chinnusamy V et al (2013) STA1, an Arabidopsis pre-mRNA processing factor 6 homolog, is a new player involved in miRNA biogenesis. Nucleic Acids Res 41:1984–1997

Billi AC, Alessi AF, Khivansara V et al (2012) The Caenorhabditis elegans HEN1 ortholog, HENN-1, methylates and stabilizes select subclasses of germline small RNAs. PLoS Genet 8: e1002617

Blevins T, Pontvianne F, Cocklin R et al (2014) A two-step process for epigenetic inheritance in Arabidopsis. Mol Cell 54:30–42

Blevins T, Podicheti R, Mishra V et al (2015) Identification of Pol IV and RDR2-dependent precursors of 24 nt siRNAs guiding de novo DNA methylation in Arabidopsis. elife 4:e09591

Bologna NG, Voinnet O (2014) The diversity, biogenesis, and activities of endogenous silencing small RNAs in Arabidopsis. Annu Rev Plant Biol 65:473–503

Borges F, Martienssen RA (2015) The expanding world of small RNAs in plants. Nat Rev Mol Cell Biol 16:727–741

Borsani O, Zhu J, Verslues PE et al (2005) Endogenous siRNAs derived from a pair of natural cis-antisense transcripts regulate salt tolerance in Arabidopsis. Cell 123:1279–1291

Burroughs AM, Ando Y, de Hoon MJ et al (2010) A comprehensive survey of 3′ animal miRNA modification events and a possible role for 3′ adenylation in modulating miRNA targeting effectiveness. Genome Res 20:1398–1410

Carlsbecker A, Lee JY, Roberts CJ et al (2010) Cell signalling by microRNA165/6 directs gene dose-dependent root cell fate. Nature 465:316–321

Castel SE, Martienssen RA (2013) RNA interference in the nucleus: roles for small RNAs in transcription, epigenetics and beyond. Nat Rev Genet 14:100–112

Chan CM, Zhou C, Brunzelle JS et al (2009a) Structural and biochemical insights into 2′-O-methylation at the 3′-terminal nucleotide of RNA by Hen1. Proc Natl Acad Sci USA 106:17699–17704

Chan CM, Zhou C, Huang RH (2009b) Reconstituting bacterial RNA repair and modification in vitro. Science 326:247

Chang HM, Triboulet R, Thornton JE et al (2013) A role for the Perlman syndrome exonuclease Dis3l2 in the Lin28-let-7 pathway. Nature 497:244–248

Chellappan P, Xia J, Zhou X et al (2010) siRNAs from miRNA sites mediate DNA methylation of target genes. Nucleic Acids Res 38:6883–6894

Chen X (2009) Small RNAs and their roles in plant development. Annu Rev Cell Dev Biol 25:21–44

Chen T, Cui P, Xiong L (2015) The RNA-binding protein HOS5 and serine/arginine-rich proteins RS40 and RS41 participate in miRNA biogenesis in Arabidopsis. Nucleic Acids Res 43:8283–8298

Cho SK, Ben Chaabane S, Shah P et al (2014) COP1 E3 ligase protects HYL1 to retain microRNA biogenesis. Nat Commun 5:5867

Choi K, Kim J, Muller SY et al (2016) Regulation of microRNA-mediated developmental changes by the SWR1 chromatin remodeling complex. Plant Physiol 171:1128–1143

Coruh C, Shahid S, Axtell MJ (2014) Seeing the forest for the trees: annotating small RNA producing genes in plants. Curr Opin Plant Biol 18C:87–95

Dong Z, Han MH, Fedoroff N (2008) The RNA-binding proteins HYL1 and SE promote accurate in vitro processing of pri-miRNA by DCL1. Proc Natl Acad Sci USA 105:9970–9975

Dunoyer P, Brosnan CA, Schott G et al (2010) An endogenous, systemic RNAi pathway in plants. EMBO J 29:1699–1712

Fang Y, Spector DL (2007) Identification of nuclear dicing bodies containing proteins for microRNA biogenesis in living Arabidopsis plants. Curr Biol 17:818–823

Fang X, Cui Y, Li Y et al (2015a) Transcription and processing of primary microRNAs are coupled by elongator complex in Arabidopsis. Nat Plants 1:15075

Fang X, Shi Y, Liu X et al (2015b) CMA33/XCT regulates small RNA production through modulating the transcription of dicer-like genes in Arabidopsis. Mol Plant 8:1227–1236

Fei Q, Xia R, Meyers BC (2013) Phased, secondary, small interfering RNAs in posttranscriptional regulatory networks. Plant Cell 25:2400–2415

Francisco-Mangilet AG, Karlsson P, Kim MH et al (2015) THO2, a core member of the THO/TREX complex, is required for microRNA production in Arabidopsis. Plant J 82:1018–1029

Fujioka Y, Utsumi M, Ohba Y et al (2007) Location of a possible miRNA processing site in SmD3/SmB nuclear bodies in Arabidopsis. Plant Cell Physiol 48:1243–1253

Fukunaga R, Doudna JA (2009) dsRNA with 5′ overhangs contributes to endogenous and antiviral RNA silencing pathways in plants. EMBO J 28:545–555

Garcia D, Garcia S, Pontier D et al (2012) Ago hook and RNA helicase motifs underpin dual roles for SDE3 in antiviral defense and silencing of nonconserved intergenic regions. Mol Cell 48:109–120

Gasciolli V, Mallory AC, Bartel DP et al (2005) Partially redundant functions of Arabidopsis DICER-like enzymes and a role for DCL4 in producing trans-acting siRNAs. Curr Biol 15:1494–1500

Gregory BD, O'Malley RC, Lister R et al (2008) A link between RNA metabolism and silencing affecting Arabidopsis development. Dev Cell 14:854–866

Haag JR, Ream TS, Marasco M et al (2012) In vitro transcription activities of Pol IV, Pol V, and RDR2 reveal coupling of Pol IV and RDR2 for dsRNA synthesis in plant RNA silencing. Mol Cell 48:811–818

Haag JR, Brower-Toland B, Krieger EK et al (2014) Functional diversification of maize RNA polymerase IV and V subtypes via alternative catalytic subunits. Cell Rep 9:378–390

Hajheidari M, Farrona S, Huettel B et al (2012) CDKF;1 and CDKD protein kinases regulate phosphorylation of serine residues in the C-terminal domain of Arabidopsis RNA polymerase II. Plant Cell 24:1626–1642

Han BW, Hung JH, Weng Z et al (2011) The 3′-to-5′ exoribonuclease Nibbler shapes the 3′ ends of microRNAs bound to Drosophila Argonaute1. Curr Biol 21:1878–1887

Herr AJ, Jensen MB, Dalmay T et al (2005) RNA polymerase IV directs silencing of endogenous DNA. Science 308:118–120

Horwich MD, Li C, Matranga C et al (2007) The Drosophila RNA methyltransferase, DmHen1, modifies germline piRNAs and single-stranded siRNAs in RISC. Curr Biol 17:1265–1272

Huang Y, Ji LJ, Huang QC et al (2009) Structural insights into mechanisms of the small RNA methyltransferase HEN1. Nature 461:823–U886

Ibrahim F, Rohr J, Jeong WJ et al (2006) Untemplated oligoadenylation promotes degradation of RISC-cleaved transcripts. Science 314:1893–1893

Ibrahim F, Rymarquis LA, Kim EJ et al (2010) Uridylation of mature miRNAs and siRNAs by the MUT68 nucleotidyltransferase promotes their degradation in Chlamydomonas. Proc Natl Acad Sci USA 107:3906–3911

Johnson C, Kasprzewska A, Tennessen K et al (2009) Clusters and superclusters of phased small RNAs in the developing inflorescence of rice. Genome Res 19:1429–1440

Jones-Rhoades MW, Bartel DP (2004) Computational identification of plant microRNAs and their targets, including a stress-induced miRNA. Mol Cell 14:787–799

Kamminga LM, Luteijn MJ, den Broeder MJ et al (2010) Hen1 is required for oocyte development and piRNA stability in zebrafish. EMBO J 29:3688–3700

Kamminga LM, van Wolfswinkel JC, Luteijn MJ et al (2012) Differential impact of the HEN1 homolog HENN-1 on 21U and 26G RNAs in the germline of *Caenorhabditis elegans*. PLoS Genet 8:e1002702

Karlsson P, Christie MD, Seymour DK et al (2015) KH domain protein RCF3 is a tissue-biased regulator of the plant miRNA biogenesis cofactor HYL1. Proc Natl Acad Sci USA 112:14096–14101

Katiyar-Agarwal S, Morgan R, Dahlbeck D et al (2006) A pathogen-inducible endogenous siRNA in plant immunity. Proc Natl Acad Sci USA 103:18002–18007

Khraiwesh B, Arif MA, Seumel GI et al (2010) Transcriptional control of gene expression by microRNAs. Cell 140:111–122

Kim S, Yang JY, Xu J et al (2008) Two cap-binding proteins CBP20 and CBP80 are involved in processing primary microRNAs. Plant Cell Physiol 49:1634–1644

Kim W, Benhamed M, Servet C et al (2009) Histone acetyltransferase GCN5 interferes with the miRNA pathway in Arabidopsis. Cell Res 19:899–909

Kim YJ, Zheng B, Yu Y et al (2011) The role of Mediator in small and long noncoding RNA production in *Arabidopsis thaliana*. EMBO J 30:814–822

Kirino Y, Mourelatos Z (2007) The mouse homolog of HEN1 is a potential methylase for Piwi-interacting RNAs. RNA 13:1397–1401

Koster T, Meyer K, Weinholdt C et al (2014) Regulation of pri-miRNA processing by the hnRNP-like protein AtGRP7 in Arabidopsis. Nucleic Acids Res 42:9925–9936

Lange H, Holec S, Cognat V et al (2008) Degradation of a polyadenylated rRNA maturation by-product involves one of the three RRP6-like proteins in *Arabidopsis thaliana*. Mol Cell Biol 28:3038–3044

Laubinger S, Sachsenberg T, Zeller G et al (2008) Dual roles of the nuclear cap-binding complex and SERRATE in pre-mRNA splicing and microRNA processing in *Arabidopsis thaliana*. Proc Natl Acad Sci USA 105:8795–8800

Law JA, Jacobsen SE (2010) Establishing, maintaining and modifying DNA methylation patterns in plants and animals. Nat Rev Genet 11:204–220

Law JA, Vashisht AA, Wohlschlegel JA et al (2011) SHH1, a homeodomain protein required for DNA methylation, as well as RDR2, RDM4, and chromatin remodeling factors, associate with RNA polymerase IV. PLoS Genet 7:e1002195

Law JA, Du J, Hale CJ et al (2013) Polymerase IV occupancy at RNA-directed DNA methylation sites requires SHH1. Nature 498:385–389

Li J, Yang Z, Yu B et al (2005) Methylation protects miRNAs and siRNAs from a 3′-end uridylation activity in Arabidopsis. Curr Biol 15:1501–1507

Li S, Vandivier LE, Tu B et al (2015) Detection of Pol IV/RDR2-dependent transcripts at the genomic scale in Arabidopsis reveals features and regulation of siRNA biogenesis. Genome Res 25:235–245

Liu N, Abe M, Sabin LR et al (2011) The exoribonuclease Nibbler controls 3′ end processing of microRNAs in Drosophila. Curr Biol 21:1888–1893

Machida S, Yuan YA (2013) Crystal structure of Arabidopsis thaliana Dawdle forkhead-associated domain reveals a conserved phospho-threonine recognition cleft for dicer-like 1 binding. Mol Plant 6:1290–1300

Manavella PA, Hagmann J, Ott F et al (2012) Fast-forward genetics identifies plant CPL phosphatases as regulators of miRNA processing factor HYL1. Cell 151:859–870

Mari-Ordonez A, Marchais A, Etcheverry M et al (2013) Reconstructing de novo silencing of an active plant retrotransposon. Nat Genet 45:1029–1039

Matzke MA, Mosher RA (2014) RNA-directed DNA methylation: an epigenetic pathway of increasing complexity. Nat Rev Genet 15:394–408

McCue AD, Panda K, Nuthikattu S et al (2015) ARGONAUTE 6 bridges transposable element mRNA-derived siRNAs to the establishment of DNA methylation. EMBO J 34:20–35

Montgomery TA, Rim YS, Zhang C et al (2012) PIWI associated siRNAs and piRNAs specifically require the *Caenorhabditis elegans* HEN1 ortholog henn-1. PLoS Genet 8:e1002616

Niu D, Lii YE, Chellappan P et al (2016) miRNA863-3p sequentially targets negative immune regulator ARLPKs and positive regulator SERRATE upon bacterial infection. Nat Commun 7:11324

Nozawa M, Miura S, Nei M (2012) Origins and evolution of microRNA genes in plant species. Genome Biol Evol 4:230–239

Nuthikattu S, McCue AD, Panda K et al (2013) The initiation of epigenetic silencing of active transposable elements is triggered by RDR6 and 21-22 nucleotide small interfering RNAs. Plant Physiol 162:116–131

Ohara T, Sakaguchi Y, Suzuki T et al (2007) The 3′ termini of mouse Piwi-interacting RNAs are 2′-O-methylated. Nat Struct Mol Biol 14:349–350

Onodera Y, Haag JR, Ream T et al (2005) Plant nuclear RNA polymerase IV mediates siRNA and DNA methylation-dependent heterochromatin formation. Cell 120:613–622

Peragine A, Yoshikawa M, Wu G et al (2004) SGS3 and SGS2/SDE1/RDR6 are required for juvenile development and the production of trans-acting siRNAs in Arabidopsis. Genes Dev 18:2368–2379

Piriyapongsa J, Jordan IK (2008) Dual coding of siRNAs and miRNAs by plant transposable elements. RNA 14:814–821

Pontier D, Picart C, Roudier F et al (2012) NERD, a plant-specific GW protein, defines an additional RNAi-dependent chromatin-based pathway in Arabidopsis. Mol Cell 48:121–132

Raghuram B, Sheikh AH, Rustagi Y et al (2015) MicroRNA biogenesis factor DRB1 is a phosphorylation target of mitogen activated protein kinase MPK3 in both rice and Arabidopsis. FEBS J 282:521–536

Rajagopalan R, Vaucheret H, Trejo J et al (2006) A diverse and evolutionarily fluid set of microRNAs in *Arabidopsis thaliana*. Genes Dev 20:3407–3425

Ramachandran V, Chen X (2008) Degradation of microRNAs by a family of exoribonucleases in Arabidopsis. Science 321:1490–1492

Ream TS, Haag JR, Wierzbicki AT et al (2009) Subunit compositions of the RNA-silencing enzymes Pol IV and Pol V reveal their origins as specialized forms of RNA polymerase II. Mol Cell 33:192–203

Ren G, Yu B (2012) Critical roles of RNA-binding proteins in miRNA biogenesis in Arabidopsis. RNA Biol 9:1424–1428

Ren G, Chen X, Yu B (2012a) Uridylation of miRNAs by HEN1 SUPPRESSOR1 in Arabidopsis. Curr Biol 22:695–700

Ren G, Xie M, Dou Y et al (2012b) Regulation of miRNA abundance by RNA binding protein TOUGH in Arabidopsis. Proc Natl Acad Sci USA 109:12817–12821

Ren G, Chen X, Yu B (2014a) Small RNAs meet their targets: when methylation defends miRNAs from uridylation. RNA Biol 11:1099–1104

Ren G, Xie M, Zhang S et al (2014b) Methylation protects microRNAs from an AGO1-associated activity that uridylates 5′ RNA fragments generated by AGO1 cleavage. Proc Natl Acad Sci USA 111:6365–6370

Ron M, Alandete Saez M, Eshed Williams L et al (2010) Proper regulation of a sperm-specific cis-nat-siRNA is essential for double fertilization in Arabidopsis. Genes Dev 24:1010–1021

Ronemus M, Vaughn MW, Martienssen RA (2006) MicroRNA-targeted and small interfering RNA-mediated mRNA degradation is regulated by argonaute, dicer, and RNA-dependent RNA polymerase in Arabidopsis. Plant Cell 18:1559–1574

Saito K, Sakaguchi Y, Suzuki T et al (2007) Pimet, the Drosophila homolog of HEN1, mediates 2′-O-methylation of Piwi- interacting RNAs at their 3′ ends. Genes Dev 21:1603–1608

Shen BZ, Goodman HM (2004) Uridine addition after microRNA-directed cleavage. Science 306:997–997

Shivaprasad PV, Chen HM, Patel K et al (2012) A microRNA superfamily regulates nucleotide binding site-leucine-rich repeats and other mRNAs. Plant Cell 24:859–874

Slotkin RK, Freeling M, Lisch D (2005) Heritable transposon silencing initiated by a naturally occurring transposon inverted duplication. Nat Genet 37:641–644

Smith LM, Pontes O, Searle I et al (2007) An SNF2 protein associated with nuclear RNA silencing and the spread of a silencing signal between cells in Arabidopsis. Plant Cell 19:1507–1521

Song X, Li P, Zhai J et al (2012a) Roles of DCL4 and DCL3b in rice phased small RNA biogenesis. Plant J 69:462–474

Song X, Wang D, Ma L et al (2012b) Rice RNA-dependent RNA polymerase 6 acts in small RNA biogenesis and spikelet development. Plant J 71:378–389

Speth C, Willing EM, Rausch S et al (2013) RACK1 scaffold proteins influence miRNA abundance in Arabidopsis. Plant J 76:433–445

Sun Z, Guo T, Liu Y et al (2015) The roles of Arabidopsis CDF2 in transcriptional and posttranscriptional regulation of primary microRNAs. PLoS Genet 11:e1005598

Tu B, Liu L, Xu C et al (2015) Distinct and cooperative activities of HESO1 and URT1 nucleotidyl transferases in microRNA turnover in Arabidopsis. PLoS Genet 11:e1005119

Ustianenko D, Hrossova D, Potesil D et al (2013) Mammalian DIS3L2 exoribonuclease targets the uridylated precursors of let-7 miRNAs. RNA 19:1632–1638

Wang L, Song X, Gu L et al (2013) NOT2 proteins promote polymerase II-dependent transcription and interact with multiple MicroRNA biogenesis factors in Arabidopsis. Plant Cell 25:715–727

Wang X, Zhang S, Dou Y et al (2015) Synergistic and independent actions of multiple terminal nucleotidyl transferases in the 3′ tailing of small RNAs in Arabidopsis. PLoS Genet 11: e1005091

Wu L, Zhou H, Zhang Q et al (2010) DNA methylation mediated by a microRNA pathway. Mol Cell 38:465–475

Wu L, Mao L, Qi Y (2012) Roles of dicer-like and argonaute proteins in TAS-derived small interfering RNA-triggered DNA methylation. Plant Physiol 160:990–999

Wu X, Shi Y, Li J et al (2013) A role for the RNA-binding protein MOS2 in microRNA maturation in Arabidopsis. Cell Res 23:645–657

Wyman SK, Knouf EC, Parkin RK et al (2011) Post-transcriptional generation of miRNA variants by multiple nucleotidyl transferases contributes to miRNA transcriptome complexity. Genome Res 21:1450–1461

Xie M, Yu B (2015) siRNA-directed DNA methylation in plants. Curr Genomics 16:23–31

Xie ZX, Allen E, Fahlgren N et al (2005) Expression of Arabidopsis MIRNA genes. Plant Physiol 138:2145–2154

Xie M, Zhang S, Yu B (2015) microRNA biogenesis, degradation and activity in plants. Cell Mol Life Sci 72:87–99

Yan J, Gu Y, Jia X et al (2012) Effective small RNA destruction by the expression of a short tandem target mimic in Arabidopsis. Plant Cell 24:415–427

Yang Z, Ebright YW, Yu B et al (2006) HEN1 recognizes 21-24 nt small RNA duplexes and deposits a methyl group onto the 2′ OH of the 3′ terminal nucleotide. Nucleic Acids Res 34:667–675

Yang GD, Yan K, Wu BJ et al (2012) Genomewide analysis of intronic microRNAs in rice and Arabidopsis. J Genet 91:313–324

Yang DL, Zhang G, Tang K et al (2016) Dicer-independent RNA-directed DNA methylation in Arabidopsis. Cell Res 26:66–82

Ye R, Chen Z, Lian B et al (2016) A dicer-independent route for biogenesis of siRNAs that Direct DNA methylation in Arabidopsis. Mol Cell 61:222–235

Yoshikawa M, Peragine A, Park MY et al (2005) A pathway for the biogenesis of trans-acting siRNAs in Arabidopsis. Genes Dev 19:2164–2175

Yu B, Yang Z, Li J et al (2005) Methylation as a crucial step in plant microRNA biogenesis. Science 307:932–935

Yu B, Bi L, Zheng BL et al (2008) The FHA domain proteins DAWDLE in Arabidopsis and SNIP1 in humans act in small RNA biogenesis. Proc Natl Acad Sci USA 105:10073–10078

Zhai J, Jeong DH, De Paoli E et al (2011) MicroRNAs as master regulators of the plant NB-LRR defense gene family via the production of phased, trans-acting siRNAs. Genes Dev 25:2540–2553

Zhai J, Zhao Y, Simon SA et al (2013) Plant microRNAs display differential 3′ – truncation and tailing, modifications which are ARGONAUTE1-dependent and conserved across species. Plant Cell 25:2417–2428

Zhai J, Bischof S, Wang H et al (2015a) A one precursor one siRNA model for pol IV-dependent siRNA biogenesis. Cell 163:445–455

Zhai J, Zhang H, Arikit S et al (2015b) Spatiotemporally dynamic, cell-type-dependent premeiotic and meiotic phasiRNAs in maize anthers. Proc Natl Acad Sci USA 112:3146–3151

Zhan X, Wang B, Li H et al (2012) Arabidopsis proline-rich protein important for development and abiotic stress tolerance is involved in microRNA biogenesis. Proc Natl Acad Sci USA 109:18198–18203

Zhang X, Xia J, Lii YE et al (2012) Genome-wide analysis of plant nat-siRNAs reveals insights into their distribution, biogenesis and function. Genome Biol 13:R20

Zhang H, Ma ZY, Zeng L et al (2013a) DTF1 is a core component of RNA-directed DNA methylation and may assist in the recruitment of Pol IV. Proc Natl Acad Sci USA 110:8290–8295

Zhang S, Xie M, Ren G et al (2013b) CDC5, a DNA binding protein, positively regulates posttranscriptional processing and/or transcription of primary microRNA transcripts. Proc Natl Acad Sci USA 110:17588–17593

Zhang S, Liu Y, Yu B (2014) PRL1, an RNA-binding protein, positively regulates the accumulation of miRNAs and siRNAs in Arabidopsis. PLoS Genet 10:e1004841

Zhang S, Liu Y, Yu B (2015) New insights into pri-miRNA processing and accumulation in plants. Wiley Interdiscip Rev RNA 6:533–545

Zhao Y, Yu Y, Zhai J et al (2012) The Arabidopsis nucleotidyl transferase HESO1 uridylates unmethylated small RNAs to trigger their degradation. Curr Biol 22:689–694

Zheng B, Wang Z, Li S et al (2009) Intergenic transcription by RNA polymerase II coordinates Pol IV and Pol V in siRNA-directed transcriptional gene silencing in Arabidopsis. Genes Dev 23:2850–2860

Zhu H, Hu F, Wang R et al (2011) Arabidopsis Argonaute10 specifically sequesters miR166/165 to regulate shoot apical meristem development. Cell 145:242–256

Zubko E, Meyer P (2007) A natural antisense transcript of the Petunia hybrida Sho gene suggests a role for an antisense mechanism in cytokinin regulation. Plant J 52:1131–1139

Plant Epigenetics: Non-coding RNAs as Emerging Regulators

Juan Sebastian Ramirez-Prado, Federico Ariel, Moussa Benhamed, and Martin Crespi

Abstract The term non-coding RNA (ncRNA) refers to functional RNA molecules that, despite being transcribed from DNA, are not translated into proteins. These molecules can play an important role in the regulation of gene expression in the eukaryotic cell, and they can act either as long ncRNAs or being processed into small RNAs, being globally classified by their size, function, or genomic origin. In recent years, it has been found that diverse ncRNAs participate directly or indirectly in several epigenetic phenomena controlling different phenotypes within clonal cells, and in the specificity determination of various physiological processes. Although some of their mechanisms of action have been characterized, much remains to be known to understand the highly complex processes in which most of these molecules are involved. In this chapter, we discuss and illustrate examples of different ncRNAs that can interact with the plant epigenomic machinery or intervene in its function, leading to specific epigenetic, transcriptional, and physiological states. We explore the link between chromatin compaction, histone modifications, DNA methylation, gene silencing, and these molecules, which represent a high proportion of the cellular transcriptome.

Keywords Non-coding RNAs • miRNAs • siRNAs • RNA-directed DNA methylation (RdDM) • lncRNAs • Chromatin modifications • Transcriptional regulation • Genome topology

---

J.S. Ramirez-Prado • M. Benhamed
Division of Biological and Environmental Sciences and Engineering, King Abdullah University of Science and Technology, Thuwal 23955-6900, Kingdom of Saudi Arabia

Institute of Plant Sciences Paris-Saclay (IPS2), CNRS, INRA, Universities Paris-Sud, Evry and Paris-Diderot, Sorbonne Paris-Cite, University of Paris-Saclay, Batiment 630, 91405 Orsay, France

F. Ariel
Instituto de Agrobiotecnología del Litoral, CONICET, Universidad Nacional del Litoral, ColectoraRutaNacional 168 km 0, 3000 Santa Fe, Argentina

M. Crespi (✉)
Institute of Plant Sciences Paris-Saclay (IPS2), CNRS, INRA, Universities Paris-Sud, Evry and Paris-Diderot, Sorbonne Paris-Cite, University of Paris-Saclay, Batiment 630, 91405 Orsay, France
e-mail: martin.crespi@ips2.universite-paris-saclay.fr

N. Rajewsky et al. (eds.), *Plant Epigenetics*, RNA Technologies,
DOI 10.1007/978-3-319-55520-1_7

Contents

1 Introduction

It is well known that the eukaryotic genomes are transcribed in their vast majority (up to 90%) although only a small fraction of transcripts is translated into proteins. The proportion of the protein-coding genome varies greatly and is inversely correlated to the size and complexity of the genome—50% in *Arabidopsis thaliana*, a plant with a compact genome, and ≈5% in humans (Taft et al. 2007). The vast amount of non-translated transcripts forms the group of the ncRNAs, including miRNAs, siRNAs and lncRNAs. In eukaryotes, developmental complexity clearly correlates with the content and length of non-coding regions, leading John Mattick to propose in 1994 that these sequences have evolved in order to expand the range of trans-acting RNAs that evolve new regulatory mechanisms to build complex morphological diversity in living organisms. Within his hypothesis, he stated that developmental complexity requires a versatile RNA-based regulatory system, and that genes evolve and specialize into intronic or exonic loci (Mattick 1994).

It is currently accepted that RNA molecules, besides their well-known informational functions, perform several catalytic and regulatory processes in the cell. While it is true that the first non-coding RNAs (ncRNAs) were described in 1950s—rRNAs and tRNAs—it was not until the late1990s and early 2000s that the ncRNAs boom occurred within the scientific community (Cech and Steitz 2014). The discovery of thousands of small RNAs (sRNAs) and long non-coding RNAs (lncRNAs), thanks to major technical advances in transcriptome analysis and, notably, to the increase in the depth of sequencing techniques, has permitted to elucidate the important role of such molecules as key regulators of gene expression, post-transcriptional events, and genome remodeling.

The term ncRNAs comprises a large group of diverse short and long non-coding RNAs that are involved in the regulation of target genes through various molecular mechanisms (Quan et al. 2015). Some regulatory ncRNAs display high target specificity, while others act in a genome-wide fashion, acting as master regulators of big regulatory networks (Shin and Shin 2016). Furthermore, it has been proven in plants that some of these molecules can be mobile and may have a systemic, long distance action that can vary from post-transcriptional silencing to epigenetic changes (Chitwood et al. 2009; Dunoyer et al. 2010).

Traditionally, ncRNAs have been classified into different groups according to their size, biogenesis, and molecular function. LncRNAs include all the non-translated RNAs with a size generally over 200 nt, while the small RNAs (sRNAs) group comprises shorter transcripts that can be subcategorized indifferent groups, including microRNAs (miRNAs) and small interfering RNAs (siRNAs). The latter are highly similar in their function, length, and structure, but differentiate mainly in their biogenesis (Bologna and Voinnet 2014). Both types of molecules are known for their role in the RNAi (RNA interference) pathway, a process that inhibits gene expression through the RNA-guided mRNA degradation.

2 MicroRNAs in Plants

MicroRNAs are produced from precursors called primary microRNAs (pri-miRNAs) that present a secondary structure able to form a double-stranded RNA and are subsequently cleaved. Pri-miRNAs are mainly transcribed by RNA polymerase II and present a variable length. DICER-LIKE 1 (DCL1) cleaves sequentially the majority of pri-miRNAs, which are stabilized in nuclear foci known as dicing bodies. The transcripts are further processed in the nucleus and exported to the cytoplasm by HASTY; nevertheless, there seems to be another exporting pathway that remains so far unknown (Ha and Kim 2014). miRNAs mediate post-transcriptional gene silencing (PTGS) through mRNA degradation or translational inhibition (Borges and Martienssen 2015).

The mutation of *DCL1* in different plants leads to lethal embryos or to pleiotropic developmental effects, which are attributed mainly to the significant decrease in the miRNAs levels (Liu et al. 2005; Nodine and Bartel 2010). These results indicate the importance of this group of molecules in developmental processes in Arabidopsis and other plants. The expression of the loci encoding miRNAs is also highly regulated to finely tune the accumulation of their targets. For example, the levels of miR156 and miR164 are reduced in mutants affected in several of the subunits of SWR1-C—an ATP-dependent SWR1 chromatin-remodeling complex that exchanges the histone H2A-H2B dimer with H2A.Z-H2B. Such reduction in the miRNA transcripts leads to the accumulation of their target mRNAs, thus resulting in an associated phenotype (Choi et al. 2016). Reciprocally, miRNAs also seem to regulate the epigenomic machinery since some of these molecules regulate the expression of genes involved in the rearrangement of chromatin. For instance, miR773 targets *MET2 (DNA methyltransferase 2)* transcripts, and upon pathogen infection the miRNA levels decrease, resulting in the enhanced MET2 accumulation required for an appropriate immune response. The overexpression of this miRNA suppresses *MET2* and weakens PTI (PAMP-Triggered Immunity) (Li et al. 2010).

After the processing and exporting of miRNAs to the cytoplasm, these transcripts are loaded into one of the several ARGONAUTE (AGO) complexes present in the plant cell (10 paralogs in Arabidopsis) (Voinnet 2009). Several research groups have depicted the mechanisms by which miRNA sorting into specific AGO complexes occurs (Montgomery et al. 2008; Takeda et al. 2008; Voinnet 2009). For instance, AGO1 and AGO10 tend to associate with miRNAs presenting a uridine at the 5′, whileAGO2, AGO4, AGO6, AGO7, and AGO9 are loaded mainly with miRNAs bearing an adenosine, and AGO5 prefers cytosines. There is evidence that the presence of mismatches, and imperfect complementarities between the two strands of an miRNA, can determine their loading into a certain types of AGO. Indeed, Arabidopsis miR166 presents a mismatch at position 12, a characteristic that prevents its loading into AGO1 and promotes its sorting into AGO10 (Liu et al. 2009; Zhu et al. 2011). Interestingly, the function of AGO2 is still unknown since the loading of miRNAs into this complex abolishes their silencing properties (Mi et al. 2008).

3 Small Interfering RNAs (siRNAs)

The precursors of siRNAs are long dsRNAs that can be produced through diverse mechanisms, such as the folding of an inverted sequence, a lncRNA, hybridization of sense and antisense sequences, partial complementarity between unrelated transcripts, and the activity of RNA-dependent RNA polymerases (RDRs) which in the case of Arabidopsis, are six. The processing of siRNAs occurs mainly by action of DCL2, DCL3, and DCL4, leading to the formation of different types of siRNAs that can be classified into several subgroups.

3.1 Secondary SiRNAs

Even though these molecules are relatively rare in Arabidopsis somatic cells (Ronemus et al. 2006), other plant species with larger genomes—such as rice and maize—contain thousands of loci encoding these and other ncRNAs (Fei et al. 2013). This group of molecules comprises several subgroups of siRNAs, including phased siRNAs (phasiRNAs), trans-acting siRNAs (tasiRNAs), epigenetically activated siRNAs (easiRNAs), and natsiRNAs. The precursors of these molecules are transcribed by POLII, followed by the synthesis of a complementary RNA strand by RDR6. The resultant dsRNA is processed by DCL2 or DCL4, producing strands of 22 and 21 nucleotides, respectively (Fig. 1a) (Allen et al. 2005; Ronemus et al. 2006).

The biogenesis of phasiRNAs, tasiRNAs, and easiRNAs requires the targeting of an RNA transcript by one (or two) miRNAs (Fei et al. 2013). This transcript, which can be an mRNA, or come from transposable elements and non-coding sequences,

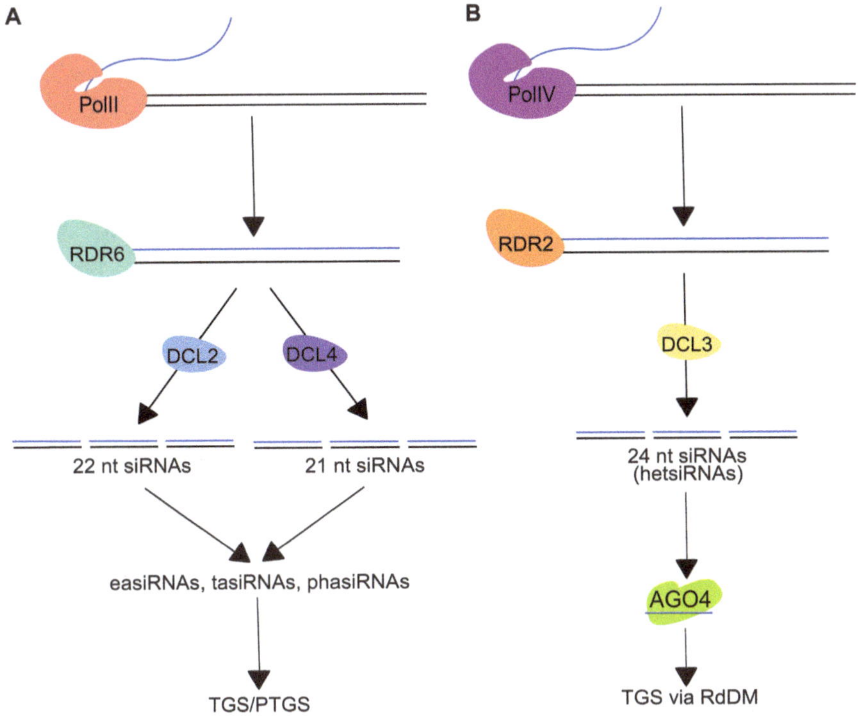

Fig. 1 Biogenesis of siRNAs in plants (**a**) secondary siRNAs are transcribed by RNA Polymerase II and transformed in dsRNA molecules by action of RDR6. Such dsRNAs are processes either DCL2 or DCL4, producing siRNAs of 22 or 21 nucleotides, respectively. The resultant molecules can be further classified into easiRNAs, tasiRNAs, and phasiRNAs, depending on their molecular function, and their activity leads mainly to PTGS, but it can also lead to TGS. (**b**) hetsiRNAs are transcribed by the plant-specific polymerase IV and processed by RDR2 and DCL3. The final products are 24-nt siRNAs that are loaded into AGO4 and recruit the RdDM machinery to their target loci in order to perform TGS

is targeted by 1 or 2 miRNA molecules, diverging into the "one-hit" or "two-hit" systems (Allen et al. 2005; Yoshikawa et al. 2005; Axtell et al. 2006). It is thought that the existence of siRNAs enables the control of a much broader range of transcripts than the activity of miRNAs per se, adding a new layer of complexity to the post-transcriptional regulation of gene expression (Morris and Mattick 2014).

In general, phasiRNAs are derived from an mRNA, turned into dsRNA by the action of RDR6 and processed by DCL4. Some phasiRNAs are further classified into the tasiRNA category, due to their capacity to function as miRNAs in a homology-dependent manner. They direct the degradation of mRNAs from genes other than those of their source (Fei et al. 2013). Arabidopsis possesses only four families of tasiRNA precursors, known as TAS genes. From these, TAS3 is conserved among different plant taxa, including bryophytes, gymnosperms, monocots, and eudicots (Fei et al. 2013). The siRNAs derived from this locus are known as tasiRNA AUXIN-RESPONSIVE FACTORS (tasi-ARFs), which are produced by the targeting of the TAS3 lncRNA by miR390 and are loaded into AGO7. TAS3 and miR390 play an important role in development since they repress the ARFs (Auxin-Responsive Factors) *ARF2, ARF3,* and *ARF4*. By repressing such factors, lateral root growth is promoted and modulated (Marin et al. 2010). Other tasiRNA of Arabidopsis, *TAS2,* is originated from a primary transcript (pri-*TAS2*) containing 11 short Open Reading Frames—ORFs— being the third of them the target of the miR173 microRNA. The mutations of this ORF, located upstream the miRNA recognition sequence leads to suppression in the accumulation of the respective tasiRNA. ORF3 was found to be translated and it plays an important role in the biogenesis of the tasiRNA (Yoshikawa et al. 2016). The 5′ and 3′ fragments of the pri-*TAS2* form a complex with miR173-programmed AGO1-RISC and SGS3. 3′-*TAS2* lacks the 5′cap, what should make it unstable; however, this association permits it to escape from degradation (Yoshikawa et al. 2013).

In Arabidopsis and other plants, some transposable elements (TEs) are inactivated by cytosine methylation and chromatin compaction, performed by DNA METHYLTRANSFERASE-1 (MET1) and DECREASED DNA METHYLATION-1 (DDM1). The activation of TEs is, nevertheless, reversible and can occur in response to developmental or environmental conditions (Tanurdzic et al. 2008; Ito et al. 2011). For this reason, plant cells count on backup mechanisms that protect the genome integrity. The activation of retrotransposons leads to the production of easiRNAs, through the targeting of transcripts by endogenous miRNAs. Several studies have been performed in the *ddm1* mutant, in order to create an artificial landscape with increased TE activation, finding that only certain families of TEs are able to induce the formation of easiRNAs (Creasey et al. 2014), which indicates the sequence-specificity of this mechanism.

Interestingly, in the *ddm1-rdr6* mutant many easiRNA-generating TEs can produce 24-nt siRNAs instead of 21- or 22-nt siRNAS. Such 24-nt molecules are produced by the activity ofDCL3 and the double strand synthesized by RDR2 (Kim and Zilberman 2014), known for competing with RDR6 for substrates (Jauvion et al. 2012). The production of 24-nt molecules leads to cytosine methylation in the target loci, resulting in heritable transcriptional gene silencing (TGS). However, this phenomenon seems to occur mainly in a mutant background since most 24-nucleotides RNAs are transcribed by Pol IV and V. This phenomenon is further discussed in the heterochromatic siRNAs (hetsiRNAs) section.

3.2 Heterochromatic SiRNAs and RNA-Directed DNA Methylation

HetsiRNAs are not classified into the secondary siRNAs category since they are transcribed by the plant exclusive DNA-directed RNA polymerases Pol IV and Pol V, their complementary strand is synthesized by RDR2, and they are processed by DCL3 into 24-nt molecules. One of their characteristic features is that their activity leads to TGS (Transcriptional Gene Silencing), via RNA-Directed DNA methylation (RdDM) (Fig. 1b), a well-characterized process in Arabidopsis (Holoch and Moazed 2015). In plants, methylation is an important covalent modification of both, DNA and histones. In the case of DNA, it is often associated to gene silencing, and it is present mainly in heterochromatic regions such as centromeres, transposable elements, and repetitive sequences, where it participates in the physical packaging of such regions. In plants, cytosine methylation occurs in three different sequence contexts, commonly referred as CG, CHG, and CHH, where H represents a different nucleotide than guanine (Xie and Yu 2015).

Generally, RdDM is initiated with the biogenesis of 24-nt siRNAs from RdDM-target loci. Pol IV transcribes an ssRNA that is converted into a double strand by RDR2. Contrary to what was thought, such transcripts, known as P4RNAs, are surprisingly short in length (30–40 nt), suggesting that one precursor RNA would produce one siRNA (Zhai et al. 2015). The RNAse III enzyme DCL3 processes such dsRNAs and the resultant 24-length products are methylated by HEN1 in order to protect them from degradation. The siRNAs are then loaded specifically into AGO4 in the cytoplasm and the complex enters the nucleus, associating itself with the KTF1 elongation factor, and recruiting the CHH DNA de novo methyltransferase DRM2. DRM2 targets the nascent transcripts produced by Pol V and catalyzes the corresponding methylation of the target loci (Borges and Martienssen 2015; Xie and Yu 2015). However, RdDM is a very complex process that includes a much higher number of proteins and processes than the formerly described, making it a highly regulated and fine-tuned mechanism (Matzke and Mosher 2014). Parallel to the recruitment of DRM2, RDM1 (RNA-DIRECTED DNA METHYLATION1) is recruited in homodimers, binding methylated ssDNA and interacting directly with Pol II, AGO4, and DRM2. Furthermore, RDM1 acts together with the chromatin remodeler DRD1 (DEFECTIVE IN RNA-DIRECTED DNA METHYLATION1) and DMS3 (DEFECTIVE IN MERISTEM SILENCING3) to form the DDR complex, which facilitates synthesis of Pol V scaffold transcripts (Sasaki et al. 2014b).

Interestingly, the methylation of DNA and histones, and the biogenesis of siRNA are highly correlated, as can be evidenced in RdDM. In this process, silencing by DNA methylation is reinforced by H3K9 histone methylation performed by KYP, SUVH5, and SUVH6 enzymes. The recruitment of KYP depends on DNA methylation (Du et al. 2014), and Pol IV is recruited by SHH1, a protein that binds specifically to unmodified H3K9 and to H3K9me2 chromatin sites, directing the generation of siRNAs from certain loci (Law et al. 2013; Holoch and Moazed 2015). It has also been found that the H3K4me3 demethylase JMJ14 participates in RdDM (Deleris et al. 2010), and its mutations affect maintenance methylation by this process (Zhang et al. 2010). On the other hand, it is also hypothesized that the histone demethylases LDL1 and LDL2 remove H3K4me2 and H3K4me3 marks during RdDM to allow SHH1 binding. It has become clearer the existence of a dependence of RdDM on histone and DNA methylations, processes linked by siRNA-directed machinery.

RdDM is not the only mechanism by which transposons and other transposable elements are silenced in Arabidopsis since mutations in *dcl3* and *rdr2* reduce siRNA levels for transposons but do not affect silencing in a significant way (Xie et al. 2004). Non-canonical RdDM mechanisms, involving DNA methylation induced by miRNAs, tasiRNAs, and 21-nt siRNAs, can occur in plants, frequently counting with the RdDM machinery plus some elements involved in PTGS (Xie and Yu 2015). Interestingly, it has been observed that some siRNAs generated through the activity of RDR6 can trigger de novo DNA methylation at some transposable loci. Such resultant methylation leads to the production of 24-nt siRNAs that participate in canonical RdDM, reinforcing DNA methylation (Nuthikattu et al. 2013). In rice, it has been observed that hairpin RNAs, transcribed by Pol II, can be processes by DCL3 instead of DCL1, to produce 24-nt miRNAs (long microRNAs) that are loaded into AGO4 and can direct DNA methylation (Chen et al. 2011). 21-nucleotide tasiRNAs, produced by the activity of DCL1, can also be loaded into AGO4 or AGO6, acting over the corresponding TAS locus through Pol V-mediated RdDM (Wu et al. 2012). A study by Sasaki and collaborators indicates that both, the Pol II- and Pol IV-dependent pathways of siRNA biogenesis can operate simultaneously in the same locus, through the recognition of different promoters for the transcription of the same template. The existence of tandem repeats with methylated CG dinucleotides seems to be important for the transcription by Pol IV, indicating an affinity of this polymerase to methylated DNA. On the other hand, Pol II recognizes traditional promoters, but the accumulation of 21-nucleotide siRNAs is affected by sequence composition (Sasaki et al. 2014a), a result that is coherent with a study by Zhang and collaborators, reporting a significant positive correlation between the GC content and the expression levels of different types of plant siRNAs (Zhang et al. 2014b). The sequence of a siRNA may represent then an intrinsic factor in the regulation of its expression and activity, which in an evolutionary context could indicate that the sequence of its targets—which determines the siRNA-target interaction—indirectly regulates the expression of its own locus.

4 Long Non-Coding RNAs

LncRNAs, in contrast to miRNAs and siRNAs, can act without being processed and cleaved by DCR proteins. Within their several functions, lncRNAs have been reported to be involved in protein and miRNAs hijack, modulation of mRNA stability and translation, and the modification of chromatin at diverse levels (Fig. 2) (Ariel et al. 2015).

LncRNAs are commonly classified according to their location regarding protein-coding genes (Rinn and Chang 2012). Long non-coding Natural Antisense transcripts (lncNATs) that start inside the coding sequence or in the 3′ of a gene are transcribed in its opposite direction and are overlapped with at least one of its exons. Intronic lncRNAs occur in intronic regions without overlapping with any

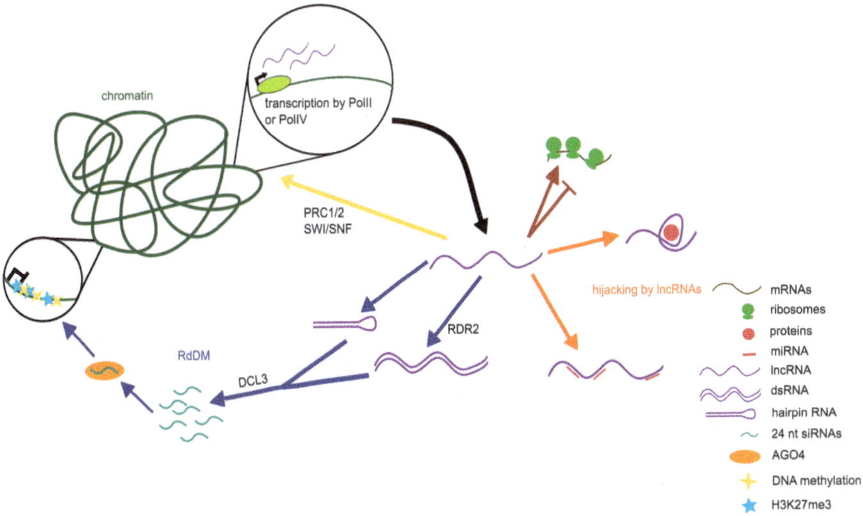

Fig. 2 Roles of lncRNAs in the eukaryotic cell. LncRNAs are either transcribed by Pol II or Pol IV. These molecules can perform as protein and miRNA hijackers, binding to them and blocking their function (*orange arrows*). NATlncRNAs can promote or prevent translation of their target mRNAs (*Red arrows*). LncRNAs can affect chromatin directly through the recruitment of the epigenomic machinery, including PRC1, PRC2, and SWI/SNF complexes (*yellow arrow*), or through the RdDM pathway, represented with *blue arrows*. RdDM can be initiated either through the spontaneous folding of an RNA molecule and the formation of a hairpin RNA, or through the action of RDR2, which synthesizes a dsRNA using the lncRNA as a template. Both latter molecules are recognized by DCL3, which processes them into 24-nt siRNAs that are loaded into AGO4. siRNA-loaded AGO4s direct the RdDM machinery to the target locus, where TGS occurs through the addition of DNA and histone methylation marks

exon, while promoter lncRNAs result from the transcription of promoter regions of protein-coding genes. Finally, Long intergenic lncRNAs (lincRNAs) occur in independent transcriptional loci, at a distance of at least 1 kb from neighbor protein-coding genes (Ariel et al. 2015).

The emergence of RNA-sequencing techniques and transcriptomic analyses permitted the scientific community to gain deeper understanding of the non-coding transcription in several plant species and other organisms. In Arabidopsis, it is well known that thousands of lncRNAs co-express simultaneously, and that some of them exhibit organ-specificity, while others are detectable only in response to environmental stimuli and stresses. On the other hand, around 70% of the Arabidopsis annotated mRNAs present coordinated or anti-regulated antisense transcription, highlighting the important regulatory function of lncRNAs (Wang et al. 2014a).

Several studies have been performed towards the identification of lncRNAs involved in stress responses. Xin et al. discovered 125 putative stress-responsive lncRNAs in wheat induced by the infection with a powdery mildew and by heat (Xin et al. 2011). In maize, the first approach, sequencing the full cDNA, led to

1900 transcript from non-protein-coding genes (Boerner and McGinnis 2012), while another integrative study, combining data obtained from databases and through sequencing, revealed 20163 putative lncRNAs in this plant, from which 1704 represent lncRNAs with high confidence (Zhang et al. 2014a). Several thousands of lncRNAs have been described in other plant species, including foxtail millet (*Setaria italica*), black cottonwood (*Populus trichocarpa*), Chinese white poplar (*Populus tomentosa*), soybean (*Glycine max*), peach (*Prunus persica*), and *Brassica rapa,* finding constantly that a big proportion of such lncRNAs present tissue- or stress-specific expression (Qi et al. 2013; Wang et al. 2013; Yu et al. 2013; Shuai et al. 2014; Ye et al. 2014; Chen et al. 2015).

While the role of lncRNAs has been extensively characterized in animals, their study in plants is relatively new and there is still a long way towards the understanding of their regulatory functions. It has been reported that many lncRNAs can be spliced, indicating that they are more than just transcriptional noise, as it has been formerly proposed. Furthermore, some of the isoforms of these molecules encode peptides (Chooniedass-Kothari et al. 2004), and inversely some mRNAs can function as trans-acting regulatory RNAs (Dinger et al. 2011; Mercer et al. 2011). Apart from being sRNAs precursors, lncRNAs have diverse cellular roles (Morris and Mattick 2014) that we further discuss in this chapter.

4.1 Plant LncRNAs, Professional Hijackers

In certain cases, miRNAs can be hijacked by "target mimickry." First described in plants, this process occurs when lncRNAs work as decoys for miRNAs (also known as miRNA kidnapping), blocking the interaction between an miRNA and its actual target, via partially complementary sequences (Fig. 2) (Franco-Zorrilla et al. 2007; Wu et al. 2013). This phenomenon has also been reported in animals, where lncRNAs involved in the sequestration of miRNAs are called competing endogenous RNAs (ceRNAs) (Kartha and Subramanian 2014). The distribution of this mechanism among different taxa indicates its importance in the regulation of many genetic networks simultaneously occurring in the eukaryotic cell.

An already classic example of miRNA kidnapping is the regulation of miR399, a major regulator of plant phosphate homeostasis. This miRNA guides the degradation of the*PHO2* (E2 ubiquitin conjugase-related protein) mRNA (Aung et al. 2006; Bari et al. 2006), which simultaneously promotes the expression of two phosphate transporters in the root and the consequent increase in phosphate uptake (Aung et al. 2006). The *IPS1* (Induced by Phosphate Starvation1) lincRNA is complementary to miR399 with a central 3-nt mismatch loop, allowing it to bind to the miRNA and decrease the repression of *PHO2*. The increase in PHO1;2 levels in response to phosphate starvation occurs together with an increase in the $cis\text{-}NAT_{pho1;2}$ transcripts, while the levels in the *PHO1*;2 mRNA remain stable. Furthermore, the downregulation of $cis\text{-}NAT_{pho1;2}$ through RNA interference affects the acropetalous transport of phosphate and leads to a reduced seed production, indicating that this antisense transcript enhances the translation of the protein. It is thought that *cis-NAT*$_{pho1;2}$ promotes the transport of the mRNA towards the polysome, positively regulating its expression, representing a novel mechanism of action for lncRNAs (Jabnoune et al. 2013). Usually, the induction of cis-lncNATs is related to the repression of target mRNAs (Zubko and Meyer 2007; Wunderlich et al. 2014); however, their function appears to be not limited to canonical antagonistic interactions (Fig. 2).

A study performed by Wu et al. predicted the existence of 20 endogenous microRNA target mimics (eTMs) conserved between Arabidopsis and rice. The overexpression of the candidate miR160 eTMin Arabidopsis led to a significant decrease in mature miR160, an increase in the levels of miR160 target transcripts (*ARF10, ARF16, and ARF17*), and developmental defects characterized by dwarfism, serrated leaves, and early flowering. Similarly, the transgenic overexpression of miR166eTM increased the levels of miR166-target expression, being these loci three HD-ZIP III genes, known as *ATHB-9, ATHB-14,* and *ATHB-15*. Besides, the transgenic line presented spoon-shaped cotyledons and abnormal leave shapes (Wu et al. 2013), evidencing the importance of the equilibrium between miRNA-directed PTGS and lncRNA-mediated miRNA target mimicking in plant development and morphology.

Surprisingly, beside mRNAs, lncRNAs can also hijack proteins (Fig. 2). This phenomenon was observed in legumes, where a peptide-coding RNA can act as an RNA regulatory molecule in development (Crespi et al. 1994). *ENOD40* (EARLY NODULIN 40) were proposed to encode short peptides involved in nodule metabolism, specifically in sucrose pathways (Rohrig et al. 2002). On the other hand, the transcript directly interacts with MtRBP1 (RNA-binding protein 1, now renamed MtNSR for Nuclear Speckle RNA-binding protein, (Bardou et al. 2014)), localized in nuclear speckles that colocalizes with the splicing machinery. This transcript promotes NSR localization into cytoplasmic granules during nodulation, independently of the encoded protein function (Campalans et al. 2004). Similarly, the expression of the *ASCO* (*ALTERNATIVE SPLICING COMPETITOR*) lincRNA *in* Arabidopsis affects alternative splicing of NSR targets by binding to these proteins, which are the Arabidopsis closest homologs to the legume MtNSR1. This hijacking phenomenon can modulate RNA processing during lateral root formation (Bardou et al. 2014), and these examples highlight the importance of RNA secondary structure in their interactions with diverse protein partners, which in general present a stable quaternary structures with defined domains. Such role of the RNA secondary structure in lncRNA biology can be evidenced in the case of rice varieties exhibiting photoperiod-sensitive male sterility (PSMS). Ding and collaborators described a lncRNA, referred as *LDMAR*, that regulates PSMS in these mutants. The existence of a spontaneous SNP in the *LDMAR* locus of the mutant line leads to a change in the secondary structure of the transcript, which simultaneously increases the DNA methylation in the putative *LDMAR* promoter, reducing its transcription levels under long day conditions. Such low levels of LDMAR result in premature programmed cell death in developing anthers, making the male organs sterile (Ding et al. 2012). The molecular mechanisms by which the described phenomenon occurs are unknown; thus, this study sets a new challenge for the understanding of the implication of lncRNAs, and other ncRNAs, in the regulation of sexual development and photoperiodic regulation in plants, since these molecules could offer a whole repertoire of options for the directed breeding and selection of specific agricultural traits.

4.2 LncRNAs Mediate Chromatin Modifications and Remodeling

Even if the study of plant lncRNAs is still in its infancy, and there are certainly plenty of undescribed processes and mechanisms in which they participate, it is already clear that they can directly or indirectly affect chromatin structure through several mechanisms (Fig. 2). The first genetic studies in the field found an association between some lncRNAs and heterochromatin and genomic imprinting in animals (Barlow et al. 1991; Brown et al. 1991). A well-known example of this is that one of the *Xist* lncRNAs from mouse, which originates from the inactive X chromosome and binds it throughout the chromosome body. Such coating recruits the PRC2 complex that performs the methylation of lysine 27 of histone 3 (H3K27) and silences the expression of local genes in a *cis*-fashion (Chaumeil et al. 2006). In humans, the case of the *HOTAIR* antisense lncRNA is a classic example of *trans* regulation mediated by lncRNA-directed epigenomic changes. *HOTAIR*, transcribed from the *HOXC* locus, is capable of regulating the expression of *HOXD* and other loci across the genome through the recruitment of PRC2 and the promotion of the deposition of H3K27me3 marks (Rinn et al. 2007; Tsai et al. 2010). In both of the latter cases, direct interaction of the methyltransferase subunit of PRC2 with the corresponding lncRNAs was proven (Zhao et al. 2008; Kaneko et al. 2010); nevertheless, the nature of such interactions is not completely understood but imposes a challenge for the comprehension of RNA–protein interactions (Heo et al. 2013). This is mainly due to our current poor understanding of the landscape of such interactions in the eukaryotic cell. The comprehension of their nature will allow predicting with accuracy other plausible interactions, integrating RNA and protein biology, further than the classical translational dogma. Among the novel mechanisms in which plant lncRNAs are known to be involved are the formation of chromatin loops and the interaction with components of the epigenomic machinery, such as histone modifiers and remodelers, and with other proteins that bind DNA (Heo et al. 2011, 2013; Saxena and Carninci 2011; Blignaut et al. 2012; Lai et al. 2013; Ariel et al. 2014; Rodriguez-Granados et al. 2016).

4.3 LncRNAs and Epigenomic Regulation of Flowering

Flowering is a highly regulated process in plants that occurs in response to environmental cues. Some plants, including Arabidopsis, flower after prolonged periods of cold, a process named vernalization (Sung and Amasino 2004). *FLC (FLOWERING LOCUS C)* is a transcriptional repressor which activity suppresses the expression of several genes required for flowering. *FLC* overexpression delays flowering, while its downregulation has the opposite effect.

When a plant is vernalized, the expression of *FLC* is gradually repressed through the action of several plant homeodomain proteins (PHD), PRC2, and two different lncRNAs, named *COOLAIR* and *COLDAIR*. The first one of these RNAs is a lncNAT produced by reverse transcription of the complete *FLC* locus, while the second is a lncRNA encoded in its first intron (Letswaart et al. 2012). *COOLAIR* is transcribed transiently under cold conditions, and its expression relies on the transcription elongation factor b (P-TEFb), a complex that integrates mRNA gene expression with histone modifications (Wang et al. 2014b). *COLDAIR* binds to CLF—a component of the PRC2 complex—and its downregulation by RNA interference produces transgenic lines displaying in sensitivity to vernalization (Heo et al. 2011). Concomitantly, they present a dramatic decrease in the enrichment of CLF and H3K27me3 in the *FLC* locus, indicating the importance of *COLDAIR* for the recruitment of PRC2.

Before vernalization, *FLC* is transcribed in a gene loop that joins the 3′and 5′ of the locus (Crevillén et al. 2013). During vernalization, BAF60, a subunit of the SWI/SNF chromatin-remodeling complex, participates in the disruption of the loop (Jégu et al. 2014), which coincides with the transcription of the *FLC* antisense lncRNA *COOLAIR*, its polyadenylation and its interaction with the locus, in order to decrease *FLC* expression (Csorba et al. 2014). This phenomenon is accomplished through the repressive activity of FCA and FPA (RNA-binding proteins), FY 3′ processing factor and FLD, a histone demethylase that removes H3K4me2 marks. The PHD–PRC2 complex performs the replacement of H3K36me3 that marks byH3K27me3 along the *FLC* locus, while VAL1 promotes histone deacetylation through its binding to the RY motifs in the nucleation region of *FLC*. The action of these elements stabilizes the repressive state, finally leading to flowering progression (Jiang et al. 2007; Qüesta et al. 2016).

ssDNA–RNA loops (R-loops) have been described in Arabidopsis and are known to have an impact on gene expression (Huertas and Aguilera 2003; El Hage et al. 2010). A recent study reported the formation of one of these structures mediated by AtNDX (a PHD protein) at the promoter region of *COOLAIR*. This RNA–DNA complex comprises the non-template ssDNA, where AtNDX binds. It is thought that AtNDX stabilizes the mentioned loop, which downregulates *COOLAIR* expression by blocking Pol II binding and elongation (Sun et al. 2013).

4.4 LncRNAs Link Hormone Signaling with Chromatin Modifications and Genome 3D-Conformation

PINOID (PID) is an auxin-inducible gene that plays an important role in the transport of this hormone and, thereby, in organ development. For this reason, the expression of this gene demands a finely tuned regulation that involves several processes, including rearrangements in chromatin architecture, the action of ncRNAs and the activity of histone-interacting proteins. Ariel and collaborators described the establishment of a gene loop encompassing the promoter of *PID*, which allows the transcription of this gene and the *APOLO (AUXIN-REGULATED PROMOTER LOOP)* lincRNA. In the absence of auxin, the expression of PID is repressed by the formation of the gene loop. This structure undergoes DNA demethylation and opens in response to increased auxin levels, allowing the divergent transcription of *PID* and *APOLO* by POLII. The accumulation of *APOLO* transcripts initiates RdDM, leading to the reestablishment of histone and DNA methylation marks, and the repressive loop—by action of the *APOLO*–LHP1 complex, PRC1 and PRC2 (Ariel et al. 2014, 2015; Rodriguez-Granados et al. 2016).

5 Concluding Remarks

The regulatory role of lncRNAs in plants and other organisms is becoming evident with recent studies that involve several state-of-the-art techniques and comprehensive approaches. As we discussed in this chapter, non-coding transcription is a highly important process in the regulation of many cellular and physiological processes occurring in plants and other eukaryotes. Apart from their classical functions, it is presently known that these molecules can intervene and participate in epigenomic processes, tuning transcription with post-transcriptional and translational regulation, and establishing a cellular network that permits fast and efficient changes in response to various stimuli,—including biotic, abiotic, and developmental cues.

Although sRNAs are known to have chromatin-related roles in *cis* and *trans*, lncRNAs have only been reported to have an epigenomic regulatory function in *cis*. Future research may elucidate the role of these molecules as regulators of complex pathways, including several genes (Ariel et al. 2015). The identification of proteins interacting with diverse ncRNAs and the increasing understanding of the genome topology of Arabidopsis and of other plants, will permit the scientific community to decipher the rules and dynamics of genome architecture modulation, and how they are associated with the regulation of gene activity.

Acknowledgements The laboratory of M.C. is funded by the "Agence Nationale de la Recherche" in France through the ANR RNAdapt project and Saclay Plant Sciences Labex

(SPS, ANR-10-LABX-40). An international KAUST (Saudi Arabia)—INRA (France) grant (EPIMMUNITY) is also acknowledged. J.S.R-P. is a KAUST Ph.D. fellow.

References

Allen E, Xie Z, Gustafson AM et al (2005) microRNA-directed phasing during trans-acting siRNA biogenesis in plants. Cell 121:207–221

Ariel F, Jegu T, Latrasse D et al (2014) Noncoding transcription by alternative rna polymerases dynamically regulates an auxin-driven chromatin loop. Mol Cell 55:383–396

Ariel F, Romero-Barrios N, Jegu T et al (2015) Battles and hijacks: noncoding transcription in plants. Trends Plant Sci 20:362–371

Aung K, Lin S-I, Wu C-C et al (2006) pho2, a phosphate overaccumulator, is caused by a nonsense mutation in a microRN399 target gene. Plant Physiol 141:1000–1011

Axtell MJ, Jan C, Rajagopalan R et al (2006) A two-hit trigger for siRNA biogenesis in plants. Cell 127:565–577

Bardou F, Ariel F, Simpson CG et al (2014) Long noncoding RNA modulates alternative splicing regulators in arabidopsis. Dev Cell 30:166–176

Bari R, Pant BD, Stitt M et al (2006) Phosphate-signaling pathway in plants 1. Plant Physiol 141:988–999

Barlow DP, Stöger R, Herrmann BG et al (1991) The mouse insulin-like growth factor type-2 receptor is imprinted and closely linked to the Tme locus. Nature 349:84–87

Blignaut M, Morris KV, Scripps T (2012) Review of non-coding RNAs and the epigenetic regulation of gene expression. A book edited by Kevin Morris. Epigenetics 7:664–666

Boerner S, McGinnis KM (2012) Computational identification and functional predictions of long noncoding RNA in Zea mays. PLoS One 7:1–16

Bologna NG, Voinnet O (2014) The diversity, biogenesis, and activities of endogenous silencing small RNAs in Arabidopsis. Annu Rev Plant Biol 65:473–503

Borges F, Martienssen RA (2015) The expanding world of small RNAs in plants. Nat Rev Mol Cell Biol 16:1–15

Brown CJ, Ballabio A, Rupert JL et al (1991) A gene from the region of the human X inactivation centre is expressed exclusively from the inactive X chromosome. Nature 349:38–44

Campalans A, Kondorosi A, Crespi M (2004) Enod40, a short open reading frame—containing mRNA, induces cytoplasmic localization of a nuclear RNA binding protein in medicago truncatula. Plant Cell 16:1047–1059

Cech TR, Steitz JA (2014) The noncoding RNA revolution—trashing old rules to forge new ones. Cell 157:77–94

Chaumeil J, Le Baccon P, Wutz A et al (2006) A novel role for Xist RNA in the formation of a repressive nuclear compartment into which genes are recruited when silenced. Genes Dev 20:2223–2237

Chen D, Meng Y, Yuan C (2011) Plant siRNAs from introns mediate DNA methylation of host genes. RNA 17(6):1012–1024

Chen J, Quan M, Zhang D (2015) Genome-wide identification of novel long non-coding RNAs in Populus tomentosa tension wood, opposite wood and normal wood xylem by RNA-seq. Planta 241:125–143

Chitwood DH, Nogueira FTS, Howell MD et al (2009) Pattern formation via small RNA mobility. Mol Cell Biol 23(5):549–554

Choi K, Kim J, Müller SY et al (2016) Regulation of microRNA-mediated developmental changes by the SWR1 chromatin remodeling complex in Arabidopsis thaliana. Plant Physiol 171:1128–1143

Chooniedass-Kothari S, Emberley E, Hamedani MK et al (2004) The steroid receptor RNA activator is the first functional RNA encoding a protein. FEBS Lett 566:43–47
Creasey KM, Zhai J, Borges F et al (2014) miRNAs trigger widespread epigenetically activated siRNAs from transposons in Arabidopsis. Nature 508:411–415
Crespi MD, Jurkevitch E, Poiret M et al (1994) enod40, a gene expressed during nodule organogenesis, codes for a non-translatable RNA involved in plant growth. EMBO J 13:5099–5112
Crevillén P, Sonmez C, Wu Z et al (2013) A gene loop containing the floral repressor FLC is disrupted in the early phase of vernalization. EMBO J 32:140–148
Csorba T, Questa JI, Sun Q et al (2014) Antisense COOLAIR mediates the coordinated switching of chromatin states at FLC during vernalization. Proc Natl Acad Sci USA 111:16160–16165
Deleris A, Greenberg MVC, Ausin I et al (2010) Involvement of a Jumonji-C domain-containing histone demethylase in DRM2-mediated maintenance of DNA methylation. EMBO Rep 11:950–955
Ding J, Lu Q, Ouyang Y et al (2012) A long noncoding RNA regulates photoperiod-sensitive male sterility, an essential component of hybrid rice. Proc Natl Acad Sci USA 109:2654–2659
Dinger ME, Gascoigne DK, Mattick JS (2011) The evolution of RNAs with multiple functions. Biochimie 93:2013–2018
Du J, Johnson LM, Groth M et al (2014) Mechanism of DNA methylation-directed histone methylation by KRYPTONITE. Mol Cell 55:495–504
Dunoyer P, Schott G, Himber C et al (2010) Small RNA duplexes function as mobile silencing signals between plant cells. Science 328:912–916
El Hage A, French SL, Beyer AL, Tollervey D (2010) Loss of topoisomerase I leads to R-loop-mediated transcriptional blocks during ribosomal RNA synthesis. Genes Dev 24:1546–1558
Fei Q, Xia R, Meyers BC (2013) Phased, secondary, small interfering RNAs in posttranscriptional regulatory networks. Plant Cell 25:2400–2415
Franco-Zorrilla JM, Valli A, Todesco M et al (2007) Target mimicry provides a new mechanism for regulation of microRNA activity. Nat Genet 39:1033–1037
Ha M, Kim VN (2014) Regulation of microRNA biogenesis. Nat Rev Mol Cell Biol 15:509–524
Heo JB, Sung S, Kim DH et al (2011) Vernalization-mediated epigenetic silencing by a long intronic noncoding RNA. Science 331:76–79
Heo JB, Lee Y-S, Sung S (2013) Epigenetic regulation by long noncoding RNAs in plants. Chromosom Res 21:685–693
Holoch D, Moazed D (2015) RNA-mediated epigenetic regulation of gene expression. Nat Rev Genet 16:71–84
Huertas P, Aguilera A (2003) Cotranscriptionally formed DNA:RNA hybrids mediate transcription elongation impairment and transcription-associated recombination. Mol Cell 12:711–721
Ietswaart R, Wu Z, Dean C (2012) Flowering time control: another window to the connection between antisense RNA and chromatin. Trends Genet 28:445–453
Ito H, Gaubert H, Bucher E et al (2011) An siRNA pathway prevents transgenerational retrotransposition in plants subjected to stress. Nature 472:115–119
Jabnoune M, Secco D, Lecampion C et al (2013) A rice cis-natural antisense RNA acts as a translational enhancer for its cognate mRNA and contributes to phosphate homeostasis and plant fitness. Plant Cell 25:4166–4182
Jauvion V, Rivard M, Bouteiller N et al (2012) RDR2 partially antagonizes the production of RDR6-dependent siRNA in sense transgene-mediated PTGS. PLoS One 7:7–11
Jégu T, Latrasse D, Delarue M et al (2014) The BAF60 subunit of the SWI/SNF chromatin-remodeling complex directly controls the formation of a gene loop at FLOWERING LOCUS C in Arabidopsis. Plant Cell 26:538–551
Jiang D, Yang W, He Y et al (2007) *Arabidopsis* relatives of the human lysine-specific demethylase1 repress the expression of *FWA* and *FLOWERING LOCUS C* and thus promote the floral transition. Plant Cell 19:2975–2987
Kaneko S, Li G, Son J et al (2010) Phosphorylation of the PRC2 component Ezh2 is cell cycle-regulated and up-regulates its binding to ncRNA. Genes Dev 24:2615–2620

Kartha RV, Subramanian S (2014) Competing endogenous RNAs (ceRNAs): new entrants to the intricacies of gene regulation. Front Genet 5:1–9
Kim MY, Zilberman D (2014) DNA methylation as a system of plant genomic immunity. Trends Plant Sci 19:320–326
Lai F, Orom UA, Cesaroni M et al (2013) Activating RNAs associate with mediator to enhance chromatin architecture and transcription. Nature 494:497–501
Law JA, Du J, Hale CJ et al (2013) Polymerase IV occupancy at RNA-directed DNA methylation sites requires SHH1. Nature 498:385–389
Li Y, Zhang Q, Zhang J et al (2010) Identification of MicroRNAs involved in pathogen-associated molecular pattern-triggered plant innate immunity. Plant Physiol 152:2222–2231
Liu B, Li P, Li X et al (2005) Loss of function of OsDCL1 affects microRNA accumulation and causes developmental defects in rice. Plant Physiol 139:296–305
Liu Q, Yao X, Pi L et al (2009) The ARGONAUTE10 gene modulates shoot apical meristem maintenance and establishment of leaf polarity by repressing miR165/166 in Arabidopsis. Plant J 58:27–40
Marin E, Jouannet V, Herz A et al (2010) miR390, Arabidopsis TAS3 tasiRNAs, and their AUXIN RESPONSE FACTOR targets define an autoregulatory network quantitatively regulating lateral root growth. Plant Cell 22:1104–1117
Mattick JS (1994) Introns: evolution and function. Curr Opin Genet Dev 4:823–831
Matzke MA, Mosher RA (2014) RNA-directed DNA methylation: an epigenetic pathway of increasing complexity. Nat Rev Genet 15:394–408
Mercer TR, Wilhelm D, Dinger ME et al (2011) Expression of distinct RNAs from 3′ untranslated regions. Nucleic Acids Res 39:2393–2403
Mi S, Cai T, Hu Y et al (2008) Sorting of small RNAs into Arabidopsis argonaute complexes is directed by the 5′ terminal nucleotide. Cell 133:116–127
Montgomery TA, Howell MD, Cuperus JT et al (2008) Specificity of ARGONAUTE7-miR390 interaction and dual functionality in TAS3 trans-acting siRNA formation. Cell 133:128–141
Morris KV, Mattick JS (2014) The rise of regulatory RNA. Nat Rev Genet 15:423–437
Nodine MD, Bartel DP (2010) MicroRNAs prevent precocious gene expression and enable pattern formation during plant embryogenesis. Genes Dev 24:2678–2692
Nuthikattu S, McCue AD, Panda K et al (2013) The initiation of epigenetic silencing of active transposable elements is triggered by RDR6 and 21–22 nucleotide small interfering RNAs. Plant Physiol 162:116–131
Qi X, Xie S, Liu Y et al (2013) Genome-wide annotation of genes and noncoding RNAs of foxtail millet in response to simulated drought stress by deep sequencing. Plant Mol Biol 83:459–473
Quan M, Chen J, Zhang D (2015) Exploring the secrets of long noncoding RNAs. Int J Mol Sci 16:5467–5496
Qüesta JI, Song J, Geraldo N et al (2016) Arabidopsis transcriptional repressor VAL1 triggers polycomb silencing at FLC during vernalization. Science 3531:485–488
Rinn JL, Chang HY (2012) Genome regulation by long noncoding RNAs. Annu Rev Biochem 81:145–166
Rinn JL, Kertesz M, Wang JK et al (2007) Functional demarcation of active and silent chromatin domains in human HOX loci by noncoding RNAs. Cell 129:1311–1323
Rodriguez-Granados NY, Ramirez-Prado JS, Veluchamy A et al (2016) Put your 3D glasses on: plant chromatin is on show. J Exp Bot 67(11):1–17
Rohrig H, Schmidt J, Miklashevichs E et al (2002) Soybean ENOD40 encodes two peptides that bind to sucrose synthase. Proc Natl Acad Sci USA 99:1915–1920
Ronemus M, Vaughn M, Martienssen RA (2006) MicroRNA-targeted and small interfering RNA-mediated mRNA degradation is regulated by Argonaute, Dicer, and RNA-dependent RNA polymerase in Arabidopsis. Plant Cell 18:1559–1574
Sasaki T, Lee TF, Liao WW et al (2014a) Distinct and concurrent pathways of Pol II- and Pol IV-dependent siRNA biogenesis at a repetitive trans-silencer locus in Arabidopsis thaliana. Plant J 79:127–138

Sasaki T, Lorković ZJ, Liang SC et al (2014b) The ability to form homodimers is essential for RDM1 to function in RNA-directed DNA methylation. PLoS One 9:1–8
Saxena A, Carninci P (2011) Long non-coding RNA modifies chromatin: epigenetic silencing by long non-coding RNAs. BioEssays 33:830–839
Shin SY, Shin C (2016) Regulatory non-coding RNAs in plants: potential gene resources for the improvement of agricultural traits. Plant Biotechnol Rep 10:35–47
Shuai P, Liang D, Tang S et al (2014) Genome-wide identification and functional prediction of novel and drought-responsive lincRNAs in Populus trichocarpa. J Exp Bot 65:4975–4983
Sun Q, Csorba T, Skourti-Stathaki K et al (2013) R-loop stabilization represses. Science 340:619–621
Sung S, Amasino RM (2004) Vernalization and epigenetics: how plants remember winter. Curr Opin Plant Biol 7:4–10
Takeda A, Iwasaki S, Watanabe T et al (2008) The mechanism selecting the guide strand from small RNA duplexes is different among Argonaute proteins. Plant Cell Physiol 49:493–500
Tanurdzic M, Vaughn MW, Jiang H et al (2008) Epigenomic consequences of immortalized plant cell suspension culture. PLoS Biol 6:2880–2895
Taft RJ, Pheasant M, Mattick JS (2007) The relationship between non-protein-coding DNA and eukaryotic complexity. Bioessays 29:288-299. doi:10.1002/bies.20544
Tsai M, Manor O, Wan Y et al (2010) Long noncoding RNA as modular Scaffold of histone modification complexes. Science 329:689–693
Voinnet O (2009) Origin, biogenesis, and activity of plant MicroRNAs. Cell 136:669–687
Wang L, Zhao S, Gu C et al (2013) Deep RNA-Seq uncovers the peach transcriptome landscape. Plant Mol Biol 83:365–377
Wang H, Chung PJ, Liu J et al (2014a) Genome-wide identification of long noncoding natural antisense transcripts and their responses to light in Arabidopsis. Genome Res 24(3):444–453
Wang Z-W, Wu Z, Raitskin O et al (2014b) Antisense-mediated FLC transcriptional repression requires the P-TEFb transcription elongation factor. Proc Natl Acad Sci USA 111:7468–7473
Wu L, Mao L, Qi Y (2012) Roles of dicer-like and argonaute proteins in TAS-derived small interfering RNA-triggered DNA methylation. Plant Physiol 160:990–999
Wu H-J, Wang Z-M, Wang M et al (2013) Widespread long noncoding RNAs as endogenous target mimics for microRNAs in plants. Plant Physiol 161:1875–1884
Wunderlich M, Grob-Hardt R, Schoffl F (2014) Heat shock factor HSFB2a involved in gametophyte development of Arabidopsis thaliana and its expression is controlled by a heat-inducible long non-coding antisense RNA. Plant Mol Biol 85:541–550
Xie M, Yu B (2015) siRNA-directed DNA methylation in plants. Curr Genomics 16:23–31
Xie Z, Johansen LK, Gustafson AM et al (2004) Genetic and functional diversification of small RNA pathways in plants. PLoS Biol 2:642–652
Xin M, Wang Y, Yao Y et al (2011) Identification and characterization of wheat long non-protein coding RNAs responsive to powdery mildew infection and heat stress by using microarray analysis and SBS sequencing. BMC Plant Biol 11:1–13
Ye C-Y, Xu H, Shen E et al (2014) Genome-wide identification of non-coding RNAs interacted with microRNAs in soybean. Front Plant Sci 5:743
Yoshikawa M, Peragine A, Park MY et al (2005) A pathway for the biogenesis of trans -acting siRNAs in Arabidopsis. Genes Dev 19:2164–2175
Yoshikawa M, Iki T, Tsutsui Y et al (2013) 3′ fragment of miR173-programmed RISC-cleaved RNA is protected from degradation in a complex with RISC and SGS3. Proc Natl Acad Sci USA 110:4117–4122
Yoshikawa M, Iki T, Numa H et al (2016) A role of a short open reading frame encompassing the microRNA173 target site of the TAS2 transcript in trans-acting small interfering RNA biogenesis. Plant Physiol 171:359–368
Yu X, Yang J, Li X et al (2013) Global analysis of cis-natural antisense transcripts and their heat-responsive nat-siRNAs in Brassica rapa. BMC Plant Biol 13:208

Zhai J, Bischof S, Wang H et al (2015) A one precursor one siRNA model for pol IV-dependent siRNA biogenesis. Cell 163:445–455

Zhang Y, Jurkowska R, Soeroes S et al (2010) Chromatin methylation activity of Dnmt3a and Dnmt3a/3L is guided by interaction of the ADD domain with the histone H3 tail. Nucleic Acids Res 38:4246–4253

Zhang Y-C, Liao J-Y, Li Z-Y et al (2014a) Genome-wide screening and functional analysis identify a large number of long noncoding RNAs involved in the sexual reproduction of rice. Genome Biol 15:1–16

Zhang Y, Yan C, Kuang H (2014b) GC content fluctuation around plant small RNA-generating sites. FEBS Lett 588:764–769

Zhao J, Sun BK, Erwin JA et al (2008) Polycomb proteins targeted by a short repeat RNA to the mouse X chromosome. Science 322:750–756

Zhu H, Hu F, Wang R et al (2011) Arabidopsis argonaute 10 specifically sequesters miR166/165 to regulate shoot apical meristem development. Cell 145:1–27

Zubko E, Meyer P (2007) A natural antisense transcript of the Petunia hybrida Sho gene suggests a role for an antisense mechanism in cytokinin regulation. Plant J 52:1131–1139

Genome-Wide Function Analysis of lincRNAs as miRNA Targets or Decoys in Plant

Guanglin Li, Zhiqiang Hao, Chunyan Fan, and Xianmiao Wu

Abstract Long intergenic noncoding RNAs (lincRNAs) are generally long transcripts of more than 200 nucleotide (nt) that lack a coding sequence (CDS) or open reading frame (ORF); now the functions of plant lincRNAs remain largely unknown. In this chapter, we describe a computational pipeline to identify the function of lincRNAs based on the relationship between miRNAs and lincRNAs. In our method, lincRNA–mRNA co-expression networks and the ceRNA hypothesis are used to infer the function of lincRNAs as miRNA targets and lincRNAs as miRNA decoys, respectively.

Keywords lincRNAs • miRNAs • miRNA targets • miRNA decoys • Co-expression

Contents

G. Li (✉) • Z. Hao • X. Wu
College of Life Science, Shaanxi Normal University, Xi'an 710119, China
e-mail: glli@snnu.edu.cn

C. Fan
College of Computer Science, Shaanxi Normal University, Xi'an 710119, China

N. Rajewsky et al. (eds.), *Plant Epigenetics*, RNA Technologies,
DOI 10.1007/978-3-319-55520-1_8

1 Introduction

Long intergenic noncoding RNAs (lincRNAs) are endogenous noncoding RNAs (ncRNAs) that are larger than 200 nucleotide (nt) in length and transcribed from intergenic regions of the genome and may regulate gene expression at transcription and/or post-transcription levels by acting as signals, decoys, guides, and scaffolds (Mercer et al. 2009; Guttman and Rinn 2012). Plant lincRNAs play critical roles in a wide range of biological processes, especially in plant reproductive development and response to stresses (Zhu and Wang 2012).

Compared with protein-coding genes, the orthologs of plant lincRNAs are less conserved in distantly related species and exhibit high rates of sequence evolution (Ponting et al. 2009; Liu et al. 2012). This makes the identification of lincRNAs and prediction of their function based only on sequence conservation infeasible. With more and more plant transcriptome data provided by next generation sequencing (NGS) technology, it is urgent to develop new methods to genome-wide function analysis of lincRNAs in plant.

Plant microRNAs (miRNAs) are approximately 21–24 nt single-stranded, small noncoding RNAs that typically form near-perfect duplexes with their targets and mediate cleavage or translation repression at the posttranscriptional level (Bartel 2004; Ding et al. 2013). Recent studies suggest that miRNAs function in a more sophisticated way than was initially assumed. In addition to protein-coding RNAs acting as miRNA targets, lincRNAs can also be directly targeted by miRNAs for cleavage (Rhoades et al. 2002; Liang et al. 2013) or serve as miRNA decoys (Franco-Zorrilla et al. 2007; Salmena et al. 2011; Wu et al. 2013).

To explore the function of plant lincRNAs acting as miRNA targets or decoys, a genome-scale network among miRNAs, lincRNAs acting as miRNA targets, lincRNAs acting as miRNA decoys, and mRNAs was first constructed. Then, the functions of plant lincRNAs acting as miRNA targets were predicted by a co-expression network between lincRNAs and mRNAs, and the functions of lincRNAs acting as miRNA decoys were predicted and annotated according to the ceRNA hypothesis (Salmena et al. 2011).

This analysis in its current form is specific for maize, but can be adapted to other species. The potential function of lincRNAs will increase our knowledge about plant lincRNAs and be useful for further experimental validation. We hope to develop many other methods in order to identify the function of lincRNAs in the future such as prediction of lincRNA–protein interaction and analysis of lincRNA synteny.

2 Materials

2.1 *Hardware Requirements*

Personal computer, preferably with a multi-core processor.

2.2 *Software Requirements*

1. perl 5.24.0: The website https://www.perl.org/ has everything you need to get started with Perl.
2. linux: There are a lot of linux distribution, including Ubuntu, CentOS, Debian, Fedora, OpenSuse, FreeBSD, ArchLinux, LinuxMint, etc. You can choose one you like best to install. The following website contains links to manuals on how to download and install linux on your machine (http://www.linuxdown.net/).
3. R 3.2.5: The following website contains links to manuals on how to download, install, and run R on your machine: https://www.r-project.org/.
4. Cytoscape 3.4.0: A software that could be used to visualize the relationships among miRNAs, lincRNAs, and mRNAs; it can be downloaded from the following website: http://www.cytoscape.org/.
5. Blast2go 3.2: A software that could be used to annotate the function of genes base on GO and KEGG pathway. Blast2go can be downloaded from the following website: https://www.blast2go.com/.

2.3 *Data Resources*

1. miRNA data: Maize miRNAs are downloaded from miRBase (Release 21: June 2014, http://www.mirbase.org/) and saved in FASTA format. A sample file is as following and named as **miRNAs.fasta**.

```
>zma-miR156e-3p
GCUCACUGCUCUCUCUGUCAUC
>zma-miR399f-3p
UGCCAAAGGAAAUUUGCCCCG
>zma-miR529-5p
AGAAGAGAGAGAGUACAGCCU
```

2. lincRNA data: Maize lincRNAs could be obtained from the published three lincRNAs papers (Li et al. 2014; Boerner and McGinnis 2012; Zhang et al. 2014) (**Note 1**) and saved in FASTA format. A sample file is as following and named as **lincRNAs.fasta**.

```
>TCONS_00011680
CGCGCCGGTCGGCAGCGTGTCCTGGGCGGGCGCGGCGCGTCGGGCAGAG-
CAACGCGGCTCCTGCGCGCGCGGTGTGTGGCTCGCTCGGCGTGCGCGCGAGCGGCCGGAG-
CAGGGCGCGCTCGGCTGGCTGCTGGGTGGCGCGGTGCAGAGAGAACGGTGGGAGGAGAGAGT-
GAGTGGGAGAGAGAAGTGAGAGAGAGAGAAGTGAGGGAGAGAGAGTGGGAGTAGCAGAGG-
GAGGCGGCGGCTGGGA
>TCONS_00094692
CTCGATCGCCGCCTTTCACCACCCTGCGCTTGGAGGTTTTGTGTCGTGCTAGTATTTGCCTC-
CATTAAAGCCGTCGACTGGAGAGAGAAAGAGAAAGGGTGTGGCCAGAGATTTGGGACCAGCC-
GATCGGGGCGCGTGGCGTGCGTAGGAGGTCGGGCGGAGTCTAGCGGGGGGAGGGGGCA-
GAGCTTTGGCATGTCCGTCGCTAAGGAAGTGTCGTTGTACGACGCCACCGTCGATCCG-
CATCCCTGGCCGCGCGGTTCATCGTGAACAGGGCAGCCGCCACATCGTAATCCGGTCGT
>TCONS_00030207
TGGCTGCACCCTTGGTAGAGCCTTAGGAAGATTTGAGTAGTTTGATTTTTTGATCCCCTTATG-
GATCTTTATTTGTATGCCTGTTTCTGGCCTCCGTAAGGAGGACTG-
TAAAACCTTTTTTGTTTTTGTTTGTTTTTTCTATTCCTCTGCTTAATATATAATGGGGCG-
CAGTTCTCCTGCGCGTTCGAGAAAAAAAAGAAGAAGAAAAGAAGCAATCCAAGCGCAAGAGCT-
CAAATGAACACAAGTCACTCTCTCACTAGCCACTATTTGATTGGGGATGATCTTTTGGCTTGG-
GAGAGAATTTGATCTCTTTGGTGTGTCTTGTATTGAATGCTATAGCTCTTGTAAGGTATA-
GAAAGTATGA
```

3. mRNA data: Maize mRNA data were firstly downloaded from http://www.plantgdb.org/XGDB/phplib/download.php?GDB=Zm and then saved in FASTA format. A sample file is as following and named as **mRNAs.fasta**.

```
>GRMZM2G007300_T01
ACTATACGGGATTGATTTCTAGTGGAGGCTAGTGCGGGGCGGAATTTTCTTTGCTTGTT-
GATTCTTGATTGTTGTAGCGGTGATGGAGGGTACTTCAATGCAATAATGAGCTTCCCA-
GAAAATTACCCAAACAGCCCGCCATCAGTCAGATTTACCTCTGAAATGTGGCATCCAAATGTT-
TATTCAGATGGGCTGGTGTGTATCTCTATTCTTCATCCACCTGGTGAAGATCCAAGTGGTTAT-
GAGCTCGCAAGTGAGCGTTGGACACCTGTGCATACAGTTGAAAGCATAGTTCTGAGCATTA-
TATCAATGCTCTCTAGTCCAAATGACGAGTCTGCAGCAAACATTGAGGCAGCTAAGGAATG-
GAGGGATAAGAGGGAGGATTTCAAGAAAAGGGTCAGGCGCATCGTGCGGAAATCACAG-
GAAATGTTATGAAGAAAATGGAATACTGGGGAAGTCAACCGTGCCATGTCAGGCTACGA-
GATTTTGCTTTCAACCTCAAAATGCTCATCTGACATCTTGTGTATTTTCCATT-
TAGCGTCTGTTCATTGGTGCACCAGAAATGCATTGGGACGTGGTGTATCGTGTGGCCTGGGCTC-
TACTAATGTTTGACAATCCTGAAGTGCCACCAGTGGTGAACCTGGTATAAGCAGGCGCCATT-
CATGTCAGCTAATCGTTGAGAAAACATATCAGTGAAATCGTGCAACTGTTAGTGTAATGAGG-
CAGTTCCTGCTTACATATTATTACTCGTGCTACCTGTCTACAGTGTCTTGAATGCATG-
GAATGTCTTGAATTTTACCTGTTTTCCTTGGTGTAAGCATGGTATGCAATAATGGGAAACG-
CAAATTTCTCAGTTAACTGTATCCT
>GRMZM2G371942_T01
CATGGCTGTCTCGATGGGTCAGCTGCGAGGCTAGCCGCGTTCGATTGATCCTCTCTT-
CAACCGCCACCCTCGATCCTCTCTCTAGACGTCGCCCTCTCCTTCTCTCTTCATCC-
GATGTCGTCGCCACCATGGCCGGCTGGCCCGTTCCCTCCCTCCTCCTCCGCCGCCGGC-
```

*GACTCCTTTCCCTTCTCCCTCTCTCCTACCGGGCTGCACTCGCCGCACCCC-
TAGCGGTCGCGCCCGGCGCATCACCGGGGCGCGTCCTCGTCGCCGCTGCCGCGCCGCCCTCTC-
CACCTGACCCCCTTCTCGGGCGGTGGCCTTTGCGATGACCGGAGGTGGTGCTCGCGGTCGTGGT-
CAAGACGGAGGCCGCTACGTCTGCAGAGGCACCGGCTTTGGGGTCTACCGTCCAGATCC-
GATGGTCTCAGGGCAGAGGACAGCGCCCGAGATGGTGGGGCACATGTCCTGTGGCTGCTGGC-
GAATGATGGCGGCGGCATCGGCAGGTGCCGGGGCCGACGACGTTGGT-
CATTTGTGGCCTCGGCGTGCGACTGTGTGAAGCTGCGGACTGCGTTGGTGGTGGCGC-
GACGCTCTGGCGTCAACGTGGCGGTGCTAGAGGTCACACGTTTGTGAGGGTGACGCCGAC-
CAGTGGCCTAGGATGGCGCAAGGTGGTGGTGGCGGTGGCAGGGTGCCCGCGCGACAGCTG-
GATTCGGTGGCCGTGTGTGCTAGTGCCATCAGTTGCCGGGTGCGGCTCTGG-
CAGGCGTCTGTGTGACACACATGGTATGGAGGACGACGATGATGGTAGCTCTTGGCCTGACGT-
TAGTGCTGGAGGCTTCTTGGGCGAGAGCCTCAGCGACGG-
CAAGGCTCGTGGGCGTCGCTTCCCCTGTTGGGGCGTCGTATTCACCTTCGATGGTTGCTTTA-
GATTATCGGCGTCTGGTATACATGTGAAGATGGCGCTTTACAACGTGAAGTCGAAGCTACTG-
CATCAGGGGTTGTGGCTAAGCAACGATGACAGCTAACGAGCCCCATTCTTTGTGGACTG-
GATTGTTTCGATGGGTCGAGCATTTGCCGGGGGTAGGATGGAGGTTCAAGCTCGTTGGTGAA-
GAATCGGAGCTGTCTCACGTGGGGTGTGTTGAGGCTTGGCACGACGAGGCATCGC-
GATCTTTCTCATTATGTGCTGGTGTTTGGTTGTGGGCGTCATCGTGTTTTGTTTGAT-
CATGTGTGGCCCGTTTTGCAAAGTCGGAGCTGCTTGTCCAGAGCTCGGCAACGATAACTCTA-
GATGTGTTGTGTATGAGGGGCTCCCATTGATGGAGATGTTTGTGCATTTTATAGGCGG-
TAAGTGTTGTGCCGCGTTGATGTCCCAATCTAACCGATGAGTTTAACGTGTTGA*

4. Data of lincRNA and mRNA abundance: RNA-seq data for *Zea mays* are downloaded from NCBI GEO database (www.ncbi.nlm.nih.gov/geo); then scripts of align_and_estimate_abundance.pl and abundance_estimates_to_matrix.pl in Trinity (https://github.com/trinityrnaseq/trinityrnaseq/tree/master/util) are used to generate the abundance of lincRNA and mRNA which is counted by RPKM. A sample file is as following and named as **Gene-expression-table.txt**.

lincRNA or mRNA ID	*Sample1*	*Sample2*	*Sample3*
lincRNA_TCONS_00030207	*0.113*	*2.691*	*0.101*
mRNA_GRMZM2G177942_T01	*0.179*	*2.775*	*0.097*
mRNA_GRMZM2G111504_T01	*0.021*	*1.663*	*0.001*

3 Methods

3.1 Identification of Unique Maize miRNAs

The same sequences that have different miRNA IDs were firstly merged and output with fasta format by the following perl script.

```
####Script name:Merged_miRNAs.pl
#!/usr/bin/perl
open IN, "miRNAs.fasta"or die;
open OUT, ">unique_miRNAs.fasta" or die;
while(<IN>){
   chomp;
   $id=$1 and next if /^>(\S+)/;
   If ($hash{$_}) {
   $hash{$_}.="/".$id; }
   else {$hash{$_} = $id; }
}
for(sort keys%hash){
   print OUT ">$hash{$_}\n$_\n";
}
```

3.2 Set up the Relationship Between Unique miRNAs and lincRNAs

The relationship between unique miRNAs and lincRNAs could be predicted by running perl script of GATAr.pl which is included in CleaveLand (Addo-Quaye et al. 2009) with the following command:

```
GSTAr.pl unique_miRNAs.fasta lincRNAs.fasta > align_ miRNAs_
and_lincRNAs.txt
```

3.3 Set up the Relationship Between Unique miRNAs and mRNAs

The relationship between unique miRNAs and mRNAs could be predicted by running perl script of GATAr.pl with the following command:

```
GSTAr.pl unique_miRNAs.fasta mRNAs.fasta > align_ miRNAs_and_
mRNAs.txt
```

3.4 Functional Prediction of lincRNAs Acting as miRNA Targets Based on the lincRNA-mRNA Co-expression Networks

3.4.1 Identification of lincRNAs Acting as miRNA Targets

LincRNAs potentially acting as miRNA targets were predicted according to the following set of rules: (a) at most, one mismatch or indel was allowed between the 9th and 12th positions of the 5′ end of miRNA sequences, (b) the total number of bulges or mismatches in the other regions was not allowed to exceed 4 nt, and (c) no continuous mismatches were allowed (Fan et al. 2015) (**Note 2**).

```
####Script name: miRNAs_target_prediction.pl
#!/usr/bin/perl
open IN,"align_ miRNAs_and_lincRNAs.txt"or die;
open OUT, ">miRNA_targets_in_lincRNAs.txt" or die;
while(<IN>){
    chomp;
    if($_=~/^5/) {
        $gene_inf=$_; @gene_infs=split" ",$gene_inf;
        $new=1;}
    if($new) {
        $align0=$1 if $_=~/^    (.*)\n/;}
    if ($_=~/^3/){
        $new=0; $miRNA_inf=$_; @miRNA_infs=split" ",$miRNA_inf;
        $miRNA_seq=reverse $miRNA_infs[1];
        @atcg=split"",$miRNA_seq;}
    if($_=~/MFE of perfect match: (\S+)/) { $mfeperfect=$1;}
    if($_=~/MFE of this site: (\S+)/) {$mfesite=$1;}
    if($_=~/MFEratio: (\S+)/) {
        $mferatio=$1;
        for($n=0;$n<scalar@atcg;$n++) {
            $m++ if $atcg[$n]=~/[AUCG]/;
            push@start_end9_12,$n if $atcg[$n]=~/[AUCG]/ and ($m==9
or $m==12);}
      $m=0;
      $align=reverse $align0;
      @aligns=split"",$align;
      for($n=$start_end9_12[0];$n<=$start_end9_12[1];$n++) {
          $indel9_12++ if $aligns[$n] eq ' ';}
      for($n=0;($n+1)<length$align;$n++)  {
          $bi=$aligns[$n].$aligns[$n+1]; $mismatch++ if $bi=~/  /;}
      $indel_all=$align=~tr/ / /;
```

```
      $condition1=1  if $indel9_12<=1;
      $condition2=1  unless $mismatch;
      $condition3=1  if $indel_all-$indel9_12<=4;
      if($condition1 and $condition2 and $condition3){
         $miRNA_nam=$miRNA_infs[4];
         ($gene_nam,$position)=split":",$gene_infs[4];
         $miRNA_target='miRNA_target: 5\".$gene_infs[1].' 3\";
         $miRNA='  miRNA: 3\".$miRNA_infs[1].' 5\";
         $result.=$miRNA_nam."\t".$gene_nam."\t".$position."\t".
$mfeperfect."\t".$mfesite."\t".$mferatio."\n".$miRNA_target."\n".
(' 'x11).$align0."\n".$miRNA."\n";}
         $indel9_12=$indel_all=$mismatch=$condition1=$condition2=
$condition3=0;}
}
$result="miRNA\tTranscript\tstart-end\tMFEperfect\tMFEsite\tMFEr-
atio\n".$result;
print OUT "$result\n";
```

3.4.2 Construction of lincRNA–mRNA Co-expression Networks

RNA-seq data were used to perform co-expression analysis between mRNAs and lincRNAs as miRNA targets; the pipeline for constructing the co-expression networks was as follows: (a) genes, including mRNAs and lincRNAs as miRNA targets, whose variances ranked in the top 75% of the expression profiles were retained; (b) the *p*-value of Pearson's correlation coefficient (*Pcc*) was calculated for each pair of genes using Fisher's asymptotic test in the *WGCNA* library of R (Langfelder and Horvath 2008), and these values were adjusted using the Bonferroni correction method; and (c) co-expression relationships showing adjusted *p*-values of less than 0.05 and ranking in the top 5% and bottom 5% of *Pcc* were selected for further analysis (Hao et al. 2015). The Bonferroni multiples test was executed using the *multtest* package from R. The co-expression networks could be visualized using Cytoscape 3.1.1 (Shannon et al. 2003). The R script is as follows:

```
#### Script name: Co-expression.R (Note 3)
library(WGCNA)
options(stringsAsFactors = FALSE)
myData = read.table("Gene-expression-table.txt", header=T)
#Gene-expression-table.txt is the expression level of gene (Note 4).
dim(myData)
names(myData)
datExpr = as.data.frame(t(myData))
adjacency =cor(datExpr,use="p")
```

```
padj<-p.adjust(corPvalueFisher(adjacency,32,twoSided = TRUE),meth-
od=c("bonferroni"))
# 32 is the number of samples, it should be replace by real sample
number
n=0
result<-c()
for(i in names(datExpr)){
   for(j in names(datExpr)){
   n<-n+1
   if (i >=j) {next}
   if(adjacency[i,j]>quantile(adjacency,0.05)&&adjacency[i,j]<quan-
tile(adjacency,0.95)){next}
   if (padj[n]>0.05){ next}
   lines<-paste(i,j,adjacency[i,j],padj[n],sep="\t")
   result<-paste(result,lines,sep="\n")
   }
}
write.table(result,file="Gene-correlation.txt",col.names=F,
quote=F,sep="\t")
```

3.4.3 Functional Prediction of lincRNAs as miRNA Targets

Based on the co-expression networks between lincRNAs and mRNAs (file of **Gene-correlation.txt** as an example), all the mRNAs ID (GRMZM2G177942_T01 and GRMZM2G111504_T01) can be firstly extracted from the file of **Gene-correlation.txt**; then the protein sequences for these mRNA ID can be obtained from the annotation of maize genome and saved in FASTA format; the file is as following and named as **Protein_seq1.fasta**.

```
>GRMZM2G177942_P01
MPPKSDSVEGIVLGFVNEQNRPLNSQNAADALQKFNLKKTAVQKALDALADSGQISFKEY
GKQKIYIARQDQFDIPNGEELEEMKKTNAKLQEELADQKKAISEVESEVRGLQSNLTLTE
ITSKKSELQSEVQEMEEKLNKLRSGVTLVKPEDKKIIENSFAEKVSQWKRRKRMFKELWD
NITEHSPKDQKEFKEELGIEYDEDVDVNVQSYSDMLASLNKRRKISR
>GRMZM2G111504_P01
MPPAPTTGGAGEEPAILKAALGSGKLRIEGYCRFISAGHPQGRKKKHRNNKYAEEWKVRA
CAGRLPSTLPPPKAPASAWRSPPWQPRRLRPGGRRRRCGRRSLPLEGSPRPSSLSLPLGF
PDGDGDAAGSVAGGVGGVRGGGKRALLDPADRAVMQRQKRMIKNRESAASSRDRKQAALE
EILYQFEEKLQAEREEAARK
```

Then the GO toolkit can be used to predict the function of lincRNAs. The commonly used GO toolkit was BLAST2GO (Conesa et al. 2005); the command is shown in the following lines:

```
blastall -p blastp -i "Protein_seq1.fasta" -d nr -b 20 -v 20 -m 7 -e 1e-3 -o "Protein_seq_blast1.out" -a 8
./blast2go_cli.run -workspace PATHWAY -nameprefix NAME -loadfasta Protein_seq1.fasta -loadblast Protein_seq_blast1.out -mapping -annotation -annex -gograph -statistics all -savereport -savedat -properties 'blast2go_cli_v1.0.2/cli.prop' -useobo 'blast2go_cli_v1.0.2/go-basic.obo' (Note 5)
```

3.5 *Functional Prediction of lincRNAs Acting as miRNA Decoys Based on CeRNA Hypothesis*

3.5.1 Identification of mRNAs Acting as miRNA Targets

Using the same script as in Sect. 3.4.1 (miRNAs_target_prediction.pl), mRNAs as miRNA targets can be predicted, except that the input file should be changed into **align_miRNAs_and_mRNAs.txt** and the output file changed into **miRNA_targets_in_mRNAs.txt**.

3.5.2 Identification of lincRNAs as miRNA Decoys

LincRNAs potentially acting as miRNA decoys were predicted according to the following set of rules: (a) the number of mismatches or indels should be larger than 1 and less than 6 between the 9th and 12th positions of the 5′ end of the miRNA sequences; (b) perfect nucleotide pairing was required between the 2nd and 8th positions of the 5′ end of miRNA sequences; and (c) the number of mismatches and indels should be no more than four in other regions (Fan et al. 2015). These rules were implemented using the following Perl scripts.

```
#!/usr/bin/perl
open IN,"align_miRNAs_and_lincRNAs.txt"or die;
open OUT, ">miRNA_decoys_in_lincRNAs.txt" or die;
while(<IN>){
   chomp;
   if($_=~/^5/) {
      $gene_inf=$_;
      @gene_infs=split" ",$gene_inf;
      $new=1; }
   if($new) {
      $align0=$1 if $_=~/^    (.*)\n/;}
   if ($_=~/^3/) {
      $new=0; $miRNA_inf=$_; @miRNA_infs=split" ",$miRNA_inf;
```

```
      $miRNA_seq=reverse $miRNA_infs[1]; @atcg=split"",$miRNA_seq;}
   if($_=~/MFE of perfect match: (\S+)/) {$mfeperfect=$1;}
   if($_=~/MFE of this site: (\S+)/) {$mfesite=$1;}
   if($_=~/MFEratio: (\S+)/) {
      $mferatio=$1;
      for($n=0;$n<scalar@atcg;$n++){
         $m++ if $atcg[$n]=~/[AUCG]/;
         push@start_end9_12,$n if $atcg[$n]=~/[AUCG]/ and ($m==9 or
$m==12);
         push@start_end2_8,$n if $atcg[$n]=~/[AUCG]/ and ($m==2 or
$m==8);}
      $m=0;
      $align=reverse $align0;
      @aligns=split"",$align;
      for($n=$start_end9_12[0];$n<=$start_end9_12[1];$n++){
         $indel9_12++ if $aligns[$n] eq ' ';}
      for($n=$start_end2_8[0];$n<=$start_end2_8[1];$n++){
         $match2_8++ if $aligns[$n] eq '|';}
      $indel_all=$align=~tr/ / /;

      $condition1=1 if $indel9_12>1 and $indel9_12<6;
      $condition2=1 if $match2_8 ==7 and $start_end2_8[0]==1 and
$start_end2_8[1]==7;
      $condition3=1  if $indel_all-$indel9_12<=4;
      if($condition1 and $condition2 and $condition3){
         $miRNA_nam=$miRNA_infs[4];
         ($gene_nam,$position)=split":",$gene_infs[4];
         $miRNA_decoy='miRNA_decoy: 5\".$gene_infs[1].' 3\";
         $miRNA=' miRNA: 3\' '.$miRNA_infs[1].' 5\";
         $result.=$miRNA_nam."\t".$gene_nam."\t".$position."\t".
$mfeperfect."\t".$mfesite."\t".$mferatio."\n".$miRNA_decoy."\n".
(' 'x 11).$align0."\n".$miRNA."\n";}
         $indel9_12=$indel_all=$match2_8=$condition1=$condition2=
$condition3=0;
      @start_end9_12=@start_end2_8=();}
}
$result="miRNA\tTranscript\tstart-end\tMFEperfect\tMFEsite
\tMFEratio\n".$result;
print OUT "$result\n";
```

3.5.3 Functional Prediction of lincRNAs as miRNA Decoys Based on CeRNA Hypothesis

To infer the function of lincRNAs as miRNA decoys, networks were firstly constructed based on the complementary pairs between miRNAs and lincRNAs (file of **miRNA_decoys_in_lincRNAs.txt** as an example) and between miRNAs and mRNAs (file of **miRNA_targets_in_mRNAs.txt** as an example). The nodes in the networks consisted of miRNAs (zma-miR399f-3p as an example), lincRNAs acting as miRNA decoys (TCONS_00094692 as an example), and mRNAs acting as miRNA targets (GRMZM2G007300_T01 and GRMZM2G371942_T01 as an example). The miRNA–lincRNA–mRNA networks were then visualized with Cytoscape. Finally, lincRNAs and mRNAs that were regulated by the same miRNA were extracted from miRNA–lincRNA–mRNA networks and saved the protein sequences for mRNAs as FASTA format; the sample file is as following and named as **Protein_seq2.fasta**.

```
>GRMZM2G007300_P01
MSFPENYPNSPPSVRFTSEMWHPNVYSDGLVCISILHPPGEDPSGYELASERWTPVHTVE
SIVLSIISMLSSPNDESAANIEAAKEWRDKREDFKKRVRRIVRKSQEML
>GRMZM2G371942_P01
MEDDDDGSSWPDVSAGGFLGESLSDGKARGRRFPCWGVVFTFDGCFRLSASGIHVKMALY
NVKSKLLHQGLWLSNDDS
```

Based on the ceRNA hypothesis and gene ontology (GO) analysis, the function of lincRNAs acting as miRNA decoys (TCONS_00094692 as example) can be predicted based on the function of mRNAs (GRMZM2G007300_P01 and GRMZM2G371942_P01 as an example). The commonly used GO toolkit was BLAST2 GO; the command is shown in the following lines:

```
  blastall -p blastp -i "Protein_seq2.fasta" -d nr -b 20 -v 20 -m
7 -e 1e-3 -o "Protein_seq_blast2.out" -a 8
  ./blast2go_cli.run -workspace PATHWAY  -nameprefix NAME  -load-
fasta  Protein_seq2.fasta   -loadblast   Protein_seq_blast2.out
-mapping -annotation -annex  -gograph -statistics all -savereport
-savedat  -properties  'blast2go_cli_v1.0.2/cli.prop'  -useobo
'blast2go_cli_v1.0.2/go-basic.obo'
```

4 Notes

Note 1 If lincRNAs in the interesting species have not been identified, CPC or other tools can be used to predict lincRNAs.

Note 2 Except for GSTAr.pl in CleaveLand, other tools can also be used to predict the miRNA targets, these tools are PsRobot, PAREsnip, SeqTar, MTide, and SoMART.

Note 3 This is a "R" script, typing "R" in Terminal of Linux can start to use R and its packages. And typing "q()" will quit.

Note 4 The expression for each gene is counted by RPKM (Reads per Kilobase per Million Reads) and listed in the following format.

Gene	Sample1	Sample2	Sample3
Gene 1	XXXXX	XXXXX	XXXXX
Gene 2	XXXXX	XXXXX	XXXXX
……			

Note 5 The interpretation for the parameter:

- workplace: the pathway to save result
- nameprefix: the prefix for output files
- other parameters are defaulted

References

Addo-Quaye C, Miller W, Axtell MJ (2009) CleaveLand: a pipeline for using degradome data to find cleaved small RNA targets. Bioinformatics 25:130–131

Bartel DP (2004) MicroRNAs: genomics, biogenesis, mechanism, and function. Cell 116:281–297

Boerner S, McGinnis KM (2012) Computational identification and functional predictions of long noncoding RNA in Zea mays. PLoS One 7:e43047

Conesa A, Gotz S, Garcia-Gomez JM et al (2005) Blast2GO: a universal tool for annotation, visualization and analysis in functional genomics research. Bioinformatics 21:3674–3676

Ding Y, Tao Y, Zhu C (2013) Emerging roles of microRNAs in the mediation of drought stress response in plants. J Exp Bot 64:3077–3086

Fan C, Hao Z, Yan J, Li G (2015) Genome-wide identification and functional analysis of lincRNAs acting as miRNA targets or decoys in maize. BMC Genomics 16:793

Franco-Zorrilla JM, Valli A, Todesco M et al (2007) Target mimicry provides a new mechanism for regulation of microRNA activity. Nat Genet 39:1033–1037

Guttman M, Rinn JL (2012) Modular regulatory principles of large non-coding RNAs. Nature 482:339–346

Hao Z, Fan C, Cheng T et al (2015) Genome-wide identification, characterization and evolutionary analysis of long intergenic noncoding RNAs in cucumber. PLoS One 10:e0121800

Langfelder P, Horvath S (2008) WGCNA: an R package for weighted correlation network analysis. BMC Bioinformatics 9:559

Li L, Eichten SR, Shimizu R et al (2014) Genome-wide discovery and characterization of maize long non-coding RNAs. Genome Biol 15:R40

Liang H, Zhang J, Zen K et al (2013) Nuclear microRNAs and their unconventional role in regulating non-coding RNAs. Protein Cell 4:325–330

Liu J, Jung C, Xu J et al (2012) Genome-wide analysis uncovers regulation of long intergenic noncoding RNAs in Arabidopsis. Plant Cell 24:4333–4345

Mercer TR, Dinger ME, Mattick JS (2009) Long non-coding RNAs: insights into functions. Nat Rev Genet 10:155–159
Ponting CP, Oliver PL, Reik W (2009) Evolution and functions of long noncoding RNAs. Cell 136:629–641
Rhoades MW, Reinhart BJ, Lim LP et al (2002) Prediction of plant microRNA targets. Cell 110:513–520
Salmena L, Poliseno L, Tay Y et al (2011) A ceRNA hypothesis: the Rosetta Stone of a hidden RNA language? Cell 146:353–358
Shannon P, Markiel A, Ozier O et al (2003) Cytoscape: a software environment for integrated models of biomolecular interaction networks. Genome Res 13:2498–2504
Wu HJ, Wang ZM, Wang M et al (2013) Widespread long noncoding RNAs as endogenous target mimics for microRNAs in plants. Plant Physiol 161:1875–1884
Zhang W, Han Z, Guo Q et al (2014) Identification of maize long non-coding RNAs responsive to drought stress. PLoS One 9:e98958
Zhu QH, Wang MB (2012) Molecular functions of long non-coding RNAs in plants. Genes 3:176–190

Plant Non-coding RNAs and the New Paradigms

Rodrigo Siqueira Reis and Jules Deforges

Abstract Non-coding RNAs vary greatly in length, shape and function. Growing interest and recent evidence have identified some of them as essential elements for life, as well as for environmental adaptation and development. Since non-coding RNAs by definition do not code for proteins, their ever-growing roles pose a paradigm shift in biology. In this chapter, we will discuss our current knowledge of two distinct plant non-coding RNAs: long non-coding RNAs (lncRNAs) and microRNAs (miRNAs). These two classes of non-coding transcripts are relatively well characterized in plants, and regulate gene expression through distinct modes of action. We thus anticipate that our current mechanistic knowledge of lncRNAs and miRNAs will provide the basis for future studies of non-coding RNAs in plant genetics and epigenetics.

Keywords LncRNA • MiRNA • Non-coding RNA • Epigenetics

Contents

R.S. Reis (✉) • J. Deforges
Département de Biologie Moléculaire Végétale, Université de Lausanne, Bâtiment Biophore, 1015 Lausanne, Switzerland
e-mail: Rodrigo.SiqueiraReis@unil.ch; Jules.Deforges@unil.ch

N. Rajewsky et al. (eds.), *Plant Epigenetics*, RNA Technologies,
DOI 10.1007/978-3-319-55520-1_9

1 Introduction

LncRNAs, by definition, are RNA transcripts with a length of >200 nt that do not encode proteins. They are found in every studied living organism, often in great numbers, and are predicted to outnumber protein-coding genes in humans (Djebali et al. 2012). Since lncRNAs have only recently been identified as a functional class with important biological roles, the great majority of them have yet unknown functions. Their conservation and functional motifs are also poorly known but clearly complex, as lncRNAs have already been implicated in a vast range of processes, through various mechanisms of action. This can be exemplified by their role in the X chromosome inactivation by Xist in humans (Penny et al. 1996) and flowering time control by COOLAIR in Arabidopsis (Swiezewski et al. 2009), in which these lncRNAs play critical but unrelated roles. Moreover, although many lncRNAs are similar to mRNAs, there are also general differences that discriminate them. The similarities often include transcription by RNA polymerase II, presence of 5′-cap and 3′-poly (A) and splicing. However, lncRNAs tend to be shorter than protein-coding transcripts, have fewer and longer exons, low expression levels and poor primary sequence conservation (Quinn and Chang 2016). LncRNA annotation is another complexity level in their study as the main curation centres and the whole community are debating the 'gold-standard' annotations (Mattick and Rinn 2015). Consequently, naming of lncRNA loci is also a topic of debate.

The efforts to set standards for lncRNAs as a single class of RNAs seem, however, an intermediate step bridging our currently poor knowledge to a yet-to-be more comprehensive understanding, as any current lncRNA commonalities are surrounded by exceptions. In plants, these so-called exceptions include nonpolyadenylated intermediate-sized ncRNAs (im-ncRNAs) (Wang et al. 2014b) and circular RNAs (cRNAs) (Franco-Zorrilla et al. 2007), and some less stable lncRNAs transcribed by Pol IV and/or Pol V (Ariel et al. 2014; Blevins et al. 2014), amongst others. Further, evidence in animals for coding potential of some lncRNAs points to a dual role, regulatory and coding (de Andres-Pablo et al. 2017). In plants, such a dual role has been reported for primary miRNA transcripts (pri-miRNAs) that encode small peptides, similar to upstream open reading frames (uORFs) in mRNAs, which regulate the expression levels of their associated miRNAs (Hellens et al. 2016; Lauressergues et al. 2015). A yet more intriguingly plant lncRNA, COLDAIR, is present within an FLC intron and is required for the vernalization-mediated epigenetic repression of FLC (Heo and Sung 2011). Indeed, we have only started to grasp the functions and mechanisms underlying the vast number of lncRNAs, and the dichotomy coding–non-coding RNAs might not be a sufficient benchmark for classification of lncRNAs.

In general, lncRNAs also exhibit unexpectedly more specific expression profiles than mRNAs, i.e. expression restricted or primarily in specific cell type, tissue, developmental stage or stress condition (Batista and Chang 2013). Such specificity has not been studied in great detail in plants; however, preliminary studies show that, in general, plant lncRNAs also are highly tissue specific and many respond to specific (a)biotic stresses (reviewed in Liu et al. 2015). Thus, although lncRNAs resemble mRNAs in many ways, they seem to have features that make them unique much beyond protein coding potential. In this chapter section, we will present and discuss some of the better understood examples of plant lncRNAs. For simplicity, these examples were broadly divided into two lncRNA regulatory activities, i.e. transcriptional and post-transcriptional.

2 LncRNAs

2.1 LncRNA Transcriptional Activity

Most of the lncRNAs characterized so far are known to regulate gene expression at the transcriptional level, through various complex mechanisms. Examples of lncRNAs involved in chromatin modification, chromosome topology dynamics, and regulation of the transcription machinery or transcription interference have been reported in different organisms and seem widespread in the eukaryotes.

LncRNAs can repress gene expression through transcriptional interference. In mammals, transcription of Igf2r non-coding RNA has been shown to be impaired by transcription of the antisense lncRNA Airn (Latos et al. 2012). Transcription of the lncRNA on the region overlapping with the promoter of Igf2r, but not the lncRNA molecule itself is essential for transcription repression. The authors suggest that transcription of the promoter region would impede recruitment of the pol II. Similar examples have been described in different organisms such as yeast and human (Hongay et al. 2006; Martianov et al. 2007).

LncRNAs can also regulate gene expression via direct interactions with the transcription machinery, modulating the binding property of transcription factors or directly interacting with the RNA polymerase. In humans, for instance, the lncRNA lnc-DC expressed in dendritic cells physically interacts with the transcription factor “signal transducer and activator of transcription 3” (STAT3). This interaction promotes the activating phosphorylation of STAT3 on tyrosine-705, stimulating dendritic cells differentiation (Wang et al. 2014a).

By contrast, a peculiar group of lncRNAs generated by Alu SINE elements has been shown to act as a transacting transcription repressor in human and mouse (Allen et al. 2004; Mariner et al. 2008). These lncRNAs are induced during heat shock and physically interact with the RNA pol II to repress a specific set of mRNAs. Interestingly, the SINE RNAs in mouse and human cells have evolved similar functions although the sequences are unrelated. Such examples have not yet been identified in plants, but are likely to emerge in the future.

Two distinct classes of lncRNAs are involved in an epigenetic silencing pathway specific to plants, the RNA-dependent DNA methylation (RdDM) (Böhmdorfer and Wierzbicki 2015). These lncRNAs are produced by pol IV and pol V. Pol IV transcripts are precursors of the 24 nucleotides siRNAs. They are converted into double stranded RNAs by the RNA-dependent RNA polymerase 2 (RDR2) and then processed into siRNA by Dicer Like 3 (DCL3) (Onodera et al. 2005; Herr et al. 2005). The siRNAs are loaded into ARGONAUTE 4 (AGO4) in the cytoplasm before being reimported into nucleus to direct de novo methylation of specific targets (Ye et al. 2012). Pol V transcripts, in the other hand, are thought to act as a scaffold, interacting with the siRNA-AGO4 complex through sequence complementarity and recruiting chromatin modifying enzymes (Wierzbicki et al. 2008, 2009; Rowley et al. 2011; Zheng et al. 2013). This class of RNA is essential for the RdDM pathway since a loss of function of pol V abolishes transcriptional gene silencing (Kanno et al. 2005).

Along with DNA methylation and histone modification, pol V transcripts are also involved in modulation of nucleosome positioning in Arabidopsis. They can interact with the lncRNA-binding protein IDN2 that recruits in turn the putative ATP-dependent chromatin-remodelling complex SWI/SNF through a direct interaction with the SWI3 subunit (Zhu et al. 2013). This interaction promotes a stabilization of the nucleosomes and thus represses transcription at the pol V transcribed loci.

A case of lncRNA regulating gene expression via chromatin topology has also been described recently in Arabidopsis (Ariel et al. 2014). Auxin Regulated Promoter Loop (APOLO) is a long intergenic non-coding RNA transcribed both by pol II and pol V. The dual transcription of APOLO locus regulates expression of its neighbouring gene PINOID (PID) via chromatin loop formation. Exogenous auxin treatment triggers the active demethylation of the APOLO-PID region as well as the opening of the loop encompassing PID promoter. This leads to the accumulation of both APOLO and PID transcripts. Pol II transcripts then start to recruit the polycomb complex (PC1), resulting in progressive loop reformation. Meanwhile, the transcripts generated by pol V associate with AGO4 and promote DNA methylation. The APOLO-PID locus is thus repressed via this feedback mechanism. Such processes of dynamic chromatin modification might be a novel layer of regulation of gene expression.

By contrast, another class of lncRNAs promotes chromatin loop formation to stimulate gene expression. These lncRNAs were identified from high-throughput studies in mammals and named enhancer RNAs (eRNAs) (Kim et al. 2010). They are bidirectionally transcribed by the RNA polymerase II from enhancer domains and help to recruit activators of transcription or to promote chromosome looping between enhancer and promoter regions (Zhang et al. 2013). Several examples of genes enhanced by eRNAs have been described in mammal cells, but this class of lncRNAs has not yet been reported in plants (Ørom et al. 2010; Melo et al. 2013; Li et al. 2013c; Zhang et al. 2013).

In Arabidopsis, flowering is probably the best understood example of biological process regulated by lncRNAs. Two distinct categories of non-coding transcripts are responsible for the vernalization-mediated repression of Flowering Locus C (FLC) by epigenetic modifications (Crevillén and Dean 2011). FLC encodes a transcription repressor that prevents the expression of genes involved in flowering transition (Michaels and Amasino 1999). COLDAIR is a lncRNA transcribed from the intron of FLC in the same orientation (Swiezewski et al. 2009; Heo and Sung 2011). It is capped but not polyadenylated. COLDAIR was shown to repress FLC expression via recruitment of PRC2 to the chromatin at FLC locus through methylation of the histone H3K27 (Csorba et al. 2014; Heo and Sung 2011). COOLAIR, by contrast, is a set of alternatively spliced and polyadenylated lncRNAs produced in antisense orientation from the 3′ end of FLC. These transcripts are induced under cold temperatures earlier than COLDAIR (Swiezewski et al. 2009) and were found to physically interact with FLC chromatin in two regulatory regions (Heo and Sung 2011). The authors propose that they would play a role in the coordinated switching of chromatin states occuring during cold. More recently, Sun et al. identified a protein, AtNDX, that regulates COOLAIR expression via chromatin loop formation (Sun et al. 2013). They showed that COOLAIR lncRNA can form an RNA-DNA heteroduplex in the promoter region, releasing a single stranded DNA region called R-loop. AtNDX binding to this single strand of DNA inhibits COOLAIR transcription, probably by preventing RNA polymerase II elongation. Similar mechanism of epigenetic repression by lncRNAs was found in different species in the eukaryotic kingdom, the best studied examples being the lncRNAs HOTAIR and Xist in mammals (Rinn et al. 2007; Cerase et al. 2015).

2.2 LncRNA Posttranscriptional Activity

LncRNAs can also modulate various processes downstream of transcription and chromatin modification, including mRNA stability, alternative splicing, miRNA decoy, protein interaction, protein sorting and translation. Most of these regulations have been identified in animals and, to a lesser extent, in plants. In animals, for instance, the lncRNA TINCR interacts with STAU1 protein to promote the stability of mRNAs containing the TINCR box motif (Kretz et al. 2013). However, STAU1 bound to half-STAU1-binding site RNAs (1/2-sbsRNAs) are guided to mRNA targets, harbouring base pairing matches to 1/2-sbsRNAs, to mediate mRNA decay (SMD) (Gong and Maquat 2011). Thus, the same STAU1 protein can either stabilize or destabilize mRNAs depending on which lncRNA is bound to it. The antisense *Uchl1*RNA, a lncRNA, interacts with the sense coding mRNA but instead of modulating its stability it leads to an increase in Uchl1 protein, in a cap-independent manner, acting like a mobile internal ribosomal entry site (IRES) (Carrieri et al. 2012). Translation inhibition guided by lncRNAs has also been reported in animals, in which lincRNA-p21, in the absence of HuR, bridges the interaction between the translational repressor Rck with the mRNAs CTNNB1 and JUNB, repressing the translation of these targeted mRNAs (Yoon et al. 2012).

In plants, some lncRNAs have been shown to have posttranscriptional activity. The lncRNA INDUCED BY PHOSPHATE STARVATION 1 (IPS1) acts as a miRNA decoy via base pair complementarity to miR399, which hijacks the miRNA from its transcript target (Franco-Zorrilla et al. 2007). miR399 target gene PHO2 posttranscriptionally represses the root-to-shoot phosphate transporter PHO1 (Liu et al. 2012). Phosphate (Pi) starvation induces IPS1 transcription to high levels, resulting in the hijacking of miR399 and higher levels of PHO2 mRNA, thus decreased PHO1 protein and root-to-shoot Pi transport (Franco-Zorrilla et al. 2007; Wu et al. 2013a). The role of IPS1 in Pi homeostasis is conserved, although there is low sequence conservation. At4 is another Arabidopsis lncRNA functionally similar to IPS1, as it is induced by Pi starvation and mimics miR399 targets (Shin et al. 2006), suggesting this process is highly posttranscriptionally regulated. Many other examples of lncRNAs acting as miRNA decoy have been predicted but are still poorly characterized (Wu et al. 2013a).

In rice, Pi starvation induces high levels of yet another lncRNA. The Pi transporter PHO1; 2 mRNA was found unexpectedly unchanged upon Pi starvation although its protein levels were elevated (Jabnoune et al. 2013). The authors then investigated the role of a *cis*-natural antisense transcript (*cis*-NAT), lncRNA produced from the antisense DNA strand in the same locus. The *cis*-NAT*pho1;2* had elevated transcript levels upon Pi starvation, and was found to enhance translation of the sense *PHO1;2* mRNA. The translational enhancement activity of *cis*-NAT*pho1;2* is critical for rice yield and grain quality. However, the mechanisms underlying this regulation are still unknown. Pi starvation and salt stress were further shown to induce expression of another *cis*-NAT. Npc536 is antisense to COG5 (AT1G67930) and its overexpression produces enhanced root growth under salt stress as well as increased primary root growth and secondary root length (Amor et al. 2009). Similar to *cis*-NAT*pho1;2*, Npc536 overexpression does not alter the transcript levels of the sense coding RNA; however, the role of Npc536 in translational enhancement was not verified.

In leguminous, EARLY NODULIN 40 (ENOD40) encodes functional peptides and has a role as lncRNA. Soybean ENOD40 encodes two peptides that bind specifically to sucrose synthase, suggesting a role in sucrose homeostasis in nitrogen-fixing nodules (Rohrig et al. 2002). ENOD40 RNA was later found to interact with a MtRBP1, a RNA-binding protein in the leguminous model *Medicago truncatula* (Campalans et al. 2004). This interaction is required for MtRBP1 proper sorting from nuclear speckles to cytoplasm during nodule development in ENOD40-expressing cells and, importantly, this role is independent of ENOD40 translation. It is possible that the role of lncRNAs in protein sorting is a general feature, particularly in plant–pathogen interactions and others interactions with their environment; however, only ENOD40 lncRNA and MtRBP1 have been described yet.

A role for lncRNAs has also been shown in alternative splicing. The nuclear speckle RNA-binding proteins (NSRs) were identified as a family of RNA-binding proteins (RBPs) that act as nuclear alternative splicing regulators in Arabidopsis. These RBPs were also shown to interact with two lncRNAs, ENOD40 and lnc351 (Amor et al. 2009), thus the authors decided to rename them Alternative Splicing Competitor RNA (ASCO-RNA) (Bardou et al. 2014). The ASCO-RNA affected alternative splicing of NSR-dependent mRNA targets by competing for the binding of NSRs, suggesting a role for lncRNAs as hijacker regulators in alternative splicing to modulate gene expression during developmental transitions.

A spontaneous mutation caused a SNP in the lncRNA LDMAR that unveiled its critical role in rice fertility and development (Ding et al. 2012). The spontaneous G→C mutation caused a potential localized change in LDMAR RNA secondary structure, leading to heritable increased methylation in its promoter region. The consequent reduced levels of LDMAR caused premature programed cell death (PCD) in another development under long days, resulting in male sterility. However, the mechanisms underlying this process are still elusive. Another elusive but striking example of lncRNA was found upon infection with *Fusarium oxysporum*. Arabidopsis infected by *F. oxysporum* revealed a large set of lncRNAs potentially involved in antifungal response (Zhu et al. 2014). Knockdown of selected lncRNAs, as well as their genetic association with coding genes involved in antifungal response, further corroborated the relevant role of lncRNAs in this process.

2.3 LncRNAs: Non-coding Transcripts or Dual RNAs?

Along with the identification of the modes of action, a question still debated is whether or not lncRNAs can be translated into peptides. Most of the functions of lncRNAs studied so far involve the RNA molecule itself rather than any encoded peptide. However, the recent finding that ribosome footprints could be detected in short Open Reading Frames (sORFs) of most lncRNAs is now challenging the 'non-coding' assumption (Ruiz-Orera et al. 2014; de Andres-Pablo et al. 2017). It has recently been proposed that translation of the sORFs could regulate lncRNA degradation via a Non-sense Mediated mRNA Decay (NMD) pathway (de - Andres-Pablo et al. 2017). Kurihara et al. reported that in plants, lncRNAs are more degraded by the NMD than coding transcripts, probably because of the presence of numerous stop triplets (Kurihara et al. 2009). This is in agreement with the recent observation that most cytoplasmic lncRNAs are bound to and degraded at ribosomes in human cells, suggesting that ribosomes could play a role in their degradation (Carlevaro-Fita et al. 2016). Andres-Pablo et al. propose the NMD to be a layer of regulation of lncRNAs expression (de Andres-Pablo et al. 2017). A particular set of lncRNAs would escape the NMD to produce the encoded peptides under specific condition. Examples of dual RNAs, with coding and regulatory functions, such as ASCO (ENOD40) have been reported in plants (Bardou et al. 2011; Ulveling et al. 2011; Sinturel et al. 2015), but the proportion of lncRNAs actually translated as well as the biological relevance of the encoded peptides remain to be determined.

3 MiRNAs

3.1 MiRNAs: Specialized Products of LncRNAs

MiRNAs are a class of small non-coding RNAs enzymatically excised from a very particular type of lncRNAs, namely primary miRNA transcripts (pri-miRNAs). Eukaryotes have evolved a diverse and complex set of miRNA-guided gene expression inhibition pathways (Chang et al. 2012; Ameres and Zamore 2013; Rogers and Chen 2013). In the canonical miRNA pathway, pri-miRNAs are transcribed by RNA polymerase II (Pol II) to produce non-coding RNAs with mRNA-like features, including a 7-methylguanosine (m^7G) cap at 5′ end and poly(A) tail at 3′ end. Due to partial self-complementarity, pri-miRNAs fold to form stem-loop structures that are cleaved in the nucleus to produce smaller sized precursor miRNA (pre-miRNA) intermediate molecules. This cleavage is performed by a Dicer protein, assisted by a dsRNA-BINDING (DRB) protein. Such protein partnerships include Drosha/DGCR8 in mammals, Drosha/Pasha in flies, and DICER-LIKE1 (DCL1)/DRB1 in plants (Gregory et al. 2004; Kurihara and Watanabe 2004; Landthaler et al. 2004). In mammals and flies, the pre-miRNA is exported to the cytoplasm and further processed into a miRNA/miRNA* duplex by a second Dicer/DRB partnership, the Dicer/TRBP and Dicer-1/Loqs interaction, respectively (Bernstein et al. 2001; Hutvágner et al. 2001; Lee et al. 2004). In plants, both the pri-miRNA and pre-miRNA precursor processing steps occur in the nucleus, in specialized nuclear bodies termed 'dicing bodies' or D-bodies, and only require the single DCL1/DRB1 protein partnership (Kurihara and Watanabe 2004). The resulting miRNA/miRNA* duplex is then loaded by an AGO protein that, in mammals and flies, removes the miRNA* passenger strand (via AGO-catalysed endonucleolytic cleavage), resulting in an active miRISC (Czech and Hannon 2010). In plants, however, the exact mechanism by which the miRNA is selected for miRISC incorporation over the corresponding miRNA* strand remains unknown, although the preferential selection and AGO loading of the miRNA guide strand has been shown to be directed by DRB1 (also called HYL1).

Our knowledge of miRNA biogenesis and activity in plants has advanced greatly in recent years. However, several of the latest findings indicate that some important mechanisms remain poorly characterized. The biogenesis of miRNA/miRNA* from miRNA-containing intermediates occurs in D-bodies, and a growing number of proteins, in addition to well-characterized core components, have been demonstrated to be also required at this stage of the miRNA pathway (Rogers and Chen 2013). Only in the past few years have approximately half of the known proteins required for D-body assembly and/or function been identified, including TOUGH (TGH), C-TERMINAL DOMAIN PHOSPHATASE-LIKE1 (CPL1), DRB2, DAWDLE (DDL), MODIFIER OF SNC1, 2 (MOS2), NEGATIVE ON TATA LESS2b (NOT2b), RECEPTOR FOR ACTIVATED C KINASE 1 (RACK1), SICKLE (SIC), STABILIZED1 (STA1) and CELL DIVISION CYCLE5 (CDC5). In fact, our knowledge of miRNA biogenesis is still largely restricted to the function of the core proteins DCL1, DRB and SE (Vazquez et al. 2004; Lobbes et al. 2006). Moreover, forward genetics (Manavella et al. 2012) and, to a lesser extent, yeast two-hybrid protein–protein interaction screens (Speth et al. 2013) are the only approaches currently used to discover novel genes relevant to miRNA biogenesis.

The recent identification of new proteins required for miRNA biogenesis suggests that this process is more complex and dynamic than previously thought. For example, NOT2b-Pol II interaction is required for efficient transcription of both protein-coding and non-coding gene transcripts, as well as to mediate the connection of *MIR* gene transcripts to core miRNA biogenesis protein machinery, including DCL1 and SE (Wang et al. 2013). The *not2a not2b* (*not2a2b*) double mutant has reduced pri-miRNA expression and mature miRNA accumulation. Furthermore, in *not2a2b* plants, DCL1 localization, but not the localization of DRB1, is affected in D-bodies, suggesting that D-body assembly is independent of DCL1. Interestingly, in the *mos2* mutant, DRB1 fails to localize to D-bodies, but MOS2 does not interact with the core proteins DRB1, DCL1 or SE (Wu et al. 2013b). However, the pri-miRNA binding affinity of DRB1, and hence pri-miRNA processing, was greatly reduced in the *mos2* mutant background. These findings led the authors to propose that the function of MOS2 is to facilitate pri-miRNA recruitment to the D-body, and that MOS2-recruited pri-miRNAs might act as scaffolding proteins for D-body formation.

CPL1 and DDL have also recently been shown to play a role in miRNA biogenesis, and their functions add another level of complexity to this silencing pathway. The identification of CPL1 revealed that DRB1 is inactive when phosphorylated, requiring CPL1 activity for its dephosphorylation and subsequent activation (Manavella et al. 2012). However, the kinase responsible for DRB1 phosphorylation remains unknown and, more importantly, the biological significance of having a pool of inactive, phosphorylated DRB1 is still unclear. Hyperphosphorylated DRB1 was readily detectable in cellular lysates, leading the authors to speculate that there is a substantial reservoir of inactive DRB1 that can be dephosphorylated and activated when required, for instance to mediate developmentally important processes, such as seed germination (Manavella et al. 2012). DDL is required in miRNA biogenesis to bind pri-miRNAs and to interact with phosphorylated DCL1 (Yu et al. 2008; Engelsberger and Schulze 2012; Machida and Yuan 2013). Interestingly, *DDL* encodes a phosphothreonine-binding forkhead-associated (FHA) domain, a domain often encoded by proteins that function in signal transduction pathways (Machida and Yuan 2013). The SMAD signal transducer, which is structurally similar to DDL (Machida and Yuan 2013), has been demonstrated to integrate the miRNA biogenesis and signal transduction pathways in humans (Davis et al. 2008). SMAD is recruited to pri-miRNA processing complexes and controls the vascular smooth muscle cell phenotype mediated by the bone morphogenetic protein (BMP) family of human growth factors. Due to their structural similarities, it is possible that DDL, like human SMAD, plays a role in integrating the miRNA biogenesis and signal transduction pathways (Machida and Yuan 2013). However, to date, no experimental evidence exists to support this proposed role for DDL in plants.

In animals, miRNAs are processed from their precursor transcripts, pri-miRNA and pre-miRNA, via a sequential two-step process in different cellular compartments. On the other hand, the vast majority of plant miRNAs require DRB1-assisted DCL1 activity for their nuclear production (Vazquez et al. 2004; Eamens et al. 2012). DRB1 is a highly characterized DCL1 partner protein, and is required by DCL1 for accurate and efficient miRNA/miRNA* processing from larger sized precursor transcripts (Kurihara et al. 2006; Dong et al. 2008). Furthermore, DRB1 has been demonstrated to mediate an additional step in the Arabidopsis miRNA pathway, the preferential selection of miRNA guide strands over its antisense paired sequence, the miRNA* strand, for loading into AGO1-catalysed miRISC (Kurihara et al. 2006; Dong et al. 2008; Eamens et al. 2009). More recently, DRB4, together with DCL4, has been shown to be required for the production of a small number of newly evolved miRNAs that are processed from precursor transcripts that, upon folding, form highly complementary stem-loop structures (Rajagopalan et al. 2006; Pélissier et al. 2011). In addition, Eamens et al. (2012) have shown that in the shoot apex and floral tissues, DRB2 is both synergistic and antagonistic to DRB1 in the biogenesis of different miRNAs. The role of DRB2 in the miRNA pathway unveiled the mechanism for sorting of miRNAs to either the transcript cleavage or translational inhibition mode of action (Reis et al. 2015b).

The dsRNA-binding domains (RBDs) are typically ~70 amino acids in length, have an αβββα fold, and are found in most proteins that recognize dsRNA. However, RBD function is not limited to the recognition and binding of dsRNA, but also involves mediation of DRB protein–protein interactions (reviewed by Daniels and Gatignol 2012). All plant DRB proteins characterized to date encode two amino-terminal RBDs, and the structure of both RBDs of Arabidopsis DRB1 has been determined (Yang et al. 2010). Surprisingly, DRB1 encodes both a canonical RBD (RBD1) and a non-canonical RBD (RBD2), whereas both DRB2 RBDs are canonical, and the RBD and C-terminal domains appear to determine their different role in the miRNA pathway (Reis et al. 2016). DRB1 RBD2 differs from the canonical structure in (1) the loop that recognizes dsRNA minor groove and (2) the α-helices that recognize both major and minor groove. Both deviations are caused by different electrostatic potential and surface shape, resulting in drastic reduction in binding affinity for pre-miRNA and dsRNA. DRB1 has also been shown to recognize and bind 21 nt dsRNA as homodimer, and Yang et al. (Yang et al. 2010) suggested that this may be mediated by the non-canonical structure of DRB1 RBD2. Amino acid sequence alignment of DRB1 and DRB2 RBDs shows that the deviation in loop structure in DRB1 RBD2 appears to be highly conserved in DRB1 orthologs found in other plant species. Interestingly, although DRB2 is required in the miRNA biogenesis pathway (Eamens et al. 2012; Reis et al. 2015b), the lack of invariant histidine that defines the non-canonical DRB1 RBD2 is not observed (Reis et al. 2016).

3.2 MiRNA Activity

Animal and plant miRNAs have distinct mechanisms of target transcript recognition and expression regulation (Ameres and Zamore 2013). In animals, the primary determinant for the binding of miRISC to the targeted mRNA(s) is a 6–8 nt domain at the 5′ end of the RISC-loaded miRNA, termed the seed region. Furthermore, the vast majority of seed-matched target sequences are located in the 3′ untranslated region (3′UTR) of the targeted mRNA(s) (Grimson et al. 2007). In animals, as a consequence of the low target homology requirements of miRNAs, a large number of unrelated genes are regulated by each miRNA (Friedman et al. 2009). The exact mechanism of miRNA-directed target gene expression regulation remains a topic of debate, but it appears to involve translation inhibition and mRNA decay, and to a lesser extent, endonuclease-catalysed mRNA cleavage (Baek et al. 2008; Guo et al. 2010). In contrast, plant miRNAs are highly complementary to their target mRNAs, and their respective target sites are usually located within the coding region of the targeted gene (German et al. 2008; Karginov et al. 2010). The high target complementarity requirements of plant miRNAs result in a small number of closely related target genes, usually a subset of genes belonging to a much larger gene family. Curiously, plant miRNAs can direct either a RISC-mediated endonucleolytic cleavage or translation inhibition to repress the expression of their target genes. Both of these mechanisms are independent of the degree of miRNA:mRNA base pairing or the position of the target site within the coding region of the targeted mRNA (Brodersen et al. 2008; Li et al. 2013b; Grant-Downton et al. 2013). The miRNA fate to cleave or inhibit translation is determined at birth by DCL1 partnering protein DRB2, as it determines translation inhibition and represses DRB1 transcription, which in the absence of DRB2 determines transcript cleavage of miRNA targets (Reis et al. 2015b).

Argonautes, the RISC effector proteins, contain several functional domains, including PAZ, MID and PIWI (Mallory and Vaucheret 2010). The MID and PAZ domains bind the 5′-monophosphorylated and 3′-nucleotide of the guide RNA, respectively, and the PIWI domain functions as the ribonucleolytic domain (Song et al. 2004). AGO slicer activity appears to be a three-step process (Wang et al. 2009). In the nucleation step, the 5′ end of the miRNA binds to the 3′ end of the miRNA target site of the mRNA. Nucleation is followed by the propagation step, characterized by rearrangement of AGO protein and extension of the miRNA: mRNA dsRNA hybrid. During propagation, PAZ domain rotation favours the correct positioning of the mRNA target site with respect to the catalytic PIWI domain. Once the mRNA target site is correctly positioned in AGO, the mRNA is cleaved at the phosphodiester bond linking mRNA nucleotides opposite to miRNA positions 10 and 11 (Wang et al. 2009). The slicer role of Arabidopsis AGO1 in miRNA pathway is well documented; however, several of the other nine Arabidopsis AGO proteins also encode a functional PAZ domain, and exhibit slicer activity, indicating that other AGOs may also perform a similar role to AGO1 in the miRNA pathway (Baumberger and Baulcombe 2005; Carbonell et al. 2012).

Interestingly, Carbonell et al. (2012) recently demonstrated that a mutated, slicer-defective AGO1 forms miRISC more efficiently with its targeted mRNA(s) than wild-type AGO1, indicating that a functional PAZ domain is not a requirement for miRISC target recognition in plants.

In plants, miRNA activity has been almost exclusively assessed at the transcript level, while translation inhibition was assumed to be a less important mechanism of silencing operating via an alternative pathway (Rogers and Chen 2013). However, rapidly growing evidence suggests that in specific plant tissues, such as floral tissues, translation inhibition, and not target mRNA slicing, is the predominant miRNA-directed silencing mechanism (Chen 2004; Grant-Downton et al. 2013). Furthermore, Li et al. recently showed that expression of amiRNAs in plants predominantly mediated highly specific translation repression and limited mRNA decay or cleavage (Li et al. 2013a). This finding has a direct impact on the design and evaluation of amiRNA efficacy in a biotechnological context, but it also reveals that translation inhibition has been largely underappreciated in plants.

Forward genetics has recently allowed identification of a number of novel genes required for miRNA-guided translation inhibition, including *KATANIN1* (*KTN1*) (Brodersen et al. 2008), *VARICOSE* (*VCS*) (Brodersen et al. 2008), *ALTERED MERISTEM PROGRAM1* (*AMP1*) (Li et al. 2013b) and *'SHUTTLE' IN CHINESE* (*SUO*) (Yang et al. 2012). The encoded proteins appear to be required for either trafficking or localization of miRISC, and/or for mRNA stability. *KTN1* encodes a microtubule severing enzyme required for the correct organization of cortical microtubules (Burk et al. 2007), suggesting that the trafficking or assembly of the cellular components required for translation inhibition may require the microtubule network. The involvement of AMP1, an integral endoplasmic reticulum (ER) enzyme (Li et al. 2013b), in translation inhibition further shows that, post assembly, RISC is transported from the nucleus to specific cytoplasmic sites. Interestingly, AGO1 is essential for the slicing activity of miRISC, and AGO1 activity has also been demonstrated to be required for miRNA-directed translation inhibition (Brodersen et al. 2008; Lanet et al. 2009). Taken together, these recent findings suggest that miRNA trafficking and miRNA complex assembly steps are crucial in the rewiring of AGO1 activity from slicer to repressor. The proteome landscape may determine the identity of additional proteins with which AGO1-catalysed miRISC interacts; these interactions in turn may mediate the ability of the complex to direct either mRNA cleavage or translation inhibition [reviewed in Reis et al. (2015a)].

RACK1 orthologs are evolutionarily conserved and contain seven WD40-β-propeller domains. These domains have been shown to be involved in mediating simultaneous interactions with multiple proteins and hence allow RACK1 proteins to act as scaffolding proteins in large and dynamic protein complexes (Adams et al. 2011). Arabidopsis RACK1 was identified in a yeast two-hybrid screen that used SE as the bait. Subsequent analyses revealed that the *rack1* mutant accumulates less mature miRNA than in wild-type plants (Speth et al. 2013), unlike the increased mature miRNA accumulation previously demonstrated in animal mutants of RACK1 orthologs (Jannot et al. 2011). The Arabidopsis

RACK1 alters miRNA accumulation and activity via distinct mechanisms: (1) it is required for efficient and precise pri-miRNA processing, possibly via its interaction with SE, in miRNA biogenesis, and (2) RACK1 is also part of the AGO1-catalysed miRISC, suggesting that RACK1 also has a role in miRNA activity (Speth et al. 2013). Although the exact role of RACK1 in the AGO1-catalysed miRISC remains unclear, RACK1 does not alter the slicer activity of AGO1. Also, in *rack1* mutants, miR398 targets CSD1, CSD2 and CCS showed increased protein accumulation without a corresponding transcript elevation, suggesting that RACK1 is involved in miR398-guided translational inhibition. However, miR398 accumulation is reduced in *rack1* mutants, which may directly explain the observed elevation in accumulation of miR398 target proteins (Speth et al. 2013). The role of RACK1 in both the biogenesis and action stages of the Arabidopsis miRNA pathway makes it experimentally challenging to assess its function only in relation to miRNA activity. Nonetheless, RACK1 is involved in protein translation in mammals and yeast (Ceci et al. 2003), and a recent report supports a similar role for Arabidopsis RACK1 (Guo et al. 2011). Hence, the observation that RACK1 scaffold protein is involved in both miRNA biogenesis and activity raises the possibility that RACK1 is required for miRISC trafficking from nucleus to the ER.

3.3 Function of MiRNA-Guided Translation Inhibition

Although widely documented in animals, insects and (more recently) plants, the biological relevance of miRNA-guided translational inhibition, over that of transcript cleavage, remains largely unknown. A plant miRNA can guide either of these modes of action to control expression of a gene, thus providing an ideal model to study the biological relevance of translation inhibition and transcript cleavage. Recently, it has been shown that these two modes of action play different roles in various plant processes, including development and environmental adaptation (Reis et al. 2015c).

In leguminous plants, the formation of nitrogen-fixing nodules (nodulation) upon bacterial infection is an important evolutionary adaptation to low nitrogen conditions [reviewed by Smith and Smith (2011)]. More recently, this process in legumes has been shown to involve miRNA activity, including miR166 and miR169 (Combier et al. 2006; Boualem et al. 2008). Furthermore, Reynoso et al. (2013) demonstrated that the interaction of *Medicago truncatula* (a model legume) with *Sinorhizobium meliloti* results in the differential accumulation of miRNAs in polysomal complexes, and that this in turn leads to differential levels of miRNA target proteins. Enrichment of miRNAs, notably miRNAs miR169 and miR172, in the polysomes revealed their association with translation machinery. Interestingly, upon infection, accumulation of miR169 is reduced in polysomal complexes and, as a consequence, the level of the miR169 target protein, HAP2-1, is elevated. The authors proposed that reduced miR169 polysome accumulation may contribute to translation derepression of HAP2-1 mRNA immediately following inoculation with

S. meliloti. miR169-guided HAP2-1 cleavage precedes its translational derepression, and may act to restrict the expression of HAP2-1 to nodule meristems for cell identity preservation (Reynoso et al. 2013).

The differential accumulation of miRNAs in polysomal complexes suggests that plants may actively control miRNA activity in order to either preserve or remove target mRNAs. Olivier Voinnet proposed that (1) miRNAs and siRNAs could operate primarily through transcript cleavage to produce irreversible gene expression changes required to establish permanent cell fates (e.g. during cell differentiation) or (2) they could guide translational repression of sRNA target transcripts in a reversible manner, thus allowing the cell to still be able to respond rapidly to environmental challenges (Voinnet 2009). Although the reversibility of translation repression can be intuitively understood, it has not been experimentally validated in plants. In animals, miRNAs induce gene silencing, with translational inhibition occurring first, which impairs the function of the eIF4F initiation complex, and being required for subsequent mRNA destabilization (Meijer et al. 2013). However, animal and plant miRNAs are distinct in that plant miRNAs do not show very clear preference towards one mode of action, while the animal counterparts appear to act preferentially through translational repression. Thus, it is reasonable to argue that plant miRNAs guide two functionally different mechanisms to regulate the expression of their target genes.

Plants need to trigger a rapid response against non-beneficial infection. If reversibility is a predominant feature of miRNA action, miRNA- and/or siRNA-guided target transcript translation inhibition may have evolved in plants to confer adaptive advantages against pathogen infection. Reversible sRNA-directed silencing response would allow for the storage of sRNA target transcripts during periods of biotic or abiotic stress. The beneficial plant–bacterial interaction in nodule development in legumes (outlined above) supports such a proposed model. Furthermore, nucleotide-binding site leucine-rich repeat (NB-LRR) genes, a major plant resistance gene (R-gene) class, are required for the recognition of specific pathogens and are encoded by very large gene families in plant genomes (Tobias and Guest 2014). Recent reports have revealed that miRNAs, and a number of classes of siRNAs, are crucial regulators of NB-LRR gene expression in plants (Zhai et al. 2011; Fei et al. 2013). The biological relevance of a RNA silencing-based mechanism to regulate NB-LRR expression remains debated, but it has been shown to be a conserved regulatory mechanism in plants. Moreover, Lucas et al. (2014) have demonstrated that siRNA-directed NB-LRR gene expression regulation is not limited to target transcript cleavage. The authors showed that upon infection of *Brachypodium distachyon* with the fungus *Fusarium culmorum*, 31 miRNAs that were predicted to target NB-LRR genes had altered accumulation. However, the authors did not observe any change to the targets at the mRNA level, suggesting that translation inhibition may play a role in the regulation of R-genes NB-LRR.

4 Conclusion

Over the years, an ever-growing number of non-coding RNAs have been characterized in the eukaryotic kingdom, regulating gene expression through a myriad of molecular mechanisms. The miRNAs are now well characterized in plants. Given their conserved structure and mechanism of action, they form a relatively homogenous class of non-coding transcripts. Such common features enable in silico prediction of putative novel miRNAs, as well as identification of targets. By contrast, lncRNAs are more loosely defined given the broad diversity of their sequence, structure or topology and little is known about the functions and modes of action in plants. Only a few examples such as COOLAIR and COLDAIR, APOLO, ASCO, or the RdDM pathway have been relatively well studied, and several mechanisms reported in mammal or yeast are yet to be identified in plants. For instance, no case of transcription interference, enhancer RNAs or lncRNA interacting directly with the transcription machinery has been reported so far. Such examples are likely to emerge in a close future with the accumulation of high-throughput data, particularly from strand-specific RNA sequencing experiments. Finally, although some lncRNAs have been shown to play a critical role in important biological processes, the biological relevance of the large majority of them is yet to be determined.

References

Adams DR, Ron D, Kiely PA (2011) RACK1, A multifaceted scaffolding protein: structure and function. Cell Commun Signal 9:22

Allen TA, Von Kaenel S, Goodrich JA et al (2004) The SINE-encoded mouse B2 RNA represses mRNA transcription in response to heat shock. Nat Struct Mol Biol 11:816–821

Ameres SL, Zamore PD (2013) Diversifying microRNA sequence and function. Nat Rev Mol Cell Biol 14:475–488

Amor BB, Wirth S, Merchan F et al (2009) Novel long non-protein coding RNAs involved in Arabidopsis differentiation and stress responses. Genome Res 19:57–69

de Andres-Pablo A, Morillon A, Wery M (2017) LncRNAs, lost in translation or licence to regulate? Curr Genet 63:29

Ariel F, Jegu T, Latrasse D et al (2014) Noncoding transcription by alternative RNA polymerases dynamically regulates an auxin-driven chromatin loop. Mol Cell 55:383–396

Baek D, Villén J, Shin C et al (2008) The impact of microRNAs on protein output. Nature 455:64–71

Bardou F, Merchan F, Ariel F et al (2011) Dual RNAs in plants. Biochimie 93:1950–1954

Bardou F, Ariel F, Simpson CG et al (2014) Long noncoding RNA modulates alternative splicing regulators in Arabidopsis. Dev Cell 30:166–176

Batista PJ, Chang HY (2013) Long noncoding RNAs: cellular address codes in development and disease. Cell 152:1298–1307

Baumberger N, Baulcombe DC (2005) Arabidopsis ARGONAUTE1 is an RNA Slicer that selectively recruits microRNAs and short interfering RNAs. Proc Natl Acad Sci USA 102:11928–11933

Bernstein E, Caudy AA, Hammond SM et al (2001) Role for a bidentate ribonuclease in the initiation step of RNA interference. Nature 409:363–366

Blevins T, Pontvianne F, Cocklin R et al (2014) A two-step process for epigenetic inheritance in Arabidopsis. Mol Cell 54:30–42

Böhmdorfer G, Wierzbicki AT (2015) Control of chromatin structure by long noncoding RNA. Trends Cell Biol 25:623–632

Boualem A, Laporte P, Jovanovic M et al (2008) MicroRNA166 controls root and nodule development in *Medicago truncatula*. Plant J 54:876–887

Brodersen P, Sakvarelidze-Achard L, Bruun-Rasmussen M et al (2008) Widespread translational inhibition by plant miRNAs and siRNAs. Science 320:1185–1190

Burk DH, Zhong R, Ye Z-H (2007) The katanin microtubule severing protein in plants. J Integr Plant Biol 49:1174–1182

Campalans A, Kondorosi A, Crespi M (2004) Enod40, a short open reading frame–containing mRNA, induces cytoplasmic localization of a nuclear RNA binding protein in *Medicago truncatula*. Plant Cell 16:1047–1059

Carbonell A, Fahlgren N, Garcia-Ruiz H et al (2012) Functional analysis of three Arabidopsis ARGONAUTES using slicer-defective mutants. Plant Cell 24:3613–2629

Carlevaro-Fita J, Rahim A, Guigó R et al (2016) Cytoplasmic long noncoding RNAs are frequently bound to and degraded at ribosomes in human cells. RNA 22:867–882

Carrieri C, Cimatti L, Biagioli M et al (2012) Long non-coding antisense RNA controls Uchl1 translation through an embedded SINEB2 repeat. Nature 491:454–457

Ceci M, Gaviraghi C, Gorrini C et al (2003) Release of eIF6 (p27BBP) from the 60S subunit allows 80S ribosome assembly. Nature 426:579–584

Cerase A, Pintacuda G, Tattermusch A et al (2015) Xist localization and function: new insights from multiple levels. Genome Biol 16:166

Chang S-S, Zhang Z, Liu Y (2012) RNA interference pathways in fungi: mechanisms and functions. Annu Rev Microbiol 66:305–323

Chen X (2004) A microRNA as a translational repressor of *APETALA2* in Arabidopsis flower development. Science 303:2022–2025

Combier J-P, Frugier F, de Billy F et al (2006) MtHAP2-1 is a key transcriptional regulator of symbiotic nodule development regulated by microRNA169 in *Medicago truncatula*. Genes Dev 20:3084–3088

Crevillén P, Dean C (2011) Regulation of the floral repressor gene FLC: the complexity of transcription in a chromatin context. Curr Opin Plant Biol 14:38–44

Csorba T, Questa JI, Sun Q et al (2014) Antisense COOLAIR mediates the coordinated switching of chromatin states at FLC during vernalization. Proc Natl Acad Sci USA 111:16160–16165

Czech B, Hannon GJ (2010) Small RNA sorting: matchmaking for Argonautes. Nat Rev Genet 12:19–31

Daniels SM, Gatignol A (2012) The multiple functions of TRBP, at the hub of cell responses to viruses, stress, and cancer. Microbiol Mol Biol Rev 76:652–666

Davis BN, Hilyard AC, Lagna G et al (2008) SMAD proteins control DROSHA-mediated microRNA maturation. Nature 454:56–61

Ding J, Lu Q, Ouyang Y et al (2012) A long noncoding RNA regulates photoperiod-sensitive male sterility, an essential component of hybrid rice. Proc Natl Acad Sci USA 109:2654–2659

Djebali S, Davis CA, Merkel A et al (2012) Landscape of transcription in human cells. Nature 489:101–108

Dong Z, Han M-H, Fedoroff N (2008) The RNA-binding proteins HYL1 and SE promote accurate in vitro processing of pri-miRNA by DCL1. Proc Natl Acad Sci USA 105:9970–9975

Eamens AL, Smith NA, Curtin SJ et al (2009) The Arabidopsis thaliana double-stranded RNA binding protein DRB1 directs guide strand selection from microRNA duplexes. RNA 15:2219–2235

Eamens AL, Kim KW, Curtin SJ et al (2012) DRB2 is required for microRNA biogenesis in *Arabidopsis thaliana*. PLoS One 7:e35933

Engelsberger WR, Schulze WX (2012) Nitrate and ammonium lead to distinct global dynamic phosphorylation patterns when resupplied to nitrogen-starved Arabidopsis seedlings. Plant J 69:978–995
Fei Q, Xia R, Meyers BC (2013) Phased, secondary, small interfering RNAs in posttranscriptional regulatory networks. Plant Cell 25:2400–2415
Franco-Zorrilla JM, Valli A, Todesco M et al (2007) Target mimicry provides a new mechanism for regulation of microRNA activity. Nat Genet 39:1033–1037
Friedman RC, Farh KK-H, Burge CB et al (2009) Most mammalian mRNAs are conserved targets of microRNAs. Genome Res 19:92–105
German MA, Pillay M, Jeong D-H et al (2008) Global identification of microRNA–target RNA pairs by parallel analysis of RNA ends. Nat Biotechnol 26:941–946
Gong C, Maquat LE (2011) lncRNAs transactivate STAU1-mediated mRNA decay by duplexing with 3′ UTRs via Alu elements. Nature 470:284–288
Grant-Downton R, Kourmpetli S, Hafidh S et al (2013) Artificial microRNAs reveal cell-specific differences in small RNA activity in pollen. Curr Biol 23:R599–R601
Gregory RI, Yan K, Amuthan G et al (2004) The microprocessor complex mediates the genesis of microRNAs. Nature 432:235–240
Grimson A, Farh KK-H, Johnston WK et al (2007) MicroRNA targeting specificity in mammals: determinants beyond seed pairing. Mol Cell 27:91–105
Guo H, Ingolia NT, Weissman JS et al (2010) Mammalian microRNAs predominantly act to decrease target mRNA levels. Nature 466:835–840
Guo J, Wang S, Valerius O et al (2011) Involvement of Arabidopsis RACK1 in protein translation and its regulation by abscisic acid. Plant Physiol 155:370–383
Hellens RP, Brown CM, Chisnall MAW et al (2016) The emerging world of small ORFs. Trend Plant Sci 21(4):317–332
Heo JB, Sung S (2011) Vernalization-mediated epigenetic silencing by a long intronic noncoding RNA. Science 331:76–79
Herr AJ, Jensen MB, Dalmay T et al (2005) RNA polymerase IV directs silencing of endogenous DNA. Science 308:118–120
Hongay CF, Grisafi PL, Galitski T et al (2006) Antisense transcription controls cell fate in *Saccharomyces cerevisiae*. Cell 127:735–745
Hutvágner G, McLachlan J, Pasquinelli AE et al (2001) A cellular function for the RNA-interference enzyme Dicer in the maturation of the let-7 small temporal RNA. Science 293:834–838
Jabnoune M, Secco D, Lecampion C et al (2013) A rice cis-natural antisense RNA acts as a translational enhancer for its cognate mRNA and contributes to phosphate homeostasis and plant fitness. Plant Cell 25:4166–4182
Jannot G, Bajan S, Giguère NJ et al (2011) The ribosomal protein RACK1 is required for microRNA function in both *C. elegans* and humans. EMBO Rep 12:581–586
Kanno T, Huettel B, Mette MF et al (2005) Atypical RNA polymerase subunits required for RNA-directed DNA methylation. Nat Genet 37:761–765
Karginov FV, Cheloufi S, Chong MMW et al (2010) Diverse endonucleolytic cleavage sites in the mammalian transcriptome depend upon microRNAs, drosha, and additional nucleases. Mol Cell 38:781–788
Kim T-K, Hemberg M, Gray JM et al (2010) Widespread transcription at neuronal activity-regulated enhancers. Nature 465:182–187
Kretz M, Siprashvili Z, Chu C et al (2013) Control of somatic tissue differentiation by the long non-coding RNA TINCR. Nature 493:231–235
Kurihara Y, Watanabe Y (2004) Arabidopsis micro-RNA biogenesis through Dicer-like 1 protein functions. Proc Natl Acad Sci USA 101:12753–12758
Kurihara Y, Takashi Y, Watanabe Y (2006) The interaction between DCL1 and HYL1 is important for efficient and precise processing of pri-miRNA in plant microRNA biogenesis. RNA 12:206–212

Kurihara Y, Matsui A, Hanada K et al (2009) Genome-wide suppression of aberrant mRNA-like noncoding RNAs by NMD in Arabidopsis. Proc Natl Acad Sci USA 106:2453–2458

Landthaler M, Yalcin A, Tuschl T (2004) The human DiGeorge syndrome critical region gene 8 and its *D. melanogaster* homolog are required for miRNA biogenesis. Curr Biol 14:2162–2167

Lanet E, Delannoy E, Sormani R et al (2009) Biochemical evidence for translational repression by Arabidopsis microRNAs. Plant Cell 21:1762–1768

Latos PA, Pauler FM, Koerner MV et al (2012) Airn transcriptional overlap, but not its lncRNA products, induces imprinted Igf2r silencing. Science 338:1469–1472

Lauressergues D, Couzigou J-M, Clemente HS et al (2015) Primary transcripts of microRNAs encode regulatory peptides. Nature 520:90–93

Lee YS, Nakahara K, Pham JW et al (2004) Distinct roles for Drosophila Dicer-1 and Dicer-2 in the siRNA/miRNA silencing pathways. Cell 117:69–81

Li J-F, Chung HS, Niu Y et al (2013a) Comprehensive protein-based artificial microRNA screens for effective gene silencing in plants. Plant Cell 25:1507–1522

Li S, Liu L, Zhuang X et al (2013b) MicroRNAs inhibit the translation of target mRNAs on the endoplasmic reticulum in Arabidopsis. Cell 153:562–574

Li W, Notani D, Ma Q et al (2013c) Functional importance of eRNAs for estrogen-dependent transcriptional activation events. Nature 498:516–520

Liu T-Y, Huang T-K, Tseng C-Y et al (2012) PHO2-dependent degradation of PHO1 modulates phosphate homeostasis in Arabidopsis. Plant Cell 24:2168–2183

Liu J, Wang H, Chua N-HH (2015) Long noncoding RNA transcriptome of plants. Plant Biotechnol J 13:319–328

Lobbes D, Rallapalli G, Schmidt DD et al (2006) SERRATE: a new player on the plant microRNA scene. EMBO Rep 7:1052–1058

Lucas SJ, Baştaş K, Budak H (2014) Exploring the interaction between small RNAs and R genes during Brachypodium response to *Fusarium culmorum* infection. Gene 536:254–264

Machida S, Yuan YA (2013) Crystal structure of *Arabidopsis thaliana* Dawdle forkhead-associated domain reveals a conserved phospho-threonine recognition cleft for Dicer-like 1 binding. Mol Plant 6:1290–1300

Mallory A, Vaucheret H (2010) Form, function, and regulation of ARGONAUTE proteins. Plant Cell 22:3879–3889

Manavella PA, Hagmann J, Ott F et al (2012) Fast-forward genetics identifies plant CPL phosphatases as regulators of miRNA processing factor HYL1. Cell 151:859–870

Mariner PD, Walters RD, Espinoza CA et al (2008) Human Alu RNA is a modular transacting repressor of mRNA transcription during heat shock. Mol Cell 29:499–509

Martianov I, Ramadass A, Serra Barros A et al (2007) Repression of the human dihydrofolate reductase gene by a non-coding interfering transcript. Nature 445:666–670

Mattick JS, Rinn JL (2015) Discovery and annotation of long noncoding RNAs. Nat Struct Mol Biol 22:5–7

Meijer HA, Kong YW, Lu W et al (2013) Translational repression and eIF4A2 activity are critical for microRNA-mediated gene regulation. Science 340:82–85

Melo CA, Drost J, Wijchers PJ et al (2013) eRNAs are required for p53-dependent enhancer activity and gene transcription. Mol Cell 49:524–535

Michaels SD, Amasino RM (1999) Flowering locus C encodes a novel MADS domain protein that acts as a repressor of flowering. Plant Cell 11:949–956

Onodera Y, Haag JR, Ream T et al (2005) Plant nuclear RNA polymerase IV mediates siRNA and DNA methylation-dependent heterochromatin formation. Cell 120:613–622

Ørom UA, Derrien T, Beringer M et al (2010) Long non-coding RNAs with enhancer-like function in human. Cell 143:46–58

Pélissier T, Clavel M, Chaparro C et al (2011) Double-stranded RNA binding proteins DRB2 and DRB4 have an antagonistic impact on polymerase IV-dependent siRNA levels in Arabidopsis. RNA 17:1502–1510

Penny GD, Kay GF, Sheardown SA et al (1996) Requirement for Xist in X chromosome inactivation. Nature 379:131–137

Quinn JJ, Chang HY (2016) Unique features of long non-coding RNA biogenesis and function. Nat Rev Genet 17:47–62

Rajagopalan R, Vaucheret H, Trejo J et al (2006) A diverse and evolutionarily fluid set of microRNAs in *Arabidopsis thaliana*. Genes Dev 20:3407–3425

Reis RS, Eamens AL, Waterhouse PM (2015a) Missing pieces in the puzzle of plant microRNAs. Trends Plant Sci 20:721–728

Reis RS, Hart-Smith G, Eamens AL et al (2015b) Gene regulation by translational inhibition is determined by Dicer partnering proteins. Nat Plants 1:14027

Reis RS, Hart-Smith G, Eamens AL et al (2015c) MicroRNA regulatory mechanisms play different roles in Arabidopsis. J Proteome Res 14:4743–4751

Reis RS, Eamens AL, Roberts TH et al (2016) Chimeric DCL1-partnering proteins provide insights into the microRNA pathway. Front Plant Sci 6:1–11

Reynoso MA, Blanco FA, Bailey-Serres J et al (2013) Selective recruitment of mRNAs and miRNAs to polyribosomes in response to rhizobia infection in *Medicago truncatula*. Plant J 73:289–301

Rinn JL, Kertesz M, Wang JK et al (2007) Functional demarcation of active and silent chromatin domains in human HOX loci by non-coding RNAs. Cell 129:1311–1323

Rogers K, Chen X (2013) Biogenesis, turnover, and mode of action of plant microRNAs. Plant Cell 25:2383–2399

Rohrig H, Schmidt J, Miklashevichs E et al (2002) Soybean ENOD40 encodes two peptides that bind to sucrose synthase. Proc Natl Acad Sci USA 99:1915–1520

Rowley MJ, Avrutsky MI, Sifuentes CJ et al (2011) Independent chromatin binding of ARGONAUTE4 and SPT5L/KTF1 mediates transcriptional gene silencing. PLoS Genet 7: e1002120

Ruiz-Orera J, Messeguer X, Subirana JA et al (2014) Long non-coding RNAs as a source of new peptides. elife 3:e03523

Shin HH-SS, Shin HH-SS, Chen R et al (2006) Loss of At4 function impacts phosphate distribution between the roots and the shoots during phosphate starvation. Plant J 45:712–726

Sinturel F, Navickas A, Wery M et al (2015) Cytoplasmic control of sense-antisense mRNA pairs. Cell Rep 12:1853–1864

Smith SE, Smith FA (2011) Roles of arbuscular mycorrhizas in plant nutrition and growth: new paradigms from cellular to ecosystem scales. Annu Rev Plant Biol 62:227–250

Song J-J, Smith SK, Hannon GJ et al (2004) Crystal structure of Argonaute and its implications for RISC slicer activity. Science 305:1434–1437

Speth C, Willing E-M, Rausch S et al (2013) RACK1 scaffold proteins influence miRNA abundance in Arabidopsis. Plant J 76:433–445

Sun Q, Csorba T, Skourti-Stathaki K et al (2013) R-loop stabilization represses antisense transcription at the Arabidopsis FLC locus. Science 340:619–621

Swiezewski S, Liu F, Magusin A et al (2009) Cold-induced silencing by long antisense transcripts of an Arabidopsis polycomb target. Nature 462:799–802

Tobias P, Guest DI (2014) Tree immunity: growing old without antibodies. Trends Plant Sci 19:367–370

Ulveling D, Francastel C, Hubé F (2011) When one is better than two: RNA with dual functions. Biochimie 93:633–644

Vazquez F, Gasciolli V, Crété P et al (2004) The nuclear dsRNA binding protein HYL1 is required for microRNA accumulation and plant development, but not posttranscriptional transgene silencing. Curr Biol 14:346–351

Voinnet O (2009) Origin, biogenesis, and activity of plant microRNAs. Cell 136:669–687

Wang Y, Juranek S, Li H et al (2009) Nucleation, propagation and cleavage of target RNAs in Ago silencing complexes. Nature 461:754–761

Wang L, Song X, Gu L et al (2013) NOT2 proteins promote polymerase II–dependent transcription and interact with multiple microRNA biogenesis factors in Arabidopsis. Plant Cell 25:715–727
Wang P, Xue Y, Han Y et al (2014a) The STAT3-binding long noncoding RNA lnc-DC controls human dendritic cell differentiation. Science 344:310–313
Wang Y, Wang X, Deng W et al (2014b) Genomic features and regulatory roles of intermediate-sized non-coding RNAs in Arabidopsis. Mol Plant 7:514–527
Wierzbicki AT, Haag JR, Pikaard CS (2008) Noncoding transcription by RNA polymerase pol IVb/Pol V mediates transcriptional silencing of overlapping and adjacent genes. Cell 135:635–648
Wierzbicki AT, Ream TS, Haag JR et al (2009) RNA polymerase V transcription guides ARGONAUTE4 to chromatin. Nat Genet 41:630–634
Wu H-J, Wang Z-M, Wang M et al (2013a) Widespread long noncoding RNAs as endogenous target mimics for microRNAs in plants. Plant Physiol 161:1875–1884
Wu X, Shi Y, Li J et al (2013b) A role for the RNA-binding protein MOS2 in microRNA maturation in Arabidopsis. Cell Res 23:645–657
Yang SW, Chen H-Y, Yang J et al (2010) Structure of *Arabidopsis* HYPONASTIC LEAVES1 and its molecular implications for miRNA processing. Structure 18:594–605
Yang L, Wu G, Poethig RS (2012) Mutations in the GW-repeat protein SUO reveal a developmental function for microRNA-mediated translational repression in Arabidopsis. Proc Natl Acad Sci USA 109:315–320
Ye R, Wang W, Iki T et al (2012) Cytoplasmic assembly and selective nuclear import of Arabidopsis ARGONAUTE4/siRNA complexes. Mol Cell 46:859–870
Yoon JH, Abdelmohsen K, Srikantan S et al (2012) LincRNA-p21 suppresses target mRNA translation. Mol Cell 47:648–655
Yu B, Bi L, Zheng B et al (2008) The FHA domain proteins DAWDLE in Arabidopsis and SNIP1 in humans act in small RNA biogenesis. Proc Natl Acad Sci USA 105:10073–10078
Zhai J, Jeong D-H, De Paoli E et al (2011) MicroRNAs as master regulators of the plant NB-LRR defense gene family via the production of phased, trans-acting siRNAs. Genes Dev 25:2540–2553
Zhang Y, Wong CH, Birnbaum RY et al (2013) Chromatin connectivity maps reveal dynamic promoter-enhancer long-range associations. Nature 504:306–310
Zheng Q, Rowley MJ, Böhmdorfer G et al (2013) RNA polymerase V targets transcriptional silencing components to promoters of protein-coding genes. Plant J 73:179–189
Zhu Y, Rowley MJ, Böhmdorfer G et al (2013) A SWI/SNF chromatin-remodeling complex acts in noncoding RNA-mediated transcriptional silencing. Mol Cell 49:298–309
Zhu QH, Stephen S, Taylor J et al (2014) Long noncoding RNAs responsive to *Fusarium oxysporum* infection in *Arabidopsis thaliana*. New Phytol 201:574–584

Epigenetic Regulation by Noncoding RNAs in Plant Development

Yu-Chan Zhang and Yue-Qin Chen

Abstract Epigenetic regulatory networks are orchestrated by various molecules, including transcription factors, mRNAs, noncoding RNAs (ncRNAs), DNA modifications, histone modifications, alterations in higher order chromatin structure, and some other signals. Noncoding RNAs constitute a substantial portion of transcribed sequences and play important roles in a wide range of biological processes in mammals and plants. ncRNAs have been further divided into small ncRNAs (sncRNAs, <200 nt), such as microRNAs (miRNAs), and long ncRNAs (lncRNAs, >200 nt). In this chapter, we focus on the epigenetic regulatory mechanisms involving ncRNAs, specifically the role of miRNAs, lncRNAs, and their regulatory networks in plant development. We also discuss future challenges of using ncRNAs in agricultural applications, including transgenic plants in crop improvement.

Keywords ncRNA • Epigenetic • microRNA • lncRNA • Plant development

Contents

Y.-C. Zhang • Y.-Q. Chen (✉)
School of Life Science, Sun Yat-sen University, Guangzhou, 510275, China
e-mail: lsscyq@mail.sysu.edu.cn

N. Rajewsky et al. (eds.), *Plant Epigenetics*, RNA Technologies,
DOI 10.1007/978-3-319-55520-1_10

1 Introduction

Cytosine methylation and post-translational histone modifications are important aspects of epigenetic gene regulation, helping to establish or maintain gene "on" or "off" states at the transcriptional level (Pikaard and Mittelsten 2014). Additionally, epigenetic regulation also occurs post-transcriptionally through targeted mRNA degradation or translational inhibition. This kind of epigenetic regulation controls the temporal or spatial distribution of developmentally important mRNAs and involves noncoding RNAs (ncRNAs).

ncRNAs are functional RNA molecules that are not translated into proteins. They constitute a majority of the genomes of both animals and plants. Many of the ncRNAs are functional and are involved in regulating gene expression at the transcriptional and post-transcriptional level. In this chapter, we discuss the epigenetic regulatory mechanisms involving ncRNAs, specifically the role of miRNAs, lncRNAs, and their regulatory networks in plant development.

2 Diverse Noncoding RNAs

ncRNAs can be divided into small ncRNAs (sncRNAs, <200 nt) and long ncRNAs (lncRNAs, >200 nt). The three major classes of sncRNAs are microRNAs (miRNAs), short interfering RNAs (siRNAs), and piwi-interacting RNAs (piRNAs) which do not exist in plants. miRNA, siRNA, and lncRNA are shown to function in heterochromatin formation, DNA methylation targeting, histone modification, and gene silencing.

As the most well known of the regulatory sncRNA classes, miRNAs are widespread sncRNAs ranging from 20 to 24 nucleotides in length, and are proved to be crucial regulators in post-transcriptional gene regulation through translational repression and/or guiding degradation of their mRNA targets (Huntzinger and Izaurralde 2011). miRNAs are transcribed by RNA Polymerase II (Pol II) into long primary miRNAs (pri-miRNAs), and the pri-miRNAs are cleaved by DCL1 into a smaller stem-loop structure called precursor miRNAs, which are subsequently processed by DCL1 to produce mature miRNA (Bologna and Voinnet 2014; Borges and Martienssen 2015).

Endogenous siRNAs in plants are primarily processed by DCL2, DCL3, and DCL4. They could be categorized into secondary siRNAs and heterochromatic siRNAs (hetsiRNAs) (Borges and Martienssen 2015). 21 and 22-nucleotide secondary siRNAs include different subclasses such as trans-acting small interfering RNAs (tasiRNAs), phased small interfering RNAs (phasiRNAs), natural antisense siRNAs (natsiRNAs), and epigenetically activated small interfering RNAs (easiRNAs), and they are produced by DCL4 and DCL2, following Pol II transcription and double-stranded RNAs (dsRNA) synthesis by RNA-Dependent RNA Polymerase 6 (RDR6) (Borsani et al. 2005; Katiyar-Agarwal et al. 2007; Ron et al. 2010). 24-nucleotide hetsiRNAs are transcribed by Pol IV followed by dsRNA synthesis by RDR2 and processing by DCL3. hetsiRNAs mediate transcriptional silencing of transposons and pericentromeric repeats through RNA-directed DNA methylation (RdDM) (Slotkin and Martienssen 2007; Matzke and Mosher 2014).

lncRNAs are greater than 200 nt in length, arise from intergenic, intronic, or coding regions in the sense and antisense directions. lncRNAs are mainly transcribed by Pol II and are polyadenylated, spliced, and mostly localized in the nucleus (Wierzbicki 2012). On the basis of their genomic origins, lncRNAs can be broadly classified as natural antisense transcripts (NATs), long intronic noncoding RNAs (lincRNAs), and long intergenic noncoding RNAs (lincRNAs).

In the following sections, we will focus on the functions of miRNAs and lncRNAs in plant and their roles in epigenetic regulation.

3 miRNAs and Epigenetics in Plant

Most plant miRNAs regulate gene expression by incorporating into the ARGONAUTE (AGO) effector complex where they cleave the target mRNAs, which are highly complementary to corresponding miRNAs. Translational inhibition is another distinct manner of gene repression based on miRNA-target complementarity, which is considered to be a more widespread method of miRNA-mediated gene suppression (Rogers and Chen 2013). miRNAs play crucial roles in a variety of plant biological processes including the developmental regulation of organs, and biotic and abiotic stress responses (Sun 2012). In Arabidopsis approximately 50% of the known miRNA targets are transcription factors, many of which modulate meristem formation and identity. Other miRNAs target mRNAs are involved in developmentally important signaling pathways. Interestingly, mature miRNAs are detected in both the nucleus and cytoplasm, suggesting that miRNAs may have functions in both cellular compartments, including directing DNA cytosine methylation.

3.1 MiRNAs and Vegetable Organ Development

Leaf is the major organ for photosynthesis and plays dominant role in plant biomass and crop plant productivity. Leaf development involves leaf primordia establishment, boundary cells separating leaf primordia from shoot apical meristems (SAM) and differentiate along leaf polarity (Byrne 2005; Takeda et al. 2011; Li and Zhang 2016).

miR319 is a conserved miRNA that regulate leaf development. Overexpression of miR319 leads to dramatically changes in the size and shape of tomato leaves (Ori et al. 2007). In rice, high expression level of miR319 displays obvious wider leaf blade (Yang et al. 2013a). miR319 targets TCP transcription factors (Schommer et al. 2008; Palatnik et al. 2003) (Fig. 1). miR396 targets another transcription

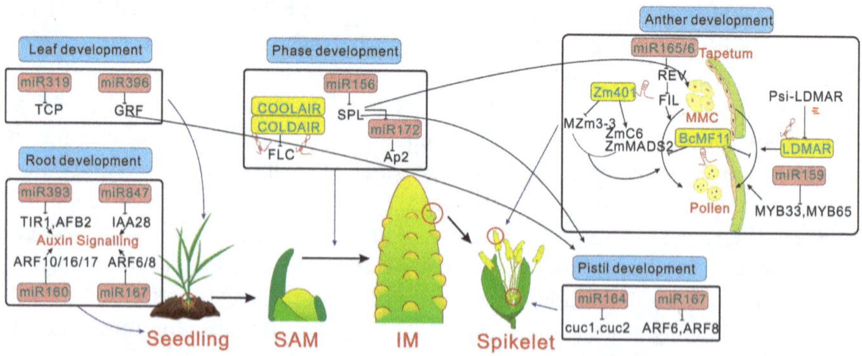

Fig. 1 miRNAs and lncRNAs in plant development. The *red* labels represent miRNAs, while the *yellow* labels indicate lncRNAs in plants. Target genes and several related genes are also shown. *SAM* shoot apical meristem, *IM* inflorescence meristem, *MMC* microspore mother cell

factors that affects leaf development. Plant Growth-Regulating Factors (GRFs) transcription factors have been reported to implicate in the regulation of leaf growth (Kim et al. 2003). miR396 targets and regulates GRFs at the transcriptional level in different species. In Arabidopsis, overexpression of miR396 down-regulates six GRFs and results in narrow leaf phenotypes, which may through attenuating cell division and proliferation during leaf growth (Liu et al. 2009; Rodriguez et al. 2010; Wang et al. 2011). This regulatory role of miR396 on leaf growth is conserved in rice (Liu et al. 2014a) (Fig. 1).

Plant root system is essential for nutrient and water uptake, hormone, and secondary metabolites production (Meng et al. 2010; Li and Zhang 2016). The root system is mainly derived from embryonic development and postembryonic development (Rogers and Benfey 2015). Embryo-derived root architecture comprises plant primary root or/and seminal root, whereas postembryonic development mostly gives rise to lateral, crown, and brace root (Rogers and Benfey 2015). Plant root system architecture is resulted from the coordination between exogenous environmental factors and endogenous signal pathways. Auxin is the dominant phytohormone in regulating root development (Gutierrez et al. 2012). Interestingly, several miRNAs play key roles in root development through regulating the auxin signaling pathway at the post-transcriptional level.

Transport Inhibitor Response Protein I (*TIR1*) is an auxin receptor in *Arabidopsis* (Dharmasiri et al. 2005a). TIR1 belongs to a small gene family that includes five other members Auxin Signaling F-BOX (*AFB1-5*) (Dharmasiri et al. 2005b). TIR1 and AFB2 both are involved in root development and regulated by miR393 (Navarro et al. 2006; Parry et al. 2009). In rice, overexpression of miR393 leads to obvious changes in root development involved altered auxin signaling, including primary root elongation and adventitious root number (Bian et al. 2012) (Fig. 1). IAA28 is an IAA/ARF transcriptional repressor involved in lateral root formation (Rogg et al. 2001). IAA28 contains partly complementary sequences of miR847 (Wang and Guo 2015). Overexpression of miR847 leads to developmental defect in lateral root formation (Wang and Guo 2015) (Fig. 1). Auxin activates signal transduction and promotes auxin mediated development mainly via Auxin Response Factor (*ARF*) family genes (Salehin et al. 2015). In the 23 members of *ARFs* in *Arabidopsis*, *ARF10*, *ARF16*, and *ARF17* are targeted by miR160 (Mallory et al. 2005; Yang et al. 2013b). Upregulation of miR160c decreases primary root and increases lateral root number and affects the gravitropism of roots, leading to curly primary root (Wang et al. 2005). The regulating role of miR160 on root development is also observed in rice (Meng et al. 2010). *ARF6* and *ARF8* are targeted by miR167. miR167 exerts positive roles in adventitious root formation (Gutierrez et al. 2009, 2012) (Fig. 1).

3.2 miRNAs and Floral Transition

After the adult plant organs formation, plants will undergo the transition from vegetative to reproductive phase, namely plant floral transition. miR156 and miR172 are well-known miRNAs involved in floral transition.

miR156 plays essential role in floral transition together with miR172. Both miRNAs always display some degree of opposite correlation. miR156 directly targets *Squamosa Promoter-binding protein-like* genes (*SPLs*, also known as *SBPs*) (Wu et al. 2009; Wang et al. 2009). miR156-targeted *SPL*s are necessary in floral transition in *Arabidopsis*. The expression level of miR156 gradually decreases from seeding stage to adult stage, and upregulation of miR156 results in delayed floral transition. In contrast, miR172 is highly expressed in plant juvenile phase and accumulating with the developmental time. Overexpressing miR172 promotes the flowering time in both monocotyledons and dicotyledons (Zhu and Helliwell 2011) (Fig. 1). miR172 targets AP2 domain transcription factors (Chen 2004). Interestingly, many SPLs which targeted by miR156 have been reported to transcriptional regulate miR172 through binding to its promoter region (Wu et al. 2009). Thus, miR156 may act upstream of miR172 in regulating juvenile-to-adult phase transition (Fig. 1). In addition, miR156 and miR172 appear to mediate the interplays between age pathway and other floral pathways.

3.3 miRNAs and Male Reproductive Development

Another development is essential for fertility and grain production in crops. Abnormal pollen development leads to male sterility, one of the most valuable genetic characteristics of hybrid production in agriculture. The formation of the tapetum provides protection and sustains early pollen development, while tapetum degeneration provides nutrition for pollen maturation (Twell 2011). Genome-wide studies have revealed that many miRNAs are preferentially transcribed in developing pollen in various crop species (Wei et al. 2011; Zhang et al. 2009; Yin and Shen 2010). Several studies have further characterized a few miRNAs as post-transcriptional regulators of male reproductive development, especially in tapetum and microspore development.

miR156 and miR159 are two conserved and well-studied miRNAs in plants, and they both regulate tapetum development (Fig. 1). miR156-targeted *SPL*s are necessary in maintaining anther fertility in *Arabidopsis*. Loss of function of miR156-targeted *SPLs* leads to male semi-sterile by disrupting primary tapetum cell and primary sporogenous cell formation. miR156 overexpression in the semi-sterile mutant background leads to full sterility, while the transgenic miR156-resistant form of *SPLs* partially attenuates the semi-sterile phenotype (Xing et al. 2010). miR156 is highly expressed in the pollen of rice (Wei et al. 2011), maize (Zhang et al. 2009), and wheat (Yin and Shen 2010). In rice, six inflorescence tissues highly expressed *SPLs* (*OsSPL2/12/13/14/16/18*) are regulated by miR156 (Yang et al. 2008; Xie et al. 2006). Three wheat *SPLs* (*SPL2/3/11*) were also identified as the targets of tae-miR156 (Yin and Shen 2010). Thus, miR156 may also be a potential regulator in crop pollen maturation although the downstream targets may not be the same. miR159 negatively regulates members of the GAMYB-like family, a group of positive regulators of gibberellin (GA) signal transduction pathways that control the programmed cell death process in anthers. High expression level of miR159 suppress the expression of *MYB33/MYB65* in *Arabidopsis* and *TaGAMYB1* in wheat, which then affects tapetum degradation and leads to male sterility due to abnormal microspore development (Wang et al. 2012; Millar and Gubler 2005; Tsuji et al. 2006).

miRNAs also participate in microspore development. miR165 and miR166 both control microspore development by down-regulating their target gene *REV*, an *HD-ZIP III* family gene in *Arabidopsis* (Fig. 1). Down-regulation of *REV* is associated with higher abundant of *FILAMENTOUS FLOWER (FIL)*. *REV* also negatively regulates the establishment of anther polarity, while *FIL* regulates the development of microsporangia and microspore mother cells (Lian et al. 2013). The biogenesis of miR165/166 requires HYPONASTIC LEAVES1 (HYL1), a crucial regulator in plant miRNAs biogenesis. *HYL1* deficiency induced down-regulated miR165/166, resulting in two microsporangia (vs. four in wild type), decreasing male fertility in *Arabidopsis* (Lian et al. 2013). The balance between *REV* and *FIL* expression regulated by HY1-dependent miR165/166 is important in maintaining the architecture of inner microsporangia and anther connectives (a small lump tissue in abaxial region) in stamen development. In rice, miR166 is highly expressed in developing rice pollen (Wei et al. 2011). Several other miRNAs (OsmiR528, OsmiR5793, OsmiR1432, OsmiR159, OsmiR812d, OsmiR2118c, OsmiR172d, miR5498) are differentially expressed in the rice anthers of a cytoplasmic male sterility (CMS) line and a maintainer line, but their roles in regulating male fertility need to be further studied (Ru et al. 2006). In maize, eight miRNA (Zma-miR601, Zma-miR602, Zma-miR603, Zma-miR604, Zma-miR605, Zma-miR606, Zma-miR607, Zma-miR397) families might be related to pollen development. 18 potential targets of these miRNAs have been predicted and are considered to be involved in a number of biological processes during pollen development (Shen et al. 2011). miR167 controls pollen development by directly targeting *ARF6* and *ARF8*, which regulates jasmonate biosynthesis by repressing downstream genes (Tabata et al. 2010) (Fig. 1). The loss of function of miR167 in *Arabidopsis* results in ARF6 and ARF8 ectopic expression in connective cells, leading to extremely large connective cells that fail to break open and release pollen (Wu et al. 2006; Ru et al. 2006). Four *ARFs (ARF6/12/17/25)* are targeted by miR167 in rice, and overexpression of miR167 in rice leads to shorter plants with fewer tillers (Liu et al. 2012a). These results demonstrate the potential of miRNAs in regulating microspore development in plants.

3.4 miRNAs and Female Reproductive Development

Female reproductive organ is composed by pistil, ovule, and stigma. Female sterility is another important subject of crop genetic engineering and is determined by female gametocyte development and morphological maturity for pollination.

Although very few ncRNAs are known to be specifically involved in female gametophyte development in crop plants, most of the miRNAs associated with male reproduction participate in female reproductive development. For example, the miR156-targeted *SPL* genes, which negatively regulate anther fertility, have an opposite regulatory response in gynoecium patterning through interference with auxin homeostasis and signaling. Down-regulation of miR156-targeted *SPLs* clearly leads to a shorter gynoecium in *Arabidopsis*. Transgenic knockdown of *SPL8*, a target of miR156, dramatically changes gynoecium shape, with a swollen upper part, an increasingly narrower basal part and severely shorter style (Xing et al. 2013) (Fig. 1). Pollen tube penetration into ovaries is also impaired by the upregulation of miR156 and down-regulation of *SPL8*. SPL8 cooperates with ETT, an auxin-responsive transcription factor (also known as ARF3), possibly facilitating auxin production or accumulation, and determining apical-basal pattering during ovary formation (Xing et al. 2013).

miR167 regulates both female and male reproduction in *Arabidopsis* by controlling *ARF6* and *ARF8* expression (Ru et al. 2006) (Fig. 1). Overexpressing *Arabidopsis* miR167 in tomatoes results in the down-regulation of tomato *ARF6* and *ARF8*, resulting in a complete absence of trichomes on the styles, failure of pollen tube formation and sterility (Liu et al. 2014b). These results demonstrate that miRNAs have essential functions in both male and female development. However, the underlying mechanisms of how a single miRNA simultaneously functions in both male and female development needs to be further investigated.

miR396 functions in female organ development (Hibara et al. 2003). In rice, miR396 controls carpel number by down-regulating its targets *GRF6/10*. The GRF6/10 complex also forms a transcription co-activator complex with GIF1. Transgenic lines overexpressing OsmiR396 have phenotypes similar to *GRF6* knockdowns, with open husks and sterile lemma (Liu et al. 2014a) (Fig. 1).

GRF6 and GRF10 also bind the GA-response element in the promoter of *OsJMJ706* (*JMJD2* family *jmjC* gene 706) and *OsCR4* (*Crinkly RECEPTOR-LIKE KINASE4*). The GRF6/10-GIF1 complex also transactivates *OsJMJ706* and *OsCR4,* which are required for floral organ development (Yuan et al. 2014). *CUP-SHAPED COTYLEDON (CUC)* genes *CUC1* and *CUC2* are targeted by miR164 and control the boundary size of meristems in *Arabidopsis* (Laufs et al. 2004). *CUC1* also activates shoot apical meristem formation, and *CUC2* controls the initiation of leaf margin development (Hibara et al. 2003; Nikovics et al. 2006). *CUC1* and *CUC2* are also required for the initiation of carpel margin meristem and are essential for septum and ovule formation. Disruption of miR164-mediated regulation of *CUC1* and *CUC2* results in extra meristematic activity at the carpel margin as well as the altered position of carpel margin meristem compared with wild type (Kamiuchi et al. 2014) (Fig. 1). In tomato, miR164 targets two *NAM* genes, *GOBLET (GOB)* and *SlNAM2* (Hendelman et al. 2013; Berger et al. 2009). Both genes are involved in boundary formation in different organs. In flowers, *SlNAM2* down-regulation leads to abnormal fusion of the sepal and whorl. Further studies of female fertility in crop plants will further the understanding of plant reproduction and will be of use in agricultural applications. The utilization of a female sterile line in hybrid breeding decreases self-fertilization of the paternal line, resulting in increased breeding efficiency.

4 lncRNAs and Epigenetics in Plant

lncRNAs are a less well-characterized group of ncRNAs. Unlike miRNAs, only a small set of lncRNAs are known to function in different developmental processes. Plant long noncoding RNAs (lncRNAs) play important roles in a wide range of biological processes, especially in plant reproductive development and response to stresses, although the detailed mechanisms remain largely unknown (Kim and Sung 2012). They are transcribed by RNA polymerase II (Pol II), Pol III and Pol V, and exert their functions by a variety of regulation pathways.

4.1 lncRNAs Discoveries in Different Plant Model Species

lncRNAs can be classified as natural antisense transcripts (NATs), long intronic noncoding RNAs and long intergenic noncoding RNAs (lincRNAs) according to their characteristics. Deep sequencing is essential for lncRNA identification and provides useful information for the characteristics of lncRNAs in different species.

In Arabidopsis, 6480 intergenic transcripts can be classified as lincRNAs by using a tiling array-based strategy, among which 2708 lincRNAs was detected by RNA sequencing experiments (Liu et al. 2012b). Interestingly, a subset of lincRNA genes shows organ-specific expression, whereas others are responsive to biotic and/or abiotic stresses. In rice, 2224 lncRNAs were identified, including 1624 lincRNAs and 600 NATs. Most of rice lncRNAs have a reproductive process preferred expressing pattern. Further functional analyze showed a set of lincRNAs could induce reproductive deficiencies (Zhang et al. 2014). These studies would provide new insight into the involvement of lncRNAs in the reproductive development of rice. Wu et al. identified a number of lncRNAs as Endogenous Target Mimics (eTM) for microRNAs (miRNAs) in both Arabidopsis and rice, in which the eTMs of several miRNAs, such as miR160, miR166, miR156, miR159, and miR172, can effectively inhibit the functions of their corresponding miRNAs, and the eTMs of miR160 and miR166 play a role in regulation of plant development (Wu et al. 2013). Wang et al. have analyzed lincRNAs and NATs from maize and rice (Wang et al. 2015). Subsequently, Boerner et al. identify the potential lncRNAs using the maize full length cDNA sequences. The results showed the noncoding transcription appears to be widespread in the maize genome, and these ncRNAs were predicted to originate from both genic and intergenic loci. Computational predictions indicated that they may function to regulate expression of other genes through multiple RNA-mediated mechanisms (Boerner and McGinnis 2012). In *Populus trichocarpa*, 2542 lincRNAs responding to drought stress have been identified by RNA-seq (Shuai et al. 2014). Recently, Xin et al. applied computational analysis and experimental approach identifying 125 putative wheat stress-responsive lncRNAs, which are not conserved among plant species (Xin et al. 2011). Among them, two lncRNAs were identified as signal recognition particle (SRP) 7S RNA variants, and three were characterized as U3 snoRNAs. Furthermore, the wheat lncRNAs also showed tissue-dependent expression patterns like the lncRNAs in Arabidopsis (Xin et al. 2011), suggesting that the highly tissue-specific expression pattern might be a general trait of lncRNAs in plant development. Zhang et al. have analyzed global patterns of allelic gene expression in developing maize endosperms from reciprocal crosses between inbreds B73 and Mo17 and found that 38 lncRNAs expressed in the endosperm are imprinted. Among them, 25 are maternally expressed transcripts, whereas 13 are paternally expressed transcripts, and transcribed in either sense or antisense orientation from intronic regions of normal protein-coding genes or from intergenic regions (Zhang et al. 2011). The growing reports of lncRNA identifications in different species indicate that lncRNAs ubiquitously exist in the plant kingdom with conserved roles.

4.2 The Regulation Pathways of lncRNAs Related to Plant Development

In contrast to small ncRNAs, much less is known about the functions and regulatory mechanisms of long ncRNAs, and only a few lncRNAs' functions are characterized. Several regulation pathways of plant lncRNAs have been identified. lncRNA COLDAIR functions through mediating the chromatin remodeling pathway.

COLDAIR participates in the repression of *FLC* after vernalization. It could bind PRC2 complex protein CURLY LEAF (CLF), with maximal interaction after 20 days of vernalization/cold exposure, and is required to recruit PRC2 to the *FLC* locus allowing deposition of the repressive H3K27me3 chromatin mark (Heo and Sung 2011). Although our knowledge of this kind of lncRNAs is still limited, there must be more lncRNAs acting in the regulation of chromatin remodeling way.

The second regulatory mechanism of lncRNAs in plants might work as a decoy of miRNAs (Wu et al. 2013), these kinds of lncRNAs exert their functions by binding miRNAs in a target mimicry mechanism to sequestrate the miRNAs' regulation roles on their target genes, such as lncRNAs IPS1 and at4. IPS1 is complementary to miR399, but contains a mismatch loop which makes it uncleavable when miR399 binding. It is induced by phosphate starvation in plants, and then allows the increased expression of miR399 targets including *PHO2*, consequently alters shoot phosphate content (Liu et al. 1997; Franco-Zorrilla et al. 2007; Martin et al. 2000). IPS1 has many family members in a number of plant species, including At4, At4-1, At4-2, and At4-3 in *Arabidopsis thaliana*, TPS11 in tomato, Mt4 in Barrel Clover and Alfalfa, Mt4-like in Soybean (Liu et al. 1997; Burleigh and Harrison 1999). At4 appears to be functionally redundant with IPS1, overexpression of one of them could result in the same phenotypic change as that when overexpressing both At4 and IPS1 (Shin et al. 2006). Moreover, the identification of a number of eTM lncRNAs by Wu et al. also supported universality of lncRNAs working as decoy of miRNAs (Wu et al. 2013).

Another function of lncRNAs in plant might serve as natural antisense transcripts (NATs) of mRNAs. An example is lncRNA COOLAIR (Swiezewski et al. 2009; Heo and Sung 2011). COOLAIR is a cold-induced antisense transcript of FLC gene and generally shows a positive correlation with FLC transcripts, except upon cold treatment. In early cold-induced vernalization, COOLAIR could silence *FLC* transcription (Swiezewski et al. 2009; Heo and Sung 2011). However, Helliwell et al. reported that abolishment of COOLAIR does not compromise vernalization-mediated *FLC* silencing; thus, the functions of these NATs needs further experimentally addressing (Helliwell et al. 2011).

The regulation pathways of most plant lncRNAs are still unclear, but the known models are all similar with that of mammal lncRNAs, suggesting that the regulation pathways of lncRNAs might be conserved in different species.

4.3 lncRNAs in Reproductive Development

Although more than 9000 plant lncRNAs have been annotated, less than 1% of lncRNAs identified have been characterized (Guttman et al. 2009; Rinn and Chang 2012; Banfai et al. 2012). Intriguingly, several well-studied lncRNAs are reported to be involved in plant reproductive process.

Some lncRNAs function in regulating pollen development. A long-day-specific male-fertility-associated long intergenic noncoding RNAs (lincRNA) named LDMAR was identified in rice (Ding et al. 2012). Sufficient expression of LDMAR transcripts is required under long-day conditions for normal pollen development. Mutational silencing of LDMAR results in premature programmed cell death in developing anthers, causing photoperiod-sensitive male sterility (PSMS) (Ding et al. 2012) (Fig. 1). In addition to the well-analyzed lncRNA, some other lncRNAs were also reported to be associated with the reproductive regulation although their regulatory mechanisms are still unclear. An example is *BcMF11*, an 828-nt lncRNA in Chinese cabbage (*Brassica campestris*) that has a crucial function in pollen development and male fertility. *BcMF11* is transcribed during most stages in pollen development. Decreasing the numbers of *BcMF11* transcripts with antisense cDNA delays tapetum degradation and, simultaneously, causes a synchronous separation of microspores and aborts the development of pollen grains. The regulation of *BcMF11* in male reproductive development is specific because the down-regulation of *BcMF11* in transgenic plants results in no significant differences compared with wild type during vegetative stages and ovule development (Song et al. 2007, 2013). Maize *Zm401* is another pollen-specific lncRNA primarily expressed in the tapetum and microspores in maize. *Zm401* expression levels significantly affect the expression of *MZm3-3, ZmC5, and ZmMADS2,* which are crucial for anther development. *Zm401* silencing upregulates *MZm3-3* and down-regulates *ZmMADS2* and *ZmC5*, inducing abnormal tapetum and microspores, which lead to male sterility (Ma et al. 2008) (Fig. 1). These data strongly indicate that specific lncRNAs are related to tapetum degeneration and the formation of microspores.

There are also several lncRNAs regulate floral transition. In Arabidopsis, COLDAIR was a cold-induced lncRNA mentioned above, which was identified from the first intron of *FLC* (Fig. 1). It is associated with PRC2 both in vitro and in vivo, and recruit PRC2 to the *FLC* locus to repress the *FLC* expression, allowing the unimpeded operation of the photoperiod pathway and starting the rapid flowering (Heo and Sung 2011). Another cold-induced lncRNA COOLAIR might also be involved in vernalization-mediated *FLC* repression although it might not be essential in this process (Helliwell et al. 2011; Heo and Sung 2011) (Fig. 1).

4.4 Stress-Responsive lncRNAs in Plants

Many lncRNAs also participate in the response to diverse stresses, including biotic and abiotic stresses. In Arabidopsis, 1832 lincRNAs have been found to significantly alter after 2 h and/or 10 h of drought, cold, high-salt, and/or abscisic acid (ABA) treatments, whereas treatment by elf18 (EF-Tu), which triggers pathogen-associated molecular pattern responses, could also increase the expression level of one of the representative stress-responsive lincRNAs (Liu et al. 2012b). 125 putative stress (powdery mildew infection and heat stress)-responsive lncRNAs were identified in wheat (Xin et al. 2011), suggesting that diverse sets of plant lncRNA were responsive to pathogen infection and stress induction, and could function in plant responses to both biotic and abiotic stresses, which would provide a starting point to understand their functions and regulatory mechanisms in the future.

In addition to the large-scale identification of stress-responsive lncRNAs, several stress-responsive lncRNAs have been further studied. IPS1 and At4 are induced by phosphate starvation, for they could block the repressing role of miR399 on its target gene *PHO2*, which then regulate the dynamic balance of shoot Pi content (Liu et al. 1997; Franco-Zorrilla et al. 2007; Martin et al. 2000; Shin et al. 2006). COLDAIR and COOLAIR are positively responding to cold weather and function in inducing flowering of Arabidopsis under appropriate season (Swiezewski et al. 2009; Heo and Sung 2011). Npc536 is another NATs lncRNA which upregulated in roots and leaves of Arabidopsis submitted to stress by phosphate starvation and salt stress, and overexpressing npc536 could promote the root growth under salt stress conditions, with increased primary root growth and secondary root length (Ben Amor et al. 2009). Thus, lncRNAs might be the important regulators in the biotic and abiotic stress response in plants.

5 Conclusion and Prospects

ncRNAs have crucial roles in regulating various developmental processes in both plants and animals. However, knowledge of the biological functions of these molecules in plant development is limited, especially for lncRNAs. Therefore, more efforts should be made to systematically analyze the regulatory roles and pathways of ncRNAs that are involved in plant development. The research approaches for lncRNAs is limited until now which has limited the functional analysis of lncRNAs. In the near future, it would be necessary to further investigate the functional motifs and secondary or tertiary structure of lncRNAs, to fully elucidate the diverse gene regulatory mechanisms of lncRNAs, and to develop the new and effective method to investigate target genes of lncRNAs.

Recently, more ncRNAs have been proven to play important roles in reproductive processes through the regulation of related genes in various species. Sexual reproduction is one of the most essential biological processes and occurs in a vast number of species. In crops, reproductive development is crucial for agronomic traits regulating and crop breeding. It is a challenge to use ncRNAs in agricultural applications, including transgenic plants in hybrid breeding, for novel genetic trait selection, for rapid character screening, and for genetic modification for crop improvement.

References

Banfai B, Jia H, Khatun J et al (2012) Long noncoding RNAs are rarely translated in two human cell lines. Genome Res 22:1646–1657

Ben Amor B, Wirth S, Merchan F et al (2009) Novel long non-protein coding RNAs involved in Arabidopsis differentiation and stress responses. Genome Res 19:57–69

Berger Y, Harpaz-Saad S, Brand A et al (2009) The NAC-domain transcription factor GOBLET specifies leaflet boundaries in compound tomato leaves. Development 136:823–832

Bian H, Xie Y, Guo F et al (2012) Distinctive expression patterns and roles of the miRNA393/TIR1 homolog module in regulating flag leaf inclination and primary and crown root growth in rice (*Oryza sativa*). New Phytol 196:149–161

Boerner S, Mcginnis KM (2012) Computational identification and functional predictions of long noncoding RNA in Zea mays. PLoS One 7:e43047

Bologna NG, Voinnet O (2014) The diversity, biogenesis, and activities of endogenous silencing small RNAs in Arabidopsis. Annu Rev Plant Biol 65:473–503

Borges F, Martienssen RA (2015) The expanding world of small RNAs in plants. Nat Rev Mol Cell Biol 16:727–741

Borsani O, Zhu J, Verslues PE et al (2005) Endogenous siRNAs derived from a pair of natural cis-antisense transcripts regulate salt tolerance in Arabidopsis. Cell 123:1279–1291

Burleigh SH, Harrison MJ (1999) The down-regulation of Mt4-like genes by phosphate fertilization occurs systemically and involves phosphate translocation to the shoots. Plant Physiol 119:241–248

Byrne ME (2005) Networks in leaf development. Curr Opin Plant Biol 8:59–66

Chen X (2004) A microRNA as a translational repressor of APETALA2 in Arabidopsis flower development. Science 303:2022–2025

Dharmasiri N, Dharmasiri S, Estelle M (2005a) The F-box protein TIR1 is an auxin receptor. Nature 435:441–445

Dharmasiri N, Dharmasiri S, Weijers D et al (2005b) Plant development is regulated by a family of auxin receptor F box proteins. Dev Cell 9:109–119

Ding J, Lu Q, Ouyang Y et al (2012) A long noncoding RNA regulates photoperiod-sensitive male sterility, an essential component of hybrid rice. Proc Natl Acad Sci USA 109:2654–2659

Franco-Zorrilla JM, Valli A, Todesco M et al (2007) Target mimicry provides a new mechanism for regulation of microRNA activity. Nat Genet 39:1033–1037

Gutierrez L, Bussell JD, Pacurar DI et al (2009) Phenotypic plasticity of adventitious rooting in Arabidopsis is controlled by complex regulation of AUXIN RESPONSE FACTOR transcripts and microRNA abundance. Plant Cell 21:3119–3132

Gutierrez L, Mongelard G, Flokova K et al (2012) Auxin controls Arabidopsis adventitious root initiation by regulating jasmonic acid homeostasis. Plant Cell 24:2515–2527

Guttman M, Amit I, Garber M et al (2009) Chromatin signature reveals over a thousand highly conserved large non-coding RNAs in mammals. Nature 458:223–227

Helliwell CA, Robertson M, Finnegan EJ et al (2011) Vernalization-repression of Arabidopsis FLC requires promoter sequences but not antisense transcripts. PLoS One 6:e21513

Hendelman A, Stav R, Zemach H et al (2013) The tomato NAC transcription factor SlNAM2 is involved in flower-boundary morphogenesis. J Exp Bot 64:5497–5507

Heo JB, Sung S (2011) Vernalization-mediated epigenetic silencing by a long intronic noncoding RNA. Science 331:76–79

Hibara K, Takada S, Tasaka M (2003) CUC1 gene activates the expression of SAM-related genes to induce adventitious shoot formation. Plant J 36:687–696

Huntzinger E, Izaurralde E (2011) Gene silencing by microRNAs: contributions of translational repression and mRNA decay. Nat Rev Genet 12:99–110

Kamiuchi Y, Yamamoto K, Furutani M et al (2014) The CUC1 and CUC2 genes promote carpel margin meristem formation during Arabidopsis gynoecium development. Front Plant Sci 5:165

Katiyar-Agarwal S, Gao S, Vivian-Smith A et al (2007) A novel class of bacteria-induced small RNAs in Arabidopsis. Genes Dev 21:3123–3134

Kim ED, Sung S (2012) Long noncoding RNA: unveiling hidden layer of gene regulatory networks. Trends Plant Sci 17:16–21

Kim JH, Choi D, Kende H (2003) The AtGRF family of putative transcription factors is involved in leaf and cotyledon growth in Arabidopsis. Plant J 36:94–104

Laufs P, Peaucelle A, Morin H et al (2004) MicroRNA regulation of the CUC genes is required for boundary size control in Arabidopsis meristems. Development 131:4311–4322

Li C, Zhang B (2016) MicroRNAs in control of plant development. J Cell Physiol 231:303–313

Lian H, Li X, Liu Z et al (2013) HYL1 is required for establishment of stamen architecture with four microsporangia in Arabidopsis. J Exp Bot 64:3397–3410

Liu C, Muchhal US, Raghothama KG (1997) Differential expression of TPS11, a phosphate starvation-induced gene in tomato. Plant Mol Biol 33:867–874

Liu D, Song Y, Chen Z et al (2009) Ectopic expression of miR396 suppresses GRF target gene expression and alters leaf growth in Arabidopsis. Physiol Plant 136:223–236

Liu H, Jia SH, Shen DF et al (2012a) Four AUXIN RESPONSE FACTOR genes downregulated by microRNA167 are associated with growth and development in *Oryza sativa*. Funct Plant Biol 39:736–744

Liu J, Jung C, Xu J et al (2012b) Genome-wide analysis uncovers regulation of long intergenic noncoding RNAs in Arabidopsis. Plant Cell 24:4333–4345

Liu H, Guo S, Xu Y et al (2014a) OsmiR396d-regulated OsGRFs function in floral organogenesis in rice through binding to their targets OsJMJ706 and OsCR4. Plant Physiol 165:160–174

Liu N, Wu S, Van Houten J et al (2014b) Down-regulation of AUXIN RESPONSE FACTORS 6 and 8 by microRNA 167 leads to floral development defects and female sterility in tomato. J Exp Bot 65:2507–2520

Ma J, Yan B, Qu Y et al (2008) Zm401, a short-open reading-frame mRNA or noncoding RNA, is essential for tapetum and microspore development and can regulate the floret formation in maize. J Cell Biochem 105:136–146

Mallory AC, Bartel DP, Bartel B (2005) MicroRNA-directed regulation of Arabidopsis AUXIN RESPONSE FACTOR17 is essential for proper development and modulates expression of early auxin response genes. Plant Cell 17:1360–1375

Martin AC, Del Pozo JC, Iglesias J et al (2000) Influence of cytokinins on the expression of phosphate starvation responsive genes in Arabidopsis. Plant J 24:559–567

Matzke MA, Mosher RA (2014) RNA-directed DNA methylation: an epigenetic pathway of increasing complexity. Nat Rev Genet 15:394–408

Meng Y, Ma X, Chen D et al (2010) MicroRNA-mediated signaling involved in plant root development. Biochem Biophys Res Commun 393:345–349

Millar AA, Gubler F (2005) The Arabidopsis GAMYB-like genes, MYB33 and MYB65, are microRNA-regulated genes that redundantly facilitate anther development. Plant Cell 17:705–721

Navarro L, Dunoyer P, Jay F et al (2006) A plant miRNA contributes to antibacterial resistance by repressing auxin signaling. Science 312:436–439

Nikovics K, Blein T, Peaucelle A et al (2006) The balance between the MIR164A and CUC2 genes controls leaf margin serration in Arabidopsis. Plant Cell 18:2929–2945

Ori N, Cohen AR, Etzioni A et al (2007) Regulation of LANCEOLATE by miR319 is required for compound-leaf development in tomato. Nat Genet 39:787–791

Palatnik JF, Allen E, Wu X et al (2003) Control of leaf morphogenesis by microRNAs. Nature 425:257–263

Parry G, Calderon-Villalobos LI, Prigge M et al (2009) Complex regulation of the TIR1/AFB family of auxin receptors. Proc Natl Acad Sci USA 106:22540–22545

Pikaard CS, Mittelsten SO (2014) Epigenetic regulation in plants. Cold Spring Harb Perspect Biol 6:a019315

Rinn JL, Chang HY (2012) Genome regulation by long noncoding RNAs. Annu Rev Biochem 81:145–166

Rodriguez RE, Mecchia MA, Debernardi JM et al (2010) Control of cell proliferation in Arabidopsis thaliana by microRNA miR396. Development 137:103–112

Rogers ED, Benfey PN (2015) Regulation of plant root system architecture: implications for crop advancement. Curr Opin Biotechnol 32:93–98

Rogers K, Chen X (2013) Biogenesis, turnover, and mode of action of plant microRNAs. Plant Cell 25:2383–2399

Rogg LE, Lasswell J, Bartel B (2001) A gain-of-function mutation in IAA28 suppresses lateral root development. Plant Cell 13:465–480

Ron M, Alandete Saez M, Eshed Williams L et al (2010) Proper regulation of a sperm-specific cis-nat-siRNA is essential for double fertilization in Arabidopsis. Genes Dev 24:1010–1021

Ru P, Xu L, Ma H et al (2006) Plant fertility defects induced by the enhanced expression of microRNA167. Cell Res 16:457–465

Salehin M, Bagchi R, Estelle M (2015) SCFTIR1/AFB-based auxin perception: mechanism and role in plant growth and development. Plant Cell 27:9–19

Schommer C, Palatnik JF, Aggarwal P et al (2008) Control of jasmonate biosynthesis and senescence by miR319 targets. PLoS Biol 6:e230

Shen Y, Zhang Z, Lin H et al (2011) Cytoplasmic male sterility-regulated novel microRNAs from maize. Funct Integr Genomics 11:179–191

Shin H, Shin HS, Chen R et al (2006) Loss of At4 function impacts phosphate distribution between the roots and the shoots during phosphate starvation. Plant J 45:712–726

Shuai P, Liang D, Tang S et al (2014) Genome-wide identification and functional prediction of novel and drought-responsive lincRNAs in *Populus trichocarpa*. J Exp Bot 65:4975–4983

Slotkin RK, Martienssen R (2007) Transposable elements and the epigenetic regulation of the genome. Nat Rev Genet 8:272–285

Song JH, Cao JS, Yu XL et al (2007) BcMF11, a putative pollen-specific non-coding RNA from *Brassica campestris* ssp. chinensis. J Plant Physiol 164:1097–1100

Song JH, Cao JS, Wang CG (2013) BcMF11, a novel non-coding RNA gene from *Brassica campestris*, is required for pollen development and male fertility. Plant Cell Rep 32:21–30

Sun G (2012) MicroRNAs and their diverse functions in plants. Plant Mol Biol 80:17–36

Swiezewski S, Liu F, Magusin A et al (2009) Cold-induced silencing by long antisense transcripts of an Arabidopsis Polycomb target. Nature 462:799–802

Tabata R, Ikezaki M, Fujibe T et al (2010) Arabidopsis auxin response factor6 and 8 regulate jasmonic acid biosynthesis and floral organ development via repression of class 1 KNOX genes. Plant Cell Physiol 51:164–175

Takeda S, Hanano K, Kariya A et al (2011) CUP-SHAPED COTYLEDON1 transcription factor activates the expression of LSH4 and LSH3, two members of the ALOG gene family, in shoot organ boundary cells. Plant J 66:1066–1077

Tsuji H, Aya K, Ueguchi-Tanaka M et al (2006) GAMYB controls different sets of genes and is differentially regulated by microRNA in aleurone cells and anthers. Plant J 47:427–444

Twell D (2011) Male gametogenesis and germline specification in flowering plants. Sex Plant Reprod 24:149–160

Wang JJ, Guo HS (2015) Cleavage of INDOLE-3-ACETIC ACID INDUCIBLE28 mRNA by microRNA847 upregulates auxin signaling to modulate cell proliferation and lateral organ growth in Arabidopsis. Plant Cell 27:574–590

Wang JW, Wang LJ, Mao YB et al (2005) Control of root cap formation by MicroRNA-targeted auxin response factors in Arabidopsis. Plant Cell 17:2204–2216

Wang JW, Czech B, Weigel D (2009) miR156-regulated SPL transcription factors define an endogenous flowering pathway in Arabidopsis thaliana. Cell 138:738–749

Wang L, Gu X, Xu D et al (2011) miR396-targeted AtGRF transcription factors are required for coordination of cell division and differentiation during leaf development in Arabidopsis. J Exp Bot 62:761–773

Wang Y, Sun F, Cao H et al (2012) TamiR159 directed wheat TaGAMYB cleavage and its involvement in anther development and heat response. PLoS One 7:e48445

Wang H, Niu QW, Wu HW et al (2015) Analysis of non-coding transcriptome in rice and maize uncovers roles of conserved lncRNAs associated with agriculture traits. Plant J 84:404–416

Wei LQ, Yan LF, Wang T (2011) Deep sequencing on genome-wide scale reveals the unique composition and expression patterns of microRNAs in developing pollen of *Oryza sativa*. Gen Biol 12:R53

Wierzbicki AT (2012) The role of long non-coding RNA in transcriptional gene silencing. Curr Opin Plant Biol 15:517–522

Wu MF, Tian Q, Reed JW (2006) Arabidopsis microRNA167 controls patterns of ARF6 and ARF8 expression, and regulates both female and male reproduction. Development 133:4211–4218

Wu G, Park MY, Conway SR et al (2009) The sequential action of miR156 and miR172 regulates developmental timing in Arabidopsis. Cell 138:750–759

Wu HJ, Wang ZM, Wang M et al (2013) Widespread long noncoding RNAs as endogenous target mimics for microRNAs in plants. Plant Physiol 161:1875–1884

Xie K, Wu C, Xiong L (2006) Genomic organization, differential expression, and interaction of SQUAMOSA promoter-binding-like transcription factors and microRNA156 in rice. Plant Physiol 142:280–293

Xin M, Wang Y, Yao Y et al (2011) Identification and characterization of wheat long non-protein coding RNAs responsive to powdery mildew infection and heat stress by using microarray analysis and SBS sequencing. BMC Plant Biol 11:61

Xing S, Salinas M, Hohmann S et al (2010) miR156-targeted and nontargeted SBP-box transcription factors act in concert to secure male fertility in Arabidopsis. Plant Cell 22:3935–3950

Xing S, Salinas M, Garcia-Molina A et al (2013) SPL8 and miR156-targeted SPL genes redundantly regulate Arabidopsis gynoecium differential patterning. Plant J 75:566–577

Yang Z, Wang X, Gu S et al (2008) Comparative study of SBP-box gene family in Arabidopsis and rice. Gene 407:1–11

Yang C, Li D, Mao D et al (2013a) Overexpression of microRNA319 impacts leaf morphogenesis and leads to enhanced cold tolerance in rice (*Oryza sativa* L.) Plant Cell Environ 36:2207–2218

Yang J, Tian L, Sun MX et al (2013b) AUXIN RESPONSE FACTOR17 is essential for pollen wall pattern formation in Arabidopsis. Plant Physiol 162:720–731

Yin ZJ, Shen FF (2010) Identification and characterization of conserved microRNAs and their target genes in wheat (*Triticum aestivum*). Genet Mol Res 9:1186–1196

Yuan JH, Yang F, Wang F et al (2014) A long noncoding RNA activated by TGF-beta promotes the invasion-metastasis cascade in hepatocellular carcinoma. Cancer Cell 25:666–681

Zhang L, Chia JM, Kumari S et al (2009) A genome-wide characterization of microRNA genes in maize. PLoS Genet 5:e1000716

Zhang M, Zhao H, Xie S et al (2011) Extensive, clustered parental imprinting of protein-coding and noncoding RNAs in developing maize endosperm. Proc Natl Acad Sci USA 108: 20042–20047

Zhang YC, Liao JY, Li ZY et al (2014) Genome-wide screening and functional analysis identify a large number of long noncoding RNAs involved in the sexual reproduction of rice. Genome Biol 15:512

Zhu QH, Helliwell CA (2011) Regulation of flowering time and floral patterning by miR172. J Exp Bot 62:487–495

RNAi Suppressors: Biology and Mechanisms

Neeti Sanan-Mishra, Supriya Chakraborty, Dinesh Gupta, and Sunil Kumar Mukherjee

Abstract Plant small RNAs, namely si- and miRNAs, control a gamut of biological functions by regulating gene expressions. One of the major functions is to protect the host genome from molecular parasites, including the viruses. The virus-infected plants allow generating the siRNAs from all over the viral genomes that, in turn, control viral gene expressions post-transcriptionally leading to inhibition of viral growth and spread. In the case of DNA viruses, the siRNAs also exert transcriptional control of viral gene expression in an epigenetic manner by promoting methylation of the promoter of viral genes. Further, transcriptional gene silencing (TGS) mechanism has also been shown to be involved in symptom remission. DNA viruses also interfere with the methyl cycle to prevent the availability of methyl donor (S-adenosyl methionine) for methylating viral DNAs. However, in the battles between the host and viruses, the viruses have also evolved to encode few proteins from their genomes that counteract the RNAi-mediated host defense reactions. Such group of proteins is collectively known as RNAi suppressors which also participate in viral life cycle in manifold ways besides thwarting the host RNAi activities towards the viruses. In addition, these virus-encoded proteins also manipulate the components of TGS machinery such as histone and/or DNA methyl transferases, to combat the antiviral silencing mechanism. These are also called the pathogenicity factors as they principally govern the disease symptoms in the host. The mechanistic action of a few of the viral-encoded suppressors has been dealt in some detail within the text. These proteins deregulate the host miRNAs during the expression of disease. Several studies have now shown that transgenic

N. Sanan-Mishra
Plant RNAi Biology Group, International Center for Genetic Engineering and Biotechnology, New Delhi, India

S. Chakraborty
School of Life Sciences, Jawaharlal Nehru University, New Delhi, India

D. Gupta
Translational Bioinformatics Group, International Center for Genetic Engineering and Biotechnology, New Delhi, India

S.K. Mukherjee (✉)
North Eastern Region-Biotechnology Programme Management Cell, New Delhi, India
e-mail: sunilmukherjee@hotmail.com

N. Rajewsky et al. (eds.), *Plant Epigenetics*, RNA Technologies,
DOI 10.1007/978-3-319-55520-1_11

expression of viral suppressors can alter the accumulation and/or functioning of miRNAs leading to developmental abnormalities. Molecules like HC-Pro, P19, etc. were shown to affect the processing and activity of miRNAs. Hence the antiviral strategies could be developed by silencing these viral suppressors. Our laboratories have developed tomato transgenics expressing miRNAs and tasiRNAs which can efficiently silence the RNAi suppressors of tomato leaf curl viruses and offer a high degree of tolerance towards the viruses. The future direction of research including the biotechnological usages of the viral suppressors has been discussed.

Keywords RNAi • RNA silencing suppression • Suppressors • Artificial miRNA • Artificial tasiRNA

Contents

1 Introduction

RNA interference (RNAi) or RNA silencing is the natural process of switching off gene expression during fundamental processes like development, genome maintenance, and defense against foreign molecules like viruses. As a counter defense, viruses have also evolved to encode proteins to suppress RNA silencing mechanisms that are known as RNAi suppressors. With the rapid advancement in science, a lot of information has emerged regarding the mechanisms and machinery of RNA silencing and its suppression (Agrawal et al. 2003; Roth et al. 2004). These are being exploited as a new tool for developing antiviral products, which have large applications in field of medicine, agriculture, and basic biology. In this review, we have discussed the virus triggered RNAi response and the mechanisms evolved by viruses to suppress this pathway for their own advantage.

2 RNAi and the Suppressors

The RNAi science evolved with the serendipitous as well as the famous story of transgenic petunia flowers in 1990 (Napoli et al. 1990). Now the various forms and associated mechanisms of the effectors of gene silencing are well known and are still being represented in the literature. The three major forms of small RNAs, namely the small interfering RNAs (siRNAs), microRNAs (miRNAs), and picoRNAs (piRNAs), are well described in almost all eukaryotic creatures including many non-model organisms like parasites, fungi, etc. (Perrimon et al. 2010; Nicolas et al. 2013). Besides these three forms, many other forms of siRNAs like rasiRNA, tasiRNA, natsiRNAs, etc. are also reported in the literature. The biogenesis, functions, and cross-talks of these small RNAs requires the participation of many silencing factors, known collectively as the RNAi factors. The functions of these factors are well conserved across evolution; the characteristics motifs underlying the functions of many factors are well recognized. However, there are also many reported factors that exert their gene silencing effects in a system- and tissue-specific manner.

An evolutionarily conserved function of a subset of RNAi factors is to safeguard the host and its genome from invading molecular parasites like viruses and transposons. Following viral entry in the host, a pathogen-triggered immunity (PTI) will be invoked in the host. If the RNAi factors provide the PTI function, the pathogen effector-triggered sensitivity (ETS) will also come in play following the Z-model of PTI-ETI scheme of the host–pathogen arms race (Jones and Dang 2006). The viral-encoded suppressor of RNAi can straightforwardly fit the criterion of ETS. These are also known as RNA silencing suppressors (RSSs) or viral suppressors of RNA silencing (VSRs) and were initially brought into the limelight through a report by Voinnet et al. (1999). Till now more than 80 VSRs of plant, animal, and insect origins are documented; however, the mechanistic details of a few of VSRs are reported in atomistic details. These suppressors generally do not have common motifs but a subset of them has GW/WG repeats and RNA binding (RBS) motifs (Bivalkar-Mehla et al. 2011). It probably entails that the VSRs have multiple independent origins leading to high divergence in function and thus intercept at various steps of RNAi pathway.

3 Antiviral RNAi

3.1 Viral SiRNA Generation

All viruses with either RNA or DNA genomes present genomic or sub-genomic forms of intracellular double-stranded (ds) RNA which are eventual sources of viral siRNAs (vi-siRNAs). The vi-siRNAs are commonly produced from three distinct processes in which the dsRNA precursors are formed and further subjected to Dicer or Dicer-like (DCL) mediated cleavage to switch on the RNA silencing mechanisms. The entire process starts with the utilization of available sources of dsRNA to yield the primary vi-siRNA, structure-associated vi-siRNA, and secondary vi-siRNA. Primary vi-siRNAs are the derivative of an intermediate of genome replication formed either due to the activity of virus-encoded RNA polymerases (encoded by RNA viruses) or through transcription of the viral genome in the case of DNA viruses. Apart from this, these structures are also produced by convergent transcription. Another class of siRNAs are structure-associated vi-siRNA, which are in fact the defectively base-paired viral transcripts forming an imperfect secondary structure. The third class includes the secondary vi-siRNA, which are produced from the ssRNA by the active participation of host RNA-dependent RNA polymerase (RDR) family (Ahlquist 2006). Biogenesis of vi-siRNA includes transcription, processing, modification and finally these vi-siRNAs load on to the RNA Induced Silencing complex (RISC) to silence the viral transcripts. Figure 1 displays the biogenesis and functions of the vi-siRNAs along with interference from the VSRs.

These viral dsRNAs are processed by endonucleolytic activity of DCL4 and DCL2 to produce 21 nt and 22 nt vi-siRNA, respectively. DCL2 acts as a substitute of DCL4 and its antiviral activity is initiated only in Arabidopsis plants lacking DCL4. Further, the duplex of vi-siRNA is stabilized by methylation at 2′ OH of 3′ terminal nucleotides by Hua Enhancer 1 (HEN1; Yu et al. 2005) to protect sRNA molecules against uridylation (Li et al. 2005) and against the exoribonuclease activity of small RNA degrading nucleases (SDN1-3) (Ramachandran and Chen 2008). Further, amplification and systematic RNA silencing occur through the activities of secondary siRNAs. RDR proteins facilitate production of elongated complementary viral RNAs (transitive RNA) which are subsequently subjected to DCL processing (Vance and Vaucheret 2001). *Arabidopsis thaliana* RDR6 contributes largely to the process of amplification. In case of DNA viruses, the 24 nt vi-siRNAs (product of DCL3) are also generated. These 24 nt vi-siRNAs initiate

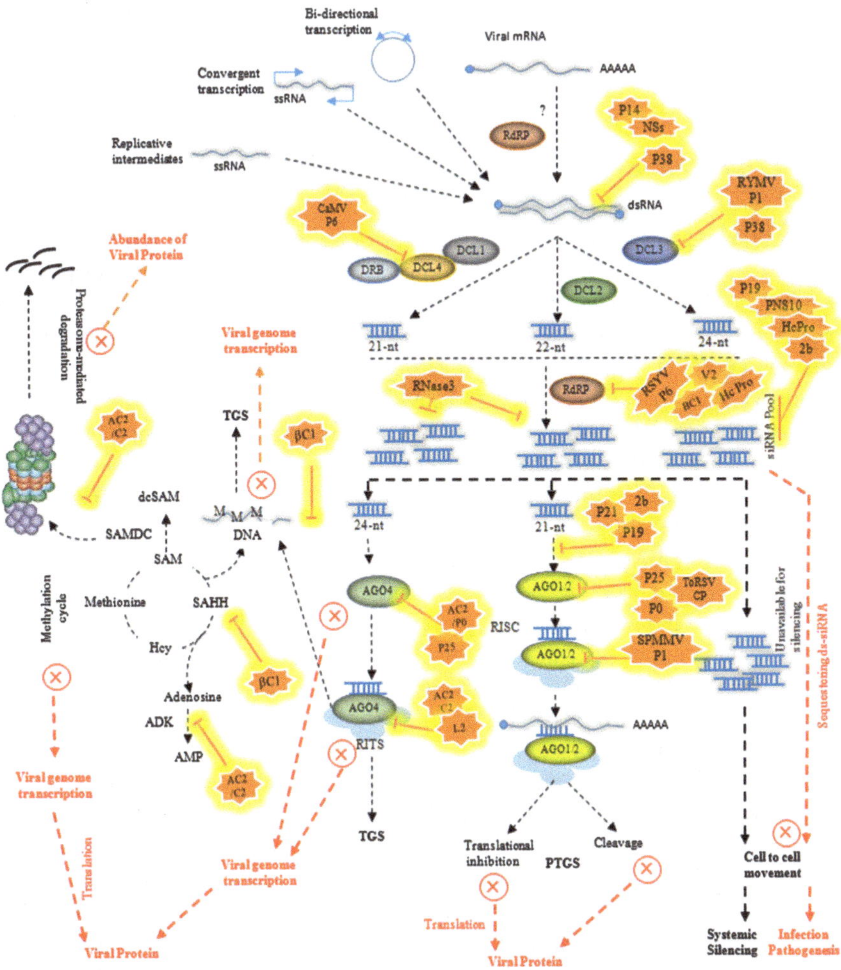

Fig. 1 A schematic representation of vi-siRNA biogenesis and function. Double-stranded RNAs generated from RNA (viral mRNAs mainly) and DNA viruses (convergent, bidirectional, and replicative intermediate transcriptions) are cleaved with the specialized proteins DCLs and produce a variety of siRNA species (21, 22, and 24 nt). VSRs target these proteins to hinder siRNAs generation and subsequently impede both TGS and PTGS pathways. VSRs are shown in *red stars*. *Black-dashed lines* indicate the normal steps of silencing pathway, while *red-dashed lines* designate the process manipulated by the suppressors. CaMV P6, *Cauliflower mosaic virus* P6 protein; RYSV P6, *Rice yellow stunt virus* P6 protein; RYMV P1, *Rice yellow mottle virus* P1 protein; SPMMV P1, *Sweet potato mild mottle ipomovirus* P1 protein; ToRSV CP, *Tomato ringspot virus* coat protein

transcriptional gene silencing by inducing cytosine methylation of target DNA sequence (Lister et al. 2008). This phenomenon is discussed in detail in Sect. 3.2.

The formation, accumulation, and functional stages of vi-siRNAs are also subject to the inhibitory activities of VSRs as described in Sect. 4. The accumulation and activities of vi-siRNAs should, thus be viewed as the host defense response towards the viruses.

3.2 Transcriptional Control of Viral Genes

Transcriptional gene silencing (TGS) involves the production of siRNA homologous to the non-coding region of a target virus, which is also linked with the corresponding methylation of the virus genome, an event that controls viral gene expression. This process consists of three key steps: initiation, effector, and amplification/spreading of silencing. The 24 nt vi-siRNAs corresponding to viral non-coding regions are the key players in this process. In the RNA-directed DNA methylation (RdDM), vi-siRNAs are recruited to the RNA-Induced Transcriptional Silencing (RITS) complex leading to chromatin remodelling (Huang et al. 2007). Argonaute 4 (AGO4) protein promotes chromatin modification through cytosine as well as histone methylation. The downstream processing requires the contribution of various enzymes responsible for either de novo methylation (Domains Rearranged Methyltransferase 2, DRM2) or maintaining the methylation (Chromomethylase 3, CMT3; Methyltransferase 1, MET1; and Kryptonite, KYP2). Apart from this, RdDM entails the action of some chromatin remodelers such as Defective in RNA-Directed DNA Methylation1 (DRD1) and Decrease in DNA Methylation1 (DDM1), which are necessary to ensure viral DNA availability to RNA signals and the maintenance of symmetric methylation, respectively (Raja et al. 2010).

Production of vi-siRNAs and the consequent DNA methylation in various plant–virus interactions have been recently highlighted. These reports advocate that upon infection by DNA virus, the host activates TGS to suppress the transcription of the viral genome. For example, generation of a wide range of vi-siRNAs have been reported in *Nicotiana–African cassava mosaic virus* (ACMV) interaction as well as during Arabidopsis–*Cabbage leaf curl virus* (CaLCuV) interaction (Akbergenov et al. 2006; Vanderschuren et al. 2007). These results put forward that TGS is the key pathway, which is implicated in the plant defense against geminivirus. Later on, the association of TGS pathways with biological functions such as symptom remission phenotypes was revealed. For example, pepper–*Pepper golden mosaic virus* (PepGMV) interaction showed a recovery phenotype provided by the presence of vi-siRNAs (Carrillo-Tripp et al. 2007).

Moreover, it has been substantiated that the expression of virus genes is frequently targeted for the DNA methylation through the RdDM pathway (Yadav and Chattopadhyay 2011; Sahu et al. 2014). Methylation in the promoter region which is essential for the viral transcription can inhibit the accumulation of viral transcripts, thus reducing the infectivity in the infected plant. The 24 nt vi-siRNAs also cause methylation of the intergenic region of *Mungbean yellow mosaic India virus* (MYMIV; Yadav and Chattopadhyay 2011) as well as *Tomato leaf curl New Delhi virus* (ToLCNDV; Sahu et al. 2014). Scanning of the viral genome producing a higher level of siRNAs revealed that there was a strong correlation between the accumulation of small RNAs and genome methylation processes. Taken together, these studies suggest that the TGS and viral genome methylation act as a key regulatory process to minimize or limit the viral gene expression. The detailed study of the viral DNA methylation suggested that the expression of genes encoding enzymes linked with the cytosine methylation occurs in specific patterns (Yadav and Chattopadhyay 2011; Sahu et al. 2014). Upon ToLCNDV infection in a tolerant cultivar of tomato, higher expression of *DRM1* had been observed leading to enhanced de novo methylation; moreover, higher expression in level of methylation maintenance genes *CMT3* was also reported. Hence, we may infer that the change in the level of key methylation maintenance enzymes might be linked with RdDM, which is plausibly involved in the progression of siRNA-directed silencing pathway in a tolerant response against geminiviruses.

3.3 Post-transcriptional Control of Viral Proteins

3.3.1 Post-transcriptional Gene Silencing

Post-transcriptional gene silencing (PTGS) is one of the most efficient defense strategies that plants have devised against viral pathogens (Baulcombe 1999; Waterhouse et al. 2001). It is regarded as a form of immune system that operates at the nucleic acid level and can act against any cytoplasmic RNA species homologous with the small RNA molecules (Voinnet 2001). This defense is not host-programmed but depends on the genome sequence of the invading DNA or RNA virus (Ruiz et al. 1998; Matthew 2004), hence it can remarkably silence the expression of potentially any virus. The RNA silencing signals can propagate to distant parts of the plant, thus conferring immunity to non-infected parts of the plant (Palauqui et al. 1997; Voinnet and Baulcombe 1997; Palauqui and Vaucheret 1998; Voinnet et al. 1998; Sonoda and Nishiguchi 2000).

RNA silencing involves diversity in its mode of action as well as its components. Besides the vi-siRNAs, the host miRNAs also exert post-transcriptional control of viral transcripts. The biogenesis and function of these forms of small RNAs requires a number of different proteins. However, the two main players in the pathways are DCL and AGO, which require attention due to their commonality to all the small RNA pathways.

DCL belongs to a family of RNase III-like endoribonucleases which act on dsRNAs and cleave them into smaller fragments in a sequence-independent manner (Bernstein et al. 2001). In general, it contains a helicase-C, DExD-helicase, PAZ, Duf283, RNaseIII, and dsRNA-binding domain. All the DCLs contain two RNaseIII domains, which act simultaneously to cleave the dsRNA (Finnegan et al. 2003; Margis et al. 2006). The number of DCLs varies in organisms from single in humans and mice (Zhang et al. 2004) to four in *A. thaliana* (Finnegan et al. 2003; Liu et al. 2005). Mutation analysis of the four *A. thaliana* DCL (AtDCL) genes showed that the species and the corresponding functions of a small RNA depend on the type of DCL enzyme involved in its biogenesis. For example, AtDCL1 has been shown to generate miRNAs, while AtDCL2 is implicated in the production of siRNAs linked with virus defense and also the production of siRNAs from natural *cis*-acting antisense transcripts. On the other hand, AtDCL3 takes part in the siRNA generation that guides chromatin modification. AtDCL4 is essentially required to produce tasiRNAs which regulates variations associated with vegetative phase (Hutvágner et al. 2001; Llave et al. 2002). In rice, 8 DCL coding genes have been identified so far; however, their distinct roles and their effect in rice development are still unclear (Kapoor et al. 2008). Mutation analysis of DCL1 in rice showed great reduction in number of miRNAs as well as developmental arrest at seedling stage (Liu et al. 2005); however, in the same mutations, the production of siRNAs was not affected.

AGO is an evolutionarily conserved protein and the main slicer element of the RISC in plants and animals (Wu et al. 2009b). Its number significantly differs in each organism and 10 different AGO proteins are known in Arabidopsis (Vaucheret 2008), whereas 19 AGO proteins have been identified in rice. AGO1 is the major effector protein of miRNA-induced silencing (Mallory and Vaucheret 2010; Wu et al. 2009a; Wang et al. 2009). When a correct pairing of 2–8 nt between the miRNA and an RNA strand is detected, the catalytic machinery of RISC-AGO complex proceeds to the silencing of the target by either cleavage of the target or translational repression. The vi-siRNAs silence the viral transcripts in a similar post-transcriptional process using the siRNA-RISC pathways. The relative weights of TGS and PTGS are virus—as well as host tissue-specific. DNA viruses are more prone to TGS while the RNA viruses are subject more to PTGS processes.

3.3.2 Host MiRNA Control of Viral Genes

Animal miRNAs are well known to control viral genes but plant miRNAs doing the same job are not reported yet. However, bioinformatic predictions about plant miRNAs have shown that they have a role in plant–virus interactions by targeting the genomes of plant infecting viruses (Naqvi et al. 2010) and they are also thought to regulate the tissue tropism of virus in the host to some extent (Ghosh et al. 2009). In order to highlight the probable geminivirus targets for miRNAs encoded by the six plant genomes, we have carried out bioinformatics analysis in detail and the same is presented in tabular (Table 1) as well as figure form (Fig. 2).

Table 1 Putative viral targets of plant miRNAs

No.	Plant Species (miRNA)	miRNA used	Targets found in begomoviruses (336)	Targets found in potyviruses (101)
1	*Arabidopsis thaliana*	427	2679	3997
2	*Glycine max*	639	5413	5962
3	*Oryza sativa*	713	6644	8234
4	*Sorghum bicolor*	241	1739	2132
5	*Vitis vinifera*	186	1651	2312
6	*Zea mays*	321	2731	3956

A total of 2527 miRNA sequences were downloaded from the miRBASE release 21 from six plants (*Arabidopsis thaliana*, *Glycine max*, *Oryza sativa*, *Sorghum bicolor*, *Vitis vinifera*, and *Zea mays*) (Griffiths-Jones et al. 2008). Complete genome sequences for two major families of plant infecting viruses, namely, geminiviruses and potyviruses, were obtained from Genbank (Benson et al. 2005) and these sequences were used to look for the targets of the above- mentioned miRNAs using a certain set of rules. A modified version of miRanda (ver. September 2008) was essentially used for target predictions (Enright et al. 2003). The miRanda scoring matrix allows G=U "wobble" pairs, important for the detection of RNA:RNA hybrid duplexes. The folding algorithm was based on the Vienna 1.3 RNA secondary structure programming protocols (Hofacker et al. 1994). Although miRanda was originally developed to look for animal miRNA targets, it can be modified and used to search for targets in other systems like viruses and plants (Hsu et al. 2007; Maziere and Enright 2007). The other criteria to consider a sequence as a putative miRNA target were: four or fewer mismatches overall, only one or none mismatches in the 5′ region of the miRNA (positions 1–12), no more than two consecutive mismatches in positions 13–21, and no mismatches in positions 10 and 11. Additionally, the miRNA:target pair should have low free-energy of bonding (maximum −20 kcal/mol) and parameter "strict" was also used to ensure no mismatches in seed region (Lin et al. 2009; Zhang et al. 2006; Schwab et al. 2005). The resultant hits in the viral targets have been summarized in the Table 1.

The number of targets found in each of the viral genomes has been displayed in Fig. 2a (for potyviruses) and Fig. 2b (for geminiviruses).

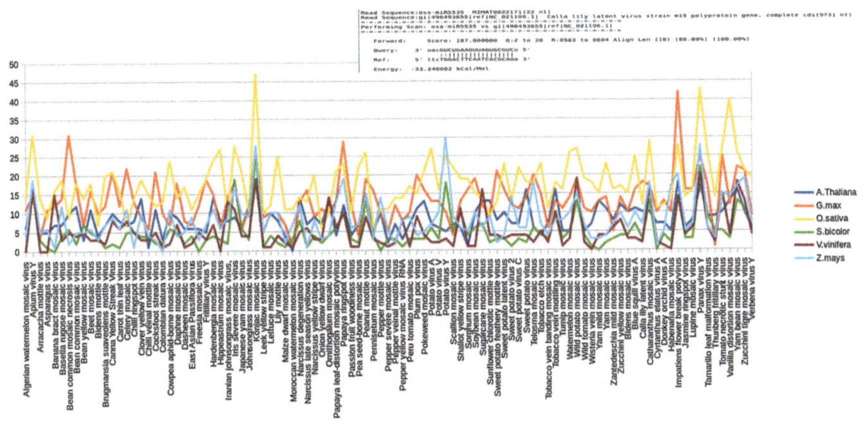

Fig. 2 (**a**) Number of potyvirus targets predicted for miRNAs encoded in the six plant genomes. The *inset box* shows one of the best representative miRNA:target alignment, between osa-miR5535 and Calla lily latent virus polyprotein gene (L594_gp1). (**b**) Number of geminivirus targets predicted for miRNAs encoded in the six plant genomes. The *inset box* shows one of the best representative miRNA:target alignments, osa-miR396f-5p:Sida golden

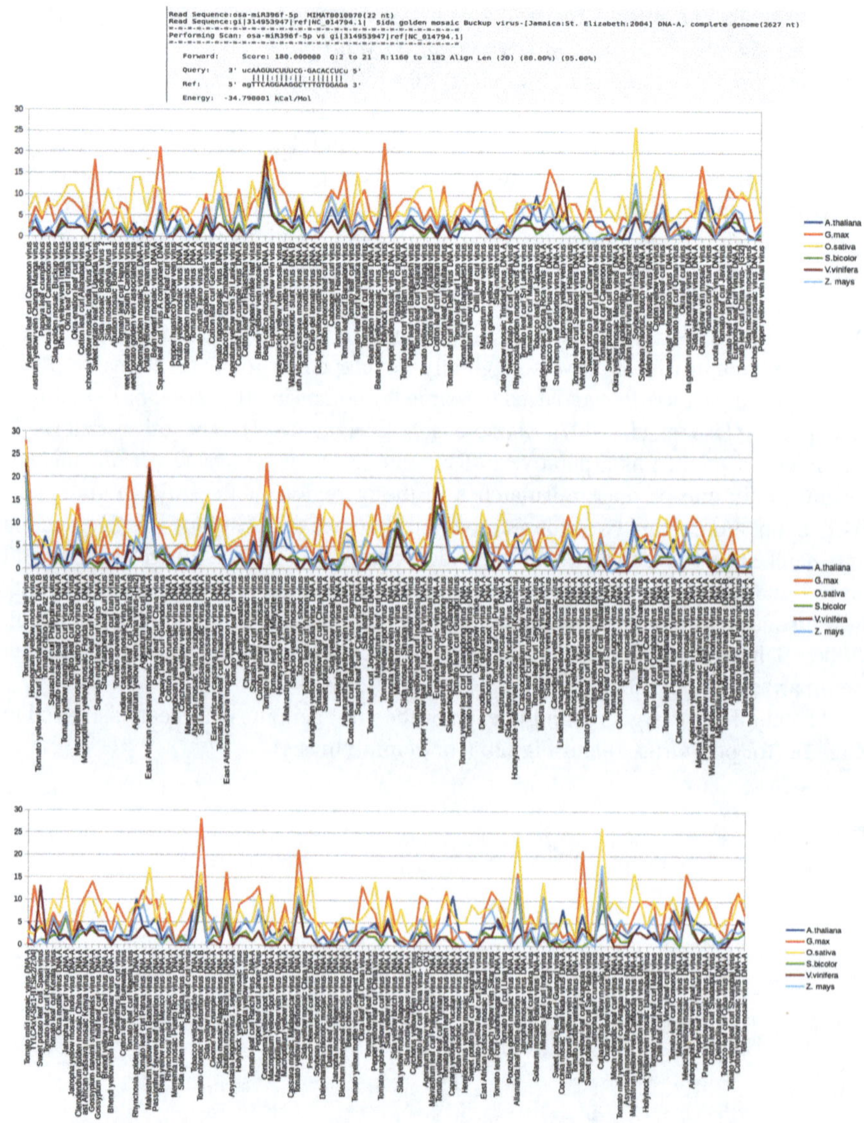

Fig. 2 (continued)

The viral targets include mostly the viral proteins and occasionally the intergenic regions. The targets are predicted to undergo slicing as well as translational repression.

4 Viral Counterstrategy

Viruses fight back the mechanism of host RNA silencing by encoding protein molecules known as RNA silencing suppressors (RSSs), and the RSSs encoded by viruses are also known as VSRs. The VSRs are known to interfere at different stages of RNA silencing pathways, thus helping in efficient infection and replication of virus in the host cell and spreading the infection systemically (Voinnet et al. 1999; Shi et al. 2002). These VSR molecules are generally usual viral proteins such as coat protein (CP), movement protein (MP), or proteases that carry the suppressor activity in the form of their secondary function (Hartitz et al. 1999). As a result, there is extensive assortment in the VSRs documented from the distinct viruses.

4.1 Earlier Experiments to Confirm RNA Silencing Suppression

The indications on the existence of the VSRs came from the early observations that certain specific proteins expressed by viruses played a significant role in their virulence. Subsequently it was highlighted that coinfection with a combination of viruses resulted in enhanced symptom severity rather than the single virus infection. One of the classical examples is *Potato virus X* (PVX) which, by itself, causes mild symptoms but multiplies vigorously during coinfection with the *Potato virus Y* (PVY) and *Tobacco etch virus* (TEV) (Pruss et al. 1997). This phenomenon was referred as synergism (Damirdagh and Ross 1967) and it is now implicit that the enhanced synergism is mainly due to the weakening of host defense by VSR targeting the silencing pathway at multiple points (Pruss et al. 1997; Mlotshwa et al. 2005).

In the year 1998, the initial report on identification of virus-mediated RSS came exclusively on a potyvirus-encoded helper component proteinase (Hc-Pro). This protein was identified as a major component involved in the enhancement of replication of unrelated viruses. In one such report, it was shown that P1/Hc-Pro suppressed the PTGS of *uidA* gene coding for β-glucuronidase (GUS) reporter on a highly expressed locus (Kasschau and Carrington 1998). In a similar but independent study, *Nicotiana tabacum* post-transcriptionally silenced for *uidA* was crossed with four independent transgenic plants expressing TEV P1/HC-Pro. It was identified that silencing efficiency was boosted in the progenies (Anandalakshmi et al. 1998).

Another study by Brigneti and coworkers (1998) revealed that PTGS of a green fluorescent protein (GFP) transgene was repressed in *Nicotiana benthamiana* infected with *Cucumber mosaic virus* (CMV); however, this suppression was not evident with PVX infection. In the same experiment, they expressed HC-Pro of PVY and 2b protein of CMV-encoded proteins in a PVX vector and demonstrated that they act as VSRs. Their study also anticipated that HC-Pro acts by hindering the maintenance of PTGS process in the tissues where silencing had previously been established. On the other hand, the 2b protein had prevented the commencement of gene silencing at the growing parts of the plants (Brigneti et al. 1998). So the identification and understanding of VSRs provided evidence for reversal of silencing of RNA as a natural antiviral defense response (Voinnet 2001). Besides, the VSRs can suppress silencing in both animal and plant cells, regardless of their host preference due to the conserved nature of the silencing phenomenon.

Apart from this, there are few reports which confirm that single virus may code for multiple VSR proteins. For example, *Citrus tristeza virus* (CTV) was reported to code for three different proteins p20, p23, and the coat protein (CP) exhibiting RSS activity (Lu et al. 2004). These preliminary findings provided a novel insight to find more VSRs as it seemed to be a universal strategy used by viruses against one of the most potent induced immune system of plants. Since then, a large number of viral proteins have been discovered which show RSS activity of dissimilar potency depending upon the host.

4.2 Assays to Detect RNA Silencing Suppressors

Following the discovery of HC-Pro as a VSR, many other virus proteins exhibited the capability of inhibiting the host antiviral mechanism. This establishment was greatly accelerated due to the availability of several simple and efficient functional assays to detect the RSS activity. Identification and functional characterization of RSS in turn facilitated the understanding of the intricacies of the RNA silencing pathway. It also provided insight into the evolutionary arms race between the host and the pathogens during pathogenicity.

During the formative period of VSR concepts, the major bottleneck in the identification of RSS was probably the unavailability of large array of screening systems. In plants, however, a number of strategies have been exploited to analyze the RSS activities of a candidate viral protein. These are mainly based on monitoring the role of the viral protein in suppressing the RNA-mediated silencing of a reporter gene. The reporter gene may be silenced constitutively (Elmayan and Vaucheret 1996) or locally by infiltrating through the Ti plasmid via infection by Agrobacterium (Voinnet and Baulcombe 1997; Voinnet et al. 1998). In the subsequent sections, we have listed and briefly explained the commonly used assays.

4.2.1 Agrobacterium-Mediated Transient Assay

In this method, transgenic tobacco plants stably silenced for a reporter gene like GFP or GUS are used. The candidate RSS is locally introduced into the transgenic silenced plant through infiltration of an *Agrobacterium* strain carrying the putative VSR gene. This method is called agro-infiltration. If the ectopically expressed protein has capabilities of suppressing the RNA silencing, then localized reversal of silencing will lead to expression of the reporter gene in the infiltrated zone. This is one of the most widely used assays for RSS analysis, due to its simple protocol and rapid generation of result (Karjee et al. 2008). A modification of this method involves coinfiltrating the reporter gene into wild-type tobacco plants along with the VSR using two *Agrobacterium* strains and monitoring the reporter gene expression (Llave et al. 2000; Voinnet et al. 2000; Johansen and Carrington 2001). The infiltration of the reporter gene will eventually initiate RNA silencing and the reporter will be silenced after three to five days. In the presence of candidate RSS protein, there will be suppression of silencing and the reporter gene expression is retained to a high level or may even increase after 6 days. By means of different reporter constructs, for instance genes organized as inverted repeats, it is possible to evaluate at which step of RNA silencing the suppressor protein acts (Takeda et al. 2002).

4.2.2 Reversal of Transgene Induced Silencing

In this method, plants expressing a reporter gene are systemically silenced through introducing the Agrobacterium expressing the reporter gene or a fragment of it. The reporter gene expression is monitored after the infection of single or multiple viral constructs. Re-establishment of reporter gene expression designates that the tested virus construct contains a RSS activity. Nevertheless, PVX encoding a RSS, which is incapable to restore the reporter gene expression, has been utilized as a vector to evaluate the RSS capability of other viral proteins (Brigneti et al. 1998).

4.2.3 Crossing Assay

This assay exploits a cross between a silenced transgenic plant and a second transgenic plant expressing a candidate viral protein (Anandalakshmi et al. 1998; Kasschau and Carrington 1998). A substantial disadvantage of this method is that higher VSR activity develops various abnormal phenotypic defects in the plants (Anandalakshmi et al. 1998). Though, the assay had been successfully utilized in various studies (Kasschau et al. 2003; Chapman et al. 2004; Dunoyer et al. 2004). A better alternate to mitigate this risk would be the ectopic expression of target gene through a heterologous viral vector system inoculated onto the silenced transgenic plants.

4.2.4 Grafting Assay

The principle of this assay is based on the fact that silencing molecules or signals display systemic movement from a silenced rootstock to a non-silenced scion in a grafted plant. The silencing signal spreads *via* RNA-mediated processes and has been extensively studied and reviewed (Chitwood and Timmermans 2010; Kalantidis et al. 2008). Grafting experiment itself has been one of the most reliable strategies to study suppression of silencing in plants. For rootstock, a line silenced for a reporter transgene is selected and the candidate RSS is introduced in it with the help of genetic crossing experiments. Further, a scion expressing the same reporter transgene is grafted on this rootstock. If the candidate is not a RSS, the silencing signal will systemically spread from rootstock to scion and the reporter will be silenced in the scion. However, if the candidate has RSS activity, the reporter will be expressed in the scion. This assay is time consuming and needs raising transgenics as well as breeding experiments. The grafting itself requires lot of practice and expertise. However, the reliability of the assay compensates for its time. This assay has been quite helpful in the identification of suppressors with specific activity on local and systemic silencing.

4.2.5 Specific Biochemical Assays

There are specific biochemical assays in the RNA silencing pathway, namely the Dicing and RISC assays (Carbonell et al. 2012; Wang et al. 2015). These are stage-specific RNA silencing assays with specific final readouts. Dicing assays convert the dsRNA substrates in siRNAs and RISC assays produce sliced RNA transcripts from the input mRNA molecules. Exogenous additions of proteins in such assays impede the formation of the final readable products. In this way, proteins with RSS activities at the defined steps could be identified.

5 Functional Mechanism of Viral Suppressors of RNAi

RNAi-based immunity in plant against viruses entails a cascade of well-established molecular processes that enhances siRNA/miRNA production and promotes cleavage of targeted transcripts. As counter-defensive mechanisms, viruses may interfere and inhibit each steps of RNAi pathway, affecting the normal sRNA biogenesis. The RNAi suppressors can suppress the pathway of RNAi at different steps as mentioned below.

5.1 Interaction Between DsRNA-VSRs

Many VSRs have attributes of dsRNA-binding proteins. This possibly imitates the fact that every RNAi-mediated antiviral reaction consistently commence with DCL-mediated processing of virus-derived dsRNAs. Hence, targeting dsRNA which acts as a DCL substrate for protection would serve as common strategy for many VSRs. The VSRs are also known to bind siRNAs and consequently inhibit downstream activities of siRNAs.

5.2 Viral Suppressors Target RNAi Effectors

It has been revealed that VSRs may target key RNAi components such as targets AGO1 and DCLs for degradation. Since AGO1 is recognized as a prime component for miRNA function in plants, suppression of these RNAi effectors by VSRs leads to the inhibition of miRNA function. A few VSRs target DCL4 and suppress dicing. VSRs also target RDR proteins of the host to lessen biogenesis of dsRNA and amplification of siRNAs.

5.3 Suppression of Systemic RNAi by VSRs

Additionally, as a counter-defense mechanism, few VSRs are proficient in precise targeting of systemic silencing signal. For example, *PVX*-encoded P25 and CTV-encoded coat protein are well distinguished for their suppression action on systemic silencing.

5.4 Epigenetic Modifications

As mentioned earlier, cytosine methylation in DNA and histone methylation are the common epigenetic marks that could be brought in by small RNAs, mostly si- and miRNAs. These marks keep DNA unavailable for transcription by various mechanisms. The genomes of the viral DNAs are also known to be subject of this transcriptional control leading to TGS, an account of which is nicely dealt in a recent review (Pooggin 2013). However, a substantial portion of TGS can be reversed by VSRs. The AC2/C2 homologs of begomovirus and Curtovirus genera and the C1 protein of beta-satellite of some begomoviruses can cause reversal of TGS by various mechanisms. The AC2 protein of *Mungbean yellow mosaic virus* (MYMV), CaLCuV, *Tomato golden mosaic virus* (TGMV), etc. inactivate Adenosine Kinase (ADK), reducing production of SAM, the methyl donor, and thus cause release of TGS (Trinks et al. 2005; Buchmann et al. 2009). The C1 protein of beta-satellite of *Tomato yellow leaf curl china virus* (TYLCCNV) inactivates S-Adenosyl Homocysteine Hydrolase (SAHH), an enzyme required for synthesis of SAM, and thus reduces the level of cytosine methylation of viral DNA (Yang et al. 2011). The C2 protein of *Beet severe curly top virus* (BSCTV) causes reduction of vi-siRNA and decreases methylation of defense response genes so that defense proteins of salicylic acid pathway, the GST superfamily, etc. could be turned on (Yang et al. 2013). In a separate report, it has been shown that the same protein increases the life span of SAMDC1 and thus suppresses DNA methylation-mediated gene silencing in Arabidopsis (Zhang et al. 2011).

Besides DNA methylation, histone methylation is also targeted by the VSR, namely the AC2 protein (Sun et al. 2015). The AC2 protein of *Indian cassava mosaic virus* (ICMV) upregulates RAV2, which acts as a transcriptional repressor, inhibiting transcription of KYP, a histone methyl transferase. In this way, AC2 dampens TGS and allows viral survival in the infected host. Not only the DNA viruses, but the RNA viruses are also known to relieve TGS. The 2b protein, an RNAi suppressor of severe Shan-Dong (SD) isolate of *Cucumber mosaic virus*, suppresses RdDM by binding and sequestering siRNAs in a process involving AGO proteins in the nucleolus (Duan et al. 2012).

6 Few Representative VSRs

Even though many VSRs have been described, extensive research has been focused on a few following selected proteins.

6.1 HC-Pro of Potyviruses

The foremost-described VSR is the potyviral HC-Pro protein (Anandalakshmi et al. 1998). This protein is mainly found to affect the processes associated with vector transmission, polyprotein processing, replication of viral genome, and the systemic movement of the virus (Kasschau et al. 1997). It was also well characterized as a comprehensive pathogenicity enhancer assisting in the enhancement in the viral RNA accumulation and development of severe symptoms of virus infection during many distinct virus infections (Pruss et al. 1997), thus representing a direct and robust influence on the maintenance of RNA silencing.

Systemic infection by PVX carrying HC-Pro was capable of reversing the expression of GUS in the reporter gene silenced transgenic plants. It was demonstrated that cross between the GUS silenced lines and HC-Pro expressing plant may possibly reinstate GUS expression. This restoration was due to the action of HC-Pro which contributed to prevent the degradation of the *gus* mRNA (Anandalakshmi et al. 1998; Brigneti et al. 1998; Hamilton et al. 2002). This suggested that HC-Pro may perhaps impede an RNase III-like enzyme involved in the generation of siRNAs from dsRNA or an active component of the RISC. Interestingly, it was revealed later that HC-Pro did not interrupt the silencing signal cascade within a plant, albeit all siRNAs were eliminated (Mallory et al. 2001). Moreover, HC-Pro was displayed to capably avert the plant from retorting the silencing signal in a grafting experiment (Hamilton et al. 2002). Furthermore, there are few contradictory reports, which suggested the possible involvement of HC-Pro in the DNA methylation at the silenced transgene locus of genome (Llave et al. 2000; Mallory et al. 2001).

A step forward discovery on the mechanism of silencing suppression was the identification of interacting partner of P1/HC-Pro of TEV (Anandalakshmi et al. 2000). In this study, transgenic plants overexpressing rgs-CaM (regulator of gene silencing-calmodulin-like protein) showed phenotypic variations, which were found to be similar to HC-pro transgenic plants. Apart from these characteristics of HC-Pro, it has also been implicated in stimulating the miRNA-mediated gene regulation, thus supporting the previous observation of developmental defects detected in the transgenic plants (Mallory et al. 2002; Kasschau et al. 2003). Molecular structure of this protein revealed that the domain of HC-Pro possesses RNA-binding properties which is essential and prerequisite for silencing suppression (Kasschau and Carrington 2001). Further studies showed that it has the highly conserved FRNK box, which apparently provides a site of interaction with siRNA and miRNA duplex. This directly influences the miRNA abundance and associated regulatory functions, leading to the symptom development (Shiboleth et al. 2007). Overall, these studies suggested that HC-Pro potentially suppresses the RNA silencing downstream of dsRNA and miRNA generation. Conversely, it also alters the upstream process of the siRNA accumulation and probably impedes the systemic spread of silencing signal.

6.2 Cucumoviruses 2b (CMV-2b)

Previous studies have suggested that CMV-2b regulates systemic viral movement, and deficiency of this protein may reduce the pathogenicity of the virus (Ding et al. 1995a, b). Apart from this, CMV-2b protein was found to manipulate viral cell to cell movement in plants (Soards et al. 2002). First report of functional characterization of CMV-2b as a RSS revealed that the inhibition of 2b protein translation of the mild Q strain (Q-D2b) caused attenuation in *Nicotiana glutinosa*, along with the deficiency of systemic infection in cucumber plant (Ding et al. 1994). Several studies aimed to understand the function of the CMV-2b protein in virulence have been carried out, in last decade. In this context, Diaz-Pendon et al. (2007) identified that the Q-D2b mutant was competent of causing disease in Arabidopsis *DCL2-4* mutants. These genes are the key components of RNA silencing component, hence provided a strong correlation between RNA silencing and CMV-2b function. Moreover, mutation in the D2b of severe Fny strain resulted in the restoration of virulence in the *rdr1/6*, *ago1*, and *ago2* mutants of Arabidopsis (Wang et al. 2011). Interestingly, it was suggested that expression of CMV-2b protein from a mild strain may harmonize the infectivity in the developing tissues in response to the synergistic effect of *Tobacco mosaic virus* (Siddiqui et al. 2011).

CMV-2b was shown to avert the initiation of RNA silencing in newly emerging tissue but it cannot reverse established RNA silencing (Beclin et al. 1998; Brigneti et al. 1998). This result advocated that 2b might be potentially required for preventing the cell to cell spread of the silencing signal, from the locally infected parts to the rest of the plant to promote further virus spread (Goto et al. 2007). The

CMV-2b exhibits dual cellular localization in the cytoplasm as well as nuclear foci (Lucy et al. 2000). Additionally, it was also revealed that CMV-2b possesses a monopartite nuclear localization signal (Lucy et al. 2000), hence may interfere with the restoration of transgene methylation, indicating functioning of 2b in the nucleus (Guo and Ding 2002).

CMV-2b has also been shown to affect the PTGS pathway by directly binding to siRNAs or long dsRNA (Guo and Ding 2002; Mitter et al. 2003), an activity, which differed from strain to strain of CMV. Exhaustive study done by Goto and coworkers (2007) revealed that 2b of severe strain (CM95R) of CMV binds in vitro to both chemically synthesized siRNAs and dsRNAs. Alternatively, 2b suppressor of an attenuated strain of CMV (CM95), which differs in single amino acid from the 2b CM95R, could barely bind to siRNAs. It signifies that the reduction in substantial RSS activity of the CM95 due to the single amino acid change may be responsible for the loss of siRNAs binding property of 2b.

It was also demonstrated that CMV-2b protein could inhibit the function of the siRNAs by directly interacting with AGO1. This interaction was studied in vitro and in vivo, and was found to be predominantly on one surface of the PAZ encompassing unit and part of the PIWI-box (Zhang et al. 2006; Ruiz-Ferrer and Voinnet 2007). This suggested that 2b specifically inhibited AGO1 cleavage activity in RISC reconstitution assays, thereby interfering with miRNA pathway and causing development abnormalities moderately phenocopying AGO1 mutant alleles.

Furthermore, 2b was revealed to be unable to inhibit the initiation of signal-independent RNA silencing of transgene and virus, by obstructing the RDR1-dependent viral siRNAs generation process (Diaz-Pendon et al. 2007). This stipulates that different mechanisms possibly will be involved in overcoming the antiviral defense by the infecting virus.

6.3 Tombusviruses P19

One of the robust VSRs is P19 of the tombusvirus, such as *Cymbidium ringspot virus*. It has the characteristic of recognizing the 2 nt extension at the 3′ end of 21 nt RNA duplexes for siRNA binding and thus inhibiting them from spreading systemically through the plant. It may also impede the activity of siRNA-primed RDR complex, which is assumed to modulate the establishment of the systemic signal (Voinnet 2001). Few reports also suggest that it has the capacity to interact and efficiently bind to variety of siRNA molecules, such as ss-siRNAs, long dsRNAs, and blunted 21 nt dsRNAs (Silhavy et al. 2002). Biochemical characterization of P19 in *Drosophila* cell extracts revealed that it might hinder the siRNAs loading into RISC effectors complexes (Lakatos et al. 2006).

Later on, the elucidation of the crystal structure of P19 binding a 21 nt siRNA duplex confirmed the physical interaction in between P19 and siRNAs. It helped the biologist to advance their understanding about how dimers of this protein are proficient in distinguishing RNA duplexes of 21 nt and also overhanging 3′ nt, which is a hallmark of the siRNAs (Vargason et al. 2003). Moreover, this siRNA binding characteristic of P19 was conserved among all the organisms containing silencing machinery, which also provided a base to develop P19 as widespread and potent tool to study RNA silencing process, Recently, inhibition of 3′ modification of small RNAs in *Carnation Italian ringspot virus* infected plants was studied and it was found that P19 binds to both 3′ modified and non-modified small RNAs in vivo. In general, 3′ modifications of viral siRNAs take place in cytoplasm, whereas in the case of miRNAs, this modification occurs in the nucleus. Hence, the P19 facilitated inhibition of the 3′ si/miRNAs alteration would entail spatial and sequential expression of both P19 and small RNAs. Finally, their data revealed that Hen1-like methyltransferase might account for the small RNA modification of their 3′-terminal nucleotide in *N. benthamiana* (Lozsa et al. 2008). Similar to HC-Pro, P19 has also been shown to interfere with the processing and activity of miRNAs by modulating the HEN-1-mediated methylation of miRNA.

Remarkably, the P19 protein of *Tomato bushy stunt virus* interacts with ALY proteins. These proteins have been shown to be associated with the export of RNAs from the nucleus and transcriptional co-activation in animal cells. P19 helps in the re-localization of a subset of these proteins from the nucleus to the cytoplasm. Co-expression of ALY proteins and P19 in *N. benthamiana* revealed that the subset of ALY proteins, which were not translocated from the nucleus significantly, altered the RNA silencing suppression ability of P19 (Canto et al. 2006).

6.4 Geminivirus AC2

Geminiviruses are characterized by small geminate particles (18–20 nm) containing either one or two single-stranded circular DNA molecules of around 2.7 kb (Stanley and Gay 1983). Based on genome organization, host range, and vector specificity, the members of the family Geminiviridae are classified into seven genera: Begomovirus, Mastrevirus, Curtovirus, Eragrovirus, Becurtovirus, Turncurtovirus, and Topocuvirus (Adams et al. 2013). The majority of begomoviruses have two components, referred to as DNA-A and DNA-B, both of which are essential for infectivity. Monopartite begomovirus such as isolates of *Tomato yellow leaf curl Sardinia virus* (Kheyr-Pour et al. 1991) has a single genomic component equivalent to DNA-A.

The protein encoded by the complementary strand of DNA-A component, named AC2, is one of the major pathogenicity factors. It is multifunctional protein encoded by all members of the genus Begomovirus. The protein has transactivation potential and is required for the expression of late viral genes AV1 and BV1 in at least some geminiviruses, thus also known as Transcriptional Activator Protein (TrAP) (Sunter and Bisaro 1991, 1992; Jeffrey et al. 1996). It binds to ssDNA in a non-specific way and only weakly to dsDNA, suggesting that it is not a canonical transcriptional factor, but probably interacts with host plant cellular proteins to trigger transcriptional activation (Hartitz et al. 1999).

In general, the AC2 protein has a modular structure consisting of three conserved domains: a basic domain with a nuclear localization signal at the N-terminus, a central DNA-binding Zn-finger motif, and C-terminal acidic activator domain (Hartitz et al. 1999). The AC2 or the C2 protein (a positional homolog of AC2 in TYLCCNV) encoded by monopartite and bipartite begomoviruses have been shown to possess strong RSS activity and are capable of suppressing TGS and or PTGS (Voinnet et al. 1999; van Wezel et al. 2002; Dong et al. 2003; Vanitharani et al. 2004; Trinks et al. 2005; Wang et al. 2005). It has been postulated that since AC2 protein of begomoviruses fails to bind any form of RNA, it thus needs to target host RNAi factors. AC2 protein was found to be directly interacting with RNA silencing pathway components like RDR6 and AGO1, which indicates its dual action site on the pathway to make the suppression more strong and effective (Kumar et al. 2015). Moreover, AC2 of CbLCV promotes the decapping activity of DCP2, which in turn accelerates mRNA turnover rate and also inhibits the siRNA accumulation (Ye et al. 2016).

AC2/C2 of TGMV (a begomovirus) and *Beet curly top virus*—BCTV (a curtovirus) have been shown to suppress PTGS by interacting and inactivating the SNF1 and adenosine kinases enzymes which appear to be involved in defense response (Hao et al. 2003; Wang et al. 2003). The adenosine kinase is known to be essential for the production of s-adenosyl methionine (SAM), an important cofactor for methyl transferases (Saze et al. 2003) and inhibition of its activity negatively affects methyl cycle (Wang et al. 2003, 2005).

In addition, C2/AC2 of the members of both begomovirus and curtovirus has been shown to be a suppressor of TGS (Buchmann et al. 2009; Zhang et al. 2011; Yang et al. 2012). Buchmann et al. (2009) first showed that Geminivirus C2 and AC2 proteins can be a TGS suppressor and demonstrated that they reduce the overall cytosine methylation. BSCTV acts as a TGS suppressor by interacting with SAM decarboxylase 1 (SAMDC1) and attenuating the degradation of SAMDC1, a key player in the methyl cycle (Zhang et al. 2011). Later, BSCTV C2 also has been shown to affect the generation of virus-derived siRNAs, a precursor for the initiation of RdDM, and thereby reducing the viral DNA methylation (Yang et al. 2012). More recently, AC2 of ICMV has been reported to inhibit kryptonite (KYP, a H3K9 methyl transferase) via the activation of transcription repressor RAV2 (RELATED TO ABI3 and VP1) (Sun et al. 2015). However, AC2 of TGMV and CbLCV has been shown to interact with the catalytic domain of KYP and further inhibits its methyl transferase activity in vitro (Castillo-Gonzalez et al. 2015). Furthermore, using TrAP protein lacking its transcription activation domain, a recent report revealed that this TrAP could reverse TGS in the reproductive plants, independent of ADK inactivation or transcription activation (Jackel et al. 2015).

6.5 Polerovirus P0

P0 adopts proteasome-mediated degradation of AGO1. Molecular analysis of Polerovirus P0 protein structure suggests that it encompasses F-box motif, which is essential to form the SCF-like complex, and also a prerequisite for P0's RSS activity. Further studies of P0 suggested that it does not essentially affect the biogenesis of primary siRNAs; however, it may target the PAZ motif and adjacent upstream sequence of AGO1 to destabilize it and subsequently lead to proteasome-mediated degradation (Baumberger et al. 2007; Bortolamiol et al. 2007).

7 Disease or Pathogenicity: Host MicroRNA Dysregulation and Affected Functions

It has been shown that the cellular miRNAs are capable of regulating viral replication. The viruses at the same time may alter the expression of cellular miRNAs through the VSR molecules. The VSR-mediated changes in the profile of host miRNA abundance and activities are well known in literature. The VSRs might treat the chemically similar duplex-miRNAs and siRNAs in a more or less similar manner, even though the former groups of molecules are processed from hairpin loop RNA precursors transcribed from endogenous genes (Ambros et al. 2003). The processing and function of miRNA pathway involve common components including DCL1 and AGO1 (Bartel 2004). The VSR-mediated deviation of the normal miRNA profile of the host following the virus infection could be a major source of viral pathogenicity.

In plants, miRNAs target a wide range of mRNAs encoding transcription factors required for development (Park et al. 2002; Rhoades et al. 2002; Palatnik et al. 2003). These include factors required for meristem identity and maintenance, patterning, cell division, hormone signalling, and developmental timing. In addition, plant miRNAs also target mRNAs encoding miRNA metabolic factors and factors of unknown function (Rhoades et al. 2002; Xie et al. 2003). Loss of miRNA biogenesis or activity in *Arabidopsis* results in pleiotropic defects during embryonic, vegetative, and reproductive development (Park et al. 2002; Schauer et al. 2002; Kasschau et al. 2003).

It is proposed that most of the developmental defects triggered by virus infection are due to interference with pathways that depend on negative regulation by miRNAs. A study with TuMV in Arabidopsis demonstrated that P1/HC-Pro is the virus-encoded factor that mediates this interference. The suppression of miRNA-directed function and RNA silencing by P1/HC-Pro is likely due to interference with a common reaction, probably involving assembly or activity of RISC-like complexes. The consequence of virus infection is ectopic expression of some mRNAs that are normally negatively regulated by miRNA-guided cleavage. Infected plants, therefore, display a range of developmental abnormalities because the aberrantly expressed target mRNAs encode proteins belonging to families that control meristem identity (NAC domain and SBP-like proteins), organ identity and separation (AP2 domain and NAC domain proteins), radial patterning (SCL-like proteins), and hormone signalling (ARF proteins). Interference with leaf and flower formation and developmental timing; ectopic induction of cell division in non-meristematic tissues; and disruption of hormone production, signalling, and response are some of the well-characterized effects of different viruses in certain susceptible host plants (Hull 2001). Given that many of the miRNA target genes are expressed or repressed in specific cell-types in meristematic and organ primordium zones, we further propose that viruses triggering the most severe developmental defects are those that (1) invade meristematic and dividing cells and (2) encode potent RNA silencing suppressors. Indeed, although many viruses are known to be excluded from meristematic zones, in situ analysis revealed that meristems and organ primordia are effectively invaded by TuMV in *Arabidopsis*.

8 VSR-Targeted Antiviral Strategy

The VSRs are the pathogenicity factors and hence are very good targets for antiviral strategy. Many RNA viruses failed to cause disease in plants expressing siRNAs targeted to silence the VSRs of the infecting viruses. Similar strategy also works in mammalian systems. The non-human primates have been found protected against the deadly Ebola viruses when the animals are systemically injected with the siRNAs meant to silence the Ebola-VSR (Thi et al. 2016). The artificial miRNAs have also been used to silence the VSRs of RNA and DNA viruses of plants, and the transgenic plants expressing the miRNAs have been found tolerant/resistant against the viruses. The literature is replete with the information on siRNAs silencing the VSRs but the corresponding reports of artificial miRNAs (amiRNAs) are few. In the following, we give an account of the amiRNAs and tasiRNAs providing the antiviral strategy.

8.1 Artificial MiRNA Strategy

The amiRNA technology is being utilized to target the invading viral gene transcripts. In this regard, the VSR transcripts have been widely subjected to degradation (Tiwari et al. 2014). It was reported that miR156 and miR393 may inhibit the invasion of foreign genetic elements like plant viruses (Xing and Zhang 2010; Zhang et al. 2011). The ath-miR-159 based amiRNAs were designed to target viral sequences encoding P69, aVSR of *Turnip yellow mosaic virus* (TYMV) and HC-Pro of *Turnip mosaic virus* (TuMV). Transgenic *Arabidopsis* lines expressing amiR-P69 and amiR-HCPro were specifically resistant to TYMV and TuMV (Niu et al. 2006). The amiRNA sequences targeting the VSR, 2b of CMV, can efficiently confer effective resistance to CMV infection (Qu et al. 2007). Later amiRNA technology was used to confer virus resistance in transgenic tobacco and tomato (Ai et al. 2011; Zhang et al. 2011). The amiRNA targeting overlapping regions of geminiviruses genes AC1, AC2, and AC4 were used to generate transgenic tomato plants, that could resist infection by begomovirus, ToLCNDV (Yadava et al. 2010; Tien et al. 2013). There are also reports in literature on using the amiRNAs for generating resistance against *Watermelon silver mottle virus* in tobacco (Kung et al. 2012).

8.2 *Artificial TasiRNA Strategy*

Besides amiRNA, artificial tasiRNA technology has also been used to generate virus tolerant plants. A binary vector has been designed incorporating control elements such as the 5′ and 3′ binding sites of miR390 and keeping the VSR sequences sandwiched between the control elements. This vector when introduced in plants produces artificial tasiRNAs from the VSR sequences. These tasiRNAs slice the VSRs of the infecting ToLCVs. Thus the transgenics producing the artificial tasiRNAs was tolerant against the invading ToLCVs (Singh et al. 2015). Such strategy could in principle be adopted to develop plants tolerant for all viruses whose VSR sequences are known.

9 Future Perspectives

RNAi has been used extensively as a tool to study gene functions. The efficiencies of these processes are presumed to be subjects of several degrees and layers of modifications. The VSRs or RSSs can contribute largely to the modification processes. The VSRs, when overexpressed, can influence the outcomes of RNAi in several systems. In this connection it is important to reveal the identities of RSSs in all of the RNAi-competent organisms. A few of these are reported in host plants like tobacco and tomato but these class of RSSs from plant sources or other organisms have remained elusive so far. Hence appropriate assays need to be devised to trap RSSs from several nonviral pathogens and their hosts. Recently a class of proteins, namely RNase III-like proteins (RTLs), have been described from plants that act as general RNAi suppressors, which are induced in response to virus infection but are functionally repressed by plant VSRs (Shamandi et al. 2015). On the other hand, the β-C1 suppressor of TYLCCN virus collaborates with tomato rgs-CaM RNAi suppressor for efficient viral growth (Li et al. 2014). Thus along with the identification of RSSs from nonviral sources, the cross-talks between the RSSs are also very important to reveal the overall biology of RNAi.

VSRs could be used for multiple purposes, namely, reversal of siRNA-mediated disease, overcoming transgene silencing, enhancing expression of viral vectors and vaccine production, etc. Tobacco plants infected with TMV bearing the pathogenic satellite RNA show darkening effects in the leaves due to loss of chlorophyll biosynthesis encoding protein CHL1 which gets silenced by the siRNAs produced from the satellite RNA. This silencing effect is strongly inhibited by the P1/HC-PRO VSR of the potyvirus (Ricaño-Rodríguez et al. 2016). Many VSRs have been used to overcome transgene- or siRNA-mediated silencing (Rahman et al. 2012, 2014). However, there is an inherent difficulty in reversing such kind of RNAi as the presence of VSRs also interfere in the biogenesis and function of the plants hosting the VSRs, making the host plant developmentally retarded. Hence either the VSRs need to be modified or these should be chosen carefully such that the selected VSRs do not interfere in the miRNA pathways. A mutant form of HC-PRO has been used by Mallory et al. to enhance transgene expression in tobacco showing no developmental anomaly (Mallory et al. 2002). Similarly the VSR proteins of *Beet yellow closterovirus* like p64, p21, etc. might have minimal impact on the miRNA pathways (Til'kunova et al. 2004). A few VSRs when expressed in the heterologous systems remove the restrictions of RNAi but do not cause perturbations in the miR pathways. The VSR B2 of insect *Flock House Virus* (FHV) suppresses RNAi in *C. elegans* and also facilitates natural infection of *Orsay virus* in *C. elegans* but is inactive against miRNA-mediated silencing (Guo and Lu 2013). Thus FHV-B2 does not harm the *C. elegans* hosts. VSRs have been extensively used in improving replication and transcription of viral vectors used for gene therapy and vaccine production. Recently, P19 VSR from *Tomato bushy stunt virus* was stably expressed in human embryonic kidney cells (B6 cells) and the replication of Adenovirus shot up 100-fold in these cells. Adenoviruses are widely used viral vectors and along with p19 the oncolysis potential of the vector is increased five- to six fold in the tumor cells, raising the hope of translating these results in preclinical and clinical trials (Rauschhuber et al. 2012). Hence the selective usages of VSRs are very beneficial to remove the undesirable restrictions of RNAi.

The intertwined and multiple-layered arms race between host and pathogen must be interpretable in terms of "molecular arms race." Both the host and viral components along with their cross-talks have been adequately described in literature (Ding and Voinnet 2007; Csorba et al. 2015). Viral genes evolve faster than host genes as the viruses want to combat host with novel winning designs, and in response, the antiviral silencing factors also evolve faster than other host genes to gain upper hand of the battle. Amidst all these, the VSRs evolve faster than any other known genes (Murray et al. 2013). Such changes impact strongly both on viruses and hosts. The diversity of VSRs' structure and functions are partly accounted by such changes. Besides silencing RNAi, the VSRs also participate in other important aspects of viral life cycle (Csorba et al. 2015). So it would be important to assess how much of the viral life processes as well as their pathogenicity has changed over the evolutionary time scale. It would also be worthwhile to watch what new functions, besides RNAi silencing, like interfering with host hormone signalling, relocalizations of interacting host factors in subcellular structures etc., are being gained by the VSRs. The vi-siRNAs and VSRs interact directly as well as indirectly with many of the host factors that are involved in antiviral silencing pathways including rgs-CaM and RAV2 (Moissiard and Voinnet 2006; Nakahara et al. 2012; Endres et al. 2010). When VSRs undergo evolution, interacting host factors might also change, thus causing hosts to evolve. It would be very interesting to study the profile of changes in host evolutionary pattern in response to the evolution of VSRs. Besides RNAi factors, hosts also offer resistance to viruses by other antiviral pathways like R-gene-mediated hypersensitive response, hormone (SA/JA) mediated SAR pathways, etc. Another interesting area of research would be to follow how the evolving VSRs intersect these pathways.

References

Adams MJ, King AMQ, Carstens EB (2013) Ratification vote on taxonomic proposals to the International Committee on Taxonomy of Viruses. Arch Virol 158:2023–2030

Agrawal N, Dasarathi PVN, Mohmed A et al (2003) RNA interference biology, mechanism and applications. Microbiol Mol Biol Rev 67:657–685

Ahlquist P (2006) Parallels among positive-strand RNA viruses, reverse-transcribing viruses and double-stranded RNA viruses. Nat Rev Micro Biol 4:371–382

Ai Y, Wang J, Johnson RE et al (2011) A novel ubiquitin binding mode in the *S. cerevisiae* translesion synthesis DNA polymerase? Mol BioSyst 7:1874–1882

Akbergenov R, Si-Ammour A, Blevins T et al (2006) Molecular characterization of geminivirus-derived small RNAs in different plant species. Nucleic Acids Res 34:462–471

Ambros V, Bartel B, Bartel DP (2003) A uniform system for microRNA annotation. RNA 9:277–279

Anandalakshmi R, Pruss GJ, Ge X et al (1998) A viral suppressor of gene silencing in plants. Proc Natl Acad Sci USA 95:13079–13084

Anandalakshmi R, Marathe R, Ge X et al (2000) A calmodulin-related protein that suppresses post-transcriptional gene silencing in plants. Science 290:142–144

Bartel DP (2004) MicroRNAs: genomics, biogenesis, mechanism and function. Cell 116:281–297

Baulcombe DC (1999) Fast forward genetics based on virus induced gene silencing. Curr Opin Plant Biol 2:109–113

Baumberger N, Tsai CH, Lie M et al (2007) The polerovirus silencing suppressor P0 targets Argonaute proteins for degradation. Curr Biol 17:1609–1614

Beclin C, Berthome R, Palauqui J et al (1998) Infection of tobacco or Arabidopsis plants by CMV counteracts systemic post-transcriptional silencing of nonviral (trans)genes. Virology 252:313–317

Benson DA, Karsch-Mizrachi I, Lipman DJ et al (2005) GenBank. Nucleic Acids Res 33(S1): D34–D38

Bernstein E, Caudy AA, Hammond SM et al (2001) Role for a bidentate ribonuclease in the initiation step of RNA interference. Nature 409:363–366

Bivalkar-Mehla S, Vakharia J, Mehla R et al (2011) Viral RNA silencing suppressors (RSS): novel strategy of viruses to ablate the host RNA interference (RNAi) defense system. Virus Res 155:1–9

Bortolamiol D, Pazhouhandeh M, Marrocco K et al (2007) The poleovirus F Box protein P0 targets Argonaute1 to suppress RNA silencing. Curr Biol 17:1615–1621

Brigneti G, Voinnet O, Li WX et al (1998) Viral pathogenicity determinants are suppressors of transgene silencing in *Nicotiana benthamiana*. EMBO J 17:6739–6746

Buchmann RC, Asad S, Wolf JN et al (2009) Geminivirus AL2 and L2 proteins suppress transcriptional gene silencing and cause genome-wide reductions in cytosine methylation. J Virol 83:5005–5013

Canto T, Uhrig JF, Swanson M et al (2006) Translocation of tomato bushy stunt virus P19 protein into the nucleus by ALY proteins compromises its silencing suppressor activity. J Virol 80:9064–9072
Carbonell A, Fahlgren N, Garcia-Ruiz H et al (2012) Functional analysis of three Arabidopsis ARGONAUTES using slicer-defective mutants. Plant Cell 24:3613–3629
Carrillo-Tripp J, Lozoya-Gloria E, Rivera-Bustamante RF (2007) Symptom remission and specific resistance of pepper plants after infection by pepper golden mosaic virus. Phytopathology 97:51–59
Castillo-González C, Liu X, Huang C et al (2015) Geminivirus-encoded TrAP suppressor inhibits the histone methyltransferase SUVH4/KYP to counter host defense. eLife 4:e06671
Chapman EJ, Prokhnevsky AI, Gopinath K et al (2004) Viral RNA silencing suppressors inhibit the microRNA pathway at an intermediate step. Genes Dev 18:1179–1186
Chitwood DH, Timmermans MC (2010) Small RNAs are on the move. Nature 467:415–419
Csorba T, Kontra L, Burgyán J (2015) Viral silencing suppressors: tools forged to fine-tune host-pathogen coexistence. Virology 479–480:85–103
Darnirdagh IS, Ross AF (1967) A marked synergistic interaction of potato viruses X and Y in inoculated leaves of tobacco. Virology 31:296–307
Diaz-Pendon JA, Li F, Li W-X et al (2007) Suppression of antiviral silencing by Cucumber mosaic virus 2b protein in Arabidopsis is associated with drastically reduced accumulation of three classes of viral small interfering RNAs. Plant Cell 19:2053–2063
Ding SW, Voinnet O (2007) Antiviral immunity directed by small RNAs. Cell 130:413–426
Ding SW, Anderson BJ, Haase HR et al (1994) New overlapping gene encoded by the Cucumber mosaic virus genome. Virology 198:593–601
Ding S-W, Li W-X, Symons RH (1995a) A novel naturally occurring hybrid gene encoded by a plant RNA virus facilitates long distance virus movement. EMBO J 14:5762–5772
Ding SW, Rathjen JP, Li WX et al (1995b) Efficient infection from cDNA clones of cucumber mosaic cucumovirus RNAs in a new plasmid vector. J Gen Virol 76:459–464
Dong X, van Wezel R, Stanley J et al (2003) Functional characterization of the nuclear localization signal for a suppressor of posttranscriptional gene silencing. J Virol 77:7026–7033
Duan CG, Fang YY, Zhou BJ et al (2012) Suppression of Arabidopsis ARGONAUTE1-mediated slicing, transgene-induced RNA silencing, and DNA methylation by distinct domains of the cucumber mosaic virus 2b protein. Plant Cell 24:259–274
Dunoyer P, Lecellier CH, Parizotto EA et al (2004) Probing the microRNA and small interfering RNA pathways with virus-encoded suppressors of RNA silencing. Plant Cell 16:1235–1250
Elmayan T, Vaucheret H (1996) Expression of single copies of a strongly expressed 35S transgene can be silenced post-transcriptionally. Plant J 9:787–797
Endres MW, Gregory BD, Gao Z et al (2010) Two plant viral suppressors of silencing require the ethylene-inducible host transcription factor RAV2 to block RNA silencing. PLoS Pathog 6: e1000729
Enright AJ, John B, Gaul U et al (2003) MicroRNA targets in Drosophila. Genome Biol 5:1
Finnegan EJ, Margis R, Waterhouse PM (2003) Posttranscriptional gene silencing is not compromised in the arabidopsis CARPEL FACTORY (DICER-LIKE1) mutant, a homolog of dicer-1 from drosophila. Curr Biol 13:236–240
Ghosh Z, Mallick B, Chakrabarti J (2009) Cellular versus viral microRNAs in host-virus interaction. Nucleic Acids Res 37:1035–1048
Goto K, Kobori T, Kosaka Y et al (2007) Characterization of silencing suppressor 2b of cucumber mosaic virus based on examination of its small RNA-binding abilities. Plant Cell Physiol 48:1050–1060
Griffiths-Jones S, Saini HK, van Dongen S et al (2008) miRBase: tools for microRNA genomics. Nucleic Acids Res 36(S1):D154–D158
Guo HS, Ding SW (2002) A viral protein inhibits the long range signaling activity of the gene silencing signal. EMBO J 21:398–407

Guo X, Lu R (2013) Characterization of virus-encoded RNAi suppressors in Caenorhabditis elegans. J Virol 87:5414–5423
Hamilton A, Voinnet O, Chappell L et al (2002) Two classes of short interfering RNA in RNA silencing. EMBO J 21:4671–4679
Hao L, Wang H, Sunter G et al (2003) Geminivirus AL2 and L2 proteins interact with and inactivate SNF1 kinase. Plant Cell 15:1034–1048
Hartitz MD, Sunter G, Bisaro DM (1999) The tomato golden mosaic virus transactivator (TrAP) is a single-stranded DNA and zinc-binding phosphoprotein with an acidic activation domain. Virology 263:1–14
Hofacker IL, Fontana W, Stadler PF et al (1994) Fast folding and comparison of RNA secondary structures. Monatsh Chem 125:167–188
Hsu PWC, Lin LZ, Hsu SD et al (2007) ViTa: prediction of host microRNAs targets on viruses. Nucleic Acids Res 35(S1):D381–D385
Huang Y, Greene E, Murray Stewart T et al (2007) Inhibition of lysine-specific demethylase 1 by polyamine analogues results in reexpression of aberrantly silenced genes. Proc Natl Acad Sci USA 104:8023–8028
Hull R (2001) Matthew's plant virology, 4th edn. Academic, San Diego
Hutvágner G, McLachlan J, Pasquinelli AE et al (2001) A cellular function for the RNA-interference enzyme Dicer in the maturation of the let-7 small temporal RNA. Science 293:834–838
Jackel JN, Buchmann RC, Singhal U et al (2015) Analysis of geminivirus AL2 and L2 proteins reveals a novel AL2 silencing suppressor activity. J Virol 89:3176–3187
Jeffrey JL, Pooma W, Petty ITD (1996) Genetic requirements for local and systemic movement of tomato golden mosaic virus in infected plants. Virology 223:208–218
Johansen LK, Carrington JC (2001) Silencing on the spot induction and suppression of RNA silencing in the Agrobacterium-mediated transient expression system. Plant Physiol 126:930–938
Jones JDG, Dang JL (2006) The plant immune system. Nature 444:323–329
Kalantidis K, Schumacher HT, Alexiadis T et al (2008) RNA silencing movement in plants. Biol Cell 100:13–26
Kapoor M, Arora R, Lama T et al (2008) Genome-wide identification, organization and phylogenetic analysis of Dicer-like, Argonaute and RNA-dependent RNA polymerase gene families and their expression analysis during reproductive development and stress in rice. BMC Genomics 9:451
Karjee S, Islam MN, Mukherjee SK (2008) Screening and identification of virus encoded silencing suppressors. Methods Mol Biol 442:187–203
Kasschau KD, Carrington JC (1998) A counter defensive strategy of plant viruses: suppression of posttranscriptional gene silencing. Cell 95:461–470
Kasschau KD, Carrington JC (2001) Long-distance movement and replication maintenance functions correlate with silencing suppression activity of potyviral HC-Pro. Virology 285:71–81
Kasschau KD, Cronin S, Carrington JC (1997) Genome amplification and long-distance movement functions associated with the central domain of Tobacco etch potyvirus helper component-proteinase. Virology 228:251–262
Kasschau KD, Xie Z, Allen E et al (2003) P1/HC-Pro a viral suppressor of RNA silencing interferes with Arabidopsis development and miRNA function. Dev Cell 4:205–217
Kheyr-Pour A, Bendahmane M, Matzeit V et al (1991) Tomato yellow leaf curl virus from Sardinia is a whitefly-transmitted monopartite geminivirus. Nucleic Acids Res 19:6763–6769
Kumar V, Mishra SK, Rahman J et al (2015) Mungbean yellow mosaic Indian virus encoded AC2 protein suppresses RNA silencing by inhibiting Arabidopsis RDR6 and AGO1 activities. Virology 486:158–172

Kung Y-J, Lin S-S, Huang Y-L et al (2012) Multiple artificial microRNAs targeting conserved motifs of the replicase gene confer robust transgenic resistance to negative-sense single-stranded RNA plant virus. Mol Plant Pathol 13:303–317

Lakatos L, Csorba T, Pantaleo V et al (2006) Small RNA binding is a common strategy to suppress RNA silencing by several viral suppressors. EMBO J 25:2768–2780

Li J, Yang Z, Yu B et al (2005) Methylation protects miRNAs and siRNAs from a 3′ end uridylation activity in Arabidopsis. Curr Biol 15:1501–1507

Li R, Weldegergis BT, Li J et al (2014) Virulence factors of geminivirus interact with MYC2 to subvert plant resistance and promote vector performance. Plant Cell 26:4991–5008

Lin SS, Wu HW, Elena SF et al (2009) Molecular evolution of a viral non-coding sequence under the selective pressure of amiRNA-mediated silencing. PLoS Pathog 5:e1000312

Lister R, O'Malley RC, Tonti-Filippini J et al (2008) Highly integrated single-base resolution maps of the epigenome in Arabidopsis. Cell 133:523–536

Liu B, Li P, Li X et al (2005) Loss of function of OsDCL1 affects microRNA accumulation and causes developmental defects in rice. Plant Physiol 139:296–305

Llave C, Kasschau KD, Carrington JC (2000) Virus-encoded suppressor of posttranscriptional gene silencing targets a maintenance step in the silencing pathway. Proc Natl Acad Sci USA 97:13401–13406

Llave C, Kasschau KD, Rector MA et al (2002) Endogenous and silencing-associated small RNAs in plants. Plant Cell 14:1605–1619

Lozsa R, Csorba T, Lakatos L et al (2008) Inhibition of 3′ modification of small RNAs in virus-infected plants requires spatial and temporal co-expression of small RNAs and viral silencing-suppressor proteins. Nucleic Acids Res 36:4099–4107

Lu R, Folimonov A, Shintaku M et al (2004) Three distinct suppressors of RNA silencing encoded by a 20-kb viral RNA genome. Proc Natl Acad Sci USA 101:15742–15747

Lucy AP, Guo HS, Li WX et al (2000) Suppression of post-transcriptional gene silencing by a plant viral protein localized in the nucleus. EMBO J 19:1672–1680

Mallory A, Vaucheret H (2010) Form, functionand regulation of ARGONAUTE proteins. Plant Cell 22:3879–3889

Mallory AC, Ely L, Smith TH et al (2001) HC-Pro suppression of transgene silencing eliminates the small RNAs but not transgene methylation or the mobile signal. Plant Cell 13:571–583

Mallory AC, Reinhart BJ, Bartel D et al (2002) A viral suppressor of RNA silencing differentially regulates the accumulation of short interfering RNAs and micro-RNAs in tobacco. Proc Natl Acad Sci USA 99:15228–15233

Margis R, Fusaro AF, Smith NA et al (2006) The evolution and diversification of Dicers in plants. FEBS Lett 580:2442–2450

Matthew L (2004) RNAi for plant functional genomics. Comp Funct Genomics 5:240–244

Maziere P, Enright AJ (2007) Prediction of microRNA targets. Drug Discov Today 12:452–458

Mitter N, Sulistyowati E, Dietzgen RG (2003) Cucumber mosaic virus infection transiently breaks dsRNA-induced transgenic immunity to Potato virus Y in tobacco. Mol Plant-Microbe Interact 16:936–944

Mlotshwa S, Schauer SE, Smith TH et al (2005) Ectopic DICER-LIKE1 expression in P1/HC-Pro Arabidopsis rescues phenotypic anomalies but not defects in microRNA and silencing pathways. Plant Cell 17:2873–2288

Moissiard G, Voinnet O (2006) RNA silencing of host transcripts by cauliflower mosaic virus requires coordinated action of the four Arabidopsis Dicer-like proteins. Proc Natl Acad Sci USA 103:19593–19598

Murray GGR, Kosakovsky Pond SL, Obbard DJ (2013) Suppressors of RNAi from plant viruses are subject to episodic positive selection. Proc R Soc B Biol Sci 280:20130965

Nakahara KS, Masuta C, Yamada S et al (2012) Tobacco calmodulin-like protein provides secondary defense by binding to and directing degradation of virus RNA silencing suppressors. Proc Natl Acad Sci USA 109:10113–10118

Napoli C, Lemieux C, Jorgensen R (1990) Introduction of a chimeric chalcone synthase gene into petunia results in reversible co-suppression of homologous genes in trans. Plant Cell 2:279–289
Naqvi AR, Haq QM, Mukherjee SK (2010) MicroRNA profiling of tomato leaf curl new delhi virus (ToLCNDV) infected tomato leaves indicates that deregulation of mir159/319 and mir172 might be linked with leaf curl disease. Virol J 7:1
Nicolas FE, Torres-Martínez S, Ruiz-Vazquez RM (2013) Loss and retention of RNA interference in fungi and parasites. PLoS Pathog 9:e1003089
Niu QW, Lin SS, Reyes JL et al (2006) Expression of artificial microRNAs in transgenic *Arabidopsis thaliana* confers virus resistance. Nat Biotechnol 24:1420–1428
Palatnik JF, Allen E, Wu X et al (2003) Control of leaf morphogenesis by microRNAs. Nature 425:257–263
Palauqui J-C, Vaucheret H (1998) Transgenes are dispensable for the RNA degradation step of cosuppression. Proc Natl Acad Sci USA 95:9675–9680
Palauqui JC, Elmayan T, Pollien JM et al (1997) Systemic acquired silencing: transgene-specific post-transcriptional silencing is transmitted by grafting from silenced stocks to non-silenced scions. EMBO J 16:4738–4745
Park W, Li J, Song R et al (2002) CARPEL FACTORY, a Dicer homolog, and HEN1, a novel protein, act in microRNA metabolism in *Arabidopsis thaliana.* Curr Biol 12:1484–1149
Perrimon N, Ni JQ, Perkins L (2010) In vivo RNAi: today and tomorrow. Cold Spring Harb Perspect Biol 2:a003640
Pooggin MM (2013) How can plant DNA viruses evade siRNA-directed DNA methylation and silencing? Int J Mol Sci 14:15233–15259
Pruss G, Ge X, Shi XM et al (1997) Plant viral synergism: the potyviral genome encodes a broad-range pathogenicity enhancer that transactivates replication of heterologous viruses. Plant Cell 9:859–868
Qu J, Ye J, Fang RX (2007) Artificial microRNA-mediated virus resistance in plants. J Virol 81:6690–6699
Rahman J, Karjee S, Mukherjee SK (2012) MYMIV-AC2, a geminiviral RNAi suppressor protein, has potential to increase the transgene expression. Appl Biochem Biotechnol 167:758–775
Rahman J, Sanan-Mishra N, Mukherjee SK (2014) MYMIV-AC2 protein suppresses hairpin-induced gene silencing in *Nicotiana tabacum* cv. Xanthi. Plant Biotechnol Rep 8:337–347
Raja P, Wolf JN, Bisaro DM (2010) RNA silencing directed against geminiviruses: post-transcriptional and epigenetic components. Biochim Biophys Acta 1799:337–351
Ramachandran V, Chen X (2008) Degradation of microRNAs by a family of exoribonucleases in Arabidopsis. Science 321:1490–1492
Rauschhuber C, Noske N, Ehrhardt A (2012) New insights into stability of recombinant adenovirus vector genomes in mammalian cells. Eur J Cell Biol 91:2–9
Rhoades MW, Reinhart BJ, Lim LP et al (2002) Prediction of plant microRNA targets. Cell 110:513–520
Ricaño-Rodríguez J, Adame-García J, Portilla-Vázquez S et al (2016) The extraordinary nature of RNA interference in understanding gene downregulation mechanism in plants. In: Abdurakhmonov IY (ed) Plant genomics. InTech
Roth BM, Pruss GJ, Vance VB (2004) Plant viral suppressors of RNA silencing. Virus Res 102:97–108
Ruiz MT, Voinnet O, Baulcombe DC (1998) Initiation and maintenance of virus-induced gene silencing. Plant Cell 10:937–946
Ruiz-Ferrer V, Voinnet O (2007) Viral suppression of RNA silencing: 2b wins the Golden Fleece by defeating Argonaute. Bioessays 29:319–323
Sahu PP, Sharma N, Puranik S, Prasad M (2014) Post-transcriptional and epigenetic arms of RNA silencing: a defense machinery of naturally tolerant tomato plant against tomato leaf curl New Delhi virus. Plant Mol Biol Rep 32:1015–1029

Saze H, Mittelsten Scheid O, Paszkowski J (2003) Maintenance of CpG methylation is essential for epigenetic inheritance during plant gametogenesis. Nat Genet 34:65–69

Schauer SE, Jacobsen SE, Meinke DW et al (2002) DICER-LIKE1: blind men and elephants in Arabidopsis development. Trends Plant Sci 7:487–491

Schwab R, Palatnik JF, Riester M et al (2005) Specific effects of microRNAs on the plant transcriptome. Dev Cell 8:517–527

Shamandi N, Zytnicki M, Charbonnel C et al (2015) Plants encode a general siRNA suppressor that is induced and suppressed by viruses. PLoS Biol 13:e1002326

Shi BJ, Palukaitis P, Symons RH (2002) Differential virulence by strains of Cucumber mosaic virus is mediated by the 2b gene. Mol Plant-Microbe Interact 15:947–955

Shiboleth YM, Haronsky E, Leibman D et al (2007) The conserved FRNK box in HC-Pro a plant viral suppressor of gene silencing is required for small RNA binding and mediates symptom development. J Virol 81:13135–13148

Siddiqui SA, Valkonen JP, Rajamäki M et al (2011) The 2b silencing suppressor of a mild strain of Cucumber mosaic virus alone is sufficient for synergistic interaction with Tobacco mosaic virus and induction of severe leaf malformation in 2b-transgenic tobacco plants. Mol Plant-Microbe Interact 24:685–693

Silhavy D, Molnar A, Lucioli A et al (2002) A viral protein suppresses RNA silencing and binds silencing-generated 21- to 25-nucleotide double-stranded RNAs. EMBO J 21:3070–3080

Singh RK, Rai N, Kumar P et al (2015) Inheritance study in tomato (*Solanum lycopersicum*) for Tomato leaf curl virus (ToLCV) resistance. Indian J Agri Sci 85:896–901

Soards AJ, Murphy AM, Palukaitis P et al (2002) Virulence and differential local and systemic spread of cucumber mosaic virus in tobacco are affected by the CMV 2b protein. Mol Plant-Microbe Interact 15:647–653

Sonoda S, Nishiguchi M (2000) Graft transmission of post-transcriptional gene silencing: target specificity for RNA degradation is transmissible between silenced and non-silenced plants, but not between silenced plants. Plant J 21:1–8

Stanley J, Gay MR (1983) Nucleotide sequence of cassava latent virus DNA. Nature 301:260–262

Sun YW, Tee CS, Ma YH et al (2015) Attenuation of histone methyltransferase KRYPTONITE-mediated transcriptional gene silencing by Geminivirus. Sci Rep 5:16476

Sunter G, Bisaro DM (1991) Transactivation in a geminivirus: AL2 gene product is needed for coat protein expression. Virology 180:416–419

Sunter G, Bisaro DM (1992) Transactivation of geminivirus AR1 and BR1 gene expression by the viral AL2 gene product occurs at the level of transcription. Plant Cell 4:1321–1331

Takeda A, Sugiyama K, Nagano H et al (2002) Identification of a novel RNA silencing suppressor, NSs protein of Tomato spotted wilt virus. FEBS Lett 532:75–79

Thi EP, Lee ACH, Geisbert JB et al (2016) Rescue of non-human primatesfrom advanced Sudan ebolavirus infection with lipid encapsulated siRNA. Nat Microbiol 1:16142

Tien VV, Choudhury NR, Mukherjee SK (2013) Transgenic tomato plants expressing artificial microRNAs for silencing the pre-coat and coat proteins of a begomovirus, Tomato leaf curl New Delhi virus, show tolerance to virus infection. Virus Res 172:35–45

Til'kunova T, Zinovkin RA, Agranovskiĭ AA (2004) RNA-binding properties of proteins of the beet yellows closterovirus. Mol Biol (Mosk) 38:553–558

Tiwari M, Sharma D, Trivedi PK (2014) Artificial microRNA mediated gene silencing in plants: progress and perspectives. Plant Mol Biol 86:1

Trinks D, Rajeswaran R, Shivaprasad PV et al (2005) Suppression of RNA silencing by a geminivirus nuclear protein, AC2, correlates with transactivation of host genes. J Virol 79:2517–2527

Vance V, Vaucheret H (2001) RNA silencing in plants – defense and counter defense. Science 292:2277–2280

Vanderschuren H, Akbergenov R, Pooggin MM et al (2007) Transgenic cassava resistance to African cassava mosaic virus is enhanced by viral DNA-A bidirectional promoter-derived siRNAs. Plant Mol Biol 64:549–557

Vanitharani R, Chellappan P, Pita JS et al (2004) Differential roles of AC2 and AC4 of cassava geminiviruses in mediating synergism and suppression of posttranscriptional gene silencing. J Virol 78:9487–9498
Vargason JM, Szittya G, Burgyan J et al (2003) Size selective recognition of siRNA by an RNA silencing suppressor. Cell 115:799–811
Vaucheret H (2008) Plant argonautes. Trends Plant Sci 13:350–358
Voinnet O (2001) RNA silencing as a plant immune system against viruses. Trends Genet 17:449–459
Voinnet O, Baulcombe DC (1997) Systemic signalling in gene silencing. Nature 389:553
Voinnet O, Vain P, Angell S et al (1998) Systemic spread of sequence-specific transgene RNA degradation in plants is initiated by localized introduction of ectopic promoter less DNA. Cell 95:177–187
Voinnet O, Pinto YM, Baulcombe DC (1999) Suppression of gene silencing: a general strategy used by diverse DNA and RNA viruses of plants. Proc Natl Acad Sci USA 96:14147–14152
Voinnet O, Lederer C, Baulcombe DC (2000) A viral movement protein prevents spread of the gene silencing signal in *Nicotiana benthamiana*. Cell 103:157–167
Wang H, Hao L, Shung CY et al (2003) Adenosine kinase is inactivated by geminivirus AL2 and L2 proteins. Plant Cell 15:3020–3032
Wang H, Buckley KJ, Yang X et al (2005) Adenosine kinase inhibition and suppression of RNA silencing by geminivirus AL2 and L2 proteins. J Virol 79:7410–7418
Wang B, Li S, Qi HH et al (2009) Distinct passenger strand and mRNA cleavage activities of human Argonaute proteins. Nat Struct Mol Biol 16:1259–1266
Wang X-B, Jovel J, Udomporn P et al (2011) The 21-nucleotide, but not 22-nucleotide, viral secondary small interfering RNAs direct potent antiviral defense by two cooperative argonautes in *Arabidopsis thaliana*. Plant Cell 23:1625–1638
Wang T, Castillo-González C, You L et al (2015) In vitro reconstitution assay of miRNA biogenesis by Arabidopsis DCL1. Bio-protocol 5:e1454
Waterhouse PM, Wang MB, Lough T (2001) Gene silencing as an adaptive defence against viruses. Nature 411:834–842
van Wezel R, Liu H, Tien P et al (2002) Mutation of three cysteine residues in tomato yellow leaf curl virus-China C2 protein causes dysfunction in pathogenesis and posttranscriptional gene silencing-suppression. Mol Plant-Microb Interact 15:203–208
Wu G, Park MY, Conway SR et al (2009a) The sequential action of miR156 and miR172 regulates developmental timing in Arabidopsis. Cell 138:750–759
Wu L, Zhang Q, Zhou H et al (2009b) Rice microRNA effector complexes and targets. Plant Cell 21:3421–3435
Xie Z, Kasschau KD, Carrington JC (2003) Negative feedback regulation of Dicer-Like1 in Arabidopsis by microRNA-guided mRNA degradation. Curr Biol 13:784–789
Xing Y, Zhang Q (2010) Genetic and molecular bases of rice yield. Annu Rev Plant Biol 61:421–442
Yadav RK, Chattopadhyay D (2011) Enhanced viral intergenic region-specific short interfering RNA accumulation and DNA methylation correlates with resistance against a geminivirus. Mol Plant-Microbe Interact 24:1189–1197
Yadava P, Suyal G, Mukherjee SK (2010) Begomovirus DNA replication and pathogenicity. Curr Sci 98:360–368
Yang X, Xie Y, Raja P et al (2011) Suppression of methylation-mediated transcriptional gene silencing by βC1-SAHH protein interaction during geminivirus-betasatellite infection. PLoS Pathog 7:e1002329
Yang L, Wu G, Poethig RS (2012) Mutations in the GW-repeat protein SUO reveal a developmental function for microRNA-mediated translational repression in *Arabidopsis*. Proc Natl Acad Sci USA 109:315–320

Yang LP, Fang YY, An CP et al (2013) C2-mediated decrease in DNA methylation, accumulation of siRNAsand increase in expression for genes involved in defense pathways in plants infected with beet severe curly top virus. Plant J 73:910–917

Ye R, Chen Z, Lian B et al (2016) A Dicer-independent route for biogenesis of siRNAs that direct DNA methylation in Arabidopsis. Mol Cell 61:222–235

Yu B, Yang Z, Li J et al (2005) Methylation as a crucial step in plant microRNA biogenesis. Science 307:932–935

Zhang H, Kolb FA, Jaskiewicz L et al (2004) Single processing center models for human Dicer and bacterial RNase III. Cell 118:57–68

Zhang X, Yuan Y-R, Pei Y et al (2006) Cucumber mosaic virus-encoded 2b suppressor inhibits Arabidopsis Argonaute1 cleavage activity to counter plant defense. Genes Dev 20:3255–3268

Zhang Z, Chen H, Huang X et al (2011) BSCTV C2 attenuates the degradation of SAMDC1 to suppress DNA methylation-mediated gene silencing in Arabidopsis. Plant Cell 2(3):273–288

Analysis of Nucleic Acids Methylation in Plants

Bi-Feng Yuan

Abstract 5-methylcysine (5-mC) and N^6-methyladenosine (m^6A) are important epigenetic marks occurring in nucleic acids of plants with regulatory roles in a broad range of biological processes. Recently, some novel modifications with potential regulatory roles such as 5-hydroxymethylcytosine (5-hmC), 5-formylcytosine (5-foC), and 5-carboxylcytosine (5-caC) have also been discovered in plants. Systematic investigation of the functions of nucleic acid modifications will promote the understanding of the mechanism underlying association of epigenetic modifications with plant development and response to environmental stresses. In this respect, great advances have been made in the development of methods for investigation of the occurrence and localization of these epigenetic modifications in nucleic acids of plants. Here, we focus on the recent methodological advances for the analysis of the global levels of DNA and RNA methylation. In addition, we will discuss the mostly used methods for mapping the genome-wide distribution of DNA and RNA methylation.

Keywords DNA methylation • RNA methylation • 5-methylcysine • 5-hydroxymethylcytosine • 5-formylcytosine • 5-carboxylcytosine • N^6-methyladenosine • global detection • mapping

Contents

B.-F. Yuan (✉)
Key Laboratory of Analytical Chemistry for Biology and Medicine (Ministry of Education), Department of Chemistry, Wuhan University, Wuhan 430072, P. R. China
e-mail: bfyuan@whu.edu.cn

N. Rajewsky et al. (eds.), *Plant Epigenetics*, RNA Technologies,
DOI 10.1007/978-3-319-55520-1_12

1 General Functions of DNA and RNA Methylation in Plants

Methylation modifications, notably in the forms of 5-methylcytosine (5-mC) and N^6-methyladenosine (m^6A) in both DNA and RNA, perform important regulatory functions in various biological processes (Chen et al. 2016; Shen et al. 2014; Wu and Zhang 2014). DNA and RNA methylation occurs in almost all living organisms, from bacteria to fungi, plants, and mammals (Motorin and Helm 2011; Zemach and Zilberman 2010). The fine-tuning of chromatin structure by DNA and RNA methylation is one of the major hallmarks of gene regulation during cellular development (He 2010; Jones 2012).

1.1 DNA Cytosine Methylation in Plants

Although the levels of 5-mC are relatively low in mammalian genomes (3–8% of total cytosine), 5-mC presents a much abundant level in plant genomes (5–25% of total cytosine) (Rangwala and Richards 2004). Plants have more complicated and sophisticated system of the genome methylation compared to animals. In plants, cytosine can be methylated at CG, CHG, and CHH sites (H is A, T, C) (Matzke et al. 2015). DNA methylation in plants is predominantly controlled by domains rearranged as methyltransferases 2 (DRM2) via the RNA-directed DNA methylation pathway and maintained by DNA methyltransferases 1 (MET1), chromomethylase 3 (CMT3), and DRM2 (Chan et al. 2005). CG methylation is mediated by MET1; CHH methylation is controlled by DRM2, while the plant-specific CMT3 regulates CHG methylation (Chan et al. 2005). DNA methylation in plants is involved in the control of genetic functions including transcription, replication, gene transposition, and cell differentiation (Law and Jacobsen 2010).

1.2 RNA Cytosine Methylation in Plants

In addition to occurring in DNA, 5-mC has also been identified in different RNA species from all kingdoms of life (Motorin and Helm 2011; Motorin et al. 2010). 5-mC residues in tRNAs are known to influence their secondary structural, stabilization, and codon recognition (Helm 2006; Squires and Preiss 2010). 5-mC sites are also found in rRNA where they play critical roles in recognition of tRNA and translational fidelity (Chow et al. 2007). And internal 5-mC in mRNA was also identified (Edelheit et al. 2013). In plants, the level of 5-mC in total RNA is about 0.88% in *Lepidium sativum*. When plant was exposed to abiotic stress, such as Cd (II) or Se(IV), 5-mC in both DNA and RNA changed (Yanez Barrientos et al. 2013). The advances in the field of epigenetics suggest that RNA cytosine methylation might play a similar role in the modulation of genetic information as DNA cytosine methylation in plants (Mattick et al. 2009).

1.3 RNA Adenine Methylation in Plants

Recent discovery of reversible m^6A modification on mRNA and mapping of m^6A in mammals revealed potential regulatory functions of this RNA adenine modification. In *Arabidopsis thaliana*, m^6A content in mRNA varies across tissues with a high ratio of m^6A/A found in flower buds, and defects in m^6A methyltransferase cause an embryo-lethal phenotype, suggesting a critical role of m^6A in plant development (Zhong et al. 2008). Recently, m^6A mapping analysis showed that m^6A is a highly conserved modification in mRNA of *Arabidopsis thaliana* (Luo et al. 2014). m^6A in mRNA of *Arabidopsis thaliana* is enriched around both the stop codon and the start codon. A positive correlation between m^6A deposition and mRNA levels indicates a regulatory role of m^6A in plant gene expression (Luo et al. 2014).

2 Global Detection of DNA and RNA Methylation in Plants

Established methods for the determination of global DNA and RNA methylation in plants mainly include liquid chromatography (LC), liquid chromatography-mass spectrometry (LC-MS), capillary electrophoresis (CE), thin layer chromatography (TLC), and immuno-based detection. The global detection of DNA and RNA methylation requires the liberation of DNA/RNA components with enzymatic/ chemical treatments followed by determination of the components with various methods.

2.1 Liquid Chromatography

The analysis of global DNA and RNA methylation by LC is based on the chromatographic separation of the components by enzymatic or chemical hydrolysis of DNA/RNA. Therefore, the baseline separation of the DNA/RNA components is necessary to avoid co-elution of analytes.

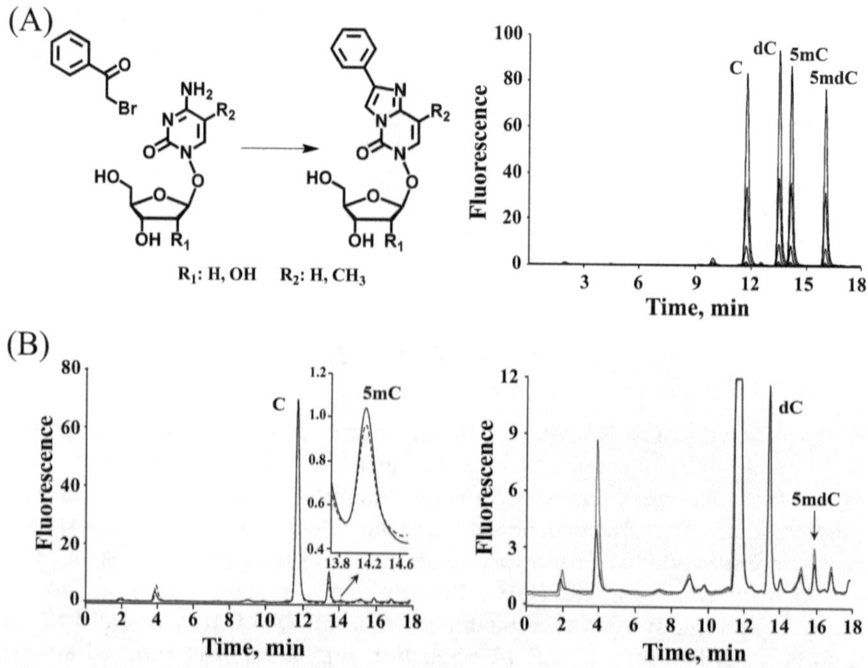

Fig. 1 (**a**) Schematic illustration for selective derivatization of cytosinemoieties with 2-bromoacetophenone for the determination of global DNA methylation by reversed phase liquid chromatography with spectrofluorimetric detection (Reprinted with permission from Torres AL, Barrientos EY, Wrobel K. *Anal Chem*, 2011, 83: 7999–8005). (**b**) LC-FLD chromatograms of control plant (*solid line*) and plant exposed to 1.0 mgL^{-1} Se(IV) (*dashed line*). *Left panel*: Y-scale and insert adjusted to visualize 5-mC and C signals (RNA methylation); *right panel*: Y-scale adjusted to visualize 5-mdC and dC signals (DNA methylation) (Reprinted with permission from Yanez Barrientos E, Wrobel K, Lopez Torres A, Gutierrez Corona F, Wrobel K. *Anal Bioanal Chem*, 2013, 405: 2397–2404)

Johnston et al. (2005) examined the methodological factors in LC analysis of plant DNA methylation using in vitro cultures of *Ribes ciliatum*. The results demonstrated that complete removal of RNA from plant DNA is difficult using RNase digestions and LiCl precipitation, suggesting that nucleobases analysis should be avoided as nucleobases from residual RNA fragments will interfere DNA-derived nucleobases. Nucleoside or nucleotide analysis is therefore recommended as a more suitable option. Liquid chromatographic techniques generally are quantitative, reproducible, but less sensitive. In this respect, a relatively large amount of genomic DNA (~1–50 μg) is normally needed.

To increase the detection sensitivity of 5-mC by LC-based detection, selective derivatization of cytosine moieties with 2-bromoacetophenone followed by reversed phase LC with spectrofluorimetric detection was developed (Torres et al. 2011; Yanez Barrientos et al. 2013) (Fig. 1). The proposed method was capable for the detection of as low as 0.06% of methylation in 80 ng of DNA and can be used for the evaluation of RNA methylation at the same time. With this method, Barrientos et al. (2013) analyzed the global DNA and total RNA methylation in *Lepidium sativum* and further assessed the effect of Cd(II) and Se(IV) exposure on DNA and RNA methylation of *Lepidium sativum*.

2.2 Liquid Chromatography-Mass Spectrometry

Due to the good selectivity and sensitivity, LC-MS has been widely used in the analysis of nucleic acid modifications.

Our group developed various methods for the detection of DNA and RNA methylation by LC-MS (Chen et al. 2013; Huang et al. 2016a, b, 2015; Jiang et al. 2015, 2016; Shen et al. 2015; Tang et al. 2013, 2014, 2015; Wang et al. 2013; Xiong et al. 2015; Yuan 2014; Yuan and Feng 2014; Yuan et al. 2011; Zhang et al. 2016). Specifically, we recently established a chemical derivatization strategy combined with liquid chromatography-electrospray ionization tandem mass spectrometry (LC/ESI-MS/MS) method to determine 5-formyl-2′-deoxycytidine (5-fodC) and 5-carboxyl-2′-deoxycytidine (5-cadC) in plants (Tang et al. 2014) (Fig. 2). Derivatization of 5-fodC and 5-cadC by Girard's reagents significantly increased the detection sensitivities of 5-fodC and 5-cadC by 52 to 260folds. Using this method, we demonstrated the widespread existence of 5-fodC and 5-cadC in genomic DNA of various plant tissues, with contents of 5-fodC ranging from 2.1 to 4.7 modifications per 10^6dG and 5-cadC ranging from 0.2 to 3.6 modifications per 10^6dG. Moreover, we found that environmental stresses of drought and salinity can change the contents of 5-fodC and 5-cadC in plant genomes, suggesting the functional roles of 5-fodC and 5-cadC in response to environmental stresses.

In addition, Liu et al. (2013) reported the use of a reversed-phase LC coupled with tandem mass spectrometry method and stable isotope-labeled standards for assessing the levels of the oxidized 5-mC nucleosides in DNA of *Arabidopsis*

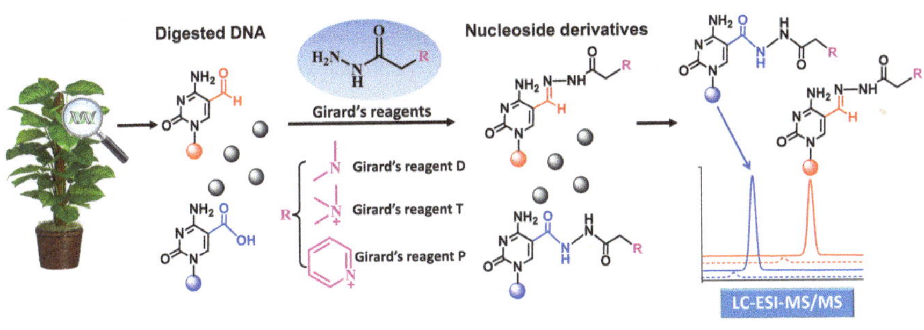

Fig. 2 Schematic diagram for determination of 5-fodC and 5-cadC in genomic DNA of plant samples by chemical derivatization using Girard's reagents coupled with LC/ESI-MS/MS analysis (Reprinted with permission from Tang Y, Xiong J, Jiang HP, Zheng SJ, Feng YQ, Yuan BF. *Anal Chem*, 2014, 86: 7764–7772)

thaliana. The quantitative results showed that the occurrence of 5-hmC, 5-foC, and 5-caC in DNA of *Arabidopsis thaliana* were 0.79, 0.79, and 0.78 modifications per 10^6nucleosides, respectively. This method is involved in the pre-enrichment of these modified nucleosides by LC, which can minimize the interference from other abundant nucleosides during subsequent mass spectrometry analysis but will increase the analytical time.

Recently, Magana et al (2016) investigated the effect of CuO nanoparticles on global DNA and RNA methylation in *Lepidium sativum* by liquid chromatography/ion trap mass spectrometry. Enhanced selectivity toward cytosine-containing nucleosides was achieved using proton-bound dimers formed in positive electrospray ionization as precursor ions for multiple reaction monitoring (MRM) quantification. The quantitative results showed that 13.03% and 0.92% methylated cytosines were found in DNA and RNA, respectively. Upon CuO nanoparticles treatment, DNA hypomethylation was observed, but RNA methylation did not present significant changes.

2.3 Capillary Electrophoresis

Several studies have developed capillary electrophoresis methods for the detection of genomic DNA methylation content in plant tissues. Fraga et al. (2000) used open-tube capillary electrophoresis system to separate acid hydrolyzed genomic DNA (nucleobases) from *Pinus radiate* trees for the evaluation of genomic DNA methylation. The problem is that the contamination of RNA may also contribute to the acid hydrolyzed products of nucleobases. In this respect, later they developed micellar capillary electrophoresis with UV-Vis detection to analyze the nucleosides of enzymatic products of DNA from *Pinus radiate* trees (Fraga et al. 2002). The detection and quantification of nucleosides through enzymatic hydrolyses notably increases the specificity and allows its exploitation in the analysis of poorly purified and/or concentrated DNA samples.

The capillary electrophoresis method offers the advantages of high resolution and cost-effective separations, providing an efficient approach to quantify nucleic acids methylation. The drawback of capillary electrophoresis method is that sample loading volume is limited and separation reproducibility can be affected by slight variations, which requires further improvements.

2.4 Thin Layer Chromatography

TLC-based method normally requires ribo- and deoxyribonucleotides that are generally distinguishable. After isolation of nucleic acids, the digested nucleosides were labeled with radioactive phosphate which enables sensitive determination of the contents of global DNA and RNA methylation. With the TLC method, global DNA methylation was successfully detected and quantified in various plant samples, including *Arabidopsis* and *Cardaminopsis arenosa* (Madlung et al. 2002),

Artemisia annua (Pandey and Pandey-Rai 2015), *Pyruscommunis* (Michalak et al. 2013), and *Quercus robur* (Michalak et al. 2015).

TLC method is cost-effective, and there is no need for sophisticated instrumentation. So TLC method is frequently used in biological lab for discovery and quantification of modified DNA and RNA. However, TLC-based method involves radioactive isotope labeling, and the analytical procedure is relatively tedious.

2.5 *Immuno-Based Detection*

Immunostaining is a technique widely used to evaluate the presence of DNA and RNA methylation. This technique relies antibodies that can selectively recognize the corresponding modified DNA and RNA inside cells for cell-based visualization.

Zluvova et al. (2001) employed immunostaining technique to examine the global changes of DNA methylation during seed germination and shoot apical meristem development in *Silene latifolia*. The data showed that a rapid decrease in global DNA methylation during seed germination occurred first in endosperm tissue and subsequently in the hypocotyl. To reveal the dynamics of the methylation pattern, correct epitope retrieval sometimes is essential and a deep denaturation step is needed.

In addition to 5-mC in DNA, Yao et al. (2012) found that DNA in leaves and flowers of *Arabidopsis thaliana* contains low level of 5-hmC by immuno-based dot-blot technique (Fig. 3). Using in vitro binding assays, the authors observed that full-length VIM1 protein binds preferentially to hemi-methylated DNA. However, when 5-hmC replaces one or both cytosine residues at CpG site, VIM1 binds with tenfold lower affinity. These results suggest that 5-hmC may contribute to VIM-mediated passive loss of cytosine methylation in vivo during DNA replication in *Arabidopsis thaliana*.

In the past several years, many commercially available kits, such as enzyme linked immunosorbent assay (ELISA) with chemiluminescence detection-based analysis, were also produced and employed for the detection of global 5-mC in *Phelipanche ramosa* (Lechat et al. 2015) and rice (Ferreira et al. 2015) and global 5-mC and 5-hmC in *B. oleracea* and *C. sativus* (Moricova et al. 2013).

3 Location Analysis of DNA and RNA Methylation in Plants

The quantitative distribution information of DNA and RNA methylation is crucial to understand their biological functions. The advance in sequencing technologies accelerates and revolutionizes the genome-wide distribution studies of DNA and RNA methylation (Mardis 2013). Two major strategies, including affinity

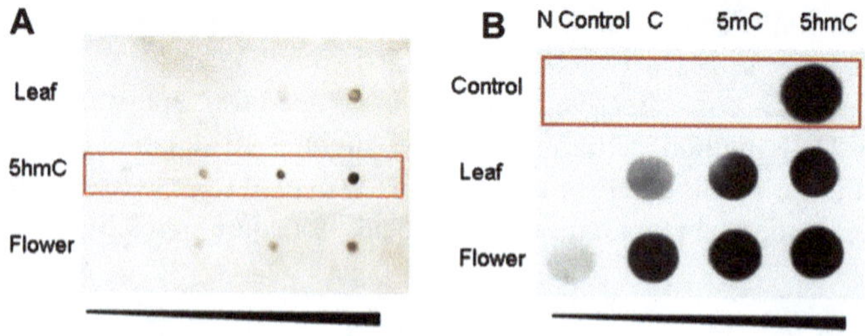

Fig. 3 Dot-blot detection of 5-hmC in DNA of *Arabidopsis thaliana*. (**a**) Dot-blot assay of synthetic DNA and *Arabidopsis thaliana* genomic DNA containing biotin-N3-5-gmC. The amount of loaded DNA increased from left to right. Detection *top row*: leaf DNA (25, 50, 100, 200 ng); *middle row*: 5-hmCcontaining DNA standard (boxed) (1–8 ng); *bottom row*: flower DNA (25, 50, 100, 200 ng). (**b**) Dot-blot assay using antibody to 5-hmC to detect 5-hmC in synthetic DNA or *Arabidopsis thaliana* genomic DNA. Spotted samples in *top row* (boxed): N Control: H_2O; C: Synthetic DNA with C; 5-mC: Synthetic DNA with 5mC; 5-hmC: Synthetic DNA with 5-hmC; 0.5 ng of each synthetic DNA were loaded; *middle row* and *bottom rows* (leaf and flower DNA, respectively): The amount of *Arabidopsis thaliana* DNA (0.5, 1, 2, 4 ng) increased from left to right (Reprinted with permission from Yao Q, Song CX, He C, Kumaran D, Dunn JJ. *Protein Expr Purif*, 2012, 83:104–111)

enrichment and bisulfite conversion followed by next-generation sequencing, have been widely used to profile the location of 5-mC in DNA and m^6A in RNA of plants.

3.1 *Affinity Enrichment-Sequencing Analysis*

Affinity enrichment of modified DNA and RNA using antibodies or affinity binding proteins has been proved to be a powerful tool for comprehensive profiling of 5-mC and its derivatives (Thu et al. 2010). The enriched DNA-antibody/protein complex can be analyzed using sequencing-based technologies (Down et al. 2008). However, affinity enrichment-sequencing methods do not provide location information at single-base resolution.

Methylated DNA immunoprecipitation sequencing (MeDIP-seq) relies on the use of antibody to precipitate fragments of DNA containing methylated cytosine (Down et al. 2008). Related methods include MBD-seq (Serre et al. 2010) and MethylCap-seq (Brinkman et al. 2010), which employ methyl CpG binding domain protein to precipitate DNA fragments containing methylated CpG sites. These approaches have the advantage that a greatly reduced portion of the genome needs to be sequenced. However, the major drawback of these approaches is the low resolution for identifying methylation sites.

Genome-wide profiles of DNA methylation for maize (*Zea mays*) inbred lines by MeDIP-seq demonstrated that DNA methylation variation is influenced by genetic and epigenetic changes that are often stably inherited and can affect the expression of nearby genes (Eichten et al. 2013). Hu et al. (2015) used MeDIP-seq to profile DNA methylation in the rice PTGMS line PA64S under two different phenotypes (sterility and fertility). The results revealed that hypermethylation was observed in PA64S (sterility), and 1258 differentially methylated regions were found between PA64S (sterility) and PA64S (fertility).

In addition to 5-mC in DNA, Luo et al. (2014) established m^6A-targeted antibody enrichment coupled with next-generation sequencing to map transcriptome-wide m^6A in *Arabidopsis thaliana*. The results showed that m^6A is a highly conserved modification of mRNA in plants. Distinct from mammals, m^6A in *Arabidopsis thaliana* is enriched around both the stop codon and the start codon. The distribution pattern of m^6A in *Arabidopsis thaliana* is associated with plant-specific pathways involving the chloroplast. A positive correlation between m^6A deposition and mRNA abundance was observed, suggesting a regulatory role of m^6A in plant gene expression. The m^6A transcriptome-wide study of *Arabidopsis thaliana* provides a starting roadmap for uncovering m^6A functions that may regulate plant metabolism.

Recently, Shen et al. (2016) employed anti-m^6A polyclonal antibody enrichment coupled with next-generation sequencing that further map the m^6A sites in mRNA of *Arabidopsis thaliana*. The results suggested an indispensable role of FIP37 (a core component of the m^6A methyltransferase complex) in mediating m^6A mRNA modification, which is required for maintaining the shoot meristem as a renewable source for continuously producing aerial organs in plants.

3.2 *Bisulfite Conversion-Sequencing Analysis*

The discovery that treatment of DNA with sodium bisulfite revolutionized DNA methylation analysis since 1990s (Clark et al. 1994; Frommer et al. 1992). And various methodologies have been developed based on bisulfite treatment that leads to the conversion of unmethylated cytosine to uracil, while methylated cytosine remains unchanged in DNA and RNA (Plongthongkum et al. 2014; Schaefer 2015). Amplification by polymerase chain reaction of converted DNA followed by sequencing can reveal positions of 5-mC in DNA and RNA. Since bisulfite conversion-sequencing strategy can provide single-base resolution for DNA and RNA cytosine methylation, the technique has been widely utilized in various plant samples, including *Arabidopsis thaliana* (Cokus et al. 2008; Feng et al. 2010; Ibarra et al. 2012; Lister et al. 2008; Shen et al. 2012; Stroud et al. 2013; Yu et al. 2013), rice (Li et al. 2012), soybean (Song et al. 2013), and *Marchantia polymorpha* (Takuno et al. 2016).

With applications of bisulfite conversion-sequencing to wild-type *Arabidopsis thaliana* and mutants defective in DNA methyltransferase or demethylase activity, Lister et al. (2008) observed local sequence effects upon methylation state and revealed a direct relationship between the location of smRNAs and DNA methylation. Cokus et al. (2008) identified sequence motifs that associate with high and low methylation for each different context of methylation in *Arabidopsis thaliana*.

Using the bisulfite conversion-sequencing, Stroud et al. (2013) substantially extended and refined the characterization of regulatory factors of the methylome by examining 86 *Arabidopsis thaliana* mutants, suggesting that individual sites of methylation may be regulated by novel RNA-directed pathways. Feng et al. (2010) carried out bisulfite conversion-sequencing in the flowering plants rice and *Arabidopsis thaliana* and found that the patterns of methylation were similar in flowering plants with methylated cytosines detected in all sequence contexts. Shen et al. (2012) found that genome-wide remodeling of DNA methylation mediated by the RNA-directed DNA methylation pathway in *Arabidopsis thaliana* may play a role in heterosis. Yu et al. (2013) found some transposable elements are demethylated and transcriptionally reactivated during antibacterial defense in *Arabidopsis thaliana*, which provides evidence that DNA demethylation is part of a plant-induced immune response and can potentially act to transcriptional activation of some defense genes linked to transposable elements and repeats. Ibarra et al. (2012) demonstrated that demethylation in companion cells reinforces transposon methylation in *Arabidopsis thaliana* gametes and contributes to stable silencing of transposable elements across generations.

In addition to the most studied model plant of *Arabidopsis thaliana*, Li et al. (2012) generated single-base resolution DNA methylome maps by bisulfite conversion-sequencing for Asian cultivated rice *Oryza sativa* ssp. *japonica*, *indica*, and their wild relatives, *Oryza rufipogon* and *Oryza nivara*. The overall methylation level of rice genomes is four times higher than that of *Arabidopsis thaliana*. Interestingly, the authors discovered that methylation in gene transcriptional termination regions can significantly repress gene expression, and the effect is stronger than that of promoter methylation.

Song et al. (2013) also analyzed the DNA methylation status in soybean roots, stems, leaves, and cotyledons of developing seeds at single-base resolution. Profiling of DNA methylation in different organs revealed 2162 differentially methylated regions among organs. Recently, Takuno et al. (2016) studied the single-base resolution methylome that span the phylogenetic breadth of land plants using bisulfite conversion-sequencing. The results showed that a basal land plant, *Marchantia polymorpha*, lacks evident signal of gene-body methylation within exons, but conifers have high methylation levels in both CG and CHG sites in expressed genes, which indicated the evolutionary forces acting on DNA methylation vary substantially across species, genes, and methylation contexts.

The advantage of bisulfite-converted strategy is that it can provide single-base resolution for DNA and RNA methylation analysis. However, the sample preparation associated with bisulfite sequencing can be time-consuming, and the conversion process may result in DNA and RNA degradation and reduce sequence complexity. As bisulfite analysis depends on the complete conversion of unmethylated cytosines to uracil, incomplete conversion will cause error or inaccurate results. In addition, discrimination between dC, 5-mC, and 5-hmC cannot be accomplished by bisulfite sequencing. Therefore, these exiting issues need to be further addressed.

3.3 *Single-Molecule Detection*

DNA methylation analysis by single-molecule, real-time (SMRT) sequencing without bisulfite conversion was first established in 2010 (Flusberg et al. 2010). The SMRT sequencing is considered to be the third-generation sequencing technology and can realize the direct distribution study of 5-mC in DNA. In SMRT sequencing, DNA polymerases catalyze the incorporation of fluorescently labeled nucleotides into complementary DNA strands. The recording of nucleotide incorporations generates the sequence readout and information about the polymerase kinetics, which are used to discriminate different nucleotides.

In addition to SMRT method, nanopore sequencing technology also has been established to single-molecule detection of modified nucleosides (Branton et al. 2008). Nanopore analysis uses a voltage to drive molecules through a nanoscale pore and monitors how the ionic current through the nanopore changes as single molecules pass through it (Venkatesan and Bashir 2011). Different nucleotides passing through nanopores generate different electric currents, which can be measured and designated to the corresponding nucleotides or modified nucleotides. The methodology is successfully used to distinguish methylated from unmethylated cytosines without bisulfite conversion (Clarke et al. 2009; Mirsaidov et al. 2009).

Using the SMRT sequencing, Kim et al. (2014) obtained high-coverage SMRT sequence datasets from five organisms including *Arabidopsis thaliana*. Later, Berlin et al. (2015)introduced the MinHash Alignment Process (MHAP) and integrated MHAP with the Celera Assembler, enabling reference-grade de novo assemblies of *Arabidopsis thaliana* from SMRT sequencing. Although the SMRT sequencing and nanopore sequencing techniques haven't been employed for DNA methylation analysis in plants so far, these powerful techniques are qualified for mapping of DNA methylation and we expect the technique will be used in plant samples in the near future.

4 Conclusions and Perspectives

DNA and RNA methylation are important epigenetic modifications in eukaryotes to maintain genome integrity and regulate gene expression. DNA methylation in plants is species-, tissue-, organelle- and age-specific. Although DNA cytosine methylome and RNA adenine methylome have been profiled in plants, however, these previous studies mainly used the mixture of different cell types for methylome analysis, which may shade somemethylation patterns in specific cell types. Therefore, the improvement on the isolation of cell type-specific nucleic acids and methylome sequencing in a single cell would facilitate to understand the dynamics of DNA and RNA methylation in plants.

The application of next-generation sequencing technology in DNA and RNA methylation studies has greatly contributed to our knowledge of DNA and RNA methylation. Future applications of some newly developed sequencing approaches such as single-molecule sequencing approaches are particularly well suited for the location study of DNA and RNA methylation. SMRT and nanopore sequencing offer the potential for direct sequencing of nucleic acid modifications without complicated pretreatment. The advancement of new technologies and methods may also lead to the discovery of novel epigenetic modifications in both DNA and RNA that will enhance our understanding of the fundamental issues in cellular developmental processing in plants.

Acknowledgment The author thanks the financial support from the National Natural Science Foundation of China (21522507, 21672166).

References

Alcazar Magana A, Wrobel K, Corrales Escobosa AR et al (2016) Application of liquid chromatography/electrospray ionization ion trap tandem mass spectrometry for the evaluation of global nucleic acids: methylation in garden cress under exposure to CuO nanoparticles. Rapid Commun Mass Spectrom 30:209–220

Berlin K, Koren S, Chin CS et al (2015) Assembling large genomes with single-molecule sequencing and locality-sensitive hashing. Nat Biotechnol 33:623–630

Branton D, Deamer DW, Marziali A et al (2008) The potential and challenges of nanopore sequencing. Nat Biotechnol 26:1146–1153

Brinkman AB, Simmer F, Ma K et al (2010) Whole-genome DNA methylation profiling using MethylCap-seq. Methods 52:232–236

Chan SW, Henderson IR, Jacobsen SE (2005) Gardening the genome: DNA methylation in Arabidopsis thaliana. Nat Rev Genet 6:351–360

Chen ML, Shen F, Huang W et al (2013) Quantification of 5-methylcytosine and 5-hydroxymethylcytosine in genomic DNA from hepatocellular carcinoma tissues by capillary hydrophilic-interaction liquid chromatography/quadrupole TOF mass spectrometry. Clin Chem 59:824–832

Chen K, Zhao BS, He C (2016) Nucleic acid modifications in regulation of gene expression. Cell Chem Biol 23:74–85

Chow CS, Lamichhane TN, Mahto SK (2007) Expanding the nucleotide repertoire of the ribosome with post-transcriptional modifications. ACS Chem Biol 2:610–619

Clark SJ, Harrison J, Paul CL et al (1994) High sensitivity mapping of methylated cytosines. Nucleic Acids Res 22:2990–2997

Clarke J, Wu HC, Jayasinghe L et al (2009) Continuous base identification for single-molecule nanopore DNA sequencing. Nat Nanotechnol 4:265–270

Cokus SJ, Feng S, Zhang X et al (2008) Shotgun bisulphite sequencing of the Arabidopsis genome reveals DNA methylation patterning. Nature 452:215–219

Down TA, Rakyan VK, Turner DJ et al (2008) A Bayesian deconvolution strategy for immunoprecipitation-based DNA methylome analysis. Nat Biotechnol 26:779–785

Edelheit S, Schwartz S, Mumbach MR et al (2013) Transcriptome-wide mapping of 5-methylcytidine RNA modifications in bacteria, archaea, and yeast reveals m5C within archaeal mRNAs. PLoS Genet 9:e1003602

Eichten SR, Briskine R, Song J et al (2013) Epigenetic and genetic influences on DNA methylation variation in maize populations. Plant Cell 25:2783–2797

Feng S, Cokus SJ, Zhang X et al (2010) Conservation and divergence of methylation patterning in plants and animals. Proc Natl Acad Sci U S A 107:8689–8694

Ferreira LJ, Azevedo V, Maroco J et al (2015) Salt tolerant and sensitive rice varieties display differential methylome flexibility under salt stress. PLoS One 10:e0124060

Flusberg BA, Webster DR, Lee JH et al (2010) Direct detection of DNA methylation during single-molecule, real-time sequencing. Nat Methods 7:461–465

Fraga MF, Rodriguez R, Canal MJ (2000) Rapid quantification of DNA methylation by high performance capillary electrophoresis. Electrophoresis 21:2990–2994

Fraga MF, Uriol E, Borja Diego L et al (2002) High-performance capillary electrophoretic method for the quantification of 5-methyl 2′-deoxycytidine in genomic DNA: application to plant, animal and human cancer tissues. Electrophoresis 23:1677–1681

Frommer M, McDonald LE, Millar DS et al (1992) A genomic sequencing protocol that yields a positive display of 5-methylcytosine residues in individual DNA strands. Proc Natl Acad Sci U S A 89:1827–1831

He C (2010) Grand challenge commentary: RNA epigenetics? Nat Chem Biol 6:863–865

Helm M (2006) Post-transcriptional nucleotide modification and alternative folding of RNA. Nucleic Acids Res 34:721–733

Hu J, Chen X, Zhang H et al (2015) Genome-wide analysis of DNA methylation in photoperiod- and thermo-sensitive male sterile rice Peiai 64S. BMC Genomics 16:102

Huang W, Xiong J, Yang Y et al (2015) Determination of DNA adenine methylation in genomes of mammals and plants by liquid chromatography/mass spectrometry. RSC Adv 5:64046–64054

Huang W, Lan MD, Qi CB et al (2016a) Formation and determination of the oxidation products of 5-methylcytosine in RNA. Chem Sci 7:5495–5502

Huang W, Qi CB, Lv SW et al (2016b) Determination of DNA and RNA methylation in circulating tumor cells by mass spectrometry. Anal Chem 88:1378–1384

Ibarra CA, Feng X, Schoft VK et al (2012) Active DNA demethylation in plant companion cells reinforces transposon methylation in gametes. Science 337:1360–1364

Jiang HP, Qi CB, Chu JM et al (2015) Profiling of cis-diol-containing nucleosides and ribosylated metabolites by boronate-affinity organic-silica hybrid monolithic capillary liquid chromatography/mass spectrometry. Sci Rep 5:7785

Jiang HP, Chu JM, Lan MD et al (2016) Comprehensive profiling of ribonucleosides modification by affinity zirconium oxide-silica composite monolithic column online solid-phase microextraction—mass spectrometry analysis. J Chromatogr A 1462:90–99

Johnston JW, Harding K, Bremner DH et al (2005) HPLC analysis of plant DNA methylation: a study of critical methodological factors. Plant Physiol Biochem 43:844–853

Jones PA (2012) Functions of DNA methylation: islands, start sites, gene bodies and beyond. Nat Rev Genet 13:484–492

Kim KE, Peluso P, Babayan P et al (2014) Long-read, whole-genome shotgun sequence data for five model organisms. Sci Data 1:140045

Law JA, Jacobsen SE (2010) Establishing, maintaining and modifying DNA methylation patterns in plants and animals. Nat Rev Genet 11:204–220

Lechat MM, Brun G, Montiel G et al (2015) Seed response to strigolactone is controlled by abscisic acid-independent DNA methylation in the obligate root parasitic plant, Phelipanche ramosa L. Pomel. J Exp Bot 66:3129–3140

Li X, Zhu J, Hu F et al (2012) Single-base resolution maps of cultivated and wild rice methylomes and regulatory roles of DNA methylation in plant gene expression. BMC Genomics 13:300

Lister R, O'Malley RC, Tonti-Filippini J et al (2008) Highly integrated single-base resolution maps of the epigenome in Arabidopsis. Cell 133:523–536

Liu S, Dunwell TL, Pfeifer GP et al (2013) Detection of oxidation products of 5-methyl-2-′-deoxycytidine in Arabidopsis DNA. PLoS One 8:e84620

Luo GZ, MacQueen A, Zheng G et al (2014) Unique features of the m6A methylome in Arabidopsis thaliana. Nat Commun 5:5630

Madlung A, Masuelli RW, Watson B et al (2002) Remodeling of DNA methylation and phenotypic and transcriptional changes in synthetic Arabidopsis allotetraploids. Plant Physiol 129:733–746

Mardis ER (2013) Next-generation sequencing platforms. Annu Rev Anal Chem (Palo Alto Calif) 6:287–303

Mattick JS, Amaral PP, Dinger ME et al (2009) RNA regulation of epigenetic processes. Bioessays 31:51–59

Matzke MA, Kanno T, Matzke AJ (2015) RNA-directed DNA methylation: the evolution of a complex epigenetic pathway in flowering plants. Annu Rev Plant Biol 66:243–267

Michalak M, Barciszewska MZ, Barciszewski J et al (2013) Global changes in DNA methylation in seeds and seedlings of Pyrus communis after seed desiccation and storage. PLoS One 8: e70693

Michalak M, Plitta-Michalak BP, Naskret-Barciszewska M et al (2015) Global 5-methylcytosine alterations in DNA during ageing of Quercus robur seeds. Ann Bot 116:369–376

Mirsaidov U, Timp W, Zou X et al (2009) Nanoelectromechanics of methylated DNA in a synthetic nanopore. Biophys J 96:L32–L34

Moricova P, Ondrej V, Navratilova B et al (2013) Changes of DNA methylation and hydroxymethylation in plant protoplast cultures. Acta Biochim Pol 60:33–36

Motorin Y, Helm M (2011) RNA nucleotide methylation. Wiley Interdiscip Rev RNA 2:611–631

Motorin Y, Lyko F, Helm M (2010) 5-methylcytosine in RNA: detection, enzymatic formation and biological functions. Nucleic Acids Res 38:1415–1430

Pandey N, Pandey-Rai S (2015) Deciphering UV-B-induced variation in DNA methylation pattern and its influence on regulation of DBR2 expression in Artemisia annua L. Planta 242:869–879

Plongthongkum N, Diep DH, Zhang K (2014) Advances in the profiling of DNA modifications: cytosine methylation and beyond. Nat Rev Genet 15:647–661

Rangwala SH, Richards EJ (2004) The value-added genome: building and maintaining genomic cytosine methylation landscapes. Curr Opin Genet Dev 14:686–691

Schaefer M (2015) RNA 5-methylcytosine analysis by bisulfite sequencing. Methods Enzymol 560:297–329

Serre D, Lee BH, Ting AH (2010) MBD-isolated genome sequencing provides a high-throughput and comprehensive survey of DNA methylation in the human genome. Nucleic Acids Res 38:391–399

Shen H, He H, Li J et al (2012) Genome-wide analysis of DNA methylation and gene expression changes in two Arabidopsis ecotypes and their reciprocal hybrids. Plant Cell 24:875–892

Shen L, Song CX, He C, Zhang Y (2014) Mechanism and function of oxidative reversal of DNA and RNA methylation. Annu Rev Biochem 83:585–614

Shen F, Huang W, Huang JT et al (2015) Decreased N(6)-methyladenosine in peripheral blood RNA From diabetic patients is associated with FTO expression rather than ALKBH5. J Clin Endocrinol Metab 100:E148–E154

Shen L, Liang Z, Gu X et al (2016) N(6)-methyladenosine RNA modification regulates shoot stem cell fate in Arabidopsis. Dev Cell 38:186–200

Song QX, Lu X, Li QT et al (2013) Genome-wide analysis of DNA methylation in soybean. Mol Plant 6:1961–1974

Squires JE, Preiss T (2010) Function and detection of 5-methylcytosine in eukaryotic RNA. Epigenomics 2:709–715

Stroud H, Greenberg MV, Feng S et al (2013) Comprehensive analysis of silencing mutants reveals complex regulation of the Arabidopsis methylome. Cell 152:352–364

Takuno S, Ran JH, Gaut BS (2016) Evolutionary patterns of genic DNA methylation vary across land plants. Nat Plants 2:15222

Tang Y, Chu JM, Huang W et al (2013) Hydrophilic material for the selective enrichment of 5-hydroxymethylcytosine and its liquid chromatography-tandem mass spectrometry detection. Anal Chem 85:6129–6135

Tang Y, Xiong J, Jiang HP et al (2014) Determination of oxidation products of 5-methylcytosine in plants by chemical derivatization coupled with liquid chromatography/tandem mass spectrometry analysis. Anal Chem 86:7764–7772

Tang Y, Zheng SJ, Qi CB et al (2015) Sensitive and simultaneous determination of 5-methylcytosine and its oxidation products in genomic DNA by chemical derivatization coupled with liquid chromatography-tandem mass spectrometry analysis. Anal Chem 87:3445–3452

Thu KL, Pikor LA, Kennett JY et al (2010) Methylation analysis by DNA immunoprecipitation. J Cell Physiol 222:522–531

Torres AL, Barrientos EY, Wrobel K (2011) Selective derivatization of cytosine and methylcytosine moieties with 2-bromoacetophenone for submicrogram DNA methylation analysis by reversed phase HPLC with spectrofluorimetric detection. Anal Chem 83:7999–8005

Venkatesan BM, Bashir R (2011) Nanopore sensors for nucleic acid analysis. Nat Nanotechnol 6:615–624

Wang ST, Huang W, Lu W et al (2013) TiO2-based solid phase extraction strategy for highly effective elimination of normal ribonucleosides before detection of 2′-deoxynucleosides/low-abundance 2′-O-modified ribonucleosides. Anal Chem 85:10512–10518

Wu H, Zhang Y (2014) Reversing DNA methylation: mechanisms, genomics, and biological functions. Cell 156:45–68

Xiong J, Jiang HP, Peng CY et al (2015) DNA hydroxymethylation age of human blood determined by capillary hydrophilic-interaction liquid chromatography/mass spectrometry. Clin Epigenetics 7:72

Yanez Barrientos E, Wrobel K, Lopez Torres A et al (2013) Application of reversed-phase high-performance liquid chromatography with fluorimetric detection for simultaneous assessment of global DNA and total RNA methylation in Lepidium sativum: effect of plant exposure to Cd (II) and Se(IV). Anal Bioanal Chem 405:2397–2404

Yao Q, Song CX, He C et al (2012) Heterologous expression and purification of Arabidopsis thaliana VIM1 protein: in vitro evidence for its inability to recognize hydroxymethylcytosine, a rare base in Arabidopsis DNA. Protein Expr Purif 83:104–111

Yu A, Lepere G, Jay F et al (2013) Dynamics and biological relevance of DNA demethylation in Arabidopsis antibacterial defense. Proc Natl Acad Sci U S A 110:2389–2394

Yuan BF (2014) 5-methylcytosine and its derivatives. Adv Clin Chem 67:151–187

Yuan BF, Feng YQ (2014) Recent advances in the analysis of 5-methylcytosine and its oxidation products. TrAC-Trend Anal Chem 54:24–35

Yuan BF, Zhang J, Wang H et al (2011) 6-Thioguanine reactivates epigenetically silenced genes in acute lymphoblastic leukemia cells by facilitating proteasome-mediated degradation of DNMT1. Cancer Res 71:1904–1911

Zemach A, Zilberman D (2010) Evolution of eukaryotic DNA methylation and the pursuit of safer sex. Curr Biol 20:R780–R785

Zhang HY, Xiong J, Qi BL et al (2016) The existence of 5-hydroxymethylcytosine and 5-formylcytosine in both DNA and RNA in mammals. Chem Commun (Camb) 52:737–740

Zhong S, Li H, Bodi Z et al (2008) MTA is an Arabidopsis messenger RNA adenosine methylase and interacts with a homolog of a sex-specific splicing factor. Plant Cell 20:1278–1288

Zluvova J, Janousek B, Vyskot B (2001) Immunohistochemical study of DNA methylation dynamics during plant development. J Exp Bot 52:2265–2273

DNA Methylation in Plants by microRNAs

Sachin Teotia, Deepali Singh, and Guiliang Tang

Abstract DNA methylation is one of the key processes that regulate gene expression by epigenetic silencing of transcription. RNA-directed DNA methylation (RdDM) is one such epigenetic process that involves both short and long non-coding RNAs. In plants, most DNA methylation takes place through classical RdDM pathway triggered by siRNAs. Contrary to this, miRNAs have been shown to play a little role in DNA methylation. However, several recent studies have provided the evidence for miRNA-directed DNA methylation in plants. These miRNA-directed DNA methylation includes the roles of not only the canonical small (20–22 nt) miRNAs but all those of long miRNAs, siRNAs generated from miRNA precursors, and also of siRNAs generated from various miRNA cleavage products. The working mechanism of many of these diverse pathways is still not clear. In this chapter, we overview the salient features of miRNA-directed DNA methylation in plants and discuss several intricacies of such an event.

Keywords Small RNA • miRNA • lmiRNA • siRNA • ta-siRNA • DNA methylation • Epigenetics

S. Teotia (✉)
College of Agronomy, Henan Agricultural University, Zhengzhou, 450002, China

School of Biotechnology, Gautam Buddha University, Greater Noida, 201312, India

Department of Biological Sciences, Michigan Technological University, Houghton, MI, 49931, USA
e-mail: steotia@mtu.edu; sachin@gbu.ac.in

D. Singh
School of Biotechnology, Gautam Buddha University, Greater Noida, 201312, India

G. Tang (✉)
College of Agronomy, Henan Agricultural University, Zhengzhou, 450002, China

Department of Biological Sciences, Michigan Technological University, Houghton, MI, 49931, USA
e-mail: gtang1@mtu.edu

N. Rajewsky et al. (eds.), *Plant Epigenetics*, RNA Technologies,
DOI 10.1007/978-3-319-55520-1_13

Contents

1 Introduction

DNA methylation is one of the key processes which induce epigenetic changes in plants (Matzke et al. 2015). DNA methylation involves the addition of a methyl (CH_3) group covalently at the 5th-carbon of the cytosine ring by methyltransferases forming a 5-methylcytosine (m5C). This reaction is carried upon by enzymes like MET1 methyltransferase, CHROMOMETHYLASE 3 (CMT3) and DOMAINS REARRANGED METHYLTRANSFERASEs (DRMs) (Gehring and Henikoff 2007). In plants, cytosine can be methylated at CG, CHG, and CHH sites, where H represents any nucleotide but guanine (Matzke et al. 2015; Castel and Martienssen 2013). Cytosine DNA methylation is a heritable modification which can induce transcriptional repression.

Small RNAs (sRNAs) have emerged as key players in regulating DNA methylation (at different cytosine residues of CG, CHG, and CHH motifs) or histone modifications, thereby controlling transcriptional networks in organisms (Holoch and Moazed 2015). Depending upon their structures and biogenesis, sRNAs are classified into small interfering RNAs (siRNAs), microRNA (miRNAs), piwi-interacting RNAs (piRNAs) (Carthew and Sontheimer 2009; Ghildiyal and Zamore 2009). In plants, heterochromatic siRNAs (hc-siRNAs) are the most abundant sRNA species which are mainly 24 nt in length and are dependent upon DNA polymerase IV (Pol IV) and RNA-dependent RNA polymerase (RDR2) for their biogenesis. To date, among sRNAs, siRNAs have been implicated to play a major role in DNA methylation, but mainly of transposons and repetitive DNA (Xie and Yu 2015). They direct de novo DNA methylation at cytosine residues in all sequence contexts through the recruitment of DNA methyl transferase, DRM2, at the target gene loci (Chan et al. 2005; Zhang and Zhu 2011; Zhong et al. 2014). This process is called, RNA-directed DNA methylation (RdDM) and the resulting methylation can be spread at a loci from several hundred to several thousand nucleotides. Contrary to that, role of miRNAs in DNA methylation have been reported for a much lesser number of cases (Jia et al. 2011). The biogenesis of

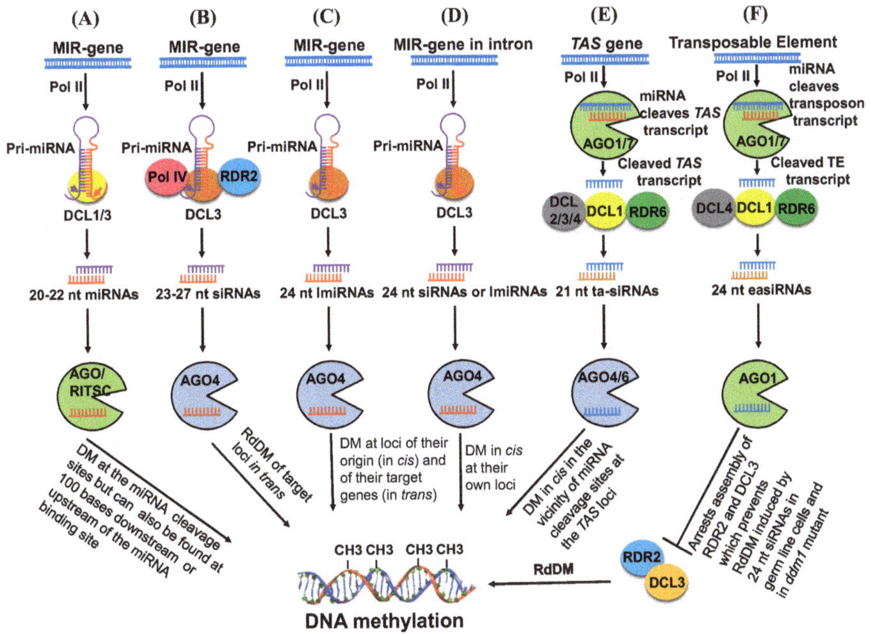

Fig. 1 microRNA-mediated DNA methylation pathways in plants: (**a**) Through canonical miRNAs in Arabidopsis and moss (Bao et al. 2004; Khraiwesh et al. 2010); (**b**) through siRNAs produced from miRNA loci in Arabidopsis (Chellappan et al. 2010); (**c**) through long miRNAs (lmiRNAs) in rice (Wu et al. 2010); (**d**) through siRNAs or lmiRNAs originating from miRNA loci located in the introns in rice (Chen et al. 2011); (**e**) through ta-siRNAs generated from miRNA-cleaved *TAS* gene transcripts in Arabidopsis (Wu et al. 2012); (**f**) through easiRNAs generated from miRNA-cleaved transposon transcripts in Arabidopsis (Creasey et al. 2014). *AGO* ARGONAUTE, *DCL* Dicer-like, *DM* DNA methylation, *Pol II* RNA polymerase II, *RDR* RNA-dependent RNA polymerase, *RITSC* RNA-induced transcriptional silencing complex

miRNAs takes place through the transcription of small non-coding genes by RNA-pol-II (Fig. 1). The single-stranded precursors of primary-miRNAs and precursor-miRNAs are then processed into mature miRNAs by the action of type III endoribonuclease Dicer-like 1 (DCL1) (Bartel 2004; Rogers and Chen 2013). The canonical miRNAs (20–22 nt) regulate the transcription of the target genes by mRNA cleavage or translational repression (Tang et al. 2003; Chen 2004; Voinnet 2009). 20–22 nt small miRNAs have been shown to trigger DNA methylation in a few cases (Bao et al. 2004). But another class of longer miRNAs (23–27 nt) have been shown to induce DNA methylation in more number of cases (Vazquez et al. 2008). These long miRNAs (lmiRNAs) are generated by DCL3 and specifically associates with AGO4 protein. These lmiRNAs can induce cytosine methylation of a loci, both in *cis* and in *trans*, which may result in transcriptional repression of the target loci (Jia et al. 2011). This chapter discusses various examples of miRNA-mediated DNA methylation in plants.

2 DNA Methylation by 20–22 nt Canonical miRNAs

2.1 *In Arabidopsis*

miR165/166 mediates DNA methylation of its target genes *PHABULOSA* (*PHB*) and *PHAVOLUTA* (*PHV*). In *PHB* and *PHV* genes, the exons are found to be heavily methylated downstream of the miRNA-binding site in the differentiated tissues, but not in the undifferentiated and reproductive tissues (Bao et al. 2004). Methylation is reduced in gain-of-function mutants *phv-1d* and *phb-1d* which are resistant to miR165/166 cleavage. This reduction is mainly in cytosines in CHH and CHG sites (Bao et al. 2004). Contrary to that, hypomethylation was observed at CG sites in *serrate* mutant, *se-3*. (Grigg et al. 2005). SERRATE interacts with DCL1 and helps in processing of miRNAs (Lobbes et al. 2006). Whether the expression of miR165/166 target genes is regulated by mRNA cleavage or DNA methylation or both, remains to be determined. The nascent transcripts of *PHV* and *PHB* might lie in close proximity of the DNA of these genes and miR165/166 might induce methylation on DNA by binding to the mRNAs of *PHB* and *PHV*. In the heterozygous *PHB/phb* or *PHV/phv* plants, carrying both the WT allele and the mutant allele, only the WT allele is methylated, suggesting that DNA methylation depends on the ability of the miRNA to bind to the transcribed mRNAs of *PHB* and *PHV* (Bao et al. 2004). The miR165/166 site in *PHB* and *PHV* loci is interrupted by an intron. If miR165/166 do not bind the DNA of these genes and miRNA biogenesis mutants such as *dcl1* and *ago1* do not affect DNA methylation, then it remains a mystery that how miRNAs induce methylation of DNA (Ronemus and Martienssen 2005). It could be possible through longer isoforms of miR165/166. Another population of 24 nt miR165/166 has been reported which is processed by DCL3 (Vazquez et al. 2008), and it is plausible that the methylation of target genes, *PHV* and *PHB*, could be brought by 24 nt miR165/166 just like in the case of rice (Wu et al. 2010) and Arabidopsis (Chellappan et al. 2010). This is consistent with the fact that the methylation directed by miR165/166 is not affected in *dcl1* and *ago1* mutants, which are predominantly responsible for the biogenesis of 21 nt miRNA isoforms (Bao et al. 2004). However, the exact mechanism of miR165/166-based DNA methylation is still unclear.

2.2 *In Moss*

miRNA-directed DNA methylation was also discovered in *Physcomitrella patens* (Khraiwesh et al. 2010). The orthologous Dicer-like proteins in moss are PpDCL1a and PpDCL1b. PpDCL1a contributes to miRNA biogenesis in moss and PpDCL1b helps in target cleavage. In *ΔPpDCL1b* mutant, miRNA processing and accumulation took normally but miRNA targets were uncleaved, indicating requirement of PpDCL1b for miRNA-guided mRNA cleavage (Khraiwesh et al. 2010).

Surprisingly, the accumulation of target mRNA transcripts reduced in *ΔPpDCL1b*, indicating toward another pathway of transcriptional control. This pathway was proven to be epigenetic in nature (Khraiwesh et al. 2010). The promoters of the target genes were found to be unmethylated in the WT, whereas in *ΔPpDCL1b* they were methylated. Apart from the promoters, the sequences flanking the miR166-binding sites in the target genes, *PpC3HDZIP1* and *PpHB10*, were found to be methylated in *ΔPpDCL1b* mutant but not in the WT. Another example is of hypermethylation of *TAS4* gene in *ΔPpDCL1b* mutant but not in the WT. This hypermethylation of target genes in *ΔPpDCL1b* mutant resulted in their transcriptional silencing in comparison to the WT. A possible explanation for this phenomenon is that PpDCL1b is critical for cleavage and that in *ΔPpDCL1b* miRNAs are not loaded to RISC but into a DNA-modifying RNA-induced transcriptional silencing (RITS) complex that induces methylation and subsequent repression of the corresponding target genes expression (Fig. 1a). Similar to *PHV* and *PHB*, the miRNA-binding sites in *PpC3HDZIP1* and *PpHB10* are intervened by introns, so a miRNA:DNA hybrid is less likely to form. Instead, the RITS complex loaded with the miRNA may first interact with target mRNAs forming miRNA:mRNA duplexes, which could then be directed to interact with the DNA of target genes to trigger methylation. In WT plants expressing artificial miRNAs (amiRNAs) against *PpGNT1*, the promoter of *PpGNT1* gene was found to be methylated. This extent of methylation was dependent on expression levels of amiRNAs; higher the dosage of amiRNA, higher the methylation. Another miRNA in moss, miR1026, is induced by ABA and cleaves the mRNA of target gene *PpbHLH*. Under induction by ABA, the methylation of *PpbHLH* gets triggered, indicating that miR1026 controls the expression of *PpbHLH* via two pathways—mRNA cleavage and methylation. Like the dosage of amiRNA, methylation by endogenous miRNA is also miRNA-dosage dependent. DNA methylation mediated by canonical miRNAs in moss indicates dual roles of canonical miRNAs in this species. This dual function of miRNAs in moss is probably emanating from two distinct forms of DCL1 proteins, PpDCL1a and PpDCL1b. In all genes analyzed in *ΔPpDCL1b*, methylation was found at cytosines at CG sites, but the possibility of other cytosine sites cannot be ruled out for other genes. Although 23–27 nt miRNAs were also found in moss, their role in DNA methylation has not been established (Chellappan et al. 2010).

3 DNA Methylation by siRNAs Produced from miRNA Loci

siRNAs generated from the miRNAs sites have also been reported to induce DNA methylation. Certain siRNAs are produced from normal miRNA sites. These novel miRNA sites produce long (23–27 nt) sRNA species in addition to canonical (20–22 nt) miRNAs (Chellappan et al. 2010). These 23–27 nt sRNA species resemble more to heterochromatic siRNA (hc-siRNA) and follow DCL3, RDR2, and Pol IV pathway for their biogenesis (Fig. 1b). hc-siRNAs are sorted into AGO4 in the cytoplasm and trigger RdDM (Ye et al. 2012). The biogenesis of hc-siRNAs is dependent on DCL3, RDR2 and Pol IV (Vaucheret 2006). Hairpin-folded single-stranded RNAs transcribed from miRNA genes form miRNAs (Bartel 2004), while long double-stranded RNAs (dsRNAs) usually form siRNAs (Hamilton et al. 2002). miRNAs 2883, 2831, and 2328 form two sRNA species, one of the 21 nt long canonical miRNA and the other of the 23–27 nt long sRNAs. The canonical miRNAs biogenesis from these sites is routed through classical pathway involving DCL1, independent of RDRs and Pol IV, and are sorted into AGO1 (Vaucheret 2008). In contrast, the longer sRNAs are formed by typical hc-siRNA pathway involving DCL3, RDR2, and Pol IV. Like hc-siRNAs, these 23–27 nt long siRNAs are loaded into AGO4 and direct RdDM of the target loci in *trans* (Chellappan et al. 2010). To much lesser extent, these siRNAs were also found to be sorted into AGO1, AGO2, AGO5, and AGO7. The 23–27 nt siRNAs are produced from the positive strand of precursor-miRNAs and originate from the same miRNA-generating sites, which also produce canonical miRNAs. These longer siRNA-like sRNAs could trigger the DNA methylation around the binding sites of their target genes, same way like miR165/166.

These siRNAs were absent in *rdr2* and *nrdp1-3* mutants, which distinguishes them from long miRNAs (lmiRNAs). DNA methylation of At4g16580 (miR2328 target) and At5g08490 (miR2831 target) was dramatically reduced at the siRNA-binding sites in *nrpd1-3*. This suggests that siRNAs derived from miRNA sites direct RdDM of target loci *in trans*. This DNA methylation also reduced the accumulation of the transcripts of these target genes. Different from the classical siRNAs, which originate from both DNA strands via DCL3/RDR2/Pol IV pathway, the MIR gene-derived siRNAs originate from the positive strand of the miRNA genes. In addition to the canonical miRNAs, both miR165 and miR166 form 23–26 nt siRNAs, which are mainly sorted into AGO4- and AGO7-guided RISC complex. As mentioned before, these siRNAs are the likely players to induce DNA methylation at the target sites in *PHB* and *PHV*, as in *dcl1* and *ago1* mutants, this methylation was unaffected.

A significant number of MIR genes in Arabidopsis, rice, and moss have been shown to produce two kinds of sRNA species: 20–22 nt miRNA-like and 23–27 nt siRNA-like from the same gene loci. Deep sequencing data revealed that 42% of Arabidopsis, 36% of moss, and 43% of rice miRNA loci produced 23–26 nt sRNA species from miRNA sites, with those loci producing different ratios of 23–26 nt sRNAs to 21 nt miRNA species (Chellappan et al. 2010). Similar to the dual mode of target gene expression control in moss, these miRNA sites which produce two kinds of sRNA species control the expression of the target genes by different modes: the smaller species via mRNA cleavage and/or translational inhibition and the longer species via DNA methylation. Two kinds of populations generated from a miRNA locus have also been found in Medicago plants (Lelandais-Briere et al. 2009).

4 DNA Methylation by lmiRNAs

Other than the canonical miRNAs, DNA methylation has been reported to be triggered by lmiRNAs in some cases. In rice, many miRNA loci are detected which produce both kinds of miRNAs: the 21 nt canonical miRNAs (which cleaves the target transcripts) and the 24 nt lmiRNAs (which directs DNA methylation). Some loci only produce the lmiRNAs (Wu et al. 2010). Like hc-siRNAs in Arabidopsis, these lmiRNAs are processed by DCL3 and loaded into AGO4 and direct cytosine DNA methylation at their target genes in *trans*, but also at their own loci in *cis* (Fig. 1c). In contrast to siRNAs emanating from the miRNA sites (Chellappan et al. 2010), these lmiRNAs do not need RDR2-dependent pathway for their biogenesis. In rice, the pri-miRNA820.1 forms two miRNA species, the 21 nt miRNA820.1, which is loaded into AGO1-guided RISC complex and the 24 nt miRNA820.2, which is loaded into AGO4-guided RISC complex. Some 24-nucleotide miRNAs, like miRNA163, which are processed by DCL1 and are sorted into AGO1 complex, may not induce DNA methylation. Some miRNA precursors like miR1850, miR168, miR396, and miR820 can produce both short (21 nt) and lmiRNAs by the coordinated actions of DCL1 and DCL3. In miR1863, CHG and CHH methylation takes place not only in sequences located in miRNA and miRNA* regions but also in the regions outside stem-loop of miRNA precursor. lmiRNAs, like miR1863, do not cleave their target genes but direct methylation within ~80 nt region around their binding sites which represses the expression of the target genes. Further studies in rice showed that out of 325 targets targeted by 24 nt miRNAs, 65 genes showed hypermethylation at CHH around miRNA-binding sites (Hu et al. 2014). The targets of miR812, miR1862c, miR1863b, miR1867, miR2121b, miR5150, and miR5831 were found to be hypermethylated around the miRNA-binding sites. One gene can be targeted by many miRNAs.

The above examples show that miRNA genes with dual function are present in all plant species which regulate DNA methylation as a conserved function. Furthermore, different plant species are likely to have different modes of biogenesis and functions for long miRNAs. In rice, 24 nt lmiRNAs do not require RDR2 for their biogenesis, and these miRNAs can direct DNA methylation at loci of their origin (in *cis*) and of their target genes (in *trans*). In contrast, in Arabidopsis, longer miRNAs/siRNAs (23–27 nt) require RDR2 for their biogenesis and mostly induce DNA methylation at the target genes loci in *trans* (Chellappan et al. 2010). In this context, analyses of sRNA species of different size proportions (originating from a miRNA loci) and their impact upon DNA methylation of different loci in *cis* and *trans*, is important to establish the role of miRNAs in DNA methylation.

5 DNA Methylation by siRNAs or lmiRNAs Originating from miRNA Genes Located in the Introns

Recent studies have shown that long hairpin introns can generate 21-, 22-, and 24-nucleotide siRNAs (Chen et al. 2011). These siRNA-generating introns are termed as sirtrons. Five miRNA genes in rice, namely, *MIR*1863b, *MIR*1863c, *MIR*1863d, *MIR*1863e, and *MIR*437b are tandemly arranged within the intron in the locus *LOC_Os12g09290.1* (Chen et al. 2011). The miRNAs processed from these miRNA genes are 24 nt in length and are generated by DCL3, independent of RDR2, and sorted into AGO4-associated RISCs (Chen et al. 2011). Thus, these intron-derived miRNAs are similar to lmiRNAs described in rice (Wu et al. 2010). This rice intron as well as some others form long hairpin structures and also generate 21-, 22-, and 24-nucleotide siRNAs from long stem regions. In contrast, miRNAs are produced from the short stem-loop regions of the same intron. Thus, siRNAs and miRNAs share the same long hairpin precursor for their biogenesis. The 24 nt siRNAs and 24 nt miRNAs originating from these hairpin structures are loaded into AGO4-based RISC complex to trigger DNA methylation at their own loci in *cis* (Fig. 1d). The long complementary regions of the long hairpins were found to be hypermethylated in comparison to the corresponding flanking sites, inferring that this methylation could be triggered by siRNAs or miRNAs produced from this loci. This methylation in *cis* may control the expression of the host gene.

6 DNA Methylation by miRNA-Triggered *TAS*/*PHAS* Loci-Derived siRNAs

Endogenous siRNAs in plants are categorized into three: hc-siRNAs, natural antisense transcript-derived siRNAs (nat-siRNAs), and *trans*-acting-siRNAs (ta-siRNAs) (Ghildiyal and Zamore 2009). ta-siRNAs are generated from miRNA-cleaved *TAS* gene transcripts which in turn degrade target mRNAs. miRNA guides cleavage of non-coding *TAS* transcripts, and the cleaved products are stabilized by SUPPRESSOR OF GENE SILENCING3 (SGS3) and converted to dsRNA by RDR6 (Vazquez et al. 2004; Axtell et al. 2006; Yoshikawa et al. 2005). These dsRNA are processed into 21 nt ta-siRNAs by DCL4 which are then loaded into AGO1 complex to cleave the target mRNA (Yoshikawa et al. 2005; Gasciolli et al. 2005). Apart from the miRNA-directed cleavage, *TAS* loci are also shown to be methylated (Lister et al. 2008). While *TAS2* locus showed less methylation, high levels of cytosine methylation at CG, CHG, and CHH were detected at *TAS1a*, *TAS1c*, and *TAS3a* loci, respectively. The methylation sites were in the vicinity of miRNAs cleavage sites at the *TAS* loci (Wu et al. 2012). To confirm if the methylation at the *TAS* loci is via classical hc-siRNA-mediated RdDM pathway involving Pol IV, RDR2, and DCL3, DNA methylation was analyzed in *rdr2-2* and *dcl3-1* mutants (Wu et al. 2012). In these mutants, even though the abundances of hc-siRNAs and 24 nt ta-siRNAs was greatly reduced, the methylation was not compromised for the *TAS* loci examined. But in the Pol V mutant (*nrpd1b-11*), DNA methylation at all three cytosine contexts was greatly reduced or absent, suggesting that the association of Pol V with the *TAS3a* and *TAS1c* scaffold transcripts is necessary for DNA methylation at their loci. Since the role of 24 nt ta-siRNAs was ruled out in *TAS* loci DNA methylation, it was questioned if 21 nt ta-siRNAs are involved in the same process. It has been shown that *TAS* transcripts are also processed into 21 nt ta-siRNAs by the action of DCL1. The abundance of *TAS3a* and *TAS1c* 21 nt ta-siRNAs was greatly reduced in the mutants related to the biogenesis pathway components, such as *dcl1-7*, *zip-1*, *rdr6-11*, *sgs3-11*, and *dcl4-2* (Wu et al. 2012; Yoshikawa et al. 2005). Except in *dcl4-2* mutants, cytosine methylation at all contexts was greatly reduced in *dcl1-7, zip-1, rdr6-11*, and *sgs3-11* mutants, indicating that 21 nt ta-siRNAs trigger DNA methylation at the *TAS* loci. DCL1, along with the combinatory roles of DCL2/3/4, play a key role in DNA methylation at *TAS3a* and *TAS1c* loci in Arabidopsis. Different DCL combinations were required in DNA methylation of different TAS loci. Furthermore, it was found that these 21 nt ta-siRNAs are loaded into AGO4/6 to induce DNA methylation at *TAS* loci (Wu et al. 2012) (Fig. 1e). But this DNA methylation at *TAS* loci does not regulate the expression of *TAS* transcripts; therefore, the functional importance of this phenomenon is not yet clear.

Another recent example of DNA methylation by siRNAs comes from phased small interfering RNAs (phasiRNAs). PhasiRNAs are derived from *PHAS* loci, which can occur at ta-siRNA loci, loci of protein-coding genes or of non-coding RNAs (Fei et al. 2013). Their biogenesis is triggered by miRNAs and is very similar to those of ta-siRNAs, involving DCL4 for 21 nt phasiRNAs and DCL5 for 24 nt phasiRNAs. PhasiRNAs occur in 21 or 24 nt phased segments and their phasing is guided by either one or two miRNA-binding sites. Their processing can happen downstream or upstream of the cleaved target site (Fei et al. 2013). Recently, it has been shown that in maize male meiocytes, miR2118 and miR2275 trigger phasiRNA formation. PhasiRNA loci in meiocytes show relatively higher cytosine DNA methylation than found in anthers and seedlings in all three contexts with the highest increase in the CHH context (Dukowic-Schulze et al. 2016). This hypermethylation in meiocytes, in comparison to the anthers and seedlings, is more pronounced in 21 nt phasiRNA loci than in 24 nt phasiRNA loci. The possible AGO proteins involved in phasiRNA-triggered DNA methylation are AGO104 or AGO18b (Dukowic-Schulze et al. 2016). But whether this DNA methylation is putatively induced by phasiRNAs in *cis*, at their own loci of origin, requires more evidence. If true, this methylation is in the line of the DNA methylation triggered by 21 nt ta-siRNA in *cis*, as mentioned before.

7 miRNA-Triggered easiRNA Biogenesis to Prevent RDR2-Dependent RdDM

A number of transposon transcripts are specifically targeted and cleaved by miRNAs bound by AGO1 and AGO7. Once cleaved, they form 21-nucleotide ta-siRNA-like "epigenetically activated" small interfering RNAs (easiRNAs) whose biogenesis is dependent upon RDR6, DCL4, or DCL1. This phenomenon is found to be happening in certain cells like vegetative nuclei of pollen grains and in dedifferentiated plant cell cultures and in the background of certain mutants, such as *DECREASED DNA METHYLATION 1* (*ddm1*) and *DNA METHYLTRANSFERASE 1* (*met1*) (Slotkin et al. 2009; Tanurdzic et al. 2008). Methylation directs transcriptional silencing of transposable element (TE) promoters located near long terminal repeat (LTR) of retrotransposons and terminal inverted repeats (TIR) of most DNA transposons (Slotkin et al. 2005). Thus, in *ddm1* mutants, where methylation is reduced, the TEs are actively transcribed and are targeted by miRNAs. Various transposable elements (TEs) which gave rise to easiRNAs include *ATHILA6*, *ATGYPSY, ATCOPIA93,* ATMU5, *CACTA*, and *VANDAL* located in pericentromeric and euchromatic regions (Creasey et al. 2014). Genome-wide miRNA-target prediction indicated that 3,662 TEs are potentially targeted by miRNAs and PARE-sequencing reveals that out of those, 2371 TEs are detected to be cleaved by miRNAs. Most of the miRNAs targeting these TEs are well known such as miR156, miR159, miR169, miR172, miR319, miR390, miR399, miR823, and miR859. But not all of the cleaved TEs transcripts formed easiRNAs. Some TEs are targeted by many miRNAs. Similar to easiRNAs, two new miRNAs, ea-miR1 and ea-miR2, were found to be originating from TEs in pollen and also in the *ddm1* mutants. These miRNAs were epigenetically active and hence named eamiRNAs (Creasey et al. 2014). In *ddm1 rdr6* and *ddm1 dcl1* mutants, the accumulation of 21 nt easiRNAs reduced and those of 24 nt heterochromatic siRNAs (hc-siRNAs) increased. Through RdDM, 24 nt hc-siRNAs guide RDR2-dependent asymmetric CHH methylation at TEs, while symmetric CG and CHG methylation states are maintained by DNA methyltransferases and histone modifications (Law and Jacobsen 2010). When transcription of TEs is activated due to loss of DNA methylation or reprogramming of the germ line in gametes, the TE transcripts are targeted and cleaved by miRNAs. The cleaved transposon products produce 21 nt easiRNAs via the actions of RDR6 and DCL4. These easiRNAs are then loaded into AGO1 and prevents assembly of RDR2 and DCL3 which arrests RdDM induced by 24 nt siRNAs (Fig. 1f). Thus for a given transposon, the actions by RDR6 and RDR2 are antagonistic. Thus, miRNA-directed easiRNAs prevent RDR2-induced RdDM and help in long-term evasion of silencing by RDR2-mediated RdDM in germ line cells.

8 miRNAs Directly Regulating Players of Methylation

Some miRNAs directly control the expression of target genes involved in methylation. In Arabidopsis, miR823 has been reported to target *CMT3* (Duarte et al. 2013) and miR-773a targets *DRM2* (Jha and Shankar 2014). *INVOLVED IN* DE NOVO *2* (*IDN2*), *IDNl1* and *IDNl2* are required for de novo DNA methylation in *Arabidopsis thaliana* (Ausin et al. 2012). miR781a and miR837 have been reported to target *IDNl1*, and miR413 and miR169 to target *IDN2* in Arabidopsis (Jha and Shankar 2014). These miRNAs, whose target genes are involved in DNA methylation, are themselves being methylated by their own target genes, suggesting a negative feedback regulation. But these target genes mediate DNA methylation through DRM2 (Jha and Shankar 2014). In mammals, miRNAs, such as miRNA153, target methyltransferases, such as *DNMT1*, and negatively regulate DNA methylation of DNMT1-targeted genes (Das et al. 2010). This indicates that this feature of miRNAs is conserved across kingdoms and miRNAs regulate overall DNA methylation in an organism by controlling key components of DNA methylation pathway.

9 Conclusions and Perspectives

miRNA-derived siRNAs or lmiRNAs are produced from the same sites as the mature miRNAs and share a common precursor formed by RNA Pol II. Their biogenesis depends upon the Dicer protein and action upon the AGO complex in which they are sorted to trigger DNA methylation (Table 1). The methylation triggered by miRNAs is mainly confined within close proximity of miRNA-binding sites on their targets, except in the case of miR165/166 and in case of moss where methylation can spread even to the promoter regions of the target genes (Khraiwesh et al. 2010). This miRNA-triggered methylation can occur both in *cis* and *trans*.

The complexity of the various pathways involving miRNA-induced DNA methylation is quite intriguing. These pathways show interesting combinations with respect to miRNAs sizes, involved loci, biogenesis (through different dicers and polymerases), and recruitment to different AGO-complexes to induce methylation. Beyond that, several questions are still present before us for which specific answers are still awaiting. The proportion of longer miRNAs over canonical miRNAs is higher in rice than in Arabidopsis, are they doing some specialized functions in rice, which is absent or less common in Arabidopsis? When most of the methylation induced by miRNAs is confined in the vicinity of binding sites, how can the methylation triggered by moss miRNAs spread to the promoter regions of the target genes and that of *PHB* and *PHV* spread to hundreds of bases downstream of the cleavage site? Which factors control the spread of DNA methylation and which factors decide to make them work in *cis* and in *trans*? What is the exact mechanism governing the miR165/166 triggered DNA methylation at *PHB* and *PHV*? Why

Table 1 Various examples showing miRNAs-triggered DNA methylation (DM) in plants

Methylation inducing small RNAs	Examples	Species	Methylated loci	Dicer involved for biogenesis	Polymerase involved for biogenesis	Proteins involved in DM	DM pattern	References
Canonical miRNA (20–22 nt)	miR165/166	Arabidopsis	*PHB* and *PHV*	DCL1	Not RDR2 dependent	ND	Several 100 bases downstream of the miRNA-binding site	Bao et al. (2004)
	miR166 miR1026	Moss	*PpC3HDZIP1* and *PpHB10* *PpbHLH*	PpDCL1a for miRNA biogenesis PpDCL1b negatively regulates DM	ND	miRNAs loaded into RITS complex	Promoters of target genes and sequences flanking the miRNA-binding sites in target genes	Khraiwesh et al. (2010)
siRNAs (23–27 nt) produced from miRNA loci	miR2328 miR2831	Arabidopsis	*At4g16580* *At5g08490*	DCL3	Pol IV, RDR2	Mainly AGO4	RdDM of target loci *in trans* which also reduced the accumulation of the transcripts of the target genes	Chellappan et al. (2010)
lmiRNAs (24 nt)	miR1863 miR820.2 miR1873.1 miR1876	Rice	*Os06g38480* *Os03g02010* *Os05g01790* *Os07g41090* and *Os02g05890*	Mainly DCL3, sometimes both DCL1 and DCL3 from dual-coding pri-miRNAs	RDR2 not needed	AGO4	Can direct DNA methylation at loci of their origin (in *cis*) and of their target genes (in *trans*). DM within ~80 nt region around their binding sites which represses the expression of the target genes.	Wu et al. (2010), Hu et al. (2014)

siRNAs (24 nt) or lmiRNAs (24 nt) originating from miRNA loci located in the introns	miR1863b miR1863c miR1863d miR1863e miR437b	Rice	Intron in the locus *LOC_Os12g09290.1*	DCL3	RDR2 not needed	AGO4	Trigger DM in *cis* at their own loci	Chen et al. (2011)
ta-siRNAs (21 nt) generated from miRNA-cleaved TAS gene transcripts	miR173 miR390	Arabidopsis	*TAS1* and *TAS2* *TAS3*	DCL1 with DCL2/3/4	RDR6	AGO4/6 and Pol V	DM in *cis* in the vicinity of miRNA cleavage sites at the TAS loci	Wu et al. (2012)
phasiRNAs (mainly 21 nt and to lesser extent 24 nt) generated from miRNA-cleaved *PHAS* loci transcripts	miR2118 miR2275	Maize	*PHAS* loci	DCL1 or DCL5	RDR6	AGO104 or AGO18b	DM in *cis* at the *PHAS* loci of origin	Dukowic-Schulze et al. (2016)
miRNA-directed easiRNAs (21 nt) produced from transposons	miRNAs such as miR156, miR159, miR169, and many more	Arabidopsis	Many transposon transcripts such as *VANDAL*, *ATHILA*, and many more	DCL4 and DCL1	RDR6	AGO1	Arrests assembly of RDR2 and DCL3 which prevents RdDM induced by 24 nt siRNAs. This happens only in *ddm1* mutant or in germ line cells of WT tissues	Creasey et al. (2014)
hc-siRNAs (24 nt) originating from transposons or other repetitive DNA sequences	Various	Various	Various	DCL3	Pol IV, RDR2	AGO4 and Pol V	Classical RdDM pathway triggers de novo DM in *cis* spanning several hundred to several thousand nucleotides	Matzke et al. (2015), Zhang and Zhu (2011)

ND Not determined

different AGOs and DCLs are required in the biogenesis of miRNAs involved in DNA methylation of various pathways? In moss, DNA methylation seems to be a stand-by phenomenon when target cleavage fails to happen, what factors are governing it? Why in moss, dosage of DNA methylation-inducing miRNA is important than in the other species? Why some miRNA genes have dual functions, generating both miRNAs and siRNAs? Why subsets of pri-miRNAs produce both small miRNAs and lmiRNAs through the coordinated actions of DCL1 and DCL3? What decides the specificity of cytosine methylation among the three contexts, and is there any biological relevance for that?

DNA methylation in many cases reduces the expression of target genes in question, but in how many cases it affects physiological or developmental processes? Answers to these questions will help solving the mysteries of miRNA-directed DNA methylation in plants.

Acknowledgments This study was partially supported by funds from Henan Agricultural University (HAU) and NSFC (31571679), China. G.T. is supported by the National Science Foundation (NSF; grants IOS-1048216 and IOS-1340001). S.T. is supported by a postdoctoral scholarship from HAU.

References

Ausin I, Greenberg MV, Simanshu DK et al (2012) INVOLVED IN DE NOVO 2-containing complex involved in RNA-directed DNA methylation in Arabidopsis. Proc Natl Acad Sci USA 109:8374–8381

Axtell MJ, Jan C, Rajagopalan R et al (2006) A two-hit trigger for siRNA biogenesis in plants. Cell 127:565–577

Bao N, Lye KW, Barton MK (2004) MicroRNA binding sites in Arabidopsis class III HD-ZIP mRNAs are required for methylation of the template chromosome. Dev Cell 7:653–662

Bartel DP (2004) MicroRNAs: genomics, biogenesis, mechanism, and function. Cell 116:281–297

Carthew RW, Sontheimer EJ (2009) Origins and mechanisms of miRNAs and siRNAs. Cell 136:642–655

Castel SE, Martienssen RA (2013) RNA interference in the nucleus: roles for small RNAs in transcription, epigenetics and beyond. Nat Rev Genet 14:100–112

Chan SW, Henderson IR, Jacobsen SE (2005) Gardening the genome: DNA methylation in *Arabidopsis thaliana*. Nat Rev Genet 6:351–360

Chellappan P, Xia J, Zhou X et al (2010) siRNAs from miRNA sites mediate DNA methylation of target genes. Nucleic Acids Res 38:6883–6894

Chen X (2004) A microRNA as a translational repressor of *APETALA2* in Arabidopsis flower development. Science 303:2022–2025

Chen D, Meng Y, Yuan C, Bai L, Huang D, Lv S, Wu P, Chen LL, Chen M (2011) Plant siRNAs from introns mediate DNA methylation of host genes. RNA 17:1012–1024

Creasey KM, Zhai J, Borges F et al (2014) miRNAs trigger widespread epigenetically activated siRNAs from transposons in Arabidopsis. Nature 508:411–415

Das S, Foley N, Bryan K et al (2010) MicroRNA mediates DNA demethylation events triggered by retinoic acid during neuroblastoma cell differentiation. Cancer Res 70:7874–7881

Duarte GT, Matiolli CC, Pant BD et al (2013) Involvement of microRNA-related regulatory pathways in the glucose-mediated control of Arabidopsis early seedling development. J Exp Bot 64:4301–4312

Dukowic-Schulze S, Sundararajan A, Ramaraj T et al (2016) Novel meiotic miRNAs and indications for a role of PhasiRNAs in meiosis. Front Plant Sci 7:762
Fei Q, Xia R, Meyers BC (2013) Phased, secondary, small interfering RNAs in posttranscriptional regulatory networks. Plant Cell 25:2400–2415
Gasciolli V, Mallory AC, Bartel DP et al (2005) Partially redundant functions of Arabidopsis DICER-like enzymes and a role for DCL4 in producing trans-acting siRNAs. Curr Biol 15:1494–1500
Gehring M, Henikoff S (2007) DNA methylation dynamics in plant genomes. Biochim Biophys Acta 1769:276–286
Ghildiyal M, Zamore PD (2009) Small silencing RNAs: an expanding universe. Nat Rev Genet 10:94–108
Grigg SP, Canales C, Hay A, Tsiantis M (2005) SERRATE coordinates shoot meristem function and leaf axial patterning in Arabidopsis. Nature 437:1022–1026
Hamilton A, Voinnet O, Chappell L, Baulcombe D (2002) Two classes of short interfering RNA in RNA silencing. EMBO J 21:4671–4679
Holoch D, Moazed D (2015) RNA-mediated epigenetic regulation of gene expression. Nat Rev Genet 16:71–84
Hu W, Wang T, Xu J, Li H (2014) MicroRNA mediates DNA methylation of target genes. Biochem Biophys Res Commun 444:676–681
Jha A, Shankar R (2014) MiRNAting control of DNA methylation. J Biosci 39:365–380
Jia X, Yan J, Tang G (2011) MicroRNA-mediated DNA methylation in plants. Front Biol 6:133–139
Khraiwesh B, Arif MA, Seumel GI et al (2010) Transcriptional control of gene expression by microRNAs. Cell 140:111–122
Law JA, Jacobsen SE (2010) Establishing, maintaining and modifying DNA methylation patterns in plants and animals. Nat Rev Genet 11:204–220
Lelandais-Briere C, Naya L, Sallet E et al (2009) Genome-wide *Medicago truncatula* small RNA analysis revealed novel microRNAs and isoforms differentially regulated in roots and nodules. Plant Cell 21:2780–2796
Lister R, O'Malley RC, Tonti-Filippini J et al (2008) Highly integrated single-base resolution maps of the epigenome in Arabidopsis. Cell 133:523–536
Lobbes D, Rallapalli G, Schmidt DD et al (2006) SERRATE: a new player on the plant microRNA scene. EMBO Rep 7:1052–1058
Matzke MA, Kanno T, Matzke AJ (2015) RNA-directed DNA methylation: the evolution of a complex epigenetic pathway in flowering plants. Annu Rev Plant Biol 66:243–267
Rogers K, Chen X (2013) Biogenesis, turnover, and mode of action of plant microRNAs. Plant Cell 25:2383–2399
Ronemus M, Martienssen R (2005) RNA interference: methylation mystery. Nature 433:472–473
Slotkin RK, Freeling M, Lisch D (2005) Heritable transposon silencing initiated by a naturally occurring transposon inverted duplication. Nat Genet 37:641–644
Slotkin RK, Vaughn M, Borges F et al (2009) Epigenetic reprogramming and small RNA silencing of transposable elements in pollen. Cell 136:461–472
Tang G, Reinhart BJ, Bartel DP, Zamore PD (2003) A biochemical framework for RNA silencing in plants. Genes Dev 17:49–63
Tanurdzic M, Vaughn MW, Jiang H et al (2008) Epigenomic consequences of immortalized plant cell suspension culture. PLoS Biol 6:2880–2895
Vaucheret H (2006) Post-transcriptional small RNA pathways in plants: mechanisms and regulations. Genes Dev 20:759–771
Vaucheret H (2008) Plant ARGONAUTES. Trends Plant Sci 13:350–358
Vazquez F, Vaucheret H, Rajagopalan R et al (2004) Endogenous trans-acting siRNAs regulate the accumulation of Arabidopsis mRNAs. Mol Cell 16:69–79
Vazquez F, Blevins T, Ailhas J et al (2008) Evolution of Arabidopsis MIR genes generates novel microRNA classes. Nucleic Acids Res 36:6429–6438

Voinnet O (2009) Origin, biogenesis, and activity of plant microRNAs. Cell 136:669–687
Wu L, Zhou H, Zhang Q et al (2010) DNA methylation mediated by a microRNA pathway. Mol Cell 38:465–475
Wu L, Mao L, Qi Y (2012) Roles of dicer-like and argonaute proteins in TAS-derived small interfering RNA-triggered DNA methylation. Plant Physiol 160:990–999
Xie M, Yu B (2015) siRNA-directed DNA methylation in plants. Curr Genomics 16:23–31
Ye R, Wang W, Iki T et al (2012) Cytoplasmic assembly and selective nuclear import of Arabidopsis Argonaute4/siRNA complexes. Mol Cell 46:859–870
Yoshikawa M, Peragine A, Park MY, Poethig RS (2005) A pathway for the biogenesis of trans-acting siRNAs in Arabidopsis. Genes Dev 19:2164–2175
Zhang H, Zhu JK (2011) RNA-directed DNA methylation. Curr Opin Plant Biol 14:142–147
Zhong X, Du J, Hale CJ et al (2014) Molecular mechanism of action of plant DRM de novo DNA methyltransferases. Cell 157:1050–1060

DNA Methylation in Plants and Its Implications in Development, Hybrid Vigour, and Evolution

Annalisa Varriale

Abstract Cytosine DNA methylation is a key, heritable, epigenetic modification widespread among major eukaryotic groups and involved in main cellular processes such as integrity of DNA structure, control of transposable elements, and regulation of gene expression. In plants, its level can be influenced and modified by a number of biotic and abiotic stresses, and its variations are prone to increase, in turn, the rate of genetic mutations in DNA regions. For this reason, these mechanisms are proposed to improve the adaptability of DNA in complex environments. DNA methylation and epigenetic marks do have a fundamental role also in fine-tuning the pattern of expressed genes, during embryogenesis and seed development, and in the heterosis process, that is, the amelioration following the crosses between individuals belonging to different variants or species.

This chapter will review the actual knowledge of these topics in plants, with a focus on the importance of DNA methylation in angiosperms. Finally, it will assess opportunities and challenges for epigenetic research to advance the molecular understanding of hybrid vigour.

Keywords Epigenetics • Angiosperms • Adaptability • Development • Heterosis

Contents

A. Varriale (✉)
School of Medicine and Surgery, University of Naples Federico II, Naples, Italy
e-mail: avarrial@libero.it

N. Rajewsky et al. (eds.), *Plant Epigenetics*, RNA Technologies,
DOI 10.1007/978-3-319-55520-1_14

1 Introduction

The term epigenetics refers to processes beyond genetics, including modifications on histones, small RNAs, and DNA methylation, that are known to play a vital role in many biological processes in eukaryotes such as development (Reik 2007) and imprinting (Rodrigues and Zilberman 2015 for a review). Several years ago, the methylated cytosine (5mC) was found in all plant DNA including ferns, mosses, and higher vascular plants (Vanyushin and Belozersky 1959), and its content was found to be higher in flowering plant DNAs than in lower plants and in monocots than in dicots (Vanyushin and Belozersky 1959), suggesting a taxonomic significance for 5mC. Now, it is known that DNA methylation is spread widely among eukaryotes and that its level can be considered species specific. In fact, it can change from one taxon to another, ranging from 3% to 8% in vertebrates and from 6% to 30% in plants (Chen and Li 2004). In spite of that, it is still quite unclear whether methylation varies among plant tissues and, if so, whether it might contribute to tissue-specific gene expression. Some evidence suggests that most plant tissues do not vary substantially in DNA methylation apart from highly specialized tissues, such as the endosperm and the pollen vegetative nucleus (Lauria et al. 2004; Hsieh et al. 2009; Zemach et al. 2010; Ibarra et al. 2012), where DNA methylation might contribute to genetic imprinting and transgenerational silencing of TEs (Choi et al. 2002; Gehring et al. 2009; Slotkin et al. 2009). Genome-wide profiling identified few DNA methylation differences also between shoot and root in rice (*Oryza sativa* L.) and only a few additional differences in CHH methylation between these two tissues and the embryo (Zemach et al. 2010).

1.1 The Machinery of DNA Methylation and Demethylation

Cytosine methylation in higher plants is regulated by two distinct yet complementary enzymatic activities known as "de novo" and "maintenance" DNA methyltransferases (MTases), together with DNA demethylases, histone-modifying enzymes, chromatin remodelling factors, and the RNA interference (RNAi) machinery (Vanyushin and Ashapkin 2011 and references therein). *Arabidopsis*, for example, the best studied plant up to now, has at least ten genes encoding DNA methyltransferases, more than any other eukaryote sequenced so far (Martienssen and Colot 2001). Furthermore, differently from animals, in angiosperms, cytosine DNA methylation occurs in three sequence contexts and needs different enzymes to establish: methylated CG (mCG) is maintained by METHYLTRANSFERASE

1 (MET1), mCHG by CHROMOMETHYLASE 3 (CMT3, which can also rescue DNA methylation defects; Chan et al. 2006), and mCHH by DOMAINS REARRANGED METHYLTRANSFERASE 2 (DRM2) or CHROMOMETHYLASE 2 (CMT2, which also performs de novo establishment of mCHH). Transposons, repetitive elements, and some genes are silenced when methylated in all three sequence contexts being targets of the RNA-directed DNA methylation pathway (RdDM) guided by 24 nucleotide short interfering RNAs (siRNAs; Law and Jacobsen 2010). In other cases, the erasure of DNA methylation in plants is carried out by a demethylation pathway, which includes the DNA glycosylases and AP lyases ROS1 (repressor of silencing 1), DME (Demeter), DML2 (Demeter-like 2), and DML3 (Demeter-like 3). It has been observed that the demethylation pathway affects genome-wide hypomethylation in the *Arabidopsis* endosperm, especially within transposable elements (Gehring et al. 2009), whereas mutation of DME partially restores methylation to the amount found in other tissues (Hsieh et al. 2009). The expression of DME in maternal-specific cells of the endosperm results in demethylation, and consequently changes in gene expression, at specific genes (Huh et al. 2008). Intriguingly, MET1-defective *A. thaliana*, in spite of lacking 99% of all genome-wide CG methylation, still produce viable offspring (Lister et al. 2008). Reinders et al. (2009), crossing MET1-deficient individuals with wild-type plants, obtained direct descendants containing numerous epialleles. Epialleles are known to cause widespread phenotypic variation in *A. thaliana* (Weigel and Colot 2012), and some well-documented epialleles were first identified in spontaneous variants discovered in natural populations or within agricultural fields (Richards 2008). Analysis of mutants deficient in some aspect of DNA methylation maintenance, such as *ddm1* and *met1* mutants, has highlighted that chromatin landscape incompatibility might disrupt crucial associations between histone modifications and DNA methylation, ultimately leading to the creation of altered epigenetic states (Deleris et al. 2012; Ito et al. 2015).

1.2 Features and Distribution of DNA Methylation in Plants

Classical experiments to study the 5mC percentage in plant DNA were carried out analysing pyrimidine sequences obtained from wheat DNA and demonstrated that methylated cytosines (5mC) in plant genome were located in the sequences Pu–m5C–Pu, Pu–m5C–T–Pu, Pu–m5C–C–Pu, and Pu–m5C–m5C–Pu (Kirnos et al. 1981). These findings were consistent with another report on methylation of cytosine residues in plant DNA where authors used nearest-neighbour analysis (Gruenbaum et al. 1981). Nowadays, with respect to the past, it has become exceptionally feasible to obtain high-resolution genome-wide mapping of cytosine methylation landscapes thank to the development of high-throughput methodologies. Sequencing of bisulphite-converted DNA has become the gold standard for several recent studies where authors have mapped the distribution of cytosine methylation in the entire genome of the model plant *Arabidopsis thaliana* at a single base-pair resolution (Zhang et al. 2006; Vaughn et al. 2007; Zilberman et al. 2007; Cokus et al. 2008; Lister et al. 2008). The results obtained using these techniques have confirmed that in plants, methylation of cytosines occurs in three sequence contexts: CG, CHG, and CHH (where H = A, C or T; see also previous section). More in detail, genome methylation levels of 24%, 6.7%, and 1.7% are attributed to CG, CHG, and CHH sites, respectively (Cokus et al. 2008), and the relative distribution of m5C residues between different sequence contexts was 55% in CG, 23% in CHG, and 22% in CHH (Lister et al. 2008). Interestingly, it has also been demonstrated that pseudogenes and non-expressed genes show higher methylation levels than actively expressing genes. This observation is in line with the prevalent view that correlates higher methylation level with a lower transcription level (Vanyushin and Ashapkin 2011 for a review). Another important finding was that most of the extensively methylated fraction of the genome (up to 80%) is composed of inactive, heterochromatic regions including clusters of tandem, inverted and interspersed repeats, and transposons frequently located within or around the centromeric regions. Euchromatic regions, including genes and non-repetitive intergenic regions, by contrast, show lower but still significant levels of cytosine methylation. It was also highlighted the mosaic nature of the distribution of 5mC within transposons and genes in plants: on one side, transposons are usually heavily methylated along their entire length, whereas on the other side, methylation within genes is often distributed away from the 3′ and 5′ ends (Gehring and Henikoff 2007).

In addition to DNA methylation, histone modifications provide another dynamic and reversible mechanism to regulate gene expression through changes in chromatin state and recruitment of protein complexes involved in the transcription process. Among the numerous modifications of histones described up to now, acetylation and methylation of K residues are the most widely studied ones (Kouzarides 2007). Whereas K acetylation is generally linked to chromatin accessibility and gene activation, K methylation can be associated with either transcriptional activation or repression, depending on the position of the K residue and the nature of methylation (Berger 2007; Li et al. 2007). Recent studies discovered that both activating and repressive histone modifications correlate with gene activity, indicating that a combinatorial interplay between opposing modifications regulates, dynamically, gene expression (Bernstein et al. 2006; Roh et al. 2006; Barski et al. 2007; Mikkelsen et al. 2007; Wang et al. 2009). Finally, small RNAs, including microRNAs (miRNAs) and small interfering RNAs (siRNAs), provide another intriguing epigenetic RNA-mediated regulation of genome stability (Ghildiyal and Zamore 2009), as in the case of transposon silencing (see previous section).

1.3 General Aspects of the Possible Roles of DNA Methylation in Plants

The level and distribution of DNA methylation in the genome is known to give rise to epigenetic mutations that have the potential to vary during development (Bird 1997; Richards 1997) and under the influence of environmental factors, promoting, in turn, genetic and, finally, phenotypic changes (Law and Jacobsen 2010; Dowen et al. 2012; Varriale 2014). For example, it has been observed that within several plant species, it is possible to find populations displaying alternative phenotypes and little or no genetic variation but increased variation in DNA methylation (Lukens and Zhan 2007; Gao et al. 2010; Lira-Medeiros et al. 2010). In their recent review, Niederhuth and Schmidt (2016) pointed out that methylation variation is extensive in plant populations (Vaughn et al. 2007; Eichten et al. 2013; Schmitz et al. 2013b) and the rate at which epimutations occur is several fold higher than the normal rate of mutation (Becker et al. 2011; Schmitz et al. 2011). However, a subset of epialleles are closely associated with genetic variation, both locally (*cis*) and distantly (*trans*), segregating with this variation in a Mendelian fashion (Vaughn et al. 2007; Eichten et al. 2013; Regulski et al. 2013; Schmitz et al. 2013a, b). This challenges the common view of DNA methylation variation being independent of the genome even if a significant number of epialleles are not associated with genotype and may be considered pure epialleles (Regulski et al. 2013; Schmitz et al. 2013a, b). Another well-characterized example of an adaptive and naturally occurring epimutation was previously found in *Linaria vulgaris*: this mutation, changing flower symmetry from bilateral to radial, was first described about 250 years ago by Linnaeus, but only much later, it was discovered that it results from an epigenetic modification on the flower morphology control gene *Lcyc* (Cubas et al. 1999). In the epimutant, it is extensively methylated and silent, and, importantly, it is transmitted in this state over generations. Epimutations (especially DNA methylation patterns) in fact, once occurred, can be easily transferred to the next generation, since germ cells are derived from somatic tissues at a late developmental stage (Grant-Downton and Dickinson 2006; Molinier et al. 2006; Mirouze and Paszkowski 2011) and become fixed. The tendency to be inherited through multiple generations seems to be one of the major differences between DNA methylation and other epigenetic modifications, DNA methylation appearing to be the most permanent kind of epigenetic marks. Considering that, it is possible to speculate that an epigenetic mutation, being heritable for hundreds generations, might play a significant role in evolution and adaptation of species (Varriale 2014 and see next section and the scheme in Fig. 1).

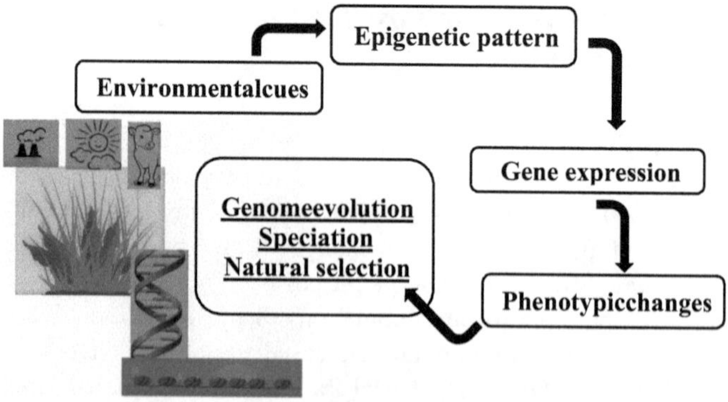

Fig. 1 Scheme representing the proposed relationship between environment, genetics, epigenetics, and evolution

2 Patterns of DNA Methylation Are Proposed to Change in Response to Environmental Stresses

When studying plant genomics, it is useful to remind that in contrast to the majority of animal species, plants are sessile organisms. They are, therefore, more strictly subjected to environmental biotic and abiotic perturbations such as changes in temperature, light intensity, nutrient and water availability, various pathogens, and insects (Richards 2008; Meyer 2015). To survive these difficulties, plants have needed to develop elaborate sensing mechanisms mediated by signalling cascades and gene transcription networks to respond to environmental cues (Meyer 2015 for a review). In these last years, it has been highlighted, even if debated (Weigel and Colot 2012 for a review), how continue exposition to environmental stress can represent a major force behind the evolutionary creation of new species through effects on epigenome (Jablonka and Lamb 2007; Mirouze and Paszkowski 2011), DNA methylation (Goll and Bestor 2005), and related mechanisms such as RNA interference (Lukens and Zhan 2007). The general conclusion was that under challenging conditions, epigenome represents a source of adaptive potential increasing the rate of evolution more rapidly than genetic recombination, through changes of transcriptional status (Jaenisch and Bird 2003) or transposon activity (Bossdorf et al. 2008; Rebollo et al. 2010). These two different authors speculated that environmental changes promote epigenetic instability and the activity of transposable elements generating, as a result, sexual isolation and speciation. The renewal of gene networks, in fact, allows the arousal of new species establishing a link between environmental changes, natural selection, and evolution. In this respect, phenotypic plasticity is currently considered a ubiquitous property in plants and animals that enables a population to achieve phenotypic variability in response to environmental change and despite genetic uniformity (West-Eberhard 2003; Boyko et al. 2007; Verhoeven et al. 2010). Its potential value is embodied directly by the influence of variations in DNA methylation on important agricultural traits such as yield, flowering time and plant height (Fieldes et al. 2005), and pathogen resistance (Akimoto et al. 2007; Boyko et al. 2007). Thereby these observations represented relevant insights for further studies aimed at clarifying how epigenetic regulation affects natural population variations (Bossdorf et al. 2008; Johannes et al. 2008; Richards et al. 2010).

2.1 *Some Examples of Responses to Biotic and Abiotic Stresses*

Among abiotic cues, since climate change has an impact on plant growth and yield, the mechanisms of stress response to temperature on methylomes and differentially methylated regions (DMRs) in plants have gained much attention (Hedhly et al. 2009). Cold has been evaluated through studies on vernalization, an epigenetic regulatory system that prevents flowering in unfavourable time. In *Arabidopsis*, premature flowering is prevented by active synthesis of a special transcription factor, FLC, that represses floral integrator genes (Kim et al. 2009 for a review). The exposure to prolonged cold at winter inactivates FLC through a mechanism involving DNA methylation. Indeed, it was showed that undermethylation of DNA, caused by antisense MET1 transgene or ddm1 mutation, induces early flowering in *Arabidopsis* plants, the magnitude of the effect being more less proportional to undermethylation level (Finnegan et al. 1998). The complete mechanism was later elucidated, and it was demonstrated that it involves multiple factors such as the level of acetylated and methylated histones, long noncoding RNAs, and the Polycomb repressive complex 2 (Finnegan et al. 2005; Swiezewski et al. 2009; Heo and Sung 2011; Turck and Coupland 2011). Vernalization is not, however, the only process under the impact of temperature: other studies, for example, focused on the influence of temperature on retrotransposons (Li et al. 2014). Steward et al. (2000), instead, observed that low-temperature stress decreases the amount of methyltransferase in corn (*Zea mays* L.), which, in turn, might reduce the level of genomic methylation. Such a decrease was found to be over 10% (Steward et al. 2002), and a quantitatively similar reduction was discovered to occur in the tobacco genome under abiotic stress (Choi and Sano 2007). In another study, Nicotra et al. (2015) examined local differentiation and adaptive phenotypic plasticity in response to elevated temperature in the alpine herb, *Wahlenbergia ceracea*, using Amplified Fragment Length Polymorphism (AFLP). They observed an increase in epigenetic variation among the seeds and found that seedlings originating from low elevations grew faster and showed stronger temperature responses than those from medium and high elevations. Other abiotic factors proposed to influence the epigenetic status of plant genomes are heavy metal pollutants (Aina et al. 2004; Yang et al. 2007; Filek et al. 2008), herbicides (Boscolo et al. 2003; Apel and Hirt 2004), water availability (Yamaguchi-Shinozaki and Shinozaki 2005), rounds of subcultures (Stroud et al. 2013), and soil salinity. About this last one, Lira-Medeiros et al. (2010) compared two populations of mangrove plants from two nearby habitats, riverside and a salt marsh neighbourhood, and found that the two populations were different in both phenotype and DNA methylation pattern. Another well-studied factor is Pi starvation (Yong-Villalobos et al. 2015). The authors found that loss of DNA methylation in specific contexts alters expression of a subset of low Pi-responsive genes as well as, in turn, a number of morphological and physiological responses to Pi starvation. Influence of biotic factors was also studied, and a few years ago, Dowen et al. (2012), looking at the effects of bacterial infection, a virulent bacteria and salicylic acid on methylome of tomato plants revealed numerous stress-induced differentially methylated regions (DMRs). Many of these were found to be associated with differentially expressed genes and/or transposons and, phenotypically, presented an increased resistance to the pathogenic bacterium *Pseudomonas syringae*. In spite of these results, Eichten and Springer (2015), analysing cold, heat, and UV-stressed maize seedlings, noticed that it was not possible to observe a consistent methylome change possibly because of technical reasons or because the methylation/demethylation machinery could be targeted to specific loci that could have been not analysed.

A last important aspect of the epigenetic plasticity is the "memory": as environmental perturbations may occur repeatedly, it is advantageous to plants to be able to remember past incidents and to use this stored knowledge to adapt to new challenges. Such memory system is termed defence priming and controls the response to a pathogen or to a herbivore attack (Pastor et al. 2013), drought stress (Ding et al. 2012), vernalization and stress response, and parasitic elements (Chinnusamy and Zhu 2009; Gómez-Díaz et al. 2012; Kinoshita and Seki 2014). In defence priming, the plant that has already encountered one of these threats displays a more rapid and robust response in the subsequent hits, thereby increasing its chances of survival and preparing future generations to better withstand biotic and abiotic stresses. This area of investigation will clearly grow in the upcoming years, and precise efforts will be needed to better analyse the capacity of the environment to create heritable phenotypes via altering DNA methylation states.

3 DNA Methylation During Plant Development

The correct development in plants, as in any other multicellular organism, requires coordinated cellular differentiation based on the proper establishment of cell type-specific gene expression and a correct epigenetic program (Goldberg et al. 2007), involving DNA methylation, Polycomb group proteins, and their associated chromatin modifications. These mechanisms, in fact, are much relevant to preserve the epigenetic status of developmental genes and in controlling proliferation in embryo and endosperm in plants (reviewed in Kohler and Grossniklaus 2002; Baroux et al. 2007) and animal embryos (Simon and Tamkun 2002). An important point is when, during developmental stages, the epigenetic patterns begin to establish. Several studies indicate that large-scale epigenetic modifications take place during gametophyte development (Berger and Twell 2011). Interestingly, Wollmann and Berger (2012) pointed out that in flowering plants, during gametogenesis and the early phase of seed development, DNA methylation needs to be erased from one generation to the next to allow developmental reprogramming. On the other side, however, the silencing of TEs must be maintained during the transition between generations and must escape reprogramming. Male game to genesis produces two identical gametic sperm cells and one large vegetative cell, which supports pollen tube growth but does not genetically contribute to the next generation. Female gametogenesis, instead, produces a megagametophyte or embryo sac with its two gametes, the egg cell and the central cell, and five accessory cells that provide support during the double fertilization (Berger and Twell 2011). The epigenetic landscape of chromatin organization and DNA methylation in the egg and central cell gametes differs from that of accessory cells, suggesting that distinct epigenetic features are established prior to or during the cellularization of the female gametophyte (reviewed in Wollmann and Berger 2012). Epigenetic changes in the central cells could be pivotal for marking genes prior to fertilization, establishing later the asymmetric gene expression observed at a significant number of imprinted gene loci in plants (reviewed by Ikeda 2012).It has been observed also that transposable elements are expressed and mobile in pollen (Slotkin et al. 2009), whereas they are methylated and silenced in most tissues. However, within the pollen grain, transposon reactivation appears to be restricted to the vegetative nucleus. This is a key distinction, as the sperm cells, but not the vegetative nucleus, provide genetic information to subsequent generations and thus their genome integrity must be protected. The epigenetic scenario of endosperm has also been described: while previous studies documented decreased DNA methylation at discrete imprinted loci in endosperm (Huh et al. 2008), two later studies showed that endosperm DNA methylation is reduced genome wide, such loss likely originating from demethylation in female central cell (Hsieh et al. 2009; Gehring et al. 2009). These findings are in line with observations that chromatin appears less condensed (and then presumably less methylated) in endosperm nuclei (Baroux et al. 2007). Despite this global decrease in DNA methylation, Hsieh et al. (2009) found increased CHH methylation in both the endosperm and embryo tissues in comparison with adult shoot tissue and suggested that this hypermethylation could result from enhanced RNA-directed DNA methylation (RdDM).

3.1 Does DNA Methylation Change During Development and Among Plant Tissues?

Even if at present the role of DNA methylation during development and its precise regulation remain not fully understood, their study represents an open field for fascinating discoveries. Earlier insights measured cytosine methylation levels in immature tomato tissues like stems, leaves, and roots and compared them with those of mature tissues like older leaves, fruits, and seeds (Messeguer et al. 1991). The results revealed that the older tissues are more methylated than the younger ones, consistently with another study where the highest methylation level was observed in senescence tissues (Brown et al. 2008). The authors hypothesized that this changing trend in cytosine methylation probably could have a role in preventing premature DNA degradation during senescence. Looking together at these results, it is possible to speculate that methylation levels in plants begin to change during development and probably continue to change during the entire lifetime of the plant. Last but not least, Wagner (2003) mentioned cases of viable mutants for chromatin remodelling factors in plants, whereas metazoan mutants for the same factors are not compatible with life. Interestingly, the author reported also examples of cloning experiments from donor differentiated cells: these experiments in animals are often unsuccessful. The cause could be attributed to genome restriction in animal differentiated cells: such restriction, probably due to epigenetic factors, might reduce the developmental potential of such nuclei. Plants, instead, possess a higher grade of plasticity and can easily regenerate structures from a number of tissues. The fact that many differentiated plant cells retain totipotency, then, possibly reflects the presence of a higher chromatin flexibility due to a different epigenetic program.

4 DNA Methylation and Its Suggested Role for Evolution in Plants

The potential role of DNA methylation in evolution has been studied by cross-species comparisons of methylomes in the model grass *Brachypodium distachyon* and *Oryza sativa* ssp. *japonica* (Takuno and Gaut 2013) and across eukaryotes, including multiple plant species (Feng et al. 2010; Zemach et al. 2010). Results demonstrated that methylation and silencing in plants have very ancient origins and that methylation in CG and non-CG contexts was found in all plants and green algae, in both gene bodies and transposons. Differences, however, appeared for CG-gene-body methylation (gbM) that was abundant in angiosperms, whereas it was very scarce in two species that diverged early from the angiosperms: *Selaginella moellendorffii* and *Physcomitrella patens* (Zemach et al. 2010). Comparison between species revealed also that slow-evolving orthologs display a surprising conservation of methylation level and that methylation and silencing of transposons are ancient processes in plants (Feng et al. 2010; Zemach et al. 2010; Takuno and Gaut 2013). Another important process for genome evolution, that is, genome duplication, was studied in soybeans, which are known to have undergone two recent whole-genome duplications during their life history (Schmutz et al. 2010). Compared to the *Arabidopsis* methylome, a much greater percentage of cytosines were methylated in soybean, and far more protein-coding genes were targeted by RdDM (Schmitz et al. 2013a). Further analysis showed that more recent gene copies were preferentially methylated, suggesting that methylation and targeting by RdDM might be a mechanism for coping with the effects of whole-genome duplication. It could act by silencing the new genes until they undergo a change towards sub- or neo-functionalization otherwise they are purged from the genome (Schmitz et al. 2013b).

Another recent study (Niederhuth et al. 2016) compared single-base resolution methylomes of 34 angiosperms. Extensive changes, both in levels and distribution of methylation, were found between species, with the greatest variation being observed in non-CG contexts. Among the families investigated, the Brassicaceae showed overall reduced mCHG levels and reduced numbers of methylator genes. In the Poaceae (a family characterized by a peculiar epigenomic architecture with mCHH often depleted in heterochromatin and enriched in genic regions), instead, mCHH levels were found to be typically lower than in other species. The authors observed also that many species with a history of clonal propagation (a technique known for its agricultural and economic importance) have lower mCHH levels. Furthermore, although gbM genes do show many conserved features, it has been observed that gbM is absent in the basal plant species *Marchantia polymorpha* (Takuno et al. 2016) and *Selaginella moellendorffii* (Zemach et al. 2010) and from the angiosperm *E. salsugineum*, which has lost also the CMT3 enzyme from its genome indicating its dispensability (Bewick et al. 2016). These authors identified an additional angiosperm, *Conringia planisiliqua*, which independently lost both CMT3 and gbM, supporting the idea that CMT3 is required for the establishment of gbM. The authors analysed also gene expression and various histone modifications in *E. salsugineum* and in *Arabidopsis thaliana* epigenetic recombinant inbred lines in order to recognize a role for gbM in regulating transcription or affecting the composition. Their results led them to propose however that gbM might be dispensable, at least in some cases.

5 Proposed Function and Evidences for the Influence of the Epigenetic State in Heterosis

Heterosis refers to the increased vigour of crosses between species (or between distantly related variants within a species) compared with the parents. Although heterosis, or hybrid vigour, has been recognized for well over a century, its molecular basis has remained a matter of debate and is surprisingly poorly understood (reviewed in Birchler et al. 2010; Chen 2010). Recent progress has pointed to a role for epigenetics in heterosis (Groszmann et al. 2011; Greaves et al. 2015). One study showed that altering the transcription of a few regulatory genes through epigenetic variations is associated with growth vigour in hybrids (Ni et al. 2009). Interestingly, Shen et al. (2012), crossing *Arabidopsis thaliana* C24 and *Landsberg erecta* ecotypes, obtained F1 hybrids with increased growth and more siliques and also a higher DNA methylation level than either parent. In addition, the growth of both hybrids was more sensitive to a chemical inhibitor of methylation than was that of the parents, supporting a role for DNA methylation in growth vigour. Shen et al. (2012) used high-throughput sequencing to examine the global sRNA expression profiles of the parents and the hybrids. They found that regions of the genome that give rise to sRNAs had increased DNA methylation in all lines, and the hybrids had even higher levels of methylation in those regions. Overall, their data support a model in which regions that give rise to sRNAs and show differential methylation between the parents account for the increased methylation in the hybrids. Shen et al. (2012) also examined the transcriptomes of the parents and hybrids, finding more genes downregulated in the hybrids than upregulated. Integration of the DNA methylome, the sRNAome, and the transcriptome data supports the idea that increased methylation of the circadian clock genes CIRCADIAN CLOCK ASSOCIATED1 and LATE ELONGATED HYPOCOTYL, which have been reported to be involved in heterosis, leads to their downregulated expression in the hybrids. Important results were obtained also recently in *Arabidopsis* by Kawanabe et al. (2016) who used mutant genes having roles in DNA methylation in order to discover a possible role for epigenetics in the heterosis process. More in particular, they found that hybrids between C24 and Columbia-0 without RNA polymerase IV or methyltransferase I function did not reduce the level of biomass heterosis, whereas hybrids with a mutation in ddm1 showed a decreased heterosis level.

Taken all together, the results, obtained until now, lead to think that exploring the inherited natural variation of epigenetic patterns between genetically diverse strains at a genome-wide level will allow to assess their importance in phenotypic plasticity and might have major consequences for biological research and agriculture (Richards 2008).

6 Conclusions and Perspectives

To date, genome-wide approaches to analyse DNA methylation have provided huge body of data on its general features throughout plant genome. Less investigated, however, remains the question on epigenome changes during plant development and in response to the action of environmental or internal stimuli. These data, in fact, are still scarce, especially if one considers to compare global levels, instead of loci or DMRs. The complexity in plants is increased by the fact that they possess methylation in three different contexts and that the massive extraction of DNA from one kind of cell or tissue is technically challenging, so in many cases, studies are conducted with DNA extracted from a mix of tissues. Anyway, the interplay of different kinds of DNA methylation and other epigenetic marks is starting to be highlighted. Two important things to take into account for future research will be "remember the past", that is, to put together the large body of data on plant DNA methylation obtained for about 60 years, and "look at the wild", that is, to extend studies on more and more plant species including wild ones that might have retained a more powerful epigenetic potential than crop lines. In this way, it will be possible to get a basic tool in order to unravel the role of epigenetics in evolution and adaptation of species to environment. It will be very useful also for new discoveries on genome reprogramming during cell differentiation, plant regeneration, and reproduction. On a practical point of view, there is no doubt that the study of epigenetics and DNA methylation will serve also as an important biotechnological tool to improve quantity, quality of crops, and their productivity.

Last but not least, the increasing emphasis on stress-induced epigenetic alterations and transgenerational phenomena should push epigenetic researchers to consider the questions on the light of ecological and evolutionary perspectives (Richards 2011; Latzel et al. 2013) since biological diversity within species can be an important driver of evolution of population and ecosystems. It will be useful to integrate and compare the knowledge acquired for animals, human, and plants to unravel the depth of epigenetic mechanisms and better understand all their biological roles. Considering the highly diverse developmental and metabolic behaviour of different crops and their importance in human usage, it is evident that the observation and analysis of genomics and epigenomics of multiple plant models will not only help to answer several intriguing questions, including those related to plant development, regulation of metabolism, evolution, and regeneration, but also to enhance the economic value of these crops.

References

Aina R, Sgorbati S, Santagostino A et al (2004) Specific hypomethylation of DNA is induced by heavy metals in white clover and industrial hemp. Physiol Plant 121:472–480

Akimoto K, Katakami H, Kim HJ et al (2007) Epigenetic inheritance in rice plants. Ann Bot (Lond) 100:205–217

Apel K, Hirt H (2004) Reactive oxygen species: metabolism, oxidative stress, and signal transduction. Annu Rev Plant Biol 55:373–399

Baroux C, Pecinka A, Fuchs J et al (2007) The triploid endosperm genome of *Arabidopsis* adopts a peculiar, parental-dosage-dependent chromatin organization. Plant Cell 19:1782–1794

Barski A, Cuddapah S, Cui K et al (2007) High-resolution profiling of histone methylations in the human genome. Cell 129:823–837

Becker C, Hagmann J, Muller J et al (2011) Spontaneous epigenetic variation in the *Arabidopsis thaliana* methylome. Nature 480:245–249

Berger SL (2007) The complex language of chromatin regulation during transcription. Nature 447:407–412

Berger F, Twell D (2011) Germline specification and function in plants. Annu Rev Plant Biol 62:461–484

Bernstein BE, Mikkelsen TS, Xie X et al (2006) A bivalent chromatin structure marks key developmental genes in embryonic stem cells. Cell 125:315–326

Bewick AJ, Ji L, Niederhuth CE et al (2016) On the origin and evolutionary consequences of gene body DNA methylation. Proc Natl Acad Sci U S A 113:9111-9116

Birchler JA, Yao H, Chudalayandi S et al (2010) Heterosis. Plant Cell 22:2105–2112

Bird A (1997) Does DNA methylation control transposition of selfish elements in the germline? Trends Genet 13:469–472

Boscolo PR, Menossi M, Jorge RA (2003) Aluminum-induced oxidative stress in maize. Phytochemistry 62:181–189

Bossdorf O, Richards CL, Pigliucci M (2008) Epigenetics for ecologists. Ecol Lett 11:106–115

Boyko A, Kathiria P, Zemp FJ et al (2007) Transgenerational changes in the genome stability and methylation in pathogen-infected plants: (virus-induced plant genome instability). Nucleic Acids Res 35:1714–1725

Brown JCL, De Decker MM, Fieldes MA (2008) A comparative analysis of developmental profiles for DNA methylation in 5-azacytidine-induced early-flowering flax lines and their control. Plant Sci 175:217–225

Chan SW-L, Henderson IR, Zhang X et al (2006) RNAi, DRD1, and histone methylation actively target developmentally important non-CG DNA methylation in *Arabidopsis*. PLoS Genet 2: e83

Chen ZJ (2010) Molecular mechanisms of polyploidy and hybrid vigor. Trends Plant Sci 15:57–71

Chen T, Li E (2004) Structure and function of eukaryotic DNA methyltransferases. Curr Top Dev Biol 60:55–89

Chinnusamy V, Zhu JK (2009) Epigenetic regulation of stress responses in plants. Curr Opin Plant Biol 12:133–139

Choi CS, Sano H (2007) Abiotic-stress induces demethylation and transcriptional activation of a gene encoding a glycerophosphodiesterase-like protein in tobacco plants. Mol Genet Genomics 277:589–600

Choi Y, Gehring M, Johnson L et al (2002) DEMETER, a DNA glycosylase domain protein, is required for endosperm gene imprinting and seed viability in *Arabidopsis*. Cell 110:33–42

Cokus SJ, Feng S, Zhang X et al (2008) Shotgun bisulphite sequencing of the *Arabidopsis* genome reveals DNA methylation patterning. Nature 452:215–219

Cubas P, Vincent C, Coen E (1999) An epigenetic mutation responsible for natural variation in floral symmetry. Nature 401:157–161

Deleris A, Stroud H, Bernatavichute Y et al (2012) Loss of the DNA methyltransferase MET1 induces H3K9 hypermethylation at PcG target genes and redistribution of H3K27 trimethylation to transposons in *Arabidopsis thaliana*. PLoS Genet 8:e1003062

Ding Y, Fromm M, Avramova Z (2012) Multiple exposures to drought 'train' transcriptional responses in *Arabidopsis*. Nat Commun 3:740

Dowen RH, Pelizzola M, Schmitz RJ et al (2012) Widespread dynamic DNA methylation in response to biotic stress. Proc Natl Acad Sci U S A 109:E2183–E2191

Eichten SR, Springer N (2015) Minimal evidence for consistent changes in maize DNA methylation patterns following environmental changes. Front Plant Sci 6:308

Eichten SR, Briskine R, Song J et al (2013) Epigenetic and genetic influences on DNA methylation variation in maize populations. Plant Cell 25:2783–2797

Feng S, Cokus SJ, Zhang X et al (2010) Conservation and divergence of methylation patterning in plants and animals. Proc Natl Acad Sci U S A 107:8689–8694

Fieldes MA, Schaeffer SM, Krech MJ et al (2005) DNA hypomethylation in 5-azacytidine-induced early-flowering lines of flax. Theor Appl Genet 111:136–149

Filek M, Keskinen R, Hartikainen H et al (2008) The protective role of selenium in rape seedlings subjected to cadmium stress. J Plant Physiol 165:833–844

Finnegan EJ, Genger RK, Kovac KA et al (1998) DNA methylation and the promotion of flowering by vernalization. Proc Natl Acad Sci U S A 95:5824–5829

Finnegan EJ, Kovac KA, Jaligot E et al (2005) The downregulation of FLOWERING LOCUS C (FLC) expression in plants with low levels of DNA methylation and by vernalization occurs by distinct mechanisms. Plant J 44:420–432

Gao L, Geng Y, Li B et al (2010) Genome-wide DNA methylation alterations of *Alternanthera philoxeroides* in natural and manipulated habitats: implications for epigenetic regulation of rapid responses to environmental fluctuation and phenotypic variation. Plant Cell Environ 33:1820–1827

Gehring M, Henikoff S (2007) DNA methylation dynamics in plant genomes. Biochim Biophys Acta 1769:276–286

Gehring M, Bubb KL, Henikoff S (2009) Extensive demethylation of repetitive elements during seed development underlies gene imprinting. Science 324:1447–1451

Ghildiyal M, Zamore PD (2009) Small silencing RNAs: an expanding universe. Nat Rev Genet 10:94–108

Goldberg AD, Allis CD, Bernstein E (2007) Epigenetics: a landscape takes shape. Cell 128:635–638

Goll MG, Bestor TH (2005) Eukaryotic cytosine methyltransferases. Annu Rev Biochem 74:481–514

Gómez-Díaz E, Jordà M, Peinado MA et al (2012) Epigenetics of host–pathogen interactions: the road ahead and the road behind. PLoS Pathog 8:e1003007

Grant-Downton RT, Dickinson HG (2006) Epigenetics and its implications for plant biology 2. The "epigenetic epiphany": epigenetics, evolution and beyond. Ann Bot 97:11–27

Greaves IK, Gonzalez-Bayon R, Wang L et al (2015) Epigenetic changes in hybrids. Plant Physiol 168:1197–1205

Groszmann M, Greaves IK, Albertyn ZI et al (2011) Changes in 24-nt siRNA levels in *Arabidopsis* hybrids suggest an epigenetic contribution to hybrid vigor. Proc Natl Acad Sci U S A 108:2617–2622

Gruenbaum Y, Naveh-Many T, Cedar H et al (1981) Sequence specificity of methylation in higher plant DNA. Nature 292:860–862

Hedhly A, Hormaza JI, Herrero M (2009) Global warming and sexual plant reproduction. Trends Plant Sci 14:30–36

Heo JB, Sung S (2011) Vernalization-mediated epigenetic silencing by a long intronic noncoding RNA. Science 331:76–79

Hsieh T-F, Ibarra CA, Silva P et al (2009) Genome-wide demethylation of *Arabidopsis* endosperm. Science 324:1451–1454

Huh JH, Bauer MJ, Hsieh T-F et al (2008) Cellular programming of plant gene imprinting. Cell 132:735–744

Ibarra CA, Feng X, Schoft VK et al (2012) Active DNA demethylation in plant companion cells reinforces transposon methylation in gametes. Science 337:1360–1364

Ikeda Y (2012) Plant imprinted genes identified by genome-wide approaches and their regulatory mechanisms. Plant Cell Physiol 53:809–816

Ito T, Tarutani Y, To TK et al (2015) Genome-wide negative feedback drives transgenerational DNA methylation dynamics in *Arabidopsis*. PLoS Genet 11:e1005154

Jablonka E, Lamb MJ (2007) Precis of evolution in four dimensions. Behav Brain Sci 30:353–365

Jaenisch R, Bird A (2003) Epigenetic regulation of gene expression: how the genome integrates intrinsic and environmental signals. Nat Genet 33:245–254

Johannes F, Colot V, Jansen RC (2008) Epigenome dynamics: a quantitative genetics perspective. Nat Rev Genet 9:883–890

Kawanabe T, Ishikura S, Miyaji N et al (2016) Role of DNA methylation in hybrid vigor in Arabidopsis thaliana. Proc Natl Acad Sci U S A 113:E6704–E6711

Kim DH, Doyle MR, Sung S et al (2009) Vernalization: winter and the timing of flowering in plants. Annu Rev Cell Dev Biol 25:277–299

Kinoshita T, Seki M (2014) Epigenetic memory for stress response and adaptation in plants. Plant Cell Physiol 55:1859–1863

Kirnos MD, Aleksandrushkina NI, Vanyushin BF (1981) 5-Methylcytosine in pyrimidine DNA sequences of plants and animals: specificity of methylation. Biokhimiya Moscow 46:1458–1474

Kohler C, Grossniklaus U (2002) Epigenetic inheritance of expression states in plant development: the role of Polycomb group proteins. Curr Opin Cell Biol 14:773–779

Kouzarides T (2007) Chromatin modifications and their function. Cell 128:693–705

Latzel V, Allan E, Bortolini Silveira A et al (2013) Epigenetic diversity increases the productivity and stability of plant populations. Nat Commun 4:2875

Lauria M, Rupe M, Guo M et al (2004) Extensive maternal DNA hypomethylation in the endosperm of Zea mays. Plant Cell 16:510–522

Law JA, Jacobsen SE (2010) Establishing, maintaining and modifying DNA methylation patterns in plants and animals. Nat Rev Genet 11:204–220

Li B, Carey M, Workman JL (2007) The role of chromatin during transcription. Cell 128:707–719

Li S, Liu J, Liu Z et al (2014) HEAT-INDUCED TAS1 TARGET mediates thermotolerance via HEAT STRESS TRANSCRIPTION FACTOR A1a-directed pathways in Arabidopsis. Plant Cell 26(4):1764–1780

Lira-Medeiros CF, Parisod C, Fernandes RA et al (2010) Epigenetic variation in mangrove plants occurring in contrasting natural environment. PLoS ONE 5:e10326

Lister R, O'Malley RC, Tonti-Filippini J et al (2008) Highly integrated single-base resolution maps of the epigenome in *Arabidopsis*. Cell 133:523–536

Lukens LN, Zhan S (2007) The plant genome's methylation status and response to stress: implications for plant improvement. Curr Opin Plant Biol 10:317–322

Martienssen RA, Colot V (2001) DNA methylation and epigenetics inheritance in plants and filamentous fungi. Science 293:1070–1074

Messeguer R, Ganal MW, Steffens JC et al (1991) Characterization of the level, target sites and inheritance of cytosine methylation in tomato nuclear DNA. Plant Mol Biol 16:753–770

Meyer P (2015) Epigenetic variation and environmental change. J Exp Bot 66:3541–3548

Mikkelsen TS, Ku M, Jaffe DB et al (2007) Genome-wide maps of chromatin state in pluripotent and lineage-committed cells. Nature 448:553–556

Mirouze M, Paszkowski J (2011) Epigenetic contribution to stress adaptation in plants. Curr Opin Plant Biol 14:267–274

Molinier J, Ries G, Zipfel C et al (2006) Transgeneration memory of stress in plants. Nature 442:1046–1049

Ni Z, Kim ED, Ha M, Lackey E et al (2009) Altered circadian rhythms regulate growth vigour in hybrids and allopolyploids. Nature 457:327–331

Nicotra AB, Segal DL, Hoyle GL et al (2015) Adaptive plasticity and epigenetic variation in response to warming in an Alpine plant. Ecol Evol 5:634–647

Niederhuth CE, Schmidt RJ (2016) Covering your bases: Inheritance of DNA methylation in plant genomes. Mol Plant 7:472–480

Niederhuth CE, Bewick A et al (2016) Widespread natural variation of DNA methylation within angiosperms. Genome Biol 17:194–213

Pastor V, Luna E, Ton J et al (2013) Fine tuning of reactive oxygen species homeostasis regulates primed immune responses in *Arabidopsis*. Mol Plant Microbe Interact 26:1334–1344

Rebollo R, Horard B, Hubert B et al (2010) Jumping genes and epigenetics: towards new species. Gene 454:1–7

Regulski M, Lu Z, Kendall J et al (2013) The maize methylome influences mRNA splice sites and reveals widespread paramutation-like switches guided by small RNA. Genome Res 23:1651–1662

Reik W (2007) Stability and flexibility of epigenetic gene regulation in mammalian development. Nature 447:425–432

Reinders J, Wulff BBH, Mirouze M et al (2009) Compromised stability of DNA methylation and transposon immobilization in mosaic *Arabidopsis* epigenomes. Genes Dev 23:939–950

Richards EJ (1997) DNA methylation and plant development. Trends Genet 13:319–323

Richards EJ (2008) Population epigenetics. Curr Opin Genet Dev 18:221–226

Richards EJ (2011) Natural epigenetic variation in plant species: a view from the field. Curr Opin Plant Biol 14:204–209

Richards CL, Bossdorf O, Verhoeven KJF (2010) Understanding natural epigenetic variation. New Phytol 187:562–564

Rodrigues JA, Zilberman D (2015) Evolution and function of genomic imprinting in plants. Genes Dev 29:2517–2531

Roh TY, Cuddapah S, Cui K et al (2006) The genomic landscape of histone modifications in human T cells. Proc Natl Acad Sci U S A 103:15782–15787

Schmitz RJ, Schultz MD, Lewsey MG et al (2011) Transgenerational epigenetic instability is a source of novel methylation variants. Science 334:369–373

Schmitz RJ, He Y, Valdes-Lopez O et al (2013a) Epigenome-wide inheritance of cytosine methylation variants in a recombinant inbred population. Genome Res 23:1663–1674

Schmitz RJ, Schultz MD, Urich MA et al (2013b) Patterns of population epigenomic diversity. Nature 495:193–198

Schmutz J, Cannon SB, Schlueter J et al (2010) Genome sequence of the palaeopolyploid soybean. Nature 463:178–183

Shen H, HeH LJ et al (2012) Genome-wide analysis of DNA methylation and gene expression changes in two *Arabidopsis* ecotypes and their reciprocal hybrids. Plant Cell 24:875–892

Simon JA, Tamkun JW (2002) Programming off and on states in chromatin: mechanisms of Polycomb and trithorax group complexes. Curr Opin Genet Dev 12:210–218

Slotkin RK, Vaughn M, Borges F et al (2009) Epigenetic reprogramming and small RNA silencing of transposable elements in pollen. Cell 136:461–472

Steward N, Kusano T, Sano H (2000) Expression of ZmMET1, a gene encoding a DNA methyltransferase from maize, is associated not only with DNA replication in actively proliferating cells, but also with altered DNA methylation status in cold-stressed quiescent cells. Nucleic Acids Res 28:3250–3259

Steward N, Ito M, Yamaguchi Y et al (2002) Periodic DNA methylation in maize nucleosomes and demethylation by environmental stress. J Biol Chem 277:37741–37746

Stroud H, Ding B, Simon SA et al (2013) Plants regenerated from tissue culture contain stable epigenome changes in rice. Elife 2:e00354

Swiezewski S, Liu F, Magusin A, Dean C (2009) Cold-induced silencing by long antisense transcripts of an *Arabidopsis* Polycomb target. Nature 462:99–802

Takuno S, Gaut BS (2013) Gene body methylation is conserved between plant orthologs and is of evolutionary consequence. Proc Natl Acad Sci U S A 110:1797–1802

Takuno S, Ran J-H, Gaut BS (2016) Evolutionary patterns of genic DNA methylation vary across land plants. Nature Plants 2:15222

Turck F, Coupland G (2011) When vernalization makes sense. Science 331:6–37

Vanyushin BF, Ashapkin V (2011) DNA methylation in higher plants: past, present and future. Bioch Biophys Acta 1809:360–368

Vanyushin BF, Belozersky AN (1959) Nucleotide composition of deoxyribonucleic acid in higher plants. Dokl Akad Nauk SSSR 129:944–946

Varriale A (2014) DNA methylation, epigenetics and evolution in vertebrates: facts and challenges. Int J Evol Biol 475981

Vaughn MW, Tanurdzic M, Lippman Z et al (2007) Epigenetic natural variation in *Arabidopsis thaliana*. PLoS Biol 5:1617–1629

Verhoeven KJF, Jansen JJ, Van Dijk PJ et al (2010) Stress induced DNA methylation changes and their heritability in asexual dandelions. New Phytol 185:1108–1118

Wagner D (2003) Chromatin regulation of plant development. Curr Opinion Plant Dev 6:20–28

Wang X, Elling AA, Li X et al (2009) Genome-wide and organ-specific landscapes of epigenetic modifications and their relationships to mRNA and small RNA transcriptomes in maize. Plant Cell 21:1053–1069

Weigel D, Colot V (2012) Epialleles in plant evolution. Genome Biol 13:249

West-Eberhard MJ (2003) Developmental plasticity and evolution. Oxford University Press, New York, NY

Wollmann H, Berger F (2012) Epigenetic reprogramming during plant reproduction and seed development. Curr Opinion Plant Biol 15:63–69

Yamaguchi-Shinozaki K, Shinozaki K (2005) Organization of cis-acting regulatory elements in osmotic- and cold-stress-responsive promoters. Trends Plant Sci 10:88–94

Yang JL, Liu LW, Gong YQ et al (2007) Analysis of genomic DNA methylation level in radish under cadmium stress by methylation sensitive amplified polymorphism technique. J Plant Physiol Mol Biol 33:219–226

Yong-Villalobos L, González-Morales SI, Wrobel K et al (2015) Methylome analysis reveals an important role for epigenetic changes in the regulation of the *Arabidopsis* response to phosphate starvation. Proc Natl Acad Sci U S A 112:E7293–E7302

Zemach A, Kim MY, Silva P et al (2010) Local DNA hypomethylation activates genes in rice endosperm. Proc Natl Acad Sci U S A 107:18729–18734

Zhang X, Yazaki J, Sundaresan A et al (2006) Genome-wide high-resolution mapping and functional analysis of DNA methylation in *Arabidopsis*. Cell 126:1189–1201

Zilberman D, Gehring M, Tran RK et al (2007) Genome-wide analysis of *Arabidopsis thaliana* DNA methylation uncovers an interdependence between methylation and transcription. Nat Genet 39:61–69

Dynamic DNA Methylation Patterns in Stress Response

Luigi Viggiano and Maria Concetta de Pinto

Abstract Plants are continuously exposed to a vast multiplicity of stressful factors, both of biotic or abiotic nature, that negatively affect their development, yield, and reproductive success. To withstand environmental changes plants have developed complex and sophisticated strategies, among which sensitive detection systems and complex signal transduction pathways. These intricate mechanisms ultimately lead to transcriptional induction of genes encoding proteins enabling adaptation to environmental challenge. Epigenetic modifications, among these DNA methylation, represent potentially robust mechanisms contributing to gene expression regulation during periods of environmental stress. The presence of enzymes involved in DNA demethylation, namely Repressor of Silencing 1, DEMETR, and DEMETR-like, makes modulation of DNA methylation highly ductile in plants. Indeed, cytosine methylation and demethylation within the promoter sequence have been shown to cause gene downregulation and upregulation, respectively, in response to different environmental stress. Due to the sessile nature of plants, this epigenetic mechanism is crucial to permit a suitable plant reaction to stress, resulting in short-term acclimation. However, plants should also be able to reset the stress-induced epigenetic alterations in order to restart normal growth when favorable environmental conditions come back.

In this chapter, dynamics and biologic significance of changes in DNA methylation patterns in plant responses to changing environment will be discussed.

Keywords Abiotic stress • Biotic stress • DNA demethylation • DNA methylation • Stress-responsive genes

Contents

L. Viggiano • M.C. de Pinto (✉)
Department of Biology, University of Bari "Aldo Moro", via E. Orabona 4, 70125 Bari, Italy
e-mail: mariaconcetta.depinto@uniba.it

N. Rajewsky et al. (eds.), *Plant Epigenetics*, RNA Technologies,
DOI 10.1007/978-3-319-55520-1_15

1 Introduction

Plants, as sessile organisms, are continuously exposed to a vast multiplicity of stress, both of biotic or abiotic nature. These stressful factors negatively affect growth, development, and reproductive success of plants and may finally lead to plant death. To withstand adverse environmental conditions, plants have developed complex and sophisticated strategies to sense the stress and to trigger defense responses. For this purpose, plants activate numerous signal transduction pathways, which permit to alter the expression of several stress-responsive genes, which ultimately cause morphological, physiological, and biochemical changes that allow plants to adapt to the adverse environment (Hirayama and Shinozaki 2010; Zhang et al. 2011).

In the last years, it has become more and more clear that epigenetic modifications, among these DNA methylation, may significantly contribute to plant's acclimation to environmental stress (Henderson and Jacobsen 2007). The functions of DNA methylation depend on the nature of the target sequence. DNA methylation greatly takes place in transposable elements (TEs) and repetitive sequences and often is related to the silencing of these regions, therefore having a significant role in genome stability (Goll and Bestor 2005; Henderson and Jacobsen 2007). DNA methylation may be also linked to gene regulation, although in this case the effects depend on the gene regions where DNA methylation occurs. The stress-induced DNA methylation can be reset to the basal level, once that the stress is relieved, resulting in short-term acclimation. On the other hand, the DNA methylation of specific regions can also be stable and inherited trough mitosis or, even, meiosis, functioning as "stress memory," associated with cell fate decisions, developmental switches, or stress responses, which can help plants to deal with successive stresses (Chinnusamy and Zhu 2009a; Angers et al. 2010; Zhang and Hsieh 2013).

There is an extensive literature that relates the stressor events with the epigenetic response of the plants. In the next paragraphs, we will try to highlight the different aspects related to DNA methylation/demethylation changes in response to stress, with particular emphasis to the mechanisms involved in the regulation of dynamics of this response.

2 DNA Methylation in Plants

DNA methylation consists in the covalent bound of a methyl group at 5′ position of cytosine DNA bases. In most eukaryotes, cytosines are methylated in the context of CG dinucleotides; in plants cytosine methylation also occurs in CNG and CNN (N = A, C, or T) sites (Gruenbaum et al. 1981).

The genome-wide methylation level significantly differs between diverse plant species, ranging from 5% of total cytosines in the plant model system *Arabidopsis thaliana* to more than 20% in wheat. This difference among species is principally due to the different content of repetitive DNA, that, as previously said, is the most important target of DNA methylation (Gehring and Henikoff 2007). Moreover, in the same genome the level of cytosine methylation in the different contexts also significantly varies. For instance, in Arabidopsis, DNA methylation occurs at 24%, 6.7%, and 1.7% at CG, CNG, and CNN sites, respectively (Law and Jacobsen 2010). In rice, it has been shown that CG methylation is typical of genic regions, while CNG and CNN methylation is abundant in TEs (Zemach et al. 2010). In *met1*, a mutant of Arabidopsis for CG methylation, the expression of genes normally methylated in wild-type plants clearly increases (Zilberman et al. 2007). CG methylation of genes, occurring in the 5′ portion, including the promoter and part of the transcribed region, and in the 3′ portion, including part of the transcribed region and the 3′ flanking sequences, may inhibit gene expression (Zhang et al. 2006; Gehring and Henikoff 2007; Zilberman et al. 2007). Two hypotheses have been proposed that can explain this inhibition: on one hand, the methylation of cytosines in promoters and enhancer regions may avoid the binding of transcription factors (TFs); on the other hand, methylated cytosines may recruit methylcytosine-binding proteins, attracting histone deacetylases and chromatin remodeling proteins, which compacting chromatin structure do not make available the access of the transcription apparatus (Cao and Jacobsen 2002). Gene body methylation also occurs in plants, but in this case DNA methylation has a less clear function (Zemach et al. 2010; Li et al. 2012; Takuno and Gaut 2012). Three different hypotheses on the role of gene body methylation have been proposed: it can repress the expression from cryptic promoters within coding regions, it can improve the correct splicing of primary transcripts, or it can be a by-product of transcription with no functional significance (Zilberman et al. 2007; Teixeira and Colot 2009; Luco et al. 2010).

In Arabidopsis, approximately one-third of genes are at least partially methylated; moreover, while DNA methylation in TEs and DNA repeats is very similar in different accessions, 50% of methylated sequences in genes change between different ecotypes (Vaughn et al. 2007). An analysis on Vancouver and Columbia ecotypes shows that 10% of CG sequences are differentially methylated. The methylation polymorphism is more often found in the upstream or downstream gene regions and it is inversely correlated with the level of gene expression. On the other hand, methylation polymorphism within the gene does not show a clear correlation with changes in gene expression (Zhang et al. 2008). In contrast, in

Populus trichocarpa, gene body methylation has a more repressive effect on transcription than promoter methylation (Vining et al. 2012).

Dynamics of DNA methylation patterns is regulated by three processes: de novo methylation, maintenance methylation, and DNA demethylation.

Methylation in the symmetrical CG and CNG sites can take place, during DNA replication via a semiconservative mechanism; conversely, the nonsymmetrical CNN methylation depends entirely on de novo methylation, occurring after the DNA replication cycle. In plants, three different classes of methyltransferases have been identified, namely methyltransferases (METs), chromomethylases (CMTs), and domains rearranged DNA methylases (DRMs) (Finnegan and Kovac 2000). Methylation in CG dinucleotide sites is performed by MET1, a homologue of mammalian DNMT1 (Kankel et al. 2003). The maintenance of DNA methylation at CNG sites is established by CMT3, which is a plant-specific DNA methylase (Lindroth et al. 2001). The functioning of these two kinds of maintenance methylases requires the activity of the chromatin remodeling factor DDM1 (decrease in DNA methylation 1) (Zemach et al. 2013). On the other hand, de novo CNN methylation is mainly established by DRM1 and DRM2, with a little contribution of CMT2. Methylation of cytosine residues in the CNN sequences of genes is frequently associated with their silencing (Wassenegger 2000). De novo DNA methylation activity in plants is regulated by the RNA-directed DNA methylation (RdDM) pathway, which using siRNA, targets DRMs in a sequence-specific manner to definite loci (Matzke and Mosher 2014).

DNA demethylation in plants can occur passively or actively. In the passive mechanism, the substitution of methylated cytosines with non-modified cytosines takes place during DNA replication; whereas, the active removal of the methyl group from cytosines is mediated by DNA 5-methylcytosine glycosylases. In Arabidopsis, DNA glycosylases are encoded by four genes: Repressor of silencing 1 (ROS1), Demeter (DME), DME-like 2 (DML2), and DML3 (Morales-Ruiz et al. 2006; Ikeda and Kinoshita 2009; Zhu 2009). DME is preferentially expressed in the cells of female gametophytes, where it is involved in gene imprinting, and in the vegetative cells of pollen, therefore, carrying out a significant role in the establishment of epigenetic state of gametophytes (Schoft et al. 2011; Ibarra et al. 2012). Conversely, ROS1, DML2, and DML3 are more ubiquitous and are expressed in different organs (Penterman et al. 2007).

3 Genome-Wide DNA Methylation Under Stress

While the genetic variations are the main source of long-term adaptation and evolution of new characters, induced epigenetic changes are able to facilitate rapid adaptation to quick environmental oscillations (Franks and Hoffmann 2012). Genome-wide DNA methylation patterns are deeply influenced during plant development, but environmental factors also impact on methylation diversity (Zhang et al. 2015). Consistently, many scientific works report that different environmental stresses, both of abiotic and biotic nature, cause changes in the genome-wide DNA methylation. The changes in methylation patterns are, in turn, involved in the control of plant adaptive response.

In *Jatropha curcas*, salt stress induces changes in the DNA methylation patterns of specific genomic fragments of leaves and roots. The authors suggest that these specific alterations in DNA methylation could play a significant function in the acclimation responses that occur under salinity stress (Mastan et al. 2012). Cold stress causes a strong genome-wide DNA demethylation in maize seedlings. The DNA demethylation is principally associated to genomic regions involved in transposon activation, hormone regulation, photosynthesis, and cold response. Coherently, an increase in the transcription of five demethylated genes occurs, thus indicating that the specific demethylation of genes is an active and rapid epigenetic response to cold (Shan et al. 2013). In grapevine plants stressed by in vitro cultivation, variation in DNA methylation occurs. When the stress is relieved, 40% of the observed methylation changes are reverted, thus, acting as a temporary and reversible stress acclimation mechanism, whereas 60% of DNA methylation diversity is maintained and most likely corresponds to mitotically inherited epimutations (Baránek et al. 2015).

DNA methylation seems to be also involved in the control of plant responses to pathogen infection. The age-dependent increase in resistance against the blight pathogen *Xanthomonas oryzae* in the rice cultivar Wase Aikoku 3 seems to depend on an overall higher level of methylation in adult plants than in the seedlings, and on a specific differential cytosine methylation between the two different developmental stages (Sha et al. 2005). In *Arabidopsis thaliana*, DNA methylation status is altered by an active demethylation mechanism in response to the attack of *Pseudomonas syringae*. This response is precocious and largely precedes plant cell death, due to pathogen attack (Pavet et al. 2006). Various Arabidopsis hypomethylated mutants display enhanced resistance to the biotrophic pathogen *Hyaloperonospora arabidopsidis*, while hypermethylated mutants are more susceptible to this pathogen. These opposite resistance phenotypes are associated with changes in cell wall defense and salicylic acid-dependent gene expression (Lopez Sanchez et al. 2016).

Alterations in DNA methylation patterns also occur after virus and viroid infection. In tomato plants infected with Tomato yellow leaf curl Sardinia virus, the majority of the identified polymorphism are related to genomic regions implicated in defense and stress response (Mason et al. 2008). Cucumber plants infected with Hop stunt viroid (HSVd) have alterations in the DNA methylation pattern of normally silenced rRNA genes, which are, therefore, transcriptionally reactivated during infection; in this case, the pathogenesis could be related to the disruptions of transcriptional machinery negatively regulated by epigenetic modifications of plant DNA (Martinez et al. 2014). HSVd RNA accumulation in *Nicotiana benthamiana* mimics the alteration induced by viroid infection in cucumber. This phenomenon seems to be linked to a loss of symmetric cytosine methylation and related to the accumulation of rRNA precursors, and it may be not restricted to a specific host but may occur in other viroid–plant interactions (Castellano et al. 2015).

3.1 Correlation of DNA Methylation Patterns with Stress Tolerance of Different Genotypes

A link between changes in the genome-wide methylation and various degrees of stress tolerance in different genotypes of the same species has been often observed. Many examples of this correlation regard the response of plants to salt and drought stress conditions. For instance, it has been shown that the different sensitivity to salt and drought of three rice cultivars correlates with the existence of several differentially methylated regions, many of which associated with the differential expression of genes involved in abiotic stress response (Garg et al. 2015). A study of DNA methylation polymorphisms in response to drought stress in rice shows that under drought conditions the drought-susceptible genotypes are more hypermethylated than the drought-tolerant ones (Gayacharan and Joel 2013). A genotypic specificity in DNA methylation patterns has been also found in the rice drought-tolerant variety DK151 and its recurrent parent IR64. Interestingly, about 30% of the sites in which epigenetic changes occur are maintained even after the stress recovery, indicating that plants may have a mechanism to remember the previous adverse environmental conditions experienced during the life cycle (Wang et al. 2011). Under salt stress, a diverse methylome flexibility has been found in salt-resistant (Pokkali) and salt-sensitive (IR29) rice varieties. The salt-tolerant rice variety has a greater capability to induce DNA methylation changes in response to salt stress than the salt-sensitive one. These differences are explained with the increased expression of the only DNA demethylases in Pokkali and the induction of both DNA demethylases and methyltransferases in the IR29 (Ferreira et al. 2015). Consistently, the analysis of the rice introgression line IL177-103 and its recurrent parent IR64, differing in salinity tolerance, shows that about 10% of the salt-induced DNA methylation changes are genotype specific and they provide epigenetic markers for salt tolerance (Wang et al. 2015). A study with two cultivars of wheat with different level of salt tolerance shows that in control conditions the global DNA methylation level was higher in the salt-tolerant cultivar than in the sensitive one; under salt stress, a clear genome-wide hypomethylation occurs in both the cultivars and the effect is dose-dependent (Zhong et al. 2009). Consistently, it has been shown that in wheat, the salinity-tolerant cultivar SR3 and its progenitor parent JN177 have a different DNA methylation level; a reduction in the DNA methylation occurs in both the cultivars following salt stress. Interestingly, methylation changes induced in 13 loci of non-stressed SR3, as result of genetic stress triggered by an arrangement of different genomes in somatic hybridization, also occur in salt-stressed JN177. Moreover, SR3 and JN177 have a different methylation pattern of some salt-responsive genes, which guarantees the stress tolerance in SR3. These results underline that changes in DNA methylation patterns can be considered as a common response of plants to stress; furthermore, methylation changes induced by somatic hybridization may participate to the greater salinity tolerance of SR3 (Wang et al. 2014).

Relationship between genotypes, DNA methylation patterns, and stress responses has been also correlated with the ability of plants to tolerate changing temperatures. The study of two collections of *Arabidopsis thaliana* accessions has permitted to found a natural CMT2 variation that is associated with genome-wide DNA methylation changes and temperature seasonality. It has been demonstrated that *cmt2* mutants are more tolerant to heat stress. These data suggest that genetic regulation of epigenetic modifications could represent a possible mechanism for natural adaptation to variable temperatures (Shen et al. 2014). Similarly, differences in DNA methylation have been found in Swedish *Arabidopsis thaliana* accessions grown at two different temperatures. A wide CNN methylation difference is associated with genetic variants. On the other hand, CG gene body methylation is not influenced by temperature, but it is correlated with the latitude of origin. Arabidopsis accessions from colder regions have a significant increase in gene body methylation, associated with an increase in the transcription for the affected genes (Dubin et al. 2015). In chickpea cells, in response to cold stress the methylation changes are higher compared to demethylation ones; however, after long-term cold stress, the demethylated regions are larger in the tolerant genotype than in the susceptible ones, demonstrating a higher potential of the tolerant genotype for activation of cold-stress responsive genes (Rakei et al. 2016). The study of wheat near-isogenic lines shows that winter wheat is more methylated as compared to spring wheat; moreover, a cold treatment, known as vernalization, results in a general DNA demethylation that is linked, not exclusively, to sequences related to flower induction (Sherman and Talbert 2002).

Finally, also the plant response to pathogens may be different depending on cultivars and their methylation patterns. For instance, resistant and sensitive chickpea genotypes show extensive cytosine methylation alterations following inoculation with *Fusarium oxysporum*. Interestingly, most of the genomic regions differentially methylated among the two genotypes have homology with disease-related genes, suggesting that the significant differences in DNA methylation between resistant and sensitive genotypes may be implicated in the chickpea resistance to Fusarium wilt (Mohammadi et al. 2015).

3.2 *Inheritable Changes in DNA Methylation Patterns in Plants Subjected to Stress*

The different stress tolerance observed among the cultivars of the same species may be due to the trans-generational acquisition of the epigenetic changes. It is believed that plants can effectively integrate the signals from the environment into a stress memory transmittable to offspring. This gained information can enable the population to effectively respond to repetitive exposures of the same stress (Conrath et al. 2006; Sani et al. 2013; Slaughter et al. 2012; Crisp et al. 2016).

Genome-wide studies show that environmental stresses are able to modify the chromatin landscape, creating new patterns of gene expression (Hauser et al. 2011; Ito et al. 2011; Sani et al. 2013). Heritable characters caused by environmental stress have been associated with changes in DNA methylation level of promoters, gene bodies, transgenes, and transposable elements (Lang-Mladek et al. 2010; Bilichak et al. 2012; Jiang et al. 2014; Secco et al. 2015). Here we report some examples of inheritable changes in DNA methylation occurring in plants subjected to stressful conditions.

The exposure of Arabidopsis plants to different kinds of stress leads to a parallel increase in global genome methylation and in stress tolerance, that persists also in the non-stressed progeny (Boyko et al. 2010). Accordingly, the progeny of Arabidopsis plants subjected to salt stress shows hypermethylation in most of the genes that change their methylation state (Bilichak et al. 2012).

Moderate and severe nitrogen deficiency triggers locus-specific methylation changes in leaf tissue of rice plants. Interestingly, 50% of the changed methylation patterns are stably inherited. Moreover, the descendent plants that inherited the altered methylation patterns are more tolerant to nitrogen deficiency (Kou et al. 2011).

The progenies of tobacco plants infected with the tobacco mosaic virus (TMV) show hypermethylated genomes and, at the same time, deep hypomethylation in several LRR-containing loci that improves their resistance to TMV. The authors suggest that the global genome hypermethylation of the progeny is part of a general protection mechanism against stress, whereas locus-specific hypomethylation is associated with a high frequency of recombination. These epigenetic changes, together, may represent an adaptive response of plants to stress (Boyko et al. 2007). Similarly, the treatment of rice seeds with the DNA methylation inhibitor, 5-azacytidine that causes genomic hypomethylation in the progeny, induces a hypomethylation at the Xa21G locus that is stably inherited and that confers an adaptive advantage against the pathogen *Xanthomonas oryzae* (Akimoto et al. 2007). Pathogens, as well as plant defense signalling molecules, such as jasmonic acid or salicylic, can induce methylation changes that lead to epiallelic variation in the Arabidopsis genome. The DNA methylation-related mutants, which have hypomethylation of genomic DNA, show an enhanced resistance to bacteria. These observations indicate that transgenerational priming of SA-dependent defense may be based on reduced DNA methylation of some regulatory genes (Luna et al. 2012).

4 Involvement of DNA Methylation Changes in the Control of Stress-Responsive Genes

Dynamics of methylation/demethylation may control the expression of genes that play a key role in acclimation responses. The expression of an antisense construct for DNA methyltransferase in tobacco plants has permitted to identify 31 genes specifically hypomethylated that accumulate their transcripts; ten of these genes are related to biotic and abiotic stress responses (Wada et al. 2004). Similarly, mutations in the Arabidopsis DNA methyltransferases lead to genome-wide hypomethylation and upregulation of defense-related genes correlated to an increased resistance to pathogen attack (Dowen et al. 2012). The association of stress-induced DNA hypomethylation with enhanced transcriptional activity of stress-responsive genes suggests that under optimal growth conditions, stress genes are maintained in a repressed state by DNA methylation, in order to not constrain normal plant growth and development (Vriet et al. 2015).

4.1 Changes in DNA Methylation of Specific Stress-Responsive Genes

Numerous studies demonstrate that dynamic changes of DNA methylation are directly involved in the regulation of specific genes in response to different kinds of stresses.

An example of activation of defense genes due to DNA demethylation regards the plant pathogen interaction occurring between tobacco plants and TMV. During the infection, specific hypomethylation of the NtAlix1 enhances the expression of this pathogen-responsive gene (Wada et al. 2004). In the same plant–pathogen interaction, a strong CG hypomethylation at the leucine-rich repeat region of the N gene for resistance to TMV occurs (Boyko et al. 2007).

In the last years, many examples of parallel changes in DNA methylation level and expression of specific stress-responsive genes, in response to abiotic stress, have been reported (Table 1). In tobacco plants, the exposure to different stress, including aluminum, salt, low temperature, and reactive oxygen species, causes changes in the methylation level of a gene coding for a glycerophosphodiesterase-like protein (NtGPDL). In particular, the GC sites of NtGPDL are selectively demethylated in the coding regions and completely demethylated in the promoter; moreover, NtGPDL transcripts are early induced after stress exposure. These results suggest a clear association between methylation and expression of NtGPDL upon abiotic stresses, which could be attributable to oxidative stress occurring in these stress conditions (Choi and Sano 2007). Consistently, the upregulation of oxidative stress-related genes in rice plants treated with the DNA methylation inhibitor 5-azacytidine leads to an increased tolerance to salt stress (Zhong et al. 2010). Similarly, the overexpression of the Arabidopsis ROS1 in tobacco decreases

Table 1 Stress-responsive genes with differential DNA methylation and expression under abiotic stress

Gene	Function of encoded protein	Organism	Stress	DNA methylation change	References
NtGPDL	Glycerophosphodiesterase-like protein; involved in abiotic stress response	*Nicotiana tabacum*	Different abiotic stress	Hypomethylation	Choi and Sano (2007)
ZmPP2c	Protein phosphatase 2C; negative regulator of the stress response	*Zea mays*	Salinity stress	Hypermethylation	Tan (2010)
ZmGST	Glutathione S-transferases; regulator of the stress response	*Zea mays*	Salinity stress	Hypomethylation	Tan (2010)
TaFLS1	Flavonol synthase	*Triticum aestivum*	Salinity stress	Hypomethylation	Wang et al. (2014)
TaWRSI5	Bowman-Birk-type protease inhibitor	*Triticum aestivum*	Salinity stress	Hypomethylation	Wang et al. (2014)
TaGAPC1	Cytosolic glyceraldehyde-3-phosphate dehydrogenase; involved in stress response	*Triticum aestivum*	Osmotic and salinity stress	Hypomethylation	Fei et al. (2016)
ASR1	Abscisic Acid Stress Ripening1; a protein of the LEA superfamily	*Solanum lycopersicum*	Abscisic acid treatment	Hypomethylation	González et al. (2011)
ERD5	Early Responsive to Dehydration 5; involved in proline degradation	*Arabidopsis thaliana*	Menadione bisulfite treatment	Hypermethylation	Borges et al. (2014)
P5CS1	Pyrroline-5-Carboxylate Synthetase 1; involved in proline biosynthesis	*Arabidopsis thaliana*	Menadione bisulfite treatment	Hypomethylation	Borges et al. (2014)
DRB2	Artemisinin aldehyde Delta 11(13) double bond reductase; involved in artemisinin biosynthesis	*Artemisia annua*	UV exposure	Hypomethylation	Pandey and Pandey-Rai (2015)

methylation levels of promoters and coding regions of genes involved in flavonoid and antioxidant pathways. Due to the higher expression level of these genes, the transgenic tobacco plants are more tolerant to salt stress (Bharti et al. 2015). In maize seedlings, salt stress induces differential methylation in two regions homologues to the first intron of protein phosphatase 2C (zmPP2C) and to glutathione S-transferases (zmGST). The salt-induced intron hypermethylation of zmPP2C, which is a negative regulator of stress response, considerably, downregulates its expression; on the other hand, salt stress-induced demethylation of zmGST, which is a positive effector of stress response, upregulates its expression (Tan 2010). In wheat, salt stress induces a shift in DNA methylation in both the promoter and coding regions of 24 genes, but only the DNA methylation changes in the promoters are associated with alterations of gene expression. Two of the differential methylated and expressed genes under salt stress, namely TaFLS1, which encodes a flavonol synthase and TaWRSI5, encoding a protease inhibitor, are able to enhance salt tolerance in Arabidopsis plants (Wang et al. 2014).

In two wheat cultivars with different degrees of drought tolerance, salt stress induces the expression of cytosolic glyceraldehyde-3-phosphate dehydrogenase (GAPC), which encodes a protein that has a positive role in plant stress response (McLoughlin et al. 2013; Fei et al. 2016). However, the increase in the transcript level is higher in the tolerant cultivar than in the susceptible one. The stress induces demethylation of CG and CNG sites in the promoter region of GAPC of the drought-tolerant cultivar; on the other hand, the stress does not affect methylation state of the CG contexts and increases the methylation of CNG and CNN sites in the promoter of GAPC of the drought-susceptible cultivar. The different types and levels of methylation, found in the promoters of GAPC of the two cultivars, underline the relationship between promoter methylation and gene expression occurring in stress response (Fei et al. 2016).

In tomato plants under drought conditions, CNN hypomethylation of Abscisic Acid Stress Ripening1 induces a contemporary increase in its expression, which can permit plant acclimation to stress (González et al. 2011). Abscisic acid enhances resistance to cold stress in the duckweed *Spirodela polyrhiza*, affecting the methylation status and consequently the expression of an ATPase gene (Zhao et al. 2015). The treatment of Arabidopsis plants with menadione sodium bisulfite, a water-soluble compound derived from vitamin K3, which is able to induce priming against biotic and abiotic stress (Borges et al. 2014), leads to changes in the methylation state of genes involved in the proline metabolism. Methylation changes in CNG and CNN contexts of P5CS1 and ERD5, involved in the biosynthesis and degradation of proline, respectively, alter the expression of these two genes and permit the accumulation of proline, which in turn play a role in stress tolerance (Jiménez-Arias et al. 2015).

In *Artemisia annua*, UV-B exposure can mediate demethylation of 4 CG, 4 CNN, and 2 CNG sites of the promoter region of the DBR2 gene, which encodes for a key enzyme of artemisinin biosynthetic pathway. *In silico* analysis has permitted to reveal that the changes in methylation state of this promoter involve seven putative TFs binding sites, including those for WRKYs, which are positive regulators of artemisinin biosynthesis. Contemporary, UV-B induces overexpression of DBR2 and accumulation of artemisinin. These data suggest that DNA demethylation is an important epigenetic plant response to UV-B radiation (Pandey and Pandey-Rai 2015).

4.2 *Hierarchic Control of DNA Methylation in the Induction of Stress-Related Genes*

Dynamic changes in DNA methylation patterns in response to environmental conditions represent a potentially strong mechanism able to regulate gene expression networks. In addition to genome-wide regulation, active DNA demethylation also functions extensively in epigenetic regulation at discrete genetic loci. However, the epigenetic alterations at different loci may be the direct outcome of stress or a derived consequence of other alterations induced by stress (Fig. 1). Indeed, some environmental stimuli alter the expression of epigenetic regulators, which in turn could cause epigenetic modifications at loci susceptible to quantitative changes of methylation (Meyer 2015). For instance, in *Nicotiana benthamiana*, the Rep

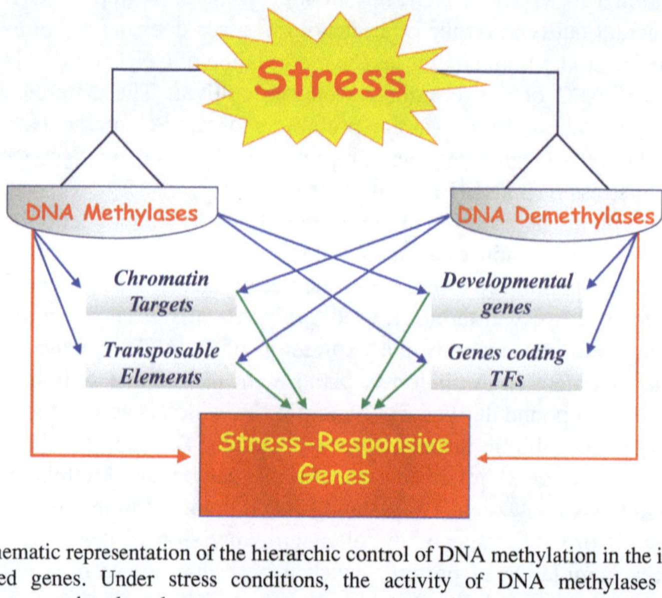

Fig. 1 Schematic representation of the hierarchic control of DNA methylation in the induction of stress-related genes. Under stress conditions, the activity of DNA methylases and DNA demethylases may be altered; as a consequence, methylase/demethylases can directly modify the methylation state of stress-responsive genes, leading to changes in their expression (*red arrows*). On the other hand, other targets, such as chromatin loci, transposable elements, or genes coding for transcription factors (TFs) or other proteins involved in the plant development, can be the target of methylase/demethylases (*blue arrows*). The alterations in the methylation state of these targets may alter, directly or indirectly, the expression of stress-responsive genes (*green arrows*). More details are given in the text

protein of geminivirus, in order to suppress transcriptional gene silencing, represses MET1 and CMT3 causing a strong reduction in the level of DNA methylation (Rodríguez-Negrete et al. 2013). The treatment of rice seedlings with high concentration of the nitric oxide donor sodium nitroprusside, which induces growth inhibition, is associated with hypomethylation at the CNG sites and transcriptional activation of various TEs and genes involved in chromatin remodeling and DNA methylation homeostasis (Ou et al. 2015).

In plants exposed to stress conditions, changes in cytosine methylation, in the CNG and CNN sites, also regulate transposons and repeat regions through chromatin remodeling (Matzke et al. 2009; Furner and Matzke 2011). The specificity of the contexts in which methylation occurs has been shown in tobacco cell cultures exposed to osmotic stress. In two heterochromatic loci, defined with repetitive DNA sequences HRS60 and GRS, reversible hypermethylation in response to stress only occurs in the CNG contexts (Kovarik et al. 1997).

Under environmental stress, DNA methylation can lead to differential gene expression by altering the chromatin structure. In maize seedlings, cold stress induces severe demethylation in nucleosomes and leaves linker regions relatively unchanged. The reduction of DNA methylation in the nucleosome core induces the expression of the ZmMI1 fragment, containing part of the coding region of a putative protein and part of a retrotransposon-like sequence (Steward et al. 2002). In *Mesembryanthemum crystallinum* plants, salt stress induces cytosine methylation in the CNG contexts of a satellite DNA. This hypermethylation is possibly linked to the formation of a specific chromatin structure that can regulate, at the same time, the expression of diverse genes involved in the switch from C3 to CAM metabolism and in the adaptation to salt stress (Dyachenko et al. 2006).

Changes in the methylation state of TEs may represent an important mechanism during plant response to stress. Stress conditions can activate TEs modifying their methylation state, as occur in the *Antirrhinum majus* Tam3 undergoing low temperature-dependent transposition (Hashida et al. 2006). Although TEs are often mutagenic, they, nevertheless, were tamed to be potentially useful for regulating gene expression in response to a wide assortment of stress. Activation of TEs can cause specific or widespread alterations in gene expression. Genes affected by activated TEs can be both the contiguous genes and the genes adjacent to the new integration site (Meyer 2015). In Arabidopsis, the ONSEN retroelement, which possesses a heat-responsive element, is activated by the exposure to high temperature and in turn induces heat-responsive genes in the new insertion sites (Ito et al. 2011; Cavrak et al. 2014). On the other hand, in Arabidopsis, in response to salicylic acid, TE-associated methylation is coupled to transcriptional changes, not only of the TEs, but also of proximal genes (Dowen et al. 2012). During the antibacterial defense in Arabidopsis, demethylation and reactivation of TEs correlates with the decreased expression of key transcriptional gene silencing factors, thus acting as an important component of plant-induced immune response (Yu et al. 2013).

A temporal hierarchy of transcriptional and epigenomic changes may be responsible for the right response occurring during stress conditions. For instance, phosphate starvation in rice induces hypermethylation in TEs localized in proximity of highly induced genes. These methylation alterations take place after the changes in the transcription of the close genes (Secco et al. 2015).

Environmental stress can also cause changes in the DNA methylation patterns of genes coding for TFs, whose expression permits downstream the regulation of a number of stress-responsive genes. In soybean, salinity stress induces the expression of 45 genes coding for TFs, ten of which are also upregulated in seedlings exposed to 5-azacytidine. Four up-regulated genes, encoding for a *MYB*, a *b-ZIP* and two *AP2/DREB* TFs, show DNA hypomethylation in promoter or coding region (Song et al. 2012).

In some cases, stress conditions can change methylation level, and consequently expression, of genes involved in normal development processes. In Arabidopsis, low relative humidity induces DNA methylation and transcriptional suppression of two positive regulator genes for stomatal development, causing a reduction of the numbers of stomata on the leaf epidermis (Tricker et al. 2012). In order to slow down normal cell metabolism and to activate stress-responsive genes, heat stress in tobacco cells alters the methylation state of genes involved in the control of cell cycle and cell expansion (Centomani et al. 2015; Sgobba et al. 2015).

5 Regulation of Dynamics of DNA Methylation Under Stress

DNA demethylation is important in defining the methylation patterns of the genome. It has been shown that in Arabidopsis, many of the CNG and CNN sites with low levels of methylation in the wild-type become heavily methylated in the *ros1* mutant; this increase in DNA methylation is associated with decreased expression. Thus, the dynamic regulation of DNA methylation patterns plays a key role in maintaining the plasticity of epigenome, which may promptly react to developmental and environmental stimuli (Zhu et al. 2007).

Methylation and demethylation have to perform two tasks apparently irreconcilable: to provide the memory of a decision and to respond quickly to any stressor event. Many RdDM targets, such as TEs, are localized close to genes; in Arabidopsis, 44% of genes have a TE within 2 kb of the transcribed region (Wang et al. 2013; Mao et al. 2015); thus de novo methylation may give rise to unwanted methylation of adjacent genes. Indeed, although RdDM directs methylation in a target DNA sequence matching with the siRNAs, methylation may reach the adjacent sequences. Unwanted methylation may also occur if a transcriptionally active locus has sequence homology with the newly created siRNAs. Consequently, plants need to balance the suppression of TEs with the maintenance of gene expression. Among the DNA glycosylases, ROS1 seems to be preferentially involved in counteracting DNA methylation established by the RdDM pathway (Gong et al. 2002; He et al. 2009; Zhu 2009; Gao et al. 2010). Interestingly, although DNA methylation is mainly involved in transcriptional repression (Zilberman et al. 2007), ROS1 is under-expressed in diverse DNA methylation mutants (Mathieu et al. 2007; Gao et al. 2010; Li et al. 2012), suggesting that DNA methylation and demethylation are somehow coordinated. In *Arabidopsis thaliana*, the decreased expression of ROS1, in the RNA-dependent polymerase 2 (*rdr2)* mutant, which is impaired in the RdDM pathway, can be restored by DNA methylation of its 5′ sequence, suggesting that the decreased expression of ROS1 in *rdr2* is directly dependent on the impaired methylation (Williams et al. 2015). The regulation of ROS1 expression is due to an RdDM target sequence in its promoter, named methylation monitoring sequence (MEMS), which is also target for ROS1. ROS1 expression is allowed when MEMS is methylated; the activation of ROS1 leads to MEMS demethylation, which consequently downregulates ROS1 (Lei et al. 2015). Thus, MEMS in the ROS1 promoter is able to regulate ROS1 expression by a negative feedback mechanism, functioning like a methylstat, able to sense DNA methylation levels and to control DNA methylation by regulating ROS1 expression (Lei et al. 2015). The existence of methylation-sensitive ROS1 expression also in *Arabidopsis lyrata* and *Zea mays* indicates that the regulation of DNA demethylases by RdDM might be a general feature of angiosperms, and thus likely adaptive. An intriguing possibility is that the methylstat has evolved to counterbalance a positive feedback between DNA methylation and RdDM activities (Johnson et al. 2014; Zhong et al. 2014).

This regulative mechanism enables plants to balance a number of extremely effective, potent, and self-reinforcing silencing mechanisms with the maintaining of active gene transcription (Williams et al. 2015). In plants, ROS1 expression is constantly balanced by its autoregulation, and this homeostatic balance may be dynamically changed in response to environmental stimuli, that require active demethylation of stress-related genes. In Arabidopsis, the triple DNA demethylase mutant *rdd* (*ros1 dml2 dml3*) has high susceptibility to the fungal pathogen *Fusarium oxysporum*. In the *rdd* mutant, a number of 348 differentially expressed genes, with a role in stress response, are downregulated. Interestingly, these genes are enriched for short TE sequences in their promoters. The reduction in CHH methylation in these TEs and their close sequences, occurring in the *rdd* mutant, indicates that the RdDM pathway may contribute to DNA demethylase-mediated regulation of stress-responsive genes. Many of the *rdd*-downregulated stress-responsive genes show decreased expression also in other mutants with impaired RdDM. Thus, a main function of DNA demethylases may be to control the expression of genes involved in stress response by targeting TE sequences present in their promoters (Le et al. 2014). Similarly, DNA demethylation occurring in antibacterial defense is essential to prime the transcriptional activation of immune-responsive genes linked to TEs and/or repeats. Also in this case, the DNA demethylation induced by the pathogen is obtained by a drastic reduction in the levels of some key components of RdDM pathway. Indeed, enhanced RdDM in *ros* mutants lowers plant resistance to bacteria (Yu et al. 2013). In Arabidopsis, the TF AtMYB74 is transcriptionally regulated by RdDM. In the promoter of AtMYB74 a DNA region, 500 bp upstream of the transcription initiation site, which is the target region for 24-nt siRNAs, is significantly demethylated under salt stress (Xu et al. 2015).

Once the stress is overcome, methylation and demethylation activities also must reset all the induced modifications in order to restore the methylation patterns existing prior of the stressful event. An example of synchronized but not simultaneous interaction between DNA demethylation and methylation has been highlighted in tobacco BY-2 cells. The exposure of these cells to moderate heat stress induces rapid changes in the methylation pattern of the upstream region of the CycD3-1 gene. At the first day after heat stress, this region shows a marked hypomethylation, while its methylation state returns to the same level of control cells after three days of heat exposure (Centomani et al. 2015). Rapid and cyclical variations in the methylation pattern of the promoter regions of some genes or TEs have been previously observed in human (Kangaspeska et al. 2008; Baccarelli et al. 2009). Based on our knowledge, the rapid and successive changes of the methylation state, observed in the CycD3-1 promoter of BY-2 cells, is the first reported case of a cyclic modification of DNA methylation pattern in plant cells.

6 Conclusions

Dynamics of DNA methylation in response to environmental stimuli is a topic of growing interest, as witnessed by the hundreds of scientific publications in the field. While in the last years significant advancements on the comprehension of stress-induced DNA methylation changes have been reached, there are yet many issues that remained unresolved and that need further studies.

DNA methylation, as part of the epigenome, represents an extra level of complexity through which the genome is interpreted. The DNA methylation-dependent control of plant response to stress is a very multifaceted event. Indeed, the significance of DNA methylation patterns may vary upon the specificity of the cytosine methylation context and the genomic regions in which methylation occurs. This uniqueness indicates that DNA methylation is not a simple on/off switch, but a flexible mechanism that ultimately is able to modify gene expression and to permit acclimation to changing environment. However, it is still unknown what is the nature of the signals that drives DNA methylation.

Methylation changes may be rapid and dynamic and can be reset to the basal level, once the stress is relieved, thus permitting short-term acclimation responses. DNA methylome flexibility is due to the presence of different demethylases. However, plants need to balance a number of silencing mechanisms with the maintaining of active gene transcription. Since pathways of de novo DNA methylation and DNA demethylation share some components (Chinnusamy and Zhu 2009b), it is plausible that DNA demethylases could exploit siRNA to target the right DNA sites.

Nevertheless, changes in DNA methylome can be also retained through the entire plant lifetime, contributing to the plant ability to be successful in changing environments. The mechanisms that decide whether and how a genomic region, or a genetic locus, becomes a DNA methylation/demethylation target are still unknown.

Environmental-responsive methylation changes may be involved in the expression of regulatory factors implicated in stimuli perception and signal transduction, and this adds a further level of complexity in the comprehension of this epigenetic modification. Indeed, it is unknown to what extent the DNA methylation changes are directly involved in the control of gene expression, or, indirectly, alter the cellular potential of gene expression.

Furthermore, some DNA methylation-mediated changes can be involved in transgenerational stress memories, leading to the creation of new adaptive epialleles that can remain in the population for many generations, driving divergence of plant ecotypes and influencing the adaptive capability of plant species. Also in this case, some questions remain unanswered; for instance, it is unclear which of the DNA methylation targets are able to create heritable epialleles and contribute to epigenetic diversity.

Due to differences in DNA methylation patterns among individuals and populations, the reaction of different plants to environmental changes may be very different, implying the importance of this epigenetic mark for phenotypical plasticity and adaptation to environment.

References

Akimoto K, Katakami H, Kim HJ et al (2007) Epigenetic inheritance in rice plants. Ann Bot-London 100:205–217

Angers B, Castonguay E, Massicotte R (2010) Environmentally induced phenotypes and DNA methylation: how to deal with unpredictable conditions until the next generation and after. Mol Ecol 19:1283–1295

Baccarelli A, Wright RO, Bollati V et al (2009) Rapid DNA methylation changes after exposure to traffic particles. Am J Respir Crit Care Med 179:572–578

Baránek M, Cechova J, Raddova J et al (2015) Dynamics and reversibility of the DNA methylation landscape of grapevine plants (*Vitis vinifera*) stressed by in vitro cultivation and thermotherapy. PLoS One 10:e0126638

Bharti P, Mahajan M, Vishwakarma AK et al (2015) AtROS1 overexpression provides evidence for epigenetic regulation of genes encoding enzymes of flavonoid biosynthesis and antioxidant pathways during salt stress in transgenic tobacco. J Exp Bot 66:5959–5969

Bilichak A, Ilnystkyy Y, Hollunder J et al (2012) The progeny of *Arabidopsis thaliana* plants exposed to salt exhibit changes in DNA methylation, histone modifications and gene expression. PLoS One 7:e30515

Borges AA, Jimenez-Arias D, Exposito-Rodriguez M et al (2014) Priming crops against biotic and abiotic stresses: MSB as a tool for studying mechanisms. Front Plant Sci 5:642

Boyko A, Kathiria P, Zemp FJ et al (2007) Transgenerational changes in the genome stability and methylation in pathogen-infected plants (Virus-induced plant genome instability). Nucleic Acids Res 35:1714–1725

Boyko A, Blevins T, Yao YL et al (2010) Transgenerational adaptation of Arabidopsis to stress requires DNA methylation and the function of Dicer-Like proteins. PLoS One 5:e9514

Cao XF, Jacobsen SE (2002) Role of the Arabidopsis DRM methyltransferases in de novo DNA methylation and gene silencing. Curr Biol 12:1138–1144

Castellano M, Martinez G, Pallas V et al (2015) Alterations in host DNA methylation in response to constitutive expression of Hop stunt viroid RNA in *Nicotiana benthamiana* plants. Plant Pathol 64:1247–1257

Cavrak VV, Lettner N, Jamge S, Kosarewicz A et al (2014) How a retrotransposon exploits the plant's heat stress response for its activation. PLoS Genet 10:e1004115

Centomani I, Sgobba A, D'Addabbo P et al (2015) Involvement of DNA methylation in the control of cell growth during heat stress in tobacco BY-2 cells. Protoplasma 252:1451–1459

Chinnusamy V, Zhu JK (2009a) Epigenetic regulation of stress responses in plants. Curr Opin Plant Biol 12:133–139

Chinnusamy V, Zhu JK (2009b) RNA-directed DNA methylation and demethylation in plants. Sci China C Life Sci 52:331–343

Choi CS, Sano H (2007) Abiotic-stress induces demethylation and transcriptional activation of a gene encoding a glycerophosphodiesterase-like protein in tobacco plants. Mol Genet Genomics 277:589–600

Conrath U, Beckers GJ, Flors V et al (2006) Priming: getting ready for battle. Mol Plant Microbe Interact 19:1062–1071

Crisp PA, Ganguly D, Eichten SR et al (2016) Reconsidering plant memory: intersections between stress recovery, RNA turnover, and epigenetics. Sci Adv 2:e1501340

Dowen RH, Pelizzola M, Schmitz RJ et al (2012) Widespread dynamic DNA methylation in response to biotic stress. Proc Natl Acad Sci USA 109:E2183–E2191

Dubin MJ, Zhang P, Meng DZ et al (2015) DNA methylation in Arabidopsis has a genetic basis and shows evidence of local adaptation. Elife 4:e05255

Dyachenko OV, Zakharchenko NS, Shevchuk TV et al (2006) Effect of hypermethylation of CCWGG sequences in DNA of *Mesembryanthemum crystallinum* plants on their adaptation to salt stress. Biochemistry (Moscow) 71:461–465

Fei Y, Xue Y, Du P, Yang S et al (2016) Expression analysis and promoter methylation under osmotic and salinity stress of TaGAPC1 in wheat (*Triticum aestivum* L). Protoplasma. doi:10.1007/s00709-016-1008-5

Ferreira LJ, Azevedo V, Maroco J et al (2015) Salt tolerant and sensitive rice varieties display differential methylome flexibility under salt stress. PLoS One 10:e0124060

Finnegan EJ, Kovac KA (2000) Plant DNA methyltransferases. Plant Mol Biol 43:189–201

Franks SJ, Hoffmann AA (2012) Genetics of climate change adaptation. Ann Rev Genet 46:185–208

Furner IJ, Matzke M (2011) Methylation and demethylation of the Arabidopsis genome. Curr Opin Plant Biol 14:137–141

Gao Z, Liu HL, Daxinger L et al (2010) An RNA polymerase II- and AGO4-associated protein acts in RNA-directed DNA methylation. Nature 465:106–109

Garg R, Chevala VVSN, Shankar R et al (2015) Divergent DNA methylation patterns associated with gene expression in rice cultivars with contrasting drought and salinity stress response. Sci Rep 5:14922

Gayacharan A, Joel J (2013) Epigenetic responses to drought stress in rice (*Oryza sativa* L.) Physiol Mol Biol Plants 19:379–387

Gehring M, Henikoff S (2007) DNA methylation dynamics in plant genomes. Biochim Biophys Acta—Gene Struct Expr 1769:276–286

Goll MG, Bestor TH (2005) Eukaryotic cytosine methyltransferases. Ann Rev Biochem 74:481–514

Gong Z, Morales-Ruiz T, Ariza RR et al (2002) ROS1, a repressor of transcriptional gene silencing in Arabidopsis, encodes a DNA glycosylase/lyase. Cell 111:803–814

González RM, Ricardi MM, Iusem ND (2011) Atypical epigenetic mark in an atypical location: cytosine methylation at asymmetric (CNN) sites within the body of a non-repetitive tomato gene. BMC Plant Biol 11:94

Gruenbaum Y, Naveh-Many T, Cedar H et al (1981) Sequence specificity of methylation in higher plant DNA. Nature 292:860–862

Hashida SN, Uchiyama T, Martin C et al (2006) The temperature-dependent change in methylation of the Antirrhinum transposon Tam3 is controlled by the activity of its transposase. Plant Cell 18:104–118

Hauser MT, Aufsatz W, Jonak C et al (2011) Transgenerational epigenetic inheritance in plants. Biochim Biophys Acta 1809:459–468

He XJ, Hsu YF, Zhu S et al (2009) An effector of RNA-directed DNA methylation in arabidopsis is an ARGONAUTE 4- and RNA-binding protein. Cell 137:498–508

Henderson IR, Jacobsen SE (2007) Epigenetic inheritance in plants. Nature 447:418–424

Hirayama T, Shinozaki K (2010) Research on plant abiotic stress responses in the post-genome era: past, present and future. Plant J 61:1041–1052

Ibarra CA, Feng XQ, Schoft VK et al (2012) Active DNA demethylation in plant companion cells reinforces transposon methylation in gametes. Science 337:1360–1364

Ikeda Y, Kinoshita T (2009) DNA demethylation: a lesson from the garden. Chromosoma 118:37–41

Ito H, Gaubert H, Bucher E et al (2011) An siRNA pathway prevents transgenerational retrotransposition in plants subjected to stress. Nature 472:115–119

Jiang C, Mithani A, Belfield EJ et al (2014) Environmentally responsive genome-wide accumulation of de novo *Arabidopsis thaliana* mutations and epimutations. Genome Res 24:1821–1829

Jiménez-Arias D, Borges AA, Luis JC et al (2015) Priming effect of menadione sodium bisulphite against salinity stress in Arabidopsis involves epigenetic changes in genes controlling proline metabolism. Environ Exp Bot 120:23–30

Johnson LM, Du JM, Hale CJ et al (2014) SRA- and SET-domain-containing proteins link RNA polymerase V occupancy to DNA methylation. Nature 507:124–128

Kangaspeska S, Stride B, Metivier R et al (2008) Transient cyclical methylation of promoter DNA. Nature 452:112–115

Kankel MW, Ramsey DE, Stokes TL et al (2003) Arabidopsis MET1 cytosine methyltransferase mutants. Genetics 163:1109–1122

Kou HP, Li Y, Song XX et al (2011) Heritable alteration in DNA methylation induced by nitrogen-deficiency stress accompanies enhanced tolerance by progenies to the stress in rice (*Oryza sativa* L.) J Plant Physiol 168:1685–1693

Kovarik A, Koukalova B, Bezdek M et al (1997) Hypermethylation of tobacco heterochromatic loci in response to osmotic stress. Theor Appl Genet 95:301–306

Lang-Mladek C, Popova O, Kiok K et al (2010) Transgenerational inheritance and resetting of stress-induced loss of epigenetic gene silencing in Arabidopsis. Mol Plant 3:594–602

Law JA, Jacobsen SE (2010) Establishing, maintaining and modifying DNA methylation patterns in plants and animals. Nat Rev Genet 11:204–220

Le TN, Schumann U, Smith NA et al (2014) DNA demethylases target promoter transposable elements to positively regulate stress responsive genes in Arabidopsis. Genome Biol 15:458

Lei MG, Zhang HM, Julian R et al (2015) Regulatory link between DNA methylation and active demethylation in Arabidopsis. Proc Natl Acad Sci USA 112:3553–3557

Li X, Qian W, Zhao Y et al (2012) Antisilencing role of the RNA-directed DNA methylation pathway and a histone acetyltransferase in Arabidopsis. Proc Natl Acad Sci USA 109:11425–11430

Lindroth AM, Cao XF, Jackson JP et al (2001) Requirement of CHROMOMETHYLASE3 for maintenance of CpXpG methylation. Science 292:2077–2080

Lopez Sanchez A, Stassen JH, Furci L et al (2016) The role of DNA (de)methylation in immune responsiveness of Arabidopsis. Plant J. 10.1111/tpj.13252

Luco RF, Pan Q, Tominaga K et al (2010) Regulation of alternative splicing by histone modifications. Science 327:996–1000

Luna E, Bruce TJA, Roberts MR et al (2012) Next-generation systemic acquired resistance. Plant Physiol 158:844–853

Mao H, Wang H, Liu S et al (2015) A transposable element in a NAC gene is associated with drought tolerance in maize seedlings. Nat Commun 6:8326

Martinez G, Castellano M, Tortosa M et al (2014) A pathogenic non-coding RNA induces changes in dynamic DNA methylation of ribosomal RNA genes in host plants. Nucleic Acids Res 42:1553–1562

Mason G, Noris E, Lanteri S et al (2008) Potentiality of methylation-sensitive amplification polymorphism (MSAP) in identifying genes involved in tomato response to Tomato yellow leaf curl Sardinia virus. Plant Mol Biol Rep 26:156–173

Mastan SG, Rathore MS, Bhatt VD et al (2012) Assessment of changes in DNA methylation by methylation-sensitive amplification polymorphism in *Jatropha curcas* L. subjected to salinity stress. Gene 508:125–129

Mathieu O, Reinders J, Caikovski M et al (2007) Transgenerational stability of the Arabidopsis epigenome is coordinated by CG methylation. Cell 130:851–862

Matzke MA, Mosher RA (2014) RNA-directed DNA methylation: an epigenetic pathway of increasing complexity. Nat Rev Genet 15:394–408

Matzke M, Kanno T, Claxinger L et al (2009) RNA-mediated chromatin-based silencing in plants. Curr Opin Cell Biol 21:367–376

McLoughlin F, Arisz SA, Dekker HL et al (2013) Identification of novel candidate phosphatidic acid-binding proteins involved in the salt-stress response of *Arabidopsis thaliana* roots. Biochem J 450:573–581

Meyer P (2015) Epigenetic variation and environmental change. J Exp Bot 66:3541–3548

Mohammadi P, Bahramnejad B, Badakhshan H et al (2015) DNA methylation changes in fusarium wilt resistant and sensitive chickpea genotypes (*Cicer arietinum* L.) Physiol Mol Plant P 91:72–80

Morales-Ruiz T, Ortega-Galisteo AP, Ponferrada-Marin MI et al (2006) DEMETER and REPRESSOR OF SILENCING 1 encode 5-methylcytosine DNA glycosylases. Proc Natl Acad Sci USA 103:6853–6858

Ou XF, Zhuang TT, Yin WC et al (2015) DNA Methylation changes induced in rice by exposure to high concentrations of the nitric oxide modulator, sodium nitroprusside. Plant Mol Biol Rep 33:1428–1440

Pandey N, Pandey-Rai S (2015) Deciphering UV-B-induced variation in DNA methylation pattern and its influence on regulation of DBR2 expression in *Artemisia annua* L. Planta 242:869–879

Pavet V, Quintero C, Cecchini NM et al (2006) Arabidopsis displays centromeric DNA hypomethylation and cytological alterations of heterochromatin upon attack by *Pseudomonas syringae*. Mol Plant Microbe Interact 19:577–587

Penterman J, Zilberman D, Huh JH et al (2007) DNA demethylation in the Arabidopsis genome. Proc Natl Acad Sci USA 104:6752–6757

Rakei A, Maali-Amiri R, Zeinali H et al (2016) DNA methylation and physio-biochemical analysis of chickpea in response to cold stress. Protoplasma 253:61–76

Rodriguez-Negrete E, Lozano-Duran R, Piedra-Aguilera A et al (2013) Geminivirus Rep protein interferes with the plant DNA methylation machinery and suppresses transcriptional gene silencing. New Phytol 199:464–475

Sani E, Herzyk P, Perrella G et al (2013) Hyperosmotic priming of Arabidopsis seedlings establishes a long-term somatic memory accompanied by specific changes of the epigenome. Genome Biol 14:R59

Schoft VK, Chumak N, Choi Y et al (2011) Function of the DEMETER DNA glycosylase in the *Arabidopsis thaliana* male gametophyte. Proc Natl Acad Sci USA 108:8042–8047

Secco D, Wang C, Shou HX et al (2015) Stress induced gene expression drives transient DNA methylation changes at adjacent repetitive elements. Elife 4:e09343

Sgobba A, Paradiso A, Dipierro S et al (2015) Changes in antioxidants are critical in determining cell responses to short- and long-term heat stress. Physiol Plant 153:68–78

Sha AH, Lin XH, Huang JB et al (2005) Analysis of DNA methylation related to rice adult plant resistance to bacterial blight based on methylation-sensitive AFLP (MSAP) analysis. Mol Genet Genomics 273:484–490

Shan XH, Wang XY, Yang G et al (2013) Analysis of the DNA methylation of maize (*Zea mays* L.) in response to cold stress based on methylation-sensitive amplified polymorphisms. J Plant Biol 56:32–38

Shen X, De Jonge J, Forsberg SKG et al (2014) Natural CMT2 variation is associated with genome-wide methylation changes and temperature seasonality. PLoS Genet 10:e1004842

Sherman JD, Talbert LE (2002) Vernalization-induced changes of the DNA methylation pattern in winter wheat. Genome 45:253–260

Slaughter A, Daniel X, Flors V et al (2012) Descendants of primed Arabidopsis plants exhibit resistance to biotic stress. Plant Physiol 158:835–843

Song YG, Ji DD, Li S et al (2012) The dynamic changes of DNA methylation and histone modifications of salt responsive transcription factor genes in soybean. PLoS One 7:e41274

Steward N, Ito M, Yamaguchi Y et al (2002) Periodic DNA methylation in maize nucleosomes and demethylation by environmental stress. J Biol Chem 277:37741–37746

Takuno S, Gaut BS (2012) Body-methylated genes in *Arabidopsis thaliana* are functionally important and evolve slowly. Mol Biol Evol 29:219–227

Tan MP (2010) Analysis of DNA methylation of maize in response to osmotic and salt stress based on methylation-sensitive amplified polymorphism. Plant Physiol Bioch 48:21–26

Teixeira FK, Colot V (2009) Gene body DNA methylation in plants: a means to an end or an end to a means? EMBO J 28:997–998

Tricker PJ, Gibbings JG, Rodriguez Lopez CM et al (2012) Low relative humidity triggers RNA-directed de novo DNA methylation and suppression of genes controlling stomatal development. J Exp Bot 63:3799–3813

Vaughn MW, Tanurdzic M, Lippman Z et al (2007) Epigenetic natural variation in *Arabidopsis thaliana*. PLoS Biol 5:1617–1629

Vining KJ, Pomraning KR, Wilhelm LJ et al (2012) Dynamic DNA cytosine methylation in the *Populus trichocarpa* genome: tissue-level variation and relationship to gene expression. BMC Genomics 13:27

Vriet C, Hennig L, Laloi C (2015) Stress-induced chromatin changes in plants: of memories, metabolites and crop improvement. Cell Mol Life Sci 72:1261–1273

Wada Y, Miyamoto K, Kusano T et al (2004) Association between up-regulation of stress-responsive genes and hypomethylation of genomic DNA in tobacco plants. Mol Genet Genomics 271:658–666

Wang WS, Zhao XQ, Pan YJ et al (2011) DNA methylation changes detected by methylation-sensitive amplified polymorphism in two contrasting rice genotypes under salt stress. J Genet Genomics 38:419–424

Wang X, Weigel D, Smith LM (2013) Transposon variants and their effects on gene expression in Arabidopsis. PLoS Genet 9:e1003255

Wang M, Qin LM, Xie C et al (2014) Induced and constitutive DNA methylation in a salinity-tolerant wheat introgression line. Plant Cell Physiol 55:1354–1365

Wang WS, Huang F, Qin Q et al (2015) Comparative analysis of DNA methylation changes in two rice genotypes under salt stress and subsequent recovery. Biochem Biophys Res Commun 465:790–796

Wassenegger M (2000) RNA-directed DNA methylation. Plant Mol Biol 43:203–220

Williams BP, Pignatta D, Henikoff S et al (2015) Methylation-sensitive expression of a DNA demethylase gene serves as an epigenetic rheostat. PLoS Genet 11:e1005142

Xu R, Wang Y, Zheng H et al (2015) Salt-induced transcription factor MYB74 is regulated by the RNA-directed DNA methylation pathway in Arabidopsis. J Exp Bot 66:5997–6008

Yu A, Lepere G, Jay F et al (2013) Dynamics and biological relevance of DNA demethylation in Arabidopsis antibacterial defense. Proc Natl Acad Sci USA 110:2389–2394

Zemach A, McDaniel IE, Silva P et al (2010) Genome-wide evolutionary analysis of eukaryotic DNA methylation. Science 328:916–919

Zemach A, Kim MY, Hsieh PH et al (2013) The Arabidopsis nucleosome remodeler DDM1 allows DNA methyltransferases to access H1-containing heterochromatin. Cell 153:193–205

Zhang C, Hsieh TF (2013) Heritable epigenetic variation and its potential applications for crop improvement. Plant Breeding Biotechnol 1:307–319

Zhang XY, Yazaki J, Sundaresan A et al (2006) Genome-wide high-resolution mapping and functional analysis of DNA methylation in Arabidopsis. Cell 126:1189–1201

Zhang X, Shiu S, Cal A et al (2008) Global analysis of genetic, epigenetic and transcriptional polymorphisms in *Arabidopsis thaliana* using whole genome tiling arrays. PLoS Genet 4: e1000032

Zhang H, Jin JP, Tang LA et al (2011) PlantTFDB 2.0: update and improvement of the comprehensive plant transcription factor database. Nucleic Acids Res 39:D1114–D1117

Zhang PY, Wang JG, Geng YP et al (2015) MSAP-based analysis of DNA methylation diversity in tobacco exposed to different environments and at different development phases. Biochem Syst Ecol 62:249–260

Zhao Z, Shi HJ, Wang ML et al (2015) Analysis of DNA methylation of *Spirodela polyrhiza* (Grater Duckweed) in response to abscisic acid using methylation-sensitive amplied polymorphism. Russ J Plant Physiol 62:127–135

Zhong L, Xu YH, Wang JB (2009) DNA-methylation changes induced by salt stress in wheat *Triticum aestivum.* Afr J Biotechnol 8:6201–6207

Zhong L, Xu YH, Wang JB (2010) The effect of 5-azacytidine on wheat seedlings responses to NaCl stress. Biol Plant 54:753–756

Zhong XH, Du JM, Hale CJ et al (2014) Molecular mechanism of action of plant DRM de novo DNA methyltransferases. Cell 157:1050–1060

Zhu JK (2009) Active DNA demethylation mediated by DNA glycosylases. Annu Rev Genet 43:143–166

Zhu JH, Kapoor A, Sridhar VV et al (2007) The DNA glycosylase/lyase ROS1 functions in pruning DNA methylation patterns in Arabidopsis. Curr Biol 17:54–59

Zilberman D, Gehring M, Tran RK et al (2007) Genome-wide analysis of *Arabidopsis thaliana* DNA methylation uncovers an interdependence between methylation and transcription. Nat Genet 39:61–69

Locus-Specific DNA Methylation Analysis and Applications to Plants

Alexandre How-Kit, Emeline Teyssier, Jean-François Deleuze, and Philippe Gallusci

Abstract In contrast to mammals where DNA methylation occurs near exclusively at CG dinucleotides, all cytosines of plant genomes can be methylated irrespective to the sequence context, including the symmetrical CG and CHG sequence contexts, and the nonsymmetrical CHH sequence context. Plant genomes do not present CG islands as found in mammalian genomes where a high frequency of CG can be observed at some promoter regions. So far, the methylome of several plants has been described showing variations of both methylation levels between plants, which ranged from 5% in *Arabidopsis thaliana* to more than 30% in corn, but also of the proportion of methylated cytosines in the CG, CHG, and CHH sequence contexts. DNA methylation was shown to have important roles in the development of many organisms including plants. In this later case, DNA methylation is involved in the regulation of flowering time, flower sex determination, seed development, or fruit ripening. To date, a wide range of low to high resolution methods initially developed for animal genomes allow the analysis of DNA methylation at specific loci or globally of whole genomes. Indeed, part of these methods is not applicable to plants due to the specificities of their methylome; however, some have been successfully modified to overcome the complexity of plant DNA methylation. In this chapter, we describe an exhaustive list of methods devoted to the locus-specific analysis of DNA methylation which have proven to be applicable to plant genomes and their potential use in high throughput screening of plant population.

Keywords Plant DNA methylation • DNA methylation analysis methods • Pyrosequencing • Locus-specific methods

---

A. How-Kit (✉)
Laboratory for Genomics, Foundation Jean Dausset – CEPH, 75010 Paris, France
e-mail: ahowkit@cephb.fr

E. Teyssier • P. Gallusci
EGFV, UMR 1287, Bordeaux Science Agro, INRA, University of Bordeaux, 33140 Villenave d'Ornon, France

J.-F. Deleuze
Laboratory for Genomics, Foundation Jean Dausset – CEPH, 75010 Paris, France

Centre National de Génotypage, CEA-Institut de Génomique, 91000 Evry, France

N. Rajewsky et al. (eds.), *Plant Epigenetics*, RNA Technologies,
DOI 10.1007/978-3-319-55520-1_16

Contents

1 Introduction

The discovery of epigenetic mechanisms during the last decades allowed the identification of new levels of regulation of gene expression mediated through ATP-dependent chromatin remodeling, posttranslational histone modifications, small and other noncoding RNAs, and DNA methylation. These epigenetic mechanisms play regulatory roles during development and allow the adaptation of the organisms to their environment. Furthermore, in a number of studies, intraspecies phenotypic variations could be correlated to epigenetic differences (Cortijo et al. 2014a; Lang et al. 2016; Schmitz and Ecker 2012). The knowledge of the epigenome appears therefore important for the understanding of phenotypic variation. Thus, different molecular tools have been developed to allow the analysis of epigenetic modifications from locus-specific regions to genome-wide, and many of these methods were developed for the analysis of the DNA methylation which is the most studied epigenetic mechanisms.

In Eukaryotes, DNA methylation occurs mainly on cytosine, although, with the development of highly sensitive methods, methyladenines have also been recently detected in a few species (Luo et al. 2015) including the unicellular algae, *Chlamydomonas reinhardtii* (Fu et al. 2015) and mouse (Wu et al. 2016), in low amount though. Indeed, the molecular tools developed for the analysis of DNA cytosine methylation rely on different approaches based on the bisulfite conversion of genomic DNA, affinity-based approaches, or on the use of methylation-specific or methylation-sensitive restriction enzymes. These methods have been initially widely developed for mammalian genomes including humans where DNA methylation occurs near exclusively at CG dinucleotides which are clustered in CpG islands, in contrary to plants where all cytosines can be methylated which includes the symmetrical CG and CHG (where H corresponds to A, T, or C) and the nonsymmetrical CHH sequence contexts (Feng et al. 2010). Moreover, plant genomes do not present CpG islands as found in mammalian genomes, and the methylome of several plants has been described showing variations of the average methylation levels between plants, which ranged from 5% in *Arabidopsis thaliana* (Becker et al. 2011; Zhang et al. 2006; Zilberman et al. 2007) to more than 30% in corn (Gent et al. 2013) and also of the proportion of methylated cytosines in the CG, CHG, and CHH sequence contexts. DNA methylation was shown to have important roles in the development of many organisms such as plants including the regulation of flowering time (Soppe et al. 2000), flower sex determination (Martin et al. 2009), seed development (Gehring et al. 2006), or fruit ripening (Liu et al. 2015; Manning et al. 2006).

In the present chapter, we describe the characteristics and distribution of DNA methylation in plants and discuss the differences which were identified between plant species. We then present an exhaustive list of the methods available for the analysis of locus-specific DNA methylation profiles applicable to plant genomes, which have been sometimes modified from the protocols initially developed for animal genomes in order to overcome the complexity of plant methylome notably due to the presence of non-CG methylation. Finally, we propose possible applications of these locus-specific methods to high throughput screening of plant population.

2 DNA Methylation in Plants

2.1 Generalities

The methylation of cytosine corresponds to the enzymatic addition of a methyl group to the fifth carbon of cytosine catalyzed by DNA methyltransferases. In plants, it occurs in the three possible sequence contexts, CG, CHG, and CHH (where H stands for any nucleotide except G). DNA methylation is usually conserved through cell divisions and eventually between generations and is therefore considered as a stable epigenetic mark. Maintenance methyltransferases which use hemi-methylated DNA molecules as substrates are responsible for cytosine methylation on newly synthesized DNA strands during DNA replication. The mechanisms that maintain DNA methylation vary between different eukaryotes (Du et al. 2015). In plants, these mechanisms have been well described in the model plant, *Arabidopsis thaliana*. Maintenance methylation is operated by two main enzymes. DNA methyltransferase 1 (MET1), a DNMT1 type enzyme, maintains the cytosine methylation in CG context and requires VIM1 and VIM2 for its activity (Kim et al. 2014; Shook and Richards 2014; Woo et al. 2008) whereas the plant-specific DNA methyltransferase CHROMOMETHYLASE 3 (CMT3) maintains cytosine methylation in CHG context (Lindroth et al. 2001). Maintenance of cytosine methylation in the CHH context requires a de novo methylation machinery. Indeed in the asymmetrical CHH context, although a cytosine on one of the two daughter DNA molecules may have to be methylated, there is no corresponding guiding methylated cytosine on the mother strand. In this case, methylation needs to be newly established in one of the two daughter DNA molecules. In *Arabidopsis thaliana*, de novo DNA methylation is established by two independent pathways: (1) the RNA-directed DNA methylation (RdDM) pathway involving the domain rearranged (DMR) methyltransferases DRM1 or DRM2, in a small RNA guided process [for a review, see Matzke et al. (2015)]. (2) Alternatively, de novo non-CG methylation may rely on the chromatin remodeler DDM1 (DECREASE IN DNA METHYLATION 1), together with the CHROMOMETHYLASE 2 or the CHROMOMETHYLASE 3 (Stroud et al. 2014; Zemach et al. 2013). Interestingly, all non-CG methylation is controlled by H3K9 methylation (Stroud et al. 2014). Indeed, the epigenetic marks $H3K9me^1$ and/or $H3K9me^2$ are essential for the binding of CHROMOMETHYLASES 2 and 3 to their target sites and for the targeting to specific genomic sites of the RdDM machinery responsible for the biogenesis of small RNAs involved in DRM1 and DRM2 guiding (Du et al. 2015).

While DNA methylation is stable, it may also be erased, either locally or globally, during development or in response to environmental signals. This process corresponds to passive DNA demethylation when maintenance methylation is not active during DNA replication. On the other hand, DNA regions may become specifically and actively demethylated through guided enzymatic activities. In plants, active demethylation relies on a family of methylcytosine DNA glycosylase–lyases, the so-called DEMETER-like proteins, which remove the methylated cytosine and cut the phosphodiester backbone at the abasic site, leaving a AP site (apuric apyrimidic) in the DNA, which is subsequently filled with an unmodified cytosine nucleotide through a base-excision repair process [see Zhu (2009) for a review].

2.2 *Distribution of DNA Methylation in Plants*

Arabidopsis thaliana's methylome was the first methylome to be deciphered (Zhang et al. 2006; Zilberman et al. 2007). Since then, an increasing number of species have been analyzed at an epigenome-wide level with BS-sequencing approaches (Niederhuth et al. 2016), which allows description of genome-wide patterns of DNA methylation at single-nucleotide resolution. Large-scale patterns of methylation are shared among flowering plants, characterized by a predominance of methylcytosines in heterochromatic transposons and repetitive sequences where they have been shown to be essential for the repression of transposons expression and mobility. Methylcytosines are also present in euchromatic regions. When genes are considered, DNA methylation can be found either in the body of genes [gene body methylation (GBM)] or in promoter regions. GBM corresponds mainly to CG methylation in the central part of genes with a clear preference for exons. In plants, GBM is prevalent in constitutively expressed genes with moderate to high level of transcription (Cokus et al. 2008; Li et al. 2012; Lister et al. 2008). But the function of gene body methylation remains largely unknown, and there is no clear evidence in support for any role in regulating gene expression (Bewick et al. 2016), although GBM is conserved in organisms from different kingdoms (Feng et al. 2010; Zemach et al. 2010). In addition, a few genes show also methylation within their promoter region (Li et al. 2012; Zhang et al. 2006). Promoter DNA methylation in plants is largely associated with transcription repression (Li et al. 2008), although the relationship between DNA methylation in the 5′ regulatory region and gene expression may not be as simple as initially thought. Indeed, the gene expression level seems to depend on additional parameters including exons and introns size and number and gene body methylation as suggested by Colicchio et al. who developed a model that predicts gene expression based on DNA methylation patterns in *Mimulus guttatus* (Colicchio et al. 2015).

DNA methylation undergoes dynamic changes during cell differentiation and in response to environmental changes. The variations in DNA methylation affect specific regions which are referred to as differentially methylated regions (DMR). Hence, different cell types are characterized by different methylomes, as illustrated for example by the analysis of the methylomes of six different cell types in the *Arabidopsis thaliana* root meristem (Kawakatsu et al. 2016b). As a consequence, care should be taken when sampling for methylation studies: more homogeneous samples will lead to more accurate DNA methylation measures which may be crucial for the detection of changes in DNA methylation level.

2.3 Differences Between Plant Genomes

The percentage and distribution of methycytosines is variable depending on the plant species and the type of sequence context (CG, CHG, and CHH). Differences in methylcytosine content between plant species had been attributed to variations in genome size and complexity that is mainly related to differences in transposons density (Seymour et al. 2014) and/or to whole genome duplication events (Springer et al. 2016). The distribution of methylcytosines between the different sequence contexts varies between plants. For example, in *Arabidopsis thaliana*, methylation occurs predominantly at the CG context (CG: 55%; CHG: 23%; CHH: 22%) (Lister et al. 2008; Zhang et al. 2006), whereas Zhong et al. found that in tomato, CHH is the major context for methylcytosines (CG: 28%; CHG: 23%; CHH: 49%) (Zhong et al. 2013). However, the average methylation level (calculated as the number of methylated sites over the total sites in a given sequence context) is always higher in CG context than in CHG and CHH context. For example in *Arabidopsis thaliana* plantlets, the average methylation levels are 24% (CG), 6.7% (CHG), and 1.7% (CHH) (Cokus et al. 2008); whereas in tomato leaves, they correspond to 85.5% (CG), 56.1% (CHG), and 8.6% (CHH) (Zhong et al. 2013). This indicates that methylation predominantly occurs in CG context compared with other contexts. Different plant species are also characterized by different epigenetic patterns, as

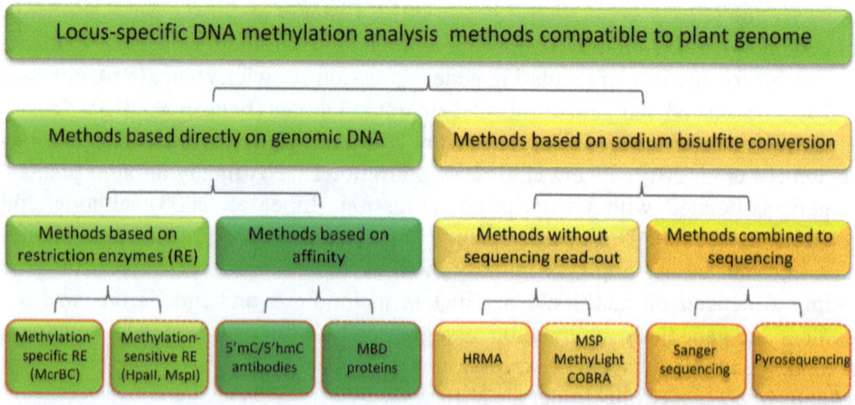

Fig. 1 Overview of locus-specific DNA methylation analysis methods compatible to plant genomes. *5′mC* 5′ methylcytosine, *5′hmC* 5′ hydroxylmethylcytosine, *MBD* methyl binding domain, *HRMA* high resolution melting analysis, MSP. *Red* and *red dotted circles* indicate highly compatible and compatible methods for the analysis of locus-specific DNA methylation in plant genomes (CG, CHG, and CHH)

illustrated for example by the comparison of *Arabidopsis thaliana* and maize DNA methylation in intergenic regions. Whereas in Arabidopsis, intergenic regions are mostly short and devoid of methylation; in maize these regions are highly enriched in transposons and exist in two major forms depending on their proximity to genes—deep intergenic chromatin is marked by dense CG and CHG methylation whereas chromatin located closer to genes is enriched in CHH methylation (Gent et al. 2013).

3 Locus-Specific DNA Methylation Analysis Methods

Locus-specific methods for the analysis of DNA methylation can be divided in two types of methods which include methods based on or not based on sodium bisulfite conversion. Methods which can be directly applied to genomic DNA without sodium bisulfite conversion either rely on the use of methylation-sensitive restriction enzymes or the anti-5-methylcytosine (5-mC) antibodies (Fig. 1).

3.1 *Methods Not Involving Sodium Bisulfite Conversion*

3.1.1 Methods Involving Methylation-Sensitive Restriction Enzyme and PCR

Restriction enzymes are endonucleases found in bacteria which cut DNA when a specific DNA sequence called restriction site is recognized by the enzyme (Kessler and Manta 1990). They were initially classified into three types of restriction enzymes: type I, II, and III, based on their enzymatic activities and compositions, the nature of the restriction sites (palindromes or not), as well as the location of the cleavage relative to the restriction site (Bickle and Kruger 1993).A fourth type of restriction enzymes has also been proposed for restriction endonucleases requiring methylated, hydroxymethylated, and/or glucosyl-hydroxymethylated DNA substrate to be active such as McrA, McrBC, and Mrr (Bickle and Kruger 1993; Roberts et al. 2003). Type II restriction enzyme can also be subclassified according to their sensitivity to methylation which includes the sensitivity/insensitivity to N^6-methyladenine, to C-5-methylcytosine, to C-5-hydroxymethylcytosine, and to C-4-methylcytosine residues which can induce an inhibition of the enzymatic activity (Kessler and Manta 1990). More than 5000 restriction and modification system enzymes are listed in REBASE which notably references information about recognition and cleavage sites, commercial availability, and methylation sensitivity (http://rebase.neb.com/rebase/rebase.html) (Roberts et al. 2015).

Due to the properties of type II and type IV restriction enzymes toward DNA methylation, different molecular tools have been developed for the analysis of locus-specific DNA methylation based on the use of restriction endonuclease. The most frequently used restriction enzymes in DNA methylation analyses are McrBc, HpaII, and MspI. McrBc is a GTP-dependent type IV restriction endonuclease which requires DNA methylation on one or both strands to be active. Its recognition sequence [5′...$R^{m}C$(N40–3000)$R^{m}C$...3′] comprises two half sites composed of a purine (R = G/A) base followed by a methylated cytosine separated by 40–3000 nucleotides with an optimal spacing of 50–100 nucleotides leading to cleavage approximately 30 nucleotides either side (Gowher et al. 2000; Roberts et al. 2003; Zhou et al. 2002). The two other endonucleases, HpaII and MspI, are isoschizomers and belong to the type II restriction enzyme which recognizes the same palindromic restriction site 5′...C^CGG...3′ and cleave it when the sequence is double stranded (Kessler and Manta 1990). HpaII is inhibited by CHG and CG methylation whereas MspI is inhibited only by CHG methylation rendering these enzymes useful for symmetrical DNA methylation analysis in plants (Johnson et al. 2002).

The most used restriction enzyme-based method for the analysis of locus-specific DNA methylation in plants is McrBC-PCR which includes a first step in which genomic DNA is digested with McrBC enzyme supplemented or not with GTP followed by a limited cycle PCR avoiding the "plateau" phase on the region of interest using the digested DNA as template subsequently revealed by agarose gel electrophoresis (Fig. 2a) (Liu et al. 2015; Vaughn et al. 2007). Consequently, the intensity of the PCR bands allows the identification of DNA methylation at the region of interest from a total absence to a strong presence where DNA methylation is associated with the disappearance of the band (Fig. 2a) (Liu et al. 2015; Vaughn et al. 2007). The reaction without GTP is a negative control of the enzymatic digestion and should always present a positive PCR amplification. Moreover McrBC digestion can also be combined to real-time PCR to measure the methylation level more precisely and to also avoid any post-PCR downstream experiments

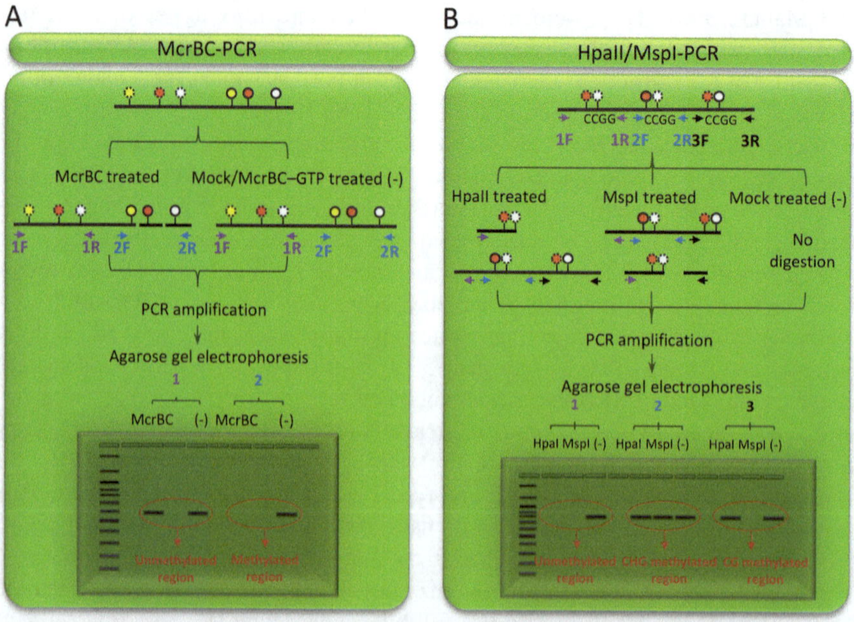

Fig. 2 Locus-specific DNA methylation analysis method based on methylation specific and methylation-sensitive restriction enzymes. (**a**) Methylation-specific *McrBC*-PCR. (**b**) Methylation-sensitive *HpaII/MspI*-PCR. *Yellow* CHH; *red* CHG; *white* CG; *dotted circle* unmethylated; *circle* methylated

such as agarose gel electrophoresis (Telias et al. 2011; Zhai et al. 2008). McrBC-PCR is thus particularly suitable to plant genomes as it allows the analysis of the methylation of cytosines in every methylation contexts (CG, CHG and CHH). However, this method only gives a blurry vision of the DNA methylation of the region of interest delimited by the PCR primers, as the methylated cytosines cannot be precisely identified and should only be used as a first screening approach.

McrBC-PCR is sometime associated with HpaII/MspI-PCR, a method that is complementary to McrBC-PCR as these methylation-sensitive enzymes cleave DNA in the restriction site when it is totally unmethylated (HpaII-PCR) or not methylated in CHG position (MspI-PCR). In HpaII- and/or MspI-PCR, genomic DNA is digested with one enzyme at a time, followed by a PCR and agarose gel electrophoresis and a PCR control reaction is performed in parallel on undigested genomic DNA for comparison to the other reactions (Fig. 2a) (Chan et al. 2008; Yin et al. 2008). The comparison of the intensities of the bands gives the level of methylation where no or low positive amplification is associated with the unmethylated DNA at the CCGG restriction sites (Chan et al. 2008; Yin et al. 2008). Regardless, at least one CCGG site must be included in the PCR amplicon and even if HpaII/MspI-PCR is more precise than McrBC-PCR, the exact methylation status of cytosines cannot be identified at base resolution and become more complex to interpret as the number of CCGG restriction sites increases in the assay.

Concerning the applications, McrBC-PCR has been successfully used in *Arabidopsis thaliana* for validation of candidate differentially methylated regions identified by tilling microarray experiments between two different ecotypes (Columbia and Landsberg *erecta*) which revealed an epigenetic natural variation of Arabidopsis (Vaughn et al. 2007). In Apple fruits from mature "Empire" trees, McrBC-real-time-PCR revealed, respectively, a positive and negative correlation of DNA methylation in *ACS1* promoter with CO_2 injury occurrence and with internal ethylene concentration suggesting a potential epigenetic regulation of ethylene biosynthesis and developmental disorder events (Gapper et al. 2013). In tomato, it has allowed the identification of genes (*RIN*, *NOR*, and *CNR*) undergoing active targeted demethylation by tomato DNA demethylase SlDML2 during fruit ripening (Liu et al. 2015). McrBC-PCR has also been combined with HpaII/MspI-PCR for the analysis 19 candidate genes as well as some class I and class II transposable elements (TE) in chromosome 4 of the rice genome and revealed DNA methylation in the candidate gene associated with H3K9 histone methylation and class I TEs and hemi-methylation in class II TEs (Yin et al. 2008). HpaII-PCR has also been used in *Selaginella moellendorffii* on five low copy sequences and five highest copy sequences and confirmed the DNA methylation of highest copy sequences in its genome (Chan et al. 2008). In rice, methylation-sensitive restriction enzymes have also been combined with Southern blotting where genomic DNAs were digested with HpaII or MspI, run in a 1% gel electrophoresis, transferred in a nylon membrane, and hybridized with different labeled probes which revealed that metal stress induced heritable locus-specific DNA hypomethylation in TEs and protein-coding genes (Ou et al. 2012).

Limitations of the use of restriction enzymes for the analysis of DNA methylation relies on their inability to distinguish methylation from hydroxymethylation and also the need of the restriction site in the sequence of interest (Nestor et al. 2010). Indeed, methylation-sensitive restriction enzymes are inhibited by both methylation and hydroxymethylation of the cytosines whereas they both activate methylation-dependent restriction enzymes. Moreover, in some cases, the region of interest does not contain the restriction sites of the enzymes, and another approach should be preferred for the analysis of DNA methylation. Therefore, these methods present low resolution and also depend on the distribution of sites within sequences.

3.1.2 Methods Involving Anti-meCytosine Antibody and PCR

An alternative to restriction enzyme for locus-specific DNA methylation analysis is the affinity-based enrichment of methylated DNA using anti-5-methyl-cytosine antibody followed by PCR (MeDIP-PCR). The great advantage of the use of a monoclonal 5-mC antibody compared to methylation-sensitive/specific restriction enzymes is its specificity for 5-mC and not 5-hmC, allowing a precise identification of only 5-methyl-cytosines (Jin et al. 2010). The MeDIP experiment typically comprises a fragmentation step of the genomic DNA by sonication which is followed by the denaturation of DNA fragments, the immunoprecipitation of single-stranded methylated DNA using the 5-mC antibody, different incubation and washing steps, and a final elution step (Cortijo et al. 2014b; Mohn et al. 2009). The enrichment efficiency in the MeDIP fraction should be evaluated by PCR on input DNA, immunoprecipitated DNA, and supernatant and always controlled using methylated and unmethylated control genes or regions, which can be an exogenous synthetic DNA added to the input DNA before the start of a MeDIP experiment as spike-in DNA (Cortijo et al. 2014b; Mohn et al. 2009). Therefore, non-treated input DNA (IN) and immunoprecipitated methylated DNA (M) can be used for locus-specific analysis by standard or real-time PCR where a positive amplification of the M fraction and its intensity compared to IN indicate different levels of methylation (Mohn et al. 2009).

The limitations of a MeDIP-PCR experiment mainly concern the antibody which must be of high quality, highly specific, and of great efficiency. These characteristics can vary greatly between different lot numbers and between suppliers, which requires much optimization. Moreover, MeDIP-PCR does not allow the precise identification of the methylated cytosines in a sequence nor give an absolute quantification of methylation. Another affinity-based enrichment of methylated DNA uses methyl-CpG-binding domain proteins (MBD) which bind symmetrically methylated CGs (Nan et al. 1993; Roloff et al. 2003; Zemach and Grafi 2003) and are also specific to 5mC and not 5hmC (Jin et al. 2010). These proteins are therefore used in the same way as 5-mC antibody for enrichment of CG methylated DNA, notably present in CpG islands which is widely found in mammals (Rauch and Pfeifer 2005), rendering this method more suitable to mammalian rather than plant genomes.

Applications using MeDIP-PCR for the analysis of locus-specific methylation in plants require in almost all cases the control of the efficiency and the quantification of DNA recovery of MeDIP experiments prior to performing MeDIP-Chip or MeDIP-Seq experiments. MeDIP therefore followed by a probe labeling and hybridized with non-treated input DNA in high density tilling arrays or followed by a library preparation followed by next generation sequencing respectively (Cortijo et al. 2014b; Vining et al. 2012; Zilberman et al. 2007). As a result MeDIP has been successfully performed on different plant species including Arabidopsis (Colome-Tatche et al. 2012; Roudier et al. 2011; Zilberman et al. 2007), poplar (Vining et al. 2012, 2013), maize (Ding et al. 2014; Eichten et al. 2011; Waters et al. 2011), rice (Hu et al. 2015) and bean (Crampton et al. 2016).

3.2 Methods Involving Sodium Bisulfite Conversion

Contrary to genome-wide approaches for DNA methylation analysis which are equally divided between methods involving or not sodium bisulfite conversion of DNA, more methods and applications are based on sodium bisulfite conversion of DNA for the analysis of locus-specific DNA methylation. They include a first common step where genomic DNA is treated with sodium bisulfite followed by PCR experiments, during which DNA methylation can be directly detected and quantified, or requiring a combination of sequencing experiments which can thus provide information on the methylation at single base resolution. Due to the specificity of the methylome of plants compared to animals, these methods have been adapted to allow the correct analysis of DNA methylation in plants for all cytosine contexts.

3.2.1 Bisulfite Conversion

The sodium bisulfite conversion of DNA is a chemical method discovered by two different groups in 1970 (Hayatsu 2008) which relies on the conversion of unmethylated cytosines into uracils leaving the methylated cytosines unaffected (Frommer et al. 1992; Wang et al. 1980). Indeed, the sodium bisulfite conversion involves three modification steps where the first step includes a reversible sulfonation of unmethylated cytosines where a HSO_3^- is added to cytosine to form cytosine sulfonate (Fig. 3a) (Clark et al. 1994). This step is followed by the irreversible hydrolytic deamination of cytosinesulfonate in uracil sulfonate which is performed when the equilibrium of cytosine sulfonate and cytosine is reached. Indeed, this equilibrium is performed rapidly at pH 7 or lower with a sodium bisulfite concentration of at least 0.5 M, and the optimum pH value for the conversion from cytosine to uracil sulfonate is between five and six (Clark et al. 1994). The third and last step is the reversible desulfonation of uracil sulfonate which is performed by an alkali treatment as uracil sulfonate is stable in neutral conditions but is converted to uracil at higher pH (Fig. 3a) (Clark et al. 1994). 5-methylcytosine is considered to not react to sodium bisulfite conversion as the kinetics of the reaction is very slow due to the presence of the protective methyl group in position 5 (Fig. 3a) (Clark et al. 1994). In contrast, 5-hydroxymethylcytosine does react rapidly with sodium bisulfite to generate cytosine-5-methyl-sulfonate which is recognized as a cytosine during PCR (Huang et al. 2010). Thus, sodium bisulfite conversion has transformed cytosine (hydroxy)methylation in a C/T polymorphism at every C nucleotide in the original genomic sequence allowing its detection and quantification by downstream experiments. After treatment, the two DNA strands are not complementary and the size of the genome to be analyzed is doubled (Fig. 3a).

The main limitations of sodium bisulfite conversion for the analysis of DNA methylation concern the inability to distinguish between methylation and hydroxymethylation of cytosines (Huang et al. 2010), the possibility of incomplete bisulfite conversion and of false positives which can happen if the DNA is not completely denatured as bisulfite conversion requires single stranded DNA (Fraga and Esteller 2002), and finally the possible degradation of the DNA during the conversion (Ehrich et al. 2007; Grunau et al. 2001). Therefore, the complete bisulfite conversion of the DNA in a plant genome should always be assessed by the analysis of the DNA methylation either of an exogenous unmethylated DNA spikes or using plastid DNA which is assumed to be totally unmethylated (Fojtova et al. 2001; Marano and Carrillo 1991).

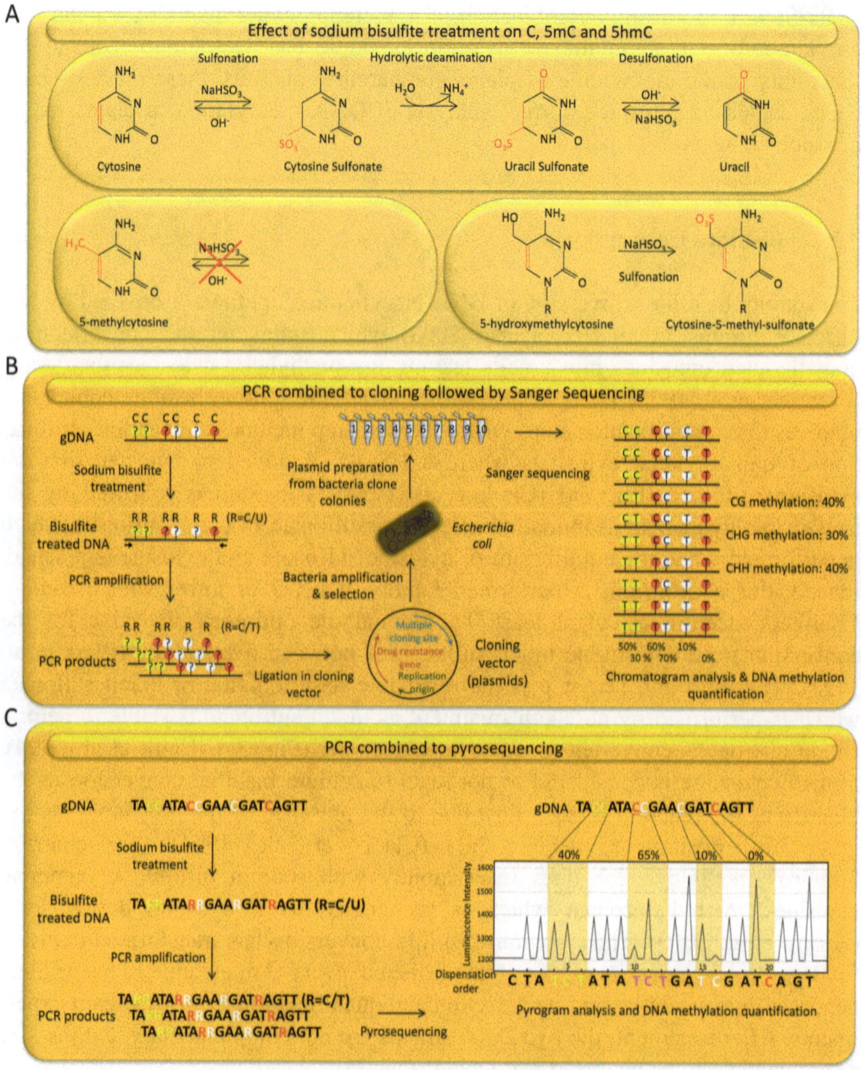

Fig. 3 Locus-specific DNA methylation analysis methods using bisulfite treated DNA and sequencing. (**a**) sodium bisulfite conversion of cytosine, 5′-methylcytosine, and 5′-hydroxymethylcytosine. (**b**) PCR combined by cloning and Sanger sequencing. (**c**) PCR combined to pyrosequencing. *Yellow* CHH; *red* CHG; *white* CG

3.2.2 Primer Design

The design of primers for the PCR amplification of bisulfite treated DNA is a particularly critical step in plants as every cytosine is potentially methylated rendering their design more complicated compared to animal genomes where only cytosine in a CG context can be methylated. As a result, primers which allow the amplification of bisulfite-treated DNA (bisulfite-specific primers) in plants also allow the amplification of untreated genomic DNA, and a complete bisulfite conversion of DNA is thus required before PCR experiments to exclude any genomic DNA contamination. Moreover, methylation-independent PCR primers are also complicated to design in plants compared to animals, which simply avoid CG dinucleotides in the primers, as they require a genomic sequence totally free of any cytosine residue which can be hard to find dependent on other constraints related to the assay. Conventional bioinformatic tools for DNA methylation analysis such as MethPrimer are not appropriate for the design of PCR primers on bisulfite-treated plant DNA (Li and Dahiya 2002). As a consequence, specific bioinformatic tools have been developed integrating the complexity of the plant methylome such as Kismeth which allow the design of bisulfite primers including some flexible parameters such as primer and amplicon length and minimal melting temperature of the primers and the analysis of methylation patterns in plants (Gruntman et al. 2008). Other conventional bioinformatic tools for PCR primer design can also be used by applying user specific parameters for the design of bisuflite-specific PCR primers in plants such as the highly flexible Primer3 (Koressaar and Remm 2007; Untergasser et al. 2012).

Different strategies have been proposed to design PCR primers and subsequent amplification of bisulfite treated DNA in plants that are all based on the use of degenerated nucleotides. A first strategy hypothesized that only CG dinucleotides are highly methylated in plants while CHH cytosines are mostly unmethylated and therefore recommends to limit to three the number of CG in each bisulfite-specific primer and replace the "C" nucleotides in CG by degenerated "Y" while replacing all other cytosines by thymines and to terminate the primer 3′ ends by one or multiple cytosines in a CHH sequence context (Henderson et al. 2010). This biased strategy aims to promote the amplification of totally converted DNA but at the expense of non-CG methylation, which can occur frequently in different plant species other than Arabidopsis. In contrast, an unbiased strategy can be to use degenerated nucleotides at all cytosines present in the primer sequence thus allowing an equally efficient PCR amplification of all types of cytosine methylation patterns (How-Kit et al. 2015). The primers should include up to five cytosine and guanine residues in the forward and reverse primers, respectively. Primers should also avoid "C" and "G" nucleotides at the 3′ end of the forward and reverse primer, respectively (How-Kit et al. 2015). The use of inosine residues instead of "Y" or "R" in the degenerated primers may also improve the assay design and efficiency (Aufsatz et al. 2002; Kanno et al. 2004).

3.2.3 Methods Without Sequencing

Different methods of PCR on bisulfite-treated DNA allow the identification of locus-specific DNA methylation without any sequencing experiments such as high resolution melting analysis (HRMA) which has been successfully performed on plants, methylation-specific PCR (MSP), Methylight, and combined bisulfite restriction analysis (COBRA).

High Resolution Melting Analysis

High resolution melting analysis (HRMA) has been initially developed for genotyping of SNPs and has been later applied to DNA methylation analysis. The method is based on the use of a saturating fluorescent dye in the PCR which is followed by a slow denaturation of the amplification products with at least 25 fluorescence acquisitions per degree and can detect slight differences between the melting curves of two samples due to polymorphisms, mutations, or differences of DNA methylation (Vossen et al. 2009; Wittwer et al. 2003). For DNA methylation analysis, HRMA is performed after a PCR on bisulfite treated DNA and can detect DNA methylation differences translated in a bisulfite conversion induced C/T polymorphism with an analytical sensitivity of 0.1% (Wojdacz and Dobrovic 2007) and where only a DNA methylation difference on one cytosine can be identified (Rodriguez Lopez et al. 2010). A control of HRMA experiments should be performed on all samples using a totally unmethylated locus and compared to a 0% DNA standard and an unconverted DNA to ensure the total bisulfite conversion of each sample which could distort the HRMA results and interpretation (Vossen et al. 2009).

Although rarely used in plants for DNA methylation analysis where HMRA is mostly used for genetic variant screening (Simko 2016), HRMA analysis has been successfully performed on Tomato *CNR* gene including CG and CHG methylated sites (Rodriguez Lopez et al. 2010) and has also revealed the induction by low relative humidity of DNA methylation of two genes in the stomatal development pathway: *SPEECHLESS* and *FAMA* in *Arabidopsis thaliana* (Tricker et al. 2012). The limitations of HRM include the inability to identify easily DNA methylation differences at base resolution and to accurately quantify DNA methylation differences.

Other Methods

Other PCR methods based on PCR amplification of bisulfite treated DNA without a sequencing read-out include methylation-specific PCR (MSP), MethyLight, and combined bisulfite restriction analysis (COBRA).

MSP allows the amplification of bisulfite treated DNA using different primers with sequences corresponding to the DNA methylation pattern whose amplification is desired so that different primer pairs are required for the different possible DNA methylation patterns (Herman et al. 1996). MethyLight relies on the use of real-time PCR and Taqman technologies dedicated to DNA methylation analysis where specific DNA methylation patterns (unmethylated or methylated) are determined by the primer and/or probe sequence where the DNA methylation discrimination is borne either by the primers, the probe, or both (Eads et al. 2000). As least two different assays for the same amplicon is required as well as a DNA methylation-independent control region assay allowing the quantification of bisulfite conversion (Eads et al. 2000). The COBRA method combines a PCR on bisulfite treated DNA using methylation-independent primers followed by the digestion of the amplicon product by restriction enzymes such as BstU1 which recognizes CGCG sites and an analysis by a gel electrophoresis (Xiong and Laird 1997). Thus, after the bisulfite treatment, the restriction site is lost in unmethylated DNA whereas it is retained in methylated DNA allowing its identification and quantification by the ratio of cleaved and un-cleaved DNA bands (Xiong and Laird 1997).

MSP is only a qualitative method which has the advantages of being simple, cheap, and rapid, COBRA is a semi-quantitative method requiring the presence of restriction sites in the analyzed sequence which is its principal limitation, and MethyLight is a highly sensitive and quantitative method. These three methods do not allow the identification of DNA methylation at base resolution and often propose to distinguish between a fully methylated and a fully unmethylated situation thus not allowing a detailed analysis as required in plants for CG, CHG, and CHH methylation.

MSP, MethyLight, and COBRA are rarely used in plants even if these methods could be easily adapted to integrate the specificities of DNA methylation in plant genomes which may be principally due to their respective limitations. MSP has been used for the analysis of DNA methylation of *Sdh1-1*, *Sdh1-2*, *Sdh2-1*, *Sdh-2*, and *Sdh2-3* in maize (Eprintsev et al. 2016).

3.2.4 Methods Including Sequencing Experiments

Two sequencing methods are used for locus-specific DNA methylation analysis in plants including Sanger sequencing which can be performed directly after a PCR or after a PCR and cloning steps and pyrosequencing which is a real-time sequencing by synthesis method performed directly after a PCR.

Cloning Combined to Sanger Sequencing

Sanger sequencing is the gold standard and most widely used sequencing method combining a PCR followed by a dye-terminator sequencing reaction in which the four types of ddNTPs (2′,3′ dideoxynucleotide triphosphates) labeled with four different fluorophores are randomly incorporated during a repeated linear enzymatic primer extension cycle including a denaturation, a primer annealing, and a primer extension step (McGinn and Gut 2013; Sanger et al. 1977). The primer extension reaction is stopped when a ddNTP is incorporated in the elongating sequence leading to a mixture of single stranded DNA of diverse lengths bearing a 3′ terminal fluorescent nucleotide (McGinn and Gut 2013; Sanger et al. 1977). The sequence of the amplicon is revealed after a capillary electrophoresis from

20–50 to 600–800 nucleotides from the sequencing primer where the length and the color of the DNA fragment give the position and the identity of the nucleotide on the reconstituted sequence, respectively (McGinn and Gut 2013; Sanger et al. 1977).

For the analysis of locus-specific DNA methylation, a PCR on bisulfite treated DNA is performed followed by the cloning of amplicon fragments in plasmids, their amplification in bacterial hosts, and their purification before Sanger sequencing experiments (Fig. 3b) (Clark et al. 1994; Foerster and Mittelsten Scheid 2010). As the bisulfite conversion potentially induces C/T polymorphisms at every cytosine, the cloning step is absolutely necessary to be able to distinguish the DNA methylation present on the cytosines of every DNA strand and also for DNA methylation quantification (Foerster and Mittelsten Scheid 2010; Henderson et al. 2010). However, the resolution of the quantification of DNA methylation by this method is relatively low, as ten clones are generally sequenced per amplicon giving a 10% resolution (Fig. 3b) (Foerster and Mittelsten Scheid 2010; Henderson et al. 2010). The analysis of the bisulfite sequencing results can be performed by Kismeth or Cymate which are software tools designed especially for plants genomes providing visualization and analysis solutions for cytosine methylation in CG, CHG, and CHH sequence contexts (Foerster et al. 2010; Gruntman et al. 2008; Hetzl et al. 2007).

As the gold standard for DNA methylation analysis, numerous studies have used Sanger sequencing for the analysis of locus-specific DNA methylation in plants at specific candidate genes or regions and some are exemplified in the following paragraph.

In Arabidopsis, the control of locus-specific asymmetric and CHG methylation by *DRM* and *CMT3* methyltransferases genes was demonstrated by bisulfite sequencing of candidate loci (*FWA* and *MEA-ISR*) where a partial to complete loss of CHH and CHG methylation is observed in *drm* and/or *cmt* mutants (Cao and Jacobsen 2002). Moreover, the role of Arabidopsis DNA Glycosylase/Lyase ROS1 in erasing CHG and CHH methylation was also demonstrated at some specific loci such as *FWA* or repeated elements (*AtMU1* and *AtLINE1*) by comparison of *ros1* mutants with WT plants (Zhu et al. 2007). Bisulfite sequencing experiments also revealed that the two imprinted genes *MEA* and *FWA* are under methylated in vegetative sperm genomes compared to sperm genome under the influence of Arabidopsis DNA glycosylase DME (Schoft et al. 2011).

In maize, bisulfite sequencing has revealed distinct CG, CHG, and CHH DNA methylation patterns of both transposable element activator ends in the wx-m9::Ac allele where the 5′ end is either unmethylated or partially methylated whereas the 3′ end is heavily methylated (Wang et al. 1996). It also revealed the DNA methylation pattern of *Fie1* and *Fie2* in different maize tissues where they are methylated and inactive except in the endosperm where the maternal allele is hypomethylated and active illustrating a paternal imprinting phenomenon (Hermon et al. 2007) and also the natural variation of DNA methylation in some genes under control of chromomethylase ZMET2 (Makarevitch et al. 2007).

In Tobacco, bisulfite sequencing allowed the identification of CG DNA methylation of auxin-binding protein gene and class I chitinase gene induced in *in vitro* plants treated with antibiotics (Schmitt et al. 1997) as well as de novo RNA-directed CG, CHG, and CHH DNA methylation induced by 30–60 bp of potato spindle tuber viroid (PSTVd) transgenes integrated in the genome at the sites of integration (Pelissier and Wassenegger 2000). It has also showed the rapid occurrence during the vegetative development of DNA methylation at P35S and NOS promoters of transgenes integrated in Tobacco plants, which is inherited by the offspring (Weinhold et al. 2013).

In Tomato, DNA methylation analysis of *colorless never ripe* (*cnr*) mutants by bisulfite sequencing revealed an epimutation occurring at CNR locus in the promoter region of an SBP-box (SQUAMOSA promoter binding protein-like) gene (*LeSPL-CNR*), controlling its expression (Manning et al. 2006) and whose inheritance is under the control of CMT3 as the silencing of *CMT3* in *cnr* mutant fruit induced a reduction of DNA methylation at *LeSPL-CNR* promoter regions. The reduction in methylation is associated to expression of the gene and fruit ripening that was also revealed by bisulfite sequencing (Chen et al. 2015).

In rice, bisulfite sequencing demonstrated the symmetrical and nonsymmetrical DNA methylation of the transposable element Kiddo at the promoter region of rice uniquitin 2 (rubq2) which regulates the transcriptional activity of rubq2 (Yang et al. 2005).

In conclusion, Sanger sequencing presents therefore the great advantage to allow the identification of strand-specific DNA methylation on every cytosine present in the PCR amplicon, but this method requires also several long and expensive steps from PCR to sequencing and presents only a low resolution for DNA methylation quantification which is typically around 10%.

Pyrosequencing

Pyrosequencing is a more recent sequencing method based on real-time sequencing by synthesis which was mainly used for SNP genotyping and mutation detection and identification. The principle of pyrosequencing is based on the cyclic incorporation from a sequencing primer of one of the four types of nucleotides complementary to a single stranded DNA template which generates a bioluminescent signal monitored in real time (Ronaghi 2001; Ronaghi et al. 1998). During each cycle, one type of each nucleotide is dispensed and incorporated to the elongating sequence by the Klenow fragment of the *Escherichia coli* DNA polymerase I, and pyrophosphate molecules (PPi) are released and used to produce ATP from adenosine phosphosulfate by the action of an ATP sulfurylase (Ahmadian et al. 2006). The released ATP molecules allow the oxidation of D-luciferin in oxyluciferin by a luciferase generating an emission of a photon which is detected by a charge-coupled device (CCD) camera (Ahmadian et al. 2006). The emission of light is proportional to the quantity of incorporated nucleotides so that the luminescence intensity of two consecutive identical nucleotides in a sequence corresponds to twice the value of one nucleotide peak rendering this method highly quantitative (Ahmadian et al. 2006). Excess ATP as well as the unincorporated nucleotides are degraded by an apyrase before the beginning of the next cycle (Ahmadian et al. 2006).

For locus-specific DNA methylation analysis, pyrosequencing has been first applied to mammalian genomes where the C/T polymorphism induced by bisulfite conversion at each CG site was accurately quantified, and the DNA methylation percentage of the cytosine present in the native genomic DNA corresponds to the proportion of C nucleotides at a CG site compared to the overall signal (C + T nucleotide) in the pyrosequenced bisulfite treated DNA (Fig. 3c) (Tost et al. 2003). The resolution and the analytical sensitivity of pyrosequencing for DNA methylation quantification is evaluated at 5% which is more precise than Sanger sequencing, and the maximum sequence length for DNA methylation analysis by pyrosequencing is evaluated at 120 nucleotides (Tost et al. 2003).

The currently available Pyrosequencing instruments for locus-specific DNA methylation analyses: PyroMark Q96ID, PyroMark Q96MD, PyroMark Q48, Autoprep, and PyroMark Q24 (Qiagen), are commercialized by Qiagen and are mainly developed for the analysis of CG methylation. Therefore, a modified protocol of pyrosequencing has been developed on the PyroMark Q96MD including a complete pipeline for the analysis of DNA methylation of cytosines in every sequence context (CG, CHG, and CHH) at base resolution using high throughput pyrosequencing which is suitable to plant genomes (How-Kit et al. 2015). The protocol includes two Microsoft Excel Visual Basic Applications (VBA) for the design and analysis of pyrosequencing assays, allowing accurate quantification of cytosine methylation in all sequence contexts from the resulting pyrosequencing raw data (How-Kit et al. 2015). Due to the specificity of pyrosequencing chemistry relative to homopolymers which are present in one peak with an intensity proportional to the number of repeated nucleotides and due to the bisulfite conversion, the quantification of the DNA methylation cannot be obtained in a homopolymer sequence such as "CC," "CCC," or "CTC" at single cytosine resolution (Fig. 3c). For these cases, the developed macro provides an average DNA methylation value for two or more cytosine sites (How-Kit et al. 2015). Such situations are not found for the majority of cytosines in plant genomes, and an unambiguous methylation value can therefore be obtained at base resolution for most cytosines.

Pyrosequencing has been successfully used for the analysis of DNA methylation in Tomato which revealed a dramatic loss of DNA methylation in *CNR* and *NOR* promoter regions during fruit ripening (How-Kit et al. 2015). Moreover, it also revealed the maintenance of DNA methylation at *CNR*, *PSY1*, and *NOR* promoters during ripening in Tomato fruits underexpressing DNA demethylase *slDML2* suggesting that the loss of DNA methylation during ripening is mediated by active demethylation (Liu et al. 2015).

Advantages of bisulfite pyrosequencing relies on its rapidity and its simplicity as it only requires a PCR followed by a fast purification and pyrosequencing with a time to results of 1–2 h after PCR and also on the possibility of accurate and reproducible quantification of DNA methylation of every cytosine at base resolution (How-Kit et al. 2015). The primary limitations of bisulfite pyrosequencing include its inability to quantify DNA methylation of some cytosines at base resolution when they are in certain sequence context and the length of the pyrosequenced sequence which is limited to 120 nucleotides (How-Kit et al. 2015).

4 Conclusion: Perspectives

Epigenomes have now been analyzed in several plants including Arabidopsis (Cokus et al. 2008; Zhang et al. 2006; Zilberman et al. 2008), rice (Li et al. 2012), maize (Gent et al. 2013), tomato (Zhong et al. 2013), and many others (Niederhuth et al. 2016). As far as DNA methylation is concerned, results indicate that the general organization of epigenomes is conserved, although variations can be observed between species depending on several factors including genome size and transposon abundance (Niederhuth et al. 2016). Polymorphism can be observed by comparing various accessions or varieties within species as shown by the recent analysis of the methylomes of 1227 different Arabidopsis accessions worldwide (Kawakatsu et al. 2016a) or by comparing the methylation distribution of rice varieties with contrasted responses to drought stress (Garg et al. 2015).

In addition, it is clearly established that variations in DNA methylation patterns can cause phenotypic changes. For example, impairing the DNA methylation/demethylation machinery in Arabidopsis and other plants result in severe alterations of plant development processes, demonstrating a direct link between methylation pattern and phenotypes (Finnegan et al. 1996; Kakutani 1997; Liu et al. 2015). DNA methylation was also shown to control traits of agronomical importance, such as yield potential in Brassica (Hauben et al. 2009). An EpiRil population characterized by plants with mosaic DNA methylation patterns but identical DNA sequences (Reinders et al. 2009) was recently used to demonstrate variation and heritability for flowering time and plant height. This indicates that DNA methylation variations contribute to the control of complex plant traits (Johannes et al. 2009; Kooke et al. 2015). Further demonstration of the importance of DNA methylation in contributing to plant phenotype diversity comes from the characterization of stable epialleles that impact flower organization in snapdragon (Cubas et al. 1999), flower sex in melon (Martin et al. 2009), or fruit development, quality, and ripening (Manning et al. 2006; Ong-Abdullah et al. 2015; Quadrana et al. 2014). Indeed, there are only few stable epialleles identified so far, and their frequency is likely to be underestimated (Richards and Wendel 2011). Yet screening for epiallele or epiQTL might be an important issue for plant improvement strategies as they can have profound impacts on plant phenotypes and agronomical characteristics.

Comparative methylome studies within crop species and in natural population may contribute to increase the number of known DNA methylation variants and to provide lists of potential epialleles of agronomical interest that could then be screened in large population using high throughput technologies to determine changes in the locus-specific methylation level and relate them to the traits of interest. An interesting example is provided by *CNR* gene which encodes a major regulator of fruit ripening. The *Cnr* epiallele that leads to inhibition of ripening is hypermethylated at site located 2.5 kb upstream of the transcriptional start site. Interestingly, this site has been shown to have different methylation levels in different tomato varieties (Manning et al. 2006). Yet the link between methylation level at the *Cnr* epimutated site, *Cnr* kinetic and level of expression, and the ripening kinetic and intensity has not been established. Hence, analyzing the variations in DNA methylation level in a wide range of tomato variety at the *Cnr* epimutable site and relating it to ripening characteristics could be a way to anticipate plant ripening properties and kinetics in breeding strategies. A similar approach could be developed to evaluate variations in methylation level at the *VitE3* promoter and relate it to changes in tomato fruit VitE content (Quadrana et al. 2014).

As a conclusion, establishing links between DNA methylation at specific loci or combination of loci and specific traits could be an important parameter to anticipate the quantitative variations of these traits in breeding population. This would require once the putative epiallelic forms have been identified to use high-throughput technology to monitor changes in methylation level at the locus of interest in large plant population. Modeling approaches based on methylation profiles could then be used to predict variations of agronomical interest.

References

Ahmadian A, Ehn M, Hober S (2006) Pyrosequencing: history, biochemistry and future. Clin Chim Acta 363:83–94

Aufsatz W, Mette MF, van der Winden J et al (2002) HDA6, a putative histone deacetylase needed to enhance DNA methylation induced by double-stranded RNA. EMBO J 21:6832–6841

Becker C, Hagmann J, Muller J et al (2011) Spontaneous epigenetic variation in the *Arabidopsis thaliana* methylome. Nature 480:245–249

Bewick AJ, Ji L, Niederhuth CE et al (2016) On the origin and evolutionary consequences of gene body DNA methylation. Proc Natl Acad Sci USA 113:9111–9116

Bickle TA, Kruger DH (1993) Biology of DNA restriction. Microbiol Rev 57:434–450

Cao X, Jacobsen SE (2002) Locus-specific control of asymmetric and CpNpG methylation by the DRM and CMT3 methyltransferase genes. Proc Natl Acad Sci USA 99(Suppl 4):16491–16498

Chan AP, Melake-Berhan A, O'Brien K et al (2008) The highest-copy repeats are methylated in the small genome of the early divergent vascular plant Selaginella moellendorffii. BMC Genomics 9:282

Chen W, Kong J, Qin C et al (2015) Requirement of CHROMOMETHYLASE3 for somatic inheritance of the spontaneous tomato epimutation Colourless non-ripening. Sci Rep 5:9192

Clark SJ, Harrison J, Paul CL et al (1994) High sensitivity mapping of methylated cytosines. Nucleic Acids Res 22:2990–2997

Cokus SJ, Feng S, Zhang X et al (2008) Shotgun bisulphite sequencing of the Arabidopsis genome reveals DNA methylation patterning. Nature 452:215–219

Colicchio JM, Miura F, Kelly JK et al (2015) DNA methylation and gene expression in Mimulus guttatus. BMC Genomics 16:507

Colome-Tatche M, Cortijo S, Wardenaar R et al (2012) Features of the Arabidopsis recombination landscape resulting from the combined loss of sequence variation and DNA methylation. Proc Natl Acad Sci USA 109:16240–16245

Cortijo S, Wardenaar R, Colome-Tatche M et al (2014a) Mapping the epigenetic basis of complex traits. Science 343:1145–1148

Cortijo S, Wardenaar R, Colome-Tatche M et al (2014b) Genome-wide analysis of DNA methylation in Arabidopsis using MeDIP-chip. Methods Mol Biol 1112:125–149

Crampton M, Sripathi VR, Hossain K et al (2016) Analyses of methylomes derived from meso-american common bean (*Phaseolus vulgaris* L.) using MeDIP-Seq and whole genome sodium bisulfite-sequencing. Front Plant Sci 7:447

Cubas P, Vincent C, Coen E (1999) An epigenetic mutation responsible for natural variation in floral symmetry. Nature 401:157–161

Ding H, Gao J, Qin C et al (2014) The dynamics of DNA methylation in maize roots under Pb stress. Int J Mol Sci 15:23537–23554

Du J, Johnson LM, Jacobsen SE et al (2015) DNA methylation pathways and their crosstalk with histone methylation. Nat Rev Mol Cell Biol 16:519–532

Eads CA, Danenberg KD, Kawakami K et al (2000) MethyLight: a high-throughput assay to measure DNA methylation. Nucleic Acids Res 28:E32–E32

Ehrich M, Zoll S, Sur S et al (2007) A new method for accurate assessment of DNA quality after bisulfite treatment. Nucleic Acids Res 35:e29

Eichten SR, Swanson-Wagner RA, Schnable JC et al (2011) Heritable epigenetic variation among maize inbreds. PLoS Genet 7:e1002372

Eprintsev AT, Fedorin DN, Karabutova LA et al (2016) Expression of genes encoding subunits A and B of succinate dehydrogenase in germinating maize seeds is regulated by methylation of their promoters. J Plant Physiol 205:33–40

Feng S, Cokus SJ, Zhang X et al (2010) Conservation and divergence of methylation patterning in plants and animals. Proc Natl Acad Sci USA 107:8689–8694

Finnegan EJ, Peacock WJ, Dennis ES (1996) Reduced DNA methylation in *Arabidopsis thaliana* results in abnormal plant development. Proc Natl Acad Sci USA 93:8449–8454

Foerster AM, Mittelsten Scheid O (2010) Analysis of DNA methylation in plants by bisulfite sequencing. Methods Mol Biol 631:1–11

Foerster AM, Hetzl J, Mullner C et al (2010) Analysis of bisulfite sequencing data from plant DNA using CyMATE. Methods Mol Biol 631:13–22

Fojtova M, Kovarik A, Matyasek R (2001) Cytosine methylation of plastid genome in higher plants. Fact or artefact? Plant Sci 160:585–593

Fraga MF, Esteller M (2002) DNA methylation: a profile of methods and applications. Biotechniques 33:632, 634, 636–649

Frommer M, McDonald LE, Millar DS et al (1992) A genomic sequencing protocol that yields a positive display of 5-methylcytosine residues in individual DNA strands. Proc Natl Acad Sci USA 89:1827–1831

Fu Y, Luo G-Z, Chen K et al (2015) N(6)-Methyldeoxyadenosine marks active transcription start sites in chlamydomonas. Cell 161:879–892

Gapper NE, Rudell DR, Giovannoni JJ et al (2013) Biomarker development for external CO_2 injury prediction in apples through exploration of both transcriptome and DNA methylation changes. AoB Plants 5:plt021

Garg R, Narayana Chevala V, Shankar R et al (2015) Divergent DNA methylation patterns associated with gene expression in rice cultivars with contrasting drought and salinity stress response. Sci Rep 5:14922

Gehring M, Huh JH, Hsieh TF et al (2006) DEMETER DNA glycosylase establishes MEDEA polycomb gene self-imprinting by allele-specific demethylation. Cell 124:495–506

Gent JI, Ellis NA, Guo L et al (2013) CHH islands: de novo DNA methylation in near-gene chromatin regulation in maize. Genome Res 23:628–637

Gowher H, Leismann O, Jeltsch A (2000) DNA of *Drosophila melanogaster* contains 5-methylcytosine. EMBO J 19:6918–6923
Grunau C, Clark SJ, Rosenthal A (2001) Bisulfite genomic sequencing: systematic investigation of critical experimental parameters. Nucleic Acids Res 29:E65–E65
Gruntman E, Qi Y, Slotkin RK et al (2008) Kismeth: analyzer of plant methylation states through bisulfite sequencing. BMC Bioinformatics 9:371
Hauben M, Haesendonckx B, Standaert E et al (2009) Energy use efficiency is characterized by an epigenetic component that can be directed through artificial selection to increase yield. Proc Natl Acad Sci USA 106:20109–20114
Hayatsu H (2008) Discovery of bisulfite-mediated cytosine conversion to uracil, the key reaction for DNA methylation analysis—a personal account. Proc Jpn Acad Ser B Phys Biol Sci 84:321–330
Henderson IR, Chan SR, Cao X et al (2010) Accurate sodium bisulfite sequencing in plants. Epigenetics 5:47–49
Herman JG, Graff JR, Myohanen S et al (1996) Methylation-specific PCR: a novel PCR assay for methylation status of CpG islands. Proc Natl Acad Sci USA 93:9821–9826
Hermon P, Srilunchang KO, Zou J et al (2007) Activation of the imprinted Polycomb Group Fie1 gene in maize endosperm requires demethylation of the maternal allele. Plant Mol Biol 64:387–395
Hetzl J, Foerster AM, Raidl G et al (2007) CyMATE: a new tool for methylation analysis of plant genomic DNA after bisulphite sequencing. Plant J 51:526–536
How-Kit A, Daunay A, Mazaleyrat N et al (2015) Accurate CpG and non-CpG cytosine methylation analysis by high-throughput locus-specific pyrosequencing in plants. Plant Mol Biol 88:471–485
Hu J, Chen X, Zhang H et al (2015) Genome-wide analysis of DNA methylation in photoperiod- and thermo-sensitive male sterile rice Peiai 64S. BMC Genomics 16:102
Huang Y, Pastor WA, Shen Y et al (2010) The behaviour of 5-hydroxymethylcytosine in bisulfite sequencing. PLoS One 5:e8888
Jin SG, Kadam S, Pfeifer GP (2010) Examination of the specificity of DNA methylation profiling techniques towards 5-methylcytosine and 5-hydroxymethylcytosine. Nucleic Acids Res 38: e125
Johannes F, Porcher E, Teixeira FK et al (2009) Assessing the impact of transgenerational epigenetic variation on complex traits. PLoS Genet 5:e1000530
Johnson L, Cao X, Jacobsen S (2002) Interplay between two epigenetic marks. DNA methylation and histone H3 lysine 9 methylation. Curr Biol 12:1360–1367
Kakutani T (1997) Genetic characterization of late-flowering traits induced by DNA hypomethylation mutation in *Arabidopsis thaliana*. Plant J 12:1447–1451
Kanno T, Mette MF, Kreil DP et al (2004) Involvement of putative SNF2 chromatin remodeling protein DRD1 in RNA-directed DNA methylation. Curr Biol 14:801–805
Kawakatsu T, Huang SS, Jupe F et al (2016a) Epigenomic diversity in a global collection of *Arabidopsis thaliana* accessions. Cell 166:492–505
Kawakatsu T, Stuart T, Valdes M et al (2016b) Unique cell-type specific patterns of DNA methylation in the root meristem. Nat Plants 2:16058–16058
Kessler C, Manta V (1990) Specificity of restriction endonucleases and DNA modification methyltransferases a review (Edition 3). Gene 92:1–248
Kim J, Kim JH, Richards EJ et al (2014) Arabidopsis VIM proteins regulate epigenetic silencing by modulating DNA methylation and histone modification in cooperation with MET1. Mol Plant 7:1470–1485
Kooke R, Johannes F, Wardenaar R et al (2015) Epigenetic basis of morphological variation and phenotypic plasticity in *Arabidopsis thaliana*. Plant Cell 27:337–348
Koressaar T, Remm M (2007) Enhancements and modifications of primer design program Primer3. Bioinformatics 23:1289–1291

Lang Z, Xie S, Zhu JK (2016) The 1001 arabidopsis DNA methylomes: an important resource for studying natural genetic, epigenetic, and phenotypic variation. Trends Plant Sci 21:906–908
Li LC, Dahiya R (2002) MethPrimer: designing primers for methylation PCRs. Bioinformatics 18:1427–1431
Li X, Wang X, He K et al (2008) High-resolution mapping of epigenetic modifications of the rice genome uncovers interplay between DNA methylation, histone methylation, and gene expression. Plant Cell 20:259–276
Li X, Zhu J, Hu F et al (2012) Single-base resolution maps of cultivated and wild rice methylomes and regulatory roles of DNA methylation in plant gene expression. BMC Genomics 13:1–15
Lindroth AM, Cao X, Jackson JP et al (2001) Requirement of CHROMOMETHYLASE3 for maintenance of CpXpG methylation. Science 292:2077–2080
Lister R, O'Malley RC, Tonti-Filippini J et al (2008) Highly integrated single-base resolution maps of the epigenome in Arabidopsis. Cell 133:523–536
Liu R, How-Kit A, Stammitti L et al (2015) A DEMETER-like DNA demethylase governs tomato fruit ripening. Proc Natl Acad Sci USA 112:10804–10809
Luo GZ, Blanco MA, Greer EL et al (2015) DNA N(6)-methyladenine: a new epigenetic mark in eukaryotes? Nat Rev Mol Cell Biol 16:705–710
Makarevitch I, Stupar RM, Iniguez AL et al (2007) Natural variation for alleles under epigenetic control by the maize chromomethylase zmet2. Genetics 177:749–760
Manning K, Tor M, Poole M et al (2006) A naturally occurring epigenetic mutation in a gene encoding an SBP-box transcription factor inhibits tomato fruit ripening. Nat Genet 38:948–952
Marano MR, Carrillo N (1991) Chromoplast formation during tomato fruit ripening. No evidence for plastid DNA methylation. Plant Mol Biol 16:11–19
Martin A, Troadec C, Boualem A et al (2009) A transposon-induced epigenetic change leads to sex determination in melon. Nature 461:1135–1138
Matzke MA, Kanno T, Matzke AJ (2015) RNA-directed DNA methylation: the evolution of a complex epigenetic pathway in flowering plants. Annu Rev Plant Biol 66:243–267
McGinn S, Gut IG (2013) DNA sequencing—spanning the generations. New Biotechnol 30:366–372
Mohn F, Weber M, Schubeler D et al (2009) Methylated DNA immunoprecipitation (MeDIP). Methods Mol Biol 507:55–64
Nan X, Meehan RR, Bird A (1993) Dissection of the methyl-CpG binding domain from the chromosomal protein MeCP2. Nucleic Acids Res 21:4886–4892
Nestor C, Ruzov A, Meehan R et al (2010) Enzymatic approaches and bisulfite sequencing cannot distinguish between 5-methylcytosine and 5-hydroxymethylcytosine in DNA. Biotechniques 48:317–319
Niederhuth CE, Bewick AJ, Ji L et al (2016) Widespread natural variation of DNA methylation within angiosperms. Genome Biol 17:194
Ong-Abdullah M, Ordway JM, Jiang N et al (2015) Loss of Karma transposon methylation underlies the mantled somaclonal variant of oil palm. Nature 525:533–537
Ou X, Zhang Y, Xu C et al (2012) Transgenerational inheritance of modified DNA methylation patterns and enhanced tolerance induced by heavy metal stress in rice (*Oryza sativa* L.) PLoS One 7:e41143
Pelissier T, Wassenegger M (2000) A DNA target of 30 bp is sufficient for RNA-directed DNA methylation. RNA 6:55–65
Quadrana L, Almeida J, Asis R et al (2014) Natural occurring epialleles determine vitamin E accumulation in tomato fruits. Nat Commun 5:3027
Rauch T, Pfeifer GP (2005) Methylated-CpG island recovery assay: a new technique for the rapid detection of methylated-CpG islands in cancer. Lab Investig 85:1172–1180
Reinders J, Wulff BB, Mirouze M et al (2009) Compromised stability of DNA methylation and transposon immobilization in mosaic Arabidopsis epigenomes. Genes Dev 23:939–950
Richards CL, Wendel JF (2011) The hairy problem of epigenetics in evolution. New Phytol 191:7–9

Roberts RJ, Belfort M, Bestor T et al (2003) A nomenclature for restriction enzymes, DNA methyltransferases, homing endonucleases and their genes. Nucleic Acids Res 31:1805–1812

Roberts RJ, Vincze T, Posfai J et al (2015) REBASE—a database for DNA restriction and modification: enzymes, genes and genomes. Nucleic Acids Res 43:D298–D299

Rodriguez Lopez CM, Guzman Asenjo B, Lloyd AJ et al (2010) Direct detection and quantification of methylation in nucleic acid sequences using high-resolution melting analysis. Anal Chem 82:9100–9108

Roloff TC, Ropers HH, Nuber UA (2003) Comparative study of methyl-CpG-binding domain proteins. BMC Genomics 4:1

Ronaghi M (2001) Pyrosequencing sheds light on DNA sequencing. Genome Res 11:3–11

Ronaghi M, Uhlen M, Nyren P (1998) A sequencing method based on real-time pyrophosphate. Science 281(363):365

Roudier F, Ahmed I, Berard C et al (2011) Integrative epigenomic mapping defines four main chromatin states in Arabidopsis. EMBO J 30:1928–1938

Sanger F, Nicklen S, Coulson AR (1977) DNA sequencing with chain-terminating inhibitors. Proc Natl Acad Sci USA 74:5463–5467

Schmitt F, Oakeley EJ, Jost JP (1997) Antibiotics induce genome-wide hypermethylation in cultured Nicotiana tabacum plants. J Biol Chem 272:1534–1540

Schmitz RJ, Ecker JR (2012) Epigenetic and epigenomic variation in *Arabidopsis thaliana*. Trends Plant Sci 17:149–154

Schoft VK, Chumak N, Choi Y et al (2011) Function of the DEMETER DNA glycosylase in the *Arabidopsis thaliana* male gametophyte. Proc Natl Acad Sci USA 108:8042–8047

Seymour DK, Koenig D, Hagmann J et al (2014) Evolution of DNA methylation patterns in the Brassicaceae is driven by differences in genome organization. PLoS Genet 10:e1004785

Shook MS, Richards EJ (2014) VIM proteins regulate transcription exclusively through the MET1 cytosine methylation pathway. Epigenetics 9:980–986

Simko I (2016) High-resolution DNA melting analysis in plant research. Trends Plant Sci 21:528–537

Soppe WJ, Jacobsen SE, Alonso-Blanco C et al (2000) The late flowering phenotype of fwa mutants is caused by gain-of-function epigenetic alleles of a homeodomain gene. Mol Cell 6:791–802

Springer NM, Lisch D, Li Q (2016) Creating order from chaos: epigenome dynamics in plants with complex genomes. Plant Cell 28:314–325

Stroud H, Do T, Du J et al (2014) Non-CG methylation patterns shape the epigenetic landscape in Arabidopsis. Nat Struct Mol Biol 21:64–72

Telias A, Lin-Wang K, Stevenson DE et al (2011) Apple skin patterning is associated with differential expression of MYB10. BMC Plant Biol 11:93

Tost J, Dunker J, Gut IG (2003) Analysis and quantification of multiple methylation variable positions in CpG islands by pyrosequencing. Biotechniques 35:152–156

Tricker PJ, Gibbings JG, Rodriguez Lopez CM et al (2012) Low relative humidity triggers RNA-directed de novo DNA methylation and suppression of genes controlling stomatal development. J Exp Bot 63:3799–3813

Untergasser A, Cutcutache I, Koressaar T et al (2012) Primer3—new capabilities and interfaces. Nucleic Acids Res 40:e115

Vaughn MW, Tanurdzic M, Lippman Z et al (2007) Epigenetic natural variation in *Arabidopsis thaliana*. PLoS Biol 5:e174

Vining KJ, Pomraning KR, Wilhelm LJ et al (2012) Dynamic DNA cytosine methylation in the *Populus trichocarpa* genome: tissue-level variation and relationship to gene expression. BMC Genomics 13:27

Vining K, Pomraning KR, Wilhelm LJ et al (2013) Methylome reorganization during in vitro dedifferentiation and regeneration of *Populus trichocarpa*. BMC Plant Biol 13:92

Vossen RH, Aten E, Roos A et al (2009) High-resolution melting analysis (HRMA): more than just sequence variant screening. Hum Mutat 30:860–866

Wang RY, Gehrke CW, Ehrlich M (1980) Comparison of bisulfite modification of 5-methyldeoxycytidine and deoxycytidine residues. Nucleic Acids Res 8:4777–4790
Wang L, Heinlein M, Kunze R (1996) Methylation pattern of activator transposase binding sites in maize endosperm. Plant Cell 8:747–758
Waters AJ, Makarevitch I, Eichten SR et al (2011) Parent-of-origin effects on gene expression and DNA methylation in the maize endosperm. Plant Cell 23:4221–4233
Weinhold A, Kallenbach M, Baldwin IT (2013) Progressive 35S promoter methylation increases rapidly during vegetative development in transgenic Nicotiana attenuata plants. BMC Plant Biol 13:99
Wittwer CT, Reed GH, Gundry CN et al (2003) High-resolution genotyping by amplicon melting analysis using LCGreen. Clin Chem 49:853–860
Wojdacz TK, Dobrovic A (2007) Methylation-sensitive high resolution melting (MS-HRM): a new approach for sensitive and high-throughput assessment of methylation. Nucleic Acids Res 35:e41
Woo HR, Dittmer TA, Richards EJ (2008) Three SRA-domain methylcytosine-binding proteins cooperate to maintain global CpG methylation and epigenetic silencing in Arabidopsis. PLoS Genet 4:e1000156
Wu TP, Wang T, Seetin MG et al (2016) DNA methylation on N6-adenine in mammalian embryonic stem cells. Nature 532:329–333
Xiong Z, Laird PW (1997) COBRA: a sensitive and quantitative DNA methylation assay. Nucleic Acids Res 25:2532–2534
Yang G, Lee YH, Jiang Y et al (2005) A two-edged role for the transposable element Kiddo in the rice ubiquitin2 promoter. Plant Cell 17:1559–1568
Yin BL, Guo L, Zhang DF et al (2008) Integration of cytological features with molecular and epigenetic properties of rice chromosome 4. Mol Plant 1:816–829
Zemach A, Grafi G (2003) Characterization of *Arabidopsis thaliana* methyl-CpG-binding domain (MBD) proteins. Plant J 34:565–572
Zemach A, McDaniel IE, Silva P et al (2010) Genome-wide evolutionary analysis of eukaryotic DNA methylation. Science 328:916–919
Zemach A, Kim MY, Hsieh PH et al (2013) The Arabidopsis nucleosome remodeler DDM1 allows DNA methyltransferases to access H1-containing heterochromatin. Cell 153:193–205
Zhai J, Liu J, Liu B et al (2008) Small RNA-directed epigenetic natural variation in *Arabidopsis thaliana*. PLoS Genet 4:e1000056
Zhang X, Yazaki J, Sundaresan A et al (2006) Genome-wide high-resolution mapping and functional analysis of DNA methylation in arabidopsis. Cell 126:1189–1201
Zhong S, Fei Z, Chen YR et al (2013) Single-base resolution methylomes of tomato fruit development reveal epigenome modifications associated with ripening. Nat Biotechnol 31:154–159
Zhou Y, Bui T, Auckland LD et al (2002) Undermethylated DNA as a source of microsatellites from a conifer genome. Genome 45:91–99
Zhu J-K (2009) Active DNA demethylation mediated by DNA glycosylases. Annu Rev Genet 43:143–166
Zhu J, Kapoor A, Sridhar VV et al (2007) The DNA glycosylase/lyase ROS1 functions in pruning DNA methylation patterns in Arabidopsis. Curr Biol 17:54–59
Zilberman D, Gehring M, Tran RK et al (2007) Genome-wide analysis of *Arabidopsis thaliana* DNA methylation uncovers an interdependence between methylation and transcription. Nat Genet 39:61–69
Zilberman D, Coleman-Derr D, Ballinger T et al (2008) Histone H2A.Z and DNA methylation are mutually antagonistic chromatin marks. Nature 456:125–129

Epigenetics in Plant Reproductive Development: An Overview from Flowers to Seeds

Antoine L.F. Gady, Cristiane S. Alves, and Fabio T.S. Nogueira

Abstract Plant development is governed by a wide variety of genetic and epigenetic events that regulate cell fate. Flower to seed developmental transition varies greatly between plants and is of importance in research programs because of its relevance for crop production and human diet. In this chapter, we review the latest research on epigenetics regulation of flower, fruit, and seed development in crop plants. We use tomato (*Solanum lycopersicum* L.) as our reference crop model while referring to *Arabidopsis thaliana* for in-depth studies and look into additional crop model plants such as maize (*Zea mays*), wheat (*Triticum* spp.), and rice (*Oryza sativa*) in order to cover a wide range of flower and fruit/seed types. Tomato is an interesting biological model thanks to its fleshy fruit. Tomato has the second natural epimutation reported, the *Colorless non-ripening* (*Cnr*), as well as newly reported studies on the paramutation SLTAB2, the role of the demethylase DML2 in fruit ripening, and the identification of two long noncoding RNAs (lncRNAs) involved in the ripening process. Altogether, these works make tomato an interesting and important epigenetic model for plants. A variety of epigenetic-based regulations are involved in each stage of the tomato fruit set, development, and ripening. Four epigenetic mechanisms are proved to be involved in flower, fruit, and developmental processes: histone modifications, DNA (de)methylation, small RNA posttranscriptional *locus* regulation, and lncRNA-associated regulatory pathways. Epigenetic mechanisms are involved at all stages of reproductive organs development, from the flower to the mature seed.

Keywords Tomato • Small RNAs • DNA methylation • Epimutation

A.L.F. Gady • F.T.S. Nogueira (✉)
Biological Sciences Department, Laboratory of Molecular Genetics of Plant Development, University of Sao Paulo (USP), 13418-900 Piracicaba, Sao Paulo, Brazil
e-mail: ftsnogue@usp.br

C.S. Alves
Biological Sciences Department, Laboratory of Molecular Genetics of Plant Development, University of Sao Paulo (USP), 13418-900 Piracicaba, Sao Paulo, Brazil

Genetics Department, Laboratory of Telomeres, São Paulo State University (Unesp), Institute of Biosciences, 18618-689 Botucatu, Sao Paulo, Brazil

N. Rajewsky et al. (eds.), *Plant Epigenetics*, RNA Technologies,
DOI 10.1007/978-3-319-55520-1_17

Contents

1 Introduction

Plasticity plays a fundamental role in adaptation and resilience of crops allowing cultivars to adjust and grow in different environments. Plasticity is responsible for homeostasis maintenance when environmental conditions change, allowing gene expression to adapt to biotic or abiotic stress or variations like altitude, soil type, seasonal, day length, rain, ambient temperature, disease, plant–plant competition, herbivory, among others (Latzel et al. 2012). To our current knowledge, epigenetic mechanisms are responsible for conferring genetic plasticity to crops by DNA methylation, histone modification, noncoding RNAs, and chromatin modulation. These variations can regulate genome expression leading to new phenotypes in response to environmental changes.

Eukaryotic chromatin is a nucleoprotein complex, where the DNA is packed and condensed enfolding histone proteins. Even being tightly condensed, the chromatin needs to be dynamic in order to allow basic functions like transcription, replication, and DNA repair to occur. For the compaction to happen, 147 base pairs of DNA are wrapped into an octamer of core histones, containing two of each H2A, H2B, H3, and H4, linked by H1 at the entry/exit point. Histones play important roles in metabolic chromatin functions such as chromatin integrity maintenance, DNA recombination, and process of DNA replication in several organisms. In fact, these proteins date back to the dawn of eukaryotic evolution, spanning protozoans, fungi, animals, and plants. Prokaryal and archaeal species are the earliest genomes known to have evolved histone-like proteins (Grove 2011; Sandman and Reeve 2006). In the last 15 years, histones became the focus of research interests thanks to the light shed on the histone code and its implications. This code is a resultant from covalent posttranslational modifications (PTM)—methylation, acetylation, phosphorylation, ubiquitination, and poly-ADP ribosylation—that takes place at the Cor N-terminal tail of histones and histone variants. There are five histones families—H1, H2A, H2B, H3, and H4—which are subject to PTM. Euchromatin is usually constituted by transcribed *loci* and heterochromatin is enriched with TEs (transposable elements) and is typically transcriptionally silenced through vegetative phases by DNA methylation and histone modifications (Bernatavichute et al. 2008; Law and Jacobsen 2010).

Contrarily to animals, in which cytosine DNA methylation in the CG context is predominant, in plant kingdom cytosine DNA methylation can occur in three contexts: CG, CHG, and CHH (where H=A, T or C) (Law and Jacobsen 2010). Plant DNA methyltransferases are distributed in four families: (1) MET1: similar to DNMT1, which is a CG maintenance DNA methyltransferase; (2) CMT: a plant-specific chromomethylase that lacks the N-terminal extension, and it is a CNG DNA methyltransferase; (3) DRM: a plant de novo DNA methylases of non-CG DNA sequences, which have an unusual arrangement of some conserved catalytic motifs; and (4) DNMT2: highly conserved but enigmatic methyltransferase (Vanyushin and Ashapkin 2009). Recent studies in *Solanum lycopersicum* (tomato) genome reveal the presence of nine genes encoding DNA methyltransferases, comprising the four DNA methyltransferase families: *MET1*, *SlDNMT2*, *SlCMT2*, *SlCMT3*, *SlCMT4*, *SlDRM5*, *SlDRM6*, *SlDRM7*, *SlDRM8* (Kumar et al. 2016).

The DNA methylation marks are reversible, which confer plasticity by facilitating the modulation of RNA transcription from genomic regions or specific *loci*. Alteration in DNA methylation can have meaningful outcomes on chromatin structure, with increases in 5mC marks connected with the arrangement of nucleosomes in plant genomes (Chodavarapu et al. 2010). Based on bisulfite sequencing—where unmodified cytosines are converted to uracil (sequenced as T) and 5mC remains sequenced as C, researchers generated methylome maps for tomato (Zhong et al. 2013), *Arabidopsis* (Zhang et al. 2006), and rice (*Oryza*; Li et al. 2012). Tissue-specific methylomes can provide evidence of epigenetic plasticity by analyzing the changing patterns in gene regulation. Euchromatin is associated with hypomethylated DNA in actively transcribed regions, while heterochromatin contains silenced genes, which were frequently hypermethylated (May 2010).

Noncoding RNAs, RNA molecules that do not code for proteins, have important roles in the regulation of plant reproductive organ development. Small interfering RNAs (siRNA) function as part of the RNA-directed DNA methylation (RdDM) machinery where they direct the RdDM complex to specific DNA locations to be methylated. MicroRNAs (miRNA) provoke transcriptional gene silencing by recruiting the RISC complex to specific target mRNAs by sequence complementarity. Finally, long noncoding RNAs (lncRNA) are transcripts over 200 bp that can act in a multitude of ways to regulate gene expression; depending on their genome location, they can act as *cis*- or *trans*-regulatory elements (Chekanova 2015; Liu et al. 2015b). The implication of ncRNAs in the development of both flower and fruit has been repeatedly proven and have unraveled a vast network of RNA molecules acting as fine regulators of gene expression during plant organ development.

In this chapter, we will highlight the main findings on epigenetic regulation affecting flower, fruit, and seed development. We will discuss the latest findings regarding epigenetic regulations of tomato flower and fruit development that might help to improve the future of tomato breeding.

2 Flowering and Pollen Development

Perennial and overwintering annual crops can respond to the length and severity of the winter thanks to their plasticity, securing that individual plants are adjusted and reaching reproductive maturity. In some cases, this environmental cue causes changes that are mitotically stable throughout the rest of development. To date, the best-characterized example of such epigenetic memory in plants is the vernalization process, namely the acceleration of flowering as a result of exposure to cold temperatures in winter. Flowering program is promoted by the perception of the vernalization signals, including chromatin-based mechanisms, as the winter passes (reviewed in Bloomfield et al. 2014).

FLOWERING LOCUS C (*FLC*) is a major floral repressor that suppresses flowering during exposure to prolonged periods of low temperature. An additive effect of histone modifications at the *FLC* locus takes place in vernalization-sensitive species when the length of winter daytime is noted as temperature degrees below a threshold (Sheldon et al. 2009). Depending on the species, the vernalization signaling can happen at different developmental stages, registered by epigenetic switches maintained during in vitro vegetative propagation and deleted through sexual reproduction (Song et al. 2012). When the winter ends, histone remodification gradually occurs allowing the derepression of *FLC locus* (Song et al. 2012). In response to cold treatment, 5mC (5-methyl cytosine) is significantly reduced both in winter and spring *B. napus* following a gradual DNA re-methylation to pretreatment levels in spring *B. napus*, but only up to 70% in winter *B. napus* (Guzy-Wrobelska et al. 2013). It seems that this primary mechanism of epigenetic regulation is conserved across plant species, as similar results from cold treatment also change 5mC patterning in different crops like cotton, maize, rice, and wheat (Steward et al. 2002; Sherman and Talbert 2002; Pan et al. 2011; Fan et al. 2013). Active repression of *FLC* or *FLC*-like *loci* is required as a standard mechanism until a target threshold is sensed (Sheldon et al. 2009; Xiao et al. 2013). *FLC* is repressed in *Arabidopsis* by decreasing H3 acetylation and demethylation of H3K9 through *REDUCED VERNALIZATION1/VERNALIZATION INSENSITIVE 3* (*VRN1/VIN3*) (Bastow et al. 2004; Sung and Amasino 2004).

Pollen development undergoes a transcriptional and translational reprogramming to promote the production of male gametes from somatic lineages (Calarco et al. 2012). Small RNAs (sRNAs), important components of the plant epigenetic machinery, play an essential role in the pollen reprogramming process, altering the transcriptional and translational dynamics characterizing the individual developmental stages (Borges et al. 2011). Pollen is notably sensitive to elevated temperatures, and little is known about the mechanisms underlying this stress response. Recently, it has been shown in heat-stressed tomatoes that the accumulation of pollen miRNAs, tRNAs, and snoRNAs is affected by heat stress in distinct pollen developmental stages, especially in post-meiotic and mature stages of male gametophyte development (Bokszczanin et al. 2015). Though still speculative, some of these sRNAs might have important roles in the epigenetic regulation of pollen development in response to environmental cues.

3 Flower and Fruit Development

The final size and shape of the fruit is the result of a multitude of developmental events that go back as far as the floral initiation. The differentiation of the inflorescence meristem into a floral meristem marks the first stage of floral initiation. The size of the floral meristem depends on the number of cells it is composed of, and therefore, it is a parameter that influences the final fruit size (van der Knaap et al. 2014; Xu et al. 2015). Five days after floral initiation, the gynoecium starts to grow. The way the gynoecium will develop pre- and post-anthesis will determine the shape and size of the final fruit. See van der Knaap et al. (2014) for a complete review of the genes influencing tomato fruit weight and shape.

3.1 Histone Acetylation Mediated Regulation

Flower architecture is molecularly determined by the reference ABC model (Bowman et al. 1991). In this model, specific combinations of gene expression and protein interactions determine the geographical limits of each flower whorl (sepals, petals, stamens, and carpel). As demonstrated in *Arabidopsis*, the A class gene *APETALA2* (*AP2*) regulates target gene expression as part of complex it forms together with *TOPLESS* (*TPL*) and *HISTONE DEACETYLASE19* (*HDA19*). That transcription repressing complex negatively regulates *AGAMOUS* (*AG*), a C class gene, and *SEPALATA3* (*SEP3*), an E class gene from the ABC flower architecture model. The gene repression is mediated by deacetylation of H4K16 in regulatory regions of both *AG* and *SEP3* (Krogan et al. 2012). Expression studies in *Arabidopsis* tissues highlighted additional HDACs expressed in reproductive tissues: *HDA5*, HDA6, HDA7, *HDA9, HDA15*, and *HDA18*, but their function in either fruit or flower development remains unidentified. Only for *HDA6* do we know that it is involved in the regulation of flowering time. *HDA6* directly interacts with *FLOWERING LOCUS D* (*FLD*), a histone H3K4 demethylase. The complex removes acetyl and methyl groups from Histone 3 at loci of three repressors of flowering: *FLD*, *MADS AFFECTING FLOWERING 4* (*MAF4*), and *MAF5*, thus repressing their expression (Yu et al. 2011). From that set of HDACs, *HDA6* is known to be involved in RNA-directed DNA methylation (RdDM), a plant-specific mechanism to regulate chromatin silencing of developmental genes as well as transposable and repetitive elements.

In plants, the major small-interfering RNA (siRNA)-mediated epigenetic pathway is RNA-directed DNA methylation (RdDM). RdDM is a complex epigenetic machinery that involves a large number of players whose activity can be broken down into a few steps (Matzke and Mosher 2014). Mainly two types of transcripts are involved in the RdDM machinery: Pol IV and Pol V transcripts. Pol IV transcribes long noncoding RNAs (lncRNAs) that are subsequently converted into double-stranded RNAs (dsRNAs) by *RDR2* (Haag et al. 2012). These dsRNAs are processed by *DICER-like3* (*DCL3*) into siRNAs. The siRNAs are exported to the cytoplasm where they are loaded into *AGO4* and reimported into the nucleus. The role of the siRNAs is to guide, by specific base pairing, *AGO4* toward nascent scaffold transcripts of Pol V. The formation of this siRNA, *AGO4*, Pol V-derived lncRNA scaffold ultimately recruits histone deacetylases (HDACs) and DNA methyltransferases that through histone deacetylation and DNA methylation silence the genomic loci transcribed by Pol V. Promoters silenced by RdDM are characterized by histone deacetylation, which in *Arabidopsis* is acted by the RPD3-type histone deacetylase *AtHDA6*, for which the tomato homologue is *SlHDA3*. *AtHDA6* activity results in the deacetylation of histone H3 lysines 9 and 14 which leads to gene expression downregulation (Aufsatz et al. 2007). Functional *AtHDA6* is required to control siRNA-dependent heterochromatin and that deacetylation is a prerequisite for subsequent methylation by HMTs (Aufsatz et al. 2007). Complete loss-of-function mutants for *AtHDA6* exhibit reactivation of RdDM-silenced promoters, despite the continuous presence of the RNA-silencing signal. Moreover, cytosine methylation is reduced, highlighting a function for *AtHDA6* in methylation maintenance. This function of *AtHDA6* might be mediated by the physical association with DMTs, *MET1*, and *CMT3* (Aufsatz et al. 2002).

The roles of histone acetylases and deacetylases in tomato flower or fruit development are poorly understood. Using the sequencing data generated by the international tomato genome sequencing consortium (Tomato Genome C 2012), Cigliano et al. (2013) identified in silico potentially all the histone modifiers of the tomato genome. Next they used the RNAseq data from that same source to look at the expression profile of each histone modifiers in 10 sample tissues. The histone acetylases *SlHAG18* and *SlHAG6* both presented peaks of expression in the flower samples, possibly indicating a role in reproductive development. A recent study (Zhao et al. 2014) identified 15 tomato histone deacetylases. *AtHDA6* tomato homologue, *SlHDA3*, was expressed in all tissues with a maximum expression at the flower stage. *SlHDA1*, the homologue to *AtHDA19*, was as well highly expressed in flowers and then repressed for all fruit samples except for a peak at the ripe fruit stage. In addition, yeast two-hybrid assays showed that *SlHDA1*, *SlHDA3*, and *SlHDA4* interact with MADS-box transcription factors TOMATO AGAMOUS1 (*TAG1)* and TOMATO MADS-BOX29 (*TM29)* (Zhao et al. 2014). *TAG1* is a transcription factor necessary to express ethylene dependent and independent ripening genes (Klee and Giovannoni 2011). *TM29* is a homologue to

SEPALATA which when silenced provokes the formation of parthenocarpic fruits and aberrant flowers (Ampomah-Dwamena et al. 2002).

3.2 DNA Methylation-Mediated Regulation

The *clark kent* hypermethylated epialleles of the *SUPERMAN (SUP)* gene are a clear illustration that DNA methylation can affect flower development through the regulation of gene expression level. The *sup-5 Arabidopsis* mutant which contains a nearly complete deletion of the *SUP* gene produces flowers with an increased number of stamens and carpels: 12 stamens compared to six and three carpels against two in wild-type *Arabidopsis* flowers. The stronger *clk3* epiallele contains an average of eight stamens and three carpels. The *clk* alleles have the exact *SUP* sequence of the wild-type accession but are extensively methylated from the start of transcription and covering most of the gene region (Jacobsen and Meyerowitz 1997). Antisense *MET1* and *ddm1* and *ddm2* mutant plants are hypomethylated but contain methylation-rich regions such as the *SUP* locus that is consistently hypermethylated in the three DNA methylases' mutant or antisense backgrounds (Jacobsen et al. 2000). In addition, *agamous* mutant-like flower phenotypes were identified in the MET1-antisense plants. Reduced levels of *AGAMOUS* mRNA were observed in the transgenic plants. An increase in methylation of the *AG* locus was measured by bisulfite genomic sequencing. Interestingly, hypermetylation of the *AG* locus only occurred in lines with a hypermethylated *SUP* locus, thus suggesting that hypermethylation of *SUP* is necessary to the hypermethylation of *AG*.

To understand the extent of the epigenetic regulation of flower development by DNA methylation, Yang et al. (2015) used high-throughput sequencing of DNA fragments obtained by MspJI digestion to obtain a whole genome profile of DNA methylation patterns. They divided the *Arabidopsis* flower development in three key stages: (1) floral meristem from *ap1 cal* double mutants, (2) wild-type early flowers at stages 1–9, and (3) wild-type late flowers at stages 10–12. In plants, DNA methylation occurs on the Cytosine nucleotide in three distinct sequence contexts: *MET1* transfers a methyl residue to CG sites, *CMT3* to CHG sites, and *DRMs* add methyl groups to CHH sites where H stands for A, T, or C nucleotides. The number of methylated cytosine increased by 8% between floral meristem and early flower stage and then decreased by 0.55% from the early to late flower stages. The initial stages of flower development are marked by numerous de novo methylation: 80,056 new methylation events reported. The study identified 3067 genes out of 24,035 that are co-differential: genes with significant variations in both methylation and gene expression. In the transition from floral meristem to early flower stage, 1048 genes were co-differential at ^{m}CG, 601 at ^{m}CHG, and 509 ^{m}CHH-containing genes. A total of 909 genes were co-differential in the early to late flower transition. Among them are important flower development regulators such as *SEP1*, *LEUNIG* (*LUG*), and *SEEDSTICK*. Moreover, 33 genes associated with flower development,

21 genes involved with pollen development, 201 transcriptional regulators, 29 genes linked to chromatin organization, and 56 to signal transduction have differential gene expression profile along flower development that is linked to DNA methylation variation. While the number of methylated cytosine increased during the meristem to early flower transition in all sequence contexts, only ^{m}CGs' number increased in the early to late flower shift, thus hinting at a role for MET1 methylations in a developmental phase characterized by organ growth. Considering the three studied developmental stages, over 1000 genes for each CHG and CHH methylation were both transcriptionally and epigenetically differentially regulated, suggesting an important role for these methylation types into reproductive organ development. Finally, Yang et al. (2015) also show that DNA methylation does not only regulate protein coding gene expression but as well that other epigenetic players are regulated in this manner: transposable elements, miRNAs, and noncoding RNA also had variation in the methylation status along flower development. This data thus hints toward the idea that the activity of additional epigenetic players is regulated by their methylation state.

3.3 MiRNA Mediated Regulation

MicroRNAs, small 21 nucleotides RNA molecules, regulate gene expression by specifically binding to mRNA with near-perfect sequence complementarity and thereof provoking their degradation. In *Arabidopsis thaliana*, at least eight miRNA families are responsible for the regulation of transcription factors involved in flower development. The miR164, miR169, and miR172 families are participating in setting boundaries between floral organs. The miR164 family regulates transcription factor of the NAC-domain family such as *CUP-SHAPED COTYLEDONS* (*CUC1* and *CUC2*) in *Arabidopsis*. Alteration to *CUC1* and *CUC2* expression results in modified sepal boundaries leading to fused sepals and fewer petals (Baker et al. 2005; Laufs et al. 2004). In addition, miR164 could play a role in carpel development (Baker et al. 2005; Sieber et al. 2007). Through the regulation of NF-YA transcription factors, the miR169 family limits the expression of the C-class family of genes to the inner two whorls of petunia and antirrhinum flowers. MiR172 on the other hand restricts the expression of *AP2* to the two outer whorls of the floral meristem (Chen 2004; Wollmann et al. 2010; Zhao et al. 2007). In addition to its role in flower development in *Arabidopsis*, miR172 was shown to be actively regulating flower formation in both rice and barley (Lee and An 2012; Nair et al. 2010; Zhu et al. 2009). MiR159 targets GAMYB-like genes such as *LEAFY*, *MYB33*, or *MYB65* and thus regulates flowering time but as well anther formation during flower development. Overexpression of miR159 downregulates *MYB33* and results in male sterility. Similarly, to what is observed in *Arabidopsis*, in rice the GAMYB gene expression is restricted to anthers. Resembling the effect of miR159, the overexpression of miR319 as well leads to defects in stamen development and male sterility, but the phenotype is the consequence of the mis-regulation of a set of TCP transcription factor genes (Palatnik et al. 2007; Schommer et al. 2012). Plants with reduced levels of miR159 and miR319 show similar floral phenotypes as *arf6/arf8* double mutants. Auxin Response Factors 6 and 8 through the regulation of Auxin level regulate the extent of cytokinin activity in the developing floral meristem. *ARF6* and *ARF8* are regulated by miR167 (Rubio-Somoza and Weigel 2011), and miR167 is upregulated by either *TCP4* or *MYB33*, the targets of miR159 and miR319, thus forming a complex floral development regulatory network. Another regulator of ARF genes is miR160 which targets *ARF10*, *ARF16*, and *ARF17* genes. Downregulation of miR160 in transgenic *Arabidopsis* plants increases its target expression. These plants have defects in fertility and in floral organ formation and floral organs appeared inside siliques (Liu et al. 2010). MiR165 and miR166 regulate HD-ZIP III genes *ATHB15*, *ATHB8*, *REVOLUTA*, *PHABULOSA*, and *PHAVOLUTA*. Downregulation of these genes through overexpression of miR165 results in plants with carpel developmental defects, enlarged apical meristem, and short sterile carpels (Kim et al. 2005). The involvement of miRNAs in *Arabidopsis* reproductive development from juvenile to the flower producing plant phases is reviewed in Hong and Jackson (2015).

The expression of *SlARF6* and *SlARF8* is also regulated by miR167 in tomato plants. Plants overexpressing AtmiR167a produce female sterile flowers with shortened sepals, stamens, and style which is a consequence of the *SlARF6* and *SlARF8* low expression levels in developing flowers (Liu et al. 2014). MiR160 is abundant in tomato ovaries. In tomato, miR160 preferentially targets ARF10A and to a lesser extent Sl*ARF10B* and Sl*ARF17* (Damodharan et al. 2016). Therefore, the use of a target mimic to sequestrate miR160 and inhibit its natural function provokes an increase in *SlARF10A* accumulation in tomato ovaries which result in perturbed ovary patterning: an excessive elongation of its proximal end and thinning of the placenta. Consequently, postfertilization, the fruit is pear shaped. This fruit shape phenotype is the result of a mis-distribution of auxin in the early stages of ovary development regulated by SlmiR160 (Damodharan et al. 2016). In rice, Huang et al. (2016) showed that OsmiR160 regulates *OsARF18* and thereby auxin signaling. mOsARF18 transgenic plants express a modified allele of *OsARF18* that is not recognized and thus not regulated by OsmiR160. These transgenic rice plants had overall growth and development defects such as dwarfism, rolled leaves, small seeds, and abnormal flowers. mOsARF18 plants were impaired in reproductive organ development: the lemma and palea did not contain flowers and stamens remained attached to developing seeds when fertilization did occur, suggesting abnormal senescence of stamens, reminiscent of the senescence phenotypes observed in tomato (Damodharan et al. 2016).

SlmiR396 targets 8 out of the 13 tomato Growth Regulating Factors (GRF): *SlGRF1*, *SlGRF2*, *SlGRF3*, *SlGRF4*, *SlGRF5*, *SlGRF7*, *SlGRF8*, and *SlGRF12*. GRFs are a class of transcription factors expressed in most developing organs. GRFs regulate cell number, and their overexpression in *Arabidopsis* results in enlarged organs (Cao et al. 2016). On the other hand, overexpression of miR396 in *Arabidopsis* plants developed flowers with a single carpel; in rice a similar approach led to similar results: altered floral organ morphology (Cao et al. 2016).

Strong downregulation of miR396a and miR396b with STTM396a/396a-88 in tomato produced plants with increased cell number and cell size in both flowers and fruits leading to larger sepals and larger fruits (Cao et al. 2016). Fruits from STTM396a/396a-88 transgenic lines were 39 and 45% larger than control fruits. This might prove to be a new way to improve yield. Overexpression of miR172 was shown as well to increase fruit size in tomato (Yao et al. 2016). MiR172 overexpressing tomato plants contain numerous flower defects such as sepal to petal transformation, poorly developed stamen that produce sterile pollen, and the development of seedless parthenocarpic fruits with ectopic ovaries inside the fruits and occasionally fruit in fruit phenotypes. Similar phenotypes were also observed by our group in Micro-Tom tomato plants overexpressing miR156 (Silva et al. 2014). The altered fruit morphology, fruit-like structures emerging from the main fruit, was correlated with accumulation of miR156 in meristematic tissues such as placenta and ovules of developing ovaries and immature fruits. miR156 overexpression plants (miR156-OE) had flower buds with extra whorls and meristem-like structures that developed into ectopic structures instead of ovaries and ovules. The overexpression of miR156 prolonged the phase of floral meristem proliferation, and when organs finally formed, they produced flowers with additional partly fused carpels that likely account for the appearance of the miR156-OE fruits (Silva et al. 2014). We have identified five miR156-targeted SQUAMOSA promoter binding protein like (*SPL*/*SBP-box*) genes that are differentially expressed in pre- and post-anthesis ovaries. Our data show that the *LeT6*/*TKn2* and *GOBLET* (*GOB*) are repressed by Sl-SBPs to control meristem maintenance and cell proliferation at the onset of flower organs' initiation and differentiation, thereby controlling proper carpel and ovule development. In addition, we showed that *MACROCALLIX* (*MC*), *FRUITFUL1* (*FUL1*), and *FALSIFLORA* (*FA*) may act under the control of the miR156/*Sl-SBP* node to regulate floral meristem identity and specification of organ whorls, but while *GOB* is controlled by the miR156/*Sl-SBP* node through miR164, it remains unknown how miR156/*Sl-SBP* controls *MC*, *FUL1*, and *FA*.

The epigenetic factors controlling the expression profile of genes involved in both flower and fruit development as reviewed in this chapter are all schematically overviewed in the Fig. 1.

4 Fruit Ripening

The ripening development process is a unique feature to plants bearring fleshy fruits. Its function is to help the dissemination of the plants seeds through animal consumption. Therefore, the fruit undergoes important changes in, for example, color: making them more visible, in metabolite composition, such as production of sugars and volatiles. In tomato, this transition from a mature green to a ripe red fruit is induced at the breaker stage by a concomitant burst in ethylene production and by a sharp increase in differential expression of transcription factors. Our

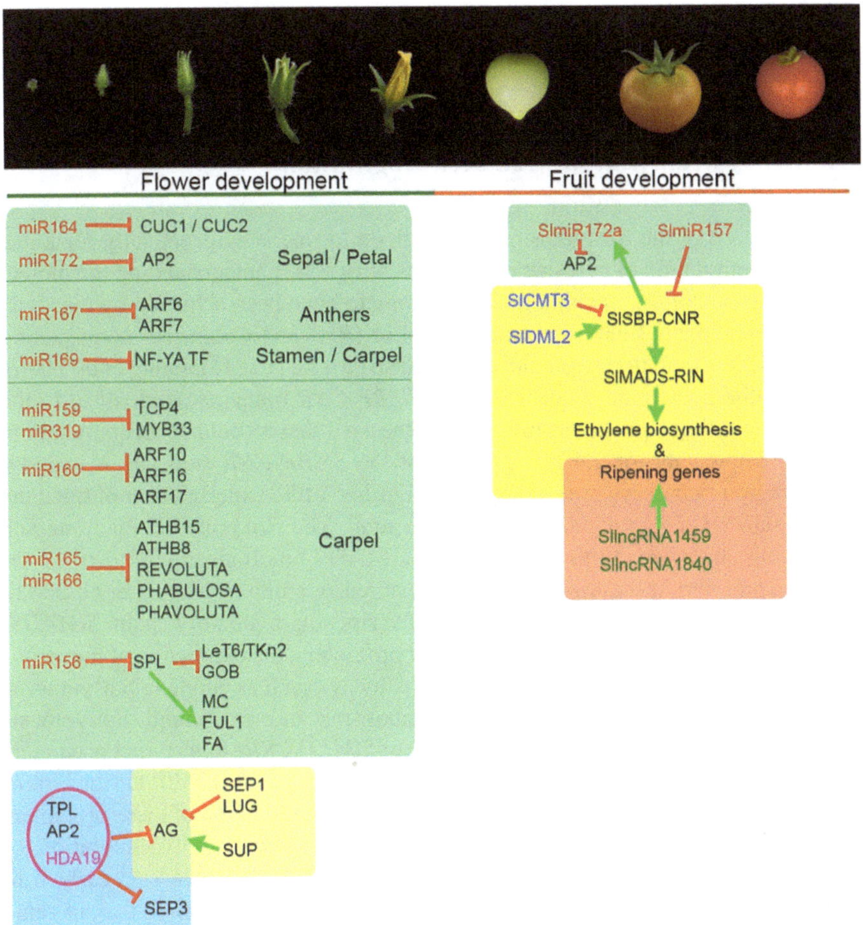

Fig. 1 Epigenetic regulation of genes governing flower and fruit development. Flower development, on the *left*, is monitored by a combination of histone acetylation, DNA methylation, and miRNA posttranscriptional gene silencing. The *green* backgrounds encapsulate regulations monitored by miRNAs. The list of miRNA directed regulations is divided according to organ location specificity. On the *yellow* backgrounds are genes whose expression is dependent on DNA methylation status: *SEPALATA1* (*SEP1*), *LEUNIG* (*LUG*), and *SUPERMAN* (*SUP*), all of which regulate *AGAMOUS* (*AG*). On the *blue* background are represented flower development genes regulated by *HISTONE DEACETYLASE 19 (HDA19)*. Regulation of the transition from *green* to maturing fruit is regulated by the pivotal gene *SLSBP-CNR* which is controlled (1) by promoter methylation status thanks to *SlCMT3* and *SlDML2* and (2) by SlmiR157 (*green* background). lncRNA1459 and lncRNA1840 that are involved in the regulation of fruit ripening genes are shown on the *red* background

understanding of the role of epigenetic players in fruit ripening regulation is a recent addition to this complex development process.

4.1 DNA Methylation-Mediated Regulation

The regulation of the fruit transition from green to ripe is epigenetically regulated by DNA methylation of cytosine nucleotides. Progress in understanding the molecular regulations of the ripening process in tomato have been achieved by the study of ripening mutants such as ripening inhibitor (*Rin*) and Colourless non-ripening (*Cnr*). The *Cnr* mutation was mapped to a SlSBP3-like gene (Manning et al. 2006), and since the nucleotide sequence of the *SlSBP-CNR* mutant was identical to the one of the wild-type plant, the authors hypothesized that it could be an epimutation. Bisulfite sequencing of 2.4 kb upstream of the *SlSBP-CNR* start site revealed a 286 bp region rich in cytosine methylation. Further VIGS experiment confirmed the role of *SlSBP-CNR* as regulator of fruit ripening. The *Rin* mutation mapping and sequencing characterized the Rin locus as a MADS box transcription factor gene whose expression in the *Rin* background restored ripening (Vrebalov et al. 2002). *SlMADS-RIN* acts upstream of ethylene in the ripening regulatory chain. *SlMADS-RIN* regulates, through interaction with their promoters, the expression of numerous genes involved in various ripening-related pathways such as ethylene biosynthesis, perception, and response; cell wall metabolism; and carotenoid biosynthesis (Martel et al. 2011). In addition, the binding of *SlMADS-RIN* to its target promoters cannot occur in the hypermethylated *Cnr* mutant (Martel et al. 2011); thus, transcription activation by *SlMADS-RIN* is impaired by methylation of these promoters and can only occur in plants with proper *SlSBP-CNR* activity. Inhibition of 5-cytosine DNA methyltransferases in green immature fruits leads to early fruit ripening, before seeds are mature. Whole genome methylome showed that, in genic regions, differentially methylated regions were located on the 5′-end of genes, therefore likely to be associated with promoter regulatory regions (Zhong et al. 2013). While whole genome methylome showed that a substancial epigenome reprogramming is occurring during fruit development, it also identified 292 genes that are regulated by the *SlRIN* MADS box transcription factor. These genes, in the hypermethylated *Cnr* background, have methylation in their promoter region that prevent binding and activation by *SlRIN*. This work confirmed 16 previously identified RIN targets. RIN-regulated genes are demethylated in the control plants, thus allowing normal production of ethylene, volatiles, sugar metabolites, carotenoids, and fruit softening. Genome methylation is thus the third key determining factor to the transition to ripening in fleshy fruit plants in addition to ethylene hormonal control and fruit-specific transcription factors (Zhong et al. 2013).

To identify which DNA methyltransferase is responsible for the methylation of the promoter of the *Cnr* locus Chen et al. (2015b) used the VIGS system to silence a range of DNA methyltransferases in *Cnr* fruits. *Cnr* fruits with silenced *SlDRM7*, *SlMET1*, *SlCMT2*, and *SlCMT3* ripened to various degrees. *SlCMT3*-silenced fruits ripened almost completely. The expression of *SlCNR* and of the ripening genes *SlRIN*, *SlAP2a*, *SlTAGL1*, of the ethylene biosynthesis, and signaling pathway were all upregulated in the red sections of the VIGS-SlCMT3 fruits. Bisulfite sequencing determined that 8 out of 18 methylated cytosine in the *Cnr* promoter were demethylated in the VIGS-SlCMT3-silenced fruits. This reduction in methylation was sufficient to alter the expression level of the CNR gene. In addition, the bisulfite sequencing data showed that hypomethylation in *SlCMT3*-silenced tissues also occurred at the promoter sites of *SlRIN* and RIN-targeted promoters directing the expression of ripening genes. Thus, *SlCMT3* is essential to the maintenance of the stable *Cnr* epiallele. In wild-type plants, at the time when the fruit is ready to switch into its ripening phase, the promoters of genes discussed above have to be demethylated. The tomato genome contains four DEMETER-like DNA demethylases (DMLs), but only *SlDML2* is expressed from the onset of fruit ripening and further on until the fruit is fully ripe (Liu et al. 2015a). *SlDML2* RNAi silencing lines showed delayed initiation of ripening, from 10 to 20 days, and the ripening was never fully completed in fruits of these lines. Thus, active demethylation is a prerequisite to tomato fruit ripening. *SlDML2* is the only demethylase expressed at the developmental stage corresponding to demethylation of ripening genes characterized in *Rin* and *Cnr* mutants such as *SlPSY1* (Liu et al. 2015a).

4.2 *MiRNA Mediated Regulation*

Deep sequencing of tomato short RNAs and comparative genomics have contributed to the identification of hundreds of miRNA expressed in tomato tissues (Din and Barozai 2014; Karlova et al. 2013; Moxon et al. 2008; Zuo et al. 2012). These large-scale projects showed that miRNAs are involved in most aspects of plant and fruit development and along all stages of fruit growth and ripening. But few of the predicted miRNA/target gene interaction have been experimentally validated. Moxon et al. (2008) predicted that miR157 and miR156 play a role during fruit ripening through an interaction with *SlSBP-CNR*. This prediction was experimentally validated, and it was demonstrated that SlmiR157 regulates the expression of *SlSBP-CNR* and thereby participates in the fine-tuning of the ripening process (Chen et al. 2015a). SlmiR156 on the other hand does not take part in ripening regulation, but on top of its role in fruit development (Silva et al. 2014), it has a function in fruit softening (Chen et al. 2015a). While *SlSBP-CNR* is to some extent regulated by SlmiR157, *SlRIN* on the other hand controls the expression of multiple miRNAs (Gao et al. 2015). Out of 33 identified miRNA families in the *rin* mutant, 14 of them were differentially regulated in ripening fruits of the mutant plants. *SlRIN* CArG-box binding sites were identified in four out of ten looked into promoters of SlmiRNA precursors. A ChiP-qPCR assay experimentally proved that indeed *SlRIN* binds to the promoter region of miR172a.

4.3 LncRNA Mediated Regulation

Besides small RNAs, long noncoding RNAs (lncRNAs) are also important for epigenetic regulation in plants. LncRNAs are defined as RNA molecules over 200 bp that do not contain an open reading frame coding for a polypeptide longer than 100 amino acids. Around 40,000 lncRNAs were identified in *Arabidopsis* (Liu et al. 2012; Wang et al. 2014). lncRNAs are transcribed by Pol II, Pol IV, and PolV polymerases. Pol II lncRNA transcripts have a 5′-cap and a 3′polyadenylated tail, similarly to mRNA. Long noncoding RNAs function as epigenetic regulators through various mechanisms. Functions as target mimics of miRNA were identified in *Arabidopsis* and Rice (Franco-Zorrilla et al. 2007; Wu et al. 2013); in this role, the lncRNA is a decoy that capture specific miRNA in place of the miRNA target gene, thus regulating the miRNA activity. Additional functions include posttranscriptional modification of transcription factors, regulation of mRNA alternative splicing, regulation of the Pol II transcription machinery, and working as enhancers or super-enhancers of mRNA transcription, and lncRNAs have a central role in the plant-specific RdDM epigenetic machinery [lncRNA roles in plants are reviewed in Chekanova (2015) and Liu et al. (2015b)]. Pol IV and Pol V transcripts are long noncoding RNAs (lncRNAs) essential to the RdDM machinery: Pol IV transcripts are transformed into double-stranded RNA (dsRNA) by RDR2 and broken down into siRNA by DCL3. They direct, through sequence complementarity to Pol V lncRNA, the DNA and histone epigenetic modifiers to their target genome location. Therefore, both types of lncRNA are involved in the direction of the RdDM complex to the genome's target sites. A G to C SNP mutation in the rice lncRNA, long-day-specific male fertility-associated RNA (LDMAR), was sufficient to affect the RNA secondary structure which in turn led to increased methylation in the promoter of LDMAR. Promoter methylation decreased the accumulation of LDMAR levels in anthers under long day condition, thus leading to male sterility (Ding et al. 2012). lncRNAs, like mRNAs, are differentially expressed depending on location and developmental stages, thus when aiming at systematic identification of expressed transcripts, it is important to carefully select the tissues under examination. From RNAseq data generated from rice anthers, pistils, seeds, and shoots, 2224 lncRNAs were identified (Zhang et al. 2014). Among them, the authors identified several lncRNAs acting as target mimics for miR160 and miR164. In addition, the T-DNA insertion mutant of lncRNA XLOC_057324 had earlier flowering and low seed set phenotypes, pointing to a role for that lncRNA in the formation of rice panicule and flower fertility. In the rice and maize cereals, lncRNAs are very probably contributing to agronomic traits because the combination of results from Genome Wide Association Studies (GWAS) with the position of about 29,000 lncRNAs in rice and maize showed that 234 SNPs associated with 34 morphological, developmental, and agronomical traits were mapping to lncRNA genomic positions (Wang et al. 2015a). In tomato, Wang et al. (2015b) generated RNAseq datasets from control and TYLC virus-infected leaves to identify lncRNAs involved with the defense of tomato against TYLCV. They predicted 1565 lncRNAs as potentially involved in TYLCV infection. Results from differential expression analysis were confirmed by qRT-PCR, and the implication of one lncRNA into the response to TYLCV virus infection was confirmed using a VIGS approach (Zhu et al. 2015), focused on the fruit ripening process. They compared lncRNAs identified in a RNAseq dataset from breaker stage (the transition from green to ripe tomato fruit) in Aisla Craig control plant to a RNAseq dataset from the same ripening stage but from the *ripening inhibitor* (*rin*) mutant. A total of 3679 lncRNAs were identified in these samples from which 677 were differentially expressed between the two conditions. The involvement of lncRNAs in the ripening transition process was validated for two lncRNAs (SllncRNA 1459 and SllncRNA1840) using the VIGS method to silence the target lncRNAs. VIGS plants for both lncRNAs showed delayed ripening compared to the control, thus validating the function of lncRNA in the ripening process. The mechanisms by which these lncRNAs operate to produce the observed phenotype remain to be deciphered. Using publicly available RNAseq datasets, Wang et al. (2016) looked into the evolution of lncRNAs by analyzing noncoding transcripts in both cultivated *Solanum lycopersicum* and the wild accessions *Solanum pimpinellifolium* and *Solanum pennellii*. The authors identified 413 lncRNAs from *S. lycopersicum* Heinz1706 and confirmed by qRT-PCR that the transcripts they identified match the expression profile of the RNAseq data. The datasets from Heinz1706 used in this study were generated by the tomato genome sequencing consortium. They cover a vast range of tomato tissues: from roots to fruits. Wang et al. (2016) analyzed the expression profile of 413 lncRNAs in developing flowers and fruits and confirmed with qRT-PCR that some lncRNAs are differentially regulated throughout fruit development, thus leading to the hypothesis that lncRNAs are involved in this process.

5 Seed Development

Seeds are the keystone of human development. Plants evolved several strategies and a wide range of adaptations to preserve successful germination of its genetic content and to conquer several different environments. Angiosperm seed development initiates with the double fertilization of the megagametophyte, where the pollen tube delivers two haploid sperm cells to the embryo sac. One sperm cell fuses with the haploid egg to generate a diploid embryo, and the other sperm cell fuses with the diploid central cell to form the triploid endosperm. The resulting embryo and endosperm are genetically identical except for their ploidy level with the endosperm having two maternal doses of the genome and one male dosage (reviewed in Bai and Settle 2014). The fertilized egg and central cell go on to form the embryo and the endosperm, respectively, by multiplying and expanding through several cell cycles. Core cell cycle factors play important roles in the regulation of the cell division cycle during seed development and its coordination with cell differentiation and maturation. Diverse aspects of the seed development such as seed dormancy and embryo and endosperm development involve epigenetic mechanisms (Kohler and Makarevich 2006; Wollmann and Berger 2012). Although seed development is regulated through physiological and transcriptional regulation, in this chapter we are going to focus only on the epigenetic aspects of seed development.

5.1 Seed Dormancy

Seed dormancy is a process that allows germination delay until a favorable environment arrives. Pieces of evidence for an epigenetic regulation of gene expression in controlling dormancy and germination in cereal seeds have emerged only recently. Genes associated with histone and chromatin structure are overrepresented among *loci* transcriptionally induced at the whole-seed level during germination of non-dormant barley seeds, specifically during the phase of late germination (An and Lin 2011). Moreover, the SET family—transcription factors that play role in histone methylation—are consistently expressed in the embryos during germination of non-dormant rice seeds (Malagnac et al. 2002; Xiao et al. 2003; Howell et al. 2008). Recent study analyzing the whole wheat seed transcription showed several genes activated during imbibition of after-ripened samples that were enriched in the chromatin assembly gene ontology (Gao et al. 2012). Such genes include those encoding for histone proteins such as H4, HTA11, HTA12, HTB11, HTB9, and FASCIATA 1 (FAS1), a histone-binding protein, which are important for nucleosome and chromatin formation, and thus gene expression regulation. Furthermore, orthologues of DNA methylation-related genes, including *CHROMOMETHYLASE 3* (*CMT3*) and *METHYLTRANSFERASE 1* (*MET1*), exhibit transcriptional induction in embedded, after-ripened dormant seeds. Such findings suggest an epigenetic role in regulating gene expression and modulating after-ripening-induced developmental switch of wheat seeds from dormant to non-dormant state (Fig. 2) (Gao and Ayele 2014). Additional studies are required to identify more dormancy-related epigenetic regulators and define how epigenetic mechanisms are involved in the control of wheat seed dormancy and germination.

In *Arabidopsis thaliana*, seed dormancy-specific genes include the *DELAY OF GERMINATION* (*DOG*) family. *DOG1* is expressed in seeds during the maturation stage; the transcript accumulates during seed maturation stage with peaks around 14–16 days after pollination (DAP) (Bentsink et al. 2006) and then is downregulated around 20% in freshly harvested seeds, vanishing during imbibition. Loss of function of *DOG1* results in the absence of dormancy (Bentsink et al. 2006). Wheat transcription factor Histone Binding Protein-1b (HBP-1b) displays the highest similarity with *Arabidopsis* DOG1 (Bentsink et al. 2006). The leucine zipper class transcription factor HBP-1b binds to the H3 hexamer motif ACGTCA in the promoter regions of wheat histone H3 loci (Mikami et al. 1989). This motif is required for transcription of wheat *H3 histone locus* (Nakayama et al. 1989).

A suitable candidate for a seed dormancy-imposing gene is *HUB1* (*histone mono-ubiquitination1*) since *hub1* seeds exhibit reduced dormancy (Liu et al.

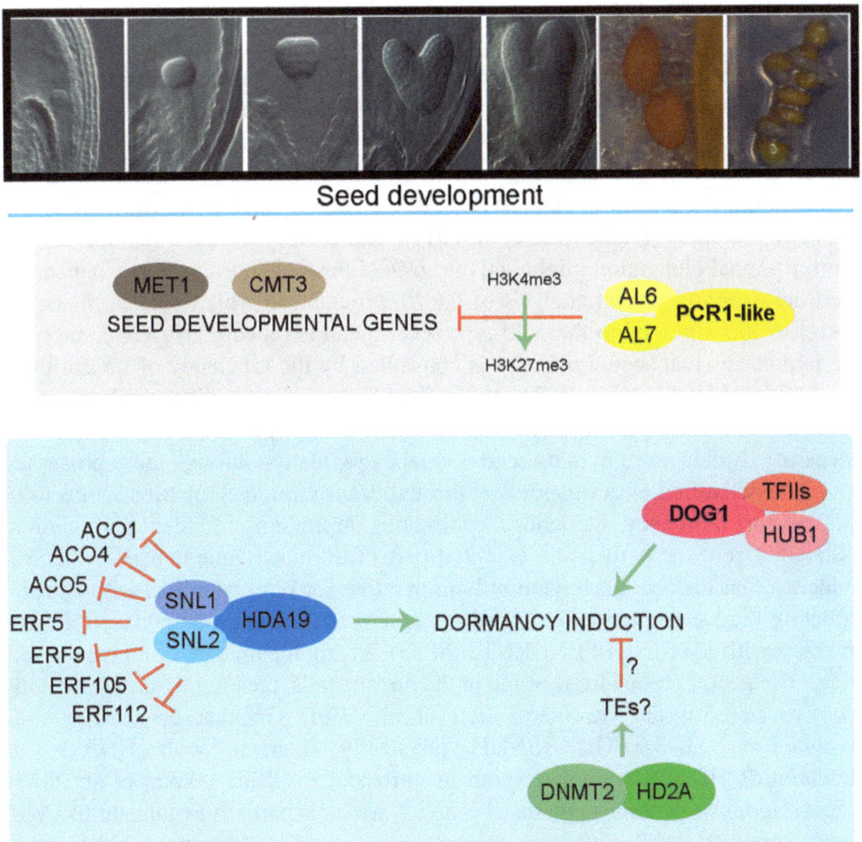

Fig. 2 Epigenetics of *Arabidopsis thaliana* seed development. Embryo developmental stages shown as octant stage, globular stage, early-heart stage, heart stage, torpedo stage, mature dry seed, and imbibed seeds, respectively. *CHROMOMETHYLASE 3* (*CMT3*) and *METHYLTRANSFERASE 1* (*MET1*) regulate developmental seeds genes by methylating DNA during embryogenesis. AL (*Alfin1-like*) proteins *AL6* and *AL7* interact with the Polycomb repressive complex 1 (PRC1)-like complex repressing seed developmental genes by switching from active H3K4me3 to inactive H3K27me3 marks of seed genes during seed germination. *DELAY OF GERMINATION1* (*DOG1*) targets *HUB1* (histone mono-ubiquitination 1) and *TFIIS* (transcription elongator factor IIS) controlling seed dormancy. *SIN3-LIKE1* (*SNL1*) and *SNL2* interact with *HDA19* (histone deacetylase19) positively regulating seed dormancy, targeting *1-AMINOCYCLOPROPANE-1-CARBOXYLATE OXIDASE 1* (*ACO1*), ACO4, and ACO5 and ethylene responsive genes, such as ETHYLENE RESPONSE FACTOR 9 (ERF9), ERF105, and ERF112, triggering seed dormancy by suppressing the ethylene pathway, affecting positively the seed germination. HISTONE DEACETYLASE2A (HD2A) and DNA METHYLTRANSFERASE2 (DNMT2) are possibly working together in the germination process affecting early embryogenesis stages

2007). *ABA INSENSITIVE 4* (*ABI4*), *DOG1*, *NINE-CIS-EPOXYCAROTENOID DIOXYGENASE 9* (*NCED9*), and other genes have been identified as possible targets of *HUB1* (Liu et al. 2007). *TFIIS* (transcription elongator factor IIS) and *HUB1*, two positive regulators of transcription, are induced during the same stages of seed maturation (~18–19 DAP). There is a significant overlap of differentially expressed genes in *tfIIs* and *hub1* mutants. This finding suggests that *TFIIS* and *HUB1* might share common targets. One of the genes commonly downregulated in both two mutants is *DOG1* (Liu et al. 2011). Therefore, chromatin remodeling and transcriptional elongation might activate *DOG1* through a primary mechanism for seed dormancy. A recent analysis of the *tfIIs* mutant, in which seed dormancy is decreased but returned to the wild-type level by an extra copy of *DOG1*, supports the hypothesis that seed dormancy is controlled by the efficiency of transcription elongation of *DOG1* (Fig. 2) (Mortensen and Grasser 2014). More analyses of the specific targets of epigenetic modification and transcriptional elongation will be necessary to shed more light on seed dormancy regulation through these processes.

While activation of dormancy *loci* through transcription elongation seems to be critical for dormancy induction, continuous repression of seed germination-associated genes is perhaps an essential part of dormancy maintenance. There is evidence that histone deacetylation is imperative for repression of loci positively affecting seed germination. In mammals and yeast, histone deacetylase (HDAC) interacts with SWI-INDEPENDENT3 (SIN3), an amphipathic helix-repeat protein, to remove acetyl groups from lysine in the histone tails, creating a transcriptionally inactive state of the chromatin (Lai et al. 2001; Grzenda et al. 2009). In *Arabidopsis*, SIN3-LIKE1 (SNL1) physically interacts with HDA19, an *Arabidopsis* HDAC orthologue, both in vitro and in planta (Wang et al. 2013). The *Arabidopsis* genome contains also *SNL2*, which is partially redundant to *SNL1*. Seeds from *snl1*/*snl2* double mutant exhibit decreased dormancy. A decreased dormancy phenotype is also observed in *hda19* mutant seeds (Wang et al. 2013). These data imply that *SNLs* and *HDA19* are seed dormancy positive regulators. It seems that suitable repression of SNL-HDA19 complex targets, which are most likely germination-inducing *loci*, through histone deacetylation, is essential for typical seed dormancy. Acetylation of H3K9/18 and H3K14 is increased in the *snl1*/*snl2* double mutant (Wang et al. 2013), corroborating that in wild-type seeds, the SIN3-HDAC complex deacetylates histones and therefore adds repressive marks on the chromatin (Richon and O'Brien 2002). Global gene expression analysis of *snl1*/*snl2* double mutant and wild-type seeds identified possible targets for *SNL-HDA19* such as *1-AMINOCYCLOPROPANE-1-CARBOXYLATE OXIDASE 1* (*ACO1*), *ACO4*, and *ACO5* and ethylene responsive genes, such as *ETHYLENE RESPONSE FACTOR 9* (*ERF9*), *ERF105*, and *ERF112* (Wang et al. 2013). Quantitative PCR combined with chromatin immunoprecipitation employing H3K9/18 acetylation-specific antibodies confirmed that *ACOs* and *ERFs* genes were indeed hyper-acetylated in the mutant, mostly in the promoter region, although hyperacetylation was also found in coding regions (Wang et al. 2013). These results indicate that *SNL-HDA19* triggers seed dormancy by suppressing the ethylene pathway, affecting positively the seed germination in *Arabidopsis* (Chiwocha et al. 2005; Arc et al. 2013) (Fig. 2).

Another family of plant histone deacetylase, the HD2s (HD2A, HD2B, HD2C, HD2D), are probably part of the seed dormancy-regulated pathways. HD2A can mediate transcriptional repression (Ueno et al. 2007) and is correlated with early stages of somatic embryo development (Zhou et al. 2004). HD2s and DNMT2 (an intriguing enzyme that holds a DNA methylation motif, but methylates specific tRNAs instead) proteins interact in *Arabidopsis* nucleus; then one of the hypothesis is that they are working together in epigenetic pathways, playing role in plant development (Song et al. 2010). Unpublished data from our group on *Arabidopsis hd2a/dnmt2* double mutants show a consistent difference at the germination and seedling growing rates in the first 48 h compared to wild-type seeds, where double mutant seeds germinate earlier and seedlings grow faster than the wild type. It seems that HD2A and DNMT2 work together, most likely as a complex (Song et al. 2010), in the germination process affecting male gametogenesis and/or early embryogenesis stages (Alves 2015). This conclusion is supported by data on mice, suggesting that DNMT2 is required for male gametogenesis (Kiani et al. 2013). Furthermore, HD2A is already known to be a seed germination and fine-tune growing regulator in *Arabidopsis* (Zhou et al. 2004; Colville et al. 2011). Further studies are necessary to unravel the epigenetic mechanisms by which DNMT2-HD2A complex regulates seed germination and early seedling development.

Histone modification may be partially inherited through cell division and epigenetic reprogramming should take place at fertilization, but these processes remain far from being completely understood. DNMT2 targets transcripts other than tRNAs (Alves 2015), and it could work as a cofactor together with HD2A to remove histone acetylation and, as a result, some *loci* that should be repressed during seed germination may be activated or partially activated in the *hd2a/dnmt2 Arabidopsis* double mutant. Moreover, methylome profiling of pollen indicates dynamic DNA methylation changes during male gametogenesis, but information regarding methylation enzymes acting at this stage is unknown. Non-CG methylation increases inn pollen vegetative cells on transposable element (TE)-rich regions, probably to regulate these elements (Borges et al. 2012). Normally, LTR (long terminal repeats) retrotransposons are regulated by RdDM pathway. Low levels of siRNAs associated with LTR retrotransposons were found in the pollen vegetative cell (Slotkin et al. 2009). However, 21-nucleotide (nt) siRNAs are found at this stage and not 24-nt siRNAs. The current model is that noncanonical RdDM pathways take place at this stage to control these elements.

5.2 Embryo–Endosperm Interaction

In *Arabidopsis* endosperm, genes and TEs are regulated by both DNA and histone (H3K37me3) methylation (Schmidt et al. 2013), with substantial variation between endosperm and embryo tissues (Gehring et al. 2009). In rice, endosperm hypomethylation occurs in all sequence contexts (CG and non-CG), although CG methylation is not similar to CHG and CHH, which are hypomethylated similarly across the genome (Zemach et al. 2010). For seed viability, MET1 and CMT3 activity is required during *Arabidopsis* embryogenesis (Xiao et al. 2006), leading to preferential maternal hypomethylation in the endosperm, while paternal methylated alleles are maintained. However, the function of the remaining methylated *loci* is largely unknown (Zhang and Xue 2013).

Epigenetic regulation is crucial for leading tissue differentiation into distinct *primordium* cell lineages and driving inheritance of each transcriptional programmed through mitosis at early stages of embryo development (Bantignies and Cavalli 2006). In the egg cell, MET1 is expressed following gametogenesis, yet the developing embryo, endosperm, and seed coat also contain its transcripts (Schmidt et al. 2013). The jmjC domain-containing histone demethylase from BcJMJ30 in *Brassica rapa* is associated with pollen development and fertilization (Li et al. 2012). Although *Helianthus LEAFY COTYLEDON1-LIKE* (*HaL1L*) is involved in early stages of zygotic and somatic embryogenesis, with multiplexed transcriptional regulation by DNA methylation, TFs (Transcription Factors), auxin, and ABA (Salvini et al. 2012).

The *MATERNALLY EXPRESSED LOCUS 1* (*MEG1*) in maize is expressed only in the basal nutrient transfer region of the endosperm (Gutiérrez-Marcos et al. 2004), where the genomic imprinting (an epigenetic event that silence one allele from one of the parents; see next section) of *MEG1* supports nutrient transfer from endosperm to the newly developing embryo (Costa et al. 2012). AL (Alfin1-like) proteins are PHD-containing proteins, and there are seven AL proteins in *Arabidopsis*, identified by AL1–AL7 (Lee et al. 2009; Molitor et al. 2014). The AL proteins are named following their homologue Alfin1 in alfalfa, which participates in salt tolerance (Winicov 2000). Functional studies have uncovered that AL6 and AL7 interact with the Polycomb repressive complex 1 (PRC1)-like complex (a complex that contains Polycomb group-like/PcG-like proteins) responsible for the methylation of Lys 27 of histone H3 (Deleris et al. 2012), in *Arabidopsis* to repress seed developmental genes by switching from active H3K4me3 to inactive H3K27me3 marks of seed genes during seed germination and early seedling growth. Also, this function depends on the interaction between AL6/AL7 and H3K4me3 (Molitor et al. 2014). A delay in seed germination under osmotic treatments but not under normal conditions is manifested in Al6 and Al7 double mutants, in agreement with the function of their homologue Alfin1 (Molitor et al. 2014; Winicov 2000). However, the single mutants of *al6* or *al7* show a normal phenotype under any conditions, indicating that AL6 and AL7 act redundantly in seed germination (Molitor et al. 2014).

5.3 Genomic Imprinting

Imprinting is an epigenetic phenomenon by which one of the alleles is silenced through methylation and histone modification mechanisms (Raissig et al. 2011). Imprinted alleles can be inherited maternally or paternally, and the imprinting takes place at the germline and is maintained through mitosis in somatic cells. It is an important mechanism to ensure the correct information transmission to the offspring. Imprinting also guarantees that TEs stay epigenetically silenced during reprogramming of plant gametogenesis, facilitating seed germination events (Wollmann and Berger 2012). Exposure to pathogens can initiate differential 5mC (5-methylcytosine) patterning, activating *NON-EXPRESSOR OF PR GENE* (*NPR1*), a defense regulatory gene (Dowen et al. 2012; Luna and Ton 2012). The transgenerational genomic imprinting of *NPR1* is probably due to posttranslational histone modifications and expression of RNA Polymerase V acting along with siRNAs to recruit methylation machinery (Luna and Ton 2012; You et al. 2013). In developing embryo genomic imprinting, sRNAs produced maternally in plant reproductive tissue can be mobile and may target specific genes, providing the first evidence for a link between genomic imprinting and RNA silencing in plants (Gutierrez-Marcos et al. 2012; Mosher et al. 2009). A variety of microRNAs, including at least four associated with nutrient homeostasis (miR169, miR395, miR398, and miR399), are indeed mobile and graft transmissible and detected in the phloem (Marín-González and Suárez-López 2012). These findings reinforced that maternally produced small RNAs (miRNAs or siRNAs) may be present in the next generation. sRNA-based regulation found in fertilization can take place during seed maturation and possibly during seed dormancy as well (Mosher et al. 2009). It is possible that the female sRNAs may influence gene regulation during germination to assist seedling establishment. This female-specific genomic imprinting mechanism may have evolved from the advantage fitness from the maternal niche germinated offspring (Gorecki et al. 2012). Throughout plant embryogenesis, hypomethylation is less prominent when compared to mammalian systems, with a higher proportion of parental DNA methylation events carried to the following generation (Reinders et al. 2009). An example is the different methylated profile in the *Arabidopsis PHE1* (*PHERES1*), in which the male *PHE1* allele is methylated, and the female allele is hypomethylated (Kohler and Makarevich 2006; Makarevich et al. 2008).

A few years back, plant imprinting was believed to occur only in the triploid endosperm, so, gymnosperms were presumed to lack imprinting mechanisms (Garnier et al. 2008). Nonetheless, more recently, it has been shown that genomic imprinting can occur in angiosperm and gymnosperm embryos (Scholten 2010). The exposure to different temperatures during embryo development can store epigenetic memory during embryogenesis, fixing epigenetic marks before seed maturation, leading to modified germination time and seedling development in the gymnosperm Norway spruce (Yakovlev et al. 2010). The epigenetic memory in long-lived plant species may confer adaptive plasticity to environmental drift in a single generation, with significant consequences for perennial and clonally propagated crops (Bloomfield et al. 2014).

6 Conclusions and Future Prospects

The epigenetic mechanism most amenable to plant breeding programs is gene expression regulation through DNA methylation. Epigenetic variability in crop plants can be either induced through chemical treatment using methyltransferase inhibitors or induced by exposure to specific stressful growing environments. Individual plants showing desired phenotypes as well as stable methylation profiles can be selected to be part of selection programs (Rodriguez Lopez and Wilkinson 2015). Thus, fixed epimutations as described for the *Cnr* mutant are not the only source of epigenetic diversity. Using a population of isogenic *Arabidopsis* lines that segregate for differentially methylated regions, Cortijo et al. (2014) showed that two agronomical traits, flowering time and primary root length, are controlled by epigenetic quantitative trait loci. Thus, forward epigenetic approaches could be actually implemented as part of crop breeding for the improvement of significant traits.

The *MutS* HOMOLOGUE1 (*MSH1*) gene is known in *Arabidopsis* to influence plant growth behavior. *msh1* mutant plants have affected vigor and development reprogramming linked to altered genome methylation. The increased plant vigor is characterized by rapid growth and earlier flowering, and greater aboveground biomass was also identified in tomato plants silenced for the *MSH1* gene (Yang et al. 2015). In tomato, the most relevant phenotype in a plant breeding perspective was the increased flower and fruit set that resulted in increased yield. *MSH1* RNAi plants were crossed with the wild-type Rutgers parent and −/− plants not carrying the RNAi transgene were selected as epi-lines. The enhanced hybrid vigor and fruit yield was increased until the epiF4 generation. Such heritable increased methylation in *msh1* plants was observed in multiple plant species (*Arabidopsis*, tomato, sorghum). This work, together with the studies demonstrating the role of methylation status of ripening genes, proves the relevance of considering the methylome as part of breeding programs.

Methylations and miRNAs are both involved in the regulation of the tomato fruit ripening process. Another group of epigenetic regulators involved in relevant traits is the lncRNAs. Zhu et al. (2015) showed that lncRNAs are another layer of regulation to the ripening process. In addition, lncRNAs are involved in tomato defense against pathogens such as TYLCV (Wang et al. 2015b) and *phytophthora infestans* (Cui et al. 2017). Together, the works on the identification of lncRNAs implicated in plant quality traits and disease resistance traits have permitted the identification of thousands of lncRNAs in a large array of tissues, developmental stages, and disease treatments. lncRNAs, as genes, are being mapped to the genome sequence and, therefore like genes, can and should be looked at in association studies, forward and reverse genetic approaches. Discovery of tomato lncRNAs is at its early stage and much of their functions remains to be identified. One particular type of lncRNA is of interest for future fundamental and applied studies: lncRNAs coding for micropeptides (miPEPs). In *Medicago truncatula* and *Arabidopsis thaliana*, miPEPs coded by pri-miRNA of miR171b and miR165a are involved in root development through positive feedback regulation of their own pri-miRNAs (Lauressergues et al. 2015). Treatment with 0.1 μM of miPEP172c increased soybean root nodulation through the stimulation of miR172c and thus *AP2* downregulation (Couzigou et al. 2016). This illustrates the potential of this novel type of epigenetic regulator to regulate agronomical traits and to be able to regulate gene networks through a simple treatment with a synthetic micropeptide.

References

Alves CS (2015) Análise funcional do papel da enzima DNA metiltransferase2 (DNMT2) no desenvolvimento e resposta à estresses e identificação e caracterização de fragmentos derivados de tRNA (tRFs) em *Arabidopsis thaliana*. Thesis dissertation, State University of Sao Paulo, Botucatu, Sao Paulo

Ampomah-Dwamena C, Morris BA, Sutherland P et al (2002) Down-regulation of TM29, a tomato SEPALLATA homolog, causes parthenocarpic fruit development and floral reversion. Plant Physiol 130:605–617

An Y-Q, Lin L (2011) Transcriptional regulatory programs underlying barley germination and regulatory functions of Gibberellin and abscisic acid. BMC Plant Biol 11:105. doi: 10.1186/1471-2229-11-105

Arc E, Sechet J, Corbineau F et al (2013) ABA crosstalk with ethylene and nitric oxide in seed dormancy and germination. Front Plant Sci 4:63

Aufsatz W, Mette MF, van der Winden J et al (2002) HDA6, a putative histone deacetylase needed to enhance DNA methylation induced by double-stranded RNA. EMBO J 21:6832–6841

Aufsatz W, Stoiber T, Rakic B et al (2007) *Arabidopsis* histone deacetylase 6: a green link to RNA silencing. Oncogene 26:5477–5488

Bai F, Settles AM (2014) Imprinting in plants as a mechanism to generate seed phenotypic diversity. Front Plant Sci 5:780

Baker CC, Sieber P, Wellmer F et al (2005) The early extra petals1 mutant uncovers a role for microRNA miR164c in regulating petal number in *Arabidopsis*. Curr Biol 15:303–315

Bantignies F, Cavalli G (2006) Cellular memory and dynamic regulation of polycomb group proteins. Curr Opin Cell Biol 18:275–283

Bastow R, Mylne JS, Lister C et al (2004) Vernalization requires epigenetic silencing of FLC by histone methylation. Nature 427:164–167

Bernatavichute YV, Zhang X, Cokus S et al (2008) Genome-wide association of histone H3 lysine nine methylation with CHG DNA methylation in Arabidopsis thaliana. PLoS One 3:e3156. doi:10.1371/journal.pone.0003156

Bentsink L, Jowett J, Hanhart CJ et al (2006) Cloning of DOG1, a quantitative trait locus controlling seed dormancy in *Arabidopsis*. Proc Natl Acad Sci USA 103:17042–17047

Bloomfield JA, Rose TJ, King GJ (2014) Sustainable harvest: managing plasticity for resilient crops. Plant Biotechnol J 12:517–533

Bokszczanin KL, Krezdorn N, Fragkostefanakis S et al (2015) Identification of novel small ncRNAs in pollen of tomato. BMC Genomics 16:714

Borges F, Calarco JP, Martienssen R (2012) Reprogramming the epigenome in Arabidopsis pollen. Cold Spring Harb Symp Quant Biol 77:1–5. doi:10.1101/sqb.2013.77.014969

Borges F, Pereira PA, Slotkin RK et al (2011) MicroRNA activity in the *Arabidopsis* male germline. J Exp Bot 62:1611–1620

Bowman JL, Smyth DR, Meyerowitz EM (1991) Genetic interactions among floral homeotic genes of *Arabidopsis*. Development 112:1–20

Calarco JP, Borges F, Donoghue MT et al (2012) Reprogramming of DNA methylation in pollen guides epigenetic inheritance via small RNA. Cell 151:194–205

Cao D, Wang J, Ju Z et al (2016) Regulations on growth and development in tomato cotyledon, flower and fruit via destruction of miR396 with short tandem target mimic. Plant Sci 247:1–12

Chekanova JA (2015) Long non-coding RNAs and their functions in plants. Curr Opin Plant Biol 27:207–216

Chen X (2004) A microRNA as a translational repressor of APETALA2 in *Arabidopsis* flower development. Science 303:2022–2025

Chen W, Kong J, Lai T et al (2015a) Tuning LeSPL-CNR expression by SlymiR157 affects tomato fruit ripening. Sci Rep 5:7852

Chen W, Kong J, Qin C et al (2015b) Requirement of CHROMOMETHYLASE3 for somatic inheritance of the spontaneous tomato epimutation colourless non-ripening. Sci Rep 5:9192

Chiwocha SDS, Cutler AJ, Abrams SR et al (2005) The etr1-2 mutation in Arabidopsis thaliana affects the abscisic acid, auxin, cytokinin and gibberellin metabolic pathways during maintenance of seed dormancy, moistchilling and germination. Plant J 42:35–48. doi:10.1111/j.1365-313X.2005.02359.x

Chodavarapu RK, Feng S, Bernatavichute YV et al (2010) Relationship between nucleosome positioning and DNA methylation. Nature 466:388–392

Cigliano RA, Sanseverino W, Cremona G, Ercolano MR, Conicella C, Consiglio FM (2013) Genome-wide analysis of histone modifiers in tomato: gaining an insight into their developmental roles. BMC Genomics 14

Colville A, Alhattab R, Hu M et al (2011) Role of HD2 genes in seed germination and early seedling growth in Arabidopsis. Plant Cell Rep 30:1969–1979. doi:10.1007/s00299-011-1105-z

Cortijo S, Wardenaar R, Colome-Tatche M, Gilly A, Etcheverry M, Labadie K, Caillieux E, Hospital F, Aury JM, Wincker P et al (2014) Mapping the epigenetic basis of complex traits. Science 343:1145–1148

Costa LM, Yuan J, Rouster J et al (2012) Maternal control of nutrient allocation in plant seeds by genomic imprinting. Curr Biol 22:160–165

Couzigou JM, Andre O, Guillotin B, Alexandre M, Combier JP (2016) Use of microRNAencoded peptide miPEP172c to stimulate nodulation in soybean. New Phytol 211:379–381

Cui J, Luan Y, Jiang N, Bao H, Meng J (2017) Comparative transcriptome analysis between resistant and susceptible tomato allows the identification of lncRNA16397 conferring resistance to Phytophthora infestans by co-expressing glutaredoxin. Plant J 89:577–589

Damodharan S, Zhao D, Arazi T (2016) A common miRNA160-based mechanism regulates ovary patterning, floral organ abscission and lamina outgrowth in tomato. Plant J 86:458–471

Deleris A, Stroud H, Bernatavichute Y et al (2012) Loss of the DNA methyltransferase MET1 Induces H3K9 hypermethylation at PcG target genes and redistribution of H3K27 trimethylation to transposons in *Arabidopsis thaliana*. PLoS Genet 8:e1003062

Din M, Barozai MY (2014) Profiling microRNAs and their targets in an important fleshy fruit: tomato (*Solanum lycopersicum*). Gene 535:198–203

Ding J, Lu Q, Ouyang Y et al (2012) A long noncoding RNA regulates photoperiod-sensitive male sterility, an essential component of hybrid rice. Proc Natl Acad Sci USA 109:2654–2659

Dowen RH, Pelizzola M, Schmitz RJ et al (2012) Widespread dynamic DNA methylation in response to biotic stress. Proc Natl Acad Sci USA 109:E2183–E2191

Fan HH, Wei J, Li TC et al (2013) DNA methylation alterations of upland cotton (*Gossypium hirsutum*) in response to cold stress. Acta Physiol Plant 35:2445–2453

Franco-Zorrilla JM, Valli A, Todesco M et al (2007) Target mimicry provides a new mechanism for regulation of microRNA activity. Nat Genet 39:1033–1037

Gao F, Ayele BT (2014) Functional genomics of seed dormancy in wheat: advances and prospects. Front Plant Sci 5:458. doi:10.3389/fpls.2014.00458

Gao F, Jordan MC, Ayele BT (2012) Transcriptional programs regulating seed dormancy and its release by after-ripening in common wheat (*Triticum aestivum* L.) Plant Biotechnol J 10:465–476

Gao C, Ju Z, Cao D et al (2015) MicroRNA profiling analysis throughout tomato fruit development and ripening reveals potential regulatory role of RIN on microRNAs accumulation. Plant Biotechnol J 13:370–382

Garnier O, Laouiellé-Duprat S, Spillane C (2008) Genomic imprinting in plants. Adv Exp Med Biol 626:89–100

Gehring M, Bubb KL, Henikoff S (2009) Extensive demethylation of repetitive elements during seed development underlies gene imprinting. Science 324:1447–1451

Gorecki MJ, Long RL, Flematti GR et al (2012) Parental environment changes the dormancy state and karrikinolide response of *Brassica tournefortii* seeds. Ann Bot 109:1369–1378

Grove A (2011) Functional evolution of bacterial histone-like HU proteins. Curr Issues Mol Biol 13:1–12

Grzenda A, Lomberk G, Zhang J-S, Urrutia R (2009) Sin3: master scaffold and transcriptional corepressor. Biochim Biophys Acta 1789:443–450. doi:10.1016/j.bbagrm.2009.05.007

Gutiérrez-Marcos JF, Costa LM, Biderre-Petit C et al (2004) Maternally expressed gene1 Is a novel maize endosperm transfer cell-specific gene with a maternal parent-of-origin pattern of expression. Plant Cell 16:1288–1301

Gutierrez-Marcos JF, Constância M, Burton GJ (2012) Maternal to offspring resource allocation in plants and mammals. Placenta 33:e3–e10

Guzy-Wrobelska J, Filek M, Kaliciak A et al (2013) Vernalization and photoperiod-related changes in the DNA methylation state in winter and spring rapeseed. Acta Physiol Plant 35:817–827

Haag JR, Ream TS, Marasco M et al (2012) In vitro transcription activities of Pol IV, Pol V, and RDR2 reveal coupling of Pol IV and RDR2 for dsRNA synthesis in plant RNA silencing. Mol Cell 48:811–818

Hong Y, Jackson S (2015) Floral induction and flower formation—the role and potential applications of miRNAs. Plant Biotechnol J 13:282–292

Howell KA, Narsai R, Carroll A et al (2008) Mapping metabolic and transcript temporal switches during germination in rice highlights specific transcription factors and the role of RNA instability in the germination process. Plant Physiol 149:961–980

Huang J, Li Z, Zhao D (2016) Deregulation of the OsmiR160 target gene OsARF18 causes growth and developmental defects with an alteration of auxin signaling in rice. Sci Rep 6:29938

Jacobsen SE, Meyerowitz EM (1997) Hypermethylated SUPERMAN epigenetic alleles in *Arabidopsis*. Science 277:1100–1103

Jacobsen SE, Sakai H, Finnegan EJ et al (2000) Ectopic hypermethylation of flower-specific genes in *Arabidopsis*. Curr Biol 10:179–186

Karlova R, van Haarst JC, Maliepaard C et al (2013) Identification of microRNA targets in tomato fruit development using high-throughput sequencing and degradome analysis. J Exp Bot 64:1863–1878

Kiani J, Rassoulzadegan M (2013) A load of small RNAs in the sperm—how many bits of hereditary information? Cell Res 23:18–19. doi:10.1038/cr.2012.181

Kim J, Jung JH, Reyes JL et al (2005) microRNA-directed cleavage of ATHB15 mRNA regulates vascular development in *Arabidopsis* inflorescence stems. Plant J 42:84–94

Klee HJ, Giovannoni JJ (2011) Genetics and control of tomato fruit ripening and quality attributes. Annu Rev Genet 45:41–59

van der Knaap E, Chakrabarti M, Chu YH et al (2014) What lies beyond the eye: the molecular mechanisms regulating tomato fruit weight and shape. Front Plant Sci 5:227

Köhler C, Makarevich G (2006) Epigenetic mechanisms governing seed development in plants. EMBO Rep 7:1223–1227

Krogan NT, Hogan K, Long JA (2012) APETALA2 negatively regulates multiple floral organ identity genes in *Arabidopsis* by recruiting the co-repressor TOPLESS and the histone deacetylase HDA19. Development 139:4180–4190

Kumar R, Chauhan PK, Khurana A (2016) Identification and expression profiling of DNA methyltransferases during development and stress conditions in Solanaceae. Funct Integr Genomics 16:513–528

Lai A, Kennedy BK, Barbie DA et al (2001) RBP1 recruits the mSIN3-histone deacetylase complex to the pocket of retinoblastoma tumor suppressor family proteins found in limited discrete regions of the nucleus at growth arrest. Mol Cell Biol 21:2918–2932

Latzel V, Zhang Y, Karlsson Moritz K et al (2012) Epigenetic variation in plant responses to defence hormones. Ann Bot 110:1423–1428. doi:10.1093/aob/mcs088

Laufs P, Peaucelle A, Morin H et al (2004) MicroRNA regulation of the CUC genes is required for boundary size control in *Arabidopsis* meristems. Development 131:4311–4322

Lauressergues D, Couzigou JM, Clemente HS, Martinez Y, Dunand C, Becard G, Combier JP (2015) Primary transcripts of microRNAs encode regulatory peptides. Nature 520:90–93

Law JA, Jacobsen SE (2010) Establishing, maintaining and modifying DNA methylation patterns in plants and animals. Nat Rev Genet 11:204–220

Lee DY, An G (2012) Two AP2 family genes, supernumerary bract (SNB) and Osindeterminate spikelet 1 (OsIDS1), synergistically control inflorescence architecture and floral meristem establishment in rice. Plant J 69:445–461

Lee WY, Lee D, Chung WI et al (2009) *Arabidopsis* ING and Alfin1-like protein families localize to the nucleus and bind to H3K4me3/2 via plant homeodomain fingers. Plant J 58:511–524

Li X, Zhu J, Hu F et al (2012) Single-base resolution maps of cultivated and wild rice methylomes and regulatory roles of DNA methylation in plant gene expression. BMC Genomics 13:300

Liu Y, Koornneef M, Soppe WJJ (2007) The absence of histone H2B monoubiquitination in the *Arabidopsis* hub1 (rdo4) mutant reveals a role for chromatin remodeling in seed dormancy. Plant Cell 19:433–444

Liu X, Huang J, Wang Y et al (2010) The role of floral organs in carpels, an *Arabidopsis* loss-of-function mutation in MicroRNA160a, in organogenesis and the mechanism regulating its expression. Plant J 62:416–428

Liu Y, Geyer R, van Zanten M et al (2011) Identification of the *Arabidopsis* reduced dormancy 2 gene uncovers a role for the polymerase associated factor 1 complex in seed dormancy. PLoS One 6:1–8

Liu J, Jung C, Xu J et al (2012) Genome-wide analysis uncovers regulation of long intergenic noncoding RNAs in *Arabidopsis*. Plant Cell 24:4333–4345

Liu N, Wu S, Van Houten J et al (2014) Down-regulation of AUXIN RESPONSE FACTORS 6 and 8 by microRNA 167 leads to floral development defects and female sterility in tomato. J Exp Bot 65:2507–2520

Liu R, How-Kit A, Stammitti L et al (2015a) A DEMETER-like DNA demethylase governs tomato fruit ripening. Proc Natl Acad Sci USA 112:10804–10809

Liu X, Hao L, Li D et al (2015b) Long non-coding RNAs and their biological roles in plants. Genomics Proteomics Bioinformatics 13:137–147

Luna E, Ton J (2012) The epigenetic machinery controlling transgenerational systemic acquired resistance. Plant Signal Behav 7:615–618

Makarevich G, Villar CBR, Erilova A et al (2008) Mechanism of PHERES1 imprinting in *Arabidopsis*. J Cell Sci 121:906–912

Malagnac F, Bartee L, Bender J (2002) An *Arabidopsis* SET domain protein required for maintenance but not establishments of DNA methylation. EMBO J 21:6842–6852

Manning K, Tor M, Poole M et al (2006) A naturally occurring epigenetic mutation in a gene encoding an SBP-box transcription factor inhibits tomato fruit ripening. Nat Genet 38:948–952

Marín-González E, Suárez-López P (2012) "And yet it moves": cell-to-cell and long-distance signaling by plant microRNAs. Plant Sci. 196:18–30

Martel C, Vrebalov J, Tafelmeyer P et al (2011) The tomato MADS-box transcription factor RIPENING INHIBITOR interacts with promoters involved in numerous ripening processes in a COLORLESS NONRIPENING-dependent manner. Plant Physiol 157:1568–1579

Matzke MA, Mosher RA (2014) RNA-directed DNA methylation: an epigenetic pathway of increasing complexity. Nat Rev Genet 15:394–408

May O (2010) DNA methylation: fingerprints of the (epi)genome. In: Epigenetics and gene regulation, vol 8. Cayman Chemical, Ann Arbor, pp 22–23

Mikami K, Takase H, Tabata T et al (1989) Multiplicity of the DNA-binding protein HBP-1 specific to the conserved hexameric sequence ACGTCA in various plant gene promoters. FEBS Lett 256:67–70

Molitor AM, Bu Z, Yu Y et al (2014) *Arabidopsis* AL PHD-PRC1 complexes promote seed germination through H3K4me3-to-H3K27me3 chromatin state switch in repression of seed developmental genes. PLoS Genet 10:e:1004091

Mortensen SA, Grasser KD (2014) The seed dormancy defect of *Arabidopsis* mutants lacking the transcript elongation factor TFIIS is caused by reduced expression of the DOG1 gene. FEBS Lett 588:47–51

Mosher RA, Melnyk CW, Kelly KA et al (2009) Uniparental expression of PolIV-dependent siRNAs in developing endosperm of *Arabidopsis*. Nature 460:283–286

Moxon S, Jing R, Szittya G et al (2008) Deep sequencing of tomato short RNAs identifies microRNAs targeting genes involved in fruit ripening. Genome Res 18:1602–1609

Nair SK, Wang N, Turuspekov Y et al (2010) Cleistogamous flowering in barley arises from the suppression of microRNA-guided HvAP2 mRNA cleavage. Proc Natl Acad Sci USA 107:490–495

Nakayama T, Ohtsubo N, Mikami K et al (1989) Cisacting sequences that modulate transcription of wheat histone H3 and 3'processing of H3 premature mRNA. Plant Cell Physiol 30:825–832

Palatnik JF, Wollmann H, Schommer C et al (2007) Sequence and expression differences underlie functional specialization of *Arabidopsis* microRNAs miR159 and miR319. Dev Cell 13:115–125

Pan Y, Wang W, Zhao X et al (2011) DNA methylation alterations of rice in response to cold stress. POJ 4:364–369

Raissig MT, Baroux C, Grossniklaus U (2011) Regulation and flexibility of genomic imprinting during seed development. Plant Cell 23:16–26

Reinders J, Wulff BBH, Mirouze M et al (2009) Compromised stability of DNA methylation and transposon immobilization in mosaic *Arabidopsis* epigenomes. Genes Dev 23:939–950

Richon VM, O'brien JP (2002) Histone deacetylase inhibitors: a new class of potential therapeutic agents for cancer treatment. Clin Cancer Res 8:662–664

Rodriguez Lopez CM, Wilkinson MJ (2015) Epi-fingerprinting and epi-interventions for improved crop production and food quality. Front Plant Sci 6:397

Rubio-Somoza I, Weigel D (2011) MicroRNA networks and developmental plasticity in plants. Trends Plant Sci 16:258–264

Salvini M, Sani E, Fambrini M et al (2012) Molecular analysis of a sunflower gene encoding an homologous of the B subunit of a CAAT binding factor. Mol Biol Rep 39:6449–6465

Sandman K, Reeve JN (2006) Archaeal histones and the origin of the histone fold. Curr Opin Microbiol 9:520–525

Schmidt A, Wöhrmann HJP, Raissig MT et al (2013) The polycomb group protein MEDEA and the DNA methyltransferase MET1 interact to repress autonomous endosperm development in *Arabidopsis*. Plant J 73:776–787

Scholten S (2010) Genomic imprinting in plant embryos. Epigenetics 5:455–459

Schommer C, Bresso EG, Spinelli LS et al (2012) Role of microRNA miR319 in plant development. In: Sunkar R (ed) MicroRNAs in plant development and stress responses. Springer, Berlin, pp 29–47

Sheldon CC, Jean Finnegan E, James Peacock W et al (2009) Mechanisms of gene repression by vernalization in *Arabidopsis*. Plant J 59:488–498

Sherman JD, Talbert LE (2002) Vernalization-induced changes of the DNA methylation pattern in winter wheat. Genome 45:253–260

Sieber P, Wellmer F, Gheyselinck J et al (2007) Redundancy and specialization among plant microRNAs: role of the MIR164 family in developmental robustness. Development 134:1051–1060

Silva GF, Silva EM, Azevedo MD et al (2014) microRNA156-targeted SPL/SBP box transcription factors regulate tomato ovary and fruit development. Plant J 78:604–618

Slotkin RK, Vaughn M, Borges F et al (2009) Epigenetic reprogramming and small RNA silencing of transposable elements in pollen. Cell 136:461–472

Song J, Angel A, Howard M et al (2012) Vernalization – a cold-induced epigenetic switch. J Cell Sci 125:3723–3731

Song Y, Wu K, Dhaubhadel S et al (2010) Arabidopsis DNA methyltransferase AtDNMT2 associates with histone deacetylase AtHD2s activity. Biochem Biophys Res Commun 396:187–192. doi:10.1016/j.bbrc.2010.03.119

Steward N, Ito M, Yamaguchi Y et al (2002) Periodic DNA methylation in maize nucleosomes and demethylation by environmental stress. J Biol Chem 277:37741–37746

Sung S, Amasino RM (2004) Vernalization in *Arabidopsis thaliana* is mediated by the PHD finger protein VIN3. Nature 427:159–164

Tomato Genome C (2012) The tomato genome sequence provides insights into fleshy fruit evolution. Nature 485:635–641

Ueno Y, Ishikawa T, Watanabe K et al (2007) Histone deacetylases and ASYMMETRIC LEAVES2 are involved in the establishment of polarity in leaves of Arabidopsis. Plant Cell 19:445–457. doi:10.1105/tpc.106.042325

Vanyushin BF, Ashapkin VV (2009) DNA methylation in plants. Nova Science, New York

Vrebalov J, Ruezinsky D, Padmanabhan V et al (2002) A MADS-box gene necessary for fruit ripening at the tomato ripening-inhibitor (rin) locus. Science 296:343–346

Wang Z, Cao H, Sun Y et al (2013) *Arabidopsis* paired amphipathic helix proteins SNL1 and SNL2 redundantly regulate primary seed dormancy via abscisic acid-ethylene antagonism mediated by histone deacetylation. Plant Cell 25:149–166

Wang H, Chung PJ, Liu J et al (2014) Genome-wide identification of long noncoding natural antisense transcripts and their responses to light in *Arabidopsis*. Genet Res 24:444–453

Wang H, Niu QW, Wu HW et al (2015a) Analysis of non-coding transcriptome in rice and maize uncovers roles of conserved lncRNAs associated with agriculture traits. Plant J 84:404–416

Wang J, Yu W, Yang Y et al (2015b) Genome-wide analysis of tomato long non-coding RNAs and identification as endogenous target mimic for microRNA in response to TYLCV infection. Sci Rep 5:16946

Wang X, Ai G, Zhang C et al (2016) Expression and diversification analysis reveals transposable elements play important roles in the origin of Lycopersicon-specific lncRNAs in tomato. New Phytol 209:1442–1455

Winicov I (2000) Alfin1 transcription factor overexpression enhaces plant root growth under normal and saline conditions and improves salt tolerance in alfafa. Planta 210:416–422

Wollmann H, Berger F (2012) Epigenetic reprogramming during plant reproduction and seed development. Curr Opin Plant Biol 15:63–69

Wollmann H, Mica E, Todesco M et al (2010) On reconciling the interactions between APETALA2, miR172 and AGAMOUS with the ABC model of flower development. Development 137:3633–3642

Wu H-J, Wang Z-M, Wang M et al (2013) Wide-spread long non-coding RNAs (lncRNAs) as endogenous target mimics (eTMs) for microRNAs in plants. Plant Physiol 161:1875–1884

Xiao B, Wilson JR, Gamblin SJ (2003) SET domains and histone methylation. Curr Opin Struct Biol 13:699–705

Xiao W, Custard KD, Brown RC et al (2006) DNA methylation is critical for *Arabidopsis* embryogenesis and seed viability. Plant Cell 18:805–814

Xiao J, Zhang H, Xing L et al (2013) Requirement of histone acetyltransferases HAM1 and HAM2 for epigenetic modification of FLC in regulating flowering in *Arabidopsis*. J Plant Physiol 170:444–451

Xu C, Liberatore KL, MacAlister CA et al (2015) A cascade of arabinosyltransferases controls shoot meristem size in tomato. Nat Genet 47:784–792

Yakovlev IA, Fossdal CG, Johnsen Ø (2010) MicroRNAs, the epigenetic memory and climatic adaptation in Norway spruce. New Phytol 187:1154–1169

Yang H, Chang F, You C et al (2015) Whole-genome DNA methylation patterns and complex associations with gene structure and expression during flower development in *Arabidopsis*. Plant J 81:268–281

Yao JL, Tomes S, Xu J et al (2016) How microRNA172 affects fruit growth in different species is dependent on fruit type. Plant Signal Behav 11:e1156833

You W, Lorkovic ZJ, Matzke AJM et al (2013) Interplay among RNA polymerases II, IV and V in RNA-directed DNA methylation at a low copy transgene locus in *Arabidopsis thaliana*. Plant Mol Biol 82:85–96

Yu CW, Liu X, Luo M et al (2011) HISTONE DEACETYLASE6 interacts with FLOWERING LOCUS D and regulates flowering in *Arabidopsis*. Plant Physiol 156:173–184

Zemach A, Kim MY, Silva P et al (2010) Local DNA hypomethylation activates genes in rice endosperm. Proc Natl Acad Sci USA 107:18729–18734

Zhang JJ, Xue HW (2013) OsLEC1/OsHAP3E participates in the determination of meristem identity in both vegetative and reproductive developments of rice. J Integr Plant Biol 55:232–249

Zhang X, Yazaki J, Sundaresan A et al (2006) Genome-wide high-resolution mapping and functional analysis of DNA methylation in *Arabidopsis*. Cell 126:1189–1201

Zhang Y-C, Liao J-Y, Li Z-Y et al (2014) Genome-wide screening and functional analysis identify a large number of long noncoding RNAs involved in the sexual reproduction of rice. Genome Biol 15:1–16

Zhao L, Kim Y, Dinh TT et al (2007) miR172 regulates stem cell fate and defines the inner boundary of APETALA3 and PISTILLATA expression domain in *Arabidopsis* floral meristems. Plant J 51:840–849

Zhao L, Lu J, Zhang J et al (2014) Identification and characterization of histone deacetylases in tomato (*Solanum lycopersicum*). Front Plant Sci 5:760

Zhong S, Fei Z, Chen YR et al (2013) Single-base resolution methylomes of tomato fruit development reveal epigenome modifications associated with ripening. Nat Biotechnol 31:154–159

Zhou C, Labbe H, Sridha S et al (2004) Expression and function of HD2-type histone deacetylases in Arabidopsis development. Plant J 38:715–724. doi:10.1111/j.1365-313X.2004.02083.x

Zhu QH, Upadhyaya NM, Gubler F et al (2009) Over-expression of miR172 causes loss of spikelet determinacy and floral organ abnormalities in rice (*Oryza sativa*). BMC Plant Biol 9:149

Zhu B, Yang Y, Li R et al (2015) RNA sequencing and functional analysis implicate the regulatory role of long non-coding RNAs in tomato fruit ripening. J Exp Bot 66:4483–4495

Zuo J, Zhu B, Fu D et al (2012) Sculpting the maturation, softening and ethylene pathway: the influences of microRNAs on tomato fruits. BMC Genomics 13:7

Epigenetic Regulation of Phase Transitions in *Arabidopsis thaliana*

Inês Trindade, Daniel Schubert, and Valérie Gaudin

Abstract In plants the adult organs are not determined during embryogenesis. Throughout their life cycle, plants undergo different developmental changes, which require deep alterations in the overall gene expression patterns. The timing at which these phase changes occur can have a strong impact on plant viability and fitness and therefore needs to be tightly regulated to assure they take place under optimal conditions. On the other hand, in *Arabidopsis thaliana*, once a transition is initiated, it is usually stable, independently of the surrounding environment, suggesting it should be memorized. Over the last years, several studies have shown that epigenetics plays an important role in the regulation of phase transitions in plants. For instance, *Arabidopsis* mutants where the repressive trimethylation of lysine 27 of histone 3 (H3K27me3) is absent were shown to germinate and further revert into calli-like structures bearing somatic embryos. Moreover, this histone mark has recently been shown to also regulate the transition from vegetative to reproductive development, as some mutants with reduced H3K27me3 levels are incapable of memorizing floral inducible stimuli, when transferred to suboptimal conditions. On the other hand, the epigenetic modifications need to be reset at the end of each generation, to ensure the normal development of the progeny. In fact, plants have developed an entire reprogramming machinery that culminates at gametogenesis, where a genome-wide resetting of gene programs takes place.

In this chapter we summarize recent findings on different layers of epigenetic regulation during *Arabidopsis* major developmental transitions: embryo to seedling, juvenile to adult and vegetative to reproductive. We present phase-specific regulatory mechanisms and highlight common aspects throughout development and the importance of resetting at the end of each generation.

I. Trindade • D. Schubert
Institute for Biology, Freie Universität Berlin, Königin-Luise-Strasse 12-16, 14195 Berlin, Germany

V. Gaudin (✉)
Institut Jean-Pierre Bourgin, INRA, AgroParisTech, CNRS, Université Paris-Saclay, 78000 Versailles, France
e-mail: valerie.gaudin@inra.fr

N. Rajewsky et al. (eds.), *Plant Epigenetics*, RNA Technologies,
DOI 10.1007/978-3-319-55520-1_18

Keywords Polycomb group (PcG) • Histone deacetylase (HDAC) • Chromatin remodellers • Juvenile phase • Germination • Vegetative development • Flowering • Reproduction • Epigenetic reprogramming

Contents

1 Introduction

In plants, most organs, including true leaves and flowers, are not established during embryogenesis, but are only generated after germination. As a consequence, the adult body is composed of a sequence of organs that were formed at different developmental phases, a feature known as heteroblasty (Goebel 1900). This characteristic is achieved through the maintenance of meristems, like the root apical meristem (RAM) and the shoot apical meristem (SAM), which keep a niche of pluripotent stem cells throughout the plant's life cycle. Changes in the identity of the SAM, induced by endogenous or environmental stimuli, lead to the emergence of morphologically and functionally distinct organs and often mark important developmental transitions in plants (reviewed in Barton 2010).

The life cycle of a plant begins with the formation of the embryo after fertilization, which in Angiosperms is enclosed inside a seed, where it can remain after maturation in a dormant state. At the onset of germination, the transition from embryonic to photoautotrophic development, the embryo has to overcome two types of barriers: a chemical barrier, in order to break dormancy induced and maintained by hormonal stimuli, and a mechanical barrier, as it must grow through the endosperm and the seed coat (Nonogaki 2014). Post-embryonic development is characterized by successive phases: juvenile vegetative, adult vegetative and reproductive development. Although morphologically distinct, these phases are often molecularly related and part of a continuum of changes in the same regulatory networks (Huijser and Schmid 2011; Wu et al. 2009). During the vegetative phase, the plant has a high photosynthetic activity and increases its biomass rapidly. The transition from juvenile to adult development is marked by the acquisition of reproductive competence, which is often accompanied by changes in leaf morphology (Huijser and Schmid 2011). At a later stage, in response to multiple stimuli, further changes in SAM identity (from vegetative to inflorescence shoot apical meristem) trigger the transition to the reproductive phase (Amasino and Michaels 2010). During the development of the reproductive organs, the somatic-to-reproductive transition is initiated by the differentiation of the spore mother cells. These

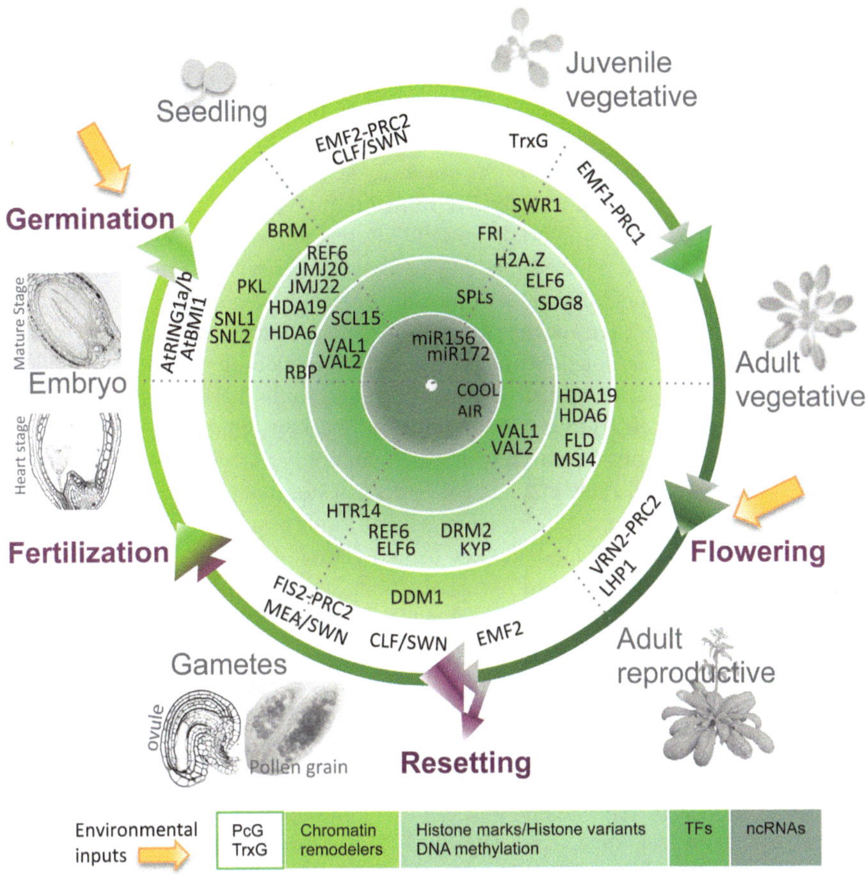

Fig. 1 The *A. thaliana* life cycle with its developmental phase transitions and different layers of epigenetic regulation. Major developmental phase transitions ensure the progression throughout the plant life cycle, which are regulated by environmental cues. Epigenetic factors involved in the control of these phase transitions which ensure optimal fitness and development can be organized in main classes: PcG and TrxG proteins, chromatin remodellers, histone variants, proteins involved in DNA methylation or histone post-translational modifications, as well as non-coding RNAs (ncRNAs). The epigenetic factors as well as transcription factors interacting with chromatin-associated proteins, presented in this chapter, are represented. The diagram is clearly non-exhaustive but displays the main (epigenetic) actors. The embryo and ovule pictures were kindly provided by Enrico Magnani (IJPB, Versailles)

cells are destined for meiosis, followed by a series of mitotic divisions to form the gametes, which contain the genetic information for the next generation (Yang and Sundaresan 2000). A scheme of the main developmental transitions in *Arabidopsis thaliana* is depicted in Fig. 1.

Phase transitions require deep changes at the molecular level, which often involve switching ON and OFF entire developmental networks. The timing at which these transitions occur needs to be tightly regulated by endogenous and environmental factors, as it can have a strong impact on plant fitness and productivity. Additionally, in several plant species, including *A. thaliana*, once a transition is initiated, it is stable and the plants do not revert back to the previous developmental phase. On the other hand, the molecular mechanisms underlining these transitions need to have a certain degree of plasticity that allows for a reprogramming of the networks at the onset of each generation.

Back in 1962, Brink had already envisioned that phase transitions in plants are regulated by "accessory material" present at the chromosomes (Brink 1962). Nowadays, it is well established that these accessories consist of chromatin factors that control phase transitions as well as their epigenetic stability. Chromatin is a dynamic structure that can occur in different states, defined on the basis of the degree of condensation, nucleosome occupancy, presence of specific histone and DNA modifications, among other features (Ernst and Kellis 2010; Sequeira-Mendes et al. 2014). Polycomb group (PcG) proteins are key regulators of chromatin structure and nuclear organization, which are commonly associated with repressive states (reviewed in Del Prete et al. 2015). Two types of plant PcG complexes have been described, which are conserved in several organisms: polycomb repressive complex 2 (PRC2), which mediates the trimethylation of lysine 27 of histone H3 (H3K27me3) (Cao et al. 2002; Muller et al. 2002; Schubert et al. 2006), and PRC1, which is associated with H2AK119 monoubiquitination (Bratzel et al. 2010; Yang et al. 2013a). Plants exhibit diverse PRC2-like complexes, which are formed by distinct *Arabidopsis* homologues of the *Drosophila*'s enhancer of zeste [E(z)] [CURLY LEAF (CLF), SWINGER (SWN) or MEDEA (MEA)] and suppressor of zeste (Su(z)12) subunits [VERNALIZATION 2 (VRN2), EMBRYONIC FLOWER 2 (EMF2) or FERTILIZATION INDEPENDENT SEED 2 (FIS2)], and function at specific developmental phases throughout plant development (reviewed in Derkacheva and Hennig 2014). The PRC1 H3K27me3 reader LIKE HETEROCHROMATIN PROTEIN 1 (LHP1) was shown to strongly co-localize to regions marked with H3K27me3, suggesting both PRC1 and PRC2 could target a similar subset of genes (Zhang et al. 2007). Nevertheless, the relation between both complexes in plants is not yet fully understood and likely varies for distinct PcG target genes, with different studies placing PRC1 either upstream, downstream or acting independently of PRC2 (reviewed in Merini and Calonje 2015). PcG silencing is counteracted by Trithorax group (TrxG) proteins, which form a functionally diverse group generally involved in gene activation (reviewed in Schuettengruber et al. 2011). Several TrxG proteins have been identified in plants, including proteins involved in the deposition of H3K4me3 (Carles and Fletcher 2009; Guo et al. 2010; Tamada et al. 2009) and H3K36me2/3 (Xu et al. 2008), two histone marks associated with active transcription. PcG silencing can be additionally counteracted by histone demethylases that actively remove H3K27me3 marks, like EARLY FLOWERING 6 (ELF6) and RELATIVE OF EARLY FLOWERING 6 (REF6) (Crevillen et al. 2014; Cui et al. 2016; Lu et al. 2011).

Arabidopsis, like many eukaryotes, contain histone variants that differ in their amino acid sequences and exhibit distinct biochemical, physical and remodelling properties (Jarillo and Piñeiro 2015; Kotlinski et al. 2016; Stroud et al. 2012;

Yelagandula et al. 2014). Reprogramming can also be achieved by the loading of histone variants into nucleosomes, which can occur dependent or independent on DNA replication, and relies on the activity of chromatin remodelling proteins and histone chaperones (reviewed in Deal and Henikoff 2011; Talbert et al. 2012; Wu et al. 2005). Chromatin remodellers form another important group of chromatin modifiers that function by inducing non-covalent changes in chromatin and affect DNA-histone interactions as well as nucleosome assembly and positioning. Chromatin remodelling proteins have also been associated with non-coding RNA (ncRNA)-mediated gene silencing in *Arabidopsis* (Zhu et al. 2013). Over the last years, ncRNAs have emerged as important epigenetic regulators, which in plants have been associated with the regulation of the RNA polymerase II machinery and DNA methylation through the RNA-directed DNA methylation pathway (RdDM) (reviewed in Chekanova 2015; Matzke et al. 2015). Additionally, chromatin remodelling proteins participate in the regulation of histone acetylation and were shown to interact with histone deacetylases (HDACs) (Buszewicz et al. 2016; Tong et al. 1998; Xue et al. 1998). The acetylation of residues at the tails of the histones H3 and H4 is associated with active transcription and is catalysed by histone acetyltransferases (HACs), which have been reported to counteract PcG proteins, thus showing TrxG-like activity. Consistently, in metazoans, H3K27me3 and H3K27ac were shown to be mutually exclusive and to form a gene expression ON/OFF switch regulated by PcG/HACs (Pasini et al. 2010; Tie et al. 2009). How chromatin remodellers are recruited to their targets is not fully understood, but some studies suggest that recognition of specific histone marks (Zhang et al. 2012) as well as interactions with transcription factors (TFs) (Smaczniak et al. 2012) might be involved in target recognition.

In this chapter, we review the current knowledge on the epigenetic regulation of the major phase transitions in *Arabidopsis thaliana*. Some of the most studied phase-specific regulatory mechanisms are presented and common aspects to the different transitions throughout development are highlighted. Additionally, we refer to the reprogramming mechanisms at the onset of the next generation, which guarantee the correct development of the progeny and the maintenance of specific requirements to assure the right timing of each transition. Finally, we briefly discuss the necessary crosstalk between environmental stimuli and the epigenetic regulators of phase transitions and how these are regulated in plants with different lifestyles.

2 Epigenetic Regulation of the Embryo-to-Seedling Transition

The first major developmental transition in the life cycle of flowering plants is the shift from the mature embryo to young seedling (Fig. 1). It comprises two important phenomena, seed dormancy termination and germination per se, and requires the repression of seed maturation and dormancy programs and the activation of genes involved in vegetative development (reviewed in Bassel 2016).

The LAFL TF network, composed of B3 domain TFs from the AFL clade [ABSCISIC ACID INSENSITIVE3 (ABI3), FUSCA3 (FUS3) and LEAFY COTYLEDON2 (LEC2)] and the LEC1-type HAP3 family, plays a central role in seed maturation. It is involved in the regulation of important processes like the accumulation of storage reserves, seed desiccation and dormancy (reviewed in Jia et al. 2014), which need to be shut down at the onset of germination. Additionally, LAFL TFs, particularly FUS3 (Gazzarrini et al. 2004), are responsible for the regulation of numerous genes involved in seed hormone metabolism (Jia et al. 2014). Phytohormones like abscisic acid (ABA) and gibberellins (GA) are key regulators of the embryo-to-seedling transition, and the balance between both determines the direction of development towards dormancy maintenance or post-embryonic growth (reviewed in Shu et al. 2016). Seed dormancy is also regulated independently of hormonal signalling pathways through *DELAY OF GERMINATION1* (*DOG1*), which encodes a protein of unknown function that, when mutated, leads to a complete loss of dormancy but no additional phenotypes (Nakabayashi et al. 2012). The repression of genes, which encode proteins like DOG1 and LAFL TFs, is therefore an essential step in the transition to post-embryonic development. VAL (VIVIPAROUS1/ABI3-LIKE) 1/2/3 proteins (also B3 domain TFs) are key repressors of the LAFL network, and, consequently, *val1 val2* seedlings exhibit embryonic traits in both shoots and roots (Suzuki et al. 2007) (Fig. 2a). The activity of VAL proteins seems to be, at least in part, dependent on epigenetic mechanisms, as VAL2/HSI2-LIKE1 (HSL1) was found to act together with HISTONE DEACETYLASE19 (HDA19), in order to repress seed maturation genes by removal of histone acetylation at these loci (Zhou et al. 2013). HDA19 likely acts in the embryo-to-seedling transition as part of a protein complex, which includes the SWI-INDEPENDENT3 (SIN3) homologues SIN3-LIKE1 (SNL1) and SNL2 (Wang et al. 2013). The *snl1* and *snl2* single and double mutants show enhanced dormancy likely due to the upregulation of ABA signalling as a result of increased H3K9/18 and H3K14 acetylation at key genes involved in these pathways (Wang et al. 2013). Interestingly, although *snl* mutants exhibit a wild-type-like post-embryonic phenotype, *hda19* mutants have more pleiotropic phenotypes, suggesting that the complex formed by SNL1/SNL2/HDA19 acts specifically during the embryo-to-seedling transition (Wang et al. 2013). Recently, the SCARECROW-like15 (SCL15) TF was also found to interact with HDA19 to fulfil a transcriptional repression function in the embryo-to-seedling transition (Gao et al. 2015). In wild-type *Arabidopsis* plants, overall HDAC activity was found to transiently increase one day after imbibition (Tai et al. 2005), suggesting the involvement of additional HDACs during germination. In fact, HDA19 seems to act redundantly with HDA6 in the repression of embryonic traits, as the ectopic emergence of embryonic characteristics was stronger in seedlings in which both genes are downregulated, compared to single knockdown mutants (Tanaka et al. 2008). Additionally, plant-specific HD2-like histone deacetylases, some of which interact with HDA6, are repressed by ABA (Luo et al. 2012), suggesting an increase

Fig. 2 Alterations in developmental phase transitions in mutants affected in chromatin-associated proteins. (**a**) In vitro seedling of the *val1 val2* double mutant (Photo kindly provided by Myriam Calonje). (**b–c**) In vitro seedlings at the two-cotyledon stage of the *pkl* mutant (**b**) with a closer view of the root apex stained with Fat Red 7B and of the wild-type plant (**c**). Photos kindly provided by Claudia Köhler. (**d**) Two-month-old in vitro plantlet of the *Atring1a Atring1b* double mutant after Fat Red 7B staining. Photo kindly provided by Wen-Hui Shen. (**e–f**) *Atbmi1a Atbmi1b* mutant at 18 (**e**) and 30 (**f**) days after germination (DAG). Photos kindly provided by Myriam Calonje. (**g–h**) *swn-3 clf-50* double mutant plants at 35 DAG with a *closer view* (**h**) of a somatic embryo formed on the callus-like tissue. Photos kindly provided by Justin Goodrich (Chanvivattana et al. 2004). (**i–j**) Stems of plants 35 days after shift from long days to short days. *clf swn pCLF::CLF-GR* (**i**). *emf2-10 vrn2-1* (**j**). *Arrowhead*: reversion nodes. *Asterisks*: pre-reversion flowers. Photos kindly provided by Ralf Müller-Xing (Müller-Xing et al. 2014). (**k**) Stems of an *emf2-10 vrn2-1* plant continuously grown in short days. Photo kindly provided by Ralf Müller-Xing. (**l**) Inflorescence of the *Atring1a Atring1b* double mutant showing fasciation and altered flowers. Photo kindly provided by Wen-Hui Shen. (**m**) Altered inflorescence from an indeterminated carpel of the *lif2* mutant (*closer view*). Photos provided by Valérie Gaudin

in expression during germination. Further proof of the key role of histone acetylation during the embryo-to-seedling transition came from treating *Arabidopsis* seedlings with trichostatin A (TSA), an HDAC inhibitor, which led to the upregulation of *LAFL* genes during germination (Tanaka et al. 2008).

The repression of the *LAFL* network during the embryo-to-seedling transition requires not only the removal of histone marks associated with active transcription but also the subsequent deposition of modifications associated with repression. Besides histone acetylation, genes like *ABI3* and *DOG1* also exhibit the active mark H3K4me3 in mature seeds (Footitt et al. 2015; Molitor et al. 2014). During germination H3K4me3 at these loci is lost, while the levels of the repressive H3K27me3 simultaneously increase (Footitt et al. 2015; Molitor et al. 2014). PcG proteins were recently shown to act in concert with VAL proteins in the repression of seed maturation genes (Yang et al. 2013a). Moreover, VAL proteins interact with the PRC1 core component AtBMI1, an E3 H2A monoubiquitin (H2Aub) ligase, which is thought to mark seed maturation genes for further PRC2-dependent silencing (Bratzel et al. 2010; Yang et al. 2013a). In fact, the *Atbmi1a/b/c* and *Atring1a/b* triple and double mutants (Fig. 2d–f, l), in which PRC1 catalytic subunits are suppressed, exhibit a phenotype that resembles the *val1 val2* double mutant and is characterized by low levels of H3K27me3 at different *LAFL* genes (Chen et al. 2010; Yang et al. 2013a). Interestingly, the H3K27me3 reader LIKE HETEROCHROMATIN PROTEIN 1 (LHP1) seems to have no impact, or at least a minor role, in the embryo-to-seedling transition (Wang et al. 2016), despite the fact that it was reported to bind *ABI3* and *DOG1* loci (Molitor et al. 2014), possibly suggesting the existence of complementary control mechanisms.

The ATP-dependent chromatin remodeller PICKLE (PKL) has also emerged as a key regulator of the embryo-to-seedling transition (Henderson et al. 2004; Ogas et al. 1999; Rider et al. 2003). The *pkl* mutant phenotype is characterized by a so-called pickle root (Fig. 2b–c), which exhibits embryonic traits, as the accumulation of seed storage proteins and lipids (Ogas et al. 1997), likely caused by the misexpression of *LEC1*, *LEC2* and *FUS3* (Aichinger et al. 2011; Ogas et al. 1999; Rider et al. 2003). The chromatin remodeller PKL belongs to the chromodomain/helicase/DNA binding domain (CHD3/CHD4) subfamily of proteins (Ogas et al. 1999), which in animals are known to function in transcription repression as part of the nucleosome remodelling and deacetylase (NuRD) complex (reviewed in Ahringer 2000). In plants, however, PKL seems to act preferentially as a monomer and not to be involved in histone deacetylation, although it still retains the ability to bind DNA and nucleosomes (Ho et al. 2013). PKL-induced repression of transcription during germination is at least partly dependent on PRC2, as *pkl* mutants show reduced H3K27me3 levels at several seed development genes (Zhang et al. 2008, 2012). Interestingly, PKL could be indirectly required for PRC2-mediated repression of embryonic genes, as it does not seem to bind *LEC1*, *FUS3* and *ABI3*, but instead promotes the expression of PRC2 genes, such as *SWN* and *EMF2* (Aichinger et al. 2009). PKL has also been implicated in the hormonal regulation of germination (Perruc et al. 2007; Zhang et al. 2014). In fact, *ABI3* and its downstream target *ABI5*, in which transcripts usually accumulate in dry seeds, are also upregulated in *pkl* seedlings, correlating with lower levels of the repressive histone marks H3K9me2 and H3K27me2 (Belin and Lopez-Molina 2008; Perruc et al. 2007). Additionally, PKL was shown to interact with DELLA proteins, which are master repressors of GA signalling, and its expression increased upon exogenous GA3 treatment (Zhang et al. 2014), suggesting it might also contribute to the increasing levels of gibberellins at the onset of germination.

Despite the important role of PKL during germination, it is not essential during post-embryonic development, as some seed-specific genes that are misexpressed in *pkl* young seedlings are expressed at wild-type levels in 14-day-old *pkl* seedlings (Rider et al. 2003; Zhang et al. 2012). On the other hand, the SWI2/SNF2 chromatin remodeller BRAHMA (BRM) seems to be at least partly required for post-embryonic development, as its suppression leads to an accumulation of seed storage proteins in vegetative tissues (Tang et al. 2008). BRM represses genes encoding for these proteins by directly binding to their promoters, likely as part of a complex that includes AtSWI3C, a homologue of the yeast SWI/SNF complex subunit SWI3, and BSH, the homologue of SNF5 (Tang et al. 2008). Additionally, several lines of evidence suggest that PcG proteins are not essential for germination itself, but play a crucial role in the maintenance of the post-embryonic developmental phase. For instance, *fie* and *clf swn* mutants, which are affected in PRC2 subunits, have extremely reduced or absent levels of the repressive mark H3K27me3 and are able to germinate. Later, they develop into callus-like structures with somatic embryos (Fig. 2g–h), likely due to the misexpression of *DOG1* and *LAFL* genes (Bouyer et al. 2011; Chanvivattana et al. 2004; Lafos et al. 2011; Makarevich et al. 2006). A similar phenotype was also described for mutants in which the levels of the RETINOBLASTOMA-RELATED (RBR) protein are reduced (Gutzat et al. 2011). RBR proteins are thought to act together with several chromatin remodellers (reviewed in Kuwabara and Gruissem 2014) and were shown to be required to maintain the PRC2-dependent repressed state of late embryogenesis genes, like *LEC2* and *ABI3* (Gutzat et al. 2011).

3 Epigenetic Regulation of the Juvenile-to-Adult and Vegetative-to-Reproductive Transitions

SQUAMOSA PROMOTER BINDING-LIKE (SPL) TFs and microRNAs (miRNAs) are key regulators of post-embryonic development for the establishment of adult traits and the acquisition of flowering competence (Fouracre and Poethig 2016; Wang et al. 2009; Xu et al. 2016b). *Arabidopsis* has 16 *SPL* genes (Cardon et al. 1999), often with overlapping functions, ten of which are post-transcriptionally regulated by miR156 (Rhoades et al. 2002; Xu et al. 2016b). The levels of miR156 peak in juvenile plants and decrease with increasing age, allowing for the accumulation of SPL proteins in adult plants (Wang et al. 2009; Wu and Poethig 2006). miR156 is itself under the control of the sugar status of the plant, which is correlated with photosynthetic activity in the leaves (Yang et al. 2013b; Yu et al. 2013). The fine-tuning of the miR156/SPL network is of utmost importance for the shift from juvenile to adult vegetative development. Consistently, transgenic *Arabidopsis* lines expressing miR156-resistant forms of several *SPL* genes show an accelerated juvenile-to-adult phase transition (Wu and Poethig 2006; Xu et al. 2016b), while plants with reduced or absent miR156 show almost no signs of juvenile development (Wu et al. 2009). miR156 is a known target of both PRC2 (Lafos et al. 2011) and LHP1 PRC1 subunit (Molitor et al. 2016), and its age-related downregulation was recently shown to be accompanied by an increase in the levels of H3K27me3 at *MIR156A*/*MIR156C*, likely related to the removal of the active histone mark H3K27ac (Xu et al. 2016a). Interestingly, this accumulation of H3K27me3 seems to require the chromatin remodeller PKL and PRC1-induced H2Aub (Picó et al. 2015; Xu et al. 2016a), resembling the downregulation of *LAFL* TFs during germination. Likewise, the expression of *MIR156A* and *MIR156B* was higher in *val1 val2* mutants (Picó et al. 2015), indicating that the pathway responsible for the repression of LAFL TFs during germination could also be involved in post-embryonic phase changes. *SPL* genes were also found to be downregulated in the *Arabidopsis* mutant *hag1*. The *hag1* mutation affects the catalytic subunit of *Arabidopsis* Spt–Ada–Gcn5–acetyltransferase (SAGA)-like complex (Kim et al. 2015), which has been implicated in histone acetylation and deubiquitination (reviewed in Moraga and Aquea 2015). Consistently with the role of SPLs in adult vegetative development, this mutant shows a delayed transition from the juvenile phase, exposing a role for histone acetylation in vegetative phase change (Kim et al. 2015). The age-dependent accumulation of some SPLs in adult plants leads to an increase in the levels of miR172, which is also associated with the establishment of adult features and reproductive competence, through the repression of *APETALA 2* (*AP2*)-like TFs (Wu et al. 2009). Accordingly, plants overexpressing miR172, or plants in which the miR172 targets *TARGET OF EAT1 1* (*TOE1*) and *TOE2* are suppressed, show a reduced juvenile developmental phase (Wu et al. 2009). Similar to miR156, the regulation of *SPL*s and *MIR172* is also dependent on PRC1 (Picó et al. 2015), and the histone acetyltransferase GCN5 was shown to target *MIR172a*, affecting the levels of H3K14ac at this locus (Kim et al. 2009). Interestingly, different PRC1 variants seem to act at different developmental stages, as *Atbmi1a/b* weak mutants (which are able to develop post-embryonically) do not show altered flowering time (Picó et al. 2015). On the other hand, mutants lacking the PRC1 member *EMF1*, a functional orthologue of *Drosophila* Posterior sex combs (Beh et al. 2012), show increased levels of *SPL3*, *SPL9* and *MIR172* and an early flowering phenotype (Picó et al. 2015), suggesting a role for an EMF1-PRC1 in the maintenance of the vegetative phase. Additionally, EMF proteins help prevent a precocious floral transition by repressing the expression of floral homeotic genes like *AGAMOUS* (*AG*), *APETALA 3* (*AP3*) and *PISTILLATA* (*PI*) (Calonje et al. 2008; Kim et al. 2010). Consistently, *emf* mutants almost completely skip the vegetative phase during their life cycle and flower upon germination (Sung et al. 1992).

Vegetative development is also accompanied by a gradual downregulation of floral repressors and consequent accumulation of flowering-inducing signals that culminate in the transition to reproductive development. The timing of this transition is of crucial importance in a plant's life cycle, as it can severely affect productivity and seed yield. Therefore, it is tightly regulated by multiple endogenous and environmental factors that converge at different floral integrators in several intricate pathways (reviewed in Andres and Coupland 2012). FLOWERING LOCUS T (FT) is a major component of the florigenic mobile signal produced in the leaves in response to inductive photoperiods. FT is then translocated to the shoot apical meristem, where it promotes the conversion from vegetative to reproductive meristem (Corbesier et al. 2007). *FT* expression in seeds and juvenile plants is strongly repressed by the MADS box complex formed by SHORT VEGETATIVE PHASE (SVP) and FLOWERING LOCUS C (FLC)), the latter directly binding the first intron of *FT* (Jang et al. 2009; Michaels and Amasino 1999; Searle et al. 2006). *FLC* and *SVP* repression are therefore key steps for the transition from vegetative to reproductive development to occur. Several *Arabidopsis* accessions, known as winter annuals, have an active *FRIGIDA* (*FRI*) allele (a major activator of *FLC*) and require a prolonged exposure to cold in order to flower (Choi et al. 2011; Michaels and Amasino 1999). This process, known as vernalization, promotes the mitotically stable epigenetic repression of *FLC* and constitutes a classical example of the role of chromatin and epigenetic modifications in plant development. The genetic and molecular basis of vernalization has been extensively studied and reviewed elsewhere (Berry and Dean 2015; Hepworth and Dean 2015; Schmitz and Amasino 2007; Zhu et al. 2015). Hence, it will not be discussed in this chapter.

In *Arabidopsis* accessions that do not contain an active *FRI* allele, *FLC* expression is usually low, and therefore the plants do not require vernalization to flower (Michaels and Amasino 2001). During early vegetative development, chromatin at the *FLC* locus is enriched in histone modifications associated with active transcription, like H3K4me3, H3K36me2/3 and histone H3 and H4 acetylation (Cao et al. 2008; He et al. 2003; Jiang et al. 2011; Pien et al. 2008; Yang et al. 2014). FLOWERING LOCUS D (FLD) and FVE/MULTICOPY SUPPRESSOR OF IRA1 4 (MSI4) are two key regulators of the autonomous pathway that play a major role in the removal of these marks, leading to *FLC* repression (Ausin et al. 2004; Gu et al. 2011; He et al. 2003; Yu et al. 2016). Mutants in both genes show increased histone acetylation and H3K4me3 and flower later than wild-type plants, exhibiting a prolonged vegetative phase, which is based on high *FLC* expression (Ausin et al. 2004; He et al. 2003). FLD is an orthologue of the human lysine-specific demethylase 1 (LSD1), which exhibits H3K4me demethylase activity (He et al. 2003), while *FVE* encodes a nuclear WD-repeat retinoblastoma-associated protein (Ausin et al. 2004). Both proteins have been shown to interact with HDA6, which binds directly to *FLC* and the *FLC* paralogous gene *MADS AFFECTING FLOWERING4* (*MAF4*) (Gu et al. 2011; Wu et al. 2008; Yu et al. 2011). Additionally, HDA6, FVE and FLD were recently shown to be involved in the regulation of a similar set of genes, suggesting they may act as a complex to promote the floral transition (Yu et al. 2016). VAL proteins also seem to be involved in this process, as VAL1 is required for vernalization-induced *FLC* repression through its function in histone deacetylation, likely by recruiting HDA19 to the *FLC* locus (Questa et al. 2016). This observation suggests that HDA19 may act redundantly with HDA6 and resembles the regulation of seed maturation genes during germination, exposing a more general mechanism of initiation of phase transitions in *Arabidopsis*. Besides its role in histone deacetylation and demethylation, FLD is additionally required for *FLC* repression mediated by the RNA-binding protein FCA, another member of the autonomous pathway (Liu et al. 2007). FCA is required for the processing of COOLAIR, a group of antisense long non-coding RNAs (lncRNAs) involved in *FLC* repression and originating from the 3' end of the *FLC* locus (Csorba et al. 2014; Liu et al. 2010), thus connecting the autonomous pathway to RNA-mediated chromatin silencing.

The removal of histone modifications associated with active transcription allows for the deposition of repressive marks, in order to ensure a stable transition to reproductive development. In rapid-cycling accessions, which do not require vernalization to flower, *FLC* is also repressed during vegetative development by a PRC2 complex containing the histone methyltransferase CLF, independent of environmental stimuli (Jiang et al. 2008). Interestingly, FVE was reported to interact with CLF (Pazhouhandeh et al. 2011), establishing a link between the autonomous pathway and PcG-mediated gene repression. Accordingly, *fve* mutants show an extended vegetative phase and have almost no H3K27me3 at *FLC* (Pazhouhandeh et al. 2011). VAL proteins are also involved in the regulation of PRC2-induced silencing of *FLC*, as the suppression of *VAL1* led to a reduction in cold-induced accumulation of H3K27me3 at this locus (Questa et al. 2016). Interestingly, H3K27me3 maintenance upon re-exposure to higher temperatures was not disturbed in the *val1* mutant, suggesting VAL proteins are required to trigger PRC2-induced *FLC* repression, likely by promoting histone deacetylation, but are not essential for the maintenance of flowering competence (Questa et al. 2016).

The nucleosomal composition and density also participate in the control of flowering. Indeed, the choice of the histone variants incorporated in the nucleosome core particle can impact nucleosome mobility, stability and remodelling, and thus, forms another control layer of transcription (Talbert et al. 2012; Talbert and Henikoff 2014). For instance, an early event in response to cold is the stabilization of the first nucleosome at the transcriptional start site (TSS) through the replacement of H3.3, the histone variant associated with active chromatin, by the H3.1 histone variant (Finnegan 2015). Additionally, the replacement of H2A by the H2A.Z histone variant, which requires the chromatin remodelling complex SWR1 formed by the PIE1, SURF3 and SEF subunits, also impacts *FLC* expression (Deal et al. 2007), but does not seem to participate to vernalization (Finnegan and Dennis 2007; Jarillo and Piñeiro 2015). Recently, it was highlighted that chromatin states defined by the presence or absence of specific histone marks might be the early events for the transcriptional commitment by modulating rates of RNA polymerase II initiation and elongation, and thus quantitatively regulate *FLC* expression (Wu et al. 2016; Yang et al. 2016).

Finally, the distance between the major *cis*-regulatory blocks of the *FT* regulatory regions involved in photoperiod control also has an influence on *FT* expression and, therefore, flowering time (Liu et al. 2014a), suggesting that the tridimensional chromatin conformation of the *FT* locus is also connected to its transcriptional regulation. This was also well exemplified by the disruption of the chromatin gene loop formed by the 5′ and 3′ flanking regions of *FLC*, which occurs within the first two weeks of cold treatment (Crevillen et al. 2013). Furthermore, the VRN2 and VRN5 PRC2 components were shown to participate to the physical clustering of *FLC* alleles in the nuclear space, which is also required for its silencing (Rosa et al. 2013; reviewed in Del Prete et al. 2015). Thus, the transcriptional control of the flowering repressor *FLC* assembles various chromatin-associated regulatory mechanisms to ensure proper expression from highly active to transiently and stably repressed states, in a quantitative and cell autonomous way. The maintenance of *FLC* repression was demonstrated to be different in the two *FLC* copies, thus dependent on *cis*-regulatory mechanisms (Berry et al. 2015).

The transition from vegetative to reproductive development in *Arabidopsis* is usually stable, independent of whether the plants are grown under inductive (long days; LD) or non-inductive (short days; SD) photoperiods after flowering has been induced (Müller-Xing et al. 2014). Despite the fact that floral induction in *Arabidopsis* has been thoroughly studied, not much is known about how the reproductive phase is maintained. PcG proteins have recently emerged as key factors for the stability of the transition to reproductive development (Müller-Xing et al. 2014), as previously reported for other developmental transitions in *Arabidopsis* (see above). Hence, mutants in genes encoding for the PcG proteins EMF2 and VERNALIZATION2 (VRN2), which show reduced levels of H3K27me3 (Lafos et al. 2011), exhibit strong floral reversion when shifted from LD to SD conditions (Müller-Xing et al. 2014) (Fig. 2j–k). This loss of memory of the new developmental stage was shown to be dependent on the derepression of the floral repressors *FLC* and *SVP*, which caused downregulation of *FT* (Müller-Xing et al. 2014). In agreement, photoperiod-independent *FT* post-fertilization expression in floral organs has been shown to be required for the maintenance of the reproductive developmental phase, as *ft* mutants revert to vegetative development when grown in non-inducing conditions (Liu et al. 2014b; Müller-Xing et al. 2014). Additionally, this function of *FT* does not require induction by CONSTANS (CO), which mediates photoperiod-dependent activation of *FT* in leaves, as *co* mutants do not exhibit floral reversion (Liu et al. 2014b). Finally, LHP1 participates to the maintenance of the shoot meristem identity, its defect leading to the conversion of the inflorescence meristem into floral meristem and thus inflorescence termination (Gaudin et al. 2001; Kotake et al. 2003). On the contrary, a defect in the LHP1-INTERACTING FACTOR2 (LIF2) causes a reversion to inflorescence structures in the carpel (Latrasse et al. 2011; Molitor et al. 2016) (Fig. 2m).

4 Epigenetic Reprogramming

Throughout the plant life cycle and in response to certain cues, the developmental programs that have been set, might have to be erased. Hence, reprogramming events occur in different contexts. One of them is associated with plant gametogenesis, which occurs late in development. Because the plant gametes are formed from somatic cells, the resetting of several programs is required, in order to assure the correct development of the subsequent generation and erasure of information that is stored during somatic development. Additionally, resetting during reproduction is required for the reacquisition of pluripotency in the embryonic cells that will develop into shoot and root apical meristems. Reprogramming of gene networks can also be induced in vitro. For ages, agricultural practices largely exploited plant plasticity by the means of grafting, layering, rooting or vegetative propagation, which also require reprogramming events.

Gametogenesis in flowering plants involves two major steps: sporogenesis, in which the haploid spores (the primordia of the gametophytes) are formed through meiosis, and gametogenesis, where the gametes are formed through a series of mitotic divisions (reviewed in Yang and Sundaresan 2000). The female gametophyte (embryo sac, also known as megagametophyte) develops in the ovules and contains several types of haploid cells, including the egg cell (female gamete) (reviewed in Drews et al. 1998). On the other hand, the male gametophyte (pollen) contains one vegetative nucleus and two sperm cells, one of which will fuse with the egg cell resulting in zygote formation, and another will fuse with the central cell to form the triploid endosperm (reviewed in McCormick 1993). Epigenetic reprogramming in *Arabidopsis* has been reported at different stages of gametogenesis: in the spore mother cells, before meiosis (She et al. 2013); after meiosis, during gametophyte development; and after fertilization, at the initial stages of seed development (Ingouff et al. 2010). Moreover, it is the basis for genomic imprinting, a phenomenon through which the allele from one of the parents is silenced in the subsequent generation. Imprinting has been reported for several genes in *Arabidopsis* and was previously reviewed elsewhere (Gehring 2013), and therefore will not be included in the scope of the present chapter.

An interesting mechanism of DNA methylation reprogramming guided by ncRNAs was identified by Slotkin et al. (2009) in the male gametophyte. The authors have analysed whole genome expression profiles from sperm cells and the vegetative nucleus of mature pollen and found that the chromatin remodelling ATPase *DECREASE IN DNA METHYLATION 1* (*DDM1*) is detected exclusively in the sperm cells (Slotkin et al. 2009). This observation coincided with a generalized decrease of DNA methylation at several loci encoding for transposable elements (TEs) in the vegetative nucleus, whose genetic information is not transmitted to the next generation. This derepression of TEs was accompanied by an increase in the formation of 21 nt ncRNAs from these loci, which, surprisingly, were detected in both vegetative nucleus and sperm cells. This work suggests a role for reprogramming during pollen maturation in the silencing of TEs in the subsequent generation, contributing to maintain genome stability (Slotkin et al. 2009). Additionally, the DNA methyltransferase DOMAINS REARRANGED METHYLASE2 (DRM2), which is a key enzyme in the RdDM pathway, was found to be downregulated in sperm cells (Calarco et al. 2012). Consequently, a decrease in the levels of de novo CHH DNA methylation in retrotransposons was registered in sperm cells and restored in the zygote upon fertilization, likely guided by maternally inherited ncRNAs (Calarco et al. 2012). Surprisingly, DNA methylation in both CG and CHG contexts is generally maintained in the sperm cells and after fertilization, suggesting it might be inherited in the next generation (Calarco et al. 2012).

As discussed before, the repression of *FLC*, either induced by cold or by endogenous factors, involves several epigenetic mechanisms and therefore constitutes a good example of the impact of epigenetics in plant development. However, the chromatin changes induced during post-embryonic development need to be reset at the beginning of the next generation, in order to enable floral transition to occur under optimal conditions in the progeny. Choi et al. (2009) have identified three stages in the reprogramming of *FLC* at the onset of a new generation: (1) repression during gametogenesis, which was found to be independent of PRC2 proteins and the autonomous pathway, (2) reactivation in the initial steps of embryogenesis (3) maintenance of the active state during late embryogenesis. *FLC* reactivation does not depend on FRI or FRI-interacting factors, like SUPRESSOR OF FRIGIDA 4 (SUF4), but does seem to require PIE1, a member of the *Arabidopsis* SWR1 complex (Choi et al. 2009). Moreover, ELF6, a jumonji-C domain-containing H3K27me2/3 demethylase, targets *FLC* in seedlings and is in part required for its reactivation (Crevillen et al. 2014). Consistently, the *elf6-5* mutant, which bears a substitution in a highly conserved alanine residue, exhibits higher *FLC* expression in globular stage embryos, when compared to the wild-type plants (Crevillen et al. 2014). Recently, ELF6 was found to physically interact with SET DOMAIN GROUP 8 (SDG8), the methyltransferase responsible for H3K36me3 deposition (Yang et al. 2016), which was previously reported to counteract H3K27me3 in *FLC* regulation (Yang et al. 2014). Thus, the work by Yang et al. (2016) unveils a new molecular mechanism for switching genes ON and OFF. In the future, it will be therefore interesting to see if this could be a general mechanism to reset PcG silencing during somatic development and in the next generation.

Changes in the loading of histone variants also seem to be a common feature to different reprogramming mechanisms. For instance, lower levels of the histones H1.1, H1.2 and H2A.Z were detected during female and male gametogenesis in *Arabidopsis* (She et al. 2013; She and Baroux 2015). Interestingly, in both gametophytes, histone eviction seems to occur differently in the gametes (egg cell and sperm cells) and the companion cells (central cell and vegetative nucleus) (Ingouff et al. 2010). At the mature female gametophyte, different histone H3 variants were detected in the central cell (which expresses H3.1, H3.3 and the unusual variant encoded by *HTR14*) compared to the egg cell, where no H3.1 and high levels of H3.3 were detected (Ingouff et al. 2010). Additionally, a study reported that a histone H3 gene was specifically expressed in the *Arabidopsis* male gametes (Okada et al. 2005). Similar results were described in in vitro dedifferentiation studies in protoplasts and protoplast-derived cells (Chupeau et al. 2013), suggesting a requirement for changes in nucleosomal conformation during the resetting of differentiation. In agreement with this idea, Chitteti et al. (2008) reported an overall decrease of the content of all histone variants in *Arabidopsis* cotyledon cells subjected to an in vitro hormonal treatment that induces dedifferentiation. The eviction of H3.1 in differentiated cells was also recently reported in the vicinity of the root apical meristem (Otero et al. 2016). Consistently, in the shoot apical meristem, the switch ON and OFF of genes during differentiation was correlated with the loading of H3.3, which is associated with actively transcribed genes (Wollmann et al. 2012). Further studies on dedifferentiated *Arabidopsis* cells support the idea that a more relaxed chromatin conformation is required for the resetting of genetic programs (Del Prete et al. 2014). Rosa et al. (2014), for instance, have used imaging techniques to show that the mobility of the histone H2B is higher in pluripotent root meristematic cells when compared to the nearby differentiated cells. Tessadori et al. (2007), on the other hand, have observed a disruption of chromocenters upon protoplast induction from vegetative tissue. Interestingly, this conformational change, which was not accompanied by changes in histone modifications or DNA methylation, was reversible after a prolonged culture time (Tessadori et al. 2007), exposing a transient nature for chromatin loosening during reprogramming.

Polycomb proteins also seem to play a role in the regulation of reprogramming mechanisms in *Arabidopsis*, as multiple genes are differentially targeted by PRCs in pluripotent cells when compared to differentiated somatic cells (He et al. 2012; Lafos et al. 2011). Additionally, leaves of the PRC2 *clf swn* and *emf2* mutants were shown to be highly recalcitrant to callus induction (He et al. 2012). In *Arabidopsis*, this phenomenon, even when initiated from aerial tissues, requires the activation of lateral root developmental programs (Sugimoto et al. 2010). Interestingly, callus formation was not affected in root explants from *clf swn* and *emf2* mutants, suggesting PRC2 might be required for the silencing of shoot-specific developmental programs (He et al. 2012). Other repressive histone modifications might play a similar role in dedifferentiation as seeds and leaf explants from mutants in the H3K9me2 methyltransferase KRYPTONITE (KYP) are also recalcitrant to callus induction (Grafi et al. 2007).

5 Conclusions and Future Prospects

As summarized in this chapter, epigenetic gene regulation provides a stable mechanism to switch and maintain ON/OFF molecular networks, as it does not involve changes in the nucleotide sequences. Because epigenetic modifications can be maintained across mitotic cell divisions, but still are reversible, they provide the regulatory plasticity required during plant development. Our current knowledge on the epigenetic regulation of the phase transitions in *Arabidopsis* has unveiled an intricate network involving multiple layers (Fig. 1). While HDACs, in concert with VAL proteins and chromatin remodellers like PKL or BRM, seem to play a general role during the transition itself, PRCs seem to be generally more relevant for the maintenance of the new developmental state. At the moment, not much is known about the molecular mechanisms involved in maintaining the stability of phase changes, and further work will be required to evaluate the possible involvement of other regulators. Additionally, although the above-mentioned key players seem to act throughout several developmental phases, the molecular mechanisms that determine their specificity at each stage remain to be elucidated. PRC2 specificity, for instance, is known to rely at least in part on the phase-specific occurrence of subunits with redundant functions (reviewed in Derkacheva and Hennig 2014). Like in metazoans, PRC1 in plants can also be composed of different subunit combinations (reviewed in Merini and Calonje 2015), which could affect PRC1 activity in a similar way. Additionally, ncRNAs have been reported to recruit chromatin modifiers and remodellers and provide nucleotide sequence-based specificity (reviewed in Holoch and Moazed 2015), which are promising candidates to mediate target recognition.

Plant development relies on the maintenance of stem cell fate at the SAM and RAM, which provide cells that differentiate and participate to the formation of mature organs. In mammals, the presence of both active H3K4me2/3 and repressive H3K27me3 histone marks in the same chromatin fibre seems to be a feature of pluripotent stem cells (Azuara et al. 2006; Bernstein et al. 2006). This bivalent state is thought to maintain genes in a poised condition and to play important roles in the regulation of developmental programs. In plants, only a few studies have reported the analysis of bivalent marks (Luo et al. 2013; Sequeira-Mendes et al. 2014), but one could speculate that a similar mechanism may be involved in the maintenance of cell fate at the apical meristems, allowing for a quick switch of specific genes during phase transitions. However, the studies conducted so far in plants have used samples resulting from a mixture of tissues and cell types, which can provide misleading results. In order to better understand the role of this chromatin state in plant development and in the ON/OFF switch of specific genetic programs in particular, the study of individual cell lines would be required. However, this places several technical difficulties when it comes to working with plant cells, as they dedifferentiate when kept in culture. Recent advances in techniques that allow for the isolation of single cell types from plant tissues (Deal and Henikoff 2010; Moreno-Romero et al. 2016) are expected to contribute to better understand the role of epigenetic regulation in plant development.

Another important aspect of phase transitions is that they have to be tightly regulated, in order to ensure that they take place at the correct time. For instance, if the transition to reproductive development occurs under suboptimal conditions, it can have a strong impact on the ability of the plant to develop seeds and thus affect fitness. Therefore, the entire molecular machinery involved in the initiation of phase transitions needs to be directly or indirectly regulated by external stimuli. Over the last years, epigenetic regulators have been widely implicated in the responses to environmental stimuli (reviewed in Kleinmanns and Schubert 2014), suggesting they might also be affected by external factors during development. However, there are only a few cases discovered in which an environmental stimulus induces expression or activity of an epigenetic regulator. One such example is VIN3, which is induced by cold and required for the vernalization response (Sung and Amasino 2004). Similarly, the JmjC domain-containing histone arginine demethylases JMJ20 and JMJ22 were shown to be activated as a consequence of phytochrome B (PHYB) light-induced activation and are required for germination to occur (Cho et al. 2012). Interestingly, this pathway culminated in an increase in the levels of gibberellins, placing epigenetic regulators in a bridge between exogenous and endogenous stimuli responsible for triggering germination (Cho et al. 2012). Environmental factors, like temperature and water availability, are known to affect not only germination, but also other developmental transitions. Further work will be required to understand whether the activity of epigenetic regulators relies on environmental cues, possibly in a tissue-specific manner.

In the current chapter we have described the role of epigenetics in the major phase transitions in *A. thaliana*, which is an annual plant, meaning every generation is completed at the end of the reproductive phase. However, alternative regulatory mechanisms are expected to occur in plants with different lifestyles. Unlike *Arabidopsis*, most perennials exhibit polycarpic growth and undergo multiple rounds of reproductive development throughout their life cycle (reviewed in Bergonzi and Albani 2011). This feature likely requires more complex mechanisms to simultaneously maintain meristems with different identities (vegetative and reproductive) and keep competence to respond to stimuli, which is likely also maintained by epigenetic factors. Increasing work with perennial species will likely unveil new interesting differentiation/reprogramming regulatory mechanisms.

As our knowledge on the epigenetic regulation of plant development increases, new networks and degrees of complexity are unveiled, which, as discussed above, raise new questions that remain to be answered. Recent technological advances open new windows to further explore epigenetic regulation in plants and are likely to be very fruitful in a near future.

Acknowledgements We are grateful to our colleagues Myriam Calonje (IBVF, Seville), Justin Goodrich (University of Edinburgh, Edinburgh), Claudia Köhler (SLU, Uppsala), Enrico Magnani (IJPB, Versailles), Ralf Müller-Xing (Northeast Forestry University Harbin, Harbin) and Wen-Hui Shen (IBMP, Strasbourg) and thank them for providing us photographs for the figures of this chapter. IT and DS were supported by the Freie Universität Berlin within the Excellence Initiative of the German Research Foundation. VG was supported by INRA and the LabEx Saclay Plant Sciences-SPS (ANR-10-LABX-0040-SPS).

References

Ahringer J (2000) NuRD and SIN3 histone deacetylase complexes in development. Trends Genet 16:351–356

Aichinger E, Villar CB, Farrona S et al (2009) CHD3 proteins and polycomb group proteins antagonistically determine cell identity in Arabidopsis. PLoS Genet 5:e1000605

Aichinger E, Villar CB, Di Mambro R et al (2011) The CHD3 chromatin remodeler PICKLE and polycomb group proteins antagonistically regulate meristem activity in the Arabidopsis root. Plant Cell 23:1047–1060

Amasino RM, Michaels SD (2010) The timing of flowering. Plant Physiol 154:516–520

Andres F, Coupland G (2012) The genetic basis of flowering responses to seasonal cues. Nat Rev Genet 13:627–639

Ausin I, Alonso-Blanco C, Jarillo JA et al (2004) Regulation of flowering time by FVE, a retinoblastoma-associated protein. Nat Genet 36:162–166

Azuara V, Perry P, Sauer S et al (2006) Chromatin signatures of pluripotent cell lines. Nat Cell Biol 8:532–538

Barton MK (2010) Twenty years on: the inner workings of the shoot apical meristem, a developmental dynamo. Dev Biol 341:95–113

Bassel GW (2016) To grow or not to grow? Trends Plant Sci 21:498–505

Beh LY, Colwell LJ, Francis NJ (2012) A core subunit of Polycomb repressive complex 1 is broadly conserved in function but not primary sequence. Proc Natl Acad Sci USA 109:E1063–E1071

Belin C, Lopez-Molina L (2008) Arabidopsis seed germination responses to osmotic stress involve the chromatin modifier PICKLE. Plant Signal Behav 3:478–479

Bergonzi S, Albani MC (2011) Reproductive competence from an annual and a perennial perspective. J Exp Bot 62:4415–4422

Bernstein BE, Mikkelsen TS, Xie XH et al (2006) A bivalent chromatin structure marks key developmental genes in embryonic stem cells. Cell 125:315–326

Berry S, Dean C (2015) Environmental perception and epigenetic memory: mechanistic insight through FLC. Plant J 83:133–148

Berry S, Hartley M, Olsson TS et al (2015) Local chromatin environment of a Polycomb target gene instructs its own epigenetic inheritance. elife 4:e07205

Bouyer D, Roudier F, Heese M et al (2011) Polycomb repressive complex 2 controls the embryo-to-seedling phase transition. PLoS Genet 7:e1002014

Bratzel F, Lopez-Torrejon G, Koch M et al (2010) Keeping cell identity in Arabidopsis requires PRC1 RING-finger homologs that catalyze H2A monoubiquitination. Curr Biol 20:1853–1859

Brink A (1962) Phase change in higher plants and somatic cell heredity. Q Rev Biol 37(1):1–22

Buszewicz D, Archacki R, Palusinski A et al (2016) HD2C histone deacetylase and a SWI/SNF chromatin remodelling complex interact and both are involved in mediating the heat stress response in Arabidopsis. Plant Cell Environ 39:2108–2122

Calarco JP, Borges F, Donoghue MT et al (2012) Reprogramming of DNA methylation in pollen guides epigenetic inheritance via small RNA. Cell 151:194–205

Calonje M, Sanchez R, Chen L et al (2008) Embryonic Flower1 participates in polycomb group-mediated AG gene silencing in Arabidopsis. Plant Cell 20:277–291

Cao R, Wang L, Wang H et al (2002) Role of histone H3 lysine 27 methylation in Polycomb-group silencing. Science 298:1039–1043

Cao Y, Dai Y, Cui S et al (2008) Histone H2B monoubiquitination in the chromatin of FLOWERING LOCUS C regulates flowering time in Arabidopsis. Plant Cell 20:2586–2602

Cardon G, Hohmann S, Klein J et al (1999) Molecular characterisation of the Arabidopsis SBP-box genes. Gene 237:91–104

Carles CC, Fletcher JC (2009) The SAND domain protein ULTRAPETALA1 acts as a trithorax group factor to regulate cell fate in plants. Genes Dev 23:2723–2728

Chanvivattana Y, Bishopp A, Schubert D et al (2004) Interaction of Polycomb-group proteins controlling flowering in Arabidopsis. Development 131:5263–5276
Chekanova JA (2015) Long non-coding RNAs and their functions in plants. Curr Opin Plant Biol 27:207–216
Chen D, Molitor A, Liu C et al (2010) The Arabidopsis PRC1-like ring-finger proteins are necessary for repression of embryonic traits during vegetative growth. Cell Res 20:1332–1344
Chitteti BR, Tan F, Mujahid H et al (2008) Comparative analysis of proteome differential regulation during cell dedifferentiation in Arabidopsis. Proteomics 8:4303–4316
Cho JN, Ryu JY, Jeong YM et al (2012) Control of seed germination by light-induced histone arginine demethylation activity. Dev Cell 22:736–748
Choi J, Hyun Y, Kang MJ et al (2009) Resetting and regulation of flowering locus C expression during Arabidopsis reproductive development. Plant J 57:918–931
Choi K, Kim J, Hwang HJ et al (2011) The FRIGIDA complex activates transcription of FLC, a strong flowering repressor in Arabidopsis, by recruiting chromatin modification factors. Plant Cell 23:289–303
Chupeau MC, Granier F, Pichon O et al (2013) Characterization of the early events leading to totipotency in an Arabidopsis protoplast liquid culture by temporal transcript profiling. Plant Cell 25:2444–2463
Corbesier L, Vincent C, Jang S et al (2007) FT protein movement contributes to long-distance signaling in floral induction of Arabidopsis. Science 316:1030–1033
Crevillen P, Sonmez C, Wu Z et al (2013) A gene loop containing the floral repressor FLC is disrupted in the early phase of vernalization. EMBO J 32:140–148
Crevillen P, Yang H, Cui X et al (2014) Epigenetic reprogramming that prevents transgenerational inheritance of the vernalized state. Nature 515:587–590
Csorba T, Questa JI, Sun Q et al (2014) Antisense COOLAIR mediates the coordinated switching of chromatin states at FLC during vernalization. Proc Natl Acad Sci USA 111:16160–16165
Cui X, Lu F, Qiu Q et al (2016) REF6 recognizes a specific DNA sequence to demethylate H3K27me3 and regulate organ boundary formation in Arabidopsis. Nat Genet 48:694–699
Deal RB, Henikoff S (2010) A simple method for gene expression and chromatin profiling of individual cell types within a tissue. Dev Cell 18:1030–1040
Deal RB, Henikoff S (2011) Histone variants and modifications in plant gene regulation. Curr Opin Plant Biol 14:116–122
Deal RB, Topp CN, McKinney EC et al (2007) Repression of flowering in Arabidopsis requires activation of flowering locus C expression by the histone variant H2A.Z. Plant Cell 19:74–83
Del Prete S, Arpon J, Sakai K et al (2014) Nuclear architecture and chromatin dynamics in interphase nuclei of *Arabidopsis thaliana*. Cytogenet Genome Res 143:28–50
Del Prete S, Mikulski P, Schubert D et al (2015) One, two, three: polycomb proteins hit all dimensions of gene regulation. Genes (Basel) 6:520–542
Derkacheva M, Hennig L (2014) Variations on a theme: polycomb group proteins in plants. J Exp Bot 65:2769–2784
Drews GN, Lee D, Christensen CA (1998) Genetic analysis of female gametophyte development and function. Plant Cell 10:5–17
Ernst J, Kellis M (2010) Discovery and characterization of chromatin states for systematic annotation of the human genome. Nat Biotechnol 28:817–825
Finnegan EJ (2015) Time-dependent stabilization of the +1 nucleosome is an early step in the transition to stable cold-induced repression of FLC. Plant J 84:875–885
Finnegan EJ, Dennis ES (2007) Vernalization-induced trimethylation of histone H3 lysine 27 at FLC is not maintained in mitotically quiescent cells. Curr Biol 17:1978–1983
Footitt S, Muller K, Kermode AR et al (2015) Seed dormancy cycling in Arabidopsis: chromatin remodelling and regulation of DOG1 in response to seasonal environmental signals. Plant J 81:413–425
Fouracre JP, Poethig RS (2016) The role of small RNAs in vegetative shoot development. Curr Opin Plant Biol 29:64–72

Gao MJ, Li X, Huang J et al (2015) Scarecrow-like15 interacts with histone deacetylase19 and is essential for repressing the seed maturation programme. Nat Commun 6:7243

Gaudin V, Libault M, Pouteau S et al (2001) Mutations in like heterochromatin protein 1 affect flowering time and plant architecture in Arabidopsis. Development 128:4847–4858

Gazzarrini S, Tsuchiya Y, Lumba S et al (2004) The transcription factor FUSCA3 controls developmental timing in Arabidopsis through the hormones gibberellin and abscisic acid. Dev Cell 7:373–385

Gehring M (2013) Genomic imprinting: insights from plants. Annu Rev Genet 47:187–208

Goebel K (1900) Organography of plants. Oxford, Clarendon Press

Grafi G, Ben-Meir H, Avivi Y et al (2007) Histone methylation controls telomerase-independent telomere lengthening in cells undergoing dedifferentiation. Dev Biol 306:838–846

Gu X, Jiang D, Yang W et al (2011) Arabidopsis homologs of retinoblastoma-associated protein 46/48 associate with a histone deacetylase to act redundantly in chromatin silencing. PLoS Genet 7:e1002366

Guo L, Yu YC, Law JA et al (2010) Set domain group2 is the major histone H3 lysine 4 trimethyltransferase in Arabidopsis. Proc Natl Acad Sci USA 107:18557–18562

Gutzat R, Borghi L, Futterer J et al (2011) Retinoblastoma-related protein controls the transition to autotrophic plant development. Development 138:2977–2986

He Y, Michaels SD, Amasino RM (2003) Regulation of flowering time by histone acetylation in Arabidopsis. Science 302:1751–1754

He CS, Chen XF, Huang H et al (2012) Reprogramming of H3K27me3 is critical for acquisition of pluripotency from cultured Arabidopsis tissues. PLoS Genet 8:e1002911

Henderson JT, Li HC, Rider SD et al (2004) PICKLE acts throughout the plant to repress expression of embryonic traits and may play a role in gibberellin-dependent responses. Plant Physiol 134:995–1005

Hepworth J, Dean C (2015) Flowering locus C's lessons: conserved chromatin switches underpinning developmental timing and adaptation. Plant Physiol 168:1237–1245

Ho KK, Zhang H, Golden BL et al (2013) PICKLE is a CHD subfamily II ATP-dependent chromatin remodeling factor. Biochim Biophys Acta 1829:199–210

Holoch D, Moazed D (2015) RNA-mediated epigenetic regulation of gene expression. Nat Rev Genet 16:71–84

Huijser P, Schmid M (2011) The control of developmental phase transitions in plants. Development 138:4117–4129

Ingouff M, Rademacher S, Holec S et al (2010) Zygotic resetting of the HISTONE 3 variant repertoire participates in epigenetic reprogramming in Arabidopsis. Curr Biol 20:2137–2143

Jang S, Torti S, Coupland G (2009) Genetic and spatial interactions between FT, TSF and SVP during the early stages of floral induction in Arabidopsis. Plant J 60:614–625

Jarillo JA, Piñeiro M (2015) H2A.Z mediates different aspects of chromatin function and modulates flowering responses in Arabidopsis. Plant J 83:96–109

Jia H, Suzuki M, McCarty DR (2014) Regulation of the seed to seedling developmental phase transition by the LAFL and VAL transcription factor networks. Wiley Interdiscip Rev Dev Biol 3:135–145

Jiang D, Wang Y, He Y (2008) Repression of flowering locus C and flowering locus T by the Arabidopsis Polycomb repressive complex 2 components. PLoS One 3:e3404

Jiang D, Kong NC, Gu X et al (2011) Arabidopsis COMPASS-like complexes mediate histone H3 lysine-4 trimethylation to control floral transition and plant development. PLoS Genet 7:e1001330

Kim W, Benhamed M, Servet C et al (2009) Histone acetyltransferase GCN5 interferes with the miRNA pathway in Arabidopsis. Cell Res 19:899–909

Kim SY, Zhu T, Sung ZR (2010) Epigenetic regulation of gene programs by EMF1 and EMF2 in Arabidopsis. Plant Physiol 152:516–528

Kim JY, Oh JE, Noh YS et al (2015) Epigenetic control of juvenile-to-adult phase transition by the Arabidopsis SAGA-like complex. Plant J 83:537–545

Kleinmanns JA, Schubert D (2014) Polycomb and Trithorax group protein-mediated control of stress responses in plants. Biol Chem 395:1291–1300

Kotake T, Takada S, Nakahigashi K et al (2003) Arabidopsis terminal flower 2 gene encodes a heterochromatin protein 1 homolog and represses both flowering locus T to regulate flowering time and several floral homeotic genes. Plant Cell Physiol 44:555–564
Kotlinski M, Rutowicz K, Knizewski L et al (2016) Histone H1 variants in Arabidopsis are subject to numerous post-translational modifications, both conserved and previously unknown in histones, suggesting complex functions of H1 in plants. PLoS One 11:e0147908
Kuwabara A, Gruissem W (2014) Arabidopsis retinoblastoma-related and Polycomb group proteins: cooperation during plant cell differentiation and development. J Exp Bot 65:2667–2676
Lafos M, Kroll P, Hohenstatt ML et al (2011) Dynamic regulation of H3K27 trimethylation during Arabidopsis differentiation. PLoS Genet 7:e1002040
Latrasse D, Germann S, Houba-Herin N et al (2011) Control of flowering and cell fate by LIF2, an RNA binding partner of the polycomb complex component LHP1. PLoS One 6(1):e16592
Liu F, Quesada V, Crevillen P et al (2007) The Arabidopsis RNA-binding protein FCA requires a lysine-specific demethylase 1 homolog to downregulate FLC. Mol Cell 28:398–407
Liu F, Marquardt S, Lister C et al (2010) Targeted 3′ processing of antisense transcripts triggers Arabidopsis FLC chromatin silencing. Science 327:94–97
Liu L, Adrian J, Pankin A et al (2014a) Induced and natural variation of promoter length modulates the photoperiodic response of flowering locus T. Nat Commun 5:4558
Liu L, Farrona S, Klemme S et al (2014b) Post-fertilization expression of flowering locus T suppresses reproductive reversion. Front Plant Sci 5:164
Lu FL, Cui X, Zhang SB et al (2011) Arabidopsis REF6 is a histone H3 lysine 27 demethylase. Nat Genet 43:715–U144
Luo M, Wang YY, Liu X et al (2012) HD2C interacts with HDA6 and is involved in ABA and salt stress response in Arabidopsis. J Exp Bot 63:3297–3306
Luo CY, Sidote DJ, Zhang Y et al (2013) Integrative analysis of chromatin states in Arabidopsis identified potential regulatory mechanisms for natural antisense transcript production. Plant J 73:77–90
Makarevich G, Leroy O, Akinci U et al (2006) Different Polycomb group complexes regulate common target genes in Arabidopsis. EMBO Rep 7:947–952
Matzke MA, Kanno T, Matzke AJM (2015) RNA-directed DNA methylation: the evolution of a complex epigenetic pathway in flowering plants. Annu Rev Plant Biol 66:243–267
McCormick S (1993) Male gametophyte development. Plant Cell 5:1265–1275
Merini W, Calonje M (2015) PRC1 is taking the lead in PcG repression. Plant J 83:110–120
Michaels SD, Amasino RM (1999) Flowering locus C encodes a novel MADS domain protein that acts as a repressor of flowering. Plant Cell 11:949–956
Michaels SD, Amasino RM (2001) Loss of flowering locus C activity eliminates the late-flowering phenotype of FRIGIDA and autonomous pathway mutations but not responsiveness to vernalization. Plant Cell 13:935–941
Molitor AM, Bu Z, Yu Y et al (2014) Arabidopsis AL PHD-PRC1 complexes promote seed germination through H3K4me3-to-H3K27me3 chromatin state switch in repression of seed developmental genes. PLoS Genet 10:e1004091
Molitor A, Latrasse D, Zytnicki M et al (2016) The Arabidopsis hnRNP-Q protein LIF2 and the PRC1 subunit LHP1 function in concert to regulate the transcription of stress-responsive genes. Plant Cell. doi:10.1105/tpc.16.00244
Moraga F, Aquea F (2015) Composition of the SAGA complex in plants and its role in controlling gene expression in response to abiotic stresses. Front Plant Sci 6:865
Moreno-Romero J, Jiang H, Santos-Gonzalez J et al (2016) Parental epigenetic asymmetry of PRC2-mediated histone modifications in the Arabidopsis endosperm. EMBO J 35:1298–1311
Muller J, Hart CM, Francis NJ et al (2002) Histone methyltransferase activity of a Drosophila Polycomb group repressor complex. Cell 111:197–208
Müller-Xing R, Clarenz O, Pokorny L et al (2014) Polycomb-group proteins and flowering locus T maintain commitment to flowering in *Arabidopsis thaliana*. Plant Cell 26:2457–2471

Nakabayashi K, Bartsch M, Xiang Y et al (2012) The time required for dormancy release in Arabidopsis is determined by delay of germination1 protein levels in freshly harvested seeds. Plant Cell 24:2826–2838

Nonogaki H (2014) Seed dormancy and germination—emerging mechanisms and new hypotheses. Front Plant Sci 28:233

Ogas J, Cheng JC, Sung ZR et al (1997) Cellular differentiation regulated by gibberellin in the *Arabidopsis thaliana* pickle mutant. Science 277:91–94

Ogas J, Kaufmann S, Henderson J et al (1999) PICKLE is a CHD3 chromatin-remodeling factor that regulates the transition from embryonic to vegetative development in Arabidopsis. Proc Natl Acad Sci USA 96:13839–13844

Okada T, Endo M, Singh MB et al (2005) Analysis of the histone H3 gene family in Arabidopsis and identification of the male-gamete-specific variant AtMGH3. Plant J 44:557–568

Otero S, Desvoyes B, Peiro R et al (2016) Histone H3 dynamics reveal domains with distinct proliferation potential in the Arabidopsis root. Plant Cell 28:1361–1371

Pasini D, Malatesta M, Jung HR et al (2010) Characterization of an antagonistic switch between histone H3 lysine 27 methylation and acetylation in the transcriptional regulation of Polycomb group target genes. Nucleic Acids Res 38:4958–4969

Pazhouhandeh M, Molinier J, Berr A et al (2011) MSI4/FVE interacts with CUL4-DDB1 and a PRC2-like complex to control epigenetic regulation of flowering time in Arabidopsis. Proc Natl Acad Sci USA 108:3430–3435

Perruc E, Kinoshita N, Lopez-Molina L (2007) The role of chromatin-remodeling factor PKL in balancing osmotic stress responses during Arabidopsis seed germination. Plant J 52:927–936

Picó S, Ortiz-Marchena MI, Merini W et al (2015) Deciphering the role of polycomb repressive complex1 variants in regulating the acquisition of flowering competence in Arabidopsis. Plant Physiol 168:1286–1297

Pien S, Fleury D, Mylne JS et al (2008) Arabidopsis trithorax1 dynamically regulates flowering locus C activation via histone 3 lysine 4 trimethylation. Plant Cell 20:580–588

Questa JI, Song J, Geraldo N et al (2016) Arabidopsis transcriptional repressor VAL1 triggers Polycomb silencing at FLC during vernalization. Science 353:485–488

Rhoades MW, Reinhart BJ, Lim LP et al (2002) Prediction of plant microRNA targets. Cell 110:513–520

Rider S Jr, Henderson JT, Jerome RE et al (2003) Coordinate repression of regulators of embryonic identity by PICKLE during germination in Arabidopsis. Plant J 35:33–43

Rosa S, De Lucia F, Mylne JS et al (2013) Physical clustering of FLC alleles during Polycomb-mediated epigenetic silencing in vernalization. Genes Dev 27:1845–1850

Rosa S, Ntoukakis V, Ohmido N et al (2014) Cell differentiation and development in Arabidopsis are associated with changes in histone dynamics at the single-cell level. Plant Cell 26:4821–4833

Schmitz RJ, Amasino RM (2007) Vernalization: a model for investigating epigenetics and eukaryotic gene regulation in plants. Biochim Biophys Acta 1769:269–275

Schubert D, Primavesi L, Bishopp A et al (2006) Silencing by plant Polycomb-group genes requires dispersed trimethylation of histone H3 at lysine 27. EMBO J 25:4638–4649

Schuettengruber B, Martinez AM, Iovino N et al (2011) Trithorax group proteins: switching genes on and keeping them active. Nat Rev Mol Cell Biol 12:799–814

Searle I, He Y, Turck F et al (2006) The transcription factor FLC confers a flowering response to vernalization by repressing meristem competence and systemic signaling in Arabidopsis. Genes Dev 20:898–912

Sequeira-Mendes J, Araguez I, Peiro R et al (2014) The functional topography of the Arabidopsis genome is organized in a reduced number of linear motifs of chromatin states. Plant Cell 26:2351–2366

She WJ, Baroux C (2015) Chromatin dynamics in pollen mother cells underpin a common scenario at the somatic-to-reproductive fate transition of both the male and female lineages in Arabidopsis. Front Plant Sci 6:294

She W, Grimanelli D, Rutowicz K et al (2013) Chromatin reprogramming during the somatic-to-reproductive cell fate transition in plants. Development 140:4008–4019

Shu K, Liu XD, Xie Q et al (2016) Two faces of one seed: hormonal regulation of dormancy and germination. Mol Plant 9:34–45

Slotkin RK, Vaughn M, Borges F et al (2009) Epigenetic reprogramming and small RNA silencing of transposable elements in pollen. Cell 136:461–472

Smaczniak C, Immink RGH, Muino JM et al (2012) Characterization of MADS-domain transcription factor complexes in Arabidopsis flower development. Proc Natl Acad Sci USA 109:1560–1565

Stroud H, Otero S, Desvoyes B et al (2012) Genome-wide analysis of histone H3.1 and H3.3 variants in *Arabidopsis thaliana*. Proc Natl Acad Sci USA 109:5370–5375

Sugimoto K, Jiao YL, Meyerowitz EM (2010) Arabidopsis regeneration from multiple tissues occurs via a root development pathway. Dev Cell 18:463–471

Sung SB, Amasino RM (2004) Vernalization in *Arabidopsis thaliana* is mediated by the PHD finger protein VIN3. Nature 427:159–164

Sung ZR, Belachew A, Shunong B et al (1992) EMF, an Arabidopsis gene required for vegetative shoot development. Science 258:1645–1647

Suzuki M, Wang HH, McCarty DR (2007) Repression of the leafy cotyledon 1/B3 regulatory network in plant embryo development by VP1/abscisic acid insensitive 3-like B3 genes. Plant Physiol 143:902–911

Tai HH, Tai GC, Beardmore T (2005) Dynamic histone acetylation of late embryonic genes during seed germination. Plant Mol Biol 59:909–925

Talbert PB, Henikoff S (2014) Environmental responses mediated by histone variants. Trends Cell Biol 24:642–650

Talbert PB, Ahmad K, Almouzni G et al (2012) A unified phylogeny-based nomenclature for histone variants. Epigenetics Chromatin 5:7

Tamada Y, Yun JY, Woo SC et al (2009) Arabidopsis trithorax-related7 is required for methylation of lysine 4 of histone H3 and for transcriptional activation of flowering locus C. Plant Cell 21:3257–3269

Tanaka M, Kikuchi A, Kamada H (2008) The Arabidopsis histone deacetylases HDA6 and HDA19 contribute to the repression of embryonic properties after germination. Plant Physiol 146:149–161

Tang X, Hou A, Babu M et al (2008) The Arabidopsis BRAHMA chromatin-remodeling ATPase is involved in repression of seed maturation genes in leaves. Plant Physiol 147:1143–1157

Tessadori F, Chupeau MC, Chupeau Y et al (2007) Large-scale dissociation and sequential reassembly of pericentric heterochromatin in dedifferentiated Arabidopsis cells. J Cell Sci 120:1200–1208

Tie F, Banerjee R, Stratton CA et al (2009) CBP-mediated acetylation of histone H3 lysine 27 antagonizes Drosophila Polycomb silencing. Development 136:3131–3141

Tong JK, Hassig CA, Schnitzler GR et al (1998) Chromatin deacetylation by an ATP-dependent nucleosome remodelling complex. Nature 395:917–921

Wang JW, Czech B, Weigel D (2009) miR156-regulated SPL transcription factors define an endogenous flowering pathway in *Arabidopsis thaliana*. Cell 138:738–749

Wang Z, Cao H, Sun Y et al (2013) Arabidopsis paired amphipathic helix proteins SNL1 and SNL2 redundantly regulate primary seed dormancy via abscisic acid-ethylene antagonism mediated by histone deacetylation. Plant Cell 25:149–166

Wang H, Liu C, Cheng J et al (2016) Arabidopsis flower and embryo developmental genes are repressed in seedlings by different combinations of polycomb group proteins in association with distinct sets of Cis-regulatory elements. PLoS Genet 12:e1005771

Wollmann H, Holec S, Alden K et al (2012) Dynamic deposition of histone variant H3.3 accompanies developmental remodeling of the Arabidopsis transcriptome. PLoS Genet 8: e1002658

Wu G, Poethig RS (2006) Temporal regulation of shoot development in *Arabidopsis thaliana* by miR156 and its target SPL3. Development 133:3539–3547
Wu WH, Alami S, Luk E et al (2005) Swc2 is a widely conserved H2AZ-binding module essential for ATP-dependent histone exchange. Nat Struct Mol Biol 12:1064–1071
Wu K, Zhang L, Zhou C et al (2008) HDA6 is required for jasmonate response, senescence and flowering in Arabidopsis. J Exp Bot 59:225–234
Wu G, Park MY, Conway SR et al (2009) The sequential action of miR156 and miR172 regulates developmental timing in Arabidopsis. Cell 138:750–759
Wu Z, Ietswaart R, Liu F et al (2016) Quantitative regulation of FLC via coordinated transcriptional initiation and elongation. Proc Natl Acad Sci USA 113:218–223
Xu L, Zhao Z, Dong AW et al (2008) Di- and tri- but not monomethylation on histone H3 lysine 36 marks active transcription of genes involved in flowering time regulation and other processes in *Arabidopsis thaliana*. Mol Cel Biol 28:1348–1360
Xu M, Hu T, Smith MR et al (2016a) Epigenetic regulation of vegetative phase change in Arabidopsis. Plant Cell 28:28–41
Xu M, Hu T, Zhao J et al (2016b) Developmental functions of miR156-regulated squamosa promoter binding protein-like (SPL) Genes in *Arabidopsis thaliana*. PLoS Genet 12:e1006263
Xue YT, Wong JM, Moreno GT et al (1998) NURD, a novel complex with both ATP-dependent chromatin-remodeling and histone deacetylase activities. Mol Cell 2:851–861
Yang WC, Sundaresan V (2000) Genetics of gametophyte biogenesis in Arabidopsis. Curr Opin Plant Biol 3:53–57
Yang C, Bratzel F, Hohmann N et al (2013a) VAL- and AtBMI1-mediated H2Aub initiate the switch from embryonic to postgerminative growth in Arabidopsis. Curr Biol 23:1324–1329
Yang L, Xu ML, Koo Y et al (2013b) Sugar promotes vegetative phase change in *Arabidopsis thaliana* by repressing the expression of MIR156A and MIR156C. elife 2:e00260
Yang H, Howard M, Dean C (2014) Antagonistic roles for H3K36me3 and H3K27me3 in the cold-induced epigenetic switch at Arabidopsis FLC. Curr Biol 24:1793–1797
Yang H, Howard M, Dean C (2016) Physical coupling of activation and derepression activities to maintain an active transcriptional state at FLC. Proc Natl Acad Sci USA 113:9369–9374
Yelagandula R, Stroud H, Holec S et al (2014) The histone variant H2A.W defines heterochromatin and promotes chromatin condensation in Arabidopsis. Cell 158:98–109
Yu CW, Liu X, Luo M et al (2011) Histone deacetylase6 interacts with flowering locus D and regulates flowering in Arabidopsis. Plant Physiol 156:173–184
Yu S, Cao L, Zhou CM et al (2013) Sugar is an endogenous cue for juvenile-to-adult phase transition in plants. elife 2:e00269
Yu CW, Chang KY, Wu K (2016) Genome-wide analysis of gene regulatory networks of the FVE-HDA6-FLD complex in Arabidopsis. Front Plant Sci 7:555
Zhang X, Germann S, Blus BJ et al (2007) The Arabidopsis LHP1 protein colocalizes with histone H3 Lys27 trimethylation. Nat Struct Mol Biol 14:869–871
Zhang H, Rider SD Jr, Henderson JT et al (2008) The CHD3 remodeler PICKLE promotes trimethylation of histone H3 lysine 27. J Biol Chem 283:22637–22648
Zhang H, Bishop B, Ringenberg W et al (2012) The CHD3 remodeler PICKLE associates with genes enriched for trimethylation of histone H3 lysine 27. Plant Physiol 159:418–432
Zhang D, Jing Y, Jiang Z et al (2014) The chromatin-remodeling factor pickle integrates brassinosteroid and gibberellin signaling during skotomorphogenic growth in Arabidopsis. Plant Cell 26:2472–2485
Zhou Y, Tan B, Luo M et al (2013) Histone deacetylase19 interacts with HSL1 and participates in the repression of seed maturation genes in Arabidopsis seedlings. Plant Cell 25:134–148
Zhu YY, Rowley MJ, Bohmdorfer G et al (2013) A SWI/SNF chromatin-remodeling complex acts in noncoding RNA-mediated transcriptional silencing. Mol Cell 49:298–309
Zhu D, Rosa S, Dean C (2015) Nuclear organization changes and the epigenetic silencing of FLC during vernalization. J Mol Biol 427:659–669

Epigenetics in Plant–Pathogen Interactions

Giorgio Gambino and Vitantonio Pantaleo

Abstract Epigenetics describes phenomena associated with changes in gene expression that occur without apparent modification in genomic sequence. In plants, it explains at least in part phenotypic differences between genetically identical plants under environmental stimuli. In this manuscript, we describe and discuss studies carried out by the scientific community regarding epigenetic phenomena that are likely linked to plant–pathogen interaction.

Keywords Viruses • Bacteria • Histone modification • DNA methylation • Priming • RNA silencing • Systemic acquired resistance

Contents

G. Gambino
Institute for Sustainable Plant Protection, National Research Council (IPSP-CNR), Strada delle Cacce 73, 10135 Torino, Italy

V. Pantaleo (✉)
Institute for Sustainable Plant Protection, National Research Council (IPSP-CNR), Research Unit of Bari, Via Amendola 165/a, 70126 Bari, Italy
e-mail: vitantonio.pantaleo@cnr.it

N. Rajewsky et al. (eds.), *Plant Epigenetics*, RNA Technologies,
DOI 10.1007/978-3-319-55520-1_19

1 Introduction

Plants live in sessile conditions and therefore need to rapidly acquire and interpret environmental inputs during their life cycle. Thus, plants modulate their metabolism and growth (i.e., organ and tissue development and differentiation) to the changed environmental condition. Plant responses to stressors generally last the time of the external triggering stimuli, whereas in some instances they persist longer. Indeed, stressors of various kinds can induce transient, stable, and heritable change in gene expression that occurs without changes in DNA sequences. A widely known branch of genetics that studies such phenomena is epigenetics. The epigenetic modification of plant genomes explains, at least in part, the phenotypic plasticity between genetically identical plants observed in different environmental states. The outcome of plant adaptation to the environment is manifested at phenotypic level, but it is evident that the complexity of all these biological phenomena involves (1) signal sensors, (2) signal effectors, (3) modulation of gene expression in response to those effectors, and (4) maintenance of the modified state until the stimuli change. Epigenetic studies in plants have been developing at an exponential rate in recent years and are being discovered, as many biological phenomena have an epigenetic basis whether completely or partially. Some biological phenomena are well understood at the molecular level and are summarized as follows:

1. Control of developmental switches such as vegetative to reproductive transition
2. Silencing of transposable elements for ensuring centromeric functions and genomic integrity
3. Parental imprinting
4. Paramutations
5. Virus-induced gene silencing (VIGS)

In addition to the description of the molecular marks that characterize epigenetics, in the first part of this chapter we will provide examples and the essential traits that characterize the five phenomena of environmental-induced epigenetic regulation. Among them, particular focus will be on VIGS, given that plant–virus systems have led to the discovery and investigation of many aspects of plant epigenetics. Importantly, plants are able to respond to pathogen attack in order to restrain growth of a systemic infection. VIGS involves transgenerationally inherited epigenetic modification, thus their infections may induce heritable phenotypic variation that influences plant fitness in response to recurring pathogen invasions. Moreover, in this chapter we will illustrate studies showing that other plant pathogens such as bacteria and fungi can trigger responses that could be associated to epigenetic modifications.

2 Epigenetic Modification Marks in Plants

Epigenetic trait has been defined as a "stably heritable phenotype resulting from changes in a chromosome without alterations in the DNA sequence" (Berger et al. 2009). Epigenetic traits, such as DNA methylation and histone variants, influence the chromatin structure and transcriptional levels of genes. These traits can be inherited through mitotic and meiotic divisions and continuously modify during development or in response to environmental stimuli.

2.1 DNA Methylation

DNA cytosine methylation is one of the main epigenetic mechanisms in all eukaryotes and is produced and maintained over time by several molecular pathways. De novo methylation is mediated by a process involving small interfering (si) RNAs and RNA-directed DNA methylation (RdDM) that can induce the transfer of a methyl group to a carbon at position 5 of cytosine to produce 5-methylcytosine. In mammals, DNA methylation occurs almost exclusively on cytosines in the symmetric dinucleotides CG, whereas in plants it occurs at both symmetric sites (CG and CHG, where H is A, T, or C) and asymmetric sites (CHH). Cytosine methylation is maintained during replication by METHYLTRANSFERASE1 (MET1) in the CG context and by CHROMOMETHYLASE3 (CMT3) in the CHG sites, while CHH methylation is lost during DNA replication and could be generated de novo by the 24-nt siRNA-directed RdDM pathway (see Sect. 3.2) (Furner and Matzke 2011). MET1, for example, causes C residue methylation to be replicated if C is adjacent to a G on its 3′ side. The daughter strand is methylated on the C residue opposite to this G and, following the same pattern in a second round of replication, the methylation is added to a C in the original position. Such a mechanism does not maintain methylation at C residues that are not adjacent to G, thus the extent of DNA methylation in the maintenance phase of an epigenetic mechanism is less extensive than in the presence of the initiator. The canonical pathway of RdDM is represented in Fig. 1a.

Another important component of the methylcytosine pathway is represented by the active DNA demethylation process catalyzed by the DNA glycosylase family of DNA demethylases. At least four DNA demethylases [DEMETER (DME), REPRESSOR OF SILENCING1 (ROS1)/DEMETER-LIKE 1 (DML1), DML2, and DML3] have been identified in *Arabidopsis*. These enzymes replace methylcytosine with unmethylated cytosine through a base excision repair mechanism (Zhu 2009). Importantly, DMEs are expressed in the cell of the female gametophyte and are involved in the complex process regulating methylation inheritance parental imprinting (see Sect. 3.3) (Bauer and Fischer 2011).

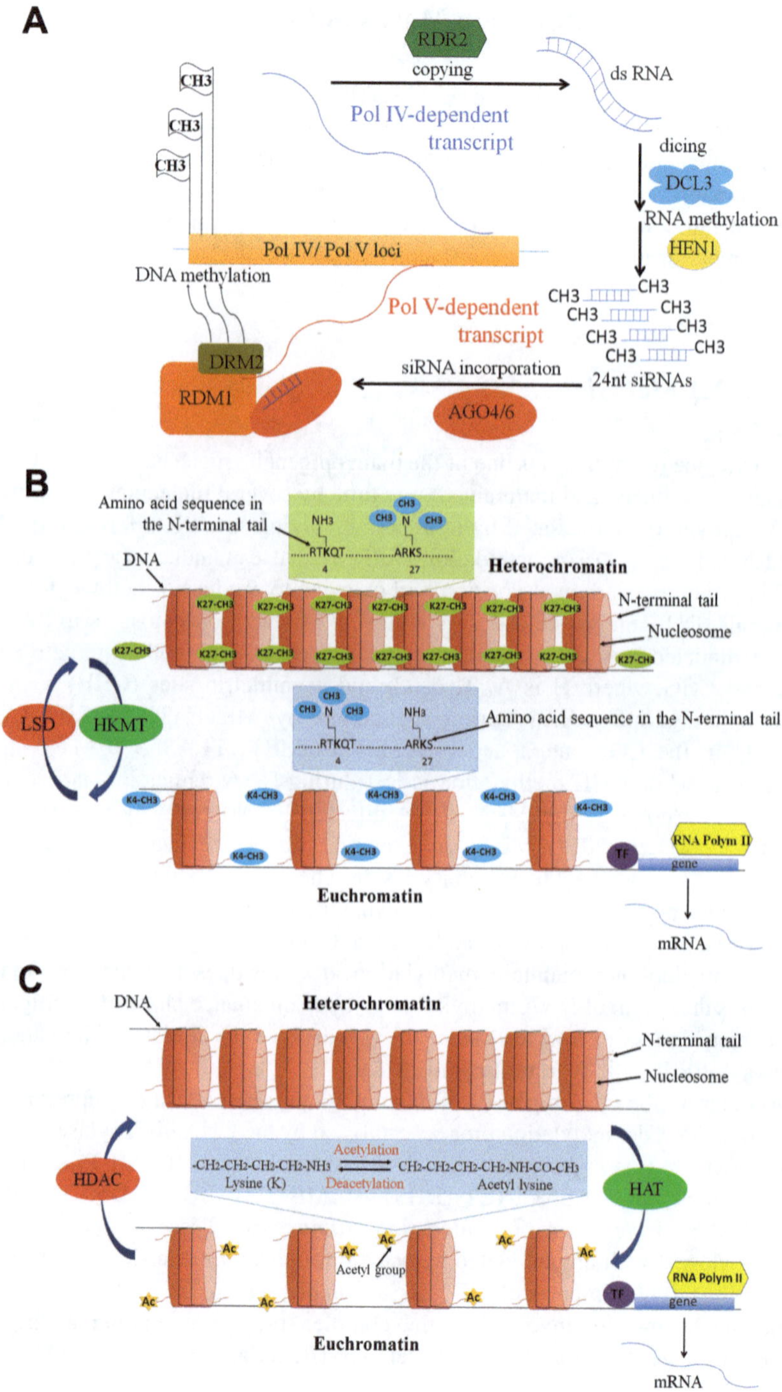

Fig. 1 Molecular marks in epigenetics. (**a**) *Canonical RdDM pathway*: DNA-DEPENDENT RNA POLIMERASE IV (Pol IV) transcribes single-stranded (ss)RNA that is copied by plant RNA-DEPENDENT RNA POLYMERASE 2 (RDR2) into double-stranded (ds) RNA. dsRNA

2.2 Histone Modifications

Histone modification is another pivotal epigenetic molecular mark present in all organisms. A core histone octamer for nucleosomes is composed of histones H2A, H2B, H3, and H4. The basic residues (amino acids such as lysine and arginine) in histone tails (N-terminal region) may be covalently modified by methylation, acetylation, phosphorylation, and ubiquitination. Thus, such variants alter the expression of genes enveloped around the histone. The relationships between histone modifications and gene activity have been studied mainly in yeast, but are highly conserved also in animals, human, and plants (Zhou et al. 2013).

The effect of histone modification on gene expression depends on the targeted amino acid residue and on the type of modifications. The acetylation of lysine residues in H3 and H4 neutralizes the positive charge of the histone tails, which alters the accessibility of transcription factors to the DNA strand. Histone acetylation tends to induce gene activation (Kuo et al. 1996; Shahbazian and Grunstein 2007), whereas the removal of acetylation can lead to gene silencing (Chen et al. 2010). For instance, acetylation at position H3K9ac in *Arabidopsis* under water stress conditions induces the up-regulation of some genes involved in drought response like RD20 and RD29A, whereas during recovery from drought the H3K9ac level decreases in conjunction with transcriptional repression of these stress-regulated genes (Kim et al. 2012). In Fig. 1b, c schematic representations of histone methylation and acetylation are shown, respectively.

Fig. 1 (continued) is substrate for DICER-like 3 (DCL3) that generates 24nt-long siRNAs. Upon HUA ENHANCER 1 (HEN1)-dependent methylation at their 3′-ends, siRNAs are incorporated into ARGONAUTE (AGO) effector proteins, i.e., AGO4 and/or 6. Incorporated siRNAs guide AGO effector complexes in a sequence-specific manner to DNA-DEPENDENT RNA POLIMERASE V (Pol V)-dependent nascent transcripts RNA-DIRECTED DNA METHYLATION 1 (DRM1) binds AGO-containing effector and DOMAINS REARRANGED METHYLTRANSFERASE1 (DRM2), which drives the de novo DNA methylation. (**b**) *Histone methylation*: methylation of Lysine (K) in histone tail is a dynamic process regulated by histone LYSINE METHYLTRANSFERASE (HKMT) and LYSINE-SPECIFIC DEMETHYLASE (LSD). The effects of histone methylation on gene expression depend on the site of the modification. The tri-methylation of the fourth lysine in the N-terminal tail of H3 (K4-CH3) induces a relaxation of chromatin structure (Euchromatin) and the DNA is more accessible to transcription factors (TF) and to RNA POLYMERASE II (RNA POLYM II). Whereas, the tri-methylation of the 27th lysine of N-terminal tail of H3 (K27-CH3) results in a denser chromatin structure (Heterochromatin), and a decrease of transcription. (**c**) *Histone acetylation*: DNA is enveloped around nucleosomes, which are composed of eight histones with two copies of histones H2A, H2B, H3, and H4. Each histone molecule has a long N-terminal tail rich in lysine residues (K), which are the sites of enzymatic modification. Acetylation of the histone tails is catalyzed by HISTONE ACETYLTRANSFERASE (HAT). Acetylation activates gene expression by making the chromatin structure less dense (Euchromatin) and accessible to transcription factors (TF) and to RNA POLYM II. Deacetylation is carried out by the enzyme HISTONE DEACETYLASE (HDAC) and results in a denser chromatin structure (Heterochromatin), and therefore reduced gene expression

The effects of histone methylation depend on the site of the modification. For example, tri-methylation of the fourth lysine in H3 (H3K4me3) induces gene expression, while tri-methylation of the 27th lysine of H3 (H3K27me3) induces decreased transcription (Cao et al. 2002; Finnegan and Dennis 2007). H3K4me3 methylation in *Arabidopsis* by the histone methyltransferase trithorax-like 1 (*ATX1*) is involved in the activation of 9-*cis*-epoxycarotenoid dioxygenase 3 (*NCED3*) under drought stress, a key enzyme in the ABA biosynthesis pathway (Ding et al. 2012). Other examples of histone modifications involved in plant–pathogen interactions are reported in Sect. 4.

3 Overview of Gene/Genomic Regulation in Plants Based on Epigenetics

3.1 Control of Developmental Switches: Vegetative to Reproductive Transition

Plants need to align flowering with spring, a more favorable environmental condition, thus ensuring effective flower formation, pollination, fruit set, and seed production to maximize reproductive success. Environmental/seasonal changes are therefore deeply monitored by plants and can act to regulate the timing of different developmental switches. One of the earliest and well-characterized processes involving epigenetic regulation in plants is vernalization: a process by which prolonged exposure to cold temperatures promotes flowering. This is due to the ability of plant cells to "remember" periods of prolonged cold they have experienced (Sung and Amasino 2004).

Vernalization involves the epigenetic silencing of a floral repressor in response to cold periods. In *Arabidopsis*, flowering locus c (*FLC*) plays the role of the floral repressor. FLC is a MADS box transcriptional factor that represses genes required to switch the meristem for flower differentiation (Michaels and Amasino 1999). Prolonged cold progressively silences expression of *FLC*, and this is epigenetically maintained during subsequent development in warm conditions so that, once plants have detected long photoperiods, the switch to flowering is activated.

Molecular studies identify changes in histone methylation as required for the maintenance of *FLC* silencing (Finnegan and Dennis 2007; Gendall et al. 2001). The degree of silencing is dependent on how much cold the plant perceives. The quantitative nature of vernalization ensures that the plant distinguishes a cold break in autumn from an entire winter. The molecular mechanisms at the basis of the quantitative feature of vernalization are at cellular level. Indeed, lengthening cold would increase the number of cells in which *FLC* is fully epigenetically repressed (Angel et al. 2011). The epigenetic mechanisms controlling vernalization are extensively studied and contain much detail that also reveals the involvement of long non-coding RNAs, which can be further investigated in reviews and recent scientific publications (Berry and Dean 2015; Questa et al. 2016).

Vernalization is one example of biological phenomena influenced by epigenetic modification that is independent on DNA methylation since it is based on histone modification and therefore is reset in each generation. In this case, resetting is biologically functional since plants must experience an appropriate cold period, which may vary yearly, to find the best flowering period.

3.2 Silencing of Transposable Elements

Transposable elements (TE) are mobile units of genomic DNA. They contribute to genomic size and impact the entire genome depending on the species, i.e., ~60% in *Zea mays*, ~50% in *Cicer arietinum*, ~14% in *Oryza sativa* and *Arabidopsis thaliana*, ~3% in *Saccharomyces cerevisiae*, ~6% in *Caenorhabditis elegans*, and ~44% in *Homo sapiens* (Kidwell 2002). In man, TE are almost completely silent (Pace and Feschotte 2007), whereas in plants they are particularly active and are involved in chromosome architecture and gene regulation, depending also on where they locate (Sigman and Slotkin 2016). Since the discovery of TE in plants in the 1950s by Barbara McClintock (Nobel prize winner in Physiology or Medicine in 1983), it is clear that they are particularly prone to epigenetic silencing. Indeed, TE possess high reactivation capacity and mutagenic potential and therefore appear as invasive DNA capable of altering genome functionality. As counter-defense, eukaryotes have evolved RNA silencing mechanisms with the primary function of inactivating TE and preserving genomic functionality (Plasterk 2002).

3.3 The RNA Silencing Pathways Involved in TE Silencing: 24nt-Long and 22nt-Long siRNAs in RNA-Directed DNA Methylation

The RNA silencing pathway that appears to be specifically associated with transposon and heterochromatic repeat silencing involves several components. In *Arabidopsis*, these include DICER-LIKE3 (DCL3) (a RNase3-like protein, formally Dicer-like, DCL), ARGONAUTE 4 (AGO4), RNA-dependent RNA polymerase 2 (RDR2), and the components of RNA polymerase IV. This pathway appears to be functionally distinct from that which is involved in micro (mi)RNA processing or antiviral RNA silencing, which also have to do with the production and use of siRNAs. Thus, although mutations in DCL3 and RDR2 eliminate siRNAs from some transposons, they have no effect on miRNA accumulation; the same is true for AGO4 (Chan et al. 2004; Chellappan et al. 2010; Herr et al. 2005; Xie et al. 2004; Zilberman et al. 2003). The available evidence suggests that these factors cooperate to maintain and/or initiate heterochromatic silencing of many endogenous TE elements through RNA-dependent DNA methylation. In *Arabidopsis*, a genomic-wide approach has revealed, for instance, that DNA methylation is particularly frequent in relation to chromosomal regions with high density of repetitive elements such as transposons. This confirms the strong involvement of RdDM in TE epigenetic control (Zhang et al. 2006).

DCL3 is known to produce 24nt-long siRNAs and, indeed, they are mainly involved in RdDM (Henderson et al. 2006). Recently, it has been discovered that 21–22nt-long siRNAs, a typical length of siRNAs involved in post-transcriptional regulation and antiviral silencing, derived from TE transcripts can be incorporated into AGO6, thus guiding it to methylate TEs at DNA levels (McCue et al. 2015). These findings elegantly demonstrate the biological bridges played by AGO6 between post-transcriptional gene silencing and RdDM. An additional link between post-transcriptional silencing and RdDM is that 24nt-long miRNAs (known to be gene regulators at post-transcriptional level) can enter into AGO4 and mediate RdDM of the loci from where they derive (i.e., miRNA genes or their targets (Wu et al. 2010; Chellappan et al. 2010). Importantly, the movement of 24nt-long sRNAs regulates patterns of TE DNA methylation at wide-genomic level (Lewsey et al. 2016). This is practically important in agriculture where the plant grafting is routinely used to combine rootstocks and scions, and it is even more important since the DNA methylation patterns in grafted plants may be heritable by the progeny (Wu et al. 2013). It is noteworthy that the mechanisms of siRNA biogenesis mediated by DCLs are temperature sensitive and have been shown to have a short-lasting effect against viruses (Szittya et al. 2003) (see also Sect. 3.6).

Most of the DNA methylation marks at TE level are transgenerationally inherited but, as other epigenetic modifications, could show instability and can be influenced by the environment. The transcription of the copia-like retrotransposon denoted as "ONSEN" and the generation of DNA copies are extremely active in *Arabidopsis* under heat stress. After stress, the transcript and DNA extra copies gradually decay and are no longer present after 20–30 days. This decay effect is associated with siRNA biogenesis pathways. Indeed, in plants impaired for siRNA biogenesis the ONSEN decay is not present and, furthermore, a high frequency of ONSEN insertion is present in the progeny of the stressed plant (Ito et al. 2011).

3.4 Parental Imprinting

Parental imprinting leads to differential allelic expression depending on whether a gene was inherited through a female or male gametophyte. The phenomenon is well known both in animals and plants; despite a different evolutionary origin, imprinting always occurs in tissues deputed to sustain embryo growth, such as endosperm in plants and placenta in mammalians. The biological function of imprinting has not yet been fully clarified (Jiang and Kohler 2012).

In angiosperm (flowering plants), parental imprinting involves the 2-celled pollen grains and the 8-celled embryo sac, the male and female gametophytes, respectively. One sperm cell mates with an egg cell (forming the 2*n* central cell) and a second cell fuses with the binucleated central cell of the embryo sac (forming the 3*n* embryonic cell), leading to development of the embryo and the triploid endosperm, respectively.

The phenotype of the progeny is imprinted by the maternal genotype: this is due to the fact that cytosine residues of the maternal genes [i.e., medea (*MEA*), fertilization-independent seed2 (*FIS2*), and flowering wageningen (*FWA*)] are hypomethylated by DME in the central cell were MET1 is suppressed since female gametogenesis. In contrast, in somatic cells cytosine methylation is maintained by MET1 [reviewed by Iwasaki and Paszkowski (2014)]. As described above, molecular marks of imprinting are DNA methylations, but not only. Indeed, the evolutionary conserved POLYCOMB GROUP PROTEIN 2 (PRC) that includes MEA and FIS2 catalyzes histone methylation of the paternal alleles when the maternal alleles are expressed [reviewed by Jiang and Kohler (2012)]. Furthermore, recent findings indicate that DNA methylation has antagonistic behavior with PRC2 binding and histone modification (Weinhofer et al. 2010); this could be a general feedback mechanism in epigenetic regulation.

3.5 Paramutation

In paramutation, a paramutagenic allele of a gene can alter the expression of a second paramutable allele of the same gene. In many cases, the altered allele can then become paramutagenic itself and alter the functionality of a locus. The phenomenon has best been studied in maize for genes (i.e., *BOOSTER* and *RED* loci) deputed to determine the antocyanic pigmentation of the kernel (Chandler et al. 2000). Paramutation is also present in tomato plants, in the case of the sulfurea (*SULF*) locus; fully paramutated *sulf* tissues are chlorotic, and paramutation is associated with reduced auxin production (Ehlert et al. 2008). Paramutation has been associated to the process of RdDM in which paramutagenic-derived siRNAs mediate the silencing of the paramutable allele (Chandler and Stam 2004; Gouil et al. 2016).

3.6 Virus-Induced Gene Silencing

RNA silencing is a conserved mechanism present in all the kingdoms of living organisms that underwent to a wide diversification in different lineages. In plants, the diversity of RNA silencing pathways is well palpable probably due to the extreme plasticity of plant genomes, i.e., through duplication, neofunctionalization, and specialization of RNA silencing factors (Zamore 2002). These variations involve endogenous genetic elements including genes, viruses, and transposable

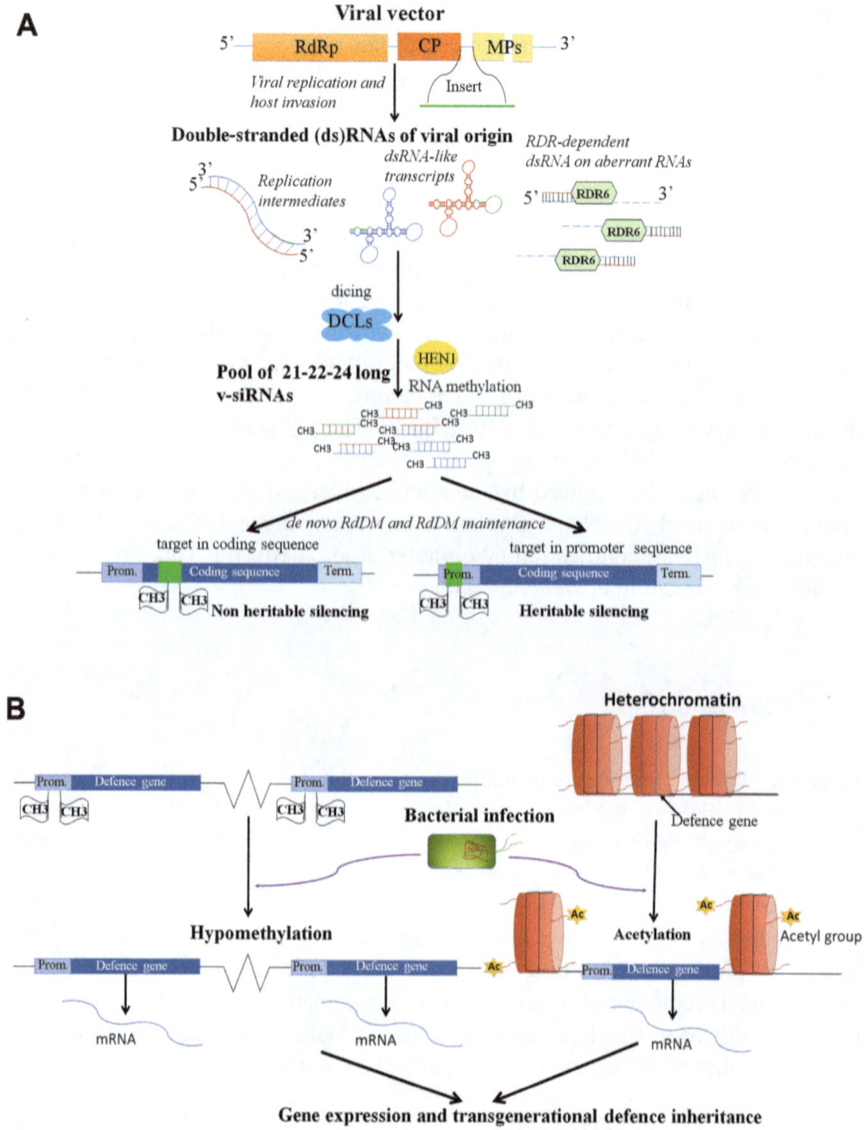

Fig. 2 Virus-induced and bacteria-induced epigenetic modifications. (**a**) *Virus-induced gene silencing*: many viral vector constructs are RNA replicating molecules that encode structural (i.e., coat protein—CP) and nonstructural viral proteins (i.e., viral RNA-dependent RNA polymerases—RdRp and movement proteins—MP). Viral vectors can be used to carry inserts (represented by *green color lines*) corresponding either to promoter (Prom.) or to coding sequences (*green boxes* in gene locus representation). Viral vectors replicate in host cells and invade host plant tissues. Double-stranded (ds) RNA of viral origin could be viral replication intermediates, folded dsRNA-like viral genomic (*blue line*) or antigenomic (*red lines*) RNAs and/or RDR6-dependent (RDR6 = host RNA-DEPENDENT RNA POLYMERASE 6) dsRNAs from viral aberrant RNAs. Viral dsRNAs are processed by plant DICER-LIKE PROTEINS (DCLs) into viral (v) siRNAs 21, 22, and 24 nucleotide (nt) long. v-siRNAs are then stabilized by HEN1 RNA METHYLASE. The pool of v-siRNAs includes those deriving from the insert region (in *green*); they drive de novo DNA methylation in a sequence-specific manner, thus inducing epigenetic

elements and indeed the process of generation and action of VIGS somehow overlaps with RdRM (see also Sect. 3.2).

RNA silencing-based antiviral defense operates when plant cells recognize the double-stranded (ds) RNAs of viral origin as foreign invaders. dsRNA of viral origin is processed mainly by DCL2, DCL3, and DCL4 into 21–24-nt siRNAs (Fig. 2a) [reviewed by Shimura and Pantaleo (2011)]. The 3′-overhang ds-siRNAs of viral origin (v) are incorporated into an effector ribonucleoprotein complex that also contains an AGO nuclease (RNA-induced silencing complex, RISC). AGO1, AGO2, AGO4, AGO5, AGO7, and AGO10 were shown to be implicated in antiviral defense plants (Carbonell and Carrington 2015). AGOs have structural similarity to ribonuclease H and the v-siRNA serves as a guide for AGO: siRNA guides RISC in a sequence-specific manner to its target for degradation or for suppression of translation. The most frequent, but not exclusively, target of v-siRNAs would, of course, be the viral RNA itself (Dunoyer and Voinnet 2005). The steady-state level of virus accumulation is therefore a reflection of the relative kinetics of silencing and the rate of virus accumulation. In complex ecosystems, antiviral RNA silencing is likely to be considered as a mechanism ensuring the coexistence of the host and the infectious entities (Csorba et al. 2015). Furthermore, the viral steady-state level in plant cells and tissues is also influenced by the extent to which the viral RNA can evade the silencing machinery during replication in protected cellular compartments, by encapsidation or through the action of viral silencing suppressors [reviewed by Burgyan and Havelda (2011) and Pantaleo (2011)].

3.6.1 RdRM-Induced by Viral and Subviral Infectious Entities

Viroids are plant infectious entities consisting of circular rod-shaped, non-coding RNAs of a few hundreds of nucleotides (Flores et al. 2005). They are taxonomically classified as subviral agents being smaller than viruses and sharing only a few of their characteristics (ICTV 2012). RdDM was first detected in *Potato spindle tuber viroid* (PSTVd)-infected tobacco plants (Wassenegger et al. 1994). In the original experiments, viroidal genomic sequence was integrated as a cDNA transgene into tobacco plants. As a result, it suddenly became de novo methylated at cytosine residues in the entire sequence with at least 30 bp in length (Pelissier et al. 1999;

Fig. 2 (continued) modification of corresponding gene region. DNA methylation within promoter regions could be maintained and therefore persist into the plant progeny. (**b**) *Bacteria-induced epigenetic modifications*: plants have evolved the ability to prime the response against pathogens promoting a fast and strong induction of defense in case of recurrent attacks. In noninfected plants, several cellular mechanisms mediate the epigenetic suppression of defense systems through the hypermethylation of cytosine in the promoters of defense genes and the heterochromatin formation. In response to bacterial infection, the cytosine hypomethylation and the histone acetylation in sites adjacent to defense genes increase the transcription of the same genes, and it contributes to the induction of a resistance response transmitted on to offspring

Pelissier and Wassenegger 2000). The same discovery of de novo methylation was shortly extended to the chalcone sintase (*CHS*) reporter transgene in tobacco (Ingelbrecht et al. 1994).

Further connections between RdRM and pathogenic viral entities were obtained shortly thereafter. Maule's group (1998) has indeed described Pea seed-borne mosaic virus (PSbMV)-induced post-transcriptional gene silencing (PTGS) in peas in association with methylation of the corresponding PSbMV transgene (Jones et al. 1998). Afterwards, an epigenetic rather than post-transcriptional effect of the virus-derived 24-nt siRNA was highlighted by the use of green fluorescent protein (*GFP*) reporter gene transgenic plants (Jones et al. 2001; Ruiz et al. 1998). Recombinant potato virus X (PVX) carrying full-length GFP, or parts thereof, was used to infect *Nicotiana benthamiana* plants transgenic for *GFP* under the control of the cauliflower mosaic virus (CaMV) promoter 35S. To induce epigenetic silencing, the insert in the virus vector had to correspond to the promoter of transgene 35S rather than the *GFP;* the silencing was indeed inherited. Furthermore, the same authors show that heritable *GFP* silencing is caused by 24-nt v-siRNA-mediated DNA methylation of the associated gene. RdDM mechanism can also target endogenous plant genomic elements if there is sequence similarity between the virus and the plant genome, and in this case the silencing is inherited as well (Jones et al. 2001). More recent works have consolidated the idea that 24nt long v-siRNAs are involved in RdDM, indeed they look able to reinforce and maintain RdDM. In addition, 21–22nt-long v-siRNAs exhibit a capacity to establish RdDM using a non-canonical pathway that involves PolV and DRM2 instead of PolIV (Bond and Baulcombe 2015). In Fig. 2a, a schematic representation of viral vector-induced RdDM is reported.

3.6.2 Influence of Viral Silencing Suppressor on Virus-Induced RdRM

After RdDM was reported for PSTVd (Wassenegger et al. 1994) for PVX vector (Jones et al. 1999) and for Tobacco rattle virus (TRV) vector (Jones et al. 2001) harboring a portion of the transgene sequence, several additional studies were carried out in order to better reveal the mechanisms regulating the phenomenon. In 2006, Kanazawa and colleagues published the setup of an additional viral vector able to induce RdDM—the tripartite plant virus Cucumber mosaic virus (CMV) (Otagaki et al. 2006, 2011). In contrast to other viral vectors, CMV was shown to remarkably succeed in inducing RdDM of several plant endogenous genes; such efficient RdDM induction is achieved by the function of the well-described viral encoded silencing suppressor 2b (Kanazawa et al. 2011a, b).

Viral silencing suppressors (VSR) have been shown to play roles as effectors in counter-defense of viruses against antiviral RNA silencing [reviewed by Burgyan and Havelda (2011)]. The main mechanism of action of VSR, though not exclusive, is siRNA duplex binding, thus preventing v-siRNA incorporation into the antiviral AGOs; CMV2b has been characterized with dsRNA binding activity (Gonzalez et al. 2010; Goto et al. 2007). Furthermore, once translated in the cytoplasm from

CMV viral genomic RNA3, the 2b protein moves for nuclear localizations (Lucy et al. 2000; Wang et al. 2004); in this manner, CMV2b facilitates virus-induced and inherited epigenetic modification infected plants through the transport of siRNAs to the nucleus (Kanazawa et al. 2011a).

VSRs target various components of the plant's silencing machinery. Plant viruses can induce specific symptoms resembling developmental anomalies and affect organs and tissues such as leaves, flowers, and fruits. These anomalies are often reconcilable with virus-induced alterations of RNA silencing-based endogenous pathways through: (1) the direct activity of VSRs on endogenous siRNAs or on silencing-related effectors; (2) the abundance of v-siRNAs in competition with endogenous sRNAs; and (3) the action of specific v-siRNAs entering into RNA silencing complexes and targeting specific host genes. It is therefore not unlikely that their presence in plants can alter the functionality of endogenous RdDM.

It is noteworthy that viral infections may induce transient modifications of RdDM leading to a hypomethylated state of the genome that, in turn, can result in reactivation of retrotransposons (Grandbastien et al. 1997).

In the case of the nuclear replicating viruses with the circular dsDNA genome, Geminivirus, the role of RdDM pathways in plant–virus interactions is more evident. Upon infection by the geminiviruses—*Cabbage leaf curl virus* (CaLCuV) and Beet curly top virus (BCTV)—AGO4 seems to contribute to the heterochromatinization of viral genomes. Indeed, *Arabidopsis* mutants of cytosine and histone methyltransferases, of methyl cycle enzymes, and of other components of RdDM, including AGO4, are hypersusceptible to viral infection (Raja et al. 2008). Thus, RdDM has a clear and direct antiviral role against geminiviruses; this is further supported by the fact that (1) most of the v-siRNAs from geminiviruses are of 24nt (Miozzi et al. 2013; Rodriguez-Negrete et al. 2009) and that (2) - geminivirus-encoded V2 protein can suppress methylation-mediated transcriptional silencing (Wang et al. 2014).

4 Epigenetic Modifications and Systemic Acquired Resistance

Besides viruses, pathogenic bacteria and fungi can also invade plant tissues. Plants have an innate defense system that is based on the ability to quickly and specifically modulate its own transcriptome and activate the defense responses against bacterial and fungal pathogens. Moreover, plants have evolved the ability to prime such an "immune" system, thus promoting a faster and stronger induction of defense in the event of recurrent attacks (Conrath 2011; Luna et al. 2012). For example, a localized infection caused by a pathogen can often induce systemic immunity termed Systemic Acquired Resistance (SAR). SAR is associated with the priming phenomenon: plants are able to respond faster and/or to a greater extent to subsequent infection [reviewed by Jaskiewicz et al. (2011)]. SAR requires the accumulation of salicylic acid (SA) hormone, the protein NON EXPRESSOR OF PR1 (NPR1), and is mediated by a multitude of expressed signaling networks that can vary according to the environment and pathogen (Durrant and Dong 2004). Jaskiewicz and collaborators (2011) showed that treatment of a plant with SA (or synthetic SA-analogues) induces the expression of several WRKY transcription factors in response to a second stress (Jaskiewicz et al. 2011). This modulation is associated with some histone modifications, such as H3 and H4 acetylation and H3K4 methylation. Such histone modifications are likely responsible for generating a memory of the primary infection that is associated with an amplified reaction to a second stress. Importantly, SA-mediated memory of infection can be inherited by the progeny in specific conditions, with yet unrevealed mechanisms (see later in the text) (Jaskiewicz et al. 2011).

The *Arabidopsis-Pseudomonas syringae* pathosystem has been studied extensively over the years by several authors to highlight the interactions between SAR, priming, and epigenetic modifications. One of the first reports dates back to 2006, when Pavet and collaborators reported the hypomethylation of several cytosines in peri/centromeric regions in *Arabidopsis* infected by *P. syringae*, likely generated by an active demethylation process (Pavet et al. 2006). This work is in line with those reported by Grandbastien (1997) years before, who observed an enhanced expression of Tnt1 retrotransposons in tobacco, *Arabidopsis*, and tomato exposed to factors of microbial origin (including viruses), by external stressors and by pathogen attacks (Grandbastien et al. 1997) (see also Sect. 3.6.2). Tnt1 expression is linked to the early steps of the metabolic pathways leading to the activation of plant defense genes. Nowadays, it well known that the regulation of transposon is under epigenetic control. In recent years, Luna et al. (2012) have demonstrated how SAR can be inherited epigenetically in *Arabidopsis* infected by *P. syringae* pv tomato DC3000 (*PstDC3000*) (Luna et al. 2012). Progeny from *PstDC3000*-inoculated plants activate SA defense genes and are more resistant to *Hyaloperonospora arabidopsidis* and recurrent infections of *PstDC3000*. This transgenerational SAR was sustained over one stress-free generation, suggesting an epigenetic basis of the phenomenon. Furthermore, this progeny displayed reduced responsiveness to jasmonic acid (JA)-inducible genes and enhanced susceptibility to a necrotrophic fungus without changes in corresponding hormone levels. Chromatin immunoprecipitation analyses revealed that SA-inducible promoters of pathogenesis-related protein 1 (*PR1*), *WRKY6*, and *WRKY53* in progeny are enriched with acetylated histone H3 (H3K9ac). Conversely, the JA-inducible promoter of plant defensin1.2 (*PDF1.2*) showed increased H3 triple methylation at lysine 27 (H3K27me3). The latter is an epigenetic mark related to repressed gene transcription. Luna and collaborators (2012) have also showed how the transgenerational acetylation of H3K9 requires an intact NPR1 protein since mutant (*npr1*)-1 failed to induce transgenerational defense phenotypes and the triple mutant *drm1drm2cmt3* that is affected in non-CpG DNA methylation mimicked the transgenerational SAR phenotype. The induction of DNA hypomethylation in Arabidopsis by *PstDC3000* suggests that transgenerational SAR is likely transmitted by hypomethylated genes that direct priming of SA-dependent defenses in the subsequent generations (Luna et al. 2012; Pavet et al. 2006). This represents an interesting phenotypically plastic mechanism for enhanced defense across generations.

Similar transgenerational resistance phenomena in response to priming-inducing stimuli have also been reported. Slaughter et al. (2012) reported a faster and greater accumulation of transcripts of defense genes in the SA signaling pathway and enhanced disease resistance upon inoculation with a virulent isolate of *P. syringae* in descendants of primed *Arabidopsis* plants with an avirulent isolate of the bacteria *P. syringae* pv tomato (PstavrRpt2) (Slaughter et al. 2012). Rasmann et al. (2012) showed that *Arabidopsis* and tomato subjected to herbivory or mechanical damage produce progeny resistance associated with transgenerational priming of JA-dependent defense responses in both species (Rasmann et al. 2012). *Arabidopsis* mutants that are deficient in jasmonate perception or in the biogenesis of siRNAs (*dcl2dcl3dcl4*, nuclear RNA polymerase d2a-*nrpd2a*, and nuclear RNA polymerase d2b-*nrpd2b*) do not exhibit inherited resistance. Thus, DNA methylation, which is impacted by PolIV- and DCL2-dependent siRNA production and can be inherited through meiosis, is a possible mechanism of transgenerational inheritance.

Interestingly, the hypomethylation state (especially at sites adjacent to defense genes) in response to bacterial infection was highlighted by other authors as well. For instance, in response to SA, DNA methylation changes within repetitive and transposable elements and can regulate neighboring genes (Dowen et al. 2012). Moreover, reactivation of gypsy-like retrotransposons was clearly observed during bacterial infection, and DNA demethylation is associated with activation of not-better identified disease resistance genes that could limit the spread of the pathogen throughout the plant xylem vessels (Yu et al. 2013). This effect is correlated with the downregulation of transcriptional gene silencing factors and is partly dependent on an active demethylation process. DNA demethylation restricts multiplication of the bacterial pathogen and, accordingly, some immune-response genes are negatively regulated by DNA methylation. These results suggest that in noninfected plants, several cellular mechanisms may mediate the epigenetic suppression of defense systems. One possible explanation for this apparent contradiction, namely, why a plant should silence resistance genes, could be linked to fitness cost to the plant to support the disease resistance (Tian et al. 2003). The silencing of the region regulating the transcription of defense genes by hypermethylation would be beneficial in environments, where infection pressure is low and substantially unnecessary. Conversely, in environments with high infection pressure the hypomethylation of sites adjacent to resistance genes would activate the defense priming system and mitigate the fitness penalty due to disease. The transgenerational transmission of this information might be especially advantageous because it would allow rapid adaptation of plant populations to changes in infection pressure (Fig. 2b) (Bond and Baulcombe 2014).

In addition to the *Arabidopsis-P. syringae* system, other plant–pathogen interactions are influenced by epigenetic changes in the plant genome, in particular by hypomethylation. In rice, the demethylation of the promoter abolished the constitutive silencing of the resistance gene Xa21G due to hypermethylation, resulting in acquisition of disease resistance. Furthermore, both hypomethylation and the resistant trait were stably inherited (Akimoto et al. 2007).

In another interesting study, Gohlke et al. (2013) provide evidence that epigenetic changes regulate transcription, physiological processes, and the development of crown gall tumors induced after integration of the T-DNA of *Agrobacterium tumefaciens* strains into the plant genome (Gohlke et al. 2013). The methylation pattern of *Arabidopsis* crown galls was analyzed by a genome-wide approach after bisulfite sequencing. Interestingly, the oncogenes isopentenyl transferase (*IPT*), tryptophan monooxygenase (*IAAM*), and indole-3-acetamide hydrolase (*IAAH*, responsible for the proliferation of transformed cells in crown galls) were unmethylated in crown galls, and siRNA-mediated promoter methylation caused transcriptional silencing of oncogenes preventing crown gall proliferation. *Arabidopsis* mutants with reduced methylation developed larger tumors than the wild-type controls. In addition, the authors demonstrate a close link between the phytohormone Abscissic Acid (ABA) and these epigenetic regulations: high ABA levels in crown galls may mediate DNA methylation and regulate expression of genes involved in drought stress (Gohlke et al. 2013).

Hypomethylation of the regulatory sequences of resistance genes observed in different pathosystems requires a pivotal role of DNA demethylases. A triple DNA demethylase mutant of *Arabidopsis* repressor of silencing 1 (*ros1*), demeter-like 2 (*dml2*), and *dml3* shows increased susceptibility to the fungal pathogen *Fusarium oxysporum* (Le et al. 2014). The same authors identified a significant proportion of genes involved in stress response downregulation in a triple mutant, suggesting that DNA demethylases maintain or positively regulate the expression of these genes. In addition, a general reduction in CHH methylation was observed, suggesting that RdDM is responsible for maintenance of CHH methylation and may participate in DNA demethylase-mediated regulation of stress response genes. Indeed, the RdDM mutants *nrpe1* and *ago4* demonstrate enhanced susceptibility to *F. oxysporum* infection. Likewise, *Arabidopsis* null- or loss-of-function mutants of *nrpd2* or *ago4*, the second largest subunit of Pol IV and V, appear more susceptible to *Botrytis cinerea* and *Plectosphaerella cucumerina* infections (Lopez et al. 2011).

5 Conclusions and Perspectives

More and more studies have shown that many host gene expressions are modulated and reprogrammed in response to various pathogenic challenges. These pathogens are known to promote the modulation of non-coding RNAs (including siRNAs), which in turn alter gene expression through post-transcriptional gene silencing by mRNA degradation, or translational inhibition, or transcriptional gene silencing by direct DNA methylation or chromatin modification. As a counter-measure, viruses and bacteria have developed strategies (including silencing suppressor) to suppress host RNAi machinery and compromise disease resistance in plants. Silencing suppressors or pathogen-derived non-coding RNAs may have an off-target effect, since they have the property to intercept RNAi pathways at any step; pathogen-associated symptoms could often be ascribable to specific alterations of endogenous epigenetic phenomena. To continuously combat evolving pathogens, plants also have evolved components, such as resistance proteins (including NB-LRR) that can recognize pathogen elicitors and trigger robust and rapid resistance responses.

The study of epigenetic-based mechanisms in plant immunity is an emerging field, and we expect that increasing and novel scientific evidence will emerge employing state-of-the-art RNA technology, such as RNA deep sequencing. We also foresee that the identification and characterization of epigenetic mechanisms, which are at the basis of plant interactions with viruses, bacteria, fungi, and oomycetes, will increase our understanding of the biodiversity and coevolution between pathogens and plant hosts. These studies will elucidate the molecular mechanisms of plant defense responses and ultimately lead to the development of efficient and effective strategies for sustainable plant protection.

References

Akimoto K, Katakami H, Kim HJ et al (2007) Epigenetic inheritance in rice plants. Ann Bot 100:205–217

Angel A, Song J, Dean C, Howard M (2011) A Polycomb-based switch underlying quantitative epigenetic memory. Nature 476:105–108

Bauer MJ, Fischer RL (2011) Genome demethylation and imprinting in the endosperm. Curr Opin Plant Biol 14:162–167

Berger SL, Kouzarides T, Shiekhattar R, Shilatifard A (2009) An operational definition of epigenetics. Genes Dev 23:781–783

Berry S, Dean C (2015) Environmental perception and epigenetic memory: mechanistic insight through FLC. Plant J 83:133–148

Bond DM, Baulcombe DC (2014) Small RNAs and heritable epigenetic variation in plants. Trends Cell Biol 24:100–107

Bond DM, Baulcombe DC (2015) Epigenetic transitions leading to heritable, RNA-mediated de novo silencing in *Arabidopsis thaliana*. Proc Natl Acad Sci USA 112:917–922

Burgyan J, Havelda Z (2011) Viral suppressors of RNA silencing. Trends Plant Sci 16:265–272

Cao R, Wang L, Wang H et al (2002) Role of histone H3 lysine 27 methylation in Polycomb-group silencing. Science 298:1039–1043

Carbonell A, Carrington JC (2015) Antiviral roles of plant ARGONAUTES. Curr Opin Plant Biol 27:111–117

Chan SW, Zilberman D, Xie Z et al (2004) RNA silencing genes control de novo DNA methylation. Science 303:1336

Chandler VL, Stam M (2004) Chromatin conversations: mechanisms and implications of paramutation. Nat Rev Genet 5:532–544

Chandler VL, Eggleston WB, Dorweiler JE (2000) Paramutation in maize. Plant Mol Biol 43:121–145

Chellappan P, Xia J, Zhou X et al (2010) siRNAs from miRNA sites mediate DNA methylation of target genes. Nucleic Acids Res 38:6883–6894

Chen M, Lv S, Meng Y (2010) Epigenetic performers in plants. Dev Growth Differ 52:555–566

Conrath U (2011) Molecular aspects of defence priming. Trends Plant Sci 16:524–531

Csorba T, Kontra L, Burgyan J (2015) viral silencing suppressors: tools forged to fine-tune host-pathogen coexistence. Virology 479–480:85–103

Ding Y, Fromm M, Avramova Z (2012) Multiple exposures to drought 'train' transcriptional responses in Arabidopsis. Nat Commun 3:740

Dowen RH, Pelizzola M, Schmitz RJ et al (2012) Widespread dynamic DNA methylation in response to biotic stress. Proc Natl Acad Sci USA 109:E2183–E2191

Dunoyer P, Voinnet O (2005) The complex interplay between plant viruses and host RNA-silencing pathways. Curr Opin Plant Biol 8:415–423

Durrant WE, Dong X (2004) Systemic acquired resistance. Annu Rev Phytopathol 42:185–209

Ehlert B, Schottler MA, Tischendorf G et al (2008) The paramutated SULFUREA locus of tomato is involved in auxin biosynthesis. J Exp Bot 59:3635–3647

Finnegan EJ, Dennis ES (2007) Vernalization-induced trimethylation of histone H3 lysine 27 at FLC is not maintained in mitotically quiescent cells. Curr Biol 17:1978–1983

Flores R, Hernandez C, Martinez de Alba AE et al (2005) Viroids and viroid-host interactions. Annu Rev Phytopathol 43:117–139

Furner IJ, Matzke M (2011) Methylation and demethylation of the Arabidopsis genome. Curr Opin Plant Biol 14:137–141

Gendall AR, Levy YY, Wilson A et al (2001) The VERNALIZATION 2 gene mediates the epigenetic regulation of vernalization in Arabidopsis. Cell 107:525–535

Gohlke J, Scholz CJ, Kneitz S et al (2013) DNA methylation mediated control of gene expression is critical for development of crown gall tumors. PLoS Genet 9:e1003267

Gonzalez I, Martinez L, Rakitina DV et al (2010) Cucumber mosaic virus 2b protein subcellular targets and interactions: their significance to RNA silencing suppressor activity. Mol Plant-Microbe Interact 23:294–303

Goto K, Kobori T, Kosaka Y et al (2007) Characterization of silencing suppressor 2b of cucumber mosaic virus based on examination of its small RNA-binding abilities. Plant Cell Physiol 48:1050–1060

Gouil Q, Novak O, Baulcombe DC (2016) SLTAB2 is the paramutated SULFUREA locus in tomato. J Exp Bot 67:2655–2664

Grandbastien MA, Lucas H, Morel JB et al (1997) The expression of the tobacco Tnt1 retrotransposon is linked to plant defense responses. Genetica 100:241–252

Henderson IR, Zhang X, Lu C et al (2006) Dissecting *Arabidopsis thaliana* DICER function in small RNA processing, gene silencing and DNA methylation patterning. Nat Genet 38:721–725

Herr AJ, Jensen MB, Dalmay T et al (2005) RNA polymerase IV directs silencing of endogenous DNA. Science 308:118–120

ICTV (2012) Virus taxonomy: ninth report of the international committee on taxonomy of viruses. Elsevier/Academic Press, New York

Ingelbrecht I, Van Houdt H, Van Montagu M, Depicker A (1994) Posttranscriptional silencing of reporter transgenes in tobacco correlates with DNA methylation. Proc Natl Acad Sci USA 91:10502–10506

Ito H, Gaubert H, Bucher E et al (2011) An siRNA pathway prevents transgenerational retrotransposition in plants subjected to stress. Nature 472:115–119

Iwasaki M, Paszkowski J (2014) Epigenetic memory in plants. EMBO J 33:1987–1998

Jaskiewicz M, Conrath U, Peterhansel C (2011) Chromatin modification acts as a memory for systemic acquired resistance in the plant stress response. EMBO Rep 12:50–55

Jiang H, Kohler C (2012) Evolution, function, and regulation of genomic imprinting in plant seed development. J Exp Bot 63:4713–4722

Jones AL, Thomas CL, Maule AJ (1998) De novo methylation and co-suppression induced by a cytoplasmically replicating plant RNA virus. EMBO J 17:6385–6393

Jones L, Hamilton AJ, Voinnet O et al (1999) RNA-DNA interactions and DNA methylation in post-transcriptional gene silencing. Plant Cell 11:2291–2301

Jones L, Ratcliff F, Baulcombe DC (2001) RNA-directed transcriptional gene silencing in plants can be inherited independently of the RNA trigger and requires Met1 for maintenance. Curr Biol 11:747–757

Kanazawa A, Inaba J, Kasai M et al (2011a) RNA-mediated epigenetic modifications of an endogenous gene targeted by a viral vector: a potent gene silencing system to produce a plant that does not carry a transgene but has altered traits. Plant Signal Behav 6:1090–1093

Kanazawa A, Inaba J, Shimura H et al (2011b) Virus-mediated efficient induction of epigenetic modifications of endogenous genes with phenotypic changes in plants. Plant J 65:156–168

Kidwell MG (2002) Transposable elements and the evolution of genome size in eukaryotes. Genetica 115:49–63

Kim JM, To TK, Ishida J et al (2012) Transition of chromatin status during the process of recovery from drought stress in *Arabidopsis thaliana*. Plant Cell Physiol 53:847–856

Kuo MH, Brownell JE, Sobel RE et al (1996) Transcription-linked acetylation by Gcn5p of histones H3 and H4 at specific lysines. Nature 383:269–272

Le TN, Schumann U, Smith NA et al (2014) DNA demethylases target promoter transposable elements to positively regulate stress responsive genes in Arabidopsis. Genome Biol 15:458

Lewsey MG, Hardcastle TJ, Melnyk CW et al (2016) Mobile small RNAs regulate genome-wide DNA methylation. Proc Natl Acad Sci USA 113:E801–E810

Lopez A, Ramirez V, Garcia-Andrade J et al (2011) The RNA silencing enzyme RNA polymerase v is required for plant immunity. PLoS Genet 7:e1002434

Lucy AP, Guo HS, Li WX et al (2000) Suppression of post-transcriptional gene silencing by a plant viral protein localized in the nucleus. EMBO J 19:1672–1680

Luna E, Bruce TJ, Roberts MR et al (2012) Next-generation systemic acquired resistance. Plant Physiol 158:844–853

McCue AD, Panda K, Nuthikattu S et al (2015) ARGONAUTE 6 bridges transposable element mRNA-derived siRNAs to the establishment of DNA methylation. EMBO J 34:20–35

Michaels SD, Amasino RM (1999) Flowering locus C encodes a novel MADS domain protein that acts as a repressor of flowering. Plant Cell 11:949–956

Miozzi L, Pantaleo V, Burgyan J et al (2013) Analysis of small RNAs derived from tomato yellow leaf curl Sardinia virus reveals a cross reaction between the major viral hotspot and the plant host genome. Virus Res 178:287–296

Otagaki S, Arai M, Takahashi A et al (2006) Rapid induction of transcriptional and post-transcriptional gene silencing using a novel Cucumber mosaic virus vector. Plant Biotechnol 23:256–265

Otagaki S, Kawai M, Masuta C et al (2011) Size and positional effects of promoter RNA segments on virus-induced RNA-directed DNA methylation and transcriptional gene silencing. Epigenetics 6:681–691

Pace JK 2nd, Feschotte C (2007) The evolutionary history of human DNA transposons: evidence for intense activity in the primate lineage. Genome Res 17:422–432

Pantaleo V (2011) Plant RNA silencing in viral defence. Adv Exp Med Biol 722:39–58

Pavet V, Quintero C, Cecchini NM et al (2006) Arabidopsis displays centromeric DNA hypomethylation and cytological alterations of heterochromatin upon attack by pseudomonas syringae. Mol Plant-Microbe Interact 19:577–587

Pelissier T, Wassenegger M (2000) A DNA target of 30 bp is sufficient for RNA-directed DNA methylation. RNA 6:55–65

Pelissier T, Thalmeir S, Kempe D, Sanger HL, Wassenegger M (1999) Heavy de novo methylation at symmetrical and non-symmetrical sites is a hallmark of RNA-directed DNA methylation. Nucleic Acids Res 27:1625–1634

Plasterk RH (2002) RNA silencing: the genome's immune system. Science 296:1263–1265

Questa JI, Song J, Geraldo N, An H, Dean C (2016) Arabidopsis transcriptional repressor VAL1 triggers Polycomb silencing at FLC during vernalization. Science 353:485–488

Raja P, Sanville BC, Buchmann RC, Bisaro DM (2008) Viral genome methylation as an epigenetic defense against geminiviruses. J Virol 82:8997–9007

Rasmann S, De Vos M, Casteel CL et al (2012) Herbivory in the previous generation primes plants for enhanced insect resistance. Plant Physiol 158:854–863

Rodriguez-Negrete EA, Carrillo-Tripp J, Rivera-Bustamante RF (2009) RNA silencing against geminivirus: complementary action of posttranscriptional gene silencing and transcriptional gene silencing in host recovery. J Virol 83:1332–1340
Ruiz MT, Voinnet O, Baulcombe DC (1998) Initiation and maintenance of virus-induced gene silencing. Plant Cell 10:937–946
Shahbazian MD, Grunstein M (2007) Functions of site-specific histone acetylation and deacetylation. Annu Rev Biochem 76:75–100
Shimura H, Pantaleo V (2011) Viral induction and suppression of RNA silencing in plants. Biochim Biophys Acta 1809:601–612
Sigman MJ, Slotkin RK (2016) The first rule of plant transposable element silencing: location, location, location. Plant Cell 28:304–313
Slaughter A, Daniel X, Flors V et al (2012) Descendants of primed Arabidopsis plants exhibit resistance to biotic stress. Plant Physiol 158:835–843
Sung S, Amasino RM (2004) Vernalization and epigenetics: how plants remember winter. Curr Opin Plant Biol 7:4–10
Szittya G, Silhavy D, Molnar A et al (2003) Low temperature inhibits RNA silencing-mediated defence by the control of siRNA generation. EMBO J 22:633–640
Tian D, Traw MB, Chen JQ et al (2003) Fitness costs of R-gene-mediated resistance in *Arabidopsis thaliana*. Nature 423:74–77
Wang Y, Tzfira T, Gaba V et al (2004) Functional analysis of the Cucumber mosaic virus 2b protein: pathogenicity and nuclear localization. J Gen Virol 85:3135–3147
Wang B, Li F, Huang C et al (2014) V2 of tomato yellow leaf curl virus can suppress methylation-mediated transcriptional gene silencing in plants. J Gen Virol 95:225–230
Wassenegger M, Heimes S, Riedel L et al (1994) RNA-directed de novo methylation of genomic sequences in plants. Cell 76:567–576
Weinhofer I, Hehenberger E, Roszak P et al (2010) H3K27me3 profiling of the endosperm implies exclusion of polycomb group protein targeting by DNA methylation. PLoS Genet 6:E10
Wu L, Zhou H, Zhang Q et al (2010) DNA methylation mediated by a microRNA pathway. Mol Cell 38:465–475
Wu R, Wang X, Lin Y et al (2013) Inter-species grafting caused extensive and heritable alterations of DNA methylation in Solanaceae plants. PLoS One 8:e61995
Xie Z, Johansen LK, Gustafson AM et al (2004) Genetic and functional diversification of small RNA pathways in plants. PLoS Biol 2:E104
Yu A, Lepere G, Jay F et al (2013) Dynamics and biological relevance of DNA demethylation in Arabidopsis antibacterial defense. Proc Natl Acad Sci USA 110:2389–2394
Zamore PD (2002) Ancient pathways programmed by small RNAs. Science 296:1265–1269
Zhang X, Yazaki J, Sundaresan A et al (2006) Genome-wide high-resolution mapping and functional analysis of DNA methylation in arabidopsis. Cell 126:1189–1201
Zhou BR, Feng H, Kato H et al (2013) Structural insights into the histone H1-nucleosome complex. Proc Natl Acad Sci USA 110:19390–19395
Zhu JK (2009) Active DNA demethylation mediated by DNA glycosylases. Annu Rev Genet 43:143–166
Zilberman D, Cao X, Jacobsen SE (2003) ARGONAUTE4 control of locus-specific siRNA accumulation and DNA and histone methylation. Science 299:716–719

Epigenetic Reprogramming During Plant Reproduction

Jer-Young Lin and Tzung-Fu Hsieh

Abstract Epigenetics is the study of heritable change in gene expression state that is independent of DNA sequence variation. Such change can occur through DNA methylation or posttranscriptional modifications of histones. Epigenetic mechanisms play critical roles in regulating gene expression during development and in response to environmental stimulation. Such epigenetic information represents the transcriptional memory associated with cell fate decisions, developmental switches, or stress responses; memory that often needs to be erased and reset during reproduction. By contrast, transgenerational epigenetic information refers to more indelible marks that can be stably transmitted through meiosis and inherited in the subsequent generation. Epigenetic reprogramming, a global change in DNA and/or histone methylation, has been reported during reproduction in mammals and in flowering plants. Such reprogramming is thought to be essential for ensuring meiosis competence, establishing genomic imprinting, and silencing transposons. In Arabidopsis, gene imprinting is a consequence of a large-scale epigenetic reprogramming via DEMETER-mediated active DNA demethylation during gametogenesis. Such reprogramming is believed to be critical for the maintenance of trans-generational epigenome integrity.

Keywords Epigenetics • Reprogramming • Gametogenesis • Reproduction • DNA Methylation • Active DNA Demethylation • Chromatin Remodeling • Transgenerational Inheritance • siRNAs

Contents

J.-Y. Lin
Department of Molecular, Cell, and Developmental Biology, University of California, Los Angeles, CA 90095, USA

T.-F. Hsieh (✉)
Department of Plant and Microbial Biology, Plants for Human Health Institute, North Carolina State University, North Carolina Research Campus, Kannapolis, NC 28081, USA
e-mail: thsieh3@ncsu.edu

N. Rajewsky et al. (eds.), *Plant Epigenetics*, RNA Technologies,
DOI 10.1007/978-3-319-55520-1_20

1 Introduction

Epigenetic marks are enzymatic modifications of DNA and histone proteins that regulate genome function and affect gene activity without altering DNA sequence. They influence higher-order organization of the chromatin, modulate accessibility of transcriptional machinery to the DNA template, and thus have important effects on the cellular transcriptional state. For example, DNA methylation and H3K9 methylation are important for silencing of transposons and repetitive sequences and the formation of heterochromatin. Epigenetic plays important roles in regulating gene expression during differentiation, developmental transition, and in response to environmental stress.

In organisms that reproduce sexually, gametes are derived from specific cell lineage that undergoes somatic-to-reproduction cell fate transition. There are two rounds of DNA methylation reprograming in mammalian life cycle. The first round occurs in the gamete after fertilization and the second round occurs during germline development. Erasing DNA methylation after fertilization occurs in the paternal genome. DNA methylation is rebuilt in the inner cell mass (ICM) of the blastocyst that later forms epiblast, from which the primordial germ cells (PGCs) inherit the DNA methylation. During PGCs migrating, global DNA methylation is erased, which is concomitant with the emerging of developmental potency. During later stages of gametogenesis, global DNA methylation is established again and developmental potency is restricted. Imprinting is established in male germ line during global DNA methylation and then later imprinting occurs in female germ line (Seisenberger et al. 2013).

By contrast, plant germline development does not occur during embryogenesis. Instead, reproductive cells in plants are specified late in development and derived from somatic cells (Walbot and Evans 2003). The life cycle of flowering plants alternates between a dominant diploid sporophytic phase and a reduced haploid gametophytic phase that gives rise to the gametes. Male gametophyte development consists of two mitotic divisions that resulted in a three-celled pollen grain, in which two sperm cells are encased within a larger vegetative cell. The two cell types have a very different nuclear organization—highly condensed chromatin in sperm cell nuclei and diffused chromatin in the vegetative cell. In the female side, the mature gametophyte is a seven-cell structure that consists of one egg cell, two synergid cells, one central cell, and three antipodal cells. The central cell nucleus is formed by the fusion of two haploid polar nuclei. Epigenetic reprogramming does not occur in plant sperm cells and egg cells, which are combined after fertilization to form embryo of the next generation. Instead, epigenetic reprogramming occurs in the companion cells in the gametophyte, the central cells in female gametophyte, and vegetative cells in male gametophyte (Feng et al. 2010).

2 Epigenetic Mechanisms Mediated by DNA Methylation

2.1 DNA Methylation by DNA Methyltransferases

DNA methylation is primarily targeted to transposons and repetitive sequences of the genome to silence and prevent unwanted transposition and to maintain genome integrity. DNA methylation is also found in gene body of some plant genes with moderate and constitutive transcription (Zilberman et al. 2007; Bewick et al. 2016). In plant, DNA methylation is found in CG, CHG, and CHH sequence contexts (where H is A, C, or T). DNA methylation is established de novo by the RNA-directed DNA methylation (RdDM) pathway that involves small RNAs (siRNAs) and maintained by three enzymes, METHYLTRANSFERASE 1 (MET1), CHROMOMETHYLASE 3 (CMT3), and CMT2 CHROMOMETHYLASE 2 (CMT2). The symmetric CG and CHG methylation in Arabidopsis are maintained by MET1 and CMT3, respectively. Asymmetric CHH context is methylated by CMT2 and RdDM pathway. CMT2 is a homologous enzyme of CMT3, and its major target loci are longer transposons in heterochromatin associated with H3K9me2. CMT2 methylation works in concert with a nucleosome remodeling protein DECREASE IN DNA METHYLATION 1 (DDM1), which can remodel compacted heterochromatin for CMT2 to access (Zemach et al. 2013; Stroud et al. 2014).

Methylation at the asymmetric CHH contexts is also methylated by the DOMAINS REARRANGED METHYLTRANSFERASE 2 (DRM2) of the RdDM pathway. Two RNA polymerases have been found to participate in the RdDM in Arabidopsis. In RNA Polymerase II (Pol II)-mediated RdDM, the Pol II-transcribed aberrant RNA is recognized by the RNA-dependent RNA polymerase 6 (RDR6) to generate double-stranded RNA (dsRNA). dsRNA is cleaved by

Dicer-like nucleases 2 (DCL2) and DCL4 to generate 21- and 22-nucleotide (nt) small RNA (siRNA) or by DCL3 to form 24-nt siRNA. The resulting siRNAs are loaded onto the Argonaute protein 4 (AGO4) and AGO6 to form the RNA-induced silencing complexes (RISC). The RISC matches with the Pol II-transcribed RNA and recruits DRM2 to mediate de novo methylation of cytosines in all cytosine contexts. In Pol IV/Pol V-mediated RdDM, Pol IV-transcribed transcripts are used as template by RDR2 to generate dsRNA. Similar to the Pol II-mediated RdDM, the dsRNA is cleaved by DCL3 into 24-nt sRNA and loaded onto AGO4 and AGO6 to target Pol V transcripts for de novo DNA methylation.

2.2 *DNA Demethylation by DEMETER Family Glycosylases*

DNA methylation can be removed by two different means. Passive DNA demethylation can occur when maintenance DNA methylation machinery is suspended and DNA methylation is not maintained in the newly synthesized DNA strands. By contrast, active DNA demthylation is the enzymatic removal of methylated cytosines mediated by the 5-methylcytosine (5mC) DNA glycosylases, which include DEMETER (DME), DEMETER-LIKE2 (DML2), DML3, and REPRESSOR OF SILENCING1 (ROS1) (Choi et al. 2002; Gong et al. 2002). ROS1, DML2, and DML3 are widely expressed in vegetable tissues whereas DME is expressed primarily in reproductive tissues (Huh et al. 2008). The DME-like proteins are bifunctional DNA glycosylases that can excise 5mC and cleave DNA backbone, resulting in a single nucleotide gap flanked with a 5′-phosphate termini and a 3′-phosphate termini or a 3′-phospho-alpha, beta-unsaturated aldehyde (3′-PUA) termini (Agius et al. 2006; Morales-Ruiz et al. 2006). The Arabidopsis apurinic/apyrimidinic endonuclease (AP endonuclease) APE1L converts 3′-PUA or the zinc finger DNA 3′-phosphoesterase (ZDP) converts 3′-phosphate to generate a 3′-OH end, allowing DNA polymerase and DNA ligase I (AtLIG1) to fill the gap with an unmethylated cytosine (Martinez-Macias et al. 2012; Li et al. 2015a, 2015b).

3 Epigenetic Mechanisms Mediated by Histone Modifications

3.1 *Histone Modification by the PRC2*

The histone tails of chromatin are subject to numerous reversible chemical modifications, including methylation, acetylation, deacetylation, phosphorylation, ubiquitinylation, sumolyation, and ADP-ribosylation. These modifications can attract and recruit different chromatin reader complexes and cause structural changes to chromatin architecture, leading to transcriptional activation or repression depending on the type of modification (Kouzarides 2007). Polycomb group proteins (PcGs) are highly conserved chromatin modifiers that regulate key developmental processes in plants and animals (Schwartz and Pirrotta 2007; Kohler and Villar 2008; Derkacheva and Hennig 2014). PcG proteins mediate repression of gene expression through methylation of histone H3 at lysine 27, which leads to chromatin remodeling and condensation (Kim et al. 2012; Bouyer et al. 2011; Ikeuchi et al. 2015; Horst et al. 2016). In Arabidopsis, the PcG Repressive Complex 2 (PRC2) were found to play a major role in regulating various developmental processes, including transition from gametophytic to sporophytic phase, embryogenesis, and organogenesis, and transition from the vegetative to reproductive phase and flower development.

3.1.1 Endosperm Development

The Arabidopsis PRC2 involved in endosperm development (also called the FIE–MEA–FIS complex) consists of the SET-domain containing protein Medea (MEA), a C2H2 zinc-finger protein Fertilization Independent Seed 2 (FIS2), and two WD-40 proteins Fertilization Independent Endosperm (FIE) and Multicopy Suppressor of IRA 1 (MSI1) (Derkacheva and Hennig 2014). The FIE–MEA–FIS complex contributes to endosperm development by mediating genome imprinting in endosperm (Hsieh et al. 2011). Mutations in any one of these genes, when maternally inherited, cause endosperm over proliferation, arrested embryo development, and seed abortion. By contrast, inheritance of a mutant paternal allele has no detectable effect on seed development. Thus, one main function of this PRC2 complex is to prevent premature activation of endosperm proliferation genes and to repress gene activity from one of parental copy resulting in parent-of-origin expression pattern (Mozgova et al. 2015; Gehring 2013).

3.1.2 Seed to Seedling Phase Transition

During seed development, master transcriptional regulators Leafy Cotyledon (LEC1) and LEC2 and Abscisic Acid Insensitive 3 (ABI3) play major roles in embryo development and seed maturation and trigger storage protein and oil accumulation. During transition from mature seed to germination, PRC2 represses key seed developmental programs in seedling by modulating multiple hormone signaling pathways, including ABA and GA (Bouyer et al. 2011; Deng et al. 2013; Kim et al. 2012). Thus, PRC2 is responsible for repressing these embryonic traits during germination. For example, PRC2 components EMF1 and FIE bind to major seed developmental genes (e.g., *LEC1*, *LEC2*, and *ABI3*) and mark them with repressive histone H3k27me3 for silencing (Bouyer et al. 2011; Suzuki et al. 2007).

3.1.3 Vegetative to Reproductive Phase Transition

In mature plant, reproductive tissues (inflorescence meristems and floral meristems that produce floral organs) are induced from shoot apical meristem (SAM). The Arabidopsis flowering inducer FLOWERING LOCUS T (FT) can induce expression of floral-meristem identity genes *APETALA1* (*AP1*) and *LEAFY* (*LFY*) whereas the flowering inhibitor FLOWERING LOCUS C (FLC) and FLOWERING LOCUS M (FLM) repress the flower inducer *FT* (Henderson and Dean 2004; Liu et al. 2009). Under noninductive conditions, FLC protein recruits PRC2 to *FT* locus to repress *FT* transcription (Wang et al. 2014). Upon induction of flowering, PRC2 is recruited to the *FLC* locus to repress *FLC* gene activity and thus derepresses *FT*, induces flowering, and maintains reproductive program in inflorescence and flower meristems (Muller-Xing et al. 2014).

3.1.4 Vernalization

In Arabidopsis, naturally occurring flowering time variations among different accessions are caused by allelic variations in two loci, *FRIGIDA* (*FRI*) and *FLOWERING LOCUS C* (*FLC*). In winter-annual accessions, *FRI* acts as a positive regulator of *FLC*, a strong suppressor of flowering, and vernalization is required for early flowering (Sung and Amasino 2005; Henderson et al. 2003). In Arabidopsis accessions that require vernalization, FLC can repress the flower inducer *FT* and vernalization can repress *FLC* to induce flowering process (Michaels and Amasino 1999). The 3′ region of the *FLC* locus A can be transcribed into a series of antisense transcripts *COOLAIR* to interact with *FLC* chromatin, resulting in the reduction of H3K36 methylation (Csorba et al. 2014). *COLDAIR*, a long noncoding transcript transcribed from the first intron of *FLC*, can recruit PRC2, leading to H3K27 methylation (Heo and Sung 2011; Sung et al. 2006). Prolonged cold exposure then induces vernalization-specific VRN PRC2 complex, which consists of VERNALIZATION 5 (VRN5), VERNALIZATION INSENSITIVE 3 (VIN3), and VERNALIZATION5/VIN3-LIKE 1 (VEL1) to spread H3K27me3 across the *FLC* locus for prolonged repression of *FLC* (Song et al. 2013).

4 Epigenetic Reprogramming During Arabidopsis Male Gametogenesis (Fig. 1)

4.1 Microgametogenesis in Arabidopsis

The Arabidopsis stamen, the male reproductive organ, contains four anther locules, each with a microsporagium where pollen grains develop. The male sexual reproduction involves the specification and differentiation of the spore mother cells

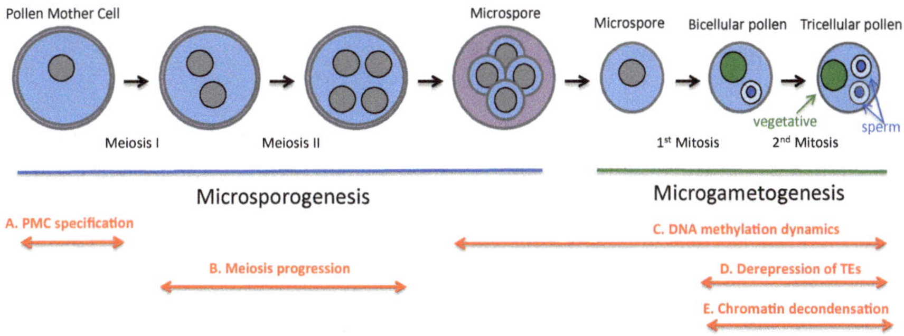

Fig. 1 Diagram of microsporogenesis and microgametogenesis stages and potential epigenetic reprogramming events. (**a**) PMC specification. PMC enlargement, H1, H2A.Z reduction, chromatin decondensation, H3K27me1 and H3K27me3 reduction, and upregulation of H3K4me3. (**b**) Meiosis progression. MEL1 (an AGO5 homolog) and MEL2 (an RRM protein) are required in rice, suggesting siRNA pathway is importing for meiosis in microsoporogenesis. (**c**) DNA methylation dynamics during microgametogenesis. The microspore and sperm cells retain CG methylation. CHH methylation is lost from retrotransposons in microspores and sperm cells and is restored in the vegetative nucleus. In the vegetative nucleus, CG methylation is lost from targets of the DNA glycosylases DME. 24-nt siRNAs are produced from the DME target loci and are required for de novo CHH methylation in sperm. (**d**) DDM1 downregulation and derepression of TEs in the vegetative nucleus. DDM1 is silenced in VN, TEs loose DNA methylation and are derepressed, 21-nt easiRNAs are produced to silence TEs in VN and in sperm cells. (**e**) Chromatin decondensation in the vegetative nucleus. The vegetative cell nuclei undergo loss of CenH3 by the ATPase CDC48/p97 molecular chaperon. One possible purpose is to activate rRNA loci in preparation for pollen tube growth and delivery of sperm cells to the ovule for double fertilization

within the anther and marks the somatic to reproductive developmental transition (Berger and Twell 2011). The male meiocyte, the microspore or pollen mother cell (PMC), then undergoes meiosis to produce four haploid microspores. Each microspore undergoes an asymmetrical cell division to generate a larger vegetative cell (male companion cell) and a smaller generative cell (germ cell). The generative cell is then engulfed by the vegetative cell to form a unique cell within cell structure. The generative cell then undergoes a second mitotic division to produce two sperm cells (Twell 2011). Thus, the mature pollen is an encapsulated tricellular male gametophype. The asymmetric division of the microspore marks the establishment of the male germline and is a critical process in male gametogenesis. Upon landing on specialized cells at the tip of the stigma of a receptive plant, the pollen grain rehydrates and the vegetative cell germinates, producing a pollen tube that protrudes and grows to the ovules and delivers two sperm cells to the female gametophyte where fertilization of egg and central cell takes place (McCormick 2004).

4.2 Chromatin Reorganization During Pollen Mother Cell Differentiation

During PMC specification and differentiation, the Arabidopsis PMC can be distinguished from the surrounding somatic cells due to its cellular and nuclear enlargement. Similar phenomena were also reported in early cytological studies in many plant species (Armstrong and Jones 2003), suggesting the presence of possible large-scale chromatin reorganization. This is recently confirmed by a more detailed study (She and Baroux 2015). In addition, the authors also reported drastic reduction of linker histone H1 and histone variant H2A.Z expression in the PMC nucleus (She and Baroux 2015). However, it remains to be investigated whether loss of H1-GFP signals was due to H1 eviction, chromatin decondensation, or being replaced with a male-specific histone variant. Decondensed chromatin with reduced heterochromatin and H1 expression is thought to result in a transcriptionally more active state in PMC, which is supported by the observation of a decrease in H3K27me1 and H3K27me3 signals (both mark repressive chromatins) and an increase in H3K4me3 permissive chromatin signal in the PMC (She and Baroux 2015). Indeed, transcription profiling of male meiocytes from various plant species indicates a large number of genes, including meiosis-specific genes, as well as transposons, are activated in the meiocytes (Zhou and Pawlowski 2014; Chen et al. 2010; Yang et al. 2011), suggesting that dynamic change in chromatin states and likely epigenetic reprogramming might be a requirement to rewire transcriptional program that is necessary for meiosis progression.

Significantly, the observed chromatin reorganization in Arabidopsis PMC is reminiscent of what happens during megaspore mother cell (MMC) differentiation (see discussion below), suggesting that flowering plants might use a common strategy to regulate somatic to reproduction cell fate transition. The rice MEIOSIS ARRESTED IN LEPTOTENE 1 and 2 (MEL1, ortholog of the Arabidopsis AGO5, and MEL2, a novel RNA-recognition-motif protein) are required for meiosis during sporogenesis, indicating that siRNAs also play a critical role in PMC differentiation (Nonomura et al. 2007, 2011).

4.3 Chromatin Remodeling During Male Gametogenesis

After meiosis, the haploid microspore undergoes an asymmetric mitotic division to differentiate a larger vegetative cell and a smaller generative cell, the male germline. The vegetative cell exits the cell cycle whereas the generative cell undergoes a second mitotic division to produce two identical sperm cells encapsulated by the vegetative cell (Berger and Twell 2011). The vegetative cell nuclei undergo loss of centromere-specific histone H3 variant (CenH3), extensive decondensation of the centromeric heterochromatin, and loss of centromere identify, which represent the most prominent evidence of epigenetic reprogramming in the plant germline (Schoft et al. 2009). CenH3 is a major component of the centromeres and is important for kinetochore assembly and chromosome segregation during cell division. Centromeres are tightly associated with heterochromatin, which is required to promote CenH3 assembly at centromeres (Folco et al. 2008; Black and Bassett 2008; Henikoff and Furuyama 2012). Loss of CenH3 and decondensation of the centromeric heterochromatin occurs in the binucleate stage of pollen development, right after the asymmetric mitotic division. Using a combination of genetic screening, nucleus cytology, and proteomic analysis, Merai et al identified that the conserved ATPase CDC48/p97 molecular chaperone is responsible for disassembling centromeres, decondensing heterchromatin and activating ribosomal RNA biosynthesis (Merai et al. 2014). The CDC48/p97 chaperon recognizes and removes sumoylated CenH3 from the vegetative centromeres, triggering heterochromatin decondensation. By contrast, the *cdc48a* vegetative cell nuclei showed robust heterichromatic H3K9me2 and CenH3 signals. Thus, the CDC48/p97 ATPase is required for the overall chromatin reconfiguration observed in the pollen vegetative cell (Merai et al. 2014).

4.4 Dynamic Changes of DNA Methylation During Male Gametogenesis

The decondensed vegetative cell (VC) nuclei resembled the phenotype of mutant nuclei defective in the *DDM1* (Probst et al. 2003; Schoft et al. 2009), which is consistent with the report that *DDM1* expression is not detectable in VC (Slotkin et al. 2009). *DDM1* is the master regulator of TE silencing in Arabidopsis, and most TEs lose DNA methylation, repressive histone modifications, and 24-nt siRNAs in *ddm1* mutants, resulting in strong transcriptional activation of TEs (Lippman et al. 2004; Zemach et al. 2013). Therefore not surprisingly, one consequence of global chromatin decondensation in the VC is the conspicuous activation of transposable elements (Slotkin et al. 2009). Interestingly, activation of TEs in the VC leads to production of a class of 21-nt siRNAs called epigenetically activated small RNAs, or easiRNAs, that accumulate at high levels in purified sperm cells (Slotkin et al. 2009).

Our understanding of epigenetic reprogramming in the plant germline is greatly advanced in recent years owing to the breakthrough in genome-wide methylation profiling techniques coupled with isolation of different male gametic cell types and their progenitor microspore, by fluorescence-activated flow cytometry (Calarco et al. 2012; Ibarra et al. 2012; Borges et al. 2012). Whole genome methylation profiles of purified microspores, sperm cells, and vegetative cells revealed that CG and CHG methylations are retained in differentiating germline, and throughout development. By contrast, CHH methylation is lost from pericentromeric retrotransposons and satellite repeats in microspores and sperm cells, but is restored in vegetative nuclei (Calarco et al. 2012). This is likely due to reduced RdDM activity during pollen meiosis and mitosis as DRM2-GFP fusion protein is visible in the VCs, but barely detectable in microspores, generative cells, and sperm cells (Schoft et al. 2009; Calarco et al. 2012).

Analysis of CG differentially methylated regions (DMRs) between vegetative cell nuclei (VN) and sperm cells indicates that local CG hypomethylation is obvious in the VN (Calarco et al. 2012; Ibarra et al. 2012). By comparing DMRs between wt and demethylase mutants in other tissues, Calarco et al. concluded that DNA demethylases (DME, ROS1, DML2, DML3) are responsible for the loss of CG in the VN (Calarco et al. 2012), consistent with previous report that they are expressed in the VCs, but not in the sperm cells (Schoft et al. 2011). Ibarra et al. compared CG methylation between wild type and *dme+/−* VN and showed that CG sites that are demethylated in wild-type vegetative cells exhibited higher methylation in *dme+/−* VN, indicating that VN demethylation requires DME (Ibarra et al. 2012).

As discussed above, activation of TE transcription leads to accumulation of 21-nt easiRNAs in VN that are likely to be transported to sperm cells (Slotkin et al. 2009). More recently it was shown that when activated, certain TE mRNAs are processed through an RNA interference mechanism that represents a posttranscriptional regulation for TE activity (McCue et al. 2012). This results in the biogenesis of the easiRNAs that requires RDR6, DCL4, and AGO1 and is mediated through miRNAs for posttranscriptional gene silencing (PTGS) of TE transcripts (Creasey et al. 2014). The accumulation of 21-nt easiRNAs in the sperm cells is postulated to be transported from the VN, as small RNAs are known to travel and communicate from cell to cell (Molnar et al. 2010; Lewsey et al. 2016). However, the possibility of siRNAs movement into germ cells remains controversial (Slotkin et al. 2009; Grant-Downton et al. 2013). To address this discrepancy, Martinez et al. used a series of transgenes with cell type-specific expression at late stages of pollen to demonstrate that an AGO1-AGO2-DCL4 pathway in VN can induce PTGS of TEs in VN, and in sperm cells, substantiating the mobility of RNA information into the sperm gametes (Martinez et al. 2016).

In addition to the accumulation of 21-nt easiRNAs, sperm cells also accumulate 24-nt siRNAs known to mediate RdDM, and some of these siRNAs overlap with genomic loci flanking the imprinted genes that lost CG methylation in the VN (Slotkin et al. 2009; Calarco et al. 2012), suggesting that these 24-nt siRNAs in sperm might be the result of active demethylation in the VN. Supporting this hypothesis, Ibarra et al. reported that CG sites demethylated by DME in VN show preferential CHH hypomethylation in *dme+/−* sperm, indicating that DME activity in the vegetative cell is required for normal methylation of a subset of sperm TEs (Ibarra et al. 2012). Although it remains to be experimentally tested, these observations imply that active demethylation in the VN may produce 24-nt siRNAs that are transported to the sperm cells to reinforce TE methylation in sperm and possibly in the zygote.

Upon fertilization, the CHH hypomethylated paternal genomes must be remethylated in the embryo and the endosperm as both tissues have relatively high levels of CHH methylation (Hsieh et al. 2009; Ibarra et al. 2012). The remethylation is likely guided by the maternal 24-nt siRNAs as previous report showed that most of the 24-nt siRNAs in seeds are of maternal origin that target transposons (Mosher et al. 2009; Lu et al. 2012). Remethylation by the maternal siRNAs might provide an opportunity to allow the maternal plants to distinguish self from non-self pollen (Creasey et al. 2014) and might have functional implications in cross-species hybridization or hybrid vigor.

5 Epigenetic Reprogramming During Arabidopsis Female Gametogenesis (Fig. 2)

5.1 Megagametogenesis in Arabidopsis

In Arabidopsis, the female gametophyte formation is initiated by the differentiation of a nucellus cell into an archesporial cell. Megaspore mother cells (MMCs) are specified in a subepidermal layer of the sporangia, and each will undergo meiosis to produce a tetrad of haploid megaspores. In Arabidopsis and most flowering species, only one functional megaspore (FMS) survives while the others degenerate. The haploid FMS undergoes mitosis to generate two syncytium nuclei that migrate to the opposite end of the megaspore. The second mitosis produces a 4-nuclei embryo sac with a pair of nuclei at either pole. The third mitosis produces an 8-nucleate embryo sac with 3 nuclei near the micropylar end, 3 at the chalazal end, and 2 polar

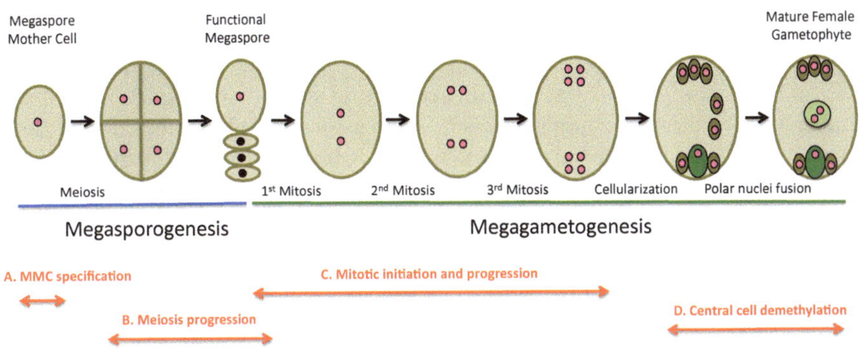

Fig. 2 Diagram of megasporogenesis and megagametogenesis stages and potential epigenetic reprogramming events. (**a**) MMC specification. MMC nuclear enlargement, chromatin decondensation, depletion of H1, reduction in heterochromatin content, possible depletion of CenH3, and a requirement of the AGO9 siRNA pathway. (**b**) Meiosis progression. The maize *AGO104*, *DMT102*, and *DMT103* are required for meiosis, suggesting the importance of DNA methylation for normal meiosis. (**c**) Mitosis initiation and progression. The AGO5-dependent siRNA pathway is required for promotion of megagametogenesis. In addition, switch-like chromatin-remodeling protein gene *CHR11* is involved in regulating gametogenesis. (**d**) Central cell DNA demethylation. Demethylation of central cell genome by DME establishes gene imprinting, promotes PRC2 complex, and likely produces siRNAs for TE silencing in the gamete and the zygote

nuclei closer to the micropylar end of the embryo sac. Distinct cell fates are specified when cellularization occurs that results in the formation of a 7-celled, 8-nuclei embryo sac consisting of 2 synergids, 1 egg, 1 central, and 3 antipodal cells (Yang et al. 2010). It has been shown that patterning and cell fate determination of the Arabidopsis female gametophyte depends on a gradient of local auxin concentration, and aberrant positioning of nuclei in the *eostra* mutant gives rise to ectopic functional egg cell in place of a synergid cell (Pagnussat et al. 2009; Sundaresan and Alandete-Saez 2010). Thus, an auxin gradient model for patterning of female gametophyte was proposed where concentration of auxin is highest at the micropylar pole and lowest at the chalazal end (Pagnussat et al. 2009; Sundaresan and Alandete-Saez 2010).

5.2 *Mobile siRNAs During Megaspore Mother Cell Differentiation and Meiosis*

Cell–cell communication through siRNA pathways appears to play important roles during female gametogenesis. Genetic evidence indicates that signaling by small RNA affects the establishment of the female germline. The Arabidopsis AGO9 protein is preferentially expressed in the epidermal (L1) layer of the ovule primordium and is absent from the MMCs (Olmedo-Monfil et al. 2010). AGO9 primarily binds a class of 24-nt siRNAs derived from TEs, and many of them are derepressed in *ago9* mutant ovules, indicating that TEs are primary targets of AGO9-dependent silencing in ovule (Duran-Figueroa and Vielle-Calzada 2010; Olmedo-Monfil et al. 2010). Mutations in *AGO9*, *RDR6*, and *SUPPRESSOR OF GENE SILENCING 3* (*SDG3*) required for the biogenesis of trans-acting siRNAs (ta-siRNAs) that move as signal molecules and cause gene silencing in a non-cell autonomous manner lead to supernumerary FMS formation in the ovule (Olmedo-Monfil et al. 2010). Thus, MMC specification and differentiation in Arabidopsis is regulated by an AGO9–RDR6–SDG3-dependent siRNA silencing pathway.

Other siRNA pathways have also been implicated in the regulation of female gametogenesis. In maize, mutation in a member of the AGRONAUT protein *AGO104* produces functional unreduced gametes due to defects in meiosis. *AGO104* is specifically expressed in somatic cells surrounding the female meiocyte, suggesting again a mobile signal that acts non-cell autonomously to regulate megasporogenesis (Singh et al. 2011). In Arabidopsis, Tucker et al. used laser capture microdissection (LCM) and transcription profiling to identify pathways involved in the transition from megasporogenesis to megagametogenesis and found expression of *AGO5* in the somatic cells surrounding the MMC and megaspores in a manner similar to that of *AGO9*. A semidominant mutant allele of *AGO5*, the *ago5-4*, shows defects in the initiation of megagametogenesis that is independent of *AGO9* (Tucker et al. 2012). Thus, at least two somatic siRNA pathways are active during female gametogenesis, the AGO9-dependent pathway that restricts multiple FMS formation, and the AGO5-dependent pathway that promotes megagametogenesis (Tucker et al. 2012).

In maize, mutations in *DMT102* (*CMT3* homolog) and *DMT103*(*DRM2* homolog) (Garcia-Aguilar et al. 2010) result in production of functional unreduced gametes due to defects in meiosis, suggesting the requirement of certain siRNA silencing pathways for normal progression of meiosis in maize, and highlight the importance of DNA methylation pathways during plant reproduction.

5.3 *Chromatin Reorganization During Megasporogenesis and Megagametogenesis*

At the cytology level, specification of MMC in Arabidopsis is accompanied by visible changes in nuclear morphology, including nuclear enlargement, chromatin decondensation, depletion of linker histone H1, reduction in heterochromatin content, and a possible depletion of CenH3 in the MMC (Schneitz et al. 1995; Sniezko 2006; Armstrong and Jones 2003; She et al. 2013). These observations highlight the presence of large-scale chromatin reorganization, likely to enable meiosis competent transcription state in the MMC. Consistent with this view, histone modification landscapes during megasporogenesis suggest a dynamic change toward a more active chromatin state (increase for H3K4me3 marks) in the MMC (She et al. 2013).

Completion of meiosis produces a tetrad of haploid spores, where one spore differentiates into the functional megaspore that proceeds through three rounds of mitotic divisions to complete gametogenesis. Although not much is known about whether epigenetic mechanism regulates the megaspore mitotic divisions, knocking down a switch-like chromatin-remodeling protein gene *CHR11*during gametogenesis caused arrest in megaspore mitotic proliferation (Huanca-Mamani et al. 2005), suggesting a potential role of chromatin remodeling during gametogenesis.

5.4 *Active DNA Demethylation by the DEMETER Glycosylase in the Gametophytes*

During fertilization, one sperm nucleus fuses with the egg cell to form the zygote, a diploid cell that goes through an asymmetric cell division to generate a smaller apical cell, the precursor of embryo proper, and a larger basal cell that gives rise to the suspensor, a terminally differentiated structure that connects the developing embryo to the maternal tissues for nutrient supply. The embryo proper develops through a series of well-programmed cell divisions and differentiates to generate a mature embryo with properly specified organs (cotyledons, meristems, and axis) and tissue types (protoderm, parenchyma, and procambium) (Wendrich and

Weijers 2013; Lau et al. 2012). A second sperm nucleus fuses with the central cell to give rise to a triploid primary endosperm nucleus. The primary endosperm nucleus replicates to produce a coenocyte of nuclei that surround the developing embryo until cellularization takes place around when the developing embryo reaches the heart stage of embryogenesis (Olsen 2004). Endosperm produces and stores starch, lipids, and storage proteins, and serves as a nutrient reservoir for the developing embryo, and is the major tissue where gene imprinting occurs in plants (Gehring 2013; Kohler et al. 2012).

Genomic imprinting is the differential expression of two alleles of a gene dependent on their parent of origins. Early studies of gene imprinting in Arabidopsis revealed that imprinted expression of several maternally expressed imprinted genes (MEGs) involves MET1-mediated DNA methylation and DME demethylation (Jullien et al. 2006; Gehring et al. 2006; Xiao et al. 2003; Kinoshita et al. 2004; Tiwari et al. 2008). For example, DME is required for the expression of *MEA*, *FIS2*, and *FWA* in the central cell prior to fertilization and in the endosperm after fertilization, while MET1 is responsible for the silencing of *FIS2* and *FWA* paternal alleles. Seeds that inherit maternal *dme* alleles abort due to failure to activate *MEA* and *FIS2* in the central cell whereas *met1* mutation suppresses *dme* seed abortion by restoring *MEA* expression, indicating that DME and MET1 antagonistically regulate *MEA* expression (Xiao et al. 2003).

DME encodes a novel DNA glycosylase that is required for removal of cytosine methylation in vitro and in vivo (Gehring et al. 2006). Since DME expression is mainly confined in the central cell within the female gametophyte, it was speculated that DME demethylates the maternal genome to establish gene imprinting. Indeed, recent genome-wide DNA methylation studies confirmed that endosperm DNA is hypomethylation compared to embryo DNA in multiple plant species, including Arabidopsis, rice, maize, and castro bean (Xu et al. 2016; Gehring et al. 2009; Hsieh et al. 2009; Lauria et al. 2004; Zemach et al. 2010). Furthermore, allele-specific analyses of DNA methylation revealed that hypomethylated alleles are uniformly of maternal original in the species examined (Zhang et al. 2014; Xu et al. 2014; Waters et al. 2011; Ibarra et al. 2012; Rodrigues et al. 2013). Since the embryo and endosperm maternal genomes derive from the egg and the central cells, respectively, it is likely that the two gametes have very distinct DNA methylation pattern prior to fertilization, presumably due to the activity of DME in the central cell.

Although demethylation of the central cell genome by DME is linked to establishment of genomic imprinting, the abundance of DME targets in gene poor heterochromatin and the shared target sites in the central and vegetative cells suggest the establishment of genomic imprinting is not the primary function of DME. It was proposed that DNA demethylation and activation of TEs in the vegetative cell generate siRNAs that would reinforce silencing of corresponding TEs in sperm (Slotkin et al. 2009). Similarly, several TEs demethylated by DME in the central cell are maternally expressed in wild type, but not in *dme* endosperm. Derepression of TEs results in production of mobile siRNAs that affects methylation of the gametes. Supporting such model, transgenic plants expressing artificial microRNA in the central cell or the vegetative cell targeting cleavage of green fluorescent protein (GFP) transcripts have been demonstrated to be capable of silencing GFP transgene expressed in the egg cell or the sperm cell (Ibarra et al. 2012; Slotkin et al. 2009). As discussed earlier, if DME-mediated companion cell sRNAs production is to reinforce gamete genome silencing, lack of DME in the companion cell would be expected to reduce RdDM of DME target sequences in gametes. Indeed, CG sites demethylated by DME in vegetative cells show preferential CHH hypomethylation in *dme+/−* sperm. Thus, DME activity in the vegetative cell is required for full methylation of some sperm TEs, indicating that demethylation in companion cells generates mobile sRNAs that protect the gametes against TE activation (Ibarra et al. 2012).

In addition to global methylation pattern difference, endosperm and embryo also inherit distinct chromatin and transcriptional states from the gametes, including enlarged nuclear volume, less condensed chromatin, and peculiar heterochromatin organization in the endosperm compared to embryo (Baroux et al. 2007; Pillot et al. 2010). The difference in chromatin state is likely maternal in origin as two sperms are thought to be functionally interchangeable (Faure et al. 2003; Russell 1991). Whether or not this chromatin state difference between egg and central cell is solely due to DME activity is not known. Demethylation of central cell by DME activates PRC2 complex that in turn repressed other genes, a process critical for seed viability in Arabidopsis (Gehring 2013). For example, the Arabidopsis maintenance CG methylation gene *VIM5* is a *PEG* whose maternal allele is silenced by the PRC2 in the endosperm. As a result, CG methylation maintenance in endosperm depends on the *VIM5* paternal allele. However, if *VIM5* is subject to the regulation of "delayed activation of paternal genome" that affects many endosperm genes, passive DNA demethylation likely occurs during early endosperm development before *VIM5* paternal allele is activated. (Nodine and Bartel 2012; Vielle-Calzada et al. 2000; Autran et al. 2011). This hypothesis is supported by the observation that in *fie* mutant endosperm, *VIM5* maternal allele is derepressed, and endosperm CG hypomethylation is restored to the level of embryo CG methylation (Ibarra et al. 2012; Hsieh et al. 2011; Hsieh et al. 2009). Thus, genome-wide CG hypomethylation in endosperm is the collective results of active DNA demethylation by DME in the central cell and passive DNA demethylation during early endosperm development.

6 Conclusions and Future Perspectives

Although certain outcome of epigenetic reprogramming is apparent (e.g., establishment of gene imprinting by DME demethylation), it is not clear whether there is a common, unified purpose for large-scale epigenetic programming. During gametogenesis, non-cell-autonomous siRNAs likely play a major role in somatic-to-reproductive transition and in the progression of meiosis. Likewise, siRNAs produced by the gamete companion cells have a direct effect on the epigenetic integrity of the gametes and the zygotes. However, understanding the functions of these mobile siRNAs is significantly restricted by the difficulty in isolating sufficient amounts of pure gametes and zygotes to identify siRNAs and to access their affect in TE silencing and DNA methylation. For example, if the purpose of demethylation in the central cell is to generate mobile siRNAs to enforce silencing in the gamete and the zygote, it would make sense to fortify the RdDM machinery in the gamete and during early zygote development, which is what was observed in Arabidopsis (Jullien et al. 2012; Belmonte et al. 2013). Our knowledge in the DNA methylation reprogramming during plant reproduction will greatly benefit from continuing technical advance in the area of specific cell type isolation using novel fluorescence activated cell sorting or INTACT techniques (Park et al. 2016; Schoft et al. 2015; Deal and Henikoff 2011; Wang and Deal 2015) coupled with single cell/ultra low input methylation profiling procedures (Smallwood et al. 2014).

Small RNA-mediated de novo DNA methylation in zygote plays an important role in transgenerational inheritance of DNA methylation patterns. In Arabidopsis *met1* or *ddm1* mutants that are defective in CG methylation, CG remethylation when outcrossed to wild-type plant is slow and depends on the RdDM pathway, raising the question that fluctuations in the RdDM activity during developing embryo might generate variation in epigenetic inheritance. Indeed, two genome-wide surveys in Arabidopsis examined spontaneously occurring variation in DNA methylation at base-pair resolution by comparing plants propagated by single-seed descent for 30 generations. Both studies found that the rate of spontaneous variation in DNA methylation was at least four orders of magnitude greater than spontaneous genetic mutations (Schmitz et al. 2011; Becker et al. 2011). More interestingly, the distribution of these differentially methylated cytosines is not random, and overlaps significantly with loci that are affected by MET1 or RdDM mutations (Schmitz et al. 2011), indicating that incomplete reinforcement of methylation is one major factor in transgenerational epigenetic variation.

In addition to developmentally wired reprogramming, large-scale epigenetic changes can occur in response to environmental stress. For example, it is well known that chromatin modifications play a role in plants' response to environmental stimuli. Recent reports have demonstrated substantial change in DNA methylation can be induced by various biotic and abiotic stresses (Lang-Mladek et al. 2010; Pecinka et al. 2010; Tittel-Elmer et al. 2010). Furthermore, stress-induced epigenetic change can also arise from activation and mobilization of transposons, generating new genetic and epigenetic variations. Thus, epigenetics is a versatile mechanism that allows the organism to modulate genome activity for specific developmental needs, to cope with environmental assaults, and to create epigenetic diversity in the progeny, all in an easily reversible manner without making extensive and permanent changes to the genetic blueprint.

References

Agius F, Kapoor A, Zhu JK (2006) Role of the Arabidopsis DNA glycosylase/lyase ROS1 in active DNA demethylation. Proc Natl Acad Sci U S A 103:1796–1801

Armstrong SJ, Jones GH (2003) Meiotic cytology and chromosome behaviour in wild-type Arabidopsis thaliana. J Exp Bot 54:1–10

Autran D, Baroux C, Raissig MT et al (2011) Maternal epigenetic pathways control parental contributions to Arabidopsis early embryogenesis. Cell 145:707–719

Baroux C, Pecinka A, Fuchs J et al (2007) The triploid endosperm genome of Arabidopsis adopts a peculiar, parental-dosage-dependent chromatin organization. Plant Cell 19:1782–1794

Becker C, Hagmann J, Muller J et al (2011) Spontaneous epigenetic variation in the Arabidopsis thaliana methylome. Nature 480:245–249

Belmonte MF, Kirkbride RC, Stone SL et al (2013) Comprehensive developmental profiles of gene activity in regions and subregions of the Arabidopsis seed. Proc Natl Acad Sci U S A 110: E435–E444

Berger F, Twell D (2011) Germline specification and function in plants. Annu Rev Plant Biol 62:461–484

Bewick AJ, Ji L, Niederhuth CE et al (2016) On the origin and evolutionary consequences of gene body DNA methylation. Proc Natl Acad Sci U S A 113:9111–9116

Black BE, Bassett EA (2008) The histone variant CENP-A and centromere specification. Curr Opin Cell Biol 20:91–100

Borges F, Gardner R, Lopes T et al (2012) FACS-based purification of Arabidopsis microspores, sperm cells and vegetative nuclei. Plant Methods 8:44

Bouyer D, Roudier F, Heese M et al (2011) Polycomb repressive complex 2 controls the embryo-to-seedling phase transition. PLoS Genet 7:e1002014

Calarco JP, Borges F, Donoghue MT et al (2012) Reprogramming of DNA methylation in pollen guides epigenetic inheritance via small RNA. Cell 151:194–205

Chen C, Farmer AD, Langley RJ et al (2010) Meiosis-specific gene discovery in plants: RNA-Seq applied to isolated Arabidopsis male meiocytes. BMC Plant Biol 10:280

Choi Y, Gehring M, Johnson L et al (2002) DEMETER, a DNA glycosylase domain protein, is required for endosperm gene imprinting and seed viability in *Arabidopsis*. Cell 110:33–42

Creasey KM, Zhai J, Borges F et al (2014) miRNAs trigger widespread epigenetically activated siRNAs from transposons in Arabidopsis. Nature 508:411–415

Csorba T, Questa JI, Sun Q et al (2014) Antisense COOLAIR mediates the coordinated switching of chromatin states at FLC during vernalization. Proc Natl Acad Sci U S A 111:16160–16165

Deal RB, Henikoff S (2011) The INTACT method for cell type-specific gene expression and chromatin profiling in Arabidopsis thaliana. Nat Protocols 6:56–68

Deng W, Buzas DM, Ying H et al (2013) Arabidopsis polycomb repressive complex 2 binding sites contain putative GAGA factor binding motifs within coding regions of genes. BMC Genomics 14:593

Derkacheva M, Hennig L (2014) Variations on a theme: polycomb group proteins in plants. J Exp Bot 65:2769–2784

Duran-Figueroa N, Vielle-Calzada JP (2010) ARGONAUTE9-dependent silencing of transposable elements in pericentromeric regions of Arabidopsis. Plant Signal Behav 5:1476–1479

Faure JE, Rusche ML, Thomas A et al (2003) Double fertilization in maize: the two male gametes from a pollen grain have the ability to fuse with egg cells. Plant J 33:1051–1062

Feng S, Jacobsen SE, Reik W (2010) Epigenetic reprogramming in plant and animal development. Science 330:622–627

Folco HD, Pidoux AL, Urano T et al (2008) Heterochromatin and RNAi are required to establish CENP-A chromatin at centromeres. Science 319:94–97

Garcia-Aguilar M, Michaud C, Leblanc O et al (2010) Inactivation of a DNA methylation pathway in maize reproductive organs results in apomixis-like phenotypes. Plant Cell 22:3249–3267

Gehring M (2013) Genomic imprinting: insights from plants. Annu Rev Genet 47:187–208

Gehring M, Huh JH, Hsieh TF et al (2006) DEMETER DNA glycosylase establishes MEDEA polycomb gene self-imprinting by allele-specific demethylation. Cell 124:495–506

Gehring M, Bubb KL, Henikoff S (2009) Extensive demethylation of repetitive elements during seed development underlies gene imprinting. Science 324:1447–1451

Gong Z, Morales-Ruiz T, Ariza RR et al (2002) ROS1, a repressor of transcriptional gene silencing in Arabidopsis, encodes a DNA glycosylase/lyase. Cell 111:803–814

Grant-Downton R, Kourmpetli S, Hafidh S et al (2013) Artificial microRNAs reveal cell-specific differences in small RNA activity in pollen. Curr Biol 23:R599–R601

Henderson IR, Dean C (2004) Control of Arabidopsis flowering: the chill before the bloom. Development 131:3829–3838

Henderson IR, Shindo C, Dean C (2003) The need for winter in the switch to flowering. Annu Rev Genet 37:371–392

Henikoff S, Furuyama T (2012) The unconventional structure of centromeric nucleosomes. Chromosoma 121:341–352

Heo JB, Sung S (2011) Vernalization-mediated epigenetic silencing by a long intronic noncoding RNA. Science 331:76–79

Horst NA, Katz A, Pereman I et al (2016) A single homeobox gene triggers phase transition, embryogenesis and asexual reproduction. Nat Plants 2:15209

Hsieh TF, Ibarra CA, Silva P et al (2009) Genome-wide demethylation of Arabidopsis endosperm. Science 324:1451–1454

Hsieh TF, Shin J, Uzawa R et al (2011) Regulation of imprinted gene expression in Arabidopsis endosperm. Proc Natl Acad Sci U S A 108:1755–1762

Huanca-Mamani W, Garcia-Aguilar M, Leon-Martinez G et al (2005) CHR11, a chromatin-remodeling factor essential for nuclear proliferation during female gametogenesis in Arabidopsis thaliana. Proc Natl Acad Sci U S A 102:17231–17236

Huh JH, Bauer MJ, Hsieh T-F et al (2008) Cellular programming of plant gene imprinting. Cell 132:735–744

Ibarra CA, Feng X, Schoft VK et al (2012) Active DNA demethylation in plant companion cells reinforces transposon methylation in gametes. Science 337(6100):1360–1364

Ikeuchi M, Iwase A, Rymen B et al (2015) PRC2 represses dedifferentiation of mature somatic cells in Arabidopsis. Nat Plants 1:15089

Jullien PE, Kinoshita T, Ohad N et al (2006) Maintenance of DNA methylation during the Arabidopsis life cycle is essential for parental imprinting. Plant Cell 18:1360–1372

Jullien PE, Susaki D, Yelagandula R et al (2012) DNA methylation dynamics during sexual reproduction in Arabidopsis thaliana. Curr Biol 22:1825–1830

Kim SY, Lee J, Eshed-Williams L et al (2012) EMF1 and PRC2 cooperate to repress key regulators of Arabidopsis development. PLoS Genet 8:e1002512

Kinoshita T, Miura A, Choi Y et al (2004) One-way control of FWA imprinting in Arabidopsis endosperm by DNA methylation. Science 303:521–523

Kohler C, Villar CB (2008) Programming of gene expression by Polycomb group proteins. Trends Cell Biol 18:236–243

Kohler C, Wolff P, Spillane C (2012) Epigenetic mechanisms underlying genomic imprinting in plants. Annu Rev Plant Biol 63:331–352

Kouzarides T (2007) Chromatin modifications and their function. Cell 128:693–705

Lang-Mladek C, Popova O, Kiok K et al (2010) Transgenerational inheritance and resetting of stress-induced loss of epigenetic gene silencing in Arabidopsis. Mol Plant 3:594–602

Lau S, Slane D, Herud O et al (2012) Early embryogenesis in flowering plants: setting up the basic body pattern. Annu Rev Plant Biol 63:483–506

Lauria M, Rupe M, Guo M et al (2004) Extensive maternal DNA hypomethylation in the endosperm of Zea mays. Plant Cell 16:510–522

Lewsey MG, Hardcastle TJ, Melnyk CW et al (2016) Mobile small RNAs regulate genome-wide DNA methylation. Proc Natl Acad Sci U S A 113:E801–E810

Li Y, Cordoba-Canero D, Qian W et al (2015a) An AP endonuclease functions in active DNA demethylation and gene imprinting in Arabidopsis [corrected]. PLoS Genet 11:e1004905

Li Y, Duan CG, Zhu X et al (2015b) A DNA ligase required for active DNA demethylation and genomic imprinting in Arabidopsis. Cell Res 25:757–760

Lippman Z, Gendrel AV, Black M et al (2004) Role of transposable elements in heterochromatin and epigenetic control. Nature 430:471–476

Liu C, Thong Z, Yu H (2009) Coming into bloom: the specification of floral meristems. Development 136:3379–3391

Lu J, Zhang C, Baulcombe DC et al (2012) Maternal siRNAs as regulators of parental genome imbalance and gene expression in endosperm of Arabidopsis seeds. Proc Natl Acad Sci U S A 109:5529–5534

Martinez G, Panda K, Kohler C et al (2016) Silencing in sperm cells is directed by RNA movement from the surrounding nurse cell. Nat Plants 2:16030

Martinez-Macias MI, Qian W, Miki D et al (2012) A DNA 3′ phosphatase functions in active DNA demethylation in Arabidopsis. Mol Cell 45:357–370

McCormick S (2004) Control of male gametophyte development. Plant Cell 16(Suppl):S142–S153

McCue AD, Nuthikattu S, Reeder SH et al (2012) Gene expression and stress response mediated by the epigenetic regulation of a transposable element small RNA. PLoS Genet 8:e1002474

Merai Z, Chumak N, Garcia-Aguilar M et al (2014) The AAA-ATPase molecular chaperone Cdc48/p97 disassembles sumoylated centromeres, decondenses heterochromatin, and activates ribosomal RNA genes. Proc Natl Acad Sci U S A 111:16166–16171

Michaels SD, Amasino RM (1999) FLOWERING LOCUS C encodes a novel MADS domain protein that acts as a repressor of flowering. Plant Cell 11:949–956

Molnar A, Melnyk CW, Bassett A et al (2010) Small silencing RNAs in plants are mobile and direct epigenetic modification in recipient cells. Science 328:872–875

Morales-Ruiz T, Ortega-Galisteo AP, Ponferrada-Marin MI et al (2006) DEMETER and REPRESSOR OF SILENCING 1 encode 5-methylcytosine DNA glycosylases. Proc Natl Acad Sci U S A 103:6853–6858

Mosher RA, Melnyk CW, Kelly KA et al (2009) Uniparental expression of PolIV-dependent siRNAs in developing endosperm of Arabidopsis. Nature 460:283–286

Mozgova I, Kohler C, Hennig L (2015) Keeping the gate closed: functions of the polycomb repressive complex PRC2 in development. Plant J 83:121–132

Muller-Xing R, Clarenz O, Pokorny L et al (2014) Polycomb-Group proteins and FLOWERING LOCUS T maintain commitment to flowering in Arabidopsis thaliana. Plant Cell 26:2457–2471

Nodine MD, Bartel DP (2012) Maternal and paternal genomes contribute equally to the transcriptome of early plant embryos. Nature 482:94–97

Nonomura K, Morohoshi A, Nakano M et al (2007) A germ cell specific gene of the ARGONAUTE family is essential for the progression of premeiotic mitosis and meiosis during sporogenesis in rice. Plant Cell 19:2583–2594

Nonomura K, Eiguchi M, Nakano M et al (2011) A novel RNA-recognition-motif protein is required for premeiotic G1/S-phase transition in rice (Oryza sativa L.) PLoS Genet 7:e1001265

Olmedo-Monfil V, Duran-Figueroa N, Arteaga-Vazquez M et al (2010) Control of female gamete formation by a small RNA pathway in Arabidopsis. Nature 464:628–632

Olsen OA (2004) Nuclear endosperm development in cereals and Arabidopsis thaliana. Plant Cell 16(Suppl):S214–S227

Pagnussat GC, Alandete-Saez M, Bowman JL et al (2009) Auxin-dependent patterning and gamete specification in the Arabidopsis female gametophyte. Science 324:1684–1689

Park K, Frost JM, Adair AJ et al (2016) Optimized methods for the isolation of Arabidopsis female central cells and their nuclei. Mol Cells 39:768–775

Pecinka A, Dinh HQ, Baubec T et al (2010) Epigenetic regulation of repetitive elements is attenuated by prolonged heat stress in Arabidopsis. Plant Cell 22:3118–3129

Pillot M, Baroux C, Vazquez MA et al (2010) Embryo and endosperm inherit distinct chromatin and transcriptional states from the female gametes in Arabidopsis. Plant Cell 22:307–320

Probst AV, Fransz PF, Paszkowski J et al (2003) Two means of transcriptional reactivation within heterochromatin. Plant J 33:743–749

Rodrigues JA, Ruan R, Nishimura T et al (2013) Imprinted expression of genes and small RNA is associated with localized hypomethylation of the maternal genome in rice endosperm. Proc Natl Acad Sci U S A 110:7934–7939

Russell SD (1991) Isolation and characterization of sperm cells in flowering plants. Annu Rev Plant Physiol Plant Mol Biol 42:189–204

Schmitz RJ, Schultz MD, Lewsey MG et al (2011) Transgenerational epigenetic instability is a source of novel methylation variants. Science 334:369–373

Schneitz K, Hulskamp M, Pruitt RE (1995) Wild-type ovule development in Arabidopsis thaliana: a light microscope study of cleared whole-mount tissue. Plant J 7:731–749

Schoft VK, Chumak N, Mosiolek M et al (2009) Induction of RNA-directed DNA methylation upon decondensation of constitutive heterochromatin. EMBO Rep 10:1015–1021

Schoft VK, Chumak N, Choi Y et al (2011) Function of the DEMETER DNA glycosylase in the Arabidopsis thaliana male gametophyte. Proc Natl Acad Sci U S A 108:8042–8047

Schoft VK, Chumak N, Bindics J et al (2015) SYBR Green-activated sorting of Arabidopsis pollen nuclei based on different DNA/RNA content. Plant Reprod 28:61–72

Schwartz YB, Pirrotta V (2007) Polycomb silencing mechanisms and the management of genomic programmes. Nat Rev Genet 8:9–22

Seisenberger S, Peat JR, Hore TA et al (2013) Reprogramming DNA methylation in the mammalian life cycle: building and breaking epigenetic barriers. Philos Trans R Soc Lond B Biol Sci 368:20110330

She W, Baroux C (2015) Chromatin dynamics in pollen mother cells underpin a common scenario at the somatic-to-reproductive fate transition of both the male and female lineages in Arabidopsis. Front Plant Sci 6:294

She W, Grimanelli D, Rutowicz K et al (2013) Chromatin reprogramming during the somatic-to-reproductive cell fate transition in plants. Development 140:4008–4019

Singh M, Goel S, Meeley RB et al (2011) Production of viable gametes without meiosis in maize deficient for an ARGONAUTE protein. Plant Cell 23:443–458

Slotkin RK, Vaughn M, Borges F et al (2009) Epigenetic reprogramming and small RNA silencing of transposable elements in pollen. Cell 136:461–472

Smallwood SA, Lee HJ, Angermueller C et al (2014) Single-cell genome-wide bisulfite sequencing for assessing epigenetic heterogeneity. Nat Methods 11:817–820

Sniezko R (2006) Meiosis in plants. In: Dashek WV, Harrison P (eds) Plant cell biology. Science Publisher, New Hampshire, pp 227–258

Song J, Irwin J, Dean C (2013) Remembering the prolonged cold of winter. Curr Biol 23:R807–R811

Stroud H, Do T, Du J et al (2014) Non-CG methylation patterns shape the epigenetic landscape in Arabidopsis. Nat Struct Mol Biol 21:64–72

Sundaresan V, Alandete-Saez M (2010) Pattern formation in miniature: the female gametophyte of flowering plants. Development 137:179–189

Sung S, Amasino RM (2005) Remembering winter: toward a molecular understanding of vernalization. Annu Rev Plant Biol 56:491–508

Sung S, He Y, Eshoo TW et al (2006) Epigenetic maintenance of the vernalized state in Arabidopsis thaliana requires LIKE HETEROCHROMATIN PROTEIN 1. Nat Genet 38:706–710

Suzuki M, Wang HH, McCarty DR (2007) Repression of the LEAFY COTYLEDON 1/B3 regulatory network in plant embryo development by VP1/ABSCISIC ACID INSENSITIVE 3-LIKE B3 genes. Plant Physiol 143(2):902–911

Tittel-Elmer M, Bucher E, Broger L et al (2010) Stress-induced activation of heterochromatic transcription. PLoS Genet 6(10):e1001175

Tiwari S, Schulz R, Ikeda Y et al (2008) MATERNALLY EXPRESSED PAB C-TERMINAL, a novel imprinted gene in Arabidopsis, encodes the conserved C-terminal domain of polyadenylate binding proteins. Plant Cell 20:2387–2398

Tucker MR, Okada T, Hu Y et al (2012) Somatic small RNA pathways promote the mitotic events of megagametogenesis during female reproductive development in Arabidopsis. Development 139:1399–1404

Twell D (2011) Male gametogenesis and germline specification in flowering plants. Sex Plant Reprod 24:149–160

Vielle-Calzada JP, Baskar R, Grossniklaus U (2000) Delayed activation of the paternal genome during seed development. Nature 404:91–94

Walbot V, Evans MM (2003) Unique features of the plant life cycle and their consequences. Nat Rev Genet 4:369–379

Wang D, Deal RB (2015) Epigenome profiling of specific plant cell types using a streamlined INTACT protocol and ChIP-seq. Meth Mol Biol 1284:3–25

Wang Y, Gu X, Yuan W et al (2014) Photoperiodic control of the floral transition through a distinct polycomb repressive complex. Dev Cell 28:727–736

Waters AJ, Makarevitch I, Eichten SR et al (2011) Parent-of-origin effects on gene expression and DNA methylation in the maize endosperm. Plant Cell 23:4221–4233

Wendrich JR, Weijers D (2013) The Arabidopsis embryo as a miniature morphogenesis model. New Phytol 199:14–25

Xiao W, Gehring M, Choi Y et al (2003) Imprinting of the MEA Polycomb gene is controlled by antagonism between MET1 methyltransferase and DME glycosylase. Dev Cell 5:891–901

Xu W, Dai M, Li F et al (2014) Genomic imprinting, methylation and parent-of-origin effects in reciprocal hybrid endosperm of castor bean. Nucleic Acids Res 42:6987–6998

Xu W, Yang T, Dong X et al (2016) Genomic DNA methylation analyses reveal the distinct profiles in castor bean seeds with persistent endosperms. Plant Physiol 171:1242–1258

Yang WC, Shi DQ, Chen YH (2010) Female gametophyte development in flowering plants. Annu Rev Plant Biol 61:89–108

Yang H, Lu P, Wang Y et al (2011) The transcriptome landscape of Arabidopsis male meiocytes from high-throughput sequencing: the complexity and evolution of the meiotic process. Plant J 65:503–516

Zemach A, Kim MY, Silva P et al (2010) Local DNA hypomethylation activates genes in rice endosperm. Proc Natl Acad Sci U S A 107:18729–18734

Zemach A, Kim MY, Hsieh PH et al (2013) The Arabidopsis nucleosome remodeler DDM1 allows DNA methyltransferases to access H1-containing heterochromatin. Cell 153:193–205

Zhang M, Xie S, Dong X et al (2014) Genome-wide high resolution parental-specific DNA and histone methylation maps uncover patterns of imprinting regulation in maize. Genome Res 24:167–176

Zhou A, Pawlowski WP (2014) Regulation of meiotic gene expression in plants. Front Plant Sci 5:413

Zilberman D, Gehring M, Tran RK et al (2007) Genome-wide analysis of Arabidopsis thaliana DNA methylation uncovers an interdependence between methylation and transcription. Nature Genet 39:61–69

Rice Epigenomics: How Does Epigenetic Manipulation of Crops Contribute to Agriculture?

Yoshiki Habu

Abstract Production of rice (*Oryza sativa*)—the staple food of over half the world's population, especially those living in poverty—must continue to increase to meet the rising demand. The availability of a wide variety of natural rice resources has enabled highly efficient breeding approaches that have successfully improved productivity as well as biotic and abiotic stress tolerances. However, recent changes in global climate tendencies are imposing additional pressures on rice production, with the need for varieties showing unprecedented characteristics to counter adverse environments calling for innovative responses from breeders and researchers. Recent developments in epigenetic research in *Arabidopsis thaliana* have provided a plethora of data on epigenetic regulation in gene expression and development, paving the way to crop improvement via epigenetic manipulation. At ~400 Mb, the rice genome is the smallest among cereal crops and is relatively tractable with current molecular genetics techniques. This chapter begins by comparing characteristics of the rice genome and epigenome with those of *Arabidopsis*, before presenting some examples of epigenetic regulation in plants, with the emphasis on agriculturally important traits including abiotic stress responses. Most molecular studies on epigenetic modifications affecting plant phenotypes have been done in *Arabidopsis*, but examples of epigenetic regulation of agriculturally important traits in rice are accumulating rapidly. Current problems and difficulties in applying epigenetic manipulation to rice and ensuring stable maintenance of the modified epigenetic states to secure given agricultural traits under natural conditions will then be discussed.

Keywords Abiotic stress • Crop breeding • Epigenome • Rice • Transgenerational epigenome inheritance

---

Y. Habu (✉)
Institute of Agrobiological Sciences, National Institute of Agriculture and Food Research Organization, Kannondai 2-1-2, Tsukuba 305-8602, Ibaraki, Japan
e-mail: habu@affrc.go.jp

N. Rajewsky et al. (eds.), *Plant Epigenetics*, RNA Technologies,
DOI 10.1007/978-3-319-55520-1_21

Contents

1 Introduction

Forecasts of world cereal production predict 2542.9 million tons in 2016, of which rice accounts for 19% of total production after maize (40%) and wheat (28%) (Food and Agriculture Organization of the United Nations 2016b). In a clear contrast to other major cereals, 96% of rice is produced in developing countries (cf. maize 50%, wheat 47%; Food and Agriculture Organization of the United Nations 2016a). These statistics indicate that technical innovation, especially when aimed at mitigating the effects of adverse culture conditions, holds the potential for dramatic increases in rice yield. Key factors in achieving high yields are tolerance to environmental stresses and resistance to pathogens. Global warming, especially, poses an unavoidable threat to world crop production, and predictions indicate that central Europe, north and south America, and south Africa will suffer severe drought and that western Russia, the Middle East, Australia, and southern China will experience moderate drought by the end of this century (Dai 2013). In addition to identifying quantitative trait loci, and their introgression to cultivated crop lines by crossing (Yamamoto and Yano 2008; Fleury and Langridge 2014), recent strategies aimed at increasing crop tolerance to environmental stress have adopted overexpression of transcription factors that govern genes directly involved in responses to stresses (Nakashima et al. 2014; Singh and Laxmi 2015; Joshi et al. 2016). However, recent studies with *Arabidopsis thaliana* have revealed that engineering of epigenetic modification of genes can provide a useful additional layer in the armory of responses to environmental stress that are tractable to manipulation (Chinnusamy et al. 2014; Kim et al. 2015).

In order to apply engineering of epigenetic states of a particular locus to an agricultural crop, stability of the altered epigenetic state is crucial, especially after removal of the initial actions implemented to induce the desired changes in the epigenetic state. Although stable maintenance of alleles carrying spontaneously, or inducibly altered, epigenetic states in plants has been reported (Boyko and Kovalchuk 2011; Hauser et al. 2011; Paszkowski and Grossniklaus 2011; Iwasaki and Paszkowski 2014a, b), active maintenance of artificially altered epigenetic states has not yet been achieved. Current directions in this field are to study the basic mechanisms of initiation, maintenance, and transgenerational inheritance of altered epigenetic states in model plants such as *Arabidopsis* and to seek appropriate procedures for application of the latter to crops. Accordingly, I begin by reviewing characteristics of the genomes and epigenetic regulation in *Arabidopsis* and in rice, the latter having the best characterized genome and epigenetic factors among the major cereal crops (Zhao and Zhou 2012; Chen and Zhou 2013).

2 Epigenome Regulation in *Arabidopsis* and Rice

The genome of cultivated rice (*Oryza sativa* L.) consists of 12 chromosomes. The latest estimate of genome size of a reference Nipponbare cultivar is 384.2–386.5 Mbp (Kawahara et al. 2013). Transposons occupy at least 35% of the genome and explain the variation in genome size among species in the genus *Oryza* (Zuccolo et al. 2007). Rice centromeres contain 155–165 bp CentO repeats and various transposons, including Ty3-*gypsy* and the centromere-specific retrotransposon *RIRE7* (International Rice Genome Sequencing Project 2005; Sasaki et al. 2008). Centromeres on chromosomes 4 and 8 have been sequenced, and active genes have been shown to be present in the centromeric regions on chromosome 8 (Nagaki et al. 2004; Zhang et al. 2004). The genome-wide epigenetic status in rice has been characterized in various ways, and its overall features are roughly similar to those of *Arabidopsis thaliana* (Li et al. 2008; Feng et al. 2010; He et al. 2010; Zemach et al. 2010). However, in contrast to *Arabidopsis*, in which transposons are highly concentrated around centromeres, rice chromosomes show a broad distribution of transposons, with higher levels of CG and CHG methylation (H = A, T, or C) around the centromere on all chromosomes (Numa et al. 2015).

Rice epigenetic regulators have been identified and characterized (reviewed in Zhao and Zhou 2012; Cen and Zhou 2013; Table 1). METHYLTRANSFERASE1 (MET1) is an *Arabidopsis* methyltransferase that maintains CG methylation upon DNA replication (Finnegan and Dennis 1993; Law and Jacobsen 2010). Two copies of genes homologous to *MET1* are present in the rice genome, with one of them (*OsMET1-2*) being an orthologous gene (Hu et al. 2014). A much larger fraction of genes was affected in the mutant *Osmet1-2* than in *Arabidopsis met1* (Hu et al. 2014). In contrast, DECREASE IN DNA METHYLATION1 (DDM1)—a chromatin remodeling protein required for the maintenance of heterochromatin characteristics in *Arabidopsis* (Vongs et al. 1993; Lippman et al. 2004)—is encoded in rice by two genes with redundant function (Tan et al. 2016). Mutants of *MET1* and *DDM1* are fertile in *Arabidopsis*, but a null mutant of *OsMET1-2* and a double mutant of the two rice *DDM1* genes are lethal at early stages of seedling growth and the mature stage, respectively (He et al. 2010; Tan et al. 2016). An antisense knockdown line of *OsDDM1* (*asDDM1*) is viable, and a genome-wide analysis of *asDDM1* indicated that deficiency of *OsDDM1* activity affects genes expressed in a tissue-specific manner (Numa et al. 2015). A striking effect of *asDDM1* on the distribution of DNA methylation was that changes in CHH methylation increased in centromeric regions (Numa et al. 2015). Since activity of *OsDDM1* genes is required for the maintenance of deposition of CENH3—a centromere-specific histone H3 variant in rice (Habu et al. 2015)—, increased CHH methylation at

Table 1 Details of epigenetic regulators in rice described in this chapter

Functional category	Gene name	Abbreviation	MSU ID	RAP ID	Homologous gene in *Arabidopsis thaliana*	TAIR ID	Function in rice	Reference
DNA methylation/ demethylation	*METHYLTRANSFERASE1*	*OsMET1-2*	LOC_Os07g08500	Os07g0182900	*MET1*	AT5G49160	A null mutant shows severe growth defects	Hu et al. (2014)
	DOMAINS REARRANGED METHYLASE2	*OsDRM2*	LOC_Os03g02010	Os03g0110800	*DRM2*	AT5G14620	Mutants show severe developmental defects	Moritoh et al. (2012), Tan et al. (2016)
		OsROS1a	LOC_Os01g11900	Os01g0218032	*ROS1*	AT2G36490	A mutant allele is not transmitted to the next generation	Ono et al. (2012)
Chromatin remodeling	*DECREASE IN DNA METHYLATION1a*	*OsDDM1a*	LOC_Os09g27060	Os09g0442700	*DDM1*	AT5G66750	OsDDM1a and OsDDM1b function redundantly	Tan et al. (2016)
	DECREASE IN DNA METHYLATION1b	*OsDDM1b*	LOC_Os03g51230	Os03g0722400	*DDM1*	AT5G66750		Tan et al. (2016)
Histone modification	*HISTONE DEACETYLASE705*	*OsHDA705*	LOC_Os08g25570	Os08g0344100	*HDA6*	AT5G63110	Involved in seed germination and abiotic stress responses	Wang et al. (2014), Zhao et al. (2016)
	HISTONE DEACETYLASE711	*OsHDA711*	LOC_Os04g33480	Os04g0409600	*HDA9*	AT3G44680	No information for its function	Fu et al. (2007), Ma et al. (2013), Wang et al. (2014)
	HISTONE DEACETYLASE702	*OsHDA702*	LOC_Os06g38470	Os06g0583400	*HDA19*	AT4G38130	Controls seedling root growth	Chung et al. (2009), Wang et al. (2014)
	HD2 FAMILY HISTONE DEACETYLASE701	*OsHDT701*	LOC_Os05g51830	Os05g0597100	*HD2C*	AT5G03740	Negatively regulates innate immunity	Ding et al. (2012), Wang et al. (2014)
	HD2 FAMILY HISTONE DEACETYLASE702	*OsHDT702*	LOC_Os01g68104	Os01g0909100	*HD2B*	AT5G22650	Involved in leaf development	Hu et al. (2009), Wang et al. (2014)
	SIRTUIN2 FAMILY HISTONE DEACETYLASE701	*OsSRT701*	LOC_Os04g20270	Os04g0271000	*SRT1*	AT5G55760	Targets stress- and metabolism-related genes and transposons	Fu et al. (2007), Ma et al. (2013), Wang et al. (2014)
	SIRTUIN2 FAMILY HISTONE DEACETYLASE702	*OsSRT702*	LOC_Os12g07950	Os12g0179800	*SRT2*	AT5G09230	Predicted to be a mitochondrial protein	Zhong et al. (2013), Wang et al. (2014)

RNA-directed DNA methylation	*DICER-LIKE3*	*OsDCL3a*	LOC_Os01g68120	Os01g0909200	*DCL3*	AT3G43920	Involved in siRNA production and hormone homeostasis	Wei et al. (2014)
Posttranscriptional gene silencing	*RNA-DEPENDENT RNA POLYMERASE6*	*OsRDR6*	LOC_Os01g34350	Os01g0527600	*RDR6*	AT3G49500	Involved in 21- and 24-nt siRNA production	Song et al. (2012)

centromeric regions might be a mechanism to compensate for loss of CG and CHG methylation in order to maintain centromeric function. Although increased CHH methylation over generations has also been observed at *rRNA* loci and pericentromeric repeat sequences in a *met1* mutant of *Arabidopsis* (Mathieu et al. 2007), *Arabidopsis ddm1* shows no increase in CHH methylation in centromeric regions (Numa et al. 2015). It should be noted that CHH methylation is concentrated in centromeric regions in wild-type *Arabidopsis* (Cokus et al. 2008; Stroud et al. 2013), but not in rice (Li et al. 2012; Numa et al. 2015; Tan et al. 2016), suggesting that regulation of centromeric DNA methylation is divergent between *Arabidopsis* and rice. Since a similar increase in CHH methylation is also observed in a double mutant of two *OsDDM1* genes in rice that is not fertile, and therefore cannot be maintained over generations (Tan et al. 2016), compensation of centromeric function by DNA methylation cannot have been a gradual change over generations, but rather a rapid and crucial requirement for faithful maintenance of centromeric function in rice.

RNA-directed DNA methylation (RdDM) is a pathway for de novo DNA methylation, including cytosines in a CHH context, in plants, and utilizes 24 -nt-small RNA for target selection (Law and Jacobsen 2010; Kanno and Habu 2011). A set of proteins functioning from precursor RNA production to DNA methylation has been characterized in detail in *Arabidopsis* (Law and Jacobsen 2010). Plant-specific RNA polymerases (Pol IV and Pol V) play pivotal roles in RdDM, but involvement of RNA polymerase II in the RdDM pathway has also been shown in *Arabidopsis* (Zheng et al. 2009). In contrast, information on RdDM pathways in rice is currently limited. A major methyltransferase required for RdDM in *Arabidopsis* is DOMAINS REARRANGED METHYLASE2 (DRM2) (Cao and Jacobsen 2002). Although a loss-of-function mutant of *DRM2* in *Arabidopsis*, even as a double mutant with an additional *DRM* gene (*DRM1*), shows normal growth and development, a rice mutant of *OsDRM2* exhibited strong morphological defects (Moritoh et al. 2012; Tan et al. 2016). DICER-LIKE3 (DCL3) is required for producing 24-nt RNAs from their precursor dsRNAs in the *Arabidopsis* RdDM pathway, and a homologue of DCL3, OsDCL3a, plays the same role in rice (Wei et al. 2014). Knockdown of OsDCL3a activity causes impaired development, including dwarfism and altered leaf angle (Wei et al. 2014). These studies suggest that, in contrast to *Arabidopsis*, RdDM functions in normal growth and development in rice.

Another RNA-dependent pathway for regulation of DNA methylation includes RNA-DEPENDENT RNA POLYMERASE6 (RDR6), which was initially thought to be involved in RNAi pathways that cleave target RNAs in a small-RNA-dependent manner (Vaucheret 2006). The RDR6-RdDM pathway induces DNA methylation of a subset of genes, including *TAS* loci producing trans-acting siRNAs and a transgene (Eamans et al. 2008; Wu et al. 2012; Kanno et al. 2013). A recent finding also indicated involvement of RDR6 in DNA methylation of transposons (Nuthikattu et al. 2013). Although mutants of *Arabidopsis RDR6* are viable, a mutant of *RDR6* in rice again shows severe developmental defects of the spikelet, suggesting an essential role for *RDR6* in normal development in rice (Song et al.

2012). Additional evidence for involvement of DNA methylation in basic rice growth came from the analysis of *REPRESSOR OF SILENCING1* (*ROS1*)—a bifunctional DNA glycosylase/lyase targeting methylated DNA that functions in active demethylation (Gong et al. 2002). A null mutation allele of the rice *OsROS1a* gene was not inherited to the next generation, indicating that demethylation activity by *OsROS1a* is crucial for gametophyte development in rice (Ono et al. 2012). These data suggest that rice epigenome regulators function not only on heterochromatic regions and transposons but also on genes involved in normal development. The higher content of transposons often present in gene-rich regions in the rice genome may explain the difference in phenotypes between rice and *Arabidopsis* mutants.

An advanced study in rice has reported the involvement of microRNAs (miRNAs) directing DNA methylation that depend on canonical RdDM components (Wu et al. 2010). miRNAs are known to function in broad aspects of plant development and tolerance to environmental stresses (Chen 2009; Shriram et al. 2016). Currently, it is not clear to what extent miRNA-mediated methylation contributes to shaping epigenome status upon environmental responses and development in plants.

3 Epigenome Regulation in Response to Abiotic Stresses

Tolerance to environmental stresses in crops is of prime importance in agriculture, and enhancement of responses upon application of stresses has been a major target in plant breeding programs. A straightforward strategy is overexpression of transcription factors that activate stress responsive genes, thereby increasing levels of stress responses and resulting in improved tolerances (Todaka et al. 2015). However, overexpression of stress-responsive transcription factors is often accompanied by negative side effects on development or yield, and therefore, strategies based on novel concepts are required for practical application of stress tolerance on crops (Cabello et al. 2014). Changes in epigenetic states in plants upon exposure to environmental stresses have been reported, and epigenetic regulators of DNA and histone modification have been shown to be involved in these processes (reviewed in Chinnusamy and Zhu 2009; Paszkowski and Grossniklaus 2011; Kim et al. 2015; Vriet et al. 2015). Epigenetic changes as responses to environmental stimuli can be divided into two overlapping sets of changes: transient and transgenerational (Mirouze and Paszkowski 2011). DNA methylation is thought to play a major role in long-term transgenerational maintenance of epigenetic changes, and histone modifications have been considered as additional components that directly or indirectly supports transgenerational maintenance of epigenetic states (Sano 2010; Mirouze and Paszkowski 2011; Hauser et al. 2011). In contrast to plants, in which DNA methylation is present on cytosines in all sequence contexts (CG, CHG, and CHH), many other organisms, including vertebrates, insects, and yeast, have restricted DNA methylation and instead utilize histone modification as the major tool regulating chromatin status. However, mechanisms connecting histone modifications to inheritance of acquired epigenetic changes in organisms other than plants currently remain unclear (Campos et al. 2014).

Histone deacetylases (HDAs) remove acetyl groups from the side chains of lysine and arginine of histones and regulate expression of genes and chromatin status in various organisms including plants (Loidl 2004). A combination of chemical modifications other than acetylation (methylation, phosphorylation, ubiquitination, etc.) on amino acids in N-terminal histone tails produces specific messages that are read by various types of modified histone-binding proteins for activating downstream signals (Jenuwein and Allis 2001; Kouzarides 2007). Transition between acetylated and deacetylated states is rapid and reversible (Taddei et al. 2001), and the flexible nature of chromatin modification without cell division is a key characteristic of environmental responses (Kim et al. 2015). Dozens of histone acetyltransferases and deacetylases have been characterized in plants (Ma et al. 2013; Wang et al. 2014), and not only their temporal response to environmental stimuli but also their regulatory functions in various processes in normal development of plants have been reported (Zhao and Zhou 2012; Ma et al. 2013). Based on sequence similarities, plant HDAs are classified into three groups (Ma et al. 2013), among which the Reduced potassium dependency3 (Rpd3)-like families are predominant in plants, and the HD2 family is plant specific (Loidl 2004; Wang et al. 2014).

Rpd3-like HDAs in *Arabidopsis* (HDA6, HDA9, and HDA19) have been shown to play critical roles in responses to abiotic stresses (Chen et al. 2010; To et al. 2011; Zhong et al. 2013; Zheng et al. 2016). HDA6 and HDA19 are also required for plant responses to exogenously applied hormones that are tightly linked to abiotic and biotic stresses (Zhou et al. 2005; Wu et al. 2008), and evidence and models for direct interaction of HDAs with promoters of stress responsive genes have been presented (Chen and Wu 2010; Zhu et al. 2011; Jung et al. 2013). Plant HDAs belonging to HD2 and SIR2 are also involved in abiotic stress responses (Sridha and Wu 2006; Ding et al. 2012; Zhong et al. 2013). However, because plant HDAs are also involved in various developmental processes (Chung et al. 2009; Jang et al. 2003; Kim et al. 2013; Hao et al. 2016; Chung and Kim 2009; Hu et al. 2009), simple application of epigenetic regulation through *HDA* mutants in agriculture would require careful consideration.

How do plant HDAs specifically regulate stress-related genes and hormones upon perception of environmental stimuli? In animals, a major role of Rpd3-like HDA is thought to be its contribution to chromatin formation through interaction with chromatin-modifying proteins (Nan et al. 1998; Xue et al. 1998; Fuks et al. 2000). Related functions of *Arabidopsis* HDAs have also been reported, especially in maintenance of silent chromatin (Probst et al. 2004; Early et al. 2010; Kim To et al. 2011), and direct binding of HDAs to various chromatin proteins has been confirmed (Gu et al. 2011; Liu et al. 2012; Lee and Cho 2016), suggesting that HDAs are recruited to target loci by their interacting proteins. Several HDA-binding proteins have been identified in *Arabidopsis*, and, strikingly, the majority of these are involved in biotic and abiotic stress responses (Ma et al.

2013). This is consistent with the abovementioned idea that histone acetylation plays a role in rapid responses to environmental stimuli. A direct link connecting HDAs to stress responses in *Arabidopsis* has been provided from a study on the interaction of a histone deacetylase complex with ABA-receptor genes (Mehdi et al. 2016). An additional molecular mechanism linking HDAs to stress responses is the direct interaction between HDAs and proteins that transduce stress hormonal signaling to downstream transcription activation (Devoto et al. 2002; Zhu et al. 2011).

4 Stable Maintenance of Altered Epigenomic State for Agricultural Applications

As described above, epigenetic modifications of a subset of genes change in response to environmental stimuli. In most cases, these changes are transient and the altered epigenetic states are reversed when the initial signals that provoked them disappear. However, in rare cases, altered epigenetic states and activities of genes are fixed and stably inherited over generations (Paszkowski and Grossniklaus 2011). Factors affecting such transgenerational stability of altered states have been elucidated in *Arabidopsis*; most involve transcriptional silencing of endogenous transposons (Mathieu et al. 2007; Iwasaki and Paszkowski 2014a, b; Blevins et al. 2014).

Activation of transposons is often observed in mutants deficient in chromatin-modifying enzymes, suggesting that epigenetic mechanisms suppress at least a subset of transposon activities in *Arabidopsis* (Miura et al. 2001; Mirouze et al. 2009; Tsukahara et al. 2009; Ito et al. 2011) and rice (Ding et al. 2007; Qin et al. 2010; Cui et al. 2013). In addition, even in wild-type plants growing under normal conditions, active states of transposons in particular lines can persist over generations in many plants (e.g., Gerats et al. 1990; Naito et al. 2006; Fedoroff 2013). Since transposons can be regarded as parasites in the host genome, the balance of epigenetic suppression and prolonged transposon activation represents a battle between transposons and their host plants. Studies on plant morphology have discovered non-transposon genes carrying altered states of epigenetic modification that persist over generations; alleles that carry no changes in their nucleotide sequence but have altered states of epigenetic modification are called epialleles (Kalisz and Purugganan 2004). Transposons are often found inserted into, or near, epialleles, suggesting that host mechanisms for silencing transposons contribute to the establishment of epialleles. Some epialleles are derived from mutants deficient in epigenetic regulation (e.g., Saze and Kakutani 2007; Stokes and Richards 2002), but others are thought to be produced spontaneously under natural conditions (e.g., Cubas et al. 1999; Bender and Fink 1995; Ong-Abdullah et al. 2015; Wang et al. 2015). Among them, epialleles directly and positively connected to agricultural and ornamental importance include those affecting floret closing (cleistogamy) in barley (Wang et al. 2015), oil yield in oil palm (Ong-Abdullah et al. 2015), and flower symmetry (Cubas et al. 1999).

Alongside maize, rice is one of the best characterized crops in terms of epialleles (Fu et al. 2007; Zhao and Zhou 2012; Chen and Zhou 2013). So far, several epialleles have been reported in rice; some are not connected to agriculturally superior phenotypes (Miura et al. 2009; Zhang et al. 2012; Chen et al. 2015; Zhang et al. 2015), but others provide direct connection to agriculturally important traits. The latter include heritable changes in DNA methylation induced by nitrogen deficiency (Kou et al. 2011), heavy metal stress (Ou et al. 2012), and resistance to *Xanthomonas oryzae* (Akimoto et al. 2007). Understanding of the mechanisms establishing and maintaining these epialleles is required if they are to be exploited in agriculture.

Naturally occurring RNA silencing is another mechanism that produces heritable phenotypic changes. A rice line containing low levels of glutelin—a seed storage protein—carries an inverted repeat structure of glutelin genes, which produces double-stranded RNAs and silences other glutelin genes in a dominant manner (Kusaba et al. 2003). Low glutelin rice has potentially important applications for patients suffering from kidney disorders who must restrict protein intake. Another example of an epigenetic change caused by RNA silencing is the pale yellow color of soybean seed coat, which was shown to be due to silencing of an inverted repeat structure in the chalcone synthase gene (Kasai et al. 2007).

5 Perspectives

This review focuses on epigenetics. In addition to the topics covered here, other important agricultural traits, whose molecular mechanisms are as yet unclear, could also be governed by epigenetic regulation. An understanding of heterosis was one of the most influential breakthroughs in twentieth-century agriculture (Gowen 1952; Duvick 2001). Heterosis is observed not only in plants but also in other organisms including livestock (Chen 2013). The molecular mechanisms of heterosis remain somewhat enigmatic, however, with several classical (dominance, overdominance, and epistasis) and epigenetic models having been proposed (Chen 2013). Many studies have suggested that, in crops, genetic components govern heterosis (Krieger et al. 2010; Huang et al. 2016), but recent studies support the possible involvement of epigenetic mechanisms in heterosis through hormonal and growth regulation (Dapp et al. 2015; Zhang et al. 2016). Another issue to be solved at the molecular level that could be applied to agriculture is regulation of meiotic recombination. Recent developments in marker-assisted breeding, which utilizes information from genome sequences, have accelerated the speed and efficiency of selecting individual plants carrying better traits, but the efficiency of mixing genomes carrying such traits that scatter in the genome has not yet been addressed. Low rates of recombination in genome regions other than meiosis hotspots remain an uncontrollable obstacle in breeding. Recent advances in epigenetic mechanisms have indicated that chromatin modifications could determine meiotic recombination points, and alteration of genome-wide epigenetic states in mutants could change the distribution of meiotic recombination sites in *Arabidopsis* (Perrella et al. 2010; Colomé-Tatché et al. 2012; Melamed-Bessudo and Levy 2012; Mirouze et al. 2012; Yelina et al. 2012) and rice (Habu et al. 2015).

As mentioned above, mutants of many genes functioning in epigenome regulation in rice are non-fertile or show developmental abnormalities, thus precluding both detailed analysis of their function and their practical application in agriculture. A possible solution would be to generate weak alleles carrying point mutations by chemical mutagenesis (Till et al. 2007). Recent rapid developments in genome editing tools available in plants could also provide a means to precisely modify essential genes (Osakabe and Osakabe 2015). Such strategies would be suitable for epigenetic regulators functioning with various binding partners; a mutation in a specific domain might disturb a function governed by that domain but not others. For example, *Arabidopsis* HDA6, which regulates various environmental stress responses, binds to MET1 and AtTRB2—a telomere-binding protein—through its C-terminal region (Liu et al. 2012; Lee and Cho 2016). Engineering of the C-terminal region of HDA6 would result in impaired genome integrity and abnormal development, thus illustrating that detailed information on the structure–function relationships of the target proteins and precise and careful manipulation of protein structure are crucial for practical applications. Clearly, the main purpose of genetic engineering of rice is to increase yield of this edible crop. In addition to considering the acceptability to consumers of engineered crops, avoiding unwanted side effects that impede yield while retaining elite grain characteristics, including eating quality and growth traits, are the challenges that must be met in the development of final products for public and commercial release. As outlined in this review, epigenetics has an important role to play in modern agriculture.

Acknowledgements I thank Helen Rothnie for comments on the manuscript and Rie Takahashi for technical assistance. This work was supported by a grant (CREST) from Japan Science and Technology Agency to YH.

References

Akimoto K, Katakami H, Kim HJ et al (2007) Epigenetic inheritance in rice plants. Ann Bot 100:205–217

Bender J, Fink GR (1995) Epigenetic control of an endogenous gene family is revealed by a novel blue fluorescent mutant of Arabidopsis. Cell 83:725–734

Blevins T, Pontvianne F, Cocklin R et al (2014) A two-step process for epigenetic inheritance in Arabidopsis. Mol Cell 54:30–42

Boyko A, Kovalchuk I (2011) Genome instability and epigenetic modification-heritable responses to environmental stress? Curr Opin Plant Biol 14:260–266

Cabello JV, Lodeyro AF, Zurbriggen MD (2014) Novel perspectives for the engineering of abiotic stress tolerance in plants. Curr Opin Biotechnol 26:62–70

Campos EI, Stafford JM, Reinberg D (2014) Epigenetic inheritance: histone bookmarks across generations. Trends Cell Biol 24:664–674

Cao X, Jacobsen SE (2002) Role of the *Arabidopsis DRM* methyltransferases in de novo DNA methylation and gene silencing. Curr Biol 12:1138–1144

Chen X (2009) Small RNAs and their roles in plant development. Annu Rev Cell Dev Biol 25:21–44

Chen ZJ (2013) Genomic and epigenetic insights into the molecular bases of heterosis. Nat Rev Genet 14:471–482

Chen LT, Wu K (2010) Role of histone deacetylases HDA6 and HDA19 in ABA and abiotic stress response. Plant Signal Behav 5:1318–1320

Chen X, Zhou DX (2013) Rice epigenomics and epigenetics: challenges and opportunities. Curr Opin Plant Biol 16:164–169

Chen LT, Luo M, Wang YY et al (2010) Involvement of *Arabidopsis* histone deacetylase HDA6 in ABA and salt stress response. J Exp Bot 61:3345–3353

Chen X, Liu X, Zhao Y et al (2015) Histone H3K4me3 and H3K27me3 regulatory genes control stable transmission of an epimutation in rice. Sci Rep 5:13251

Chinnusamy V, Zhu JK (2009) Epigenetic regulation of stress responses in plants. Curr Opin Plant Biol 12:133–139

Chinnusamy V, Dalal M, Zhu JK (2014) Epigenetic regulation of abiotic stress responses in plants. In: Jenks MA, Hasegawa PM (eds) Plant abiotic stress, 2nd edn. Wiley, Ames, pp 203–229

Chung PJ, Kim JK (2009) Epigenetic interaction of OsHDAC1 with the OsNAC6 gene promoter regulates rice root growth. Plant Signal Behav 4:675–677

Chung PJ, Kim YS, Jeong JS et al (2009) The histone deacetylase OsHDAC1 epigenetically regulates the OsNAC6 gene that controls seedling root growth in rice. Plant J 59:764–776

Cokus SJ, Feng S, Zhang X et al (2008) Shotgun bisulfite sequencing of the Arabidopsis genome reveals DNA methylation patterning. Nature 452:215–219

Colomé-Tatché M, Cortijo S, Wardenaar R et al (2012) Features of the *Arabidopsis* recombination landscape resulting from the combined loss of sequence variation and DNA methylation. Proc Natl Acad Sci U S A 109:16240–16245

Cubas P, Vincent C, Coen E (1999) An epigenetic mutation responsible for natural variation in floral symmetry. Nature 401:157–161

Cui X, Jin P, Cui X et al (2013) Control of transposon activity by a histone H3K4 demethylase in rice. Proc Natl Acad Sci U S A 110:1953–1958

Dai A (2013) Increasing drought under global warming in observations and models. Nat Climate Change 3:52–58

Dapp M, Reinders J, Bédiée A et al (2015) Heterosis and inbreeding depression of epigenetic Arabidopsis hybrids. Nat Plants 1:15092

Devoto A, Nieto-Rostro M, Xie D et al (2002) COI1 links jasmonate signalling and fertility to the SCF ubiquitin-ligase complex in *Arabidopsis*. Plant J 32:457–466

Ding Y, Wang X, Su L et al (2007) SDG714, a histone H3K9 methyltransferase, is involved in *Tos17* DNA methylation and transposition in rice. Plant Cell 19:9–22

Ding B, del Rosario BM, Ning Y et al (2012) HDT701, a histone H4 deacetylase, negatively regulates plant innate immunity by modulating histone H4 acetylation of defense-related genes in rice. Plant Cell 24:3783–3794

Duvick DN (2001) Biotechnology in the 1930s: the development of hybrid maize. Nat Rev Genet 2:69–74

Eamans A, Vaistij FE, Jones L (2008) NRPD1a and NRPD1b are required to maintain post-transcriptional RNA silencing and RNA-directed DNA methylation in Arabidopsis. Plant J 55:596–606

Early KW, Pontvianne F, Wierzbicki AT et al (2010) Mechanisms of HDA6-mediated rRNA gene silencing: suppression of intergenic Pol II transcription and differential effects on maintenance versus siRNA-directed cytosine methylation. Genes Dev 24:1119–1132

Fedoroff NV (2013) The discovery of transposition. In: Fedoroff NV (ed) Plant transposons and genome dynamics in evolution. Wiley-Blackwell, Ames, pp 3–13
Feng S, Cokus SJ, Zhang X et al (2010) Conservation and divergence of methylation patterning in plants and animals. Proc Natl Acad Sci U S A 107:8689–8694
Finnegan EJ, Dennis ES (1993) Isolation and identification by sequence homology of a putative cytosine methyltransferase from *Arabidopsis thaliana*. Nucleic Acids Res 21:2383–2388
Fleury D, Langridge P (2014) QTL and association mapping for plant abiotic stress tolerance: trait characterization and introgression for crop improvement. In: Jenks MA, Hasegawa PM (eds) Plant abiotic stress, 2nd edn. Wiley, Ames, pp 257–287
Food and Agriculture Organization of the United Nations (2016a) Crop prospects and food situation. No. 2 Junee (http://www.fao.org/giews/english/cpfs/i5710e/i5710e.html)
Food and Agriculture Organization of the United Nations (2016b) Food outlook. June 2016 (http://www.fao.org/giews/english/fo/index.htm)
Fu W, Wu K, Duan J (2007) Sequence and expression analysis of histone deacetylases in rice. Biochem Biophys Res Commun 356:843–850
Fuks F, Burgers WA, Brehm A et al (2000) DNA methyltransferase Dnmt1 associates with histone deacetylase activity. Nat Genet 24:88–91
Gerats AG, Huits H, Vrijlandt E et al (1990) Molecular characterization of a nonautonomous transposable element (*dTph1*) of petunia. Plant Cell 2:1121–1128
Gong Z, Morales-Ruiz T, Ariza RR et al (2002) ROS1, a repressor of transcriptional gene silencing in Arabidopsis encodes a DNA glycosylase/lyase. Cell 111:803–814
Gowen JW (1952) Heterosis. A record of researches directed toward explaining and utilizing the vigor of hybrids. Iowa State College Press, Ames
Gu X, Jiang D, Yang W et al (2011) *Arabidopsis* homolog of retinoblastoma-associated protein 46/48 associate with a histone deacetylase to act redundantly in chromatin silencing. PLoS Genet 7:e1002366
Habu Y, Ando T, Ito S et al (2015) Epigenomic modification in rice controls meiotic recombination and segregation distortion. Mol Breed 35:103
Hao Y, Wang H, Qiao S et al (2016) Histone deacetylase HDA6 enhances brassinosteroid signaling by inhibiting the BIN2 kinase. Proc Natl Acad Sci U S A 113:10418–10423
Hauser MT, Aufsatz W, Jonak C et al (2011) Transgenerational epigenetic inheritance in plants. Biochem Biophys Acta 1809:459–468
He G, Zhu X, Elling AA et al (2010) Global epigenetic and transcriptional trends among two rice subspecies and their reciprocal hybrids. Plant Cell 22:17–33
Hu Y, Qin F, Huang L et al (2009) Rice histone deacetylase genes display specific expression patterns and developmental functions. Biochem Biophys Res Commun 388:266–271
Hu L, Li N, Xu C et al (2014) Mutation of a major CG methylase in rice causes genome-wide hypomethylation, dysregulated genome expression, and seedling lethality. Proc Natl Acad Sci U S A 111:10642–10647
Huang X, Yang S, Gong J et al (2016) Genomic architecture of heterosis for yield traits in rice. Nature 537:629–633
International Rice Genome Sequencing Project (2005) The map-based sequence of the rice genome. Nature 436:793–800
Ito H, Gaubert H, Bucher E et al (2011) An siRNA pathway prevents transgenerational retrotransposition in plants subjected to stress. Nature 472:115–119
Iwasaki M, Paszkowski J (2014a) Epigenetic memory in plants. EMBO J 33:1987–1998
Iwasaki M, Paszkowski J (2014b) Identification of genes preventing transgenerational transmission of stress-induced epigenetic states. Proc Natl Acad Sci U S A 111:8547–8552
Jang IC, Pahk YM, Song SI et al (2003) Structure and expression of the rice class-I type histone deacetylase genes *OsHDAC1-3*: *OsHDAC1* overexpression in transgenic plants leads to increased growth rate and altered architecture. Plant J 33:531–541
Jenuwein T, Allis CD (2001) Translating the histone code. Science 293:1074–1080

Joshi R, Wani SH, Singh B et al (2016) Transcription factors and plants response to drought stress: current understanding and future directions. Front Plant Sci 7:1029

Jung JH, Park JH, Lee S et al (2013) The cold signaling attenuator HIGH EXPRESSION OF OSMOTICALLY RESPONSIVE GENE1 activates *FLOWERING LOCUS C* transcription via chromatin remodeling under short-term cold stress in *Arabidopsis*. Plant Cell 25:4378–4390

Kalisz S, Purugganan MD (2004) Epialleles via DNA methylation: consequences for plant evolution. Trends Ecol Evol 19:309–314

Kanno T, Habu Y (2011) siRNA-mediated chromatin maintenance and its function in *Arabidopsis thaliana*. Biochim Biophys Acta 1809:444–451

Kanno T, Yoshikawa M, Habu Y (2013) Locus-specific requirements of DDR complexes for gene-body methylation of *TAS* genes in *Arabidopsis thaliana*. Plant Mol Biol Rep 31:1048–1052

Kasai A, Kasai K, Yumoto S et al (2007) Structural features of *GmIRCHS*, candidate of the *I* gene inhibiting seed coat pigmentation in soybean: implications for inducing endogenous RNA silencing of chalcone synthase genes. Plant Mol Biol 64:467–479

Kawahara Y, de la Bastide M, Hamilton JP et al (2013) Improvement of the *Oryza sativa* Nipponbare reference genome using next generation sequence and optical map data. Rice 6:4

Kim To T, Kim JM, Matsui A et al (2011) *Arabidopsis* HDA6 regulates locus-directed heterochromatin silencing in cooperation with MET1. PLoS Genet 7:e1002055

Kim W, Latrasse D, Servet C et al (2013) Arabidopsis histone deacetylase HDA9 regulates flowering time through repression of AGL19. Biochem Biophys Res Commun 432:394–398

Kim JM, Sasaki T, Ueda M et al (2015) Chromatin changes in response to drought, salinity, heat, and cold stresses in plants. Front Plant Sci 6:114

Kou HP, Li Y, Song XX et al (2011) Heritable alteration in DNA methylation induced by nitrogen-deficiency stress accompanies enhanced tolerance by progenies to the stress in rice. J Plant Physiol 168:1685–1693

Kouzarides T (2007) Chromatin modifications and their function. Cell 128:693–705

Krieger U, Lippman ZB, Zamir D (2010) The flowering gene SINGLE FLOWER TRUSS drives heterosis for yield in tomato. Nat Genet 42:459–463

Kusaba M, Miyahara K, Iida S et al (2003) Low glutelin content1: a dominant mutation that suppresses the glutelin multigene family via RNA silencing in rice. Plant Cell 15:1455–1467

Law JA, Jacobsen SE (2010) Establishing, maintaining and modifying DNA methylation patterns in plants and animals. Nat Rev Genet 11:204–220

Lee WK, Cho MH (2016) Telomere-binding protein regulates the chromosome ends through the interaction with histone deacetylases in *Arabidopsis* thaliana. Nucleic Acids Res 44:4610–4624

Li X, Wang X, He K et al (2008) High-resolution mapping of epigenetic modifications of the rice genome uncovers interplay between DNA methylation, histone methylation, and gene expression. Plant Cell 20:259–276

Li X, Zhu J, Hu F et al (2012) Single-base resolution maps of cultivated and wild rice methylomes and regulatory roles of DNA methylation on plant gene expression. BMC Genomics 13:300

Lippman Z, Gendrel AV, Black M et al (2004) Role of transposable elements in heterochromatin and epigenetic control. Nature 430:471–476

Liu X, Yu CW, Duan J et al (2012) HDA6 directly interacts with DNA methyltransferase MET1 and maintains transposable element silencing in Arabidopsis. Plant Physiol 158:119–129

Loidl P (2004) A plant dialect of the histone language. Trends Plant Sci 9:84–90

Ma X, Lv S, Zhang C et al (2013) Histone deacetylases and their functions in plants. Plant Cell Rep 32:465–478

Mathieu O, Reinders J, Cáikovski M et al (2007) Transgenerational stability of the *Arabidopsis* epigenome is coordinated by CG methylation. Cell 130:851–862

Mehdi S, Derkacheva M, Ramström M et al (2016) The WD40 domain protein MSI1 functions in a histone deacetylase complex to fine-tune abscisic acid signaling. Plant Cell 28:42–54

Melamed-Bessudo C, Levy AA (2012) Deficiency in DNA methylation increases meiotic cross-over rates in euchromatic but not in heterochromatic regions in *Arabidopsis*. Proc Natl Acad Sci U S A 109:E981–E988

Mirouze M, Reinders J, Bucher E et al (2009) Selective epigenetic control of retrotransposition in *Arabidopsis*. Nature 461:427–430

Mirouze M, Paszkowski J (2011) Epigenetic contribution to stress adaptation in plants. Curr Opin Plant Biol 14(3):267–274

Mirouze M, Lieberman-Lazarovich M, Aversano R et al (2012) Loss of DNA methylation affects the recombination landscape in *Arabidopsis*. Proc Natl Acad Sci U S A 109:5880–5885

Miura A, Yonebayashi S, Watanabe K et al (2001) Mobilization of transposons by a mutation abolishing full DNA methylation in *Arabidopsis*. Nature 411:212–214

Miura K, Agetsuma M, Kitano H et al (2009) A metastable *DWARF1* epigenetic mutant affecting plant stature in rice. Proc Natl Acad Sci U S A 106:11218–11223

Moritoh S, Eun CH, Ono A et al (2012) Targeted disruption of an orthologue of *DOMAINS REARRANGED METHYLASE 2*, *OsDRM2*, impairs the growth of rice plants by abnormal DNA methylation. Plant J 71:85–98

Nagaki K, Cheng Z, Ouyang S et al (2004) Sequencing of a rice centromere uncovers active genes. Nat Gent 36:138–145

Naito K, Cho E, Yang G et al (2006) Dramatic amplification of a rice transposable element during recent domestication. Proc Natl Acad Sci U S A 103:17620–17625

Nakashima K, Yamaguchi-Shinozaki K, Shinozaki S (2014) The transcriptional regulatory network in the drought response and its crosstalk in abiotic stress responses including drought, cold, and heat. Front Plant Sci 5:170

Nan X, Ng HH, Johnson CA et al (1998) Transcriptional repression by the methyl-CpG-binding protein MeCP2 involves a histone deacetylase complex. Nature 393:386–389

Numa H, Yamaguchi K, Shigenobu S et al (2015) Gene body CG and CHG methylation and suppression of centromeric CHH methylation are mediated by DECREASE IN DNA METHYLATION1 in rice. Mol Plant 8:1560–1562

Nuthikattu S, McCue AD, Panda K et al (2013) The initiation of epigenetic silencing of active transposable elements is triggered by RDR6 and 21-22 nucleotide small interfering RNAs. Plant Physiol 162:116–131

Ong-Abdullah M, Ordway JM, Jiang N et al (2015) Loss of Karma transposon methylation underlies the mantled somaclonal variant of oil palm. Nature 525:533–537

Ono A, Yamaguchi K, Fukada-Tanaka S et al (2012) A null mutation of ROS1a for DNA demethylation in rice is not transmittable to progeny. Plant J 71:564–574

Osakabe Y, Osakabe K (2015) Genome editing with engineered nucleases in plants. Plant Cell Physiol 56:389–400

Ou X, Zhang Y, Xu C et al (2012) Transgenerational inheritance of modified DNA methylation patterns and enhanced tolerance induced by heavy metal stress in rice (*Oryza sativa* L.) PLoS One 7:e41143

Paszkowski J, Grossniklaus U (2011) Selected aspects of transgenerational epigenetic inheritance and resetting in plants. Curr Opin Plant Biol 14:195–203

Perrella G, Consiglio MF, Aiese-Cigliano R et al (2010) Histone hyperacetylation affects meiotic recombination and chromosome segregation in *Arabidopsis*. Plant J 62:796–806

Probst AV, Fagard M, Proux F et al (2004) Arabidopsis histone deacetylase *HDA6* is required for maintenance of transcriptional gene silencing and determines nuclear organization of rDNA repeats. Plant Cell 16:1021–1034

Qin FJ, Sun QW, Huang LM (2010) Rice SUVH histone methyltransferase genes display specific functions in chromatin modification and retrotransposon repression. Mol Plant 3:773–782

Sano H (2010) Inheritance of acquired traits in plants, reinstatement of Lamarck. Plant Signal Behav 5:346–348

Sasaki T, Wu J, Mizuno H et al (2008) The rice genome sequence as an indispensable tool for crop improvement. In Hirano HY et al (eds) Rice biology in the genomics era. Biotechnology in agriculture and forestry. vol 62 Springer, Berlin, pp 3–12

Saze H, Kakutani T (2007) Heritable epigenetic mutation of a transposon-flanked *Arabidopsis* gene due to lack of the chromatin-remodeling factor DDM1. EMBO J 26:3641–3652

Shriram V, Kumar V, Devarumath RM et al (2016) MicroRNAs as potential targets for abiotic stress tolerance in plants. Front Plant Sci 7:817

Singh D, Laxmi A (2015) transcriptional regulation of drought responses: a tortuous network of transcriptional factors. Front Plant Sci 6:895

Song X, Wang D, Ma L et al (2012) Rice RNA-dependent RNA polymerase 6 acts in small RNA biogenesis and spikelet development. Plant J 71:378–389

Sridha S, Wu K (2006) Identification of *AtHD2C* as a novel regulator of abscisic acid responses in Arabidopsis. Plant J 46:124–133

Stokes TL, Richards EJ (2002) Induced instability of two Arabidopsis constitutive pathogen-response alleles. Proc Natl Acad Sci U S A 99:7792–7796

Stroud H, Greenberg MVC, Feng S et al (2013) Comprehensive analysis of silencing mutants reveals complex regulation of the *Arabidopsis* methylome. Cell 152:352–364

Tan F, Zhou C, Zhou Q et al (2016) Analysis of chromatin regulators reveals specific features of rice DNA methylation pathways. Plant Physiol 171:2041–2054

Till BJ, Cooper J, Tai TH et al (2007) Discovery of chemically induced mutations in rice by TILLING. BMC Plant Biol 7:19

To TK, Nakaminami K, Kim JM et al (2011) *Arabidopsis HDA6* is required for freezing tolerance. Biochem Biophys Res Commun 406:414–419

Todaka D, Shinozaki K, Yamaguchi-Shinozaki K (2015) Recent advances in the dissection of drought-stress regulation networks and strategies for development of drought-tolerant transgenic plants. Front Plant Sci 6:84

Tsukahara S, Kobayashi A, Kawabe A et al (2009) Bursts of retrotransposition reproduced in *Arabidopsis*. Nature 461:423–426

Vaucheret H (2006) Post-transcriptional small RNA pathways in plants: mechanisms and regulations. Genes Dev 20:759–771

Vongs A, Kakutani T, Marienssen RA et al (1993) *Arabidopsis thaliana* DNA methylation mutants. Science 260:1926–1928

Vriet C, Henning L, Laloi C (2015) Stress-induced chromatin changes in plants: of memories, metabolites and crop improvement. Cell Mol Life Sci 72:1261–1273

Wang Z, Cao H, Chen F et al (2014) The roles of histone acetylation in seed performance and plant development. Plant Physiol Biochem 84:125–133

Wang N, Ning S, Wu J et al (2015) An epiallele ay *cly1* affects the expression of floret closing (cleistogamy) in barley. Genetics 199:95–104

Wei L, Gu L, Sing X et al (2014) Dicer-like 3 produces transposable element-associated 24-nt siRNAs that control agricultural traits in rice. Proc Natl Acad Sci U S A 111:3877–3882

Wu K, Zhang L, Zhou C et al (2008) HDA6 is required for jasmonate response, senescence and flowering in *Arabidopsis*. J Exp Bot 59:225–234

Wu L, Zhou H, Zhang Q et al (2010) DNA methylation mediated by a microRNA pathway. Mol Cell 38:465–475

Wu L, Mao L, Qi Y (2012) Roles of DICER-LIKE and ARGONAUTE proteins in *TAS*-derived small interfering RNA-triggered DNA methylation. Plant Physiol 160:990–999

Xue Y, Wong J, Moreno GT et al (1998) NURD, a novel complex with both ATP-dependent chromatin-remodeling and histone deacetylase activities. Mol Cell 2:851–861

Yamamoto T, Yano M (2008) Detection and molecular cloning of genes underlying quantitative phenotypic variations in rice. In: Hirano HY et al (eds) Rice biology in the genomics era, Biotechnology in agriculture and forestry, vol 62. Springer, Berlin, pp 295–308

Yelina NE, Choi K, Chelysheva L et al (2012) Epigenetic remodeling of meiotic crossover frequency in *Arabidopsis thaliana* DNA methyltransferase mutants. PLoS Genet 8:e1002844

Zemach A, McDaniel IE, Silva P et al (2010) Genome-wide evolutionary analysis of eukaryotic DNA methylation. Science 328:916–919
Zhang Y, Huang Y, Zhang L et al (2004) Structural features of the rice chromosome 4 centromere. Nucleic Acids Res 32:2023–2030
Zhang L, Cheng Z, Qin R et al (2012) Identification and characterization of an epi-allele of *FIE1* reveals a regulatory linkage between two epigenetic marks in rice. Plant Cell 24:4407–4421
Zhang X, Sun J, Cao X et al (2015) Epigenetic mutation of *RAV6* affects leaf angle and seed size in rice. Plant Physiol 169:2118–2128
Zhang Q, Li Y, Xu T et al (2016) The chromatin remodeler DDM1 promotes hybrid vigor by regulating salicylic acid metabolism. Cell Discov 2:16027
Zhao Y, Zhou DX (2012) Epigenomic modification and epigenetic regulation in rice. J Genet Genomics 39:307–315
Zhao J, Li M, Gu D et al (2016) Involvement of rice histone deacetylase HDA705 in seed germination and in response to ABA and abiotic stresses. Biochem Biophys Res Commun 470:439–444
Zheng B, Wang Z, Li S et al (2009) Intergenic transcription by RNA polymerase II coordinates Pol IV and Pol V in siRNA-directed transcriptional gene silencing in *Arabidopsis*. Genes Dev 23:2850–2860
Zheng Y, Ding Y, Sun X et al (2016) Histone deacetylase HDA9 negatively regulates salt and drought stress responsiveness in Arabidopsis. J Exp Bot 67:1703–1713
Zhong X, Zhang H, Zhao Y et al (2013) The rice NAD^+-dependent histone deacetylase OsSRT1 targets preferentially to stress- and metabolism-related genes and transposable elements. PLoS One 8:e66807
Zhou C, Zhang L, Duan J et al (2005) *HISTONE DEACETYLASE19* is involved in jasmonic acid and ethylene signaling of pathogen response in Arabidopsis. Plant Cell 17:1196–1204
Zhu Z, An F, Feng Y et al (2011) Derepression of ethylene-stabilized transcription factors (EIN3/EIL1) mediates jasmonate and ethylene signaling synergy in *Arabidopsis*. Proc Natl Acad Sci U S A 108:12539–12544
Zuccolo A, Sebastian A, Talag J et al (2007) Transposable element distribution, abundance and role in genome size variation in the genus *Oryza*. BMC Evol Biol 7:152

Epigenetic Characterization of Satellite DNA in Sugar Beet (*Beta vulgaris*)

Falk Zakrzewski and Thomas Schmidt

Abstract Satellite DNA is a major and abundant component of plant genomes and comprises important genomic regions such as heterochromatic knobs and centromeric chromatin. However, technical barriers of assembling first- and second-generation sequencing data hampered the complete arrangement of satellite DNAs in current plant genome sequences. Consequently, heterochromatic and centromeric regions possessing satellite DNA lack detailed characterization and assignment, which limits knowledge about their epigenetic status. We applied methods to overcome these limitations and to gain insight into the epigenetic modifications of satellite DNA-rich heterochromatic and centromeric regions of the sugar beet (*Beta vulgaris*) genome. Sugar beet is an important crop of temperate climate zones, which provides nearly 30% of the world's annual sugar needs. Due to the 11% of the genome consisting of satellite DNAs, sugar beet is a suitable research object for comparative investigation and epigenetic characterization of this repeat class. We analyzed the epigenetic modifications of satellite DNA by using bisulfite sequencing, chromatin immunoprecipitation followed by sequencing (ChiP-Seq) using antibodies against histone CenH3 and dimethylated H3K9me2, and small RNA-seq data. Immunostaining of methylated cytosines and histone modifications combined with fluorescent *in situ* hybridization (FISH) coupled with super-resolution fluorescence microscopy complemented the epigenetic analysis. As a result, we uncovered individual epigenetic characteristics of plant satellite DNAs at high resolution and hypothesized a model for satellite DNA-directed heterochromatization.

Keywords Satellite DNA • DNA methylation • Histone modifications • CenH3 • H3K9me2 • Centromere • Heterochromatin • Sugar beet • *Beta vulgaris*

Contents

F. Zakrzewski • T. Schmidt (✉)
Department of Biology, Technische Universität Dresden, 01062 Dresden, Germany
e-mail: Thomas.Schmidt@tu-dresden.de

N. Rajewsky et al. (eds.), *Plant Epigenetics*, RNA Technologies,
DOI 10.1007/978-3-319-55520-1_22

1 Introduction: Sugar Beet (*Beta vulgaris*) and Its Wild Relatives

Beet (*Beta vulgaris*) cultivars are very variable in morphology and grown for sucrose production (sugar beet), as vegetable (garden beet, Swiss chard) or animal feed (fodder beet). Sugar beet is a relatively young crop which was developed in the late seventh century when it has been proved that some beet accessions contain sucrose which is chemically identical to that of sugarcane. While wild sea beets such as *B. vulgaris* ssp. *maritima* are the progenitors of all cultivated forms, today's sugar beets are derived from the White Silesian beet (Weiße Schlesische Rübe) which was selected from the offspring of a cross of fodder beet with leaf beet (Fischer 1989). During the last 200 years of sugar beet breeding, the sugar content has increased from 8% to 18% in today's cultivars.

All cultivated forms of *Beta vulgaris* ssp. *vulgaris* belong to the genus *Beta* which is a member of the *Amaranthaceae* family within the *Caryophyllales* order. The genus *Beta* is divided into the following sections: *Beta*, *Corollinae*, and *Nanae*. Species of the formerly section *Procumbentes* form a separate genus, called *Patellifolia* (Kadereit et al. 2006). The section *Beta* also harbors the wild subspecies *Beta vulgaris* ssp. *maritima* and *Beta vulgaris* ssp. *adanensis* and the wild beets *Beta patula* and *Beta macrocarpa*. The sections *Corollinae* and *Nanae* and the genus *Patellifolia* consist exclusively of wild beets.

Species of the section *Beta* are widely distributed along Mediterranean and central and northern Atlantic coastlines, while wild beets of other sections have a more limited geographic distribution and are either found on European islands of the Atlantic Ocean or at coastal and inland locations from Greece to Iran (Ford-Lloyd and Williams 1975; de Bock 1986). The section *Nanae* consists of the single species *Beta nana*, which is endemic to high altitude habitats in Greece.

Because of the narrow genetic base of sugar beet, crosses with wild beets have been performed to broaden the gene pool of the cultivar. Of particular interest are introgression lines of sugar beet containing wild beet chromatin which confers resistance to fungi, viruses, or the beet cyst nematode *Heterodera schachtii* as well as tolerance to salinity or drought.

2 Genomes, Chromosomes and Satellite DNAs

Most *Beta* species are diploid; however, in the section *Corollinae* and the genus *Patellifolia*, higher ploidy levels are found. Sugar beet (*B. vulgaris* ssp. *vulgaris*) is diploid (2n = 18) and has a relatively small genome of 758 Mbp (Arumuganathan and Earle 1991). Early attempts to analyze *Beta* genomes were classical cytogenetic studies which were hampered by the small and similar size and morphology of the metacentric chromosomes and the lack of informative chromatin staining dyes. A reference fluorescent *in situ* hybridization (FISH) karyotype of sugar beet has been established for unequivocal and high-resolution identification of each chromosome arm using a set of marker-anchored BAC (Bacterial Artificial Chromosome) probes (Paesold et al. 2012). Recently, the sugar beet genome sequence has been generated (Dohm et al. 2014), which is an excellent resource for the investigation of the large-scale chromosomal and molecular organization and epigenetic modification of repetitive DNA. However, a large proportion of repeats is only partially included in the reference sequence due to assembly limitations using first- and second-generation sequence reads making it difficult to analyze their epigenetic profile (Treangen and Salzberg 2012; Zakrzewski et al. 2014). The current genome assembly spans 567 Mbp and harbors 26,923 protein coding genes (Weisshaar et al. 2016).

Reassociation kinetics studies and bioinformatic analyses revealed that the genome consists of 63–64% repetitive DNA (Flavell et al. 1974; Kowar et al. 2016). A large proportion of this repetitive DNA comprises a high amount of satellite DNA (Schmidt and Heslop-Harrison 1998; Kowar et al. 2016), which is only fragmentarily included in the reference sequence. Plant satellites consist of homologous sequence motifs (monomers) which vary in size of 150–180 bp and multiples thereof and are tandemly organized (Hemleben et al. 2007). Tandem arrays predominantly occur in large homogenous arrays of megabases in size in prominent heterochromatin, intercalary or subtelomeric regions while smaller arrays may be dispersed along chromosomes (Palomeque and Lorite 2008; Plohl et al. 2008; Hemleben et al. 2007). Satellite DNAs are important structural sequence elements in heterochromatin formation and maintenance and hence are involved in the stability of chromatin status (Teixeira and Colot 2010; Lisch 2009; Ugarkovic 2005; Martienssen 2003).

The availability of large data sets from genome sequencing projects has dramatically increased the opportunities of satellite DNA investigations using bioinformatics approaches. Comparison of thousands of copies of satellite monomers enabled the detection of subfamily structures, either within a genome or between species. The dynamics detected as sequence diversification and changes in abundance (e.g., species-specific amplification) provides insight into the evolution of sequence families and, at a larger scale, chromosomes and genomes and is often correlated with species radiation. The three sections *Beta*, *Corollinae*, and *Nanae* of the genus *Beta* and the *Patellifolia* species consist of twelve closely and more distantly related species and subspecies that are useful for comparative genome analysis and studies of sequence and chromosome evolution. A systematic search of the genomes of *Beta* and *Patellifolia* species has revealed 29 different satellite DNA families and subfamilies; an overview about satellite DNA families is given in Zakrzewski et al. (2013). Each section of the genus *Beta* as well as the genus *Patellifolia* exhibits a characteristic set of satellite families that cover large regions of several or all chromosomes. These satellite DNA families were amplified at different periods during the evolution of *Beta* and *Patellifolia* species demonstrating an enormous diversification and distribution. Although most satellites are widely distributed in both genera, they differ strongly in species-specific amplification, sequence composition, homogenization level, and chromosomal dispersal and localization.

3 Satellite DNAs Are a Major Repeat Class in Sugar Beet Heterochromatin and Centromeric Chromatin

Repetitive DNA, in particular satellite DNA, is only partially included in the sugar beet genome sequence (Dohm et al. 2014). Therefore, this incompleteness contributes to the deviation of the estimated genome size (758 Mbp) from the size of the assembled reference sequence (567 Mbp). Furthermore, heterochromatic regions possessing a high proportion of repetitive elements lack genetic and epigenetic characterization and assignment. Sugar beet chromosomes have a characteristic distribution of highly condensed chromatin knobs: in addition to large centromeric regions and smaller subtelomeric and interspersed heterochromatic sites, each chromosome arm harbors an intercalary heterochromatic knob which differs in size from arm to arm and chromosome to chromosome. Satellite DNAs, a major group of repetitive DNAs in plants, are largely amplified in the genome of sugar beet and make up 11.15% of the total nuclear genome (Kowar et al. 2016). The two most abundant satellites are pBV and pEV, which form large tandem arrays in the heterochromatin of all 18 chromosomes (Fig. 1a and b). The pBV satellite family was the first satellite repeat discovered in the genus *Beta* (Schmidt and Metzlaff 1991; Schmidt et al. 1991). It occurs exclusively in the section *Beta* indicating massive amplification after separation of the section *Beta* from the remaining sections *Corollinae*, *Nanae,* and *Patellifolia* species. The pBV satellite is the most abundant repeat family (50,000–60,000 copies) of the beet genome and occupies large regions of the pericentromeric and centromeric heterochromatin on all chromosomes (Schmidt and Heslop-Harrison 1998; Zakrzewski et al. 2013). The satellite pEV was first described in *B. vulgaris* (Schmidt et al. 1991) and is the second frequently occurring satellite repeat (Zakrzewski et al. 2010). This repeat family is ancient and exists in the sections *Beta*, *Corollinae*, *Nanae*, in *Patellifolia* species and even in the distantly related *Chenopodium quinoa* which occurs in South America (Schmidt et al. 2014). It has been most likely amplified before species and genera radiation and, because of its widespread distribution,

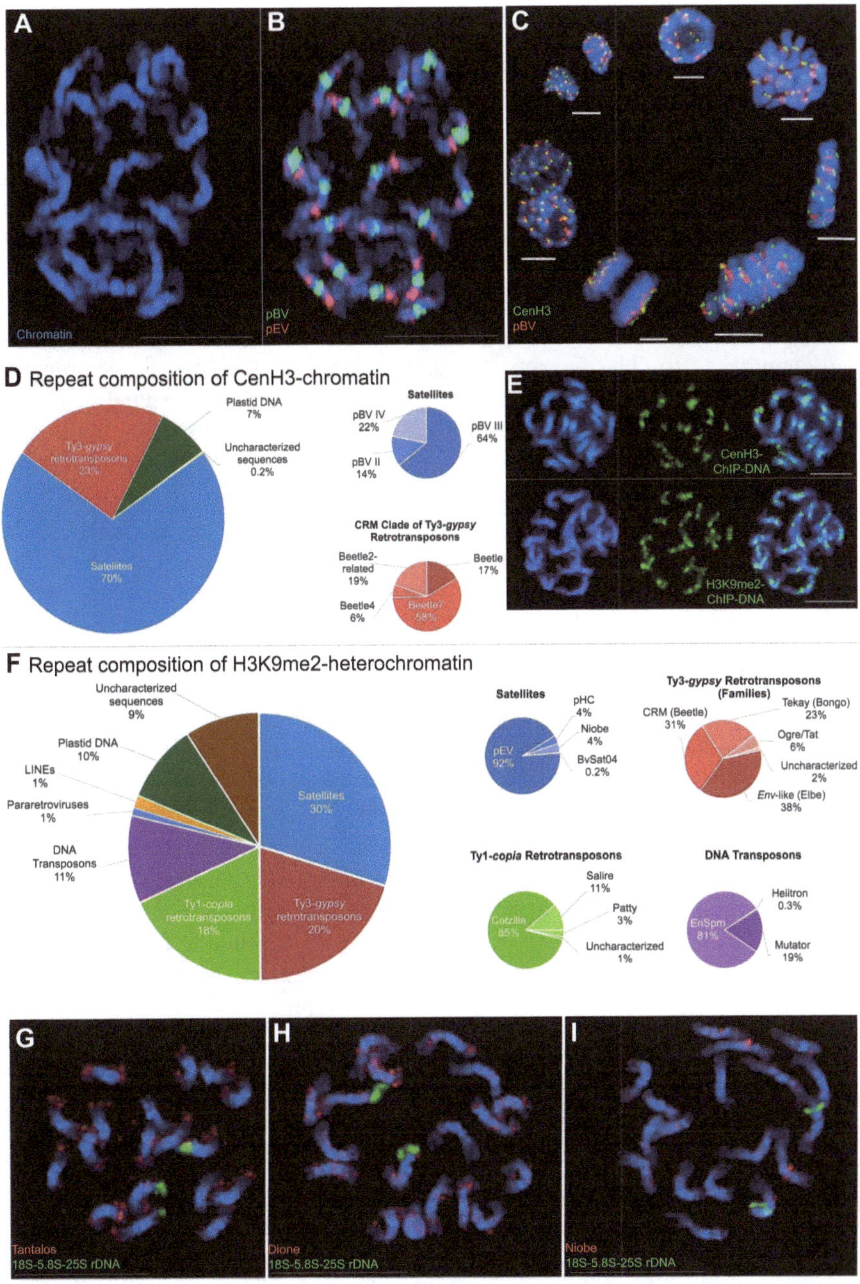

Fig. 1 Satellite DNAs are a major component of the sugar beet genome and largely contribute to CenH3-chromatin and H3K9me2-heterochromatin. Chromosomes are stained with DAPI (blue signals). Scale bars represent 5 μm. (**a**) Strong DAPI staining shows heterochromatic regions in metaphase chromosomes. (**b**) FISH of the pEV satellite (red signals) reveals the localization on both chromosome arms in intercalary heterochromatin. The pBV satellite (green signals) occupies the centromeric chromatin on all chromosomes. (**c**) Immunostaining combined with FISH shows the co-localization of the sugar beet-specific CenH3 (green signals) and the centromeric satellite

diverged species-specific subfamilies were detected. Most copies of the pEV family are concentrated in the intercalary heterochromatin knobs, present on each chromosome arm.

Kowar et al. (2016) performed a chromatin immunoprecipitation followed by sequencing (ChIP-Seq) approach to classify and characterize the repeat composition of centromeric chromatin and heterochromatic regions. The active centromere is characterized by the presence of a centromere-specific H3 histone variant CenH3, which is responsible for kinetochore formation (Heun et al. 2006; Howman et al. 2000; Lermontova et al. 2011; Régnier et al. 2005) and enables the centromere to act as early guide in cell division during mitosis and meiosis. Consequently, a sugar beet-specific antibody recognizing the sugar beet CenH3 was chosen for the ChIP-Seq approach to clarify repeat composition of the active centromere. Immunostaining using the sugar beet-specific CenH3 antibody resulted in 18 distinct and specific CenH3 signals co-localizing with pBV satellite arrays. All CenH3 signals occur in intensively DAPI-stained centromeric heterochromatin and are present in all stages of the cell cycle (Fig. 1c).

The repeats occurring in the sugar beet centromere (obtained by CenH3 ChIP-Seq) can be divided into four repeat groups (Fig. 1d), namely satellites (with 70% the most abundant repeat group),*Ty3-gypsy* retrotransposons (23%), plastid DNA (7%), and uncharacterized sequences (0.2%), which probably represent either truncated repeating units or yet unknown centromere-associated repeats. The existence of plastid DNA in sugar beet chromosomes may be due to multiple plastid DNA integration and possibly recombination events which have taken place several times independently and is therefore a frequent event rather than a single incident (Kowar et al. 2016). Within satellites, which exhibit the most abundant repeat type in centromeric chromatin, only the pBV satellite was found. The satellite pBV consists of six subfamilies varying in genome abundance and monomer size (Zakrzewski et al. 2013). In the active centromeres, only the most abundant subfamilies pBV-II, pBV-III, and pBV-IV were found. *Ty3-gypsy* retrotransposons consist exclusively of the Beetle family with Beetle7 accounting for 58% (Weber et al. 2013). All repeats detected are largely specific to centromeric chromatin, on

---

Fig. 1 (continued) pBV (red signals) on all chromosomes during all stages of the cell cycle. (**d**) Repeat composition of the CenH3-chromatin detected by CenH3-chromatin immunoprecipitation followed by sequencing (ChIP-Seq). The pBV satellite II, II, and IV variants are the most abundant repeats occurring in centromeric chromatin. (**e**) FISH using CenH3-ChIP-DNA (*upper panel*) and H3K9me2-ChIP-DNA (*lower panel*) shows strong signals in CenH3-specific and H3K9me2-specific heterochromatin. (**f**) Repeat composition of the H3K9me2-heterochromatin detected by H3K9me2-ChIP-Seq. The pEV satellite and retrotransposons are the most abundant repeats occurring in H3K9me2-heterochromatin. (**g**) The Tantalos arrays are clustered at the chromosome ends. (**h**) The Dione arrays are weakly dispersed with a tendency of enrichment in subtelomeric sites but are also detectable adjacent to pericentromeric regions. (**i**) Niobe shows a few dispersed arrays along chromosomes and a large array near the centromere on two homologues. In g, h, and i, the 18S-5.8S-25S rRNA genes (*green*) are detected at the ends of two homologues

some chromosomes extending into pericentromeric regions, as shown using the CenH3-ChIP-DNA as probe in FISH experiments (Fig. 1e, upper images).

In plants, heterochromatin is mainly characterized by the dimethylation of lysine 9 of histone H3 (H3K9me2) among other histone modifications (Zakrzewski et al. 2014; Gent et al. 2014; Roudier et al. 2011; West et al. 2014). Therefore, an antibody recognizing H3K9me2 was chosen for the ChIP-Seq experiment to detect and classify the repeats in the large intercalary heterochromatin knobs in sugar beet. Eight repeat groups occur in H3K9me2 heterochromatin. Four different satellites are the most abundant repeats, comprising 30% and among them the satellite pEV is the most amplified satellite of these chromosomal regions (92%) (Fig. 1f). *Ty1-copia* (18%) and *Ty3-gypsy* (20%) retrotransposons are highly enriched in H3K9me2-heterochromatin of which the *env*-like clade is the most abundant (Fig. 1f) containing the sugar beet-specific Elbe retrotransposons which harbor an additional ORF encoding a transmembrane-like protein domain (Wollrab et al. 2012). The CRM clade is represented in 31% of all *Ty3-gypsy* elements and contains prominent retroelements such as members of the families Beetle2, Beetle4, and Beetle7 (Weber and Schmidt 2009; Weber et al. 2013). Among the *Ty1-copia* retrotransposons, three families are enriched (Fig. 1f), namely Cotzilla, Salire, and Patty (Weber et al. 2010). Repeat clusters related to the Cotzilla family of the SIRE clade represent 85% of all *Ty1-copia* retrotransposons. These elements belong to the most abundant retrotransposon families in sugar beet comprising up to 3% of the genome. DNA transposons are represented with 11% in H3K9me2-heterochromatin, with the superfamilies En/Spm, Mutator, and Helitron identified. The vast majority of transposons belongs to the transposon superfamily En/Spm. About 9% of the sequence reads were assigned to clusters not characterized so far and represent unknown repeats specific for H3K9me2-heterochromatin. Much of the H3K9me2-ChIP-DNA is localized in the intercalary heterochromatin on both chromosome arms as demonstrated by FISH and most likely due to strong pEV satellite hybridization (Fig. 1e lower images).

The ChIP-Seq results confirm that satellites are the major repeat class in sugar beet as they make up most of the centromeric chromatin and H3K9m2-heterochromatin. The two most abundant satellites, pBV and pEV, are mutually exclusive satellites forming huge tandem arrays in active centromeres and large intercalary H3K9me2-heterochromatic blocks, respectively.

In addition to large and heavily abundant satellite arrays of pBV and pEV in CenH3-centromeric chromatin and in H3K9me2-heterochromatin, respectively, smaller and lower abundant satellite arrays occur in the sugar beet genome (Zakrzewski et al. 2013). These satellites are not restricted to specific chromosome regions or CenH3- or H3K9me2-specific chromatin states and show a relatively dispersed distribution along chromosomes with a distal tendency. Exemplarily, in FISH experiments, the three satellite families Tantalos, Dione, and Niobe show a family-specific pattern of array distribution. Tantalos is amplified in clusters that accumulate towards the chromosome ends in subtelomeric regions on most chromosomes. A few smaller arrays are dispersed on some chromosomes (Fig. 1g). Arrays of Dione are also dispersed on all chromosomes but excluded from the large heterochromatic blocks which are strongly stained by DAPI. Dione clusters are localized at distal regions of eight chromosomes while one chromosome pair contains only a small intercalary Dione cluster (Fig. 1h). Niobe is dispersed; however, only very few clusters are detectable on six chromosomes. In addition, one chromosome pair carries a large Niobe array amplified close to the centromeric heterochromatin (Fig. 1i). Green signals in Fig. 2g–i represent clusters of 18S-5.8S-25S rDNA genes at the end of chromosome pair 1.

4 Epigenetic Characterization of Satellite DNA Suggests Their Potential Function in the Establishment and Maintenance of Heterochromatin

To uncover epigenetic aspects of satellites such as the cytosine methylation, genome-wide bisulfite sequencing, analyses of transcription and occurrence of small RNAs, and immunostaining of DNA and chromatin were performed.

DNA methylation (idiom for cytosine methylation) is one of the most important contributors to the epigenetic regulation of DNA and chromatin. DNA methylation strongly influences the transcription of genes and thereby shapes the phenotypes in response to endogenous and exogenous signals such as environmental factors, developmental signals, stress, disease, and chemicals (Suzuki and Bird 2008). However, the hypothesized function of DNA methylation in transcriptional repression is not sufficient and still changing. A much more complex field has opened where DNA methylation can activate or silence genes and also repetitive DNAs with respect to varying environmental conditions (Jones 2012). Therefore, understanding the DNA methylation of the repetitive DNA, which makes up large proportions of many plant genomes (Schmidt and Heslop-Harrison 1998), is of outstanding interest to unravel the complex and expansive field of DNA methylation in whole-genome analysis. Satellite DNA as a major genome component needs to be included when analyzing the DNA methylation of plant genomes because of the high abundance and the localization in important chromosome regions.

To analyze the level of DNA methylation of satellite DNAs, the combination of immunostaining using an antibody against cytosine methylation with subsequent FISH (Zhang et al. 2008; Koo et al. 2011) targets to highlight the array-specific and chromosome-wide DNA methylation. Low methylation levels of the highly abundant sugar beet satellites pBV and pEV are detectable by application of this technique (Fig. 2a).

However, when analyzing smaller satellite loci, such as for example the dispersed sugar beet satellites Tantalos, Dione, and Niobe, the resolution of conventional UV wide-field microscopy is not sufficient for the differentiation of methylation and satellite signals. Therefore, an important step for the investigation of the chromosome-wide and array-specific DNA methylation of small satellite arrays is the adaption of super-resolution microscopy for a significant resolution

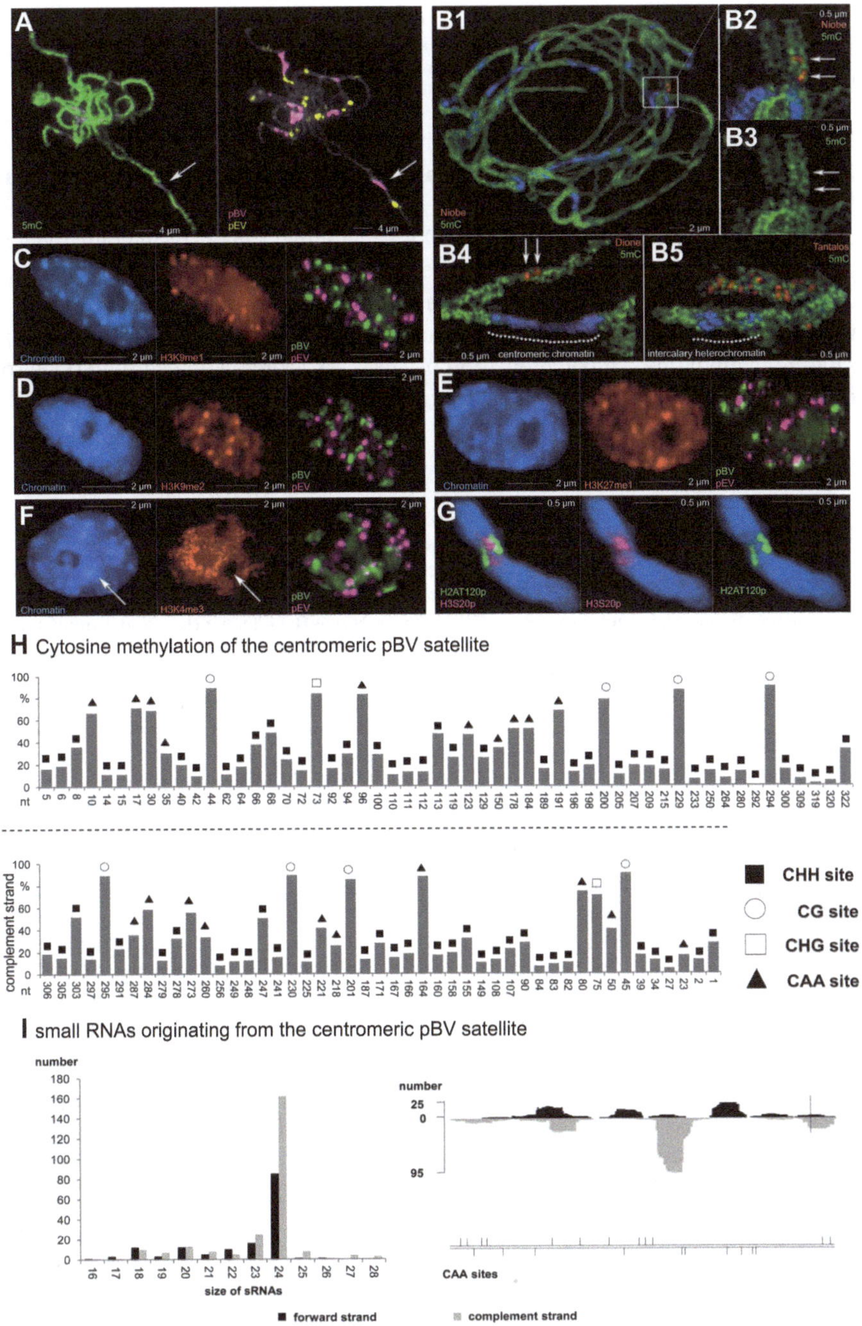

Fig. 2 Epigenetic characterization of satellite DNAs in sugar beet. (**a**) *Left*: Green signals show the distribution of the 5-methylcytosine (5mC) antibody on meiotic pachytene chromosomes (used for increased physical resolution). Large heterochromatic blocks show less signals than adjacent

enhancement. By application of this technique (SIM, Structured Illumination Microscopy), the low cytosine methylation of dispersed sugar beet satellites was determined (Fig. 2b1–b4).

The low DNA methylation of the centromeric pBV satellite is similar to the hypomethylation of centromeric satellites in *Arabidopsis thaliana* and maize (Zhang et al. 2008) observed by immunostaining and suggesting the widespread hypomethylation of plant centromeres. In contrast, the only exception are rice centromeres which are rich in satellite DNA and either hyper- or hypomethylated (Yan et al. 2010). In maize, it has been reported that satellite DNA in active centromeres is hypomethylated while inactive centromeres are elevated in DNA methylation (Koo et al. 2011), and the authors concluded that CenH3 inhibits the satellite methylation. Hypomethylation of DNA associated with centromeric chromatin may be more favorable to differential decondensation than the flanking H3

Fig. 2 (continued) euchromatic regions. *Right*: Hybridization with representative probes of the pBV satellite (magenta signals) and the pEV satellite (yellow signals) enables the detection of centromeric/pericentromeric and intercalary heterochromatin, respectively. The *arrow* points to an undermethylated centromeric region. (**b**) Chromosomal localization and low DNA methylation of the moderately abundant and dispersed satellite DNAs Niobe (**b1**), Dione (**b4**), and Tantalos (**b5**) on sugar beet meiotic pachytene chromosomes after FISH and immunostaining detected by Structured Illumination Microscopy (SIM), which significantly enhances resolution. DNA was stained with DAPI (*blue*). Centromeres are only weakly stained by the antibody against 5mC (*green*) and hence appear in *blue*. (**b1**) Niobe shows a few dispersed arrays along chromosomes and in a large array near the centromere on two homologues, shown in **b2** and with only 5mC signals (**b3**). (**b4**) An example of a Dione array (*arrows*) is located adjacent to the weakly methylated centromeric chromatin of another chromosome pair. (**b5**) Tantalos arrays clustered at chromosome ends. In addition, two homologous chromosomes below show weakly methylated intercalary heterochromatin. Detailed SIM images for the three satellites are available in Zakrzewski et al. (2014). Immunolabelling using antibodies against H3K9me1 (**c**), H3K9me2 (**d**), and H3K27me1 (**e**) on interphase nuclei (red signals). In addition, a typical euchromatic histone mark (H3K4me3) is shown in (**f**). Probes of the pBV satellite (green signals) and the pEV satellite (purple signals) were hybridized in parallel. The nuclei were stained with DAPI (blue signals). Heterochromatic regions are visible as strong DAPI positive chromocenters (exemplified by *arrows* in **f**). (**g**) The histone modifications H2AT120p (green signals) and H3S20p (red signals) are specific for active centromeres and are displayed on a representative metaphase chromosome. Antibody reactions are detectable on the outer site (H2AT120p) or the inner site (H3S20p) of the centromere. (**h**) Illustration of methylated cytosines in pBV satellite analyzed by bisulfite sequencing. Both DNA strands of satellite monomers are displayed. The relative frequency (*y*-axis) of methylated cytosines is displayed at monomer positions (*x*-axis), where a cytosine is the most frequently occurring nucleotide of the consensus monomer. CG sites are marked by an *open circle*, CHG sites with a *rectangle*, CHH sites with a *black rectangle*, and CAA sites with a *black triangle*. The images are also representative for pEV, Niobe, Dione, and Tantalos satellites (Zakrzewski et al. 2011, 2014). (**I**) *Left*: Size distribution of strand-specific small RNAs originated from the pBV satellite. The *x*-axis displays the size of the small RNAs and the *y*-axis shows the detected number of satellite homologous small RNAs. *Right*: Strand-specific distribution of 24-nt siRNAs along consensus satellite monomer sequence (the bar indicates end of one monomer). The position of CAA sites along monomers is indicated below. Clustering of 24-nt siRNAs at higher methylated CAA sites is not observable. The images are representative for pEV, Niobe, Dione, and Tantalos satellites (Zakrzewski et al. 2011, 2014)

nucleosome blocks, which may aid the loading of newly synthesized CenH3 (Koo et al. 2011).

The low satellite methylation levels observed by microscopy are the starting point for the investigation of the site-specific cytosine modification at the basepair level. Bisulfite sequencing of a large number of pBV monomers (~500) enabled to uncover the site preferences of cytosine methylation (Zakrzewski et al. 2011). Cytosines occurring in the frequently distributed asymmetric CHH (H = A, C, T) motif are mostly not methylated while cytosines in symmetric CG and CHG motifs are often methylated. Interestingly, the CAA motif (22% of all CHH sites) is higher methylated compared to other CHH sites (Fig. 2h). In total, the array-wide methylation of centromeric satellite is low because CG and CHG sites are rare and CHH sites are frequent. Furthermore, frequently occurring deamination may result in conversion of methylated cytosines in thymine bases after replication of DNA, which also explains the higher AT content of many satellites, and therefore leads to the loss of DNA methylation. A similar pattern of methylation as observed for pBV has been determined by bisulfite sequencing of thousands of other satellite monomers such as the pEV satellite in sugar beet (Zakrzewski et al. 2011) and in the wild beet *Patellifolia procumbens* (Schmidt et al. 2014) and in the moderately abundant and more dispersed smaller satellite arrays of Tantalos, Dione, and Niobe in sugar beet (Zakrzewski et al. 2014). However, the occurrence of highly methylated CG and CHG sites might explain why other subsequent epigenetic modifications, such as histone methylation or the attachment of heterochromatic proteins, can be maintained, despite the array-wide low methylation of centromeric satellites.

In animals, it has been shown that reduced levels of DNA methylation at centromeric satellite DNAs impair the localization of CENP-A, hence inhibiting centromere activity (Kim et al. 2012). The authors proposed that an altered heterochromatin status due to loss of DNA methylation and methylation of histone H3 at position lysine 9 hampers CENP-A loading. However, in animals, cytosines are mostly methylated in the CG context while in plants methylation occurs at CHH, CHG, and CG sites (Suzuki and Bird 2008). This might be the answer to the question why plant centromeres show a general and array-wide low cytosine methylation: The higher methylation at CG and CHG sites at centromeric satellite repeats might be sufficient for centromere activity, although most cytosines occur in the CHH motif and are only weakly methylated. It has been demonstrated that the most important heterochromatic histone mark, dimethylation of histone 3 at position lysine 9 (H3K9me2) (Zhang et al. 2008), is a typical marker for heterochromatin in plants (Fuchs et al. 2006). It occurs with low signal intensity at centromeric pBV repeats and strong signals at intercalary pEV arrays in sugar beet similar to H3K9me1 and H3K27me1 (Fig. 2c–e). For comparison, the plant-typical euchromatic histone modification H3K4me3 (Fuchs et al. 2006) is not enriched in pBV and pEV regions (Fig. 2f). Low levels of H3K9me2 at pBV arrays seem to be sufficient for creating an active sugar beet centromere. The enrichment of H3K9me2 clusters in the intercalary heterochromatin at pEV arrays is most likely linked to the higher frequency of CG and CHG at pEV arrays compared to pBV arrays (Zakrzewski et al. 2011). It has been reported that CHG methylation induces the methylation of H3K9 by the histone methyltransferase KRYPTONITE (KYP). CHG sites are methylated by CHROMOMETHYLASE 3 (CMT3) recognizing H3K9me2, hence generating a stably maintained heterochromatization (Law and Jacobsen 2010; Johnson et al. 2007). Furthermore, it is proposed that CMT3 interacts with the HETEROCHROMATIN PROTEIN 1 (HP1), which binds to heterochromatic regions (Jackson et al. 2002). HP1 is involved in heterochromatization and gene silencing as well as in the establishment of the epigenetic identity of centromeres in animals (Wang et al. 2000; Carroll and Straight 2006). However, the plant-specific LHP1 (HP1 in Arabidopsis) seems not to be essential for heterochromatin and centromere function (Tariq and Paszkowski 2004). Additionally, it has been demonstrated that CG methylation can also induce H3K9me2 (Johnson et al. 2007). Taken together, CG and CHG methylation present at satellite repeats while CHH methylation is low, suggesting that CG and CHG methylation at satellite DNA is sufficient for the establishment of heterochromatin. Furthermore, the histone phosphorylation of histone 2A at position threonine 120 (H2AT120p) and of histone 3 at position serine 20 (H3S20p) are centromere-specific in sugar beet (Fig. 2g) as has also been shown for other plants (Dong and Han 2012; Demidov et al. 2014).

In sugar beet, small RNAs (sRNAs), in particular of 24 nt in size, originate from the major satellites pBV and pEV (Fig. 2i, exemplarily shown for pBV) and for the dispersed satellites Niobe, Dione, and Tantalos (Zakrzewski et al. 2014). Similarly, sRNAs processed from satellite repeats were reported for several other eukaryotic species (Ghildiyal and Zamore 2009). Transcription and sRNAs of satellites act together and may fulfill essential function in the genome and are most probably involved in heterochromatization and centromere identity. Although satellite transcripts may serve autonomously as functional component in the recruitment of CenH3 to the centromeric heterochromatin (Allshire and Karpen 2008), many satellite RNAs are processed into small interfering RNAs (siRNAs) by the activity of DICER3. Subsequently, siRNAs may serve as primer for the generation of new satellite transcripts by RNA-DEPENDENT RNA POLYMERASE 2 (RDR2) which facilitate the maintenance of the siRNA pool (Martienssen 2003).

Due to the arrangement of satellite DNA in homologous monomer arrays spanning large regions of heterochromatin, each siRNA can map to multiple positions along the entire satellite array, initializing new satellite transcripts, and further siRNAs are processed. siRNAs fulfill functions in establishment and maintenance of heterochromatin by guiding DNA methylation and histone modifications at siRNA-homologous positions (Xu et al. 2013; Henderson and Jacobsen 2007; Pikaard et al. 2012). It is tempting to assume that the underlying sequence of large heterochromatic regions has to be structured as simple as possible, consisting of small and similar units (monomers) in tandem arrangement. Hence, these simple satellite structures effectively may induce heterochromatization through histone modifications and DNA methylation along large chromosomal regions.

siRNAs are also involved in the maintenance of DNA methylation patterns after replication of the DNA. The cytosine methylation at symmetric CG sites is facilitated by METHYLTRANSFERASE 1 (MET1). CHG methylation (mostly occurring on both DNA strands) is redundantly maintained by DOMAINS REARRANGED METHYLTRANSFERASE 2 (DRM2) and CMT3 (Henderson and Jacobsen 2007). However, at asymmetric CHH sites, the mother strand cannot be used as template and additional signals must serve as methylation inducer. At satellite loci, this function may be accomplished by siRNAs originating from satellite RNAs guiding DNA methyltransferase to homologous loci inducing DNA methylation. Accordingly, *de novo* methylation is established by siRNAs assisting DRM2 for *de novo* methylation of CG, CHG, and CHH sites (Chan et al. 2005).

Interestingly, in heterochromatization of repetitive DNA it has been shown that transcription and siRNAs act together with DNA and histone methyltransferases (Martienssen 2003; Law and Jacobsen 2010). Heterochromatization can take part along large chromosomal regions comprising long satellite arrays. In addition to the more or less conventional RNA polymerases I to III, plants possess also other RNA-synthesizing enzymes such as POLYMERASE IV (POL IV) and POLYMERASE V (POL V) which both are involved in the methylation of repetitive DNA (Zhang et al. 2007; Herr et al. 2005; Kanno et al. 2005; Onodera et al. 2005; Pontier et al. 2005).

5 Satellite DNA-Directed Heterochromatization

In the sections above, satellite DNAs have been linked to RNA-directed DNA methylation (RdDM). In fact, DNA methylation is one of the most important epigenetic marks and strongly involved in the regulation of the chromatin structure. In addition to DNA methylation, the methylation of histone H3 is an epigenetic modification directly acting on the level of chromatin proteins. Importantly, H3K9me2 is an H3 methylation mark necessary for heterochromatization and centromere function (Carroll and Straight 2006; Peng and Karpen 2008). Similarly, as observed for DNA methylation, small interfering RNAs (siRNAs) might induce the histone methyltransferase machinery to modify the histone H3 methylation status at the loci homologous to the corresponding siRNAs (Martienssen 2003). This may also take place at satellite repeats, and because of the organization in tandem arrays this is accomplished at a large scale (Martienssen 2003).

Taken together, the models described for RdDM and siRNA-directed guiding of histone methyltransferase, combined with the proposed mechanism of transcript and siRNA generation at satellite loci, allow to propose a simplified model for the formation and maintenance of heterochromatin along large satellite DNA arrays (Fig. 3), such for example at pEV and pBV arrays in sugar beet.

The interplay of satellite DNA transcription (POLIV), transcript processing (RDR2, DCL3, AGO4), and cytosine methylation (PolV, DRM2) is linked to

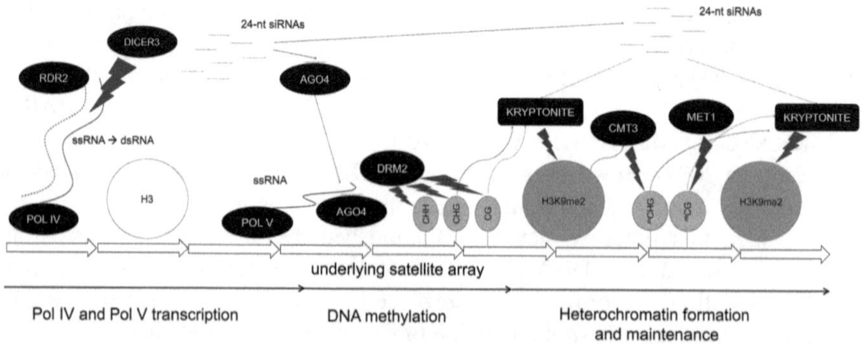

Fig. 3 A generalized and simplified model for the generation and maintenance of heterochromatin by DNA methylation and histone modifications guided by siRNAs at large satellite arrays. Transcription of satellite DNA by POLYMERASE IV (POL IV) results in single-stranded RNAs (ssRNAs) from satellite sequences. ssRNAs are synthesized into double-stranded RNAs (dsRNAs) by the activity of RNA-DEPENDENT RNA POLYMERASE (RDR2). DICER3 cuts dsRNAs into 24-nt siRNAs (homologous to satellite sequences) that are loaded onto the ARGONAUTE 4 complex (AGO4). AGO4 is guided by the 24-nt siRNA to homologous positions of ssRNAs generated by POLYMERASE V (POL V). DOMAINS REARRANGED DNA METHYLTRANSFERASE 2 (DRM2) binds to AGO4 and induces *de novo* methylation of CHH, CHG, and CG sites. The histone methyltransferase KRYPTONITE dimethylates H3K9 due to recognizing methylated CG and CHG sites. Subsequently, CHROMOMETHYLASE 3 (CMT3) methylates CHG sites induced by H3K9me2. Additionally, METHYLTRANSFERASE 1 (MET1) methylates also CG sites leading to heterochromatin formation and maintenance. siRNAs may also induce KRYPTONITE activity and hence dimethylation of H3K9 at homologous loci. Because of the tandemly arranged organization of satellite monomers this might happen at a large scale spreading along the entire satellite array leading to heterochromatization of a large region and the generation of heterochromatic knobs. Model modified according to Pikaard et al. (2012)

histone H3 modification. The 24-nt siRNAs are involved in the dimethylation of histone H3 by the KRYPTONITE methyltransferase (Martienssen 2003; Peng and Karpen 2008; Wang et al. 2000). H3K9me2 is essential and can be induced at methylated CHG and CG sites by KRYPTONITE (KYP). Interaction of KYP with CHROMOMETHYLASE 3 (CMT3) in turn results in feedback loops generating more CHG methylation, enhancing heterochromatization. Additionally, H3K27me1 (monomethylation of lysine 27 of histone 3) is typical for heterochromatin in plants and has been also observed in clusters in pEV- and pBV-rich heterochromatin in sugar beet (Zakrzewski et al. 2011). However, its establishment is not dependent on DNA methylation and is possibly mediated by other mechanisms (Liu et al. 2010).

Finally, satellite DNA is very likely a crucial sequence component for the heterochromatization due to its simplicity in structure and organization of relatively small homologous tandemly arranged monomers. Therefore, satellites serve as sequence signal for inducing and maintaining heterochromatization. Satellite transcripts and satellite 24-nt siRNAs establish and maintain DNA methylation and histone methylation at the whole satellite array, providing the epigenetic identity of heterochromatin, centromeric chromatin rich in satellite DNA and the functional centromere (Zakrzewski et al. 2014).

6 Conclusion

Satellite sequences are a major class of repetitive DNA populating large intercalary and small dispersed heterochromatic regions as well as the pericentromeric and centromeric chromatin of plant chromosomes. In this chapter, we hypothesize a model of the conservation of satellite-directed heterochromatin maintenance and centromeric chromatin formation in plants based on epigenetic mechanisms.

Numerous satellite families have been characterized in sugar beet (*Beta vulgaris*) and in other wild beets of the genus *Beta*. Satellite DNA is fast evolving, diversifying in sequence and abundance, and hence species-specific subfamilies with variable copy numbers and different chromosomal localization occur in closely related *Beta* species. Major satellites cluster in large heterochromatic blocks resulting in tandem arrays of several megabasepairs in size. Due to their localization in centromeric chromatin, major satellite families may contribute to the activity of centromeres. However, although the nucleotide sequence of centromeric satellites is highly diverged between species, monomer sizes of 150–170 bp and multiples are conserved. These similar monomer sizes most likely depend on the favored nucleosomal packaging which includes in centromeric chromatin the wrapping around nucleosomes containing CENH3, the centromere-specific variant of histone H3.

Satellite DNA has been excessively investigated on the level of the DNA sequence; however, information above this level is limited. DNA methylation is one of the most important epigenetic marks determining repeat silencing and heterochromatin identity. Therefore, a main result of this chapter was to provide comprehensive in-depth information on satellite DNA methylation. Satellite DNA is often characterized by lower levels of methylation compared to adjacent genomic regions because of the high AT content and high frequency of asymmetric CHH sites which are mostly not methylated. Symmetric CG and CHG sites are predominantly methylated but rare in sugar beet satellite monomers investigated. In particular, sugar beet centromeres are enriched of AT-rich satellite DNAs and also show reduced levels of DNA methylation. Despite this hypomethylation, satellites may serve as optimal sequence platform for the establishment and maintenance of heterochromatin due to their simple structure of monomers and their large-scale organization in tandem arrays. They provide periodically arranged CG and CHG sites along the entire satellite array which connect the DNA to repressive histone modifications, such as methylation of histone H3. Together with satellite transcripts and small interfering RNAs originating from satellites, CG, CHG, and histone methylation is established in satellite arrays leading to heterochromatin and its maintenance.

Therefore, satellite DNA can be considered as a major component of many plant genomes serving as optimal DNA sequence component for heterochromatin and centromeric chromatin, although satellite DNA is highly diverse on the level of the nucleotide sequence even between closely related species. The major point is the simplicity of tandemly arranged monomers (often with similar monomer sizes) that facilitate the satellite-directed formation and maintenance of heterochromatin and centromeric chromatin along large chromosomal regions providing the background for their epigenetic modifications.

Acknowledgments This work was supported by the BMBF grant "AnnoBeet: Annotation des Genoms der Zuckerrübe unter Berücksichtigung von Genfunktionen und struktureller Variabilität für Nutzung von Genomdaten in der Pflanzenbiotechnologie.", FKZ-0315962-C to Thomas Schmidt.

References

Allshire RC, Karpen GH (2008) Epigenetic regulation of centromeric chromatin: old dogs, new tricks? Nat Rev Genet 9:923–937

Arumuganathan K, Earle ED (1991) Nuclear DNA content of some important plant species. Plant Mol Bio Report 9:186–198

Carroll CW, Straight AF (2006) Centromere formation: from epigenetics to self-assembly. Trends Cell Biol 16:70–78

Chan SW, Henderson IR, Jacobsen SE (2005) Gardening the genome: DNA methylation in Arabidopsis thaliana. Nat Rev Genet 6:351–360

De Bock TSM (1986) The genus Beta: domestication, tayonomy and interspecific hybridization for plant breeding. Acta Hortic 82:335–344

Demidov D, Schubert V, Kumke K et al. (2014) Anti-phosphorylated histone H2AThr120: a universal microscopic marker for centromeric chromatin of mono- and holocentric plant species. Cytogenet Genome Res 143:150–156

Dohm JC, Minoche AE, Holtgräwe D et al. (2014) The genome of the recently domesticated crop plant sugar beet (Beta vulgaris). Nature 505:546–549

Dong Q, Han F (2012) Phosphorylation of histone H2A is associated with centromere function and maintenance in meiosis. Plant J 71:800–809

Fischer HE (1989) Origin of the Weiße Schlesische Rübe (white Silesian beet) and resynthesis of sugar beet. Euphytica 41:75–80

Flavell RB, Bennett MD, Smith JB et al. (1974) Genome size and proportion of repeated nucleotide-sequence DNA in plants. Biochem Genet 12:257–269

Ford-Lloyd BV, Williams JT (1975) A revision of Beta section Vulgares (Chenopodiaceae), with new light on the origin of cultivated beets. Bot J Linn Soc 71:89–102

Fuchs J, Demidov D, Houben A et al. (2006) Chromosomal histone modification patterns—from conservation to diversity. Trends Plant Sci 11:199–208

Gent JI, Madzima TF, Bader R et al. (2014) Accessible DNA and relative depletion of H3K9me2 at maize loci undergoing RNA-directed DNA methylation. Plant Cell 26:903–917

Ghildiyal M, Zamore PD (2009) Small silencing RNAs: an expanding universe. Nat Rev Genet 10:94–108

Hemleben V, Kovarik A, Torres-Ruiz RA et al. (2007) Plant highly repeated satellite DNA: molecular evolution, distribution and use for identification of hybrids. Syst Biodivers 5:277–289

Henderson IR, Jacobsen SE (2007) Epigenetic inheritance in plants. Nature 447:418–424

Herr AJ, Jensen MB, Dalmay T et al. (2005) RNA polymerase IV directs silencing of endogenous DNA. Science 308:118–120

Heun P, Erhardt S, Blower MD et al. (2006) Mislocalization of the Drosophila centromere-specific histone CID promotes formation of functional ectopic kinetochores. Dev Cell 10:303–315

Howman EV, Fowler KJ, Newson AJ et al (2000) Early disruption of centromeric chromatin organization in centromere protein A (Cenpa) null mice. Proc Natl Acad Sci U S A 97:1148–1153

Jackson JP, Lindroth AM, Cao X et al. (2002) Control of CpNpG DNA methylation by the KRYPTONITE histone H3 methyltransferase. Nature 416:556–560

Johnson LM, Bostick M, Zhang X et al. (2007) The SRA methyl-cytosine-binding domain links DNA and histone methylation. Curr Biol 17:379–384

Jones PA (2012) Functions of DNA methylation: islands, start sites, gene bodies and beyond. Nat Rev Genet 13:484–492

Kadereit G, Hohmann S, Kadereit JW (2006) A synopsis of *Chenopodiaceae* subfam. *Betoideae* and notes on the taxonomy of *Beta*. Willdenowia 36:9–19

Kanno T, Huettel B, Mette F et al. (2005) Atypical RNA polymerase subunits required for RNA-directed DNA methylation. Nat Genet 37:761–765

Kim IS, Lee M, Park KC et al. (2012) Roles of Mis18α in epigenetic regulation of centromeric chromatin and CENP-A loading. Mol Cell 46:260–273

Koo DH, Han F, Birchler JA et al. (2011) Distinct DNA methylation patterns associated with active and inactive centromeres of the maize B chromosome. Genome Res 21:908–914

Kowar T, Zakrzewski F, Macas J et al. (2016) Repeat composition of CenH3-chromatin and H3K9me2-marked heterochromatin in sugar beet (Beta vulgaris). BMC Plant Biol 16:120

Law JA, Jacobsen SE (2010) Establishing, maintaining and modifying DNA methylation patterns in plants and animals. Nat Rev Genet 11:204–220

Lermontova I, Koroleva O, Rutten T et al. (2011) Knockdown of CENH3 in Arabidopsis reduces mitotic divisions and causes sterility by disturbed meiotic chromosome segregation. Plant J 68:40–50

Lisch D (2009) Epigenetic regulation of transposable elements in plants. Annu Rev Plant Biol 60:43–66

Liu C, Lu F, Cui X et al. (2010) Histone methylation in higher plants. Annu Rev Plant Biol 61:395–420

Martienssen RA (2003) Maintenance of heterochromatin by RNA interference of tandem repeats. Nat Genet 35:213–214

Onodera Y, Haag JR, Ream T et al. (2005) Plant nuclear RNA polymerase IV mediates siRNA and DNA methylation-dependent heterochromatin formation. Cell 120:613–622

Paesold S, Borchardt D, Schmidt T et al. (2012) A sugar beet (Beta vulgaris L.) reference FISH karyotype for chromosome and chromosome-arm identification, integration of genetic linkage groups and analysis of major repeat family distribution. Plant J 72:600–611

Palomeque T, Lorite P (2008) Satellite DNA in insects: a review. Heredity 100:564–573

Peng JC, Karpen GH (2008) Epigenetic regulation of heterochromatic DNA stability. Curr Opin Genet Dev 18:204–211

Pikaard CS, Haag JR, Pontes OMF et al. (2012) A transcription fork model for Pol IV and Pol V-dependent RNA-directed DNA methylation. Cold Spring Harb Symp Quant Bio 77:205–212

Plohl M, Luchetti A, Meštrović N et al. (2008) Satellite DNAs between selfishness and functionality: structure, genomics and evolution of tandem repeats in centromeric (hetero)chromatin. Gene 409:72–82

Pontier D, Yahubyan G, Vega D et al. (2005) Reinforcement of silencing at transposons and highly repeated sequences requires the concerted action of two distinct RNA polymerases IV in Arabidopsis. Genes Dev 19:2030–2040

Régnier V, Vagnarelli P, Fukagawa T et al. (2005) CENP-A is required for accurate chromosome segregation and sustained kinetochore association of BubR1. Mol Cell Biol 25:3967–3981

Roudier F, Ahmed I, Bérard C et al. (2011) Integrative epigenomic mapping defines four main chromatin states in Arabidopsis. EMBO J 30:1928–1938
Schmidt T, Heslop-Harrison JS (1998) Genomes, genes and junk: the large-scale organization of plant chromosomes. Trends Plant Sci 3:195–199
Schmidt T, Metzlaff M (1991) Cloning and characterization of a beta-vulgaris satellite DNA family. Gene 101:247–250
Schmidt T, Jung C, Metzlaff M (1991) Distribution and evolution of two satellite DNAs in the genus Beta. Theor Appl Genet 82:793–799
Schmidt M, Hense S, Minoche AE et al. (2014) Cytosine methylation of an ancient satellite family in the wild beet *Beta procumbens*. Cytogenet Genome Res 143:157–167
Suzuki MM, Bird A (2008) DNA methylation landscapes: provocative insights from epigenomics. Nat Rev Genet 9:465–476
Tariq M, Paszkowski J (2004) DNA and histone methylation in plants. Trends Genet 20:244–251
Teixeira FK, Colot V (2010) Repeat elements and the Arabidopsis DNA methylation landscape. Heredity 105:14–23
Treangen TJ, Salzberg SL (2012) Repetitive DNA and next-generation sequencing: computational challenges and solutions. Nat Rev Genet 13:36–46
Ugarkovic D (2005) Functional elements residing within satellite DNAs. EMBO Rep 6:1035–1039
Wang G, Ma A, Chow CM et al. (2000) Conservation of heterochromatin protein 1 function. Mol Cell Biol 20:6970–6983
Weber B, Schmidt T (2009) Nested Ty3-gypsy retrotransposons of a single Beta procumbens centromere contain a putative chromodomain. Chromosom Res 17:379–396
Weber B, Wenke T, Frömmel U et al. (2010) The Ty1-copia families SALIRE and Cotzilla populating the Beta vulgaris genome show remarkable differences in abundance, chromosomal distribution, and age. Chromosom Res 18:247–263
Weber B, Heitkam T, Holtgräwe D et al. (2013) Highly diverse chromoviruses of Beta vulgaris are classified by chromodomains and chromosomal integration. Mob DNA 4:8
Weisshaar B, Himmelbauer H, Schmidt T et al. (2016) Sugar beet BeetMap-3, and steps to improve the genome assembly and genome sequence annotation. W875, Plant & Animal Genome XXIV, San Diego, USA
West PT, Li Q, Ji L et al. (2014) Genomic distribution of H3K9me2 and DNA methylation in a maize genome. PLoS One 9:e105267
Wollrab C, Heitkam T, Holtgräwe D et al. (2012) Evolutionary reshuffling in the Errantivirus lineage Elbe within the Beta vulgaris genome. Plant J 72:636–651
Xu C, Tian J, Mo B (2013) siRNA-mediated DNA methylation and H3K9 dimethylation in plants. Protein &Cell 4:656–663
Yan H, Kikuchi S, Neumann P et al. (2010) Genome-wide mapping of cytosine methylation revealed dynamic DNA methylation patterns associated with genes and centromeres in rice. Plant J 63:353–365
Zakrzewski F, Wenke T, Holtgräwe D et al. (2010) Analysis of a c0t-1 library enables the targeted identification of minisatellite and satellite families in Beta vulgaris. BMC Plant Biol 10:8
Zakrzewski F, Weisshaar B, Fuchs J et al. (2011) Epigenetic profiling of heterochromatic satellite DNA. Chromosoma 120:409–422
Zakrzewski F, Weber B, Schmidt T (2013) A molecular cytogenetic analysis of the structure, evolution, and epigenetic modifications of major DNA sequences in centromeres of *Beta* species. In: Plant centromere biology. Wiley-Blackwell, Oxford, pp 39–55
Zakrzewski F, Schubert V, Viehoever P et al. (2014) The CHH motif in sugar beet satellite DNA: a modulator for cytosine methylation. Plant J 78:937–950
Zhang X, Henderson IR, Lu C et al. (2007) Role of RNA polymerase IV in plant small RNA metabolism. Proc Natl Acad Sci U S A 104:4536–4541
Zhang W, Lee HR, Koo DH et al. (2008) Epigenetic modification of centromeric chromatin: hypomethylation of DNA sequences in the CENH3-associated chromatin in Arabidopsis thaliana and maize. Plant Cell 20:25–34

Universal and Lineage-Specific Properties of Linker Histones and SWI/SNF-Chromatin Remodeling Complexes in Plants

Andrzej Jerzmanowski and Rafal Archacki

Abstract Linker histones and SWI/SNF remodeling complexes both take part in determining the availability of chromosomal DNA to trans-acting factors and play critical roles in chromatin-mediated control and regulatory functions. Despite high degree of conservation of these universal chromatin modulators in all eukaryotes, there is increasing evidence that plants and animals have evolved a number of different ways of their use in orchestrating growth, development, and adaptation to environment. In this chapter, we discuss these lineage-specific differences from the perspective of rapidly progressing plant chromatin studies. We also refer to recent data pointing to interdependence of linker histones and SWI/SNF in controlling chromatin functions.

Keywords Heterochromatin and euchromatin • Histone H1 • SWI/SNF chromatin remodeling complexes • Transcription • Histone modifications

Contents

A. Jerzmanowski (✉) • R. Archacki (✉)
Laboratory of Systems Biology, Faculty of Biology, University of Warsaw, Warsaw, Poland

Institute of Biochemistry and Biophysics, Polish Academy of Sciences, Warsaw, Poland
e-mail: andyj@ibb.waw.pl; rafa@ibb.waw.pl

N. Rajewsky et al. (eds.), *Plant Epigenetics*, RNA Technologies,
DOI 10.1007/978-3-319-55520-1_23

1 Introduction: Chromatin in Plants and Animals: Commonalities and Differences After Over One Billion Years of Separate Evolution

The common, most likely unicellular, ancestor of plants and animals lived well over 1 billion years ago. Since then these Kingdoms of life have evolved separately, "inventing" multicellularity on the way, with its associated complex mechanisms of development and intricate systems of communication with and adaptation to the environment. In addition, both have evolved features ensuring their evolvability. In an influential paper based on whole genome sequencing studies, Elliot M. Meyerowitz concluded that while: "the logic underlying many developmental processes is similar, plants and animals have evolved development independently, and molecules that carry out the logical plan in these two major lineages are unrelated or represent novel arrangements of ancient protein domains." He then contrasted these differences to the striking conservation of chromatin proteins and chromatin-mediated mechanisms (Meyerowitz 2002).

While the above picture is in general true, it does not take into consideration more subtle differences concerning the organization and functions of chromatin that exist between these two Kingdoms. Why should this be important? Plants and animals occupy very different positions in the overall phenotypic space available to living organisms on Earth. To ensure survival and reproduction, both lineages have evolved efficient and flexible ways of adapting to the local environment. However, due to their generally sessile life habit, these strategies are far more important to plants. While numerous de novo molecular inventions have occurred independently in plants and animals in response to different selective pressures, we consider the structural and functional differences between the components of a once common chromatin regulatory system particularly interesting. Unraveling such differences and connecting them to concrete lineage-specific features may provide valuable insights into the general mechanisms underlying the complex regulatory and integratory role of chromatin in all eukaryotes. To illustrate this point, we will compare two important components of chromatin in plants and animals: linker histones and SWI/SNF nucleosome remodeling complexes.

2 Linker Histones and SWI/SNF Remodeling Complexes in Chromatin Organization and Regulatory Mechanisms: Conclusions from Studies in Yeast and Animals

2.1 Linker Histones

H1- or linker histones are basic nuclear proteins, typically about 180–200 amino acids in length, with a characteristic tripartite structure. The H1 molecule consists of a relatively short (20–35 amino acids) N-terminal domain (NTD), a centrally placed 70–80-amino acids-long globular domain (GH1), and a long and highly lysine-rich C-terminal domain (CTD) of about 80–100 amino acids (Fig. 1a). Both NTD and CTD display properties of intrinsically disordered polypeptides, which means that they are disordered in solution but capable of assuming secondary structures (including α-helix, β-structures and turns) when in contact with DNA or interacting proteins. GH1, the only natively structured domain of H1, belongs to the "winged helix" family of DNA-binding proteins. It contains a mixed α/β fold composed of three α-helices (helices I–III) and three β-strands (S1–S3) (Fig. 1b). The signature "wing" (W) motif of GH1 is placed within the domain located C-terminally to helix III and takes the form of an extended loop that joins S2 and S3 (Fig. 1a) (Bednar et al. 2016).

Linker histones are abundant components of chromatin fibers. Similarly to H2A, H2B, H3, and H4, the four highly conserved core histone types, H1 is traditionally

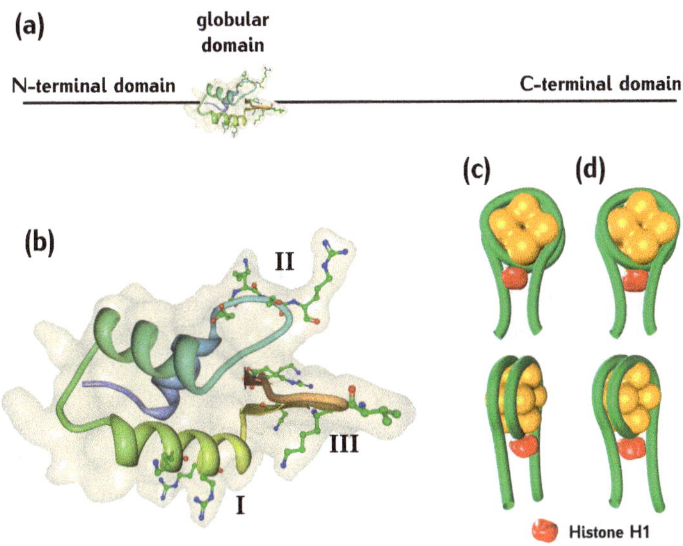

Fig. 1 Overall structure of H1 and alternative models of nucleosome binding by GH1. (**a**) Schematic view of H1's tripartite structure; (**b**) 3D structure of the GH1 domain, depicting DNA binding sites according to Cui and Zhurkin (2009); (**c**) asymmetric; and (**d**) symmetric modes of GH1 binding to nucleosomes

viewed as the fifth universally occurring histone-type component of the nucleosome: the basic structural unit of chromatin. However, despite their name, linker histones are not evolutionarily related to core histones. Phylogenetic analyses suggest that the current-type H1s had two different predecessors. CTD-like lysine-rich proteins were already present in Eubacteria, whereas GH1-like architectural proteins existed in early Eukaryotes. The latter have the ability to bind to four-way junction-like DNA structures resembling those occurring at the entry and exit points of DNA helices in the nucleosome. The fusion of these two domains at some point during the evolution of early protists gave rise to modern H1 histones. Interestingly, in some contemporary protists, the protein functionally classified as H1 still consists only of a separate CTD (Kasinsky et al. 2001).

In nucleosomal DNA, H1 is located on the outside of the Nucleosome Core Particle (NCP), a nucleoprotein complex consisting of an octamer assembly of core histones around which is wrapped a 147-bp-long DNA (almost 1.65 turns) in a left-handed helix. The NCP restricts the accessibility of this stretch of chromatin DNA to micrococcal nuclease. The presence of H1 (the NCP with H1 is called the chromatosome) also provides transient protection to approximately 20 bp of additional DNA (10 bp at each end of the wrapped strand). The way in which GH1 binds to nucleosomes has been the subject of extensive studies. Steady improvement in the resolution of structural methods has led to considerable refinement of the proposed models. This notwithstanding, the precise details of GH1 binding are still controversial. It is generally agreed that GH1 binds the nucleosome with high specificity at or near its dyad region, via two binding sites referred to as site I and site II. Both are at the surface of GH1 and encompass most of the DNA-binding amino acids, including positively charged arginine and lysine residues (Fig. 1b). In addition, a third binding site, referred to as site III and comprised mainly of the nonpolar "wing" domain, has also been postulated and is proposed to be responsible for the recognition of methyl groups in the major groove of AT-rich DNA sequences (Fig. 1b) (Cui and Zhurkin 2009). The main controversy concerns the question of whether GH1 binds (a) symmetrically, interacting with the minor groove of DNA precisely at the dyad (i.e., at the center of the nucleosome), and thus binding 10 bp DNA fragments of both the entering and exiting DNA duplexes (called linker or inter nucleosomal DNA) or (b) asymmetrically with respect to the dyad and binding an equivalent 20 bp fragment of only one of the aforementioned DNA duplexes (Fig.1c). Since the binding of H1 is critical for "locking" the DNA ends entering and exiting at the same side of the nucleosome, the importance of these seemingly minor details lies in the fact that they may have considerable consequences for the effects of H1 on the higher order organization of chromatin and its potential role as a regulator of chromatin DNA accessibility. Another important question concerns the degree of uniformity of nucleosome binding by different isoforms of H1. In the light of current evidence, it seems possible that H1 can bind in an on- or off-dyad mode depending on the isoform, which means that both of the proposed binding models could be correct (Zhou et al. 2015).

In addition to GH1, a critical role in the overall binding properties and functions of H1 is played by the CTD. This long domain with numerous alternating lysines and alanines interspersed with proline residues is important not only in the condensation and stabilization of chromatin fibers by H1 but also in assuring the cooperativity of H1 binding to chromatin. It contains two functionally important sub-domains, each about 20 amino acids in length: one located in close proximity to GH1 and the other more distant. The CTD often contains specific Ser-Pro-Lys (Arg)-Lys(Arg) (SPKK) sequence motifs which strongly enhance DNA binding and are known targets of phosphorylation. It has been shown that the CTD mediates interactions with numerous nonhistone nuclear and cytoplasmic proteins (Lu et al. 2009).

The employment of the Fluorescent Recovery after Photobleaching (FRAP) technique for measuring protein binding in vivo unexpectedly revealed that unlike core histones, linker histones are highly mobile and continuously exchanging in chromatin (Flanagan and Brown 2016), a feature that makes them open to competition by other factors with affinity for the same binding sites on the nucleosome (Postnikov and Bustin 2016). Estimation by FRAP of the in vivo binding of different H1 variants has shown that the length and amino acid composition of the CTD exerts a major influence on the chromatin residence time of H1 (Th'ng et al. 2005).

In most metazoan (animal) species of different phyla studied so far, linker histones are much more heterogeneous than core histones. For example, there are 8 H1 variants (isoforms) in *Caenorhabditis elegans* (nematode), 4 in *Strongylocentrotus purpuratus* (sea urchin) and *Danio rero* (fish), 5 in *Xenopus laevis* (amphibian), 7 in *Gallus gallus* (bird), and 11 in *Mus musculus* and human (mammals). The proportions of different H1 variants change during early development in all of these organisms (i.e., in embryogenesis). Different H1 variants have been shown to vary in their ability to bind nucleosomes, suggesting that they may exert different effects on DNA activities in vivo (Millán-Ariño et al. 2016). In general, the metazoan H1 variants can be divided into replication-coupled and replacement types, but within each category there may be differences in the spatiotemporal pattern of expression of particular variants. Some variants are abundant in almost every cell, whereas others are restricted to the cells of terminally differentiated tissues. Importantly, functional homologs, e.g., oocyte- and early embryo-specific or some terminal differentiation-specific variants, share a common evolutionary origin, suggesting that the basic sub-functionalization of H1 occurred at an early stage in metazoan evolution (Schulze and Schulze 1995). There was a long held view that the above pattern is not universally conserved among metazoans, e.g., *Drosophila melanogaster* was believed to function with just a single H1 variant. However, recent studies confirmed that this insect also contains a specialized H1 (dBigH1) that is expressed prior to the cellular blastoderm stage of embryogenesis and acts as a divergent embryonic H1 variant (Pérez-Montero et al. 2013).

In summary, studies using animal models have yet to reveal the true biological function of linker histones in multicellular eukaryotes. While it has been shown that a 50% decrease in the amount of linker histones leads to embryonic lethality in mammals (Fan et al. 2005), and that a single H1 variant is essential for development in Drosophila (Bayona-Feliu et al. 2016), no defined developmental or physiological processes that require the presence of H1 have been identified so far. This is probably due to the highly complex nature of animals, which makes the identification of H1-dependent mechanisms extremely difficult.

2.2 SWI/SNF Chromatin Remodeling Complexes

Chromatin remodeling is one of the key mechanisms ensuring chromatin reorganization during crucial nuclear processes such as transcription and replication. This activity is mediated by multi-subunit complexes assembled around a central ATPase (SNF2-type protein, named after the yeast ATPase that was the first of this class to be identified) that uses energy derived from ATP hydrolysis to alter the interactions between histone octamers and DNA. It is thought that changes in the nucleosomal organization of chromatin modulate the accessibility of DNA sequences to trans-acting factors (Arya et al. 2010). This notion is supported by the finding that the ability of many transcription factors to bind their target DNA sequence depends on active chromatin remodeling. As shown by phylogenetic analyses, multicellular organisms employ multiple proteins with chromatin remodeling ATPase activity, probably with their associated complexes. For example, 29 such proteins occur in mammals and 41 in Arabidopsis (Flaus et al. 2006; Knizewski et al. 2008), suggesting a high degree of specialization. This is consistent with the results of genetic studies indicating that mammalian and plant ATPases work mostly non-redundantly, since their mutations give rise to highly specific phenotypes (Knizewski et al. 2008; Ho and Crabtree 2010; Han et al. 2015). These diverse SNF2-type proteins can be arranged into subfamilies distinguished by structural properties of the catalytic domain and the unique composition of other domains (Flaus et al. 2006; Knizewski et al. 2008). The four main subfamilies are SWI/SNF, ISWI, INO80/SWR, and CHD. Typically, the remodeling ATPase is associated with other proteins, forming a complex that usually consists of 4–12 subunits. Complexes belonging to the SWI/SNF class were first discovered in *S. cerevisiae* in genetic screens for factors involved in the regulation of mating type switching and sugar metabolism (Stern et al. 1984; Neigeborn and Carlson 1984), hence the name (switch/sucrose non-fermentable). The SWI/SNF family is evolutionarily conserved, and since its discovery in yeast, homologous proteins have been identified and characterized in flies, mammals, and plants. In addition to the ATPase domain, SWI/SNF ATPases possess a characteristic C-terminal "bromodomain" involved in recognition of acetylated histones (Shen et al. 2007). All SWI/SNF complexes also contain accessory "core" subunits: homologs of yeast SNF5, SWI3, and SWP73, as well as other auxiliary noncore subunits [reviewed in (Clapier and Cairns 2009)]. While the ATPase provides energy for the remodeling process, the other SWI/SNF subunits are thought to participate in the assembly of the complexes, the regulation of ATPase activity, and the recruitment of SWI/SNF to target *loci* (Clapier and Cairns 2009). Native complexes consisting of 11–15 subunits have been described in yeast and mammals, with sizes of 1.14 MDa and ~2 MDa, respectively (Smith et al. 2003; Kadoch et al. 2013). Interestingly, partial complexes comprising only the core subunits have been shown to remodel chromatin in vitro with an efficiency comparable to that of the whole native complex (Phelan et al. 1999). In yeast, SWI/SNF family is represented by a pair of related complexes, ySWI/SNF and RSC. This duplication seems to be maintained at least to some extent in other eukaryotes, as exemplified by the BAP and PBAP complexes in Drosophila, and BAF and PBAF complexes in mammals (Clapier and Cairns 2009).

2.2.1 Mechanisms of Chromatin Remodeling

Initially discovered in yeast, the *swi2* and *snf2* mutations could be suppressed by mutations in the core histones H2A and H2B, indicating that the SWI/SNF complex influenced transcription by altering chromatin structure (Peterson and Herskowitz 1992). Analyses of the activity of isolated SNF2-type ATPases as well as whole complexes in vitro, using reconstituted nucleosomal templates, showed that they are able to produce various changes in chromatin structure, including the phasing or positioning of nucleosomes, exchange of nucleosomes, induction of nucleosome mobility (so-called nucleosome sliding), the eviction of nucleosomes, and changes in their composition (Narlikar et al. 2013). These activities are ATP-dependent because they require the disruption of tight histone-DNA contacts maintained by electrostatic interactions. The remodeling ATPases share similarities with DNA translocases, and the existing data support the view that DNA translocase activity is employed in the remodeling process (Narlikar et al. 2013). Several mechanisms of nucleosome remodeling have been proposed, of which the "loop recapture" model is the most favored (Kadoch et al. 2016). According to this model, the remodeler anchors to the nucleosome and conducts directional DNA translocation from the linker towards the nucleosome dyad, creating a loop of DNA on the nucleosome surface. This loop is then propagated around the nucleosome, breaking histone-DNA contacts and recreating them behind the loop (Clapier and Cairns 2009). In support of this model, imaging studies on yeast SWI/SNF using electron microscopy showed a large cavity on the remodeler surface that surrounds the nucleosome (Dechassa et al. 2008). It should be noted that despite sharing a similar basic mechanism (DNA translocase activity), separate subfamilies of chromatin remodelers differ in their substrate preferences and remodeling outcomes. For example, ISWI complexes make more limited contacts with the nucleosome than SWI/SNF and promote nucleosome spacing, while SWI/SNF complexes have been associated with nucleosome movement and eviction (Hargreaves and Crabtree 2011; Narlikar et al. 2013). The precise mechanisms of chromatin remodeling are still not fully understood, mostly due to limitations of the current experimental approaches. First, structural data for nucleosome–remodeler complexes or even isolated complex subunits are still not available. Second, all in vitro remodeling assays (like the nucleosome sliding assay), use reconstituted nucleosomes as the substrate, which only partially resemble a native chromatin template.

The development of genomic approaches has enabled genome-wide mapping of nucleosome positions, histone modifications, as well as the distribution of transcription factors and chromatin regulatory proteins (including chromatin remodelers), offering the opportunity to study their in vivo properties. The map of nucleosome positions over actively transcribed genes shows two highly characteristic 150–200 bp "nucleosome depleted regions" (NDRs), positioned upstream (5′) of the transcription start site (TSS) and downstream (3′) of the transcription termination site (TTS). These two NDRs mark distinct structural demarcation points of the transcriptional units that are especially well defined in yeast. The 5′ NDRs are flanked by two tightly bound positioned nucleosomes: the "+1" located downstream of the NDR and overlapping the TSS and the "−1" placed some distance upstream of the NDR. The first few nucleosomes downstream of the +1 position show rather strong phasing that becomes more indistinct for nucleosomes further downstream from the TSS [reviewed in (Rando and Chang 2009; Arya et al. 2010)]. It has been demonstrated that transcription factor binding sites usually co-localize with the 5′ NDRs, which makes the accessibility of these regions absolutely critical for gene expression. Recent studies have shown that the loss of yeast RSC or ySWI/SNF subunits results in an altered nucleosome profile and correlates with decreased transcriptional activity. These changes are characterized by increased nucleosome density at the NDRs of many Pol II-dependent genes and a shift in nucleosome phasing across the gene body (Tolkunov et al. 2011; Parnell et al. 2015), consistent with the notion that yeast SWI/SNF complexes act in vivo to maintain functional NDRs at target genes by nucleosome movement and ejection. Interestingly, it was recently demonstrated that this activity is probably antagonized by an ISWI-type chromatin remodeling complex, ISW1a (Parnell et al. 2015). Consistent with a role in −1 and +1 nucleosome positioning, the genome-wide distribution patterns of RSC and ySWI/SNF show high occupancy near the TSS (Yen et al. 2012; Parnell et al. 2015). In addition, the binding of RSC and ySWI/SNF is often extended into the nucleosomes of the gene body. This may support the reported role of SWI/SNF complexes in both the initiation and elongation of transcription (Brown et al. 1996; Schwabish and Struhl 2004; Tréand et al. 2006).

How mammalian SWI/SNF complexes perform their functions in vivo is much less clear. Genetic data raised some doubts as to whether the movement of nucleosomes is the main function of BAF complexes (Kadoch and Crabtree 2015). For example, mutations of some auxiliary BAF subunits result in acute phenotypes that are similar to those produced by mutations in the ATPase subunit, although these auxiliary subunits are dispensable for remodeling activity in vitro (Kadoch and Crabtree 2015). This indicates that the in vivo mechanisms of chromatin remodeling in mammals are more complex than is indicated by in vitro assays, and they depend to a large extent on the functions of the non-ATPase subunits.

2.2.2 Biological Roles of SWI/SNF Remodelers

Early studies in yeast showed that the SWI/SNF complex is required for activation of gene expression controlled by transcription factors such as GAL4 or SWI5 (Laurent and Carlson 1992; Cosma et al. 1999), and it has therefore been regarded as a general transcriptional activator. The Drosophila homolog of the yeast SWI2/SNF2 ATPase, Brahma (dBRM), was shown to co-localize with RNAPII and is similarly required for the transcriptional initiation of a large number of genes on polytene chromosomes (Armstrong et al. 2002). The *swi/snf* mutations caused reduced expression of Drosophila homeotic genes and were classified within the Thritorax group known to suppress Polycomb mutations (Tamkun et al. 1992). In mammals, SWI/SNF remodelers were shown to be essential for the transcriptional control of proliferation, differentiation, development and organogenesis, and nuclear receptor-mediated signaling [reviewed in (Clapier and Cairns 2009; Ho and Crabtree 2010)]. Pluripotency maintenance of ES cells, self-renewal of neural stem cells, dendritic morphogenesis, and cardiac development are a few examples of biological processes in mammals that are dependent on mSWI/SNF activity (Kadoch et al. 2016). In contrast to yeast, mammalian SWI/SNFs have been implicated not only in activation but also in repression of transcription (Trotter and Archer 2008; Hargreaves and Crabtree 2011). mSWI/SNF complexes consist of one of the two ATPases BRG1 or BRM, plus four other ySWI/SNF orthologs (BRM/BRG1-associated factors BAF155/170, BAF60, BAF53a/b, and BAF47), and several auxiliary subunits. Compared to their yeast homologs, BAF complexes are highly polymorphic (several of their subunits are encoded by gene families) and can be assembled in different ways depending on the tissue or developmental context. For example, BAF complexes purified from embryonic stem cells contain BRG1, BAF 155, and BAF53a, but not BRM, BAF180, or BAF53b (Ho et al. 2009). More importantly, this variation in subunit composition does not seem to be simply correlated with cell identity. Rather, as suggested by recent studies, it may play a causative role in cell fate determination. For example, during the transition from neural progenitors to post-mitotic neurons, BAF45a and BAF53a subunits are exchanged for BAF45b and BAF53b (Lessard et al. 2007), and the expression of these neuron-specific subunits in human fibroblasts can convert them to neurons (Yoo et al. 2009). The importance of individual complex subunits and combinatorial assembly of the complexes is additionally supported by the fact that their mutation or deletion often lead to developmental diseases or tumorigenesis. Recent exome sequencing studies have revealed that over 20% of human cancer lines carry a mutation in at least one of the BAF complex subunits (Kadoch et al. 2013). Mutations in tissue-specific subunits of mammalian SWI/SNF tend to give rise to specific cancers, again underscoring their tissue-specific mode of action.

2.2.3 Targeting of SWI/SNF Complexes to Specific Sites in the Genome

Their multiple roles notwithstanding, SWI/SNF complexes are specific, which means that they target only certain genes in a given pathway. Early studies in yeast indicated that the complexes are targeted to genes via recruitment by DNA binding proteins (Cosma et al. 1999). Moreover, mammalian SWI/SNF subunits were shown to co-purify with a variety of specific proteins involved in key processes regulating chromosome structure, chromatin assembly, and transcription (Euskirchen et al. 2011). Consistently, proteomic studies have demonstrated that tissue-specific BAF complexes interact with cell-type specific regulators (Ho et al. 2009). Taken together, these findings suggest that specific functions of SWI/SNF are mediated by their interaction partners. Indeed, there are reports describing transcription factor-dependent recruitment of animal SWI/SNF complexes, including the role of DAF-16/FOXO in the activation of stress responsive genes of *C. elegans* (Riedel et al. 2013), Olig2 in the regulation of myelination-associated genes (Yu et al. 2013), and SLC11A1 during macrophage differentiation of HL-60 cells (Xu et al. 2011). On the other hand, there are cases where SWI/SNF remodelers were shown to be associated with target sites prior to the binding of a specific regulator. This was documented for the transcription factor STAT3, involved in pluripotent embryonic stem cell (ESC) maintenance (Ho et al. 2011) as well as for the glucocorticoid receptor (Engel and Yamamoto 2011; Burd and Archer 2013). It is thus possible that pre-association of the remodeling machineries with particular sites in the genome establishes the chromatin state required for recruitment by a DNA-binding regulator, probably in a tissue-specific manner.

Analyses of genome-wide distribution by ChIP-chip and ChIP-seq assays demonstrated that mammalian BAF complexes have 20,000–40,000 binding sites, depending on the cell type and experimental conditions (Euskirchen et al. 2012). The distribution of the Brg1 and BAF155 subunits to a large extent overlaps with the binding sites occupied by key pluripotency factors (Ho et al. 2009). In addition, the same study demonstrated that Brg1 represses the majority of its targets in ES cells, most likely in association with pluripotency regulators such as Oct4 and Sox2 (Ho et al. 2009). The binding profiles of mammalian SWI/SNF components differ from those of their yeast counterparts. While they still show a high occupancy signal near the TSS of target genes, the number of targets bound at the TSS is quite small, with binding predominantly localized to predicted enhancers, locus control regions, and other intergenic sequences of unknown function. With regard to the impact on transcription, the binding of the remodeling complex to either the promoter or more distal sequences can lead to both activation and repression of genes. In addition, several recent reports implicate SWI/SNF complexes in the regulation of noncoding transcripts. The yeast SWI/SNF-type RSC complex appears to suppress noncoding transcripts from TTS and other genomic sites (Alcid and Tsukiyama 2014). Similarly, a human SWI/SNF-type complex specific for embryonic stem cells (esBAF) suppresses non-coding transcripts in these cells

(Hainer et al. 2015). On the other hand, SWI/SNF can activate antisense transcription from divergent promoters in yeast cells (Marquardt et al. 2014).

Binding of SWI/SNF complexes to non-promoter sequences raises questions concerning the mechanisms of their recruitment to these sites. While some could be targeted via interaction with sequence-specific transcription factors, an important role in their recruitment might also be played by recognition of specific epigenetic *loci* (Kadoch and Crabtree 2015). Different SWI/SNF subunits contain motifs capable of binding to modified histones or certain DNA structures such as cruciform DNA or AT-rich sequences. Mammalian SWI/SNF complexes contain up to eight bromodomains (one on either BRG or BRM, one on BRD7, and six on the BAF250 subunit in the PBAF complex), two PHD domains (BAF45 subunits), two chromodomains (BAF155 and BAF170), and several DNA-binding motifs (e.g., ARID, Zn finger, HMG, A/T-hook). Bromodomains are known to bind acetylated histone tails (Shen et al. 2007), while chromodomains bind methylated lysine residues (Brehm et al. 2004), and PHD domains have an affinity for different histone modifications (Sanchez and Zhou 2011). Chromatin remodeling complexes may thus be guided to sites in the genome that have specific histone modifications and DNA structures.

3 Linker Histones in Plants

3.1 Structural Features and Phylogenetic Relationships Distinguishing Plant H1s

Plant and animal H1s share a common tripartite structure consisting of N- and C-terminal domains and a conserved GH1. However, H1s from these different lineages can be readily distinguished by a characteristic difference in the "wing" domain of GH1 (see Fig. 1c). All plant H1s lack a highly conserved 5-amino-acid motif GXGAX that occurs at the tip of the "wing" in GH1 of all animal-type H1s. This fragment was proposed to constitute a third binding site in GH1 (site III), capable of recognizing (via hydrophobic interactions with methyl groups of thymine) the AT-rich tracts that are enriched at the ends of nucleosomal DNA in metazoa, and thus enhance their bending into a "stem-like" structure. Via a similar recognition mechanism, the tip of the "wing" has been proposed to account for the known preference of H1 for methylated DNA (Cui and Zhurkin 2009). Why this motif has been lost (or never acquired) by plant H1s is an intriguing question whose answer may indicate more fundamental differences between plant and animal chromatin (see below). Regarding the variability of H1s, the number of distinctive and conserved subtypes of H1 variants is considerably smaller in plants than in animals. There are several well-distinguished evolutionarily conserved subtypes within animal H1s, including the cleavage stage H1 of Echinodermata and B4 (H1M) of Xenopus, or the cycle-independent variants of H1 characteristic of the

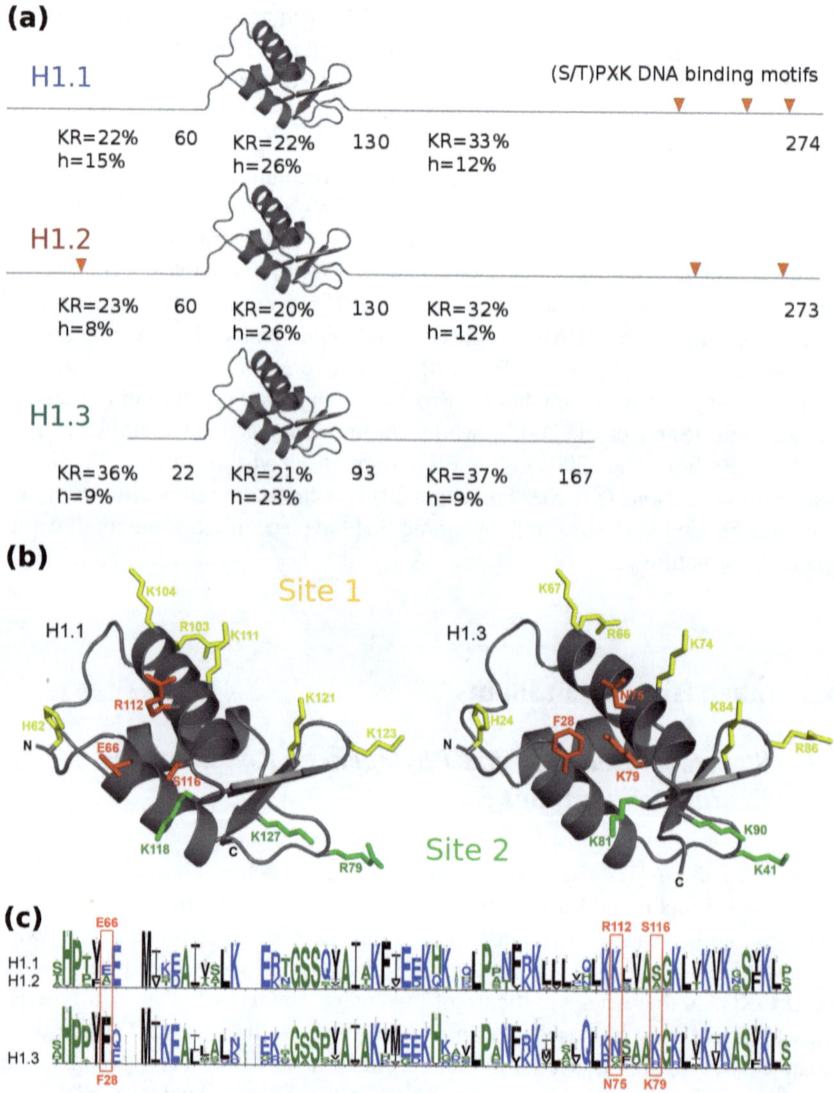

Fig. 2 Comparison of the two types of Arabidopsis linker histone. (**a**) Protein sequence features of *Arabidopsis thaliana* H1 variants. The percentages of positively charged (KR) and hydrophobic (h) residues are shown for the N-terminal domain, the central GH1 domain, and the C-terminal domain, respectively. *Red triangles* indicate (S/T)PXK motifs; (**b**) 3D models of the GH1 domains of H1.1 (*left*) and H1.3 (*right*); (**c**) amino acid sequence conservation in the plant H1.1/2-like and H1.3-like families among flowering plant species. In the models, residues corresponding to DNA-binding sites are colored *yellow* (Site I) and *green* (Site II), while those representing the most prominent differences between plant H1.1/2-like and H1.3-like variants are shown in *red*. From: (Rutowicz et al. 2015)

entire subkingdom of Deuterostomes (H1D of Echinodermata, H5 and H1° of Vertebrata). In contrast, phylogenetic analysis of plant H1s has revealed that in addition to the typical somatic variants, there is only one distinct evolutionarily conserved subtype: the so-called "stress-inducible" variants. Interestingly, the occurrence of this subtype during plant evolution coincides with the appearance of Angiosperms about 140 million years ago. The plant groups that evolved earlier, including mosses, ferns, and gymnosperms, do not have this type of H1 (Rutowicz et al. 2015). Detailed studies on the single Arabidopsis "stress-inducible" H1 (named H1.3) revealed that it significantly differs from the two main (somatic) H1 variants (H1.1 and H1.2) in both its structural (Fig. 2a) and functional characteristics. Structural differences include changes in the amino acid sequences corresponding to GH1 binding sites I and II (Fig. 2b, c). The nature of these changes in H1.3 suggests that its binding affinity to nucleosomes is decreased compared to H1.1 and H1.2. Moreover, H1.3 also has a much shorter CTD than the somatic variants, which is completely devoid of the SPXKX motifs that enhance DNA binding by H1.1 and H1.2. Collectively, the aforementioned differences probably account for the atypical mode of H1.3 binding to chromatin in vivo, as demonstrated by FRAP analysis (Rutowicz et al. 2015). While H1.1 and H1.2 bind chromatin with dynamics typical for linker histones, with full recovery 2–5 min after photobleaching and distinct stable bound fractions, the behavior of H1.3 in vivo is more similar to that of transcription factors, with superfast dynamics, characterized by full recovery from photobleaching in the range of seconds and no stable bound fraction (Rutowicz et al. 2015). The likely functional consequences of these properties of the H1.3 variant are discussed below.

Another lineage-specific feature is the occurrence in plants of a diversified group of proteins in which a typical GH1 domain is fused to other domains. In addition to the three canonical linker histones, H1.1, H1.2, and H1.3, Arabidopsis has 12 different GH1-containing proteins which also possess either AT-hook motifs or Myb domains (Kotlinski et al. manuscript in preparation). Some of the Myb-containing proteins have "telomere binding" activity, but the functions of the majority of these GH1 hybrids have not been examined. It has been proposed that the proteins with the GH1 domain plus HMG-specific AT hook motifs represent a plant-specific HMGA (formerly HMG1/Y) class (Launholt et al. 2006). We consider this unlikely because, according to our unpublished results, Arabidopsis possesses a genuine HMGA protein with no GH1 domain (Kotlinski et al. manuscript in preparation). Do these uncharacterized plant GH1-containing proteins represent highly specialized variants of H1? More detailed knowledge concerning their chromatin binding and functional properties is required to answer this question.

3.2 *Universally Conserved and Lineage-Specific Functions of Plant H1s*

Development In both animals and plants, the profile of linker histones undergoes significant modifications during early development. In animals (from sea urchin through Drosophila to vertebrates), somatic linker histones are replaced during this stage by specific oocyte/embryonic H1 variants. In Arabidopsis, neither of the two main canonical H1s can be detected in the female or male Spore Mother Cells (SMC), in the nucleus of the vegetative cell during pollen development, or in the early embryo (up to the 2–4 cell stage) (She et al. 2013). While this suggests that Arabidopsis has no H1 variant with functional homology to animal oocyte/embryonic variants, it should be remembered that, in addition to the 3 canonical H1s, its genome encodes other proteins with the GH1 domain. Thus, the possibility that some of these proteins replace canonical H1s in SMC and during early embryonic development cannot be ruled out.

A link between plant linker histones and DNA methylation was first recognized in Arabidopsis in 2005 following the analysis of changes in DNA methylation correlated with downregulation of the transcription of all three H1 variants occurring in this species (Wierzbicki and Jerzmanowski 2005). This link was soon shown not to be restricted to plants, as it was also observed in mouse (Fan et al. 2005). In plants, cytosines in DNA can undergo methylation in three different sequence contexts: CG, CHG, and CHH (where H denotes adenine, cytosine, or thymine). Regarding the possible mechanism underlying H1's role in plant DNA methylation, it was subsequently suggested that the main Arabidopsis H1 variants, H1.1 and H1.2, prevent the access of DNA methyltransferases to nucleosomal DNA, and that overcoming this restriction, especially within heterochromatin, is a major function of the ATP-dependent nucleosome remodeler DDM1 (DECREASE OF DNA METHYLATION1) (Zemach et al. 2013).

Adaptation Plant stress-inducible H1 variants were first discovered in 1990 when an *H1-D* gene identified in the wild tomato (*Solanum pennellii)* was shown to be strongly induced by drought stress and abscisic acid (Cohen and Bray 1990; Plant et al. 1991; Cohen et al. 1991; Wei and O'Connell 1996). Subsequently, Arabidopsis and tobacco were found to possess close homologs of H1-D that respond to stress in a similar way (Ascenzi and Gantt 1997; Przewloka et al. 2002). Since downregulation of the stress-inducible H1s of Arabidopsis, wild tomato and tobacco did not affect development and global chromatin organization (Ascenzi and Gantt 1999; Scippa et al. 2000, 2004; Przewloka et al. 2002), it was concluded that this group of H1 variants plays a negligible role in plant development or chromatin structure. However, the observation that—uniquely to linker histones—they can be regulated by environmental factors provided support for the notion that they are involved in some molecular mechanism of adaptation to stress conditions. This was confirmed by a recent detailed analysis of the role of Arabidopsis H1.3 in stress adaptation (Rutowicz et al. 2015). Levels of the H1.3

protein, normally absent in plant tissues except in guard cells, were shown to be massively upregulated upon exposure to complex environmental stress (concomitant limitation of water and light availability), which led to its occurrence in most tissues. Unlike the two main variants, H1.1 and H1.2, H1.3 not only binds chromatin with superfast dynamics (as mentioned above), but it has a marked preference for chromatin sites bearing histone H3 with tri-methylated lysine 4 (H3K4me3), an epigenetic memory mark of transcription. Moreover, it was also demonstrated that H1.3 is required for a substantial part of stress-related de novo CHH-type DNA methylation (Rutowicz et al. 2015). Given the aforementioned role of the main Arabidopsis H1s in limiting the accessibility of DNA to methyltransferases, this suggests that H1.3 facilitates the access of trans-acting factors to chromatin by competing with H1.1 and H1.2. Interestingly, H1.3 belongs to a narrow set of 26 proteins comprising the core network of retrograde signaling from cellular organelles (chloroplasts and mitochondria) to the nucleus: a major system connecting environmental conditions with adaptive responses at the gene level (Gläßer et al. 2014). Thus, the finding that massive upregulation of H1.3 expression occurs upon prolonged decrease in light intensity and water deprivation (conditions severely affecting photosynthesis and energy metabolism that induce strong chloroplast and mitochondrial responses) strongly suggests that this specialized H1 variant acts as an important terminal effector of stress signaling to chromatin. The fact that the evolution of stress-inducible H1 variants coincided with the appearance of Angiosperms further implies a likely role for these variants in the subsequent massive radiation of this group of plants.

Microtubules The formation of a bipolar karyo-kinetic spindle made primarily of microtubules is a *sine qua non* requirement of proper segregation of chromosomes during cell division. In animal cells, spindle microtubules grow at their "plus" end, to which the new tubulin subunits are added, while their "minus" ends remain anchored to specific structures known as centrosomes, positioned at the opposite poles of the cell. The spindle is completed when the growing "plus" ends reach kinetochores, which are protein structures assembled at the centromeres of chromosomes, capable of linking them with microtubules. Plant cells do not have centrosomes, but their spindle microtubules are known to be efficiently nucleated at the nuclear surface. Unexpectedly, it was found that the protein that promotes this process in tobacco BY-2 cells is histone H1, which is capable of forming ring-shaped complexes with tubulin dimers at the nucleus–cytoplasmic interface (Hotta et al. 2007). It is thought that these tubulin-interacting H1s are part of nucleosomes located in close proximity to the nuclear surface. While the above mechanism still requires independent confirmation by both in vitro and in vivo approaches, it is consistent with the previous demonstration that histone H1 acts as a stabilizer of sea urchin flagellar microtubules (Multigner et al. 1992) and the close evolutionary correlation between histone H1 and microtubular structures (Kaczanowski and Jerzmanowski 2001).

It is clear from the literature reviewed above that plant linker histones, apart from their universally occurring roles like the stabilization of nucleosomes or involvement in the regulation of DNA methylation, also perform functions that may be more lineage specific. Of particular interest and probably of fundamental importance, is the role of stress-inducible H1 variants in translating environmental cues into responses at the genetic level. This system may have evolved in the face of strong selective pressure to develop rapid, simple, and efficient mechanisms of adaptation to a changing environment. Such pressure could have been much more pronounced in plants, as sessile organisms, than in mobile animals. This notwithstanding, there remains a number of unanswered questions concerning the major difference in chromatin plasticity between plants and animals. In contrast to animals, plants have the potential to dedifferentiate their already differentiated cells. Further studies are required to determine whether this is due to the activity of some intricate complex mechanisms or to major structural differences between abundant chromatin components in plants and animals (e.g., the lack of "wing" tips in GH1 of H1s).

4 SWI/SNF Complexes in Plants

4.1 Composition of Plant SWI/SNF Complexes

The SWI/SNF family is represented in Arabidopsis by four chromatin remodeling ATPases: SPLAYED (SYD), BRAHMA (BRM), MINUSCULE1, and MINUSCULE2 (MINU1 and MINU2, also called CHR12 and CHR23, respectively) [reviewed in (Jerzmanowski 2007; Knizewski et al. 2008; Han et al. 2015)]. BRM and SYD are large proteins (about 250 kDa and 390 kDa, respectively), while CHR12 and CHR23 are significantly smaller. Besides the close similarity of their ATPase domains, all four ATPase proteins also share similar elements in their N-terminal regions, including the presence of a QLQ motif and HSA domain (Fig. 3), most probably involved in protein–protein interactions with other complex subunits as well as with various nuclear proteins. In contrast, large differences characterize their C-terminal regions: they are very short and lack structured motifs in MINU1 and MINU2 and are much longer in BRM and SYD. In both large ATPases, the C-terminal regions contain an A/T-hook motif, that in BRM was shown to be capable of interacting with DNA in vitro (Farrona et al. 2007). In addition, BRM is the only one of the four Arabidopsis ATPases that possesses a bromodomain, which makes it the closest homolog of yeast and animal SWI/SNF ATPases. The SYD ATPase lacks a bromodomain, but instead has a large and unstructured C-terminal sequence that was proposed to negatively affect SYD accumulation during development (Su et al. 2006) (Fig. 3).

Arabidopsis also possesses homologs of all the other major SWI/SNF complex core subunits. Similarly to animals, some of the subunits are encoded by gene families, which enables combinatorial assembly of different complexes. There are four SWI3 subunits (SWI3A, B, C, and D), two SWP73 subunits (SWP73A and B),

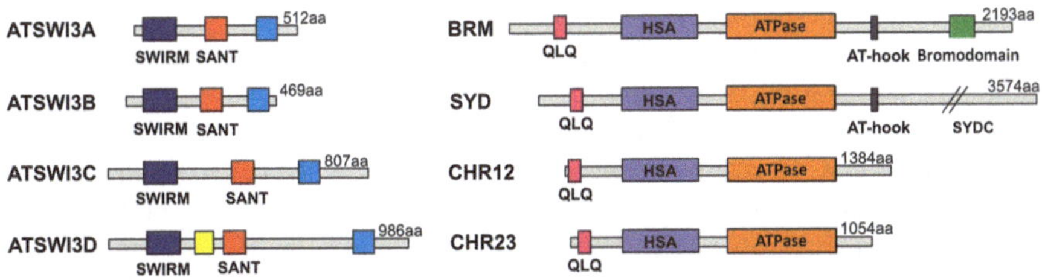

Fig. 3 Architecture of Arabidopsis SWI/SNF subunits: SWI3 (*left*) and ATPase (*right*), showing different domains facilitating protein–protein interactions and binding to modified histones or DNA. *Blue*—leucine zipper; *yellow*—zinc finger

and one SNF5 subunit (BSH) (Sarnowski et al. 2005; Sacharowski et al. 2015; Brzeski et al. 1999). As in yeast and animals, these subunits enrich the complex in additional histone- and DNA-binding domains. For example, SWI3 proteins possess SWIRM and SANT domains, leucine zippers, and a zinc finger domain (in case of SWI3D) (Fig. 3). Thus, Arabidopsis complexes, similarly to their animal counterparts, are probably also capable of recognizing chromatin modifications and local DNA architecture at their target sites.

Because of the considerable difficulties posed by the purification and isolation of intact protein complexes from plant tissues, the accurate composition of the native Arabidopsis SWI/SNF complexes remains unknown. Most of the relevant information about protein–protein interactions between core subunits has come from yeast two-hybrid and in vitro pull-down assays. It was only recently that proteomic studies using plant lines expressing tagged versions of SWI/SNF subunits or SWI/SNF-interacting proteins provided new information about complex composition and identified previously unrecognized complex subunits. In the work of Vercruyssen and colleagues, SWI/SNF complexes were co-purified from Arabidospis cell culture using GS TAP-tagged AN3 protein (Vercruyssen et al. 2014). AN3 is a homolog of human SYNOVIAL TRANSLOCATION (SYT), a transcriptional activator that was recently shown to be a stable component of BAF complexes (Kadoch et al. 2013). The AN3-associated SWI/SNF complexes contain SWI3C and SWI3D, SWP73B, ARP4, and ARP7. Other potential auxiliary subunits were identified in TAP purifications, including an ARID domain-containing LFR protein and three BRD proteins (BRD1, BRD2, BRD13) (Vercruyssen et al. 2014), each possessing one bromodomain. Surprisingly, the homolog of the core subunit SNF5 was not co-purified along with the other core subunits, despite the fact that only a single protein of this type, BSH, exists in Arabidopsis. Whether all AN3-containing complexes really lack the BSH subunit, or if it could not be detected under the experimental conditions employed, requires further clarification. AN3-containing complexes purified in the same manner from maize leaves show a very similar composition (Nelissen et al. 2015), indicating that such complexes are conserved in different plant species. SWI/SNF subunits were also purified using TAP-tagged SWI3C and SWP73B and GFP-tagged BRM (Vercruyssen et al. 2014; Li et al. 2016). BRM was the only one ATPase detected when SWI3C-TAP was used as bait, suggesting that SWI3C is a characteristic subunit of BRM-containing complexes, which is consistent with the highly similar phenotypes of *brm* and *swi3c* null mutants (Archacki et al. 2009).

In another study, SWI/SNF subunits were co-purified with GFP-tagged histone deacetylase HD2C (Buszewicz et al. 2016). Interestingly, the panel of subunits was different from that isolated in the AN3-TAP experiments. AN3-associated complexes contained SWI3C and SWI3D but not SWI3A or SWI3B. In contrast, HD2C-interacting complexes contained SWI3A and SWI3B but not SWI3C or SWI3D. Moreover, AN3 was not co-precipitated with HD2C, suggesting that AN3 and HD2C bind to different SWI/SNF complexes. This supports the notion that plants use SWI/SNF complexes with different subunit compositions to perform specialized functions. Complexes containing either SWI3A/B or SWI3C/D and one of the main ATPases (BRM or SYD) may perform separate functions, resulting from the presence of different interaction partners (HD2C repressor or AN3 activator, respectively). It should, however, be noted that to date no plant SWI/SNF complex has been purified to homogeneity. In each of the aforementioned studies, it is likely that a mixture of different complexes was obtained, as indicated by the presence of two or more ATPases in the co-precipitated fractions. Therefore, unequivocal interpretation of the data is difficult. For example, it is currently not known how many BRD proteins can be incorporated into a single complex.

4.2 Biological Roles of Plant SWI/SNF Complexes

The importance of SWI/SNF-chromatin remodeling complexes in the regulation of plant growth and development has been demonstrated by numerous genetic studies. Null mutants in the major Arabidopsis ATPases *BRM* and *SYD* show pleiotropic albeit different phenotypes during post-embryonic development, indicating that both proteins control mostly nonredundant molecular events (Wagner and Meyerowitz 2002; Hurtado et al. 2006). However, there is also some genetic evidence suggesting their partial functional overlap. The *brm syd* double null mutant is embryo-lethal (Bezhani et al. 2007), and a double hypomorphic mutant *brm-3 syd-6* shows enhanced defects during initiation of the flower primordium (Wu et al. 2015). In addition, BRM and SYD were shown to regulate some common genes in a similar way (Wu et al. 2012, 2015).

BRM and SYD seem to be involved in the control of almost every aspect of plant development. Either one or the other was found to be essential for the transcriptional control of key genes involved in cotyledon separation, repression of seed maturation, flower patterning, the promotion of cell division during leaf development, root stem cell niche maintenance, inflorescence architecture, and the initiation of flowering (Kwon et al. 2006; Tang et al. 2008; Wu et al. 2012; Vercruyssen et al. 2014; Yang et al. 2015; Zhao et al. 2015; Farrona et al. 2011; Li et al. 2015). In addition to its role in the regulation of developmental transitions, BRM is emerging as an important hub in the control of hormonal signaling pathways, as demonstrated by its involvement in orchestrating responses to abscisic acid, gibberellins, cytokinins, and auxins (Han et al. 2012; Archacki et al. 2013; Efroni et al. 2013; Wu et al. 2015) [reviewed in (Reyes 2014; Sarnowska et al. 2016)]. BRM was also shown to participate in responses to abiotic stresses, including drought and heat (Han et al. 2012; Buszewicz et al. 2016; Brzezinka et al. 2016), while SYD is involved in responses to pathogens (Walley et al. 2008).

In contrast to BRM and SYD, null mutants in genes encoding the two shorter Snf2 proteins, CHR12 and CHR23, do not exhibit any visible phenotypic defects. However, *chr12 chr23* (*minu1/minu2*) double mutant is embryo-lethal. A weak double mutant has defects in the maintenance of both shoot and the stem cell populations (Sang et al. 2012). Together, the above observations suggest a high degree of redundancy between these two smaller ATPases. In addition, overexpression of CHR12 was shown to enhance developmental growth arrest occurring in plants exposed to environmental stress (Mlynárová et al. 2007).

The other core subunits of Arabidopsis SWI/SNF complexes, including SWI3 and SWP73 proteins, were also found to be involved in numerous biological processes, mostly similar to those affected by mutations in the ATPase subunits (Sarnowski et al. 2005; Sacharowski et al. 2015; Tang et al. 2008; Han et al. 2012; Efroni et al. 2013; Sarnowska et al. 2013; Jégu et al. 2015). Analysis of the SWI3 gene family has revealed remarkable functional diversification of the four variants, strongly implying that they play specialized roles. *swi3a* and *swi3b* mutants show embryo lethality, while *swi3c* and *swi3d* mutants are viable and display pleiotropic phenotypes during vegetative development. Similarly to the SWI3 family, *swp73a* and *swp73b* mutants generally display nonoverlapping phenotypes, with *swp73b* affecting various developmental processes, while swp73a only showing defects in flowering time. The double *swp73a/b* null mutation results in embryo lethality (Sacharowski et al. 2015). Together with the lethal phenotypes of *brm/syd* and *minu1/minu2*, these genetic data suggest that at least one protein from each pair is indispensable for embryonic development and viability. Whether the severe phenotypes caused by mutation of these non-catalytic subunits of plant SWI/SNF complexes are due to mis-targeting of the complex, loss of its activity, disassembly, or some other reason is still unknown.

4.3 Mechanisms Underlying the Functions and Targeting of Plant SWI/SNF Complexes

In the studies conducted using the ChIP approach, the functions of BRM and SYD were linked to regulation and binding to specific genic loci. Similarly to its animal homologs, BRM can both activate and repress target genes. For example, two direct targets of BRM, *SCL3* and *SVP*, are downregulated, while *ABI5* is upregulated in *brm* mutants (Archacki et al. 2013; Li et al. 2015; Han et al. 2012). Available data suggest that Arabidopsis SWI/SNF is able to alter chromatin accessibility and nucleosome occupancy at its target promoters. It was shown that +1 nucleosome occupancy at the *ABI5* locus is reduced in the *brm-3* hypomorphic mutant (Han et al. 2012). In another study by the D. Wagner laboratory, accessibility of the *FIL* locus, measured by Formaldehyde Assisted Isolation of Regulatory Elements (FAIRE), was shown to be decreased in a *brm-3/syd-5* mutant. In addition, auxin-dependent destabilization of the positioned nucleosome in the *FIL* promoter region, assayed by MNase-qPCR, was impaired in the same mutant (Wu et al. 2015). Altered occupancy of selected nucleosomes was also reported for a large number of genes in *swp73* mutants (Sacharowski et al. 2015). It is likely that depending on whether the gene is activated or repressed by SWI/SNF, the complex can either decrease or increase nucleosome occupancy at the target promoter. However, it should be noted that indirect effects on nucleosome occupancy in *swi/snf* mutants are also possible, and it has yet to be ascertained whether this is a common mechanism by which Arabidopsis SWI/SNF regulates transcription.

In a few cases, the functions of BRM and SYD linked to binding and regulation of their target genes were shown to be dependent on sequence-specific transcription factors. BRM and SYD ATPases can be recruited by the transcription factors LFY and SEP3 to promoters of the flower homeotic genes *AGAMOUS* and *APETALA1* (Wu et al. 2012), the activation of which is dependent on these ATPases and required for correct floral patterning. The same ATPases are recruited by transcription factor MP upon auxin signaling to activate the *FILAMENTOUS FLOWERS* and *LEAFY loci*, encoding key regulators of flower primordium initiation (Wu et al. 2015). In turn, BP is required for the binding of BRM to the *KNAT2* and *KNAT6* genes: regulators of inflorescence architecture that are repressed by BRM (Zhao et al. 2015). Other studies have shown that plant SWI/SNF complexes are able to interact via different subunits with numerous transcription factors and regulatory proteins, possibly enabling the precise regulation of specific genes in different tissues in response to various signals. Mapping the yeast two-hybrid interactome of 6 SWI/SNF subunits, using a library of 1400 Arabidopsis TFs, identified 400 interactions with 210 transcription factors (Efroni et al. 2013). Proteins proven to interact with SWI/SNF subunits in vivo include the following: PP2C phosphatase HAB1, a key negative regulator of ABA signalling (Saez et al. 2008); TCP4, a transcription factor that promotes leaf maturation (Efroni et al. 2013); DELLA proteins, key components of gibberellin signaling (Sarnowska et al. 2013); GRF family transcription factors controlling leaf development (Vercruyssen et al. 2014), and FORGETTER1, a newly identified chromatin regulator controlling stress-induced epigenetic memory (Brzezinka et al. 2016). The interaction of SWI/SNF with long noncoding-RNA-binding protein IDN2 has also been reported, and proposed to reinforce long-ncRNA-mediated transcriptional silencing (Zhu et al. 2013).

Two studies aimed at whole genome mapping of BRM were conducted using a BRM-GFP line and specific anti-BRM antibody (Li et al. 2016; Archcki et al. 2016). The obtained global binding patterns indicate that BRM is almost exclusively associated with genes, since the vast majority of the binding sites were located within or close to gene units. The two studies identified a similar number of bound genes (approx. 5000 each) with about 3800 BRM target genes in common

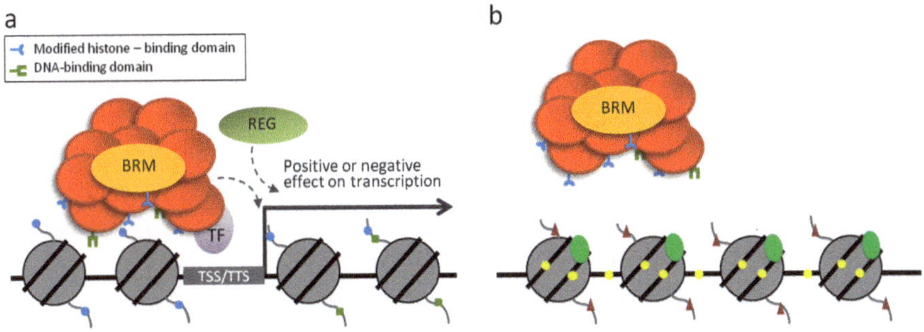

Fig. 4 Recruitment of SWI/SNF complex to different chromatin regions. (**a**) SWI/SNF complex can bind both gene promoters and terminators and mediate transcriptional activation or repression. Regulation of target genes by BRM ATPase was shown to be at least in part dependent on the protein–protein interactions with transcription factors (TF), transcriptional regulators, and chromatin regulators (REG) like REF6 histone H3K27 demethylase or histone deacetylase HD2C. SWI/SNF subunits also contain domains capable of recognizing different histone modifications and DNA structures (depicted at the surface of subunits). (**b**) Binding of BRM ATPase anti-correlates with high levels of DNA methylation (*yellow circles*) and repressive histone marks like H3K9me2 and H3K27me3 (*triangles*), as well as with the presence of major histone H1 variants H1.1 and H1.2 (*green ovals*) (Archacki et al. 2016 and our unpublished data)

(Li et al. 2016; Archcki et al. 2016). Almost 1000 of those genes were both bound by BRM and displayed altered expression in a *brm-1* null mutant, making them likely candidates for direct transcriptional targets (Archacki et al. 2016). In agreement with previous studies, gene ontology analyses of identified BRM targets showed that the regulation of genes involved in hormone, stress, and biotic responses is a substantial part of BRM function (Archacki et al. 2016; Li et al. 2016).

Unlike yeast SNF2 ATPases, which bind almost exclusively to promoters near the TSS, Arabidopsis BRM binds not only proximal, but also distal promoter regions, as well as gene bodies and gene terminators. This is more similar to the binding pattern of mammalian BRG1. Arabidopsis BRM-bound genes are mostly euchromatic active genes, characterized by a deeper nucleosome-depleted region near the TSS, strong enrichment of H3K4me3, and depletion of H3K9me2 and H3K27me3 repressive marks, as well as lower DNA methylation levels in promoters, compared to genes not bound by BRM (Archacki et al. 2016). Interestingly, BRM-bound terminators were also shown to have high H3K4me3 and low DNA methylation levels (Fig. 4a). Notably, these terminators were characterized by the presence of TATA boxes and thus resembled classical promoters. Furthermore, the genes with terminator-bound BRM showed extensive noncoding antisense transcription originating from their 3′ end regions. This transcription was either activated or repressed by BRM, similarly to sense transcription of promoter-bound genes. This suggests that the 3′-bound BRM activates antisense promoters to control sense expression of these genes (Archacki et al. 2016). Thus, the regulatory functions of SWI/SNF complexes seem to be much broader than those resulting from changes in nucleosome positioning at gene promoters. While the numerous examples are consistent with the recruitment of SWI/SNF complexes via transient interactions with sequence-specific DNA-binding proteins (see above), the results of C. Li and coworkers support a second, epigenetic-based mechanism. The binding of BRM to a substantial set of genes (~12% of all detected bound genes) was shown to require the H3K27-demethylase REF6 (Li et al. 2016).

5 Are Linker Histones and Chromatin Remodeling Structurally and Functionally Coupled?

Unless pre-associated with chromatin, ATP-dependent remodelers rarely have unrestricted access to the basic nucleosomal fiber. Before gaining such access, they must first overcome the barrier of higher order chromatin structures and deal successfully with the obstacle created by histone H1. In 2000, Imbalzano and Hill showed that the presence of H1 partially inhibits the nucleosome remodeling activity of human SWI/SNF (hSWI/SNF) complexes. Since isolated H1 had no negative effect on SWI/SNF ATPase activity in vitro, the authors concluded that the inhibition was caused by structural changes resulting from the interaction of H1 with nucleosomes (Hill and Imbalzano 2000). They also suggested that SWI/SNF is not a randomly acting nucleosome remodeling machine, but rather targets chromosomal *loci* on which H1 has been previously lost or modified. This was subsequently corroborated by a study which showed that phosphorylation of H1 by Cdc2/Cyclin B kinase can effectively rescue remodeling by SWI/SNF (Horn et al. 2002). In another in vitro study, the presence of histone H1 was shown to modulate the process of hSWI/SNF remodeling by inhibiting nucleosome movement towards the ends of nucleosomal DNA and promoting its sliding to more central positions (Ramachandran et al. 2003). Interestingly, individual H1 isoforms turned out to have different effects on SWI/SNF remodeling. On studying the role of an early embryonic variant of H1 in Xenopus, called B4, Seaki and coworkers found that in contrast to typical somatic H1 variants, it did not prevent remodeling by SWI/SNF (Saeki et al. 2005). How the removal of somatic H1 is orchestrated during the specific induction of progesterone target genes in mammals was elucidated by Vicent et al. (2011). They found that the cascade of events was started by recruitment of the chromatin remodeling complex NURF to the activated progesterone receptor, which in turn facilitated the binding of Cdk2/CyclinA required for the completion of H1 displacement. This was a prerequisite for the subsequent nucleosome remodeling by SWI/SNF. The inhibitory effect of H1 on SWI/SNF-dependent remodeling is in agreement with the structural and mechanistic properties of SWI/SNF complexes, which make extensive contacts with the nucleosome and linker DNA (see above).

To summarize, there is a general consensus that the presence of linker histones in metazoan chromatin increases the stability of more compact higher order chromatin structures. This exerts a negative effect on nucleosome mobility and thus retards chromatin structural modulation by ATP-dependent nucleosome remodeling complexes, which in turn suppresses DNA-dependent activities such as transcription and replication (see Izzo et al. 2008, for review). Interestingly, there also seems to be an inverse relationship: namely the dependence of H1 binding on the type of active nucleosome remodeling (Corona et al. 2007). In a recent study, Ocampo and colleagues noticed that genes with short internucleosomal spacing bind less H1 than genes with longer spacing. The short spacing is primarily determined by CHD1- and the long spacing by ISW1-type ATP-dependent remodelers. The authors hypothesized that CHD1-mediated short spacing leads to the eviction of H1 and partial unfolding of chromatin, whereas ISW1-mediated long spacing permits H1 binding and leads to chromatin refolding (Ocampo et al. 2016). It is possible that SWI/SNF complexes can also modulate H1 binding to chromatin. It was proposed that the BAF complex prevents the accumulation of H1 at the *Cd4* locus in lymphocytes. This in turn allows binding of the Runx1 corepressor and downregulation of *Cd4* expression (Wan et al. 2009). Together, these data strongly suggest interdependence between the occurrence of H1 in chromatin and ATP-dependent remodeling.

What are the structural and functional links between linker histones and ATP-dependent chromatin remodeling in plants? One of the most interesting and surprising connections concerns DDM1 (Decrease of DNA Methylation 1), a SNF-2 class ATPase shown to be capable of ATP-dependent nucleosome repositioning in vitro (Brzeski and Jerzmanowski 2003). Arabidopsis *ddm1* mutants are deficient in DNA methylation, mostly in heterochromatin, with the progressive loss of this modification in subsequent generations that finally leads to lethality (Tsukahara et al. 2009). It has been speculated that the maintenance of heterochromatin DNA methylation requires constant nucleosome remodeling by DDM1. Interestingly, a recent report by Zemach and coworkers showed that in an Arabidopsis mutant lacking the major H1 variants, the DDM1 ATPase is no longer required to maintain a stable level of DNA methylation at heterochromatic transposons (Zemach et al. 2013). This led to the hypothesis that DDM1-mediated nucleosome remodeling enables DNA methyltransferases to access H1-containing heterochromatin. An intriguing observation that points to the interdependence of H1 and DDM1 is that both were absent within a narrow developmental window occurring during female and male gametogenesis in Arabidopsis (Ito 2013). However, it has yet to be determined whether DDM1 directly opposes the effects of H1, or if the observed genetic interaction is due to some other unrecognized mechanism linking H1 with DNA methylation. Furthermore, the interdependence of DDM1 and H1 in controlling DNA methylation on euchromatic *loci* is less clear-cut than the situation on heterochromatic transposons (Zemach et al. 2013). One possible explanation is that on euchromatic *loci*, H1 antagonizes different chromatin remodelers, such as SWI/SNF. While there is no direct experimental evidence in support of such a mechanism, the accumulating correlative data make it plausible. First, genes bound by BRM, on either their promoters or terminators, show low levels of DNA methylation in all sequence contexts (Archacki et al. 2016), and a mutant lacking H1 is characterized by destabilization of genic methylation, which is manifested as both hypo- and hypermethylation (Zemach et al. 2013). Second, our analysis of global binding profiles of BRM and H1 variants shows clear anti-correlation along gene units suggesting possible functional antagonism (Fig. 4b). Interestingly, this anti-correlation involves the main variants H1.1 and H1.2, but not H1.3 (Archacki et al. unpublished), although it should be noted that the H1.3 profile was obtained under stress conditions. As both BRM and H1.3 have been strongly implicated in stress responses, a tempting albeit highly speculative hypothesis is that during stress signaling, H1.3, by competing with the major H1 variants, enables SWI/SNF remodelers to access previously inaccessible sites.

6 Concluding Remarks

Both linker histones and SWI/SNF-chromatin remodelers belong to an evolutionarily ancient basic molecular tool kit that mediates regulated access to genomic DNA in eukaryotes. However, over hundreds of millions of years of separate evolution, their plant and animal homologs have been exposed to different selection pressures resulting from the fundamentally different life strategies of these two lineages. The vast amount of detailed information accumulated concerning H1 and chromatin remodeling in animals and plants makes it now possible to analyze how these separate evolutionary processes have shaped their structural and functional properties. As shown in this chapter, comparative analyses based on currently available data has revealed extensive specialization in the use of linker histone variability in plants, possibly due to their much greater dependence on flexible adaptation to changing environmental conditions. The surprising richness of possibilities of combinatorial assembly of plant core SWI/SNF complexes may reflect another interesting lineage-specific specialization. Uncovering the roles of both linker histones and SWI/SNF remodelers in the adaptation of plants to a sessile lifestyle, and their ability to effectively adapt to biotic and abiotic stresses, may greatly enhance our understanding of the universal role of chromatin in the integration of environmental signals and genetic programs.

Acknowledgements This work was supported by grants from National Science Centre project no.: 2014/13/B/NZ1/00967, and the Ministry of Science and Higher Education project no.: 212/N-COST/2008/0, to A. Jerzmanowski and Lider/313/L-6/14/NCBR/2015 for R. Archacki. We thank Maciej Kotliński and Kinga Rutowicz-Kalinowska for comments and Maciej Kotliński for help in preparing Figs. 1 and 2.

References

Alcid EA, Tsukiyama T (2014) ATP-dependent chromatin remodeling shapes the long noncoding RNA landscape. Genes Dev 28:2348–2360

Archacki R, Sarnowski TJ, Halibart-Puzio J et al (2009) Genetic analysis of functional redundancy of BRM ATPase and ATSWI3C subunits of Arabidopsis SWI/SNF chromatin remodelling complexes. Planta 229:1281–1292
Archacki R, Buszewicz D, Sarnowski TJ et al (2013) BRAHMA ATPase of the SWI/SNF chromatin remodeling complex acts as a positive regulator of gibberellin-mediated responses in arabidopsis. PloS One 8:e58588
Archacki R, Yatusevich R, Buszewicz D et al (2016) Arabidopsis SWI/SNF chromatin remodeling complex binds both promoters and terminators to regulate gene expression. Nucleic Acids Res. doi:10.1093/nar/gkw1273
Armstrong JA, Papoulas O, Daubresse G et al (2002) The Drosophila BRM complex facilitates global transcription by RNA polymerase II. EMBO J 21:5245–5254
Arya G, Maitra A, Grigoryev SA (2010) A structural perspective on the where, how, why, and what of nucleosome positioning. J Biomol Struct Dyn 27:803–820
Ascenzi R, Gantt JS (1997) A drought-stress-inducible histone gene in Arabidopsis thaliana is a member of a distinct class of plant linker histone variants. Plant Mol Biol 34:629–641
Ascenzi R, Gantt JS (1999) Molecular genetic analysis of the drought-inducible linker histone variant in Arabidopsis thaliana. Plant Mol Biol 41:159–169
Bayona-Feliu A, Casas-Lamesa A, Carbonell A, Climent-Cantó P, Tatarski M, Pérez-Montero S, Azorín F, Bernués J (2016) Histone H1: lessons from Drosophila. Biochim Biophys Acta 1859:526–532. doi:10.1016/j.bbagrm.2015.09.001
Bednar J, Hamiche A, Dimitrov S (2016) H1-nucleosome interactions and their functional implications. Biochim Biophys Acta 1859:436–443
Bezhani S, Winter C, Hershman S et al (2007) Unique, shared, and redundant roles for the Arabidopsis SWI/SNF chromatin remodeling ATPases BRAHMA and SPLAYED. Plant Cell 19:403–416
Brehm A, Tufteland KR, Aasland R et al (2004) The many colours of chromodomains. BioEssays 26:133–140
Brown SA, Imbalzano AN, Kingston RE (1996) Activator-dependent regulation of transcriptional pausing on nucleosomal templates. Genes Dev 10:1479–1490
Brzeski J, Podstolski W, Olczak K et al (1999) Identification and analysis of the Arabidopsis thaliana BSH gene, a member of the SNF5 gene family. Nucleic Acids Res 27:2393–2399
Brzeski J, Jerzmanowski A (2003) Deficient in DNA methylation 1 (DDM1) defines a novel family of chromatinremodeling factors. J Biol Chem 278:823–828. doi:10.1074/jbc.M209260200
Brzezinka K, Altmann S, Czesnick H et al (2016) Arabidopsis FORGETTER1 mediates stress-induced chromatin memory through nucleosome remodeling. eLife. doi:10.7554/eLife.17061
Burd CJ, Archer TK (2013) Chromatin architecture defines the glucocorticoid response. Mol Cell Endocrinol 380:25–31
Buszewicz D, Archacki R, Palusiński A et al (2016) HD2C histone deacetylase and a SWI/SNF chromatin remodelling complex interact and both are involved in mediating the heat stress response in Arabidopsis. Plant Cell Environ 39:2108–2122
Clapier CR, Cairns BR (2009) The biology of chromatin remodeling complexes. Annu Rev Biochem 78:273–304
Cohen A, Bray EA (1990) Characterization of three mRNAs that accumulate in wilted tomato leaves in response to elevated levels of endogenous abscisic acid. Planta 182:27–33
Cohen A, Plant AL, Moses MS et al (1991) Organ-specific and environmentally regulated expression of two abscisic acid-induced genes of tomato: nucleotide sequence and analysis of the corresponding cDNAs. Plant Physiol 97:1367–1374
Corona DFV, Siriaco G, Armstrong JA, Snarskaya N, McClymont SA, Scott MP, Tamkun JW (2007) ISWI regulates higher-order chromatin structure and histone H1 assembly in vivo. PLoS Biol 5:e232. doi:10.1371/journal.pbio.0050232
Cosma MP, Tanaka T, Nasmyth K (1999) Ordered recruitment of transcription and chromatin remodeling factors to a cell cycle- and developmentally regulated promoter. Cell 97:299–311

Cui F, Zhurkin VB (2009) Distinctive sequence patterns in metazoan and yeast nucleosomes: implications for linker histone binding to AT-rich and methylated DNA. Nucleic Acids Res 37:2818–2829

Dechassa ML, Zhang B, Horowitz-Scherer R et al (2008) Architecture of the SWI/SNF-nucleosome complex. Mol Cell Biol 28:6010–6021

Efroni I, Han S-K, Kim HJ et al (2013) Regulation of leaf maturation by chromatin-mediated modulation of cytokinin responses. Dev Cell 24:438–445

Engel KB, Yamamoto KR (2011) The glucocorticoid receptor and the coregulator Brm selectively modulate each other's occupancy and activity in a gene-specific manner. Mol Cell Biol 31:3267–3276

Euskirchen GM, Auerbach RK, Davidov E et al (2011) Diverse roles and interactions of the SWI/SNF chromatin remodeling complex revealed using global approaches. PLoS Genet 7: e1002008

Euskirchen G, Auerbach RK, Snyder M (2012) SWI/SNF chromatin-remodeling factors: multiscale analyses and diverse functions. J Biol Chem 287:30897–30905

Fan Y, Nikitina T, Zhao J et al (2005) Histone H1 depletion in mammals alters global chromatin structure but causes specific changes in gene regulation. Cell 123:1199–1212

Farrona S, Hurtado L, Reyes JC (2007) A nucleosome interaction module is required for normal function of Arabidopsis thaliana BRAHMA. J Mol Biol 373:240–250

Farrona S, Hurtado L, March-Díaz R et al (2011) Brahma is required for proper expression of the floral repressor FLC in Arabidopsis. PloS One 6:e17997

Flanagan TW, Brown DT (2016) Molecular dynamics of histone H1. Biochim Biophys Acta 1859:468–475

Flaus A, Martin DMA, Barton GJ et al (2006) Identification of multiple distinct Snf2 subfamilies with conserved structural motifs. Nucleic Acids Res 34:2887–2905

Gläßer C, Haberer G, Finkemeier I et al (2014) Meta-analysis of retrograde signaling in Arabidopsis thaliana reveals a core module of genes embedded in complex cellular signaling networks. Mol Plant 7:1167–1190

Hainer SJ, Gu W, Carone BR et al (2015) Suppression of pervasive noncoding transcription in embryonic stem cells by esBAF. Genes Dev 29:362–378

Han S-K, Sang Y, Rodrigues A et al (2012) The SWI2/SNF2 chromatin remodeling ATPase BRAHMA represses abscisic acid responses in the absence of the stress stimulus in Arabidopsis. Plant Cell 24:4892–4906

Han S-K, Wu M-F, Cui S et al (2015) Roles and activities of chromatin remodeling ATPases in plants. Plant J Cell Mol Biol 83:62–77

Hargreaves DC, Crabtree GR (2011) ATP-dependent chromatin remodeling: genetics, genomics and mechanisms. Cell Res 21:396–420

Hill DA, Imbalzano AN (2000) Human SWI/SNF nucleosome remodeling activity is partially inhibited by linker histone H1. Biochemistry (Mosc) 39:11649–11656

Ho L, Crabtree GR (2010) Chromatin remodelling during development. Nature 463:474–484

Ho L, Ronan JL, Wu J et al (2009) An embryonic stem cell chromatin remodeling complex, esBAF, is essential for embryonic stem cell self-renewal and pluripotency. Proc Natl Acad Sci U S A 106:5181–5186

Ho L, Miller EL, Ronan JL et al (2011) esBAF facilitates pluripotency by conditioning the genome for LIF/STAT3 signalling and by regulating polycomb function. Nat Cell Biol 13:903–913

Horn PJ, Carruthers LM, Logie C et al (2002) Phosphorylation of linker histones regulates ATP-dependent chromatin remodeling enzymes. Nat Struct Biol 9:263–267

Hotta T, Haraguchi T, Mizuno K (2007) A novel function of plant histone H1: microtubule nucleation and continuous plus end association. Cell Struct Funct 32:79–87

Hurtado L, Farrona S, Reyes JC (2006) The putative SWI/SNF complex subunit BRAHMA activates flower homeotic genes in Arabidopsis thaliana. Plant Mol Biol 62:291–304

Ito H (2013) Small RNAs and regulation of transposons in plants. Genes Genet Syst 88:3–7

Izzo A, Kamieniarz K, Schneider R (2008) The histone H1 family: specific members, specific functions? Biol Chem 389:333–343. doi:10.1515/BC.2008.037

Jégu T, Domenichini S, Blein T et al (2015) A SWI/SNF chromatin remodelling protein controls cytokinin production through the regulation of chromatin architecture. PloS One 10:e0138276

Jerzmanowski A (2007) SWI/SNF chromatin remodeling and linker histones in plants. Biochim Biophys Acta 1769:330–345

Kaczanowski S, Jerzmanowski A (2001) Evolutionary correlation between linker histones and microtubular structures. J Mol Evol 53:19–30

Kadoch C, Crabtree GR (2015) Mammalian SWI/SNF chromatin remodeling complexes and cancer: mechanistic insights gained from human genomics. Sci Adv 1:e1500447

Kadoch C, Hargreaves DC, Hodges C et al (2013) Proteomic and bioinformatic analysis of mammalian SWI/SNF complexes identifies extensive roles in human malignancy. Nat Genet 45:592–601

Kadoch C, Copeland RA, Keilhack H (2016) PRC2 and SWI/SNF chromatin remodeling complexes in health and disease. Biochemistry (Mosc) 55:1600–1614

Kasinsky HE, Lewis JD, Dacks JB et al (2001) Origin of H1 linker histones. FASEB J 15:34–42

Knizewski L, Ginalski K, Jerzmanowski A (2008) Snf2 proteins in plants: gene silencing and beyond. Trends Plant Sci 13:557–565

Kwon CS, Hibara K, Pfluger J et al (2006) A role for chromatin remodeling in regulation of CUC gene expression in the Arabidopsis cotyledon boundary. Dev Camb Engl 133:3223–3230

Launholt D, Merkle T, Houben A et al (2006) Arabidopsis chromatin-associated HMGA and HMGB use different nuclear targeting signals and display highly dynamic localization within the nucleus. Plant Cell 18:2904–2918

Laurent BC, Carlson M (1992) Yeast SNF2/SWI2, SNF5, and SNF6 proteins function coordinately with the gene-specific transcriptional activators GAL4 and Bicoid. Genes Dev 6:1707–1715

Lessard J, Wu JI, Ranish JA et al (2007) An essential switch in subunit composition of a chromatin remodeling complex during neural development. Neuron 55:201–215

Li C, Chen C, Gao L et al (2015) The Arabidopsis SWI2/SNF2 chromatin remodeler BRAHMA regulates polycomb function during vegetative development and directly activates the flowering repressor gene SVP. PLoS Genet 11:e1004944

Li C, Gu L, Gao L et al (2016) Concerted genomic targeting of H3K27 demethylase REF6 and chromatin-remodeling ATPase BRM in Arabidopsis. Nat Genet 48:687–693

Lu X, Hamkalo B, Parseghian MH et al (2009) Chromatin condensing functions of the linker histone C-terminal domain are mediated by specific amino acid composition and intrinsic protein disorder. Biochemistry (Mosc) 48:164–172

Marquardt S, Escalante-Chong R, Pho N et al (2014) A chromatin-based mechanism for limiting divergent noncoding transcription. Cell 157:1712–1723

Meyerowitz EM (2002) Plants compared to animals: the broadest comparative study of development. Science 295:1482–1485

Millán-Ariño L, Izquierdo-Bouldstridge A, Jordan A (2016) Specificities and genomic distribution of somatic mammalian histone H1 subtypes. Biochim Biophys Acta 1859:510–519

Mlynárová L, Nap J-P, Bisseling T (2007) The SWI/SNF chromatin-remodeling gene AtCHR12 mediates temporary growth arrest in Arabidopsis thaliana upon perceiving environmental stress. Plant J Cell Mol Biol 51:874–885

Multigner L, Gagnon J, Van Dorsselaer A et al (1992) Stabilization of sea urchin flagellar microtubules by histone H1. Nature 360:33–39

Narlikar GJ, Sundaramoorthy R, Owen-Hughes T (2013) Mechanisms and functions of ATP-dependent chromatin-remodeling enzymes. Cell 154:490–503

Neigeborn L, Carlson M (1984) Genes affecting the regulation of SUC2 gene expression by glucose repression in Saccharomyces cerevisiae. Genetics 108:845–858

Nelissen H, Eeckhout D, Demuynck K et al (2015) Dynamic changes in ANGUSTIFOLIA3 complex composition reveal a growth regulatory mechanism in the maize leaf. Plant Cell 27:1605–1619

Ocampo J, Chereji RV, Eriksson PR et al (2016) The ISW1 and CHD1 ATP-dependent chromatin remodelers compete to set nucleosome spacing in vivo. Nucleic Acids Res 44:4625–4635

Parnell TJ, Schlichter A, Wilson BG et al (2015) The chromatin remodelers RSC and ISW1 display functional and chromatin-based promoter antagonism. eLife 4:e06073

Pérez-Montero S, Carbonell A, Morán T et al (2013) The embryonic linker histone H1 variant of Drosophila, dBigH1, regulates zygotic genome activation. Dev Cell 26:578–590

Peterson CL, Herskowitz I (1992) Characterization of the yeast SWI1, SWI2, and SWI3 genes, which encode a global activator of transcription. Cell 68:573–583

Phelan ML, Sif S, Narlikar GJ et al (1999) Reconstitution of a core chromatin remodeling complex from SWI/SNF subunits. Mol Cell 3:247–253

Plant AL, Cohen A, Moses MS et al (1991) Nucleotide sequence and spatial expression pattern of a drought- and abscisic acid-induced gene of tomato. Plant Physiol 97:900–906

Postnikov YV, Bustin M (2016) Functional interplay between histone H1 and HMG proteins in chromatin. Biochim Biophys Acta 1859:462–467

Przewloka MR, Wierzbicki AT, Slusarczyk J et al (2002) The "drought-inducible" histone H1s of tobacco play no role in male sterility linked to alterations in H1 variants. Planta 215:371–379

Ramachandran A, Omar M, Cheslock P et al (2003) Linker histone H1 modulates nucleosome remodeling by human SWI/SNF. J Biol Chem 278:48590–48601

Rando OJ, Chang HY (2009) Genome-wide views of chromatin structure. Annu Rev Biochem 78:245–271

Reyes JC (2014) The many faces of plant SWI/SNF complex. Mol Plant 7:454–458

Riedel CG, Dowen RH, Lourenco GF, Kirienko NV, Heimbucher T, West JA, Bowman SK, Kingston RE, Dillin A, Asara JM, Ruvkun G et al (2013) DAF-16 employs the chromatin remodeller SWI/SNF to promote stress resistance and longevity. Nat Cell Biol 15:491–501

Rutowicz K, Puzio M, Halibart-Puzio J et al (2015) A specialized histone H1 variant is required for adaptive responses to complex abiotic stress and related DNA methylation in Arabidopsis. Plant Physiol 169:2080–2101

Sacharowski SP, Gratkowska DM, Sarnowska EA et al (2015) SWP73 subunits of Arabidopsis SWI/SNF chromatin remodeling complexes play distinct roles in leaf and flower development. Plant Cell 27:1889–1906

Saeki H, Ohsumi K, Aihara H et al (2005) Linker histone variants control chromatin dynamics during early embryogenesis. Proc Natl Acad Sci U S A 102:5697–5702

Saez A, Rodrigues A, Santiago J et al (2008) HAB1-SWI3B interaction reveals a link between abscisic acid signaling and putative SWI/SNF chromatin-remodeling complexes in Arabidopsis. Plant Cell 20:2972–2988

Sanchez R, Zhou M-M (2011) The PHD finger: a versatile epigenome reader. Trends Biochem Sci 36:364–372

Sang Y, Silva-Ortega CO, Wu S et al (2012) Mutations in two non-canonical Arabidopsis SWI2/SNF2 chromatin remodeling ATPases cause embryogenesis and stem cell maintenance defects. Plant J Cell Mol Biol 72:1000–1014

Sarnowska EA, Rolicka AT, Bucior E et al (2013) DELLA-interacting SWI3C core subunit of switch/sucrose nonfermenting chromatin remodeling complex modulates gibberellin responses and hormonal cross talk in Arabidopsis. Plant Physiol 163:305–317

Sarnowska E, Gratkowska DM, Sacharowski SP et al (2016) The role of SWI/SNF chromatin remodeling complexes in hormone crosstalk. Trends Plant Sci 21:594–608

Sarnowski TJ, Ríos G, Jásik J et al (2005) SWI3 subunits of putative SWI/SNF chromatin-remodeling complexes play distinct roles during Arabidopsis development. Plant Cell 17:2454–2472

Schulze E, Schulze B (1995) The vertebrate linker histones H1 zero, H5, and H1M are descendants of invertebrate "orphon" histone H1 genes. J Mol Evol 41:833–840

Schwabish MA, Struhl K (2004) Evidence for eviction and rapid deposition of histones upon transcriptional elongation by RNA polymerase II. Mol Cell Biol 24:10111–10117

Scippa GS, Griffiths A, Chiatante D et al (2000) The H1 histone variant of tomato, H1-S, is targeted to the nucleus and accumulates in chromatin in response to water-deficit stress. Planta 211:173–181

Scippa GS, Di Michele M, Onelli E et al (2004) The histone-like protein H1-S and the response of tomato leaves to water deficit. J Exp Bot 55:99–109

She W, Grimanelli D, Rutowicz K et al (2013) Chromatin reprogramming during the somatic-to-reproductive cell fate transition in plants. Dev Camb Engl 140:4008–4019

Shen W, Xu C, Huang W et al (2007) Solution structure of human Brg1 bromodomain and its specific binding to acetylated histone tails. Biochemistry (Mosc) 46:2100–2110

Smith CL, Horowitz-Scherer R, Flanagan JF et al (2003) Structural analysis of the yeast SWI/SNF chromatin remodeling complex. Nat Struct Biol 10:141–145

Stern M, Jensen R, Herskowitz I (1984) Five SWI genes are required for expression of the HO gene in yeast. J Mol Biol 178:853–868

Su Y, Kwon CS, Bezhani S et al (2006) The N-terminal ATPase AT-hook-containing region of the Arabidopsis chromatin-remodeling protein SPLAYED is sufficient for biological activity. Plant J Cell Mol Biol 46:685–699

Tamkun JW, Deuring R, Scott MP et al (1992) Brahma: a regulator of Drosophila homeotic genes structurally related to the yeast transcriptional activator SNF2/SWI2. Cell 68:561–572

Tang X, Hou A, Babu M et al (2008) The Arabidopsis BRAHMA chromatin-remodeling ATPase is involved in repression of seed maturation genes in leaves. Plant Physiol 147:1143–1157

Th'ng JPH, Sung R, Ye M, Hendzel MJ (2005) H1 family histones in the nucleus. Control of binding and localization by the C-terminal domain. J Biol Chem 280:27809–27814

Tolkunov D, Zawadzki KA, Singer C et al (2011) Chromatin remodelers clear nucleosomes from intrinsically unfavorable sites to establish nucleosome-depleted regions at promoters. Mol Biol Cell 22:2106–2118

Tréand C, du Chéné I, Brès V et al (2006) Requirement for SWI/SNF chromatin-remodeling complex in Tat-mediated activation of the HIV-1 promoter. EMBO J 25:1690–1699

Trotter KW, Archer TK (2008) The BRG1 transcriptional coregulator. Nucl Recept Signal 6:e004

Tsukahara S, Kobayashi A, Kawabe A et al (2009) Bursts of retrotransposition reproduced in Arabidopsis. Nature 461:423–426

Vercruyssen L, Verkest A, Gonzalez N et al (2014) ANGUSTIFOLIA3 binds to SWI/SNF chromatin remodeling complexes to regulate transcription during Arabidopsis leaf development. Plant Cell 26:210–229

Vicent GP, Nacht AS, Font-Mateu J et al (2011) Four enzymes cooperate to displace histone H1 during the first minute of hormonal gene activation. Genes Dev 25:845–862

Wagner D, Meyerowitz EM (2002) SPLAYED, a novel SWI/SNF ATPase homolog, controls reproductive development in Arabidopsis. Curr Biol CB 12:85–94

Walley JW, Rowe HC, Xiao Y et al (2008) The chromatin remodeler SPLAYED regulates specific stress signaling pathways. PLoS Pathog 4:e1000237

Wan M, Zhang J, Lai D et al (2009) Molecular basis of CD4 repression by the Swi/Snf-like BAF chromatin remodeling complex. Eur J Immunol 39:580–588

Wei T, O'Connell MA (1996) Structure and characterization of a putative drought-inducible H1 histone gene. Plant Mol Biol 30:255–268

Wierzbicki AT, Jerzmanowski A (2005) Suppression of histone H1 genes in Arabidopsis results in heritable developmental defects and stochastic changes in DNA methylation. Genetics 169:997–1008

Wu M-F, Sang Y, Bezhani S et al (2012) SWI2/SNF2 chromatin remodeling ATPases overcome polycomb repression and control floral organ identity with the LEAFY and SEPALLATA3 transcription factors. Proc Natl Acad Sci U S A 109:3576–3581

Wu M-F, Yamaguchi N, Xiao J et al (2015) Auxin-regulated chromatin switch directs acquisition of flower primordium founder fate. eLife 4:e09269

Xu YZ, Thuraisingam T, Marino R et al (2011) Recruitment of SWI/SNF complex is required for transcriptional activation of the SLC11A1 gene during macrophage differentiation of HL-60 cells. J Biol Chem 286:12839–12849

Yang S, Li C, Zhao L et al (2015) The Arabidopsis SWI2/SNF2 chromatin remodeling ATPase BRAHMA targets directly to PINs and is required for root stem cell niche maintenance. Plant Cell 27:1670–1680

Yen K, Vinayachandran V, Batta K et al (2012) Genome-wide nucleosome specificity and directionality of chromatin remodelers. Cell 149:1461–1473

Yoo AS, Staahl BT, Chen L et al (2009) MicroRNA-mediated switching of chromatin-remodelling complexes in neural development. Nature 460:642–646

Yu Y, Chen Y, Kim B et al (2013) Olig2 targets chromatin remodelers to enhancers to initiate oligodendrocyte differentiation. Cell 152:248–261

Zemach A, Kim MY, Hsieh P-H et al (2013) The Arabidopsis nucleosome remodeler DDM1 allows DNA methyltransferases to access H1-containing heterochromatin. Cell 153:193–205

Zhao M, Yang S, Chen C-Y et al (2015) Arabidopsis BREVIPEDICELLUS interacts with the SWI2/SNF2 chromatin remodeling ATPase BRAHMA to regulate KNAT2 and KNAT6 expression in control of inflorescence architecture. PLoS Genet 11:e1005125

Zhou B-R, Jiang J, Wang R et al (2015) Structural mechanisms of nucleosome recognition by linker histones. J Biomol Struct Dyn 33(Suppl 1):2–3

Zhu Y, Rowley MJ, Böhmdorfer G et al (2013) A SWI/SNF chromatin-remodeling complex acts in noncoding RNA-mediated transcriptional silencing. Mol Cell 49:298–309

Abiotic Stress Induced Epigenetic Modifications in Plants: How Much Do We Know?

Sonali Bej and Jolly Basak

Abstract Epigenetics has evolved rapidly over the last two decades as a contemporary field of biology. In present day, it represents the heritable mitotic or even meiotic genetic change which does not alter the DNA sequence. Plants are considered as the masters of epigenetic regulation since they have the capability of rapid and reversible alteration of their epigenetic state and also maintaining a stable "memory" of it. Plants being sessile in nature are exposed to adverse environmental conditions which hampers their growth, development, productivity, and survival. They have developed intricate mechanisms at molecular level to withstand such stressful situations. Recent studies have documented the epigenetic control on stress-responsive mechanisms in response to various abiotic stresses. Several epigenetic mechanisms identified so far involve DNA methylation, histone modifications (acetylation, methylation, phosphorylation, ubiquitination, biotinylation, and sumoylation), chromatin remodeling, and small RNA (miRNA and siRNA) directed DNA methylation. Plants make wide use of DNA methylation as an epigenetic mark and undergo histone modifications to carry out transcriptional as well as posttranscriptional gene silencing programs. In this chapter, we have recapitulated the historical overview of the field of epigenetics followed by the various epigenetic mechanisms and lastly reviewed the studies related to various abiotic stress responses to understand the role of different epigenetic mechanisms in different plant species.

Keywords Epigenetics • Epigenetic mechanisms • DNA methylation • Histone modifications • Abiotic stress

Contents

S. Bej • J. Basak (✉)
Department of Biotechnology, Visva-Bharati, Santiniketan 731235, WB, India
e-mail: jolly.basak@visva-bharati.ac.in

N. Rajewsky et al. (eds.), *Plant Epigenetics*, RNA Technologies,
DOI 10.1007/978-3-319-55520-1_24

Abbreviations

ADA2	Alternation/deficiency in activation 2
AGO	ARGONAUTE
CAM	Crassulacean acid metabolism
CBF1	C-repeat/DRE binding factor1
CBP	CREB-binding protein
CLF	CURLY LEAF
CMT3	CHROMOMETHYLASE3
DCLs	Dicer-Like
DME	Demeter
DML	Demeter-Like
DRM	DOMAINS REARRANGED METHYLASE
dsRNAs	double-stranded RNAs
OsFIE1	*Orzya sativa* Fertilization-Independent Endosperm1
GCN5	General Control Non-derepressible5
GNAT	GCN5-related N-terminal acetyltransferase
HATs	histone acetyltransferases
HDACs	histone deacetylases
HDMs	histone demethylases
HKMTs	histone lysine methyltransferase
HOS15	High expression of osmotically responsive gene 15
HST1	HASTY
JmjC	Jumonji C
LSD1	Lysine-Specific Demethylase1
m^5C	5-methylcytosine
MEA	MEDEA
MET1	DNA METHYLTRANSFERASE1
miRNAs	microRNAs
MSAP	Methylation sensitive amplification polymorphism

ncRNAs	non-coding RNAs
NRPD1	NUCLEAR RNA POLYMERASE D1
NRPE1	Nuclear RNA Polymerase E1
pri-miRNAs	primary miRNAs
RdDM	RNA-directed DNA methylation
RDR	RNA-dependent RNA polymerases
RISC	RNA-induced silencing complex
ROS1	Repressor of Silencing1
SAHH1	S-ADENOSYL-L-HOMOCYSTEINE HYDROLASE1
SAM	S-adenosyl L-methionine
SDN	Small RNA Degrading Nuclease
siRNAs	small interfering RNAs
sRNAs	Small RNAs
ssRNAs	single stranded RNAs
SWN	SWINGER
$TAF_{II}250$	TATA binding protein-associated factors

1 Introduction

Epigenetics: The Composite Field

> "..... the tip of the iceberg is genomics....The bottom of the iceberg is epigenetics" (Weinhold 2006)

Around 75 years ago, a well-known British scientist Conard H. Waddington first introduced the term "Epigenetics" (Speybroeck 2006; Burbano 2006) merging the word "epigenesis" and "genetics." In "*The basic ideas of biology. In: Towards a Theoretical Biology,*" Waddington (1968) wrote: "Some years ago, I introduced the word 'epigenetics', derived from the Aristotelian word 'epigenesis', which had more or less passed into disuse, as a suitable name for the branch of biology which studies the causal interactions between genes and their products which bring the phenotype into being" (Waddington 1968). Since then the field of epigenetics has gained significant popularity. Over the last decade, the number of publications increased exponentially from 2500 in 2006 (Bird 2007) to 17,548 in 2016 (PubMed). This epigenetic revolution being addressed as the "next big thing" in biological sciences (Ebrahim 2012) unfortunately endured in its ambiguous definition. The word epigenetics is a polysemantic scientific term (Morange 2002), ambiguous due to the lack of a clarified definition resulting in misapplication in immensely different phenomena. The Waddingtonian view was based on developmental biology aspect of gene regulation and genotype–phenotype interactions while David Nanney and Boris Ephrussi accentuated on "steady states" and heritability of the expression (Deans and Maggert 2015). In contrary, Lederberg focused on the "epinucleic" concept (Deans and Maggert 2015). In "The dual origin of Epigenetics," Haig described how the divergence in views among scientists led to a definitional identity crisis that persists hitherto (Haig 2004; Deans and Maggert 2015). Robin Holliday's work on DNA methylation and cellular memory induced him to redefine epigenetics (Holliday 1994). He proposed two definitions encompassing all the known epigenetic processes. Although the individual definitions were incomplete, yet their combinations were highly significant in resolving most perplex phenomena (Holliday 1994). Holliday's redefinition was later delineated by Wu and Morris (2001) as "the study of changes in gene function that are mitotically and/or meiotically heritable and that do not entail change in DNA sequence."

"...not everything that is inherited is genetic"–Boris Ephrussi (1958)

In literal sense, epigenetics denotes "epi" meaning "upon or over" "genetics" i.e., "events over or beyond the gene" (Jablonka and Lamb 2002). According to Vanyushin (2014), "Epigenetics is the science of heritable properties of the organism that are not associated with changes in the DNA nucleotide sequence but can be directly encoded into the genome." In addition to the current definition, Deans and Maggert (2015) additionally emphasized epigenetic phenomena as entirely "chromosome-bound changes" to provide a more explicit definition. These heritable chromosome-bound changes include DNA methylation, histone modifications (acetylation, methylation, phosphorylation, ubiquitination, SUMOylation), and chromatin remodeling. The emerging role of non-coding RNAs (ncRNAs) has broadened our knowledge domain of epigenetics (Costa 2008). Currently, DNA methylation, histone modifications, and RNA-based mechanisms have been regarded as "the three pillars of epigenetics" (Grant-Downton and Dickinson 2005; Avramova 2011). In the last decade, epigenetics has been widely applied in diverse biological fields such as genetics, cancer genetics, aging, stem cell research, synthetic biology, evolution, species conservation, plant biotechnology, and agriculture. The advances in epigenetics research has expanded the boundaries of this field beyond expectations ranging from metabolic processes, stem cells, genomic imprinting, X chromosome inactivation to ncRNAs, prions, and polycomb mechanisms (Tollefsbol 2011). In this post-genomic era, epigenetics is believed to shed light in deciphering all the biological terra incognita phenomena.

2 The Pillars of Epigenetics

2.1 DNA Methylation

The first report of the discovery of an additional base, pyrimidine 5-methylcytosine (m^5C), was done by Johnson and Coghill in 1925 from the hydrolyzed products of tuberculinic acid of *Mycobacterium tuberculosis* (Johnson and Coghill 1925). Since then researchers tried to standardize precise techniques to detect the additional base from different organisms. Twenty-five years later, Wyatt found the occurrence of m^5C from DNA samples of animals and one higher plant by using simple chromatographic method (Wyatt 1950). Later studies discovered that the so-called notion of m^5C as "minor base" actually constitutes as high as 30% in plants (Vanyushin and Belozersky 1959; Grafi and Ohad 2013). In addition to m^5C, the existence of N6-methyladenine was also documented in mitochondrial DNA of many higher plants (Vanyushin et al. 1971; Buryanov et al. 1972).

DNA methylation mainly refers to cytosine methylation where an addition of a methyl group occurs at the 5th position of the cytosine ring. The enzymatic modification involves transfer of a methyl group from donor S-adenosyl L-methionine (SAM) while evidence of nonenzymatic DNA methylation has also been observed (Vanyushin 2014). In all organisms, cytosine methylation is majorly found in the CG dinucleotide; however in plants, the occurrence of both symmetric (CG and CHG) and asymmetric (CHH) (H stands for A, C, or T) DNA methylation has also been detected. The methylation status of *Arabidopsis* was found to be 24% in CG, 6.7% in CHG, and 1.7% in CHH sites (Dhar et al. 2014). CG and CHG methylation are simply replicated during DNA replication due to their symmetry, but CHH has to undergo de novo after each replication cycle. In plants, the enzymes required for methylation are classified into three categories: (1) DNA METHYLTRANSFERASE1 (MET1)—It is mainly involved in maintenance methylation of symmetric methylation sites (Sahu et al. 2013), (2) CHROMOMETHYLASE3 (CMT3)—Plant-specific CMT3 maintains methylation at asymmetric sites especially at centromeric repeats and transposons, and (3) DOMAINS REARRANGED METHYLASE (DRM) is involved in *de novo* methylation at asymmetric sites (Cao and Jacobsen 2002a). DRM is found to be of two types—DRM1 and DRM2. Studies have found the nonconventional role of MET1 and CMT3 in de novo methylation and role of DRM2 in maintenance methylation at symmetric sites (Cao and Jacobsen 2002b; Aufsatz et al. 2004). In addition, DRM2 was found to be involved in RNA-directed DNA methylation (RdDM) pathway with the help of small interfering RNAs (siRNAs) (Hauser et al. 2011). The global DNA methylation is regulated by events of DNA methylation and demethylation. Demethylation contributes to the reversible nature of methylation through either active or passive processes. In the passive process, loss of parental imprint or de novo methylation may be inhibited after replication (Sahu et al. 2013) while the active process is mediated by enzymes known as DNA glycosylases and AP lyases such as DEMETER (DME), Demeter-Like 2 (DML2), Demeter-Like 3 (DML3), and Repressor of Silencing1 (ROS1) (Grativol et al. 2012). It has been observed that methylation at the promoter region of gene inhibits transcription while gene body methylation may have either positive or negative gene expression. The less expressed genes are more methylated while most expressed genes are less methylated (Zemach et al. 2010). Ashapkin et al. (2002) reported that in DRM2 gene of *Arabidopsis*, not only the cytosine residues are methylated at CCGG sites but also adenines at N6 residues are methylated at GATC sites. Therefore it can be concluded that two different types of modifications coexist in plants, though it is believed that cytosine methylation might influence adenine methylation but their interaction still remains undiscovered (Vanyushin and Ashapkin 2011). Cytosine methylation plays significant role in plant development and survival by tuning up transcription regulation in certain developmental processes as well as events such as vernalization and stress adaptation.

2.2 *Histone Modifications*

Histone modifications are the most studied and variable epigenetic mechanisms. The nucleosome which is the structural unit of chromatin consists of octamer of core histone proteins (two copies each of H2A, H2B, H3, and H4). Histones are rich in basic amino acids lysine and arginine, which help to neutralize the negative charge of the DNA. The N-terminal region of histone protein, known as the histone tail, undergoes covalent modifications such as acetylation, methylation, phosphorylation, ubiquitination, biotinylation, and SUMOylation. Histone acetylation was first reported by Allfrey et al. in 1964. The acetylation and deacetylation of the histone tail are mediated by enzymes known as histone acetyltransferases (HATs) and histone deacetylases (HDACs), respectively (described in Table 1). The HATs consist of four different families in *Arabidopsis*. In many studies, HATs were found to play vital role in plant development and stress response (Luo et al. 2012). The histone deacetylases (HDACs) consist of three different families among which HD2 are plant-specific HDACs. HDACs lead to gene silencing through chromatin condensation by removal of the acetyl group of histone proteins (Chen and Tian 2007). The histone methylation and demethylation occurs at the lysine and arginine amino acid residues. In plants, the lysine methylation is catalyzed by histone lysine methyltransferases (HKMTs) containing SET domain. The SET domain is categorized into five different classes which have diverse functions in various plant development processes and stress responses. The histone demethylases (HDMs) are classified into two types, both are involved in oxidative demethylation reaction but have different cofactors and site of demethylation. Histone modifications such as acetylation, phosphorylation, and ubiquitination have been found to increase transcription (Qiao and Fan 2011). Transcriptional repression has been observed to be associated with SUMOylation and biotinylation (Qiao and Fan 2011). In general, stress-responsive gene activation is mediated by H3K4 and H3K36 methylation and H3K9 acetylation, while H3K9 and H3K27 dimethylation and H3 deacetylation mediate gene silencing (Qiao and Fan 2011; Luo et al. 2012).

2.3 *Small RNAs*

Small RNAs (sRNAs), regarded as the bioregulators of plant stress response, are ncRNAs of about 20–24 nucleotides (nt) which have recently emerged as the extended wing of the epigenetic regulation (Grativol et al. 2012). Small RNAs of plants are complex in nature and their classification is based on their biogenesis and genomic loci structure (Bej and Basak 2014). Among all the sRNAs, microRNAs

Table 1 Different enzyme families involved in histone modifications

Histone Acetylases (HATs)	Histone deacetylases (HDACs)	Histone lysine methyltransferases (HKMTs)	Histone demethylases (HDMs)
GNAT family (GCN5-related N-terminal acetyltransferase)	RPD3/HDA1 family	Class I CURLY LEAF (CLF) MEDEA (MEA) SWINGER (SWN)	Lysine-Specific Demethylase1 (KDM1/LSD1)
MYST family	SIR2 family	Class II (SET domain with AWS motif)	Jumonji C (JmjC)
CBP family (CREB-binding protein)	HD2 family	Class III ATX (Arabidopsis Trithorax-like) ATX1 to ATX5	
TAF$_{II}$250 family (TATA binding protein-associated factors)		Class IV ATXR5 and ATXR6	
		Class V SUVH (SU(VAR 3-9)	

(miRNAs) and siRNAs have crucial function in plant stress response. miRNAs mediate cleavage or translational repression of target mRNAs. In its biogenesis, the primary miRNAs (pri-miRNAs) transcribed from MIR genes are processed by the Dicer-Like (DCL1) protein, into miRNA/miRNA* duplex. On entering the cytoplasm one strand of the duplex gets degraded by Small RNA Degrading Nuclease (SDN). The mature miRNA forms a complex called RNA-induced silencing complex (RISCs) where ARGONAUTE (AGO) protein guides it to bind to the target mRNA (Bej and Basak 2014). The siRNA precursors are perfectly double-stranded RNAs (dsRNAs) that have generated by RNA-dependent RNA polymerases (RDRs) from inverted repeats or natural cis-antisense transcript pairs on conversion of single stranded to dsRNAs. siRNAs are generated by the cleavage of dsRNAs by DCL proteins, and are guided by AGO on RISC complex towards their target. The siRNAs mediate transcriptional and posttranscriptional gene silencing via the RdDM pathway (Khraiwesh et al. 2012).

3 RNA-Directed DNA Methylation Pathway

siRNAs are reported to be involved in methylation of about one-third of the methylated loci (Lister et al. 2008). The methylation of asymmetric sites is mediated by the RdDM pathway where RNA Pol II and RNA Pol IV produce substrate for siRNA biogenesis via the RNA interference pathway (Mirouze and Paszkowski 2011; Sahu et al. 2013). Plant-specific RNA Pol IV transcribes single stranded RNAs (ssRNAs) from methylated DNA loci, while RNA Pol II transcribes from inverted repeat sequences. dsRNAs are cleaved by DCL3 and loaded on AGO4 in

Table 2 Different types of chromatin remodeling factors

Chromatin Remodeling Factors (CRMs)	Functions	References
SWF (Switch)/SNF (sucrose non-fermenting) ATPases	Alteration of chromatin structure by use of ATP hydrolysis in regulation of transcription	Dhar et al. (2014), Pikaard and Scheid (2014)
CHD (chromodomain and helicase-like domain ATPases)	Important transcription regulators and have vital role in developmental processes	Dhar et al. (2014)
ISWI (Imitation Switch) ATPases	Central role in chromatin assembly	Dhar et al. (2014)

the RISC complex. The RISC complex undergoes interaction with RNA Pol V subunit at the Nuclear RNA polymerase E1 (NRPE1) which assists DRM2-mediated DNA methylation at asymmetric sites (Sahu et al. 2013).

4 Chromatin Modifications

In addition to the DNA and histone protein modification, several other processes namely the assembly/disassembly of chromatin, changes in the nucleosome occupancy, nucleosome composition and DNA–protein interactions also contribute in regulation of the transcription and other changes associated with nucleosomes and DNA (Pikaard and Scheid 2014). Chromatin remodeling factors are ATP-dependent protein complexes which mediate chromatin modifications through dissociation or relocation of nucleosomes. The chromatin remodeling ATPases can be categorized into three groups (as described in Table 2). SWF1/SNF complex was the first identified chromatin remodeling ATPase discovered in yeast for defects in mating type switching (SW1) and sucrose non-fermenting (SNF) (Sudarsanam and Winston 2000). Histone chaperones are associated in histone deposition and removal during nucleosome assembly and disassembly, respectively.

5 Abiotic Stress Directed Epigenetic Changes

Stressful conditions occur due to living (biotic) and nonliving (abiotic) factors. The disastrous effect of stressful conditions takes place due to the interactions between multiple stresses since the occurrence of an isolated stress event is usually rare. Plants being sessile autotrophs have evolved complex mechanisms to acclimatize with the changing environmental conditions. These mechanisms include interwoven genetic networks and transcriptional epigenetic regulations that are involved in both immediate and long-term stress response. The rapidness and reversibility of the epigenetic mechanisms contribute to the flexile regulation of plant stress

Table 3 Genes and transposons that are involved in DNA methylation and histone modification under different abiotic stresses

Components	Stress	Species	Function	References
DNA methylation				
ZmMI1	Cold stress	Maize	Stress-induced non-reversible demethylation	Steward et al. (2000)
Ac/Ds	Cold stress	Maize	Demethylation of transposon Ac/Ds	Steward et al. (2002)
Tam 3	Low temp	*Antirrhinum majus*	Decrease in methylation	Hashida et al. (2006)
NtGPDL	Aluminum, low temp, salt stress	Tobacco	Demethylation at coding region of gene	Choi and Sano (2007)
HRS60 and GRS	Salt, osmotic stress	Tobacco	Reversible DNA hypermethylation	Kovarˇik et al. (1997)
Histone modifications				
AtGCN5	Cold stress	*Arabidopsis*	Affect expression of COR genes	Stockinger et al. (2001), Vlachonasios et al. (2003)
Ada2b	Freezing, salt stress	*Arabidopsis*	Induces COR genes	Vlachonasios et al. (2003)
SKB1	Salt stress	*Arabidopsis*	Trimethylation of H4K3	Zhang et al. (2011)
ABO1/ ELO1	Drought stress	*Arabidopsis*	Drought tolerance	Chen et al. (2006)
ADH1 and PDC1	Submergence stress	Rice	Histone modifications of H3	Tsuji et al. (2006)
HD6	Freezing stress	*Arabidopsis*	Upregulation confer tolerance	To et al. (2011)
HOS15	Cold stress	*Arabidopsis*	Deacetylation of histone H4	Zhu et al. (2008)

response. The different genes and transposons that are discovered so far in relation to various abiotic stresses in different plant species have been summarized in Table 3. In addition to the various regulatory mechanisms to combat different stresses and to adapt to the changing environmental conditions, plants also have the ability to memorize the stress and even transmit it to the next generation.

5.1 Stress Memory

In recent times, the transgenerational inheritance of epigenetic mechanisms has become highlighted in the field of biological research which has revived the concept of Lamarckian theory of inheritance of acquired traits. In plants, certain studies have reported inheritance (both maternal and paternal) of stress-induced epigenetic changes. Stress memory are sometimes heritable through mitotic or meiotic divisions. The epigenetic modifications which are stable and irreversible are found to be inherited. Molinier et al. (2006) showed that UV-C radiation and flagellin treatment increased the frequency of the somatic homologous recombination. This increased recombination frequency was inherited by the progenies of the stress exposed plants (Chinnusamy and Zhu 2009). Feng et al. (2012) observed that the salt and alkaline stress induced DNA methylation alteration was transmitted in selfed progenies in rice. These studies suggested that the transgenerational inheritance of the epigenetic modifications contributes to the immediate and long-term stress response (Sahu et al. 2013). The best known phenomenon of stress memory system is the "defense priming" (Kinoshita and Seki 2014). In defense priming, plant memorizes the first attack of the pathogen and thereby exerts a much powerful defense during the second attack of the same pathogen. In case of abiotic stress such as drought and heat stress, such priming has also been observed. Keeping in mind the first phase of abiotic stress, plants exert strong response during the recurrence of the stress. In *Arabidopsis,* frequent dehydration stress showed upregulation of stress-responsive trainable genes. On stress removal, the trainable genes were found at basal levels, but were found to be associated with high levels of H3K4me3 and Ser5P polymerase II, which suggests stalled RNA polymerase II (Ding et al. 2012). It was the first study on stalled RNA pol II and its role in plant transcriptional memory.

6 Abiotic Stressors

6.1 Salt Stress

Salinity is the most prevalent abiotic stress that affects about 20% of the arable areas worldwide (Shrivastava and Kumar 2015). High salinity gives rise to drought-like conditions. It severely affects germination and growth, causing huge loss in crop yield and productivity. Genome-wide changes of DNA methylation in several plant species have been documented in response to various abiotic stresses (Kim et al. 2015). Methylation sensitive amplification polymorphism (MSAP) study on mangrove plants found that the riverside plants were comparatively hypermethylated than the plants grown in the salt marsh (Lira-Medeiros et al. 2010; Sahu et al. 2013). These findings suggest natural occurrence of epigenetic adaptations according to the habitat. Salinity induces water deficiency in some Crassulacean acid metabolism (CAM) plants such as the facultative halophyte *Mesembryanthemum crystallinum* which switches from the C3 to CAM pathway. This mechanism is driven by genomic methylation and specific CHG hypermethylation of the satellite DNA which results in formation of specialized chromatin structure for regulation of genes involved in the switchover. No change in the methylation pattern of the phosphoenolpyruvate carboxylase gene, key gene in the C4-CAM switchover, further establishes the role of the specialized chromatin structure (Dyachenko et al. 2006; Peng and Zhang 2009). Many global methylation studies based on MSAP technique have showed differential methylation patterns in salt-tolerant and salt-sensitive genotypes. Karan et al. (2012) studied salinity-induced methylation pattern in different rice genotypes and identified the differential expression of retrotransposons, abiotic stress responsive genes, and chromatin modifier genes. Another study on rice under both salinity and drought stress revealed the role of DRMs in differential expression of stress-responsive genes. Apart from it, sRNAs were found to be associated with hypermethylated regions (Garg et al. 2015). Wang et al. (2011a) reported demethylation of genes as an active epigenetic response in roots in response to salinity stress in rice. In tobacco, demethylation of coding region of glycerophosphodiesterase-like protein (NtGPDL) gene was observed in response to aluminum stress, low temperature, and salt stresses. Under normal conditions, the NtGPDL gene was found to be hypermethylated resulting in repression of the gene, although the gene promoter region was unaffected in both conditions (Choi and Sano 2007).

Apart from DNA methylation, the role of histone modifications in response to salinity stress has been well established. Histone acetylation, methylation, and phosphorylation have been found to be correlated with salt stress. The histone modifier enzymes histone acetyltransferases (HATs) and histone deacetylases (HDACs) have antagonistic roles in histone acetylation mechanisms. Li et al. (2014) showed that under salt stress the upregulation of ZmEXPB2 and ZmXET1 genes in maize was associated with H3K9 acetylation of the promoter and coding regions. The acetylation-induced upregulation of these genes was associated with two HAT genes—HATB and General Control Non-derepressible5 (GCN5). GCN5 is the main catalytic subunit while ADA2 (alternation/deficiency in activation 2) and SGF29 (SAGA-associated factor 29) are adaptor proteins of the GCN5-containing HAT complex (Luo et al. 2012). In a study on *Arabidopsis*, ada2b mutants were found to be hypersensitive whereas sgf29a mutants were found to be more resistant than the wild-type under salt stress (Kaldis et al. 2011) which indicates contrasting roles of the members of HAT complex. In Arabidopsis, skb1 mutant showed hypersensitivity towards salt stress. SKB1 is an arginine methyltransferase which mediates increase in trimethylation of H4R2 under normal conditions. Salt stress induces the dissociation of SKB1 from chromatin which causes activation of stress-responsive genes (Zhang et al. 2011). Sokol et al. (2007) showed that the activation of histone H3 Ser-10 phosphorylation, H3 phosphoacetylation, and H4 acetylation induces stress-specific genes in response to salinity and cold stress in *Arabidopsis* and tobacco.

6.2 Drought Stress

Drought stress or water deficient conditions occur when there is reduced availability of water levels in soil accompanied by heat and other climatic conditions. In rice genome, drought stress induces site-specific, flexible, and reversible changes in DNA methylation (Wang et al. 2011b). Labra et al. (2002) detected hypermethylation in the root tip of *Pisum sativum* which was especially specific to the inner cytosine of CCGG motif induced by water deficient conditions. Suji and Joel (2010) showed a change in methylation level among the rice cultivars under water stress and control conditions. They studied two high yielding lowland rice cultivars and two rice cultivars with drought-tolerant traits. Genomic DNA from the four cultivars was subjected to restriction digestion with methylation-sensitive isoschizomers MspI and HpaII. In all the cultivars, internal methylation (5′-CmCGG-3′) was found to be dominant, suggesting a high frequency of mCpG dinucleotide as compared to mCpC dinucleotide in the 5′-CCGG-3′ sequence. Drought susceptible cultivar under stress showed higher MspI and HpaII digestion than irrigated control, suggesting that demethylation has occurred under stress thereby altering the genome activity. In case of drought-tolerant cultivars, HpaII digestion was found to be lesser under stress conditions than irrigated control, suggesting that methylation has occurred under stress thereby altering the gene expression.

The intensity of drought is directly proportional towards the expression of drought-responsive genes (Kim et al. 2015). Kim et al. (2008) studied the histone modification status and nucleosome structure under drought stress. They determined the temporal and spatial changes in nucleosome occupancy and levels of H3K4me3, H3K9ac, H3K14ac, H3K23ac, and H3K27ac in the histone H3 N-tail on the regions of four *Arabidopsis* drought stress-inducible genes, RD29A, RD29B, RD20, and RAP2.4 under drought stress conditions. They showed that severe drought conditions cause more H3K4 trimethylation and H3K9 acetylation at RD20 and RD29A genes compared to moderate stress (Kim et al. 2008). The loss of nucleosome occurred at the RD29A gene region under severe drought whereas moderate drought caused less loss of nucleosome (Kim et al. 2015). On recovery from the stress, H3K9 acetylation on RD29A and RD20 decreased vastly (Kim et al. 2015). Studies have revealed the role of elongator HAT complex in ABA signaling, salt and drought stresses. In *Arabidopsis*, ABO1/ELO2 (ABA overly sensitive 1) was reported in drought resistant mutants. abo1 mutants showed hypersensitivity towards ABA and increase in drought resistance (Chen et al. 2006).

6.3 Heat Stress

Global warming has resulted in frequent rise of extreme temperature events worldwide. High temperature together with drought causes huge loss in crop productivity. It has been predicted that every 1 °C increase in temperature will reduce the yields of major crops like wheat, rice, maize, soybeans, barley, and sorghum by 0.6∼8.9% (Lobell and Field 2007). Many studies have found correlation between DNA methylation and heat stress although its role has not yet been known in crops. Gao et al. (2014) studied the methylation patterns between two rapeseed genotypes (heat-sensitive and heat-tolerant) in response to heat stress. They found more DNA methylation in the heat-sensitive genotype while more DNA demethylation in the heat-tolerant genotype (Gao et al. 2014). Correia et al. (2013) for the first time reported on epigenetic regulation of heat tolerance in forest trees. The study on Cork oak trees reported the interplay between DNA methylation and H3 acetylation of Cork oak trees under 55 °C temperature which might lead to its adaptation and survival. In *Arabidopsis*, heat induced the upregulation of DRM2, NRPD1 (NUCLEAR RNA POLYMERASE D1), and NRPE1 resulting in the increase of genomic methylation (Naydenov et al. 2015). In contrast, genomic hypomethylation was observed in cotton anthers resulting from downregulation of S-ADENOSYL-L-HOMOCYSTEINE HYDROLASE1 (SAHH1), DRM1, and DRM3 (Min et al. 2014). Folsom et al. (2014) reported that DNA methylation and repressive histone modification of Fertilization-Independent Endosperm1 (OsFIE1) was temperature sensitive in rice.

Acetylation and SUMOylation are known to be involved in heat stress response. The small ubiquitin-related modifier (SUMO) is posttranscriptional chromatin modifiers which is reversible in nature. H2A.Z deposition is believed to play vital role in response to thermal stress through nucleosome occupancy in *Arabidopsis* (Kumar and Wigge 2010). Change in temperature may cause replacement of core histone protein H2A in the nucleosome by H2A.Z with the help of SWR1 complex subunit encoded by ARP6 protein (March-Díaz and Reyes 2009). Boden et al. (2013) found similar role of H2A.Z deposition in *Brachypodium* under heat stress. Other studies on heat stress response have found the role of H3K56 acetylation mediated by AtASF1A/B chaperone and nucleosome loss resulting in RNA Pol II accumulation and transcription factors activation (Weng et al. 2014). SUMOylation is also observed to be an important phenomenon under thermal stress. Under heat stress, SUMOylation of H2B was found to be reduced while increased in case of GCN5 (Miller et al. 2010; Kim et al. 2015).

6.4 Submergence Stress

On one hand the world is now facing the adversities from global temperature rise resulting in severe drought while on the other hand, the sudden frequent event of floods has turned disastrous throughout the globe. Submergence causes combination of stresses such as nutrient deficiency, hypoxia, infections, and low light leading to negative effects on the yield and survival of the plant. In rice, the increased expression of submergence-induced genes—alcohol dehydrogenase 1 (ADH1) and pyruvate decarboxylase 1 (PDC1)—was found to be associated with reversible histone modifications of H3 acetylation and H3K4 trimethylation from H3K4 dimethylation (Tsuji et al. 2006).

6.5 Cold Stress

Cold stress is of two types: (1) Chilling (Temperature range: 0–18 °C) and (2) Freezing (Temperature below 0 °C) (Song et al. 2015). Cold stress is also considered as one of the major abiotic stress which restricts plant growth, productivity, and its survival. Song et al. (2015) studied an alpine subnival plant *Chorispora bungeana* to reveal the epigenetic regulation that makes it tolerant to such fluctuating chilling and freezing temperature. Based on MS-AFLP technique, the study established the frequent and flexible change in DNA modifications (methylation and demethylation) as a regulatory mechanism for adaptation of such alpine species towards the cold habitat. The demethylation event was reported more frequently under cold stress in different studies on diverse plant species (Song et al. 2015; Fan et al. 2013; Shan et al. 2013) whereas methylation event was found to be associated with transposable elements. Transposons are found to be hypermethylated in plants but under stressed conditions, they are activated via DNA demethylation which suggests that such true epigenetic changes might contribute towards better stress adaptation (Shan et al. 2013). In Maize, a 1.8 kb fragment called ZmMI1 was identified from cold stressed roots whose sequence constitutes a partial putative protein coding region and retrotransposon-like sequence (Steward et al. 2002). Cold stress induced organ-specific as well as site-specific demethylation, which was only found in roots and in the Ac/Ds transposons. The demethylation caused by downregulation of MET1 resulted in the activation of Ac/Ds transposons (Steward et al. 2000). Hashida et al. (2006) observed hypomethylation of CHH sites mediated by Tam3 transposon in *Antirrhinum majus* under low temperature. At low temperature, Tam3 undergoes low temperature dependent transposition (LTDT) mediated by DNA demethylation while under high temperature, DNA hypermethylation induces the suppression of the transposition (Qiao and Fan 2011).

Arabidopsis HATs, ADA2 and GCN5 have been found to interact with CBF1 (C-repeat/DRE binding factor 1) transcription factor to induce the COR (cold-regulated) genes. *ada2b* and *gcn5* mutants were found to have delayed induction of COR genes under low temperature (Kim et al. 2015; Vlachonasios et al. 2003). However, the non-acclimated *ada2b* mutants showed increase in freezing tolerance which suggests repression of tolerance mechanism by ADA2b caused by H3 acetylation and decrease in nucleosome occupancy at COR promoters. In *Arabidopsis*, freezing stress induced the upregulation of HDA6 which was confirmed to confer tolerance to freezing stress (To et al. 2011). Cold stress induced upregulation of HDACs was also reported to mediate H3 and H4 global deacetylation in maize (Hu et al. 2011). Zhu et al. (2008) showed that HOS15 (high expression of osmotically responsive gene 15) was involved in deacetylation of histone H4, which is crucial for repression of cold stress induced genes such as CBFs, DREBs, CORs, etc. hos15 mutants were found to be sensitive to freezing stress and were also found to have RD29A transcripts accumulation.

6.6 *Heavy Metal Stress*

Heavy metal concentration in soil depends either from natural sources such as volcanoes, weathering of rocks or from anthropogenic activities such as industrial pollution, mining, agricultural chemicals, and so on. Some heavy metals are vital trace elements (Cu^{2+}, Cr^{3+}) required for plant growth and development. However, toxic concentrations might result in cellular stress and damage which contributes to abiotic stresses to plants. On exposure to high concentrations of active heavy metals (Fe, Cu, Cd, etc.), they suffer from oxidative injury (Schützendübel and Polle 2002). Studies have proved the substantive role of hypermethylation as a defense mechanism to resist heavy metal stress. In a study, the effects of Cd, Ni, and Cr on sensitive species—Clover (*Trifolium repens* L.) and metal-tolerant species—Hemp (*Cannabis sativa* L.) have found the methylation level of hemp is three times higher that of clover. Treatment with heavy metals caused dose-dependent hypomethylation in both clover and hemp, which suggests natural methylation level might contribute to the stress tolerance in hemp (Aina et al. 2004). Studies on cadmium stress showed that the increase in methylation level is directly proportional to the dosage of stress in both radish (Yang et al. 2007) and oilseed rape (Filek et al. 2008). Ou et al. (2012) reported occurrence of specific CHG hypomethylation in rice seedlings exposed to heavy metals such as Cu^{2+}, Cd^{2+}, Cr^{3+}, and Hg^{2+}. Although most studies are based on the genome-wide patterns of DNA methylation, a limited number of studies are successful in identification of modifications of the stress-induced gene loci. Choi and Sano (2007) detected aluminum stress induced CG demethylation event in the coding region of NtGPDL soon after onset of the stress.

7 Conclusion and Future Prospectus

"The epigenetic revolution is underway"—Nessa Carrey (2012)

From the inception of agriculture till today, abiotic stresses remained a challenge to the natural environment and agriculture. Presently, with the increasing world population and decreasing arable land, we are challenged to produce more in less area under declining resources of water and global climate change severely affecting crop productivity. The challenge to engineer high yielding environmentally stable crops can be met by integrating all information about the mechanism of abiotic stress response pathways.

As stated by Nessa Carrey in her book "The Epigenetics Revolution," "*In the 21st century it is the new scientific discipline of epigenetics that is unraveling so much of what we took as dogma and rebuilding it in an infinitely more varied, more complex, and even more beautiful fashion.*" During the last decade, epigenetics has gained global attention as a vital regulator of biological mechanisms under abiotic stresses in plants. Stress-induced epigenetic changes have been evidenced in plants that play important role in their acclimatization to the changing environmental conditions. Although the epigenetic regulation of plant abiotic stress response is complex in nature, yet more in-depth studies will increase our understanding of the stress tolerance mechanisms. Analyzing the role of DNA methylation/demethylation, histone modification, small RNAs and chromatin regulation in the long-term adaptation of plants to abiotic stresses and answering the question how plants can maintain a stress memory without a nervous system will open up new avenues in crop improvement.

Acknowledgment Authors are thankful to the University Grants Commission and Visva-Bharati for providing all the facilities.

References

Aina R, Sgorbati S, Santagostino A et al (2004) Specific hypomethylation of DNA is induced by heavy metals in white clover and industrial hemp. Physiol Plant 121(3):472–480

Allfrey VG, Faulkner R, Mirsky AE (1964) Acetylation and methylation of histones and their possible role in the regulation of RNA synthesis. Proc Natl Acad Sci U S A 51:786–794

Ashapkin VV, Kutueva LI, Vanyushin BF (2002) The gene for domains rearranged methyltransferase (DRM2) in Arabidopsis thaliana plants is methylated at both cytosine and adenine residues. FEBS Lett 532:367–372

Aufsatz W, Mette MF, Matzke AJ et al (2004) The role of MET1 in RNA-directed de novo and maintenance methylation of CG dinucleotides. Plant Mol Biol 54:793–804

Avramova Z (2011) Epigenetic regulatory mechanisms in plants. In: Handbook of epigenetics: the new molecular and medical genetics, pp 251–278. doi:10.1016/B978-0-12-375709-8.00016-2

Bej S, Basak J (2014) MicroRNAs: the potential biomarkers in plant stress response. Am J Plant Sci 5:748–759

Bird A (2007) Perceptions of epigenetics. Nature 447:396–398

Boden SA, Kavanová M, Finnegan EJ et al (2013) Thermal stress effects on grain yield in Brachypodium distachyon occur via H2A.Z- nucleosomes. Genome Biol 14:R65

Burbano HA (2006) Epigenetics and genetic determinism. História, Ciências, Saúde—Manguinhos, Rio de Janeiro 13:851–863

Buryanov YI, Eroshina NV, Vagabova LM et al (1972) On the detection of 6-methylaminopurine in DNA of higher plant pollen. Dokl Akad Nauk SSSR 206:992–994

Cao X, Jacobsen SE (2002a) Role of the Arabidopsis DRM methyltransferases in de novo DNA methylation and gene silencing. Curr Biol 12:1138–1144

Cao X, Jacobsen SE (2002b) Locus-specific control of asymmetric and CpNpG methylation by the DRM and CMT3 methyltransferase genes. Proc Natl Acad Sci U S A 99:16491–16498

Carey N (2012) The epigenetics revolution: how modern biology is rewriting our understanding of genetics, disease and inheritance. Icon Books, London

Chen ZJ, Tian L (2007) Roles of dynamic and reversible histone acetylation in plant development and polyploidy. Biochim Biophys Acta 1769:295–307

Chen Z, Zhang H, Jablonowski D et al (2006) Mutations in ABO1/ELO2, a subunit of holo-Elongator, increase abscisic acid sensitivity and drought tolerance in Arabidopsis thaliana. Mol Cell Biol 26:6902–6912

Chinnusamy V, Zhu J-K (2009) Epigenetic regulation of stress responses in plants. Curr Opin Plant Biol 12:1–7

Choi CS, Sano H (2007) Abiotic-stress induces demethylation and transcriptional activation of a gene encoding a glycerophosphodiesterase-like protein in tobacco plants. Mol Genet Genomics 277:589–600

Correia B, Valledor L, Meijon M et al (2013) Is the interplay between epigenetic markers related to the acclimation of cork oak plants to high temperatures? PLoS ONE 8:e53543

Costa F (2008) Non-coding RNAs, epigenetics and complexity. Gene 410:9–17

Deans C, Maggert KA (2015) What do you mean, "Epigenetic"? Genetics 199:887–896

Dhar MK, Vishal P, Sharma R et al (2014) Epigenetic dynamics: role of epimarks and underlying machinery in plants exposed to abiotic stress. Int J Genomics 2014:187146

Ding Y, From M, Avramova Z (2012) Multiple exposures to drought 'train' transcriptional responses in Arabidopsis. Nat Commun 3:740

Dyachenko O, Zakharchenko N, Shevchuk T et al (2006) Effect of hypermethylation of CCWGG sequences in DNA of Mesembryanthemum crystallinum plants on their adaptation to salt stress. Biochemistry (Moscow) 71:461–465

Ebrahim S (2012) Epigenetics: the next big thing. Int J Epidemiol 41:1–3

Ephrussi B (1958) The cytoplasm and somatic cell variation. J Cell Comp Physiol 52:35–52

Fan HH, Wei J, Li TH et al (2013) DNA methylation alterations of upland cotton (Gossypium hirsutum) in response to cold stress. Acta Physiolgiae Plantarum 35:2445–2453

Feng Q, Yang C, Lin X et al (2012) Salt and alkaline stress induced transgenerational alteration in DNA methylation of rice (Oryza sativa). Aust J Crop Sci 6:877–883

Filek M, Keskinen R, Hartikainen H et al (2008) The protective role of selenium in rape seedlings subjected to cadmium stress. J Plant Physiol 165:833–844

Folsom JJ, Begcy K, Hao X et al (2014) Rice fertilization-independent endosperm1 regulates seed size under heat stress by controlling early endosperm development. Plant Physiol 165:238–248

Gao G, Li J, Li H et al (2014) Comparison of the heat stress induced variations in DNA methylation between heat-tolerant and heat-sensitive rapeseed seedlings. Breed Sci 64:125–133

Garg R, Chevala VVSN, Shankar R et al (2015) Divergent DNA methylation patterns associated with gene expression in rice cultivars with contrasting drought and salinity stress response. Sci Rep 5:14922

Grafi G, Ohad N (eds) (2013) Plant epigenetics: a historical perspective. In: Epigenetic memory and control in plants, signaling and communication in plants. Springer, Berlin, pp 1–19

Grant-Downton RT, Dickinson HG (2005) Epigenetics and its implications for plant biology. The epigenetic network in plants. Ann Bot (Lond) 96:1143–1164

Grativol C, Hemerly AS, Ferreira PCG (2012) Genetic and epigenetic regulation of stress responses in natural plant populations. Biochim Biophys Acta 1819:176–185

Haig D (2004) The (dual) origin of epigenetics. Cold Spring Harb Symp Quant Biol 69:67–70

Hashida SN, Uchiyama T, Martin C et al (2006) The temperature dependent change in methylation of the Antirrhinum transposon Tam3 is controlled by the activity of its transposase. Plant Cell 18:104–118

Hauser M-T, Aufsatz W, Jonak C et al (2011) Transgenerational epigenetic inheritance in plants. Biochim Biophys Acta 1809:459–468

Holliday R (1994) Epigenetics: an overview. Dev Genet 15:453–457

Hu Y, Zhang L, Zhao L et al (2011) Trichostatin A selectively suppresses the cold-induced transcription of the ZmDREB1 gene in maize. PLoSONE 6:e22132

Jablonka EVA, Lamb MJ (2002) The changing concept of epigenetics. Ann NY Acad Sci 981:82–96

Johnson TB, Coghill RD (1925) Researches on pyrimidines. C111. The discovery of 5-methyl cytosine in tuberculinic acid, the nucleic acid of the Tubercle Bacillus. J Am Chem Soc 47:2838–2844

Kaldis A, Tsementzi D, Tanriverdi O et al (2011) Arabidopsis thaliana transcriptional co-activators ADA2b and SGF29a are implicated in salt stress responses. Planta 233:749–762

Karan R, DeLeon T, Biradar H et al (2012) Salt stress induced variation in DNA methylation pattern and its influence on gene expression in contrasting rice genotypes. PLoS ONE 7:e40203

Khraiwesh B, Zhu J-K, Zhu J (2012) Role of miRNAs and siRNAs in biotic and abiotic stress responses of plants. Biochim Biophys Acta 1819:137–148

Kim J-M, To TK, Ishida J et al (2008) Alterations of lysine modifications on the histone H3 N-tail under drought stress conditions in *Arabidopsis thaliana*. Plant Cell Physiol 49:1580–1588

Kim JM, Sasaki T, Ueda M et al (2015) Chromatin changes in response to drought, salinity, heat, and cold stresses in plants. Front Plant Sci 6:114

Kinoshita T, Seki M (2014) Epigenetic memory for stress response and adaptation in plants. Plant Cell Physiol 55:1859–1863

Kovar˘ik A, Koukalová B, Bezde˘k M et al (1997) Hypermethylation of tobacco heterochromatic loci in response to osmotic stress. Theor Appl Genet 95:301–306

Kumar SV, Wigge PA (2010) H2A.Z-containing nucleosomes mediate the thermosensory response in Arabidopsis. Cell 140:136–147

Labra M, Ghiani A, Citterio S et al (2002) Analysis of cytosine methylation pattern in response to water deficit in pea root tips. Plant Biol 4:694–699

Li H, Yan S, Zhao L et al (2014) Histone acetylation associated up-regulation of the cell wall related genes is involved in salt stress induced maize root swelling. BMC Plant Biol 14:105

Lira-Medeiros CF, Parisod C, Fernandes RA et al (2010) Epigenetic variation in mangrove plants occurring in contrasting natural environment. PLoS ONE 5:e10326

Lister R, O'Malley RC, Tonti-Filippini J et al (2008) Highly integrated single-base resolution maps of the epigenome in Arabidopsis. Cell 133:523–536

Lobell DB, Field CB (2007) Global scale climate-crop yield relationships and the impacts of recent warming. Environ Res Lett 2:1–7

Luo M, Liu X, Singh P et al (2012) Chromatin modifications and remodeling in plant abiotic stress responses. Biochim Biophys Acta 1819:129–136

March-Díaz R, Reyes JC (2009) The beauty of being a variant: H2A.Z and the SWR1 complex in plants. Mol Plant 2:565–577

Miller MJ, Barrett-Wilt GA, Hua Z et al (2010) Proteomic analyses identify a diverse array of nuclear processes affected by small ubiquitin-like modifier conjugation in Arabidopsis. Proc Natl Acad Sci U S A 107:16512–16517

Min L, Li Y, Hu Q et al (2014) Sugar and auxin signaling pathways respond to high-temperature stress during anther development as revealed by transcript profiling analysis in cotton. Plant Physiol 164:1293–1308

Mirouze M, Paszkowski J (2011) Epigenetic contribution to stress adaptation in plants. Curr Opin Plant Biol 14:267–274

Molinier J, Ries G, Zipfel C, Hohn B (2006) Transgeneration memory of stress in plants. Nature 442:1046–1049

Morange M (2002) The relations between genetics and epigenetics: a historical point of view. Ann NY Acad Sci 981:50–60

Naydenov M, Baev V, Apostolova E et al (2015) High-temperature effect on genes engaged in DNA methylation and affected by DNA methylation in Arabidopsis. Plant Physiol Biochem 87:102–108

Ou X, Zhang Y, Xu C et al (2012) Transgenerational inheritance of modified DNA methylation patterns and enhanced tolerance induced by heavy metal stress in rice (Oryza sativa L.) PLoS ONE 7:e41143

Peng H, Zhang J (2009) Plant genomic DNA methylation in response to stresses: Potential applications and challenges in plant breeding. Prog Nat Sci 19:1037–1045

Pikaard CS, Scheid OM (2014) Epigenetic regulation in plants. Cold Spring Harb Perspect Biol 6: a019315

Qiao W, Fan L (2011) Epigenetics, a mode for plants to respond to abiotic stresses. Front Biol 6:477–481

Sahu PP, Pandey G, Sharma N et al (2013) Epigenetic mechanisms of plant stress responses and adaptation. Plant Cell Rep 32:1151–1159

Schützendübel A, Polle A (2002) Plant responses to abiotic stresses: heavy metal-induced oxidative stress and protection by mycorrhization. J Exp Bot 53:1351–1365

Shan XH, Wang XY, Yang G et al (2013) Analysis of the DNA methylation of maize (Zea mays L.) in response to cold stress based on methylation-sensitive amplified polymorphisms. J Plant Biol 56:32–38

Shrivastava P, Kumar R (2015) Soil salinity: a serious environmental issue and plant growth promoting bacteria as one of the tools for its alleviation. Saudi J Biol Sci 22:123–131

Sokol A, Kwiatkowska A, Jerzmanowski A et al (2007) Up-regulation of stress-inducible genes in tobacco and Arabidopsis cells in response to abiotic stresses and ABA treatment correlates with dynamic changes in histone H3 and H4 modifications. Planta 227:245–254

Song Y, Liu L, Feng Y et al (2015) Chilling- and freezing- induced alterations in cytosine methylation and its association with the cold tolerance of an alpine subnival plant, Chorispora bungeana. PLoS ONE 10:e0135485

Speybroeck L (2006) From epigenesis to epigenetics. Ann NY Acad Sci 981:61–81

Steward N, Kusano T, Sano H (2000) Expression of ZmMET1, a gene encoding a DNA methyltransferase from maize, is associated not only with DNA replication in actively proliferating cells, but also with altered DNA methylation status in cold-stressed quiescent cells. Nucleic Acids Res 28:3250–3259

Steward N, Ito M, Yamaguchi Y et al (2002) Periodic DNA methylation in maize nucleosomes and demethylation by environmental stress. J Biol Chem 277:37741–37746

Stockinger EJ, Mao Y, Regier MK et al (2001) Transcriptional adaptor and histone acetyltransferase proteins in Arabidopsis and their interactions with CBF1, a transcriptional activator involved in coldregulated gene expression. Nucleic Acids Res 29:1524–1533

Sudarsanam P, Winston F (2000) The Swi/Snf family: nucleosome-remodeling complexes and transcriptional control. Trends Genet 16:345–351

Suji KK, Joel AJ (2010) An epigenetic change in rice cultivars under water stress conditions. Elect J Plant Breed 1:1142–1143

To TK, Nakaminami K, Kim JM et al (2011) Arabidopsis HDA6 is required for freezing tolerance. Biochem Biophys Res Commun 406:414–419

Tollefsbol TO (2011) Epigenetics: the new science of genetics, Chapter 1. In: Handbook of epigenetics: the new molecular and medical genetics. Academic Press, San Diego, pp 1–6

Tsuji H, Saika H, Tsutsumi N et al (2006) Dynamic and reversible changes in histone H3-Lys4 methylation and H3 acetylation occurring at submergence-inducible genes in rice. Plant Cell Physiol 47:995–1003

Vanyushin BF (2014) Epigenetics today and tomorrow. Russ J Genet Appl Res 4(3):168–188

Vanyushin BF, Ashapkin VV (2011) DNA methylation in higher plants: past, present and future. Biochim Biophys Acta 1809:360–368

Vanyushin BF, Belozersky AN (1959) Nucleotide composition of deoxyribonucleic acid in higher plants. Dokl Akad Nauk SSSR 129:944–946

Vanyushin BF, Kadyrova DK, Karimov KK et al (1971) Minor bases in the DNA of higher plants. Biochemistry (Moscow) 36:1251–1258

Vlachonasios KE, Thomashow MF, Triezenberg SJ (2003) Disruption mutations of ADA2b and GCN5 transcriptional adaptor genes dramatically affect Arabidopsis growth, development, and gene expression. Plant Cell 15:626–638

Waddington CH (1968) The basic ideas of biology. In: Waddington CH (ed) Towards a theoretical biology, Prolegomena, vol 1. Edinburgh University Press, Edinburgh, pp 1–32

Wang W, Zhao X, Pan Y et al (2011a) DNA methylation changes detected by methylation-sensitive amplified polymorphism in two contrasting rice genotypes under salt stress. J Genet Genomics 38:419e424

Wang WS, Pan YJ, Zhao XQ et al (2011b) Drought-induced site-specific DNA methylation and its association with drought tolerance in rice (Oryza sativa L.) J Exp Bot 62:1951–1960

Weinhold B (2006) Epigenetics: the science of change. Environ Health Perspect 114:160–167
Weng M, Yang Y, Feng H et al (2014) Histone chaperone ASF1 is involved in gene transcription activation in response to heat stress in Arabidopsis thaliana. Plant Cell Environ 37:2128–2138
Wu CT, Morris JR (2001) Genes, genetics, and epigenetics: a correspondence. Science 293:1103–1105
Wyatt GR (1950) Occurrence of 5-methylcytosine in nucleic acids. Nature 166:237–238
Yang JL, Liu LW, Gong YQ et al (2007) Analysis of genomic DNA methylation level in radish under cadmium stress by methylation sensitive amplified polymorphism technique. J Plant Physiol Mol Biol 33:219–226
Zemach A, McDaniel IE, Silva P et al (2010) Genome-wide evolutionary analysis of eucaryotic DNA methylation. Science 328:916–919
Zhang Z, Zhang S, Zhang Y et al (2011) Arabidopsis floral initiator SKB1 confers high salt tolerance by regulating transcription and pre-mRNA splicing through altering histone H4R3 and small nuclear ribonucleoprotein LSM4 methylation. Plant Cell 23:396–411
Zhu J, Jeong JC, Zhu Y et al (2008) Involvement of Arabidopsis HOS15 in histone deacetylation and cold tolerance. Proc Natl Acad Sci U S A 105:4945–4950

Apple Latent Spherical Virus (ALSV) Vector as a Tool for Reverse Genetic Studies and Non-transgenic Breeding of a Variety of Crops

Ichiro Kasajima, Makoto Ito, Noriko Yamagishi, and Nobuyuki Yoshikawa

Abstract *Apple latent spherical virus* (ALSV) has recently emerged as an efficient system for reverse-genetic tool of virus-induced gene silencing (VIGS), gene expression, and potential epigenetic breeding through virus-induced transcriptional gene silencing (VITGS) of crops. ALSV does not necessarily infect all plant species, but it infects to a variety of crops such as apple, pear, cherry, soybean, pea, cucumber, watermelon, petunia, *Eustoma*, and Japanese gentian with the current protocol, at relatively high infection rates. Virus vectors are routinely constructed in binary plasmids, agro-infiltrated to *Nicotiana benthamiana*, propagated in *Chenopodium quinoa*, concentrated, RNA extracted, and introduced into crops by gold particle bombardment (biolistic inoculation). Methods and tips of virus preparation and infection are explained in detail in this chapter. We are recently testing whether VITGS technology is applicable to any gene of any plant species. Unexpected problems observed here are shown, and possible solutions to overcome these problems will be discussed.

Keywords ALSV • DNA methylation • VIGS • Virus vector • VITGS

Contents

I. Kasajima • M. Ito • N. Yamagishi • N. Yoshikawa (✉)
Laboratory of Plant Pathology, Department of Agriculture, Iwate University, Ueda 3-18-8, Morioka, Iwate 020-8550, Japan
e-mail: yoshikawa@iwate-u.ac.jp

N. Rajewsky et al. (eds.), *Plant Epigenetics*, RNA Technologies,
DOI 10.1007/978-3-319-55520-1_25

1 Introduction: Significance of ALSV Vector in Plant Reverse Genetics and Epigenetic Breeding Technology

Genetic study of higher plants and genetic breeding of crops have enjoyed enormous profits from transgenic technologies. Genetic transformation is routinely performed not only in model plants such as *Arabidopsis thaliana*, tobacco, tomato (model fruit), rice (model cereal), and torenia (model flower) but also in many other plant species. Transgenic crops, such as maize and soybean, are widely cultivated. On the other hand, we are also facing limitations of transgenic technologies. Genetic transformation is normally based on callus induction of inoculated leaf disks on solid media and dedifferentiation. The process to establish such complex system requires much labor and time. If one cultivar is successfully transformed, other cultivars may not be transformed with the same protocol. In addition, even if a brilliant transgenic cultivar is generated, we have only small chance of application. There will be essentially no proof that transgenic crops are harmful to human health or field ecosystem, but the doubts about the safety of transgenic crops does not seem to be relieved. For example, transgenic rose and carnation are the only transgenic crops developed in Japan, which are sold on the market. This quite restricted number of application is not comparable to the big population of plant scientists in Japan. Such inferiority of transgenic technology, as to commercial use, would have been a great problem restricting progress of genetic technology of plants. So, what are the possible technologies which can overcome these two serious problems (limitations in genetic transformation of cultivars and application of plant genetic technology)? *Apple latent spherical virus* (ALSV) vector is a possible means to overcome these problems. ALSV is a potent vector for gene expression and virus-induced gene silencing (VIGS). Gene expression with ALSV allows early (precautious) flowering of plants. VIGS suppresses endogenous gene expressions, enabling reverse genetic estimation of gene functions. ALSV is applicable to many plant species. Once a protocol for infection of ALSV is established for a cultivar, the same protocol is usually applicable to the other cultivars of the same crop.

Nucleotide inserts in ALSV vector for gene expression and VIGS are relatively stable. In many virus vectors, nucleotide inserts are lost (deleted) in serial infection of virus from the first infected plant to other plants. Unlike many other viruses, nucleotide inserts in ALSV and viruses in genus *Potyvirus* can be maintained even after serial infection. Tips to successfully maintain nucleotide inserts in ALSV vector will be explained elsewhere in this chapter. This relatively stable feature of nucleotide inserts in ALSV vector enables virus concentration and inoculation of concentrated viral RNA to various plant species.

"New breeding techniques" (NBT) or "new plant breeding techniques" (NPBT) enable reverse-genetic breeding of transgene-free crops. NPBT technology is represented by nuclease-supported introduction of nucleotide deletions at the targeted site of genomic DNA. Introduced mutation itself is free from transgenic sequence, although DNA sequence expressing nuclease has to be introduced by genetic transformation. Separation of this transgene from the mutation is the key to application of nuclease technology, but people may be skeptical if nuclease technology really avoids any effect of once introduced transgene. Many plant viruses including ALSV have RNA genomes. An extensively studied technology at the moment is to express nuclease gene in virus vector and introduce mutations at the targeted genomic DNA sequence, without transgene. Virus vectors are removed after introduction of the target mutations. Limitation in the size of introducible nucleotide into virus vectors is the principal problem in nuclease technology. A similar but different technology is virus-induced transcriptional gene silencing (VITGS). VITGS induces heritable cytosine methylations at the targeted promoter sequence and then suppresses gene expressions. There are already a few examples of VITGS of plant endogenous genes.

In this chapter, we will first introduce characteristics of ALSV vector as well as practical protocols for vector preparation and inoculation of plants with vector RNA. After that, examples of VIGS and VITGS with ALSV vector are explained.

2 Characteristics of the ALSV Vector

ALSV was isolated from an apple tree in an orchard of the Japanese National Institute of Fruit Trees (Koganezawa et al. 1985; Li et al. 2000). ALSV is latent (does not cause viral symptom) in apple and did not spread to nearby apple trees in orchard (Nakamura et al. 2011). Thus ALSV is not an important subject of plant pathological studies. Nevertheless, ALSV turned out to be a potent vector for gene

Fig. 1 ALSV vector. (**a**) Structure of ALSV vector (Li et al. 2000; accession 'ViralMultiSegProj15367'). ALSV-RNA1 and ALSV-RNA2 clones are expressed under the control of CaMV 35S RNA promoter in pCALSR1 and pCALSR2, respectively. PRO-co, protease cofactor; HEL, NTP-binding helicase; Vpg, viral protein genome-linked; C-PRO, cysteine protease; POL, RNA polymerase; MP, movement protein; Vp, capsid proteins. (**b**) *N. benthamiana* plants which were not inoculated (NI) or inoculated with ALSV vector with an insert of 201-base tobacco *PDS* gene fragment in XSB site (NtPDS-201; Igarashi et al. 2009)

expression and gene suppression not only in apple but also in many other plant species. Structure of ALSV vector is shown in Fig. 1a.

There are advantages of ALSV as a vector. First, ALSV uniformly penetrates to whole tissue, including leaves, stems, and shoot meristem. ALSV spreads to upper leaves and stems from inoculated leaves. Figure 1b shows VIGS of *Phytoene Desaturase* (*PDS*) gene in *Nicotiana benthamiana*, driven by ALSV vector. Virus-infected tissues are white, because of photo-bleaching by silencing of *PDS* gene. Second, ALSV has a broad host range. Infectivity of ALSV vector to various plant species is summarized in Table 1. ALSV can infect model plants (*A. thaliana* and tobacco plants), fruit trees (such as apple, pear, apricot, and sweet cherry), vegetables (such as tomato, cucumber, melon, spinach, and potato), legumes (such as soybean, pea, adzuki bean), and ornamental flowers (such as petunia, rose, *Eustoma*, and Japanese gentian). ALSV can infect only part of tomato cultivars

Table 1 List of ALSV-infected plants

Plant species	Cultivar	Infection[a]	Symptom[b]	Reference
Chenopodium quinoa		+	+	Li et al. (2004)
Apple	(Seedling)	+	–	
Nicotiana benthamiana		+	–	Yaegashi et al. (2007)
Nicotiana tabacum		+	–	
Nicotiana occidentalis		+	–	
Soybean	Jack	+	±	Yamagishi and Yoshikawa (2009)
	Enrei	+	±	
	Dewamusume	+	±	
	Tanbaguro	+	±	
	Suzukari	+	±	
	Hatayutaka	+	±	
	Chamame	+	±	
Arabidopsis thaliana	Columbia	+	–	Igarashi et al. (2009)
Nicotiana tabacum	Xanthi	+	–	
Nicotiana occidentalis		+	–	
Nicotiana glutinosa		+	–	
Nicotiana benthamiana		+	–	
Tomato	Kouju	+	–	
	Oogatafukuju	+	–	
Petunia		+	–	
Soybean	Suzukari	+	±	
Pea	Denkou	+	–	
Adzuki bean	Benidainagon	+	–	
Cowpea	Akadane-sanjaku-oonaga	+	±	
Cucumber	Tsubasa	+	±	
Muskmelon	Earl's Knight	+	±	
Cucurbita pepo	Diner	+	±	
Watermelon	Zuisyo	+	±	
Luffa	Onaga-hechima	+	±	
Bottle gourd	Onaga-yuugao	+	±	
Globe amaranth		+	–	
Plumed cockscomb		+	–	
Linaria maroccana		+	–	
Nicotiana benthamiana		+	–	Yamagishi et al. (2011)
Nicotiana tabacum	Xanthi	+	–	

(continued)

Table 1 (continued)

Plant species	Cultivar	Infection[a]	Symptom[b]	Reference
Nicotiana glutinosa		+	–	
Nicotiana occidentalis		+	–	
Arabidopsis thaliana	Columbia	+	–	
Petunia		+	–	
Apple	Fuji (seedling)	+	–	
	Orin (seedling)	+	–	
	Golden Delicious (seedling)	+	–	
Apple	(Seedling)	+	–	Sasaki et al. (2011)
Pear	(Seedling)	+	–	
Japanese pear	(Seedling)	+	–	
Rose		+	–	Ito et al. (2012)
Soybean	Harosoy	+	?	Takahashi et al. (2013)
Nicotiana benthamiana		+	–	Kon and Yoshikawa (2014)
Petunia	Red Star	+	–	
Pea	Matsushima Kinusaya	+	–	Satoh et al. (2014)
Fava bean	Kawachi-Issun	+	–	
Eustoma	Shalala blue	+	–	
Apricot	Heiwa	+	–	Kawai et al. (2014)
Japanese apricot	Nanko	–		
Japanese gentian	Polarno White	+	–	Nakatsuka et al. (2015)
Apricot	Shinyo	+	–	Kawai et al. (2016)
	Shingetsu	+	–	
	Shinshuomi	+	–	
	Nanbuhachisuke	–		
	Niigataomi	–		
Sweet cherry	Satonishiki	+	–	
Almond	Nonpareil	+	–	
	Carmel	+	–	
	Marcona	–		
Peach	Ohatsumomo	+	+	
Japanese apricot	Ryukyokoume	–		
	Benisashi	–		
	Koshinoume	–		
	Hachiro	–		
Japanese plum	Sordum	–		
European plum	Sanctus Hubertus	–		
Japanese gentian	Iwate Yume Aoi	+	–	Fekih et al. (2016)
	Iwate	+	–	

(continued)

Table 1 (continued)

Plant species	Cultivar	Infection[a]	Symptom[b]	Reference
	Alta	+	−	
Eustoma	Daburu Mikkusu	+	−	
	Newlination Pink ver. 2	+	−	
New Zealand spinach		+	−	Unpublished
Spinach	Akution	+	−	
	Active	−		
Cucumber	Aodai	+	±	
	Hokushin	+	±	
	Natsu Suzumi	+	±	
	Suzunari-Suyou	+	±	
	Tsuyatarou	+	±	
	Shakitto	+	±	
	V road	+	±	
	V summer	+	±	
	Natsunomegumi	+	±	
	Freedom series	+	±	
Muskmelon	Prince PF	+	±	
Cucumismelo var. conomon	Shirohagura	+	±	
Watermelon	Natsudaiko	+	±	
Cucurbita pepo	KZ-2	+	±	
Bottle gourd	Daijoubu	+	±	
Pea	Snack No.2	+	−	
Fava bean	Minpoh	+	−	
Tomato	Kyouryoku Beiju No.2	+	−	
	Saturn	−		
	Kiju	+	−	
	Momotaro	−		
	Komomo	−		
	Reika	−		
	Reiyou	−		
	Odoriko	−		
	Minitomato Koko	−		
	Minitomato Pepe	+	−	
Potato	Andes	+	−	
Japanese apricot	(Hybrid)	+	−	
Rice		−		

[a]+, infectious; −, noninfectious
[b]+, cause clear viral symptom; ±, latent, often with weak symptom immediately after infection; −, latent

and do not infect plum and cereals such as rice. Technical advance has to improve protocol of ALSV inoculation and let ALSV infect these plants in the future. Third, ALSV is latent in many plant species. That is, ALSV vector itself does not affect plant phenotypes. This feature allows the effect of expressed or silenced genes to be directly monitored in ALSV-infected plants. Weak viral symptoms are observed immediately after ALSV infection in cucurbits, soybean, and cowpea. In these plants, ALSV initially develop chlorotic spots in two or three leaves and then do not cause symptoms in upper leaves (i.e., latent). ALSV induces clear viral symptoms in upper leaves of *Chenopodium quinoa* and peach.

Like other virus vectors, a disadvantage of ALSV is that it does not necessarily infect to all plant species, as described above. ALSV is not highly infectious, and then virus is formed in highly competent host plant *N. benthamiana* and propagated in *C. quinoa*. Virus particles are concentrated from leaf sap, and viral RNA is extracted. High concentration of RNA is finally introduced into each plant species by particle bombardment of RNA-coated gold particles. It is rarely described in papers, but deletion of inserted sequences for gene expression and suppression is also a serious and frequently observed problem. ALSV vector with insertion sequences has lower infectivity; thus, virus is not formed even in *N. benthamiana*, depending on size, sequence, and insertion site on the vector. Things that must be considered to minimize these problems will be explained later.

ALSV is a bipartite and single-stranded RNA virus, consisting of ALSV-RNA1 and ALSV-RNA2 genome. Whole RNA1 and RNA2 genomes are cloned into binary plasmid vectors pCALSR1 and pCALSR2 (Fig. 1a). In these vectors, RNA1 and RNA2 genomes are driven by *Cauliflower mosaic virus* (CaMV) 35S RNA promoter, for expression in plant tissues. Both genomes encode single polyproteins, which are digested after translation, to form proteins necessary for virus function. Three cloning sites of exogenous nucleotide sequences are available: "SM site" at the tail of RNA1, "XSB site" at the middle of the polyprotein encoded by RNA2, and "MN site" at the tail of RNA2 (Yamagishi et al. 2016a). Artificial recognition sequences of restriction enzymes exist at each cloning site. The restriction enzymes that excise at each cloning site are *Sal*I and *Mlu*I for SM site; *Xho*I, *Sma*I, and *Bam*HI for XSB site; and *Mlu*I and *Nco*I for MN site. Among these three cloning sites, only XSB site is available for gene expression. Whole coding sequence of a gene is introduced at XSB site in frame with polyprotein (and recognition sequences of restriction enzymes). After infection to plant tissues, introduced gene is excised from polyprotein and function. For suppression of gene expression by VIGS, all three cloning sites are available. For VIGS, partial sequence of the target transcript is introduced into cloning site. After infection, defense mechanism of host plant generates small RNA targeting the insertion sequence and then degrades target transcript. A typical size of the insert for VIGS is 200 to 300 bases. Insert shorter than 200 bases may not strongly suppress gene expression (Igarashi et al. 2009). When the same insertion sequence is introduced into each cloning site, VIGS is stronger in the order XSB site > MN site > SM site (Yamagishi et al. 2016a).

3 Practical Protocol for Preparation and Infection of ALSV Vector

ALSV vector technology was recently established. Protocols for ALSV vector preparation and infection are variable between publications and still improved at present through trials of new experimental conditions, to improve infection rate and to avoid deletion of insert sequences. Here introduced are representative protocols which seem to be the best at the moment. In the order of actual experiments, protocols are introduced for vector preparation, agroinoculation to *N. benthamiana* (see also Kon and Yoshikawa 2015), propagation in *C. quinoa*, extraction of viral particles and viral RNA, and infection to target plants such as apple and Japanese gentian (see also Yamagishi and Yoshikawa 2013).

3.1 Vector Preparation

After digestion with proper restriction enzymes, introduce insertion sequences into cloning sites of pCALSR1 or pCALSR2. It is possible to introduce two separate sequences into different cloning sites, for example, insertion for gene expression at XSB site and insertion for VIGS at SM site. Vector sets with multiple inserts will have lower infectivity. Clones are first prepared in *E. coli*, and then plasmids are introduced into competent *Agrobacterium* GV3101 cells by electroporation (Gene Pulser, Bio-Rad Laboratories, Hercules, USA). *Agrobacterium* clone having pBE2113-HC-Pro is also used, to transiently express silencing suppressor (HC-Pro) to improve infection efficiency (Kon and Yoshikawa 2014). *Agrobacterium* cultures are mixed with the same volume of 50% glycerol and stored at −80 °C.

3.2 Agroinoculation of N. benthamiana

Agrobacterium cultures harboring ALSV vectors (pCALSR1 and pCALSR2) and pBE2113-HC-Pro are inoculated onto *N. benthamiana* leaves in order to express and infect ALSV.

3.2.1 Preparation of *Agrobacterium* Cultures

Three *Agrobacterium* clones, harboring ALSV-RNA1 clone, ALSV-RNA2 clone, and HC-Pro clone, respectively, are separately incubated overnight at 28 °C in liquid LB medium (tryptone, 10 g; yeast extract, 5 g; NaCl, 10 g per liter; pH adjusted to 7.0, with 5N NaOH) supplemented with suitable antibiotics (kanamycin,

50 mg/L; rifampicin, 100 mg/L). *Agrobacterium* is typically incubated in 50 mL plastic tubes. Volume of medium has to be 15 mL or less per tube, to certify aerobic growth and high bacterial concentration.

3.2.2 Pre-treatment of *Agrobacterium*

Centrifuge to collect bacterial pellet. Diffuse pellet in small volume of agroinoculation buffer ($MgCl_2$, 10 mM; MES-KOH, 10 mM, pH5.7; acetosyringone, 150 μM). 1000 times concentrated stock of acetosyringone solution in DMSO is stored at −20 °C. Agroinoculation buffer is prepared immediately before use, or otherwise acetosyringone is added immediately before use, to frozen stock of the buffer. Filter sterilization of the buffer, before use, may benefit inoculation, but not required. Measure optical density at 600 nm (OD_{600}) of bacterial solution, and then dilute to $OD_{600} = 1$. OD_{600} value of overnight culture of *Agrobacterium* is typically 0.66. Mix three *Agrobacterium* solutions (ALSV-RNA1, ALSV-RNA2, and HC-Pro) in the same ratio (1:1:1), and put in the dark (cover plastic tube with aluminum foil) at 22 °C for 2–3 h.

3.2.3 Agroinoculation

Prepare *N. benthamiana* plants (grown in growth chamber around 1 m at 25 °C, 16 h photoperiod). Fill *Agrobacterium* solution in 1 mL syringe without needle. Slowly inject bacterial solution from the backside of all expanded leaves. 3 mL of the solutions will be necessary per plant. Taking into account possible low infection rate of ALSV, 20–40 *N. benthamiana* plants had better be inoculated per ALSV construct. Incubate inoculated plants in the dark at 22 °C overnight, and then grow in growth chamber at 25 °C with proper illumination.

3.3 *RT-PCR Analysis for Detection of ALSV Infection*

To select ALSV-infected plants without deletion of the insert, from the population of inoculated plants, RNA is extracted from upper leaves and tested by RT (reverse transcription)-PCR (polymerase chain reaction).

3.3.1 Sampling of Leaves

Leaf disks (typically 10–30 mg) are hand-teared from the uppermost leaves of inoculated plants, and put in 2 mL screw-capped tubes with two stainless beads ($\phi = 4.8$ mm) inside (SUS beads, 4.8 mm, Tomy, Tokyo, Japan; Fig. 2a). RT-PCR analysis is preferably performed 3 weeks postinoculation (wpi). Analysis can be

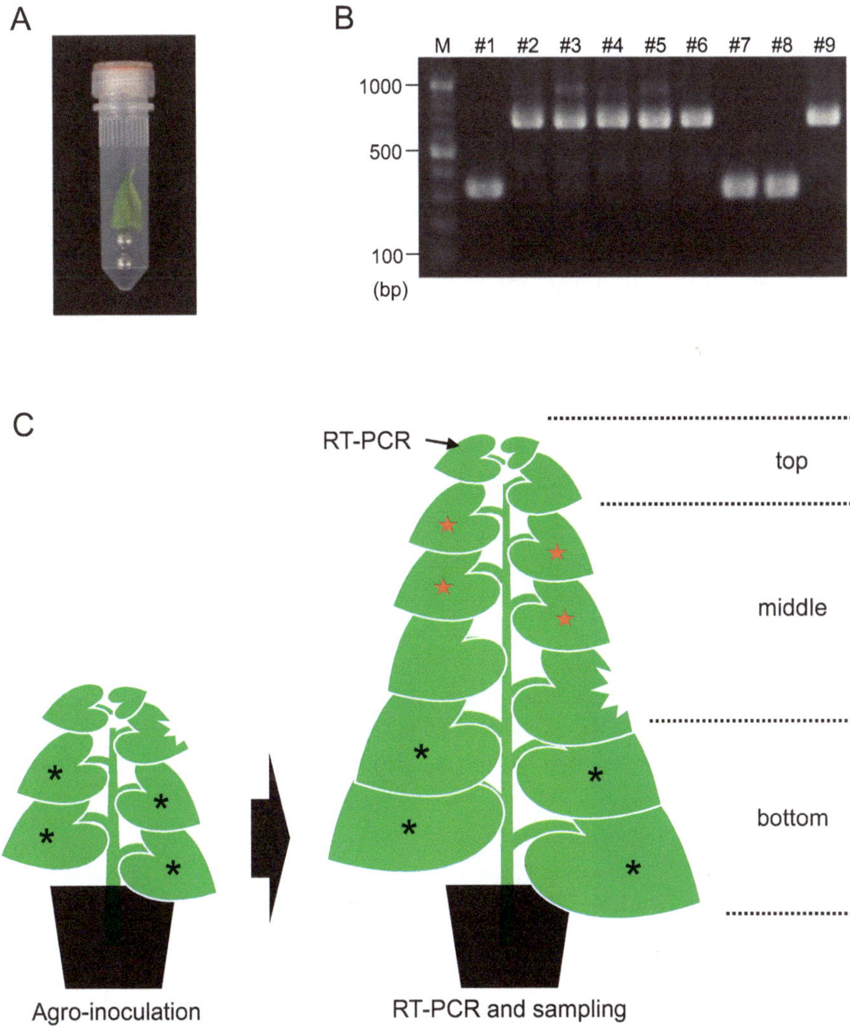

Fig. 2 Schematic representation of RT-PCR and agroinoculation. (**a**) Apple leaf disk sampled in screw-capped tube for RNA extraction. (**b**) Example of RT-PCR analysis of inoculated *N. benthamiana* plants. M, molecular weight marker. #1 through #9, independent *N. benthamiana* plants. Plants were inoculated with ALSV vector having a 400-base insert at MN site. cDNA prepared by reverse transcription was PCR amplified with ALSR2-2981(+)/ALSR2-3289(−)primers. All plants are infected by ALSV vector, but inserts are deleted in #1, #7, and #8 plants. (**c**) All expanded leaves of 1-month-old *N. benthamiana* plants are inoculated with *Agrobacterium* (marked with *black asterisks* in the left plant). After grown for 3 weeks, ALSV infection and possession/deletion of insert sequences are checked by RT-PCR, at the topmost leaves. Upper leaves in the middle part of plants are the most suitable for further virus extraction (marked with *red stars* in the right plant)

done at 2 wpi, but there remains a small possibility that ALSV has not propagated until 2 wpi and fail to be detected. Leaf samples are kept at −80 °C for at least 1 h.

3.3.2 RNA Extraction

Leaves are crushed with precooled MicroSmash (MS-100R, Tomy, Tokyo, Japan) at 2500 rpm for 30 s. Centrifuge at the maximum speed (14,000 rpm, r = 6 cm) briefly, and then add 400 μL SDS buffer (Tris-HCl, 200 mM, pH7.5; EDTA, 25 mM; NaCl, 250 mM; SDS, 0.5%; β-mercaptoethanol, 1%). PVPP+s buffer (Tris-HCl, 50 mM, pH9.5; EDTA, 10 mM; NaCl, 4M; CTAB, 1%; polyvinylpolypyrrolidone, 0.5%; β-mercaptoethanol, 1%: modified from Kasajima et al. 2013) or TRI Reagent (Molecular Research Center, Cincinnati, USA) may improve extraction efficiency according to plant species. β-Mercaptoethanol is added to buffer immediately before use. Crush with MicroSmash again, at 3000 rpm for 30 s. Centrifuge at the maximum speed for 2 min. Recover supernatant, add 100 μL PCI solution (phenol-chloroform-isoamyl alcohol, 25:25:1), mix thoroughly by vortex, centrifuge at the maximum speed for 5 min, and recover water (upper) phase to new tube. When using PVPP+s buffer, repeat this step once again and dilute water phase with the same volume of sterilized pure (deionized, distilled, or milli-Q) water. Add twice the volume of ethanol, mix, and place at −80 °C for at least 15 min. Thaw frozen sample and immediately centrifuge at the maximum speed at 4 °C for 45–60 min. Centrifugation for 30 min will yield smaller amount of RNA. Remove supernatant, dry with hair dryer, and dissolve in 50 μL of sterilized water.

3.3.3 RT-PCR Analysis

Reverse transcribe 1 μL of RNA solution in 10 μL RT reaction for 1 h with oligo-$(dT)_{20}$ primer. This cDNA solution is 5 to 10 times diluted with sterilized water and stored in freezer. ALSV is detected by PCR with relevant primer sets. Following are recommended primer sets for each cloning site:

SM site	
ALSR1-6598(+)	5′-GTACATTCCTCCCAATCAAAG-3′
ALSR1-6691(−)	5′-GGATCACGAGAACAAACTAG-3′
XSB site	
ALSR2-1213(+)	5′-ATACCACCTCATACAGGTACAC-3′
ALSR2-1484(−)	5′-CGTTCCACGACCGTGGGCCAGA-3′
MN site	
ALSR2-2981(+)	5′-TGGGAGATTCCTTCTCTGTAGATATT-3′
ALSR2-3289(−)	5′-CAAGAAACCTAACGGACCAGAGGTCAC-3′

Primers were designed by Chunjiang Li. PCR is performed for 40 cycles with annealing temperature of 55 °C and extension for 1 min, with ordinary *Taq* polymerase (such as Ex Taq, Takara, Kyoto, Japan). Nonspecific DNA amplification of different sizes can be observed in different plant species. To discriminate between specific (derived from ALSV) and nonspecific DNA amplifications, it is strongly recommended to include control RNA extracted from non-inoculated plants. Concentration of nonspecific PCR product is low, whereas concentration of specific product is usually high; thus, specific PCR product can be also discriminated by band intensity. Infection rate of wild-type (wt) ALSV, without any insertion at cloning sites, to *N. benthamiana* is 100%, whereas infection rate of ALSV with insertion is quite variable. Possession/deletion of the insertion sequences are judged by the sizes of PCR products (Fig. 2b).

3.4 Rub-inoculation of Leaf Sap onto C. quinoa

ALSV vector infected to *N. benthamiana* can be readily used for infection to various plants, such as apple. However, to increase virus concentration and infection rate, we often inoculate leaf sap to *C. quinoa*.

3.4.1 Sampling of *N. benthamiana* Leaves

Recently, we noticed that leaf positions sampled from ALSV-infected *N. benthamiana* has great effect on infection rate of ALSV and deletion rate of the inserts. Leaf positions of agro-infiltrated *N. benthamiana* are illustrated in Fig. 2c. As described above, expanded leaves of 1-month-old plants are agro-infiltrated. At the time of agroinoculation, part of the lowest non-inoculated leaves is removed as a mark of leaf positions. Three weeks after inoculation, inoculated leaves have become much wider, and many new leaves have formed. Here, leaf positions of these 3 wpi plants are classified into three: bottom (inoculated leaves), middle, and top (non-inoculated leaves). Infection rate is higher when saps are prepared from higher leaves (top > middle > bottom). On the other hand, insertion sequences are easily deleted when ALSV vectors are propagated by serial infection from the first infected plants by *Agrobacterium*. Frequency of deletion is higher in leaves at higher positions (top > middle > bottom). This is why we principally use "middle" infected leaves for further experiments. If no deleted sequence (smaller size of PCR product) is detected by RT-PCR, there is only small possibility that deletion happens in the serial infection with middle and bottom leaves, but deletion may happen with top leaves. ALSV is not detected in the lowest two or three leaves of the middle part, so these leaves are not used. Bottom *N. benthamiana* leaves may be also available, but we do not want to carry *Agrobacterium* residues to the next step. Infected leaves are reserved at −80 °C.

3.4.2 Rub-inoculation

Measure fresh weight of small amount of infected *N. benthamiana* leaf. Quickly crush with mortar and pestle. Add 2–3 volumes of ALSV buffer (Tris-HCl, 100 mM, pH7.8; NaCl, 100 mM; $MgCl_2$, 5 mM), grind again, and keep on ice. Prepare *C. quinoa* plants. *C. quinoa* prefers cool climates; then this experiment is performable only during autumn, winter, and early spring. Scatter carborundum powder through three layers of cotton mesh, on expanded *C. quinoa* leaves. Remove excessive carborundum by moderately beating leaves. Wear finger cot, and rub the upper surface of leaves with solution of leaf sap above. After rubbing all expanded leaves, rinse leaves with water. Otherwise leaves will immediately wilt. Viral symptom (yellow spots) will start to appear around 2 wpi. Select plants having ALSV vectors without deletion by RT-PCR, and sample bottom and middle leaves as soon as possible. In this sampling, weigh sum fresh weights of leaves per each sampling bag, and write these values on bags. 10–30 g of leaves is normally treated in the extraction of virus in the next step. Leaves are reserved at −80 °C.

3.5 *Bentonite Solution*

Bentonite is a clay mineral. Bentonite solution, or more correctly a phosphate buffer with diffused bentonite microcrystals, is prepared in this experiment and used for ALSV extraction in the next step.

3.5.1 Preparation of Phosphate Buffers

100 mM phosphate buffer" is prepared by adding 100 mM KH_2PO_4solution to 100 mM Na_2HPO_4solution to adjust pH to 7.4. 100 mM phosphate buffer is used to prepare "phosphate buffer 1" (phosphate 10 mM; $MgSO_4$, 10 mM) and "phosphate buffer 2" (phosphate 10 mM; $MgSO_4$, 1 mM).

3.5.2 Preparation of Bentonite Solution

Add 500 mL of phosphate buffer 1 to 25 g of bentonite. Mix for 2 h with magnetic stirrer. Centrifuge for 1 min at 1500 rpm (r = 9 cm) or 1800 rpm (r = 6 cm). Transfer supernatant to new tubes. Centrifuge for 15 min at 10,000 rpm (r = 9 cm hereafter). Transfer pellet to beaker and add 250 mL of phosphate buffer 2. Mix for 2 h with magnetic stirrer. Centrifuge for 1 min at 1500 rpm. Transfer supernatant to new tubes. Centrifuge for 15 min at 10,000 rpm. Transfer pellet to beaker, and add 100 mL of phosphate buffer 1. Mix for approximately 1 h until all pellets are diffused. Confirm that specific gravity of the solution is between 1.03 and 1.05, by weighing 1 mL solution. Bentonite solution can be reserved at 4 °C for several months or at −20 °C.

3.6 Virus Extraction

In this experiment, rough extract of ALSV is prepared from infected leaves of *C. quinoa* or *N. benthamiana*. Samples are always kept on ice. Experimental procedure of virus extraction is illustrated in Fig. 3.

3.6.1 Crushing Infected Leaves in Blender

Cool home blender (metal blade and glass cup) in refrigerator before use. Put leaves (10–30 g) which have been frozen at −80 °C and 100 mL of ALSV-β buffer (Tris-HCl, 100 mM, pH 7.8; NaCl, 100 mM; $MgCl_2$, 5 mM; β-mercaptoethanol, 1%) in blender. β-Mercaptoethanol is added to the buffer immediately before use. Wear gloves to protect your skin from β-mercaptoethanol. Blend until no large particles of leaves are left in the solution. Squeeze this dark-green solution with two layers of cotton mesh to a beaker, to remove leaf residues. Centrifuge at 9000 rpm for 10 min at 4 °C (r = 9 cm).

3.6.2 Rough Purification by Using Bentonite Solution

Recover supernatant to a new beaker. Measure the approximate volume of this extract. The volume will be around 80 mL. Mix with magnetic stirrer. Add

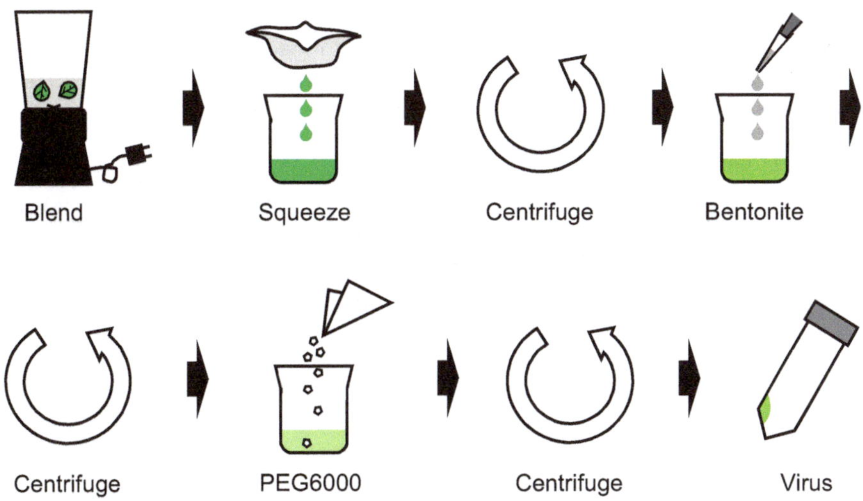

Fig. 3 Schematic representation of virus extraction

bentonite solution to stirred extract bit by bit. The total amount of bentonite solution has to be adjusted according to sample amounts. The "standard" volume of bentonite solution is 1% (0.8 mL to 80 mL extract) for *C. quinoa* and 2% for *N. benthamiana*. These percentages are for 30 g leaf samples. Percentages are reduced in proportion to the original amounts of leaves. For example, 0.5% of bentonite solution is added to extract from 15 g of *C. quinoa* leaves (i.e., 0.4 mL for 80 mL extract). Centrifuge at 9000 rpm for 10 min. Bentonite solution may be further added to the solution by observing color of extract. After adding bentonite and centrifugation, color of extract may be yellow green, light yellow green, or light yellow. Yellow-green extract still contains much impurity, so extracts have to be light yellow green or preferably slightly greenish light yellow. If the color of the extract is not yellowish enough, add bentonite and repeat centrifugation.

3.6.3 Preparation of Virus Particle

Recover supernatant to a new beaker. Measure the volume of extract. Add 8% of PEG6000 (polyethylene glycol 6000) to extract. Mix for 1 h with magnetic stirrer. Centrifuge at 9000 rpm for 10 min. Remove supernatant. Here, yellow-green pellet contains virus. When dark-green pellet is formed, that is impurity. Dissolve ALSV pellet in a small volume (1–3 mL) of ALSV buffer (Tris-HCl, 100 mM, pH7.8; NaCl, 100 mM; $MgCl_2$, 5 mM). This solution may be centrifuged again (9000 rpm, 10 min) and supernatant recovered, to increase virus purity. Store at −80 °C. This solution (virus particle) is more stable than viral RNA.

3.6.4 Extraction of Viral RNA

Add half volume of PCI solution (phenol-chloroform-isoamyl alcohol, 25:25:1) to the solution of virus particle. Mix by vortex, and centrifuge at 14,000 rpm for 5 min (r = 6 cm hereafter). Recover water (upper) phase to a new tube. Repeat PCI treatment again. Add twice the volume of ethanol, mix, place at −80 °C at least for 15 min, and centrifuge at 14,000 rpm for 45–60 min. Remove supernatant and dry. Dissolve in a small volume of sterilized water (such as 100 μL). Store at −80 °C. RNA is more stable in ethanol than in water. Dilute part of this stock solution in water, and measure RNA concentration by NanoDrop (or spectrophotometer). RNA yields are variable, but they are typically 50–200 μg from 10 g of *C. quinoa* leaves.

3.7 Preparation of Gold Particle

In preparation for particle bombardment, gold particles are coated with ALSV RNA in this step. 5 μg of viral RNA and 0.4 mg of gold particles are used per each shot of particle bombardment. Preparation for 40 shots is described here. Amounts of RNA and gold particles are adjusted according to scales of experiments.

3.7.1 Mixing Viral RNA with Gold Particle

Weigh out 16 mg of gold particle (Microcarrier, Gold, Bio-Rad, 0.6 μm, 250 mg, #165-2262) in 1.5 mL plastic tube. Add 50 μL of sterilized distilled water and vortex. Sonicate for 1 min. Mix the tube by vortex. Add RNA solution (200 μg of RNA, typically 50 μL) bit by bit into vortexed solution. Add 10 μL (1/10 volumes) of ammonium acetate (5 M). Add 220 μL (twice the volume) of isopropanol. Place at −20°C for 1 h. Gold particles coated with viral RNA will sink at the bottom of solution.

3.7.2 Preparation of RNA-coated Gold Particle

Remove supernatant with pipette. Vortex for a second. Wash gold particles with 1 mL of 100% ethanol (dehydrated with molecular sieve) for four times. Dissolve gold particles in 400 μL of 100% ethanol (dehydrated with molecular sieve). Sonicate for a second to disperse.

3.8 Particle Bombardment with NepaGene System

We recently introduced NepaGene system (GDS-80, NepaGene, Ichikawa, Japan: http://www.nepagene.jp/products_nepagene_0053.html; Fig. 4a). This system is easy, and infection rate of ALSV is relatively high. This system is also free from license. Other two systems are also available (Helios Gene Gun system, Bio-Rad; PDS-1000/He Particle Delivery System, Bio-Rad). Refer to a previous report for protocol of these systems for ALSV infection (Yamagishi and Yoshikawa 2013). An advantage of PDS-1000 system is that gold particles are shot in vacuum; then damage to plants by shots is moderate. Many plants (such as eight apple seedlings) can also be inoculated at a time with PDS-1000.

3.8.1 Setting up the Gene Gun

Connect the gene gun and gas bombe with the gas tube. Check if there is no gas leakage, following the manual. Set gas pressure. Gas pressures can be 20, 30, 40, 50, 60, or 70 psi (pounds per square inch, lb/inch^2), according to sample conditions. Gas pressure is usually set to 40 psi for bombardment of apple seedling. Rotate needle valve on the backside of the gun to the right, according to gas pressures (20 psi, 6.5 rotations; 30 psi, 6 rotations; 40 psi, 5 rotations; 50 psi,

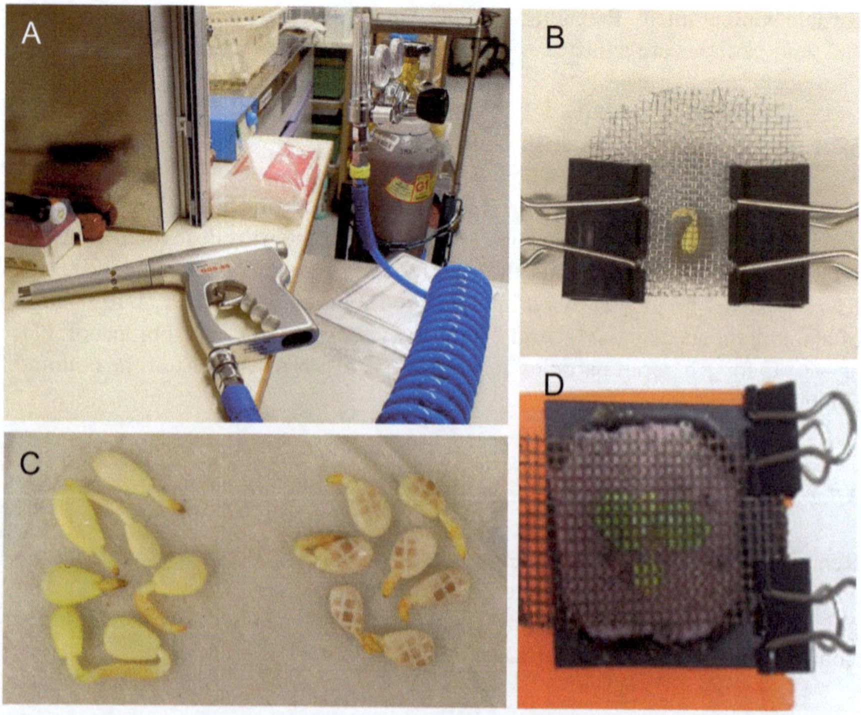

Fig. 4 Particle bombardment of the RNA of ALSV vector. (**a**) NepaGene GDS-80 gene gun, connected to gas bombe. (**b**) Apple seedling immediately after germination, stabilized between aluminum meshes. (**c**) Apple seedlings before (*left*) and after (*right*) bombardment. (**d**) Seedling of Japanese gentian stabilized with spongy and copper mesh. Photograph of Japanese gentian was taken by Rym Fekih

4 rotations; 60 psi, 3 rotations; 70 psi, 2 rotations). Trigger the gun and adjust needle valve at the bottom of flowmeter, so that the floating ball jumps to the level between 10 and 15 L/min.

3.8.2 Shooting Gold Particle

Plant samples have to be stabilized before bombardment: otherwise plants are blown off and injured. Apple seedlings immediately after germination are bound between soft aluminum mesh, and gentian seedlings are stabilized with spongy and copper mesh (Fig. 4b, d). Gentian seedlings are shot in cardboard to prevent soil particles from scattering all around. Inject 10 μL of gold particles into barrel. Do not release pipette before removing tip from the injection hole on the side of barrel. Shoot gold particles from a close distance (1–3 cm). Gold particles will come out two or three times after injection. Gold particles are visible as brown stain on the surface (Fig. 4c). After finishing all bombardments, close the gas bombe and remove the gas inside the tube, by shooting repeatedly. Close gas valves, remove gas tube, and wash barrel by sonication in water for 30 min.

3.8.3 RT-PCR Analysis

Determination of ALSV infection follows the protocols for RNA extraction and RT-PCR described above.

4 Application of ALSV Vector for Gene Expression and Gene Silencing

Foreign genes are introduced at XSB site of ALSV and expressed in infected plants. Marker gene, *GFP*, was successfully expressed by ALSV vector, but GFP fluorescence is not strong in infected plants except for *C. quinoa*, because ALSV allows only low expression level in most plant species (Li et al. 2004). Gene expression technology with ALSV vector is successfully used for early (perpetual) flowering, in several plant species (Yamagishi and Yoshikawa 2011; Yamagishi et al. 2011, 2014; Fekih et al. 2016). For early flowering of plants, *FT* gene of *A. thaliana* (*AtFT*) is introduced into ALSV vector and expressed. For some unknown reason, *AtFT* induces flowering more strongly than *FT* genes from other plants (Yamagishi et al. 2014). Flowering is further promoted by simultaneous expression of *AtFT* and suppression of *MdTFL1-1* gene with ALSV vector in apple. *MdTFL1-1* gene is one of the apple homologs of *TFL1* (*Terminal Flower 1*) gene and negatively regulates apple flowering. This ALSV system shortens the generation time of apple plant (at least 5 years) to less than 1 year (Yamagishi et al. 2014). Because ALSV is not introduced into genome sequence and removed by heat treatment (Yamagishi et al. 2016a) or generating next-generation seed (Nakamura et al. 2011; Kishigami et al. 2014), ALSV vector is expected to be a non-transgenic biotechnology for crop breeding. Genes up to 1300 bases can be introduced into XSB site (Li et al. 2004), but the introducible size may depend on DNA sequences.

In addition to *TFL1* genes, marker genes such as *PDS* (*Phytoene Desaturase*), *SU* (*Magnesium Chelatase*), and *rbcS* (*RuBisCO Small Subunit*) were also silenced by ALSV vector. Silencing of these genes is visualized by changed colors of tissues, from green to white or yellow. Transgenic *GFP* was also silenced (Ito et al. 2012; Kawai et al. 2014, 2016; Sasaki et al. 2011; Yaegashi et al. 2007). Silencing efficiency varies according to size and sequence in the inserts (Igarashi et al. 2009). *Agamous1* gene was silenced in Japanese gentian to produce double-flower strain (Nakatsuka et al. 2015). Expression of *Isoflavone Synthase* gene was silenced by ALSV vector in soybean, resulting in reduced isoflavone contents in mature seeds (Yamagishi and Yoshikawa 2009). *CPN60α*, *Actin*, and *EF-1α* were silenced in apple (Sasaki et al. 2011). *GmMYB-G20-1* was silenced in soybean to generate bluish flowers (Takahashi et al. 2013). *W15* gene was silenced by ALSV vector, to reduce overwintering survival of Japanese gentian (Hikage et al. 2016). *SVP-like* genes were silenced to generate early-flowering Japanese gentian (Yamagishi et al. 2016b). *E1-like* genes were silenced to generate early-flowering soybean (Xu et al. 2015). Thus ALSV vector can be used for downregulating expressions of specific genes through VIGS. ALSV vector also functions as vaccine against other plant viruses (Satoh et al. 2014; Taki et al. 2013; Tamura et al. 2013).

5 Application of ALSV Vector for Transcriptional Gene Silencing

After removal of ALSV by heat treatment or setting next-generation seeds, VIGS is no longer observed. To silence target genes even after removal of ALSV vector, VITGS have to be induced. In the experiment of VITGS, promoter sequence of the target gene is introduced into ALSV vector, instead of coding sequence. There has been only one report about VITGS with ALSV vector (Kon and Yoshikawa 2014). In this report, transgenic CaMV 35S RNA promoter driving *GFP* gene was targeted by ALSV vector. Cytosine residues of CaMV 35S RNA promoter were methylated, and the expression of *GFP* gene was silenced. DNA methylation and silencing was maintained in progenies, that is, even after removal of ALSV vector.

In general, VITGS of transgenic CaMV 35S RNA promoter seems easier than VITGS of endogenous genes: there are many reports about VITGS of transgenes, but the number of examples is quite limited about VITGS of endogenous genes. *Chalcone Synthase* (*CHS*)-*A* gene of petunia cultivar "Red Star" may be relatively easy to silence by VITGS. VITGS of this gene is successful with *Cucumber mosaic virus* (CMV) vector (Kanazawa et al. 2011) as well as ALSV vector (Kon and Yoshikawa 2014).

It is not clear whether or not VITGS of arbitrary endogenous gene of any plant species with ALSV vector is possible. We recently targeted promoters of *PDS2* gene of *N. benthamiana* (*NbPDS2*) and S_2-*RNase* gene of apple (*MdS*$_2$), to methylate their promoters. These original data are shown here for the first time. Promoter sequence of *NbPDS2* was introduced into MN site of ALSV vector and infected to *N. benthamiana*. Figure 5 shows methylation levels of all cytosine residues within the sequence analyzed by bisulfite treatment. The result was beyond our imagination. Large part of (proximal) promoter sequence upstream of the transcription start site was originally methylated in non-inoculated plant, and the methylation pattern is not greatly changed in infected plant with ALSV vector. Surprisingly, a similar result was obtained for *MdS*$_2$ promoter (Fig. 6). Methylation of cytosine residues in CHG array (H represents A, C, or T) may be slightly promoted in the second infected plant (MdS$_2$P-198-2), but no great difference in methylation pattern was observed between non-inoculated and infected apple plants.

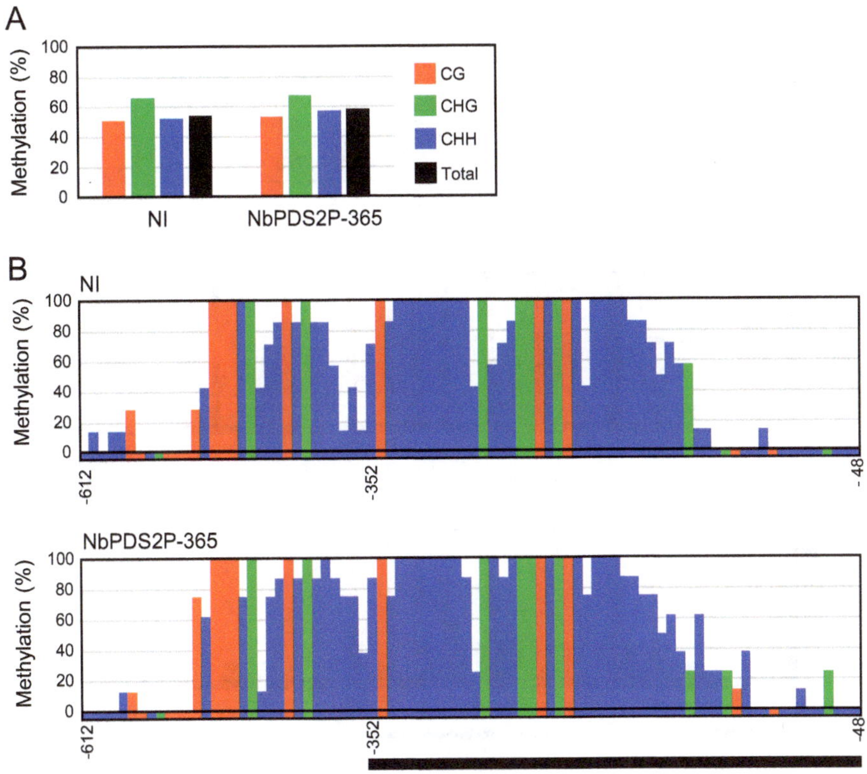

Fig. 5 Bisulfite sequencing analysis of *NbPDS2* promoter. (**a**) Methylation levels of cytosine residues in CG array, CHG array, CHH array, and all cytosine residues. Values were calculated for non-inoculated *N. benthamiana* (NI) and *N. benthamiana* infected by ALSV vector possessing 365-base insert of *NbPDS2* gene at MN site (from −367 to −3 comparative to transcription start site, NbPDS2P-365). *NbPDS2* gene (Niben101Scf14708g00023.1) was identified in database (https://solgenomics.net/organism/Nicotiana_benthamiana/genome). Expression of this gene in leaf was confirmed by RT-PCR analysis. 5′-RACE analysis identified an 862-base fragment (5′-AAGCAAGA...TCAGTAAA-3′) as 5′-UTR of *NbPDS2*. (**b**) Methylation levels of each cytosine residue in NI and NbPDS2P-365 plants. Cytosine positions are numbered relative to transcription start site. 569-base DNA sequences of the plus strand, from −614 to −46 positions, were analyzed in seven or eight replications, respectively. *Red*, *green*, and *blue bars* indicate cytosine residues in CG, CHG, and CHH arrays. Target sequences inserted into ALSV vector are indicated with *black bars* at the bottom of the graph

Through pilot experiments with *NbPDS2* and *MdS*$_2$ genes above, we learned lessons about the nature of DNA methylation and VITGS of endogenous genes. First, part of promoter (upstream) regions of endogenous genes may be originally methylated. In addition to *NbPDS2* and *MdS*$_2$, *CHS-A1* gene promoter of petunia is also partly methylated (Kon and Yoshikawa 2014). These upstream methylated sequences are likely highly repetitive transposon-like elements. The true promoter sequences will lie outside of these methylated regions, and such true promoter sequences have to be targeted by ALSV vector. Second, there may be variation in

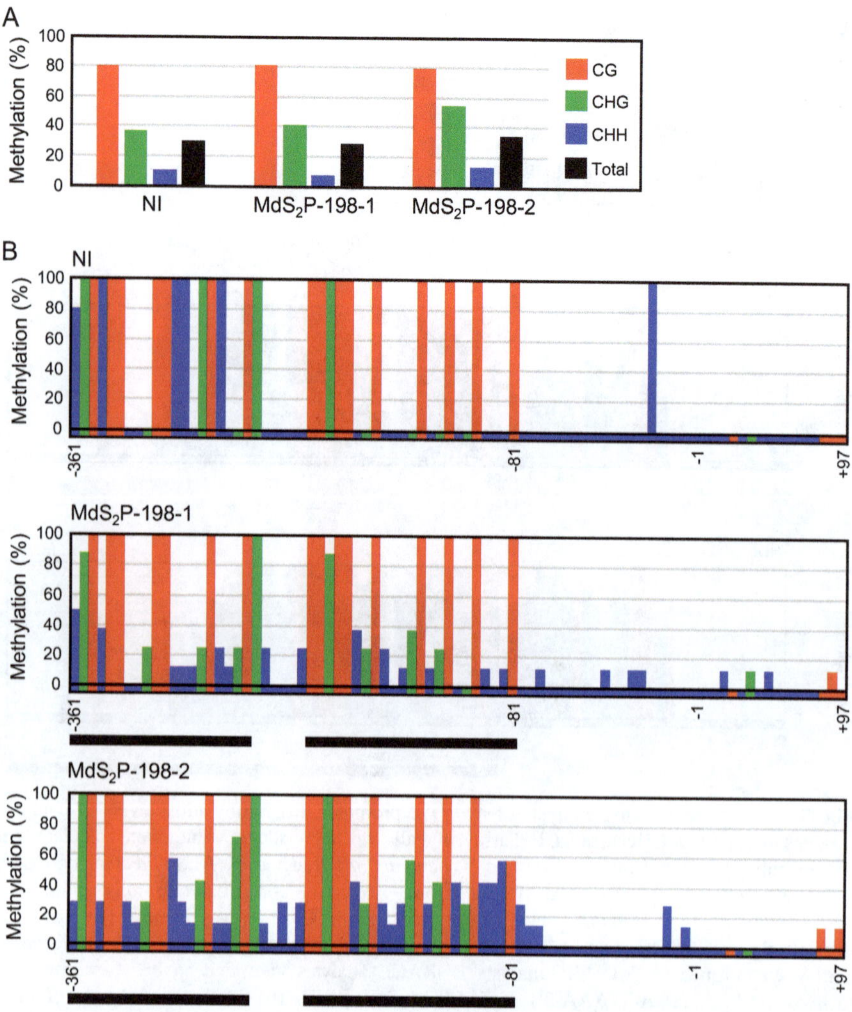

Fig. 6 Bisulfite sequencing analysis of *MdS*$_2$ promoter. (**a**) Methylation levels of cytosine residues in CG array, CHG array, CHH array, and all cytosine residues. Values were calculated for non-inoculated Orin cultivar (NI) and two Orin seedlings, having S_2 alleles (selected by genomic PCR), and infected by ALSV vector possessing 198-base insert at XSB site (combined 90-base and 108-base promoter sequences with high GC contents: MdS$_2$P-198-1,2). Four single-nucleotide mutations were introduced into insertion sequences, to eliminate stop codons. Genomic sequence of *MdS*$_2$ gene (MDC002608.619) was obtained by Blast search of apple genome (cultivar "Golden Delicious") against *MdS*$_2$ mRNA sequence (U12199.1) in Genome Database for Rosaceae (https://www.rosaceae.org/). *MdS*$_2$ promoter sequence was also consistent with a previous report (Ushijima et al. 1998). (**b**) Methylation levels of each cytosine residue in NI, MdS$_2$P-198-1, and MdS$_2$P-198-2 plants. Cytosine positions are numbered relative to transcription start site (the first T of TTCAAAA...) as "+1" and C immediately before this T as "−1." 467-base DNA sequences of the plus strand, from −366 to +101 positions, were analyzed in ten, eight, or seven replications, respectively. *Red*, *green*, and *blue bars* indicate cytosine residues in CG, CHG, and CHH arrays. Target sequences inserted into ALSV vector are indicated with *black bars* at the bottom of each graph

methylation levels between ALSV-infected plants. Remember that variation in gene expression levels are usually observed between transgenic plants expressing exogenous genes or silencing endogenous genes. Comparison of methylation levels in infected plants could be performed by "Mcr-PCR," where genomic DNA at the target site is quantified by semiquantitative PCR, after digestion with methylation-specific restriction enzymes such as McrBC and MspJI. These trials will further our technology of VITGS using ALSV vector in the future.

References

Fekih R, Yamagishi N, Yoshikawa N (2016) Apple latent spherical virus vector-induced flowering for shortening the juvenile phase in Japanese gentian and lisianthus plants. Planta 244:203–214

Hikage T, Yamagishi N, Takahashi Y et al (2016) Allelic variants of the esterase gene W14/15 differentially regulate overwinter survival in perennial gentian (*Gentiana* L.) Mol Genet Genomics 291:989–997

Igarashi A, Yamagata K, Sugai T et al (2009) Apple latent spherical virus vectors for reliable and effective virus-induced gene silencing among a broad range of plants including tobacco, tomato, *Arabidopsis thaliana*, cucurbits, and legumes. Virology 386:407–416

Ito H, Ochiai M, Kato H et al (2012) Rose *Phytoene Desaturase* gene silencing by Apple latent spherical virus vector. Hortscience 47:1278–1282

Kanazawa A, Inaba J, Shimura H et al (2011) Virus-mediated efficient induction of epigenetic modifications of endogenous genes with phenotypic changes in plants. Plant J 65:156–168

Kasajima I, Sasaki K, Tanaka Y et al (2013) Large-scale extraction of pure DNA from mature leaves of *Cyclamen persicum* Mill. and other recalcitrant plants with alkaline polyvinylpolypyrrolidone (PVPP). Sci Hortic 164:65–72

Kawai T, Gonoi A, Nitta M et al (2014) Virus-induced gene silencing in apricot (*Prunus armeniaca* L.) and Japanese apricot (*P. mume* Siebold&Zucc.) with the *Apple latent spherical virus* vector system. J Jpn Soc Hortic Sci 83:23–31

Kawai T, Gonoi A, Nitta M et al (2016) Virus-induced gene silencing in various *Prunus* species with the *Apple latent spherical virus* vector. Sci Hortic 199:103–113

Kishigami R, Yamagishi N, Ito T et al (2014) Detection of apple latent spherical virus in seeds and seedlings from infected apple trees by reverse transcription quantitative PCR and deep sequencing: evidence for lack of transmission of the virus to most progeny seedlings. J Gen Plant Pathol 80:490–498

Koganezawa H, Yanase H, Ochiai M et al (1985) An isometric virus-like particle isolated from russet ring-diseased apple (in Japanese). Ann Phytopathol Soc Jpn 51:363

Kon T, Yoshikawa N (2014) Induction and maintenance of DNA methylation in plant promoter sequences by apple latent spherical virus-induced transcriptional gene silencing. Front Microbiol 5:595

Kon T, Yoshikawa N (2015) An effective and convenient methods for the delivery of Apple Latent Spherical Virus (ALSV)-based vectors into plant cells by agroinoculation. Methods Mol Biol 1287:191–199

Li C, Yoshikawa N, Takahashi T et al (2000) Nucleotide sequence and genome organization of Apple latent spherical virus: a new virus classified into the family *Comoviridae*. J Gen Virol 81:541–547

Li C, Sasaki N, Isogai M et al (2004) Stable expression of foreign proteins in herbaceous and apple plants using Apple latent spherical virus RNA2 vectors. Arch Virol 149:1541–1558

Nakamura K, Yamagishi N, Isogai M et al (2011) Seed and pollen transmission of *Apple latent spherical virus* in apple. J Gen Plant Pathol 77:48–53

Nakatsuka T, Saito M, Yamada E et al (2015) Isolation and characterization of the C-class *MADS-box* gene involved in the formation of double flowers in Japanese gentian. BMC Plant Biol 15:182

Sasaki S, Yamagishi N, Yoshikawa N (2011) Efficient virus-induced gene silencing in apple, pear and Japanese pear using *Apple latent spherical virus* vectors. Plant Methods 7:15

Satoh N, Kon T, Yamagishi N et al (2014) *Apple latent spherical virus* vector as vaccine for the prevention and treatment of mosaic diseases in pea, broad bean, and Eustoma plants by *Bean yellow mosaic virus*. Viruses 6:4242–4257

Takahashi R, Yamagishi N, Yoshikawa N (2013) A MYB transcription factor controls flower color in soybean. J Hered 104:149–153

Taki A, Yamagishi N, Yoshikawa N (2013) Development of apple latent spherical virus-based vaccines against three tospoviruses. Virus Res 176:251–258

Tamura A, Kato T, Taki A et al (2013) Preventive and curative effects of Apple latent spherical virus vectors harboring part of the target virus genome against potyvirus and cucumovirus infections. Virology 446:314–324

Ushijima K, Sassa H, Hirano H (1998) Characterization of the flanking regions of the S-RNase genes of Japanese pear (*Pyrusserotina*) and apple (*Malus* x *domestica*). Gene 211:159–167

Xu M, Yamagishi N, Zhao C et al (2015) The soybean-specific maturity gene *E1* family of floral repressors controls night-break responses through down-regulation of *FLOWERING LOCUS T* orthologs. Plant Physiol 168:1735–1746

Yaegashi H, Yamatsuta T, Takahashi T et al (2007) Characterization of virus-induced gene silencing in tobacco plants infected with apple latent spherical virus. Arch Virol 152:1839–1849

Yamagishi N, Yoshikawa N (2009) Virus-induced gene silencing in soybean seeds and the emergence stage of soybean plants with *Apple latent spherical virus* vectors. Plant Mol Biol 71:15–24

Yamagishi N, Yoshikawa N (2011) Expression of *FLOWERING LOCUS T* from *Arabidopsis thaliana* induces precocious flowering in soybean irrespective of maturity group and stem growth habit. Planta 233:561–568

Yamagishi N, Yoshikawa N (2013) Highly efficient virus-induced gene silencing in apple and soybean by *Apple latent spherical virus* vector and biolistic inoculation. Methods Mol Biol 975:167–181

Yamagishi N, Sasaki S, Yamagata K et al (2011) Promotion of flowering and reduction of a generation time in apple seedlings by ectopical expression of the *Arabidopsis thaliana FT* gene using the *Apple latent spherical virus* vector. Plant Mol Biol 75:193–204

Yamagishi N, Kishigami R, Yoshikawa N (2014) Reduced generation time of apple seedlings to within a year by means of a plant virus vector: a new plant-breeding technique with no transmission of genetic modification to the next generation. Plant Biotechnol J 12:60–68

Yamagishi N, Li C, Yoshikawa N (2016a) Promotion of flowering by *Apple latent spherical virus* vector and virus elimination at high temperature allow accelerated breeding of apple and pear. Front Plant Sci 7:171

Yamagishi N, Kume K, Hikage T et al (2016b) Identification and functional analysis of SVP ortholog in herbaceous perennial plant *Gentiana triflora*: Implication for its multifunctional roles. Plant Sci 248:1–7

Erratum to: Growing Diversity of Plant MicroRNAs and *MIR*-Derived Small RNAs

Mariyana Gozmanova, Vesselin Baev, Elena Apostolova, Gaurav Sablok, and Galina Yahubyan

Erratum to:
© Springer International Publishing AG 2017
N. Rajewsky et al. (eds.), *Plant Epigenetics*, RNA Technologies,
DOI 10.1007/978-3-319-55520-1_3

DOI 10.1007/978-3-319-55520-1_26

The Fig. 2 in the chapter "Growing Diversity of Plant MicroRNAs and *MIR*-Derived Small RNAs" has been updated. The correct Fig. 2 is given below:

The updated original online version for this chapter can be found at
DOI 10.1007/978-3-319-55520-1_3

M. Gozmanova (✉) • V. Baev • G. Yahubyan
Department of Plant Physiology and Molecular Biology, University of Plovdiv, 24 Tsar Assen Str, Plovdiv, Bulgaria

Institute of Molecular Biology and Biotechnologies, 105, Ruski Blvd., 4000 Plovdiv, Bulgaria
e-mail: mgozmanova@gmail.com

E. Apostolova
Department of Plant Physiology and Molecular Biology, University of Plovdiv, 24 Tsar Assen Str, Plovdiv, Bulgaria

G. Sablok
Plant Functional Biology and Climate Change Cluster (C3), University of Technology, Sydney, PO Box 123, Broadway, NSW 2007, Australia

N. Rajewsky et al. (eds.), *Plant Epigenetics*, RNA Technologies,
DOI 10.1007/978-3-319-55520-1_26

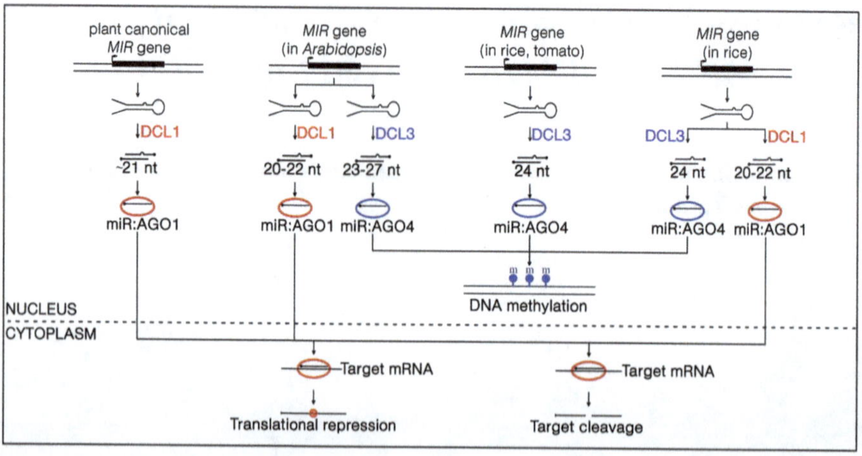

Fig. 2 Diversity of plant *MIR*-derived sRNAs. The hairpin precursors, transcribed from most plant *MIR* genes, are cut out by DCL1 to produce canonical mature ~21 nt miRNAs which associate with AGO1 and mediate target mRNA cleavage or translational repression. In addition to this classical pathway, some plant *MIR* genes can generate sRNA species that differ from the canonical miRNAs. In *Arabidopsis*, two sRNA species—canonical miRNAs and *MIR*-derived siRNAs (23–27 nt)—can be generated independently from different molecules of the same hairpin population by DCL1 and DCL3, respectively (Chellappan et al. 2010). In rice and tomato, some *MIR* genes produce only 24 ntlmiRNAs using DCL3, while other *MIR* genes can produce canonical miRNA and lmiRNA species simultaneously by coordinate activities of DCL1 and DCL3 on the same molecule (Wu et al. 2010; Kravchik et al. 2014). *MIR*-derived siRNAs and lmiRNAs associate predominantly with AGO4 and mediate DNA methylation of target genes or their own *MIR* genes. The DCL1/AGO1 pathway is depicted in *red* color, while the DCL3/AGO4 pathway is depicted in *blue* color

FSC
www.fsc.org
MIX
Papier | Fördert
gute Waldnutzung
FSC® C083411

Kolibri 360 GmbH
Ferdinand-Jühlke-Straße 7
99095 Erfurt, Deutschland
produktsicherheit@kolibri360.de